T0140326

Proceedings in Adaptation, Learning and Optimization 17

Series Editor

Meng-Hiot Lim, *Nanyang Technological University, Singapore, Singapore*

The role of adaptation, learning and optimization are becoming increasingly essential and intertwined. The capability of a system to adapt either through modification of its physiological structure or via some revalidation process of internal mechanisms that directly dictate the response or behavior is crucial in many real world applications. Optimization lies at the heart of most machine learning approaches while learning and optimization are two primary means to effect adaptation in various forms. They usually involve computational processes incorporated within the system that trigger parametric updating and knowledge or model enhancement, giving rise to progressive improvement. This book series serves as a channel to consolidate work related to topics linked to adaptation, learning and optimization in systems and structures. Topics covered under this series include:

- complex adaptive systems including evolutionary computation, memetic computing, swarm intelligence, neural networks, fuzzy systems, tabu search, simulated annealing, etc.
- machine learning, data mining & mathematical programming
- hybridization of techniques that span across artificial intelligence and computational intelligence for synergistic alliance of strategies for problem-solving.
- aspects of adaptation in robotics
- agent-based computing
- autonomic/pervasive computing
- dynamic optimization/learning in noisy and uncertain environment
- systemic alliance of stochastic and conventional search techniques
- all aspects of adaptations in man-machine systems.

This book series bridges the dichotomy of modern and conventional mathematical and heuristic/meta-heuristics approaches to bring about effective adaptation, learning and optimization. It propels the maxim that the old and the new can come together and be combined synergistically to scale new heights in problem-solving. To reach such a level, numerous research issues will emerge and researchers will find the book series a convenient medium to track the progresses made.

Indexed by INSPEC, zbMATH.

All books published in the series are submitted for consideration in Web of Science.

Harish Sharma · Apu Kumar Saha ·
Mukesh Prasad
Editors

Proceedings of International Conference on Intelligent Vision and Computing (ICIVC 2022)

Volume 1

Editors
Harish Sharma
Computer Science and Engineering
Rajasthan Technical University
Kota, Rajasthan, India

Apu Kumar Saha
Department of Mathematics
National Institute of Technology Agartala
Agartala, Tripura, India

Mukesh Prasad
School of Computer Science
University of Technology Sydney
Sydney, NSW, Australia

ISSN 2363-6084 ISSN 2363-6092 (electronic)
Proceedings in Adaptation, Learning and Optimization
ISBN 978-3-031-31166-6 ISBN 978-3-031-31164-2 (eBook)
https://doi.org/10.1007/978-3-031-31164-2

This Springer imprint is published by the registered company Springer Nature Switzerland AG
The registered company address is: Gewerbestrasse 11, 6330 Cham, Switzerland

Preface

This book contains outstanding research papers as the proceedings of the 2nd International Conference on Intelligent Vision and Computing (ICIVC 2022). ICIVC 2022 has been organized by National Institute of Technology Agartala, India, under the technical sponsorship of the Soft Computing Research Society, India. It was held on November 26–27, 2022, at National Institute of Technology Agartala, India. The conference was conceived as a platform for disseminating and exchanging ideas, concepts, and results of the researchers from academia and industry to develop a comprehensive understanding of the challenges of the advancements of intelligence in computational viewpoints. This book will help in strengthening congenial networking between academia and industry. The conference focused on collective intelligence, soft computing, optimization, cloud computing, machine learning, intelligent software, robotics, data science, data security, big data analytics, signal and natural language processing.

We have tried our best to enrich the quality of the ICIVC 2022 through a stringent and careful peer-reviewed process. ICIVC 2022 received a significant number of technical contributed articles from distinguished participants from home and abroad. After a very stringent peer-reviewing process, only 53 high-quality papers were finally accepted for presentation and for the final proceedings.

In fact, this book presents novel contributions in areas of computational intelligence, and it serves as reference material for advanced research.

Harish Sharma
Apu Kumar Saha
Mukesh Prasad

Contents

Test Pattern Modification to Minimize Test Power in Sequential Circuit

S. Asha Pon(✉), S. Priya, and V. Jeyalakshmi

Department of ECE, College of Engineering, Guindy, Chennai, Tamil Nadu, India
jpjeya@annauniv.edu

Abstract. Testing is a mandatory process to ensure the competence of Integrated circuit design. Testability measure in sequential circuits is arduous so, to alleviate the testing process, scan design is implemented. Power consumption in digital circuits surge during the testing phase due to the application of enormous test patterns to meet fault coverage. This paper proposes a power minimization technique by modifying test patterns. State skip LFSR methodology is proposed to meet test pattern compaction and Prim's algorithm to reorder compressed patterns. Prim's algorithm focuses on reduction of switching activity by decreasing hamming distance among successive test patterns. The proposed methodology is implemented in ISCAS'89 benchmark circuits. Experimental results show an improvement of 72.72% in terms of switching activity with State skip LFSR conventional LFSR and 81.81% with Prim's algorithm over conventional LFSR. Improvement in test data compression of 57.14% is observed with State skip LFSR over conventional LFSR.

Keywords: Test pattern generator · Linear feedback shift register · Prim's algorithm · State skip LFSR · test data reordering · test data compression

1 Introduction

Integrated circuits in the modern era are equipped with in-built testing set up to minimize testing cost. The addition of testing equipment ensures efficient testing with the application of larger test pattern count. Usage of larger test pattern count invokes excessive switching activity, thereby increased power consumption during the testing process is noted [1]. Hence, curtailing switching activity is mandatory to enhance the performance of digital devices.

Scan design is introduced to knock off the difficulty in testing internal storage elements in sequential circuits. Storage elements in sequential circuits are modified as Muxed D scan cell, Clocked scan cell and Level Sensitive Scan cell to intensify testing process. Muxed D scan cells are compatible with modern designs, hence, they are used abundantly in the modern digital era and so this paper also uses the same. These modification and usage of enormous test pattern count leads to drastic power consumption in sequential circuits. Reduction of testing power in sequential circuits is achieved by scan design modification, test pattern modification and test generator modification [2]. Test pattern modification is concentrated in this paper and is achieved by compressing and reordering the test pattern.

H. Sharma et al. (Eds.): ICIVC 2022, PALO 17, pp. 1–9, 2023.
https://doi.org/10.1007/978-3-031-31164-2_1

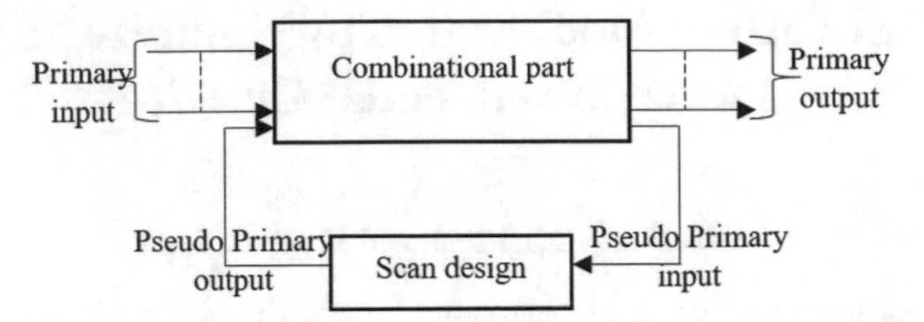

Fig. 1. Sequential circuit

Linear feedback shift register (LFSR) is used as a test pattern generator which generates 2^n-1 test patterns. 'n' represents the input size of the circuit under test. Ex OR gates are used in a feedback loop to generate all possible combinations of input. To test s27 benchmark sequential circuit two LFSR with characteristic polynomial $x^3 + x + 1$ and $x^4 + x^3 + 1$ are required. The circuit is grouped separately as combinational parts and storage elements as shown in Fig. 1. Testing of storage elements requires 3 bit LFSR to generate test patterns because 3 Muxed D scan cells are used in the circuit. The characteristic polynomial used to construct 3 bit LFSR is $x^3 + x + 1$. Figure 2 shows a 3 bit LFSR circuit which consists of 3 D flip flops and 1 Ex OR gate. Output of D flip flop 1 and 3 are tapped in feedback loop. Test pattern count obtained by this LFSR is 7 because the '000' pattern is not generated. Initial seed to the LFSR is '111'.

A new test power minimization methodology is proposed to compress and reorder test patterns to meet minimal hamming distance among consecutive test patterns. Test patterns are compressed by state skipping LFSR [3]. The compressed test patterns are reordered to minimize hamming distance among test patterns by Prim's algorithm [4].

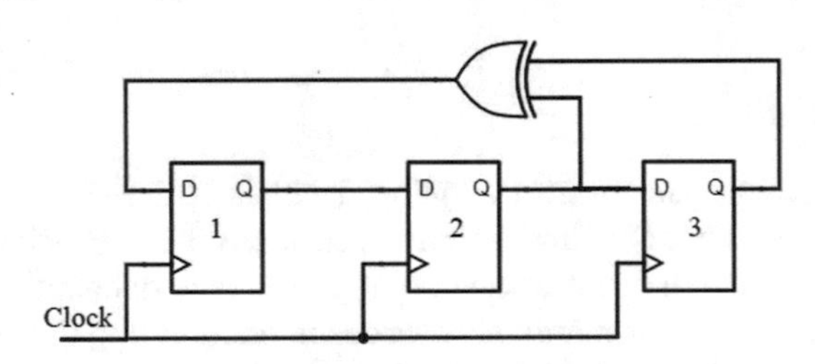

Fig. 2. 3 bit Traditional LFSR

The paper is organized as follows 2) Background review 3) Proposed methodology 4) Results and Discussion 5) Conclusion.

2 Background Review

Test power minimization can be achieved by decreasing switching activity because power directly depends on switching activity based on Eq. (1), where C is capacitance, V is supply voltage and α is switching activity.

$$P = \frac{1}{2}CV^2\alpha \tag{1}$$

Reduction of switching activity can be achieved by minimizing hamming distance among consecutive test patterns. Over past decades researchers have focused on minimizing hamming distance through test pattern reordering and test pattern compression.

Test vectors are reordered to reduce the application time and memory complexity through the Critical path tracing algorithm in [5]. Minimization of hamming distance by test vector reordering can be achieved by combining features of uniform-cost search and pure heuristic search [6]. In [7] test vector reordering is compared to traveling salesman problem and the problem of identifying minimal distance among test patterns is achieved by ant colony optimization algorithm. Minimum spanning tree based Prim's algorithm reorders test patterns to decrease hamming distance among them [8]. Test vectors are reordered to decrease switching acting activity by combining Dijkstra algorithm and Weighted transition matrix [9]. Genetic algorithm is combined with Simulated annealing to lessen switching activity by test vector reordering [10].

Test data compression also reduces test power by declining switching activity. Testing time and memory requirements are reduced by bitmask and dictionary selection based test data compression by introducing more matching patterns [11]. A better compression is achieved with test sequences by state skip LFSR and variable state skip LFSR [12]. Interrelations among test patterns have been enhanced by modifying LFSR into transition controllers to achieve better test data compression [13]. Enhancement in test compression ratio is noted with Low-Transition Generalized Linear Feedback Shift Register with bipartite (half fixed) and bit insertion techniques [14]. From the survey, it is known that to minimize test power, test data modification is mandatory. Hence, this paper concentrates on test pattern modification by compressing and reordering.

2.1 Proposed Methodology

Table 1. Test vector reordering

Unordered test pattern		Reordered test pattern	
Test pattern	α	Test pattern	α
110		110	
011	2	010	1
010	1	011	1
111	2	111	1
TSA	5	TSA	3

Test power minimization methodology proposed in this paper modifies the LFSR generated test pattern. Potent test patterns are obtained by compressing test patterns and then reordering them. Test data are compressed by state skip LFSR and are reordered by prim's algorithm. Table 1 shows the influence of test vector reordering in switching activity. It is proved that before ordering test patterns total switching activity (TSA) of

'5' is noted and after reordering it is reduced to '3'. Switching activity is calculated based on hamming distance among test patterns as shown in Eq. 2. N refers to the total number of test patterns, TP_i and TP_j are i^{th} and j^{th} test patterns.

$$HD = \sum_{i=1}^{N} \sum_{j=1}^{N} TP_i \, xor \, TP_j \tag{2}$$

3 State Skip LFSR

State skip LFSR (SS LFSR) is a revised form of traditional LFSR. In traditional LFSR test patterns are generated based on tap condition, whereas SS LFSR generates test patterns by skipping states. A N bit traditional LFSR is constructed based on characteristic polynomials with 'N' D flip flops and Ex OR gates in the feedback path. Whereas to construct N bit SS LFSR, 'N' D flip flop and 'N' Ex OR gates are required. The Ex OR gates are connected between D flip flops as shown in Fig. 3. Predetermined number of states are skipped in output when SS LFSR is used. Skipping of state occurs immediately after each pattern. N bit SS LFSR skips N-1 states compared with traditional LFSR. A SS LFSR performs successive jumps with a constant length and has more efficiency than traditional LFSR. The concept of state skipping ensures reduction of pattern count and hence test data compression is achieved.

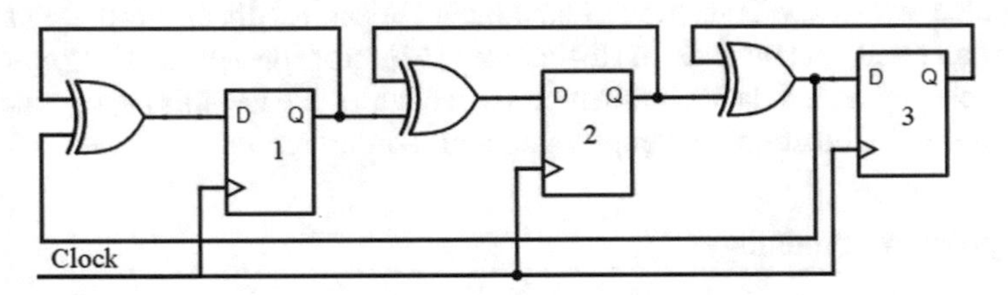

Fig. 3. 3 bit SS LFSR

3.1 Prim's Algorithm

To attain further power minimization, compressed test patterns are reordered using a greedy method called Prim's algorithm. This algorithm works based on the concept of finding minimum cost in a spanning tree. A vertex in a tree will establish connection with adjacent vertex only if the edge weight is minimum. The construction of minimum cost spanning tree primarily depends on selection of root vertex. In the test vector reordering problem, a pattern with minimum transition is chosen as the root vertex and remaining patterns are ordered with minimum hamming distance between consecutive patterns. Figure 4 with reference to Table 1 explain the concept of test vector reordering using Prim's algorithm. Unordered test patterns are shown in Fig. 4(a) whose total hamming distance amidst the patterns are noted to be '5'. Figure 4(b) portrays the ordered test patterns after application of Prim's algorithm. Here, the root node is selected as '111' because it has '0' transition and then remaining patterns are connected in such a way that the hamming distance amidst them is '1'. Figure 5 shows the flow chart of the proposed methodology.

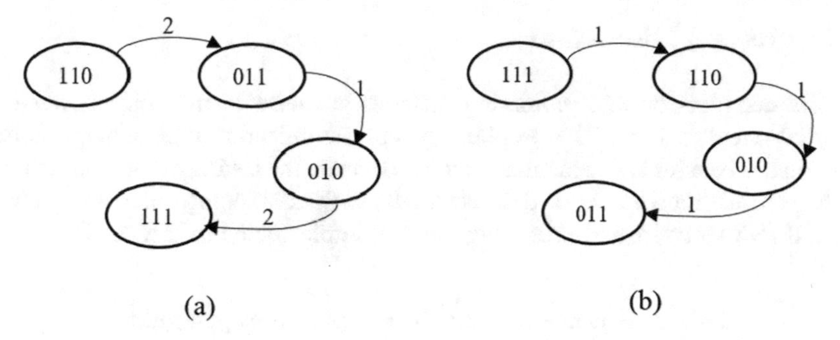

Fig. 4. (a) Unordered test patterns (b) Ordered test patterns

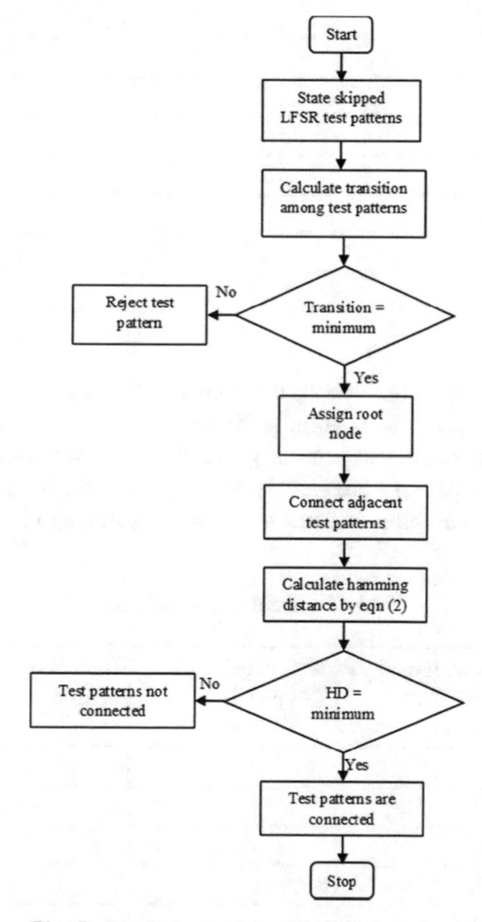

Fig. 5. Flow chart of proposed methodology

4 Results and Discussion

Test data compression and reordering methods are used to minimize power during sequential circuit testing. The proposed power minimization methodology is implemented in s27 benchmark sequential circuit. To ensure the testing of flip flops in this circuit, they are modified into Muxed D scan flip flops. State skipping LFSR is implemented using XILINX software and Prim's algorithm is implemented in MATLAB.

Table 2. Hamming distance of conventional LFSR test patterns

Test patterns	Hamming distance
111	–
011	1
101	2
010	3
001	2
100	2
110	1
THD	11

To test scan flip flops in the circuit, three bit LFSR is used. The conventional 3 bit LFSR generates 7 test patterns. Hamming distance among these patterns are shown in Table 2. Column 1 shows the test patterns generated by conventional LFSR. Hamming distance between these patterns are calculated using Eq. (2) and is indicated in column 2. THD refers to total hamming distance and it is measured as 11.

Table 3. Test data compression

Conventional LFSR Test patterns	State skip LFSR Test patterns	Hamming distance
111	111	–
011		–
101		–
010	010	2
001		–
100		–
110	110	1
	THD	3

State skipping LFSR used to achieve test data compression in s27 benchmark circuit is 3 bit. This LFSR skips 2 patterns on generating final test patterns. Hence, after generating each test pattern two patterns are skipped and a third pattern is available in the output. Test data compression and hamming distance measurement of state skip LFSR patterns are presented in Table 3. Column 1 portrays the conventional LFSR generated test patterns and State skip LFSR generated test patterns are displayed in column 2. It is noted that in test patterns generated by states skip LFSR, after each test pattern two states are skipped compared with conventional LFSR. Hence, output of SS LFSR consists of only 3 test patterns. Hamming distance of these 3 test patterns is calculated using Eq. (2) and is depicted in column 3. Total hamming distance of these patterns are calculated as 3.

The compressed test patterns are reordered using Prim's algorithm to achieve further reduction in hamming distance. Test pattern '111' is chosen as the root node because it has '0' transition among its bits. Table 4. Portray test vector reordering by Prim's algorithm. State skip LFSR test patterns are displayed in column 1 and reordered test patterns by Prim's algorithm are shown in column 2. Column 3 presents the hamming distance among reordered test patterns and total hamming distance is noted as 2.

Table 4. Test data reordering

State skip LFSR Test patterns	Prim's algorithm	Hamming distance
111	111	–
010	110	1
110	010	1
	THD	2

Test pattern count of conventional LFSR and SS LFSR are compared and depicted in Fig. 6. From the graph it is noted that the test pattern count of SS LFSR is 57.14% reduced than conventional LFSR test pattern count. Figure 7 shows the comparison of hamming distance among conventional LFSR test patterns, SS LFSR test patterns and Prim's algorithm. It is observed that SS LFSR test patterns achieves 72.72% reduction in hamming distance compared to conventional LFSR test patterns and 81.81% reduction in hamming distance is viewed with Prim's algorithm based reordered test patterns.

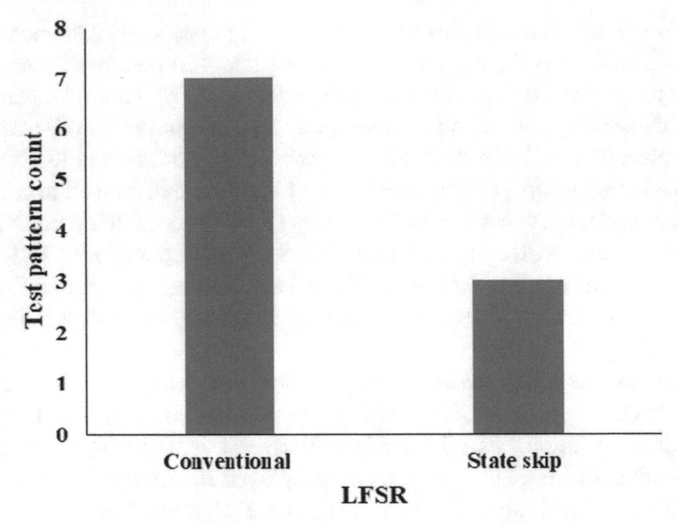

Fig. 6. Comparison of test data compression

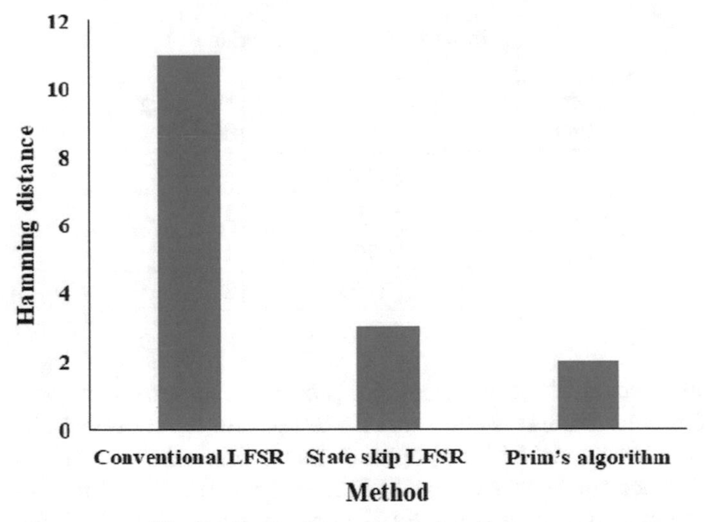

Fig. 7. Comparison of hamming distance

5 Conclusion

This paper proposes a new power minimization method based on test pattern reordering and test data compression. LFSR generated patterns are compressed by SS LFSR and reordered using Prim's algorithm. From the experimental results it is proved that the new method achieves 81.81% improvement in terms of hamming distance when compared with conventional LFSR. Reordered test patterns accomplish an improvement of 33.33% in hamming distance against compressed test patterns. Hence, it is proved that to attain minimum hamming distance LFSR generated test patterns has to be compressed and then reordered. This work can be extended further to analyze the impact on hamming distance with an increasing number of bits.

References

1. Pon, S.A., Jeyalakshmi, V.: Analysis of switching activity in various implementation of combinational circuit. In: 2020 6th International Conference on Advanced Computing and Communication Systems (ICACCS), pp. 115–121 (2020)
2. Pon, S.A., Jeyalakshmi, V.: Review of power minimization techniques in transition fault testing. IOSR J. Eng. **10**(2) (2020)
3. Manikya, D.M., Jagruthi, M., Anjum, R., Kumar, A.K.: Design of test compression for multiple scan chains circuits. In: 2021 International Conference on System, Computation, Automation and Networking (ICSCAN), pp. 1–5 (2021)
4. Maity, H., Khatua, K., Chattopadhyay, S., Sengupta, I., Patankar, G., Bhattacharya, P.: A new test vector reordering technique for low power combinational circuit testing. In: 2020 International Symposium on Devices, Circuits and Systems (ISDCS), pp. 1–6 (2020)
5. El-Maleh, A.H., Osais, Y.E.: On test vector reordering for combinational circuits. In: Proceedings. The 16th International Conference on Microelectronics, ICM 2004, pp. 772–775 (2004)
6. Roy, S., Gupta, I.S., Pal, A.: Artificial intelligence approach to test vector reordering for dynamic power reduction during VLSI testing. In: TENCON 2008 - 2008 IEEE Region 10 Conference, pp. 1–6 (2008)
7. Wang, J., Shao, J., Li, Y., Huang, Y.: Using ant colony optimization for test vector reordering. In: 2009 IEEE Symposium on Industrial Electronics & Applications, pp. 52–55 (2009)
8. Cormen, T.H., Leiserson, C.E., Rivest, R.L., Stein, C.: Introduction to algorithms. MIT press, Cambridge (2009)
9. Parmar, H., Ruparelia, S., Mehta, U.: WTM based reordering of combine test vector & output response using Dijkstra algorithm for scan power reduction. In: 2011 Nirma University International Conference on Engineering, pp. 1–6 (2011)
10. Reddy, K.V., Bharath, M.V., Suhag, A.K., Sinha, M.: Test vector reordering by using hybrid genetic algorithm-simulated annealing for lower switching activity. In: 2018 4th International Conference on Computing Communication and Automation (ICCCA), pp. 1–6 (2018)
11. Basu, K., Mishra, P.: Test data compression using efficient bitmask and dictionary selection methods. IEEE Trans. Very Large Scale Integr. (VLSI) Syst. **18**(9), 1277–1286 (2010)
12. Tenentes, V., Kavousianos, X., Kalligeros, E.: Single and variable-state-skip LFSRs: bridging the gap between test data compression and test set embedding for IP cores. IEEE Trans. Comput. Aided Des. Integr. Circuits Syst. **29**(10), 1640–1644 (2010)
13. Saraswathi, R., Manikandan, R.: Design of lfsr (linear feedback shift register) for low power test pattern generator. In: 2017 International Conference on Networks & Advances in Computational Technologies (NetACT), pp. 317–322 (2017)
14. Sakthivel, P., Kumar, A.N., Mayilsamy, T.: Low transition test pattern generator architecture for built-in-self-test. Am. J. Appl. Sci. **9**, 1396–1406 (2012)

A Social Network Approach for Improving Job Performance by Promoting Colleague Intimacy

Sreeja Velinjil Narayanan[1(✉)], Deepa Valipparambil Gopalakrishnan[1], and Maya Sreenilayam Subrayan[2]

[1] Department of Mathematics, Sree Krishna College, Guruvayur, Kerala 680104, India
drsreejavn@gmail.com

[2] Department of Statistics, Maharaja's College, Ernakulam, Kerala 682011, India

Abstract. The workplace is fast changing ins modern days. Occupational stress may be more hazardous to an employee's health than ever before. This stress has a negative impact on both employee health and job performance. Each person has distinct abilities. It really is incumbent on the organization's leaders to achieve their potential. When a group of individuals works efficiently and collaboratively toward a common objective, team members can develop great intimacy as they become more acquainted with one another, which helps them align their goals and motivates them to work harder for one another. As just a basis, the purpose of our study is to use social network analysis to determine how successful mutually supportive teamwork is in the workplace and how effective it gets as the number of people who aid each individual rises.

Keywords: Social network analysis · workplace stress · coworker closeness · job performance

1 Introduction

Recent years have seen a sharp increase in interest in social network analysis. A network is defined by a collection of nodes (actors) connected by links (relationships). Social networks are interactions between people, organizations, or other groups. They exchange social contacts, friendships, ideas, and other things just like that. Social network analysis (SNA) is a network science that uses networks and graph theory to study social systems. We are able to comprehend the relationships between nodes in great detail by using the measures used in SNA. Sociograms, which depict people as nodes with connections between them as links, are used to graphically portray networks. Recently, the field of social network analysis has developed more. Because of their pervasiveness in the world, networks have piqued the interest of researchers from a wide range of disciplines. SNA is enriched by several statistical, mathematical, economical, biological, psychological, sociological, and anthropological measures; as a result, this topic has evolved into an interdisciplinary research area. Since the middle of the 1930s, social network analysis has been used to improve research. Freeman et al. [1] showed the fundamental forms

H. Sharma et al. (Eds.): ICIVC 2022, PALO 17, pp. 10–20, 2023.
https://doi.org/10.1007/978-3-031-31164-2_2

of social structure as well as the manner in which network relational structure impacts system actors and just how actors in a social network cluster into groups. [2] will assist researchers in comprehending the theoretical assumptions underlying social network methodologies for a specific research challenge. The book written by Degenne and Fors´e is quite beneficial in understanding the excellent foundation of social network analysis. It is more extensive and discusses network methods in many societies in a precise way [3]. Those who are comfortable with the foundation can go to book [4], which covers subjects such as network data collection and evaluation, diffusion modelling, and so forth. Social network analysis techniques are still being fully investigated in this field of evaluation, despite being relatively new for modern analysis. The article [5] outlines the advantages of SNA for formative assessment. The handbook [6] is an unbeatable resource. It presents an introduction to main themes, substantiative topics, central methods, and predominant discussions in a systematic manner. As a result, both educators and learners now have access to an unparalleled resource. Rather than consulting a wide range of books and journal articles, the handbook provides a one-stop resource that readers will depend on for the foreseeable future. Book [7] primarily focuses on the social, structural, and cognitive aspects of social networks; it pays less attention to the particular problems that emerge when the structural and data-centric factors of the network interact. Accordingly, some of the prevailing opinion regarding the distinction between corporate strategic management, project management, and programme management is challenged in Book [8].

1.1 Significance of Colleague Intimacy in Reducing Workplace Stress

Stress is a state of anxiety, either emotionally or physically. Any circumstance or idea that gives you cause for annoyance, rage, or anxiety might trigger it. Your body's response to a demand or difficulty is stress. Anything that excites us and raises our level of attentiveness is what we define as stress. Stress may occasionally be beneficial, including when it enables you to prevent danger or achieve a deadline. People are comfortable and happy at work if they get along well with their coworkers; otherwise, they would have to deal with stress. While knowledge, experience, qualification, and efficiency are all influenced by job happiness, connections with coworkers and clients also play a role. Today's work pressure is increasingly intense due to industry competitiveness, the advent of new information technologies, and irrational expectations. There are several factors that make working with others harder. Every workplace has hard people; they annoy coworkers or constantly act in a manner that is against our sense of right and wrong. Employees who have poor interpersonal interactions with one another are more uncomfortable, and we are unable to leverage their expertise to develop the organization.

Now a days, many top firms treat their employees differently because they understood the value of peer group collaboration and how to inspire staff by recognising their accomplishments and rewarding them. Unfortunately, the majority of employers believe that these are silly and juvenile, which creates a stressful environment at work and prevents employers from fully utilising employees' abilities, even if they are efficient. Even when many higher authorities concur, they refuse to give it any weight since they are too preoccupied with their work and commercial success. It is conceivable for an employer to feel stressed due to troubles at work as well as any issues with their personal life, family, or friends. While we may not be able to remove the source of stress, we can provide our coworkers with peace of mind and help them complete their work effectively. For more references, one may read [9], in which they explored the managerial implications of workplace friendship potential and social aspects. Podolny et al. tackle social activity at work in a variety of contexts. They asserted that researchers may use their work to engage people and maximise involvement for the benefit of the employees, the company, and the community at large [10]. In [11], they investigated how various personality types build and benefit from social networks in workplaces. They investigated the relationship between self-monitoring attitudes and work performance. Individual job performance was independently predicted by self-monitoring and centrality in social networks. The research [12] investigates both positive and negative workplace views of social networking and presents a comprehensive assessment of the literature in the field. The relationship between social networking and workplace culture is investigated, specifically whether social networking tools are capable of energising and reshaping workplace goals and brand, which can lead to improved ways of working and enhanced levels of employee satisfaction and performance. Stress is the manner in which an individual adjusts after being exposed to internal or external issues. The organism has a number of systems that can synchronise such behavioural mechanisms at the cellular and systemic levels. We explain stress-related physical ailments in [13], as well as the mechanisms, primarily the immune and autonomic nervous systems, that take part in the link between the brain and the body. In the book [14], the key variables in the stress-strain process are identified, along with the application of multivariate techniques like structural equation modelling and multiple regression analysis. The work [15] aims to assist readers in recognising the symptoms of stress, coming up with coping mechanisms, and strengthening their own personal resilience. This study suggests ways to deal with work stress that are practical and When there is an imbalance between the demands of the job and the resources and abilities of the individual worker to meet those demands, work-related stress results. [16] addresses how to manage the stress that comes with the job. With globalisation and the current financial crisis, which is affecting almost all countries, occupations, and worker categories as well as families and societies, occupational stress has recently increased. Since employee stress has been shown to have a significant negative impact on business productivity, many organisations are working to reduce and prevent it. According to [17], individuals commonly experience pressure when they are unable to effectively handle the situation. Stress is therefore situation-specific and more probable in some individuals than in many others. It was discovered that men experience physical strain, while women are more likely to experience psychological stress. Stress can have negative effects in the field of psychology.

1.2 Terminologies

For analytical purposes, the network can be represented using graphs. $\mathbb{G} = (\mathbb{V}, \mathbb{E})$ will be a directed graph if $\mathbb{V}$ indicates the collection of vertices and $\mathbb{E}$ signifies the set of edges. Graphs are excellent tools for illustrating a variety of social situations. To indicate the total number of nodes and edges in the network, respectively, n and m will be used.

Adjacency matrix of networks is an especially useful method of representing networks, which is an $n x n$ matrix with elements

$$\mathbb{A}_{ij} = \begin{cases} 1 \text{ if node i is connected to nodej} \\ 0 \qquad\qquad \text{otherwise} \end{cases} \tag{1}$$

Directed graphs and undirected graphs are the two main categories into which the graphs or networks may be categorized. A finite directed graph $\mathbb{G}$ is formalized as consisting of a collection of vertices or actors $\mathbb{V}(\mathbb{G}) = \{\mathbb{v}_1, \mathbb{v}_2, \ldots \mathbb{v}_n\}$ and an edge set $\mathbb{E}(\mathbb{G}) \subseteq \mathbb{V}(\mathbb{G}) x \mathbb{V}(\mathbb{G})$. Each edge $(\mathbb{u}, \mathbb{v}) \in \mathbb{E}(\mathbb{G})$ may be conceptualized as a connection between the initial node $\mathbb{u}$ and final node $\mathbb{v}$; these two vertices are referred to as being adjacent to one another. The in-degree of a vertex $\mathbb{u}$, $deg_{in}(\mathbb{u})$ in a directed graph $\mathbb{G}$ is determined by the number of edges that end at $\mathbb{u}$, and the out-degree of $\mathbb{u}$, $deg_{out}(\mathbb{u})$ is decided by the number of edges that originate from $\mathbb{u}$.

2 Objective of the Study

People may experience work-related stress as a reaction to demands and pressure that are not compatible with their skills and knowledge and that test their capacity for adjustment. Although stress can arise under a variety of work conditions, it is frequently exacerbated when employees feel that they have even less support from coworkers and supervisors and little influence over work procedures. In organizations, performance management should be a primary focus as firms strategically prepare for future development. Firms always concentrate their efforts on enhancing employee performance. Executives that are driven to boost staff performance don't always need to approve large salaries, present extravagant gifts, or make new incentive offers. Instead, there are much more doable and efficient approaches to increasing labour performance. Employee involvement is the key milestone. Teamwork at the workplace is one such strategy. It's never simple to lead a team, whether there are five or five hundred employees. Grouping individuals with various demeanors may frequently result in conflicts, ambiguities, and a reduction in job efficiency. Therefore, we want to learn how to create a better team dynamic. We are attempting to comprehend that teamwork is quite beneficial, yet there is a significant difference in effectiveness when team members are willing to acknowledge themselves. Furthermore, considering the performance quotient, we are attempting to determine whether a person could perform better than previously when the team grows in size.

3 Methodology

We attempted to identify what type of teamwork is more effective by studying the spirit of teamwork in the workplace. The data set we used for this analysis included 34 workers. They came from various organisations and took part in a 21-day replenishing camp to ensure staff members did better at their jobs. They were aliens before they arrived there. These 21 days are broken into three stages of seven days each. During the first phase, each participant was given five small activities that were connected to their profession. These duties should be performed during the first week by accepting assistance from any of these 34 peers. A person was given the chance to support more than one worker if they needed it; therefore, the number of people who may be aided by an individual can range from 0 to 33. Because no one knew one another previously, they would want to seek assistance from employees seated beside one another, travelling almost the same route, or anything similar. It was noticed how many of the five tasks assigned to each of them could be nicely finished in one week. Each participant was identified as an actor in the network, and if one actor received help from another actor, we designated that as a directed arc from v_j to v_i.

Figure 1 and 2 show the resulting sociograms from phases 1 (with $n = 34$ and $m = 80$) and phase 3 (with $n = 34$ and $m = 88$) respectively. In these figures, we assigned the colour red to the node that represents the participant who has done five tasks, maroon to whom for four tasks, purple to whom for three tasks, blue to whom for two tasks, and green to whom for one task, respectively. Figure shows that compared to phase 1, the number of people who have finished two tasks only significantly reduces in phase 3. In addition, it should be noted that three people were able to successfully complete all five tasks. Participants were allowed the opportunity to meet each other better in the second phase through various mind-calming games, thought exchange, refreshments, and recreation activities together.As a result, many people become acquainted with folks who agree with their ideology. It aided such people in becoming closer. Following that, in the third phase, as in the first, each person again was assigned five tasks. They were also offered the option of hiring persons of their choosing to help them to complete the assignment. This time, the persons they enlisted for assistance were those who supported their viewpoint. Because of this development, they began to enjoy their activities and completed more of them effectively than in phase 1.

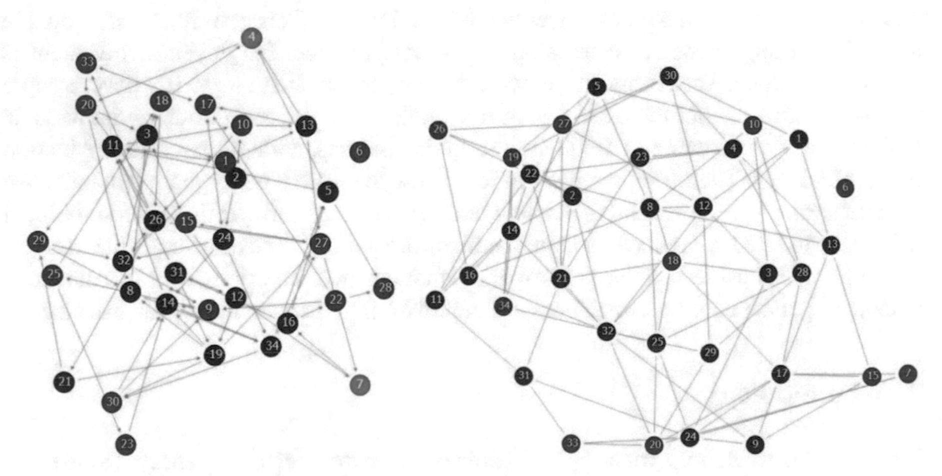

Fig. 1. Sociogram of actors in the first phase

Fig. 2. Sociogram of actors in the third phase

Performance quotient for each phase is calculated using the formula

$$PQ = \frac{\sum_{i=1}^{n} a_i t_i}{\sum_{i=1}^{n} a_i} \tag{2}$$

where a_i denotes the number of assistance received by the actor v_i and t_i denotes the number of tasks completed by the actor v_i. The performance quotient reveals how the support from colleagues affects the capability of the participant to complete the activities.

We defined some performance indicator functions below to compare the tasks completed by each participant in phases 1 and 3 with the assistance of other actors:

$$I_{a_i}I_{t_i} = \begin{cases} 1 \textit{ if the number of actors and the number of completed tasks increased} \\ 0 \textit{ otherwise} \end{cases}$$

$$I_{a_i}S_{t_i} = \begin{cases} 1 \textit{ if the number of actors increased but the number of completed tasks remains same} \\ 0 \textit{ otherwise} \end{cases}$$

$$I_{a_i}D_{t_i} = \begin{cases} 1 \textit{ if the number of actors increased but the number of completed tasks decreased} \\ 0 \textit{ otherwise} \end{cases}$$

$$S_{a_i}S_{t_i} = \begin{cases} 1 \textit{ if the number of actors and the number of completed tasks remains same} \\ 0 \textit{ otherwise} \end{cases}$$

$$S_{a_i}I_{t_i} = \begin{cases} 1 \textit{ if the number of actors remains same and the number of completed tasks increased} \\ 0 \textit{ otherwise} \end{cases}$$

$$D_{a_i}S_{t_i} = \begin{cases} 1 \textit{ if the number of actors decreased and the number of completed tasks remains same} \\ 0 \textit{ otherwise} \end{cases}$$

$$D_{a_i}I_{t_i} = \begin{cases} 1 \textit{ if the number of actors decreased but the number of completed tasks increased} \\ 0 \textit{ otherwise} \end{cases}$$

4 Results and Discussion

The number of people who assisted each actor in carrying out their tasks in the first and third phases is shown in the table below, along with the number of tasks each actor was able to complete in the allotted time. The 34 participants are represented by actors $v_1, v_2, ...v_{34}$. According to Table 1, the number of actors who assisted each participant ranges from 0 to 4 in the first and third phases. It can be observed that the number of persons assisted by each actor rose in phase 3 compared to phase 1, and hence the number of tasks completed by them increased by at least several people. In both phases, the actor v_{28} refused to accept any assistance. As a result, there was no difference in performance. Even so, the 28th actor did not seek assistance from anybody during the first phase but accepted assistance from two people during the third phase when he realised others existed during phase 2. As a result, the individual was able to finish more activities than previously.

According to Table 2, the percentage of people who were able to accomplish more activities in Phase 3 with the assistance of additional actors is $I_{a_i}I_{t_i} = 38.2352\%$. However, despite a rise in the number of actors who assisted, the number of tasks completed has declined. But the number of actors who assisted individuals and the number of their completed tasks remained the same as in phase 1 in 32.3529% of the cases. We can observe that 8.825% of the competent persons in the group were able to increase the number of tasks despite the fact that the number of actors who assisted decreased.

Table 1. Table of tasks completed successfully by the actors in first and third phase by accepting the assistance of other actors

Actor who performed tasks (v_i)	Phase 1		Phase 3		Actor who performed tasks (v_i)	Phase 1		Phase 3	
	Number of actors who assisted (a_i)	Number of tasks completed (t_i)	Number of actors who assisted (a_i)	Number of tasks completed (t_i)		Number of actors who assisted (a_i)	Number of tasks completed (t_i)	Number of actors who assisted (a_i)	Number of tasks completed (t_i)
v_1	2	2	3	3	v_{18}	2	2	2	2
v_2	2	3	3	3	v_{19}	4	4	4	5
v_3	3	3	3	3	v_{20}	2	2	2	2
v_4	1	1	3	3	v_{21}	2	3	3	3
v_5	3	3	3	4	v_{22}	3	2	4	4
v_6	0	2	0	2	v_{23}	1	2	2	3
v_7	1	1	2	2	v_{24}	3	3	3	3
v_8	3	3	3	3	v_{25}	2	2	3	3
v_9	2	2	3	3	v_{26}	4	4	3	5
v_{10}	3	2	2	2	v_{27}	2	2	2	2
v_{11}	3	3	3	3	v_{28}	0	2	2	3
v_{12}	3	3	4	4	v_{29}	2	2	2	3
v_{13}	3	3	3	3	v_{30}	2	2	3	3
v_{14}	3	3	4	4	v_{31}	3	3	2	3
v_{15}	2	2	3	2	v_{32}	3	3	4	4
v_{16}	3	3	3	3	v_{33}	2	2	2	2
v_{17}	2	2	3	3	v_{34}	4	4	3	5

Figure 3 and 4 show the number of actors who assisted each worker in completing their assigned tasks, as well as the number of tasks effectively performed by the actors in the first and third phases. According to Fig. 3, the number of people who assisted many of the performers increased in phase 3 compared to phase 1. As a result, from Fig. 4, we can observe that the majority of participants completed their tasks more efficiently.We also conclude, using the performance quotient, that there is a significant difference when a person belongs to a large team and can successfully execute their work compared to when they belong to a small team.

Performance quotient for phase 1 is $PQ_{(phase1)} = \frac{219}{80} = 2.7375$
Performance quotient for phase 3 is $PQ_{(phase3)} = \frac{306}{94} = 3.2553$

We deduced from this that an employee's performance quotient in a tiny group is 2.7375, and better performance can be accomplished successfully since the performance quotient in a bigger group is 3.2553 when team members operate with common understanding.

Table 2. Table of comparison of tasks completed successfully by individuals and the actors gave the assistance for each individual.

Actor who performed tasks (v_i)	$I_{a_i} I_{t_i}$	$I_{a_i} S_{t_i}$	$I_{a_i} D_{t_i}$	$S_{a_i} S_{t_i}$	$S_{a_i} I_{t_i}$	$D_{a_i} S_{t_i}$	$D_{a_i} I_{t_i}$
v_1	1	0	0	0	0	0	0
v_2	0	1	0	0	0	0	0
v_3	0	0	0	1	0	0	0
v_4	1	0	0	0	0	0	0
v_5	0	0	0	0	1	0	0
v_6	0	0	0	1	0	0	0
v_7	1	0	0	0	0	0	0
v_8	0	0	0	1	0	0	0
v_9	1	0	0	0	0	0	0
v_{10}	0	0	0	0	0	1	0
v_{11}	0	0	0	1	0	0	0
v_{12}	1	0	0	0	0	0	0
v_{13}	0	0	0	1	0	0	0
v_{14}	1	0	0	0	0	0	0
v_{15}	0	0	1	0	0	0	0
v_{16}	0	0	0	1	0	0	0
v_{17}	1	0	0	0	0	0	0
v_{18}	0	0	0	1	0	0	0
v_{19}	0	0	0	0	1	0	0
v_{20}	0	0	0	1	0	0	0
v_{21}	0	1	0	0	0	0	0
v_{22}	1	0	0	0	0	0	0
v_{23}	1	0	0	0	0	0	0
v_{24}	0	0	0	1	0	0	0
v_{25}	1	0	0	0	0	0	0
v_{26}	0	0	0	0	0	0	1
v_{27}	0	0	0	1	0	0	0
v_{28}	1	0	0	0	0	0	0
v_{29}	0	0	0	0	0	0	1
v_{30}	1	0	0	0	0	0	0
v_{31}	0	0	0	0	0	1	0

(*continued*)

Table 2. (*continued*)

Actor who performed tasks (v_i)	$I_{a_i} I_{t_i}$	$I_{a_i} S_{t_i}$	$I_{a_i} D_{t_i}$	$S_{a_i} S_{t_i}$	$S_{a_i} I_{t_i}$	$D_{a_i} S_{t_i}$	$D_{a_i} I_{t_i}$
v_{32}	1	0	0	0	0	0	0
v_{33}	0	0	0	1	0	0	0
v_{34}	0	0	0	0	0	0	1
Total	13	2	1	11	2	2	3
Percentage	38.2352	5.8823	2.9411	32.3529	5.88235	5.88235	8.8235

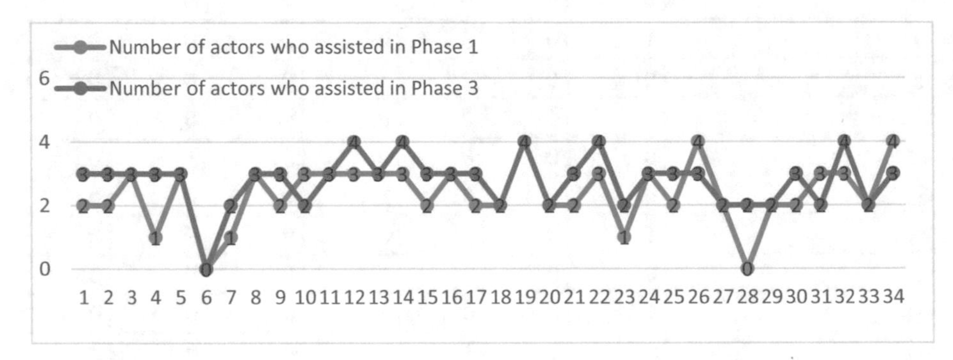

Fig. 3. Chart of number of actors assisted each worker in first and third phase

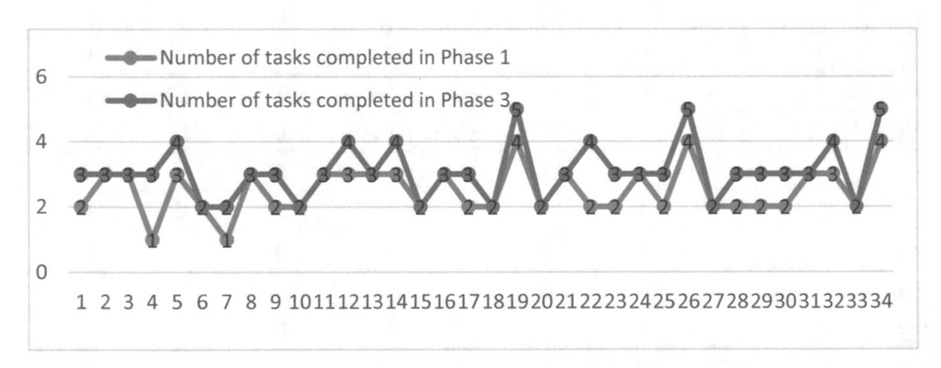

Fig. 4. Chart of number of tasks completed successfully by the actors in first and third phase

All of this suggests that an individual's job performance may be improved with the assistance of a group of people who understand one another. Each individual has unique skills. The organization's leaders should make the most of it. For a person to fully harness their potential, they must be part of a group that recognises their strengths and weaknesses. It is not feasible to complete the work perfectly simply because there are more actors in the group. If even one member of the group cannot comprehend what the others are saying, it causes tension for everyone. If a person is stressed at work, no matter how skilled they are, their abilities will be worthless to the organization. Because this study was completed in a relatively short time frame and only modest tasks were assigned, the author's desired output was not met by completing all of the activities assigned. That is not the case when it transforms into an institution. The majority of the people there will be working for many years. As a result, in addition to their own skills, they will have knowledge gained through experience. If the institution wants to transform all of them into organisational effectiveness, it would be preferable if we formed groups that could comprehend each other and work toward the same objective while reducing their stress.

5 Conclusion

Pressure at work is unavoidable due to the demands of the contemporary workplace. Stress, on the other hand, results once that situation arises or is otherwise uncontrollable. Employees' health and company performance can both be negatively impacted by stress. Employees are an institution's most crucial component since they are ultimately responsible for the success of the organisation in the first place. However, in recent years, staff performance has been declining. Since cooperation boosts effectiveness, many organisations promote its importance. They have altered their approach and now think that teamwork may lead to general advancement. In the workplace, teamwork fosters mutual understanding and support. A key component is teamwork among the members. It is advantageous to have a strong team. The team feels valuable as a whole when their work is well utilized. We thus made an effort to comprehend the significance of individuals' mutual understanding in teams and how they may function in the workplace for the development of the organization. Numerous studies have shown that teamwork is more efficient in organisations than solo effort.

Because it is an unfamiliar situation, many people will be unwilling to accept help from others, according to our study. There may be some confusion about what each person can do to help. Even when it comes to understanding certain individuals' skills, some people may believe that if more individuals seek their help, it will become harder. They are also concerned that others would believe they don't know much. When they observe individuals who perform an excellent job being commended by the authorities, they are troubled by the feeling that they too cannot achieve that level and gradually succumb to despair. When someone is unable to do their task effectively, even though they have the desire, some people feel compelled to give up. Our results show that a team that works with shared understanding performs better than a team of peers that lacks shared trust. Similarly, even though the team consists of individuals who can exchange ideas, the actual number of team members is very important. That is, even with the help of others who have similar opinions, a member of a large group can perform better than a member of a small group. Therefore, those in positions of authority who want the advancement of any organisation should pay attention to assembling a team of individuals who can cooperate with the same goals. Such groups can complete the work effectively within a short amount of time if they are given the chance to collaborate and discuss while working on it. If a project is successful in this way, the group's leadership and members will be enthusiastic to coordinate each and every future project. It guarantees the institution's consistent growth. A worker can reach their full potential and take more pleasure in their work when they collaborate with people who share their opinions than when they impose themselves in environments that are not okay for them. As a result, let our community's mental health improve as the institutions develop.

References

1. Freeman, L.C., White, D.R., Romney, A.K.: Research Methods in Social Network Analysis, Transaction Publishers. George Mason University Press, Fairfax (1989)
2. Wasserman, S., Faust, K.: Social Network Analysis: Methods and Applications. Cambridge University Press, Cambridge (1994)
3. Degenne, A., Forśe, M.: Introducing Social Networks. Sage publications Ltd., London (1999)
4. Carringtom, P.J., Scott, J., Wasserman, S.: Models and Methods in Social Network Analysis. Cambridge University Press, Cambridge (2005)
5. Durland, M., Fredericks, K.A.: An introduction to social network analysis. N. Dir. Eval. **2005**(107), 5–13 (2006)
6. Scott, J., Carrington, P.J.: The SAGE Handbook of Social Network Analysis. Sage publications Ltd., New Delhi (2014)
7. Aggerwar, C.C.: Social Network Data Analysis. Springer, New York (2011)
8. Pryke, S.: Social Network Analysis in Construction. John Wiley & Sons, Hoboken (2012)
9. Riordan, C.M., Griffeth, R.W.: The opportunity for friendship in the workplace: an under explored construct. J. Bus. Psychol. **10**, 141–154 (1995)
10. Podolny, J.M., Baron, J.N.: Resources and RELATIONSHIPS: SOCIAL NETWORKS AND MOBILITY in the workplace. Am. Sociol. Rev. **62**(5), 673–693 (1997)
11. Mehra, A., Kilduff, M., Brass, D.J.: The social networks of high and low self-monitors: Implications for workplace performance. Adm. Sci. Q. **46**(1), 121–146 (2001)
12. Bennett, J., Pitt, M., Owers, M., Tucker, M.: Workplace impact of social networking. Prop. Manag. **28**(3), 138–148 (2010)
13. Hellhammer, D.H., Hellhammer, J.: Stress: the Brain Body Connection (Key Issues in Mental Health). 174, 1st edn. Karger publishers, Switzerland (2008)
14. Kaslowsky, M.: Modeling the stress-strain relationship in work settings. Routledge progress in psychology, London and New York (1998)
15. Sahoo, S.R.: Management of stress at workplace. Glob. J. Manag. Bus. Res. Adm. Manag. **16**(16), ver 1.0 (2016)
16. Mohajan, H.K.: The occupational stress and risk of it among the employees. Int. J. Mainstream Social Sci. **2**(2), 17–34 (2012)
17. Singh, A.: Occupational stress: a comprehensive general review. Amity J. Train. Dev. ADMAA **3**(1), 36–47 (2018)

Deep Learning Model for Automated Image Based Plant Disease Classification

Akshay Dheeraj[1,2](✉) and Satish Chand[2]

[1] ICAR- Indian Institute of Soil and Water Conservation, Dehradun, India
akshaydheeraj.jmi@gmail.com, akshay.dheeraj@icar.gov.in
[2] School of Computer and Systems Sciences, Jawaharlal Nehru University, New Delhi, India

Abstract. India is a rural country where a majority of the population rely on agriculture for their living. Agriculture provides food and raw materials and also acts as a livelihood source for farmers. Farmers are bearing a big loss because of plant diseases that are difficult to be identified. These diseases can be identified by plant pathologists by visual inspection. This is a time-consuming process and needs expertise. With the improvement in digital cameras along with the evolution of machine learning and deep learning, an automated identification of plant disease is in huge demand in precision agriculture. These diseases can be identified by deep learning techniques that enable farmers to take action accordingly beforehand. In this research work, a transfer learning based deep learning model named EfficientNet B0 has been proposed to identify and classify the plant leaf disease for pepper, potato, and tomato plant. Images of these plants have been taken from a popular plant disease dataset named "PlantVillage". These images are trained on the EfficientNet B0 model and performance comparison is done with other CNN based deep learning models. The testing accuracy of the proposed EfficientNet B0 model is 99.79% which justifies the efficacy of the developed model. Therefore, the performance of the proposed model indicates the utility of an automatic plant disease recognition and its ease of application in a farmer's field at a greater scale.

Keywords: Disease Classification · Convolutional Neural Network · Transfer Learning · Deep Learning · EfficientNet B0

1 Introduction

Agriculture is playing a pivotal role in strengthening the Indian economy. It contributes around 14% of GDP. Farmers have to face various issues like climate change, weather forecasting, phenology identification, etc. Crop disease poses a big threat to agriculture as it can spoil the crop yield which creates a problem with food safety. Therefore, the identification of crop disease at an early stage is necessary to ensure a good yield and better quality of the crop. Expert knowledge about various diseases in plants is lacking among villagers. Low inter class and high intra class similarity among disease classes make the disease identification process very challenging for the farmers. This process

H. Sharma et al. (Eds.): ICIVC 2022, PALO 17, pp. 21–32, 2023.
https://doi.org/10.1007/978-3-031-31164-2_3

is time-consuming and may fail as well in a big field. In Kisan Call Centers, sometimes experts cannot see the exact problem that the farmer is facing so a wrong suggestion can also be given sometimes for plant disease identification. Hence, it is required to use modern methods like deep learning and machine learning that can assist the farmer in identifying the disease easily and provide the required solution to deal with these crop diseases.

Machine learning and deep learning are now employed in every sector for classification and prediction purposes due to the advancements in computer vision. Handcrafted and automatically extracted features can be used to detect the leaf diseases. Handcrafted features extraction is fast but generally gives poor results. On the other hand, automatic feature extraction is done by a deep learning approach that give good performances in leaf disease identification. The objective of this research study is to identify and classify the plant disease in pepper, potato, and tomato based on image processing. An automated detection of these diseases can speed up other tasks in agriculture. The images used for the proposed work have been taken from the publicly available popular dataset named "PlantVillage". These images have been trained on convolution neural network (CNN) based model EfficientNet B0 and performances are compared with another deep learning model. Contributions to the proposed research work are as follows:

- To detect disease in pepper, potato, and a tomato plant.
- To classify the disease of pepper, potato, and tomato into 15 disease types.
- To visualize the role of EfficientNet B0 in improved classification accuracy for plant disease.

Following the introduction, Sect. 2 describes the related work done in the field. Section 3 and Sect. 4 cover the proposed methodology and results of the experiment respectively. The conclusion along with future scope is mentioned in Sect. 5.

2 Literature Review

For the purpose of identifying and classifying crops and diseases, various studies have been conducted. Image processing [1] has been used to detect plant disease which assists farmers to identify diseases. Features maps are extracted by image processing [2] and sent to a neural network that classifies these plant disease images. Features are extracted manually using image processing which takes time and is less accurate. Deep Learning models have revolutionized computer vision and are being used in various research areas [3, 4] for object identification and classification. One of the computer vision problems that may be effectively handled by deep learning is the detection of plant diseases.

In research work [5], the author has used various transfer learning-based deep convolutional neural network models. Authors used pre-trained models like AlexNet [6], VGG [7], GoogleNet [8] and ResNet [9], DenseNet [10] which offer high classification accuracy for plant disease. Research work presented in [11] focuses on the identification of plant disease of 13 different types. The proposed study has an accuracy of 96.3% on the test dataset. A total of 30,880 images have been used for training the model and tested on 2589 images. In [12], the authors presented a deep convolutional neural network for disease identification for 10 commonly found rice diseases and presented model was used on 500 images of infected and healthy rice leaves. The developed model has 95.48% accuracy. In [13], research work focuses on tomato disease identification for 9 commonly occurring tomato diseases. Dataset used in research work has 14828 images of infected tomato leaves and the model offers 99.18% accuracy. CNN-based GoogleNet architecture has been used to identify the diseases. In [14], the author presented a CNN-based VGG16 model for classifying apple disease named apple black rot in four stages based on severity. The author used 1644 images for the research work and achieved an accuracy of 90.4%. In [15], the author used two deep learning frameworks AlexNet and VGG16 for tomato leaf disease classification. Data augmentation has been used to bring the variability and expand the dataset. These two proposed models achieved an accuracy of 97.49 and 96%. In [16], authors provide an overview of various classification techniques used to identify plant diseases. In agriculture, an early detection of plant diseases is very important to improve crop quality and quantity.

Research study was conducted in [17] to use a computer vision enabled Artificial Intelligence techniques for automating the yellow rust disease identification and improving the accuracy of disease identification system. In [18], the authors developed a CNN based model for tomato disease identification and used Conditional Generative Adversarial Network for creation of augmented tomato leaf images. The developed model obtains the classification accuracy of 97.11% on ten classes of tomato leaf disease. In [19], the authors have presented a VGG model combined with Random Forest and Xgboost for disease identification of corn, tomato and potato plants. The experimental results achieve the classification accuracy of 94.47% for corn, 93.91% for tomato plants and 98.74% for potato leaf images.

3 Proposed Methodology

3.1 Data Collection

The dataset used in this research study has been taken from the publicly available "PlantVillage" dataset [20]. The PlantVillage dataset has a total of 39 plant diseases along with some background images. The proposed research work is focused on 12 plant diseases of pepper, potato, and tomato leaf and three classes having healthy leaf images. The used dataset is described in Table 1.

A total of 24313 images of three plants having 12 leaf diseases along with 3 classes with healthy leaves have been used for the research work. Figure 1 displays sample diseased images of the dataset.

3.2 Data Preprocessing

Before these images are fed to CNN models, they are preprocessed and normalized to the same size and remove noises. These images are resized into 224*224 so that they are of the same sizes as the input of the proposed model. Figure 2 represents the workflow of the proposed work.

Table 1. Dataset for plant disease classification

Disease Type	Class Name	Number of Images	Total Training Image	Total Testing Image	Total Validation Image
Pepper_Bacterial spot	Pepper_BS	1000	800	100	100
Pepper_Healthy	Pepper_H	1478	1182	148	148
Potato_Early blight	Potato_EB	1000	800	100	100
Potato_Late blight	Potato_LB	1000	800	100	100
Potato_healthy	Potato_H	1000	800	100	100
Tomato_Early blight	Tomato_EB	1000	800	100	100
Tomato_Bacterial spot	Tomato_BS	2127	1702	213	212
Tomato_Late blight	Tomato_LB	1909	1527	191	191
Tomato_Leaf mold	Tomato_LM	1000	800	100	100
Tomato_mosaic virus	Tomato_MV	1000	800	100	100
Tomato_septoria leaf spot	Tomato_SLS	1771	1417	177	177
Tomato_target spot	Tomato_TS	1404	1123	140	141
Tomato_two spotted spider mite	Tomato_TSSM	1676	1341	168	167
Tomato_Yellow Leaf Curl Virus	Tomato_YLCV	5357	4285	536	536
Tomato_healthy	Tomato_H	1591	1273	159	159
Total Images		24313	19450	2432	2431

3.3 Convolutional Neural Network

A convolution Neural Network (CNN), one of the types of deep learning frameworks, is used for object detection and classification in vision tasks. It consists of a number of layers which are densely connected that extracts the features automatically. In these layers, different size filters are used for extraction of features like color, edge, curve, etc. After each layer, there is a pooling layer that is used to reduce the dimension of feature space so that computation is decreased. At last, this feature vector is fed to a fully connected layer that is connected with the output layer. This whole process is automated where features are extracted automatically unlike machine learning where feature engineering is done manually which is time-consuming. With a large dataset, machine learning generally does not give good results. Deep Learning models outperform machine learning models so we have used this approach in the proposed research. Transfer learning is an application of a model trained on different tasks to a similar task like object classification.

A well-known CNN-based Deep Learning architecture named EfficientNet B0 [21] has been used in the proposed work. This model has been selected based on its great results on ImageNet Large Scale Visual Recognition Challenge ILSVRC [22].

EfficientNet B0 Architecture

EfficientNet was proposed by Google and has eight models ranging from B0 to B7. The accuracy of these models increases gradually while the number of parameters increases slowly. The objective of EfficientNet architecture is to provide good accuracy results with appropriate scaling of width, depth of the deep network, and improvement in resolution of an image. This architecture uses compound scaling by incorporating scaling of depth, width, and image resolution so that their appropriate dimensions are determined. These dimensions can be calculated by the following formulas.

$$\begin{aligned} d &= \alpha^{\phi}(depth) \\ w &= \beta^{\phi}(width) \\ r &= \gamma^{\phi}(resolution) \\ \text{Such that } \alpha.\beta^2.\gamma^2 &\approx 2 \text{ and } \alpha \geq 1,\ \beta \geq 1,\ \gamma \geq 1 \end{aligned} \quad (1)$$

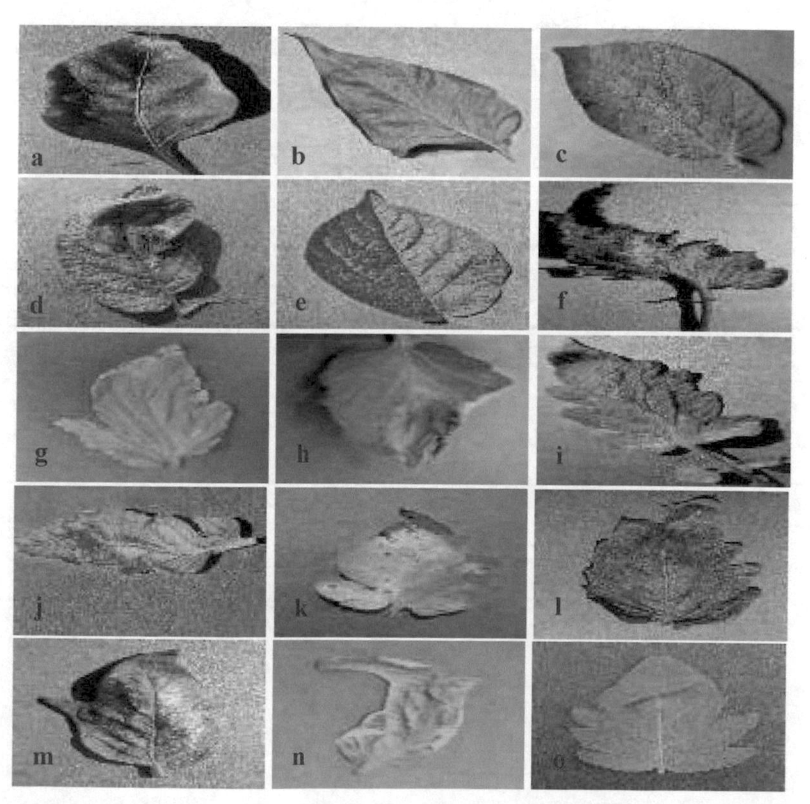

Fig. 1. a) Pepper_BS, b) Pepper_H, c) Potato_EB, d) Potato_LB, e) Potato_H, f) Tomato_EB, g) Tomato_BS, h) Tomato_LB, i) Tomato_LM, j) Tomato_MV, k) Tomato_SLS, l) Tomato_TS, m) Tomato_TSSM, n) Tomato_YLCV, o) Tomato_H

where α, β, *and* γ are the constants. Among all these parameters, the value of ϕ can be used to determine the optimum dimension of width, depth and resolution. Research work proposed in [21] found the optimum value of α, β, and γ as 1.2, 1.1, and 1.15 respectively for EfficientNet B0 architecture. Inverted bottleneck MBConv is the basic building block of this architecture that was presented in MobileNetV2 [23]. In this layer, channels are expanded and compressed. Parameters of EfficientNet B0 have been shown in Table 2. EfficientNct B0 architecture is shown in Fig. 3.

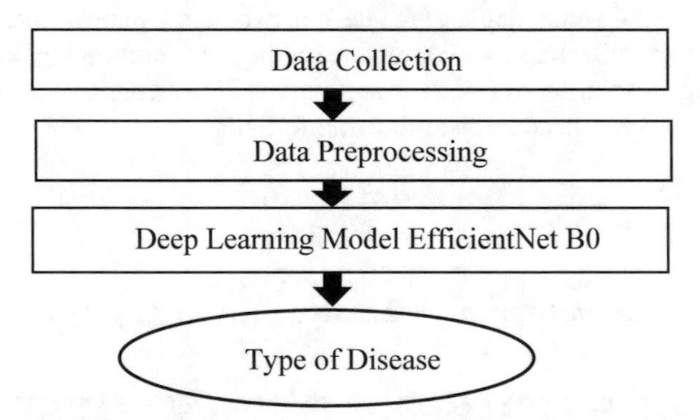

Fig. 2. Proposed methodology of research work

Table 2. Parameter of EfficientNet B0 architecture

Stage i	Operator *fi*	Resolution $\widehat{H} \times \widehat{W}$	#Channels $\widehat{C}\ i$	#Layers $\widehat{L}_i$
1	Conv3 × 3	224 × 224	32	1
2	MBConv1, k3 × 3	112 × 112	16	1
3	MBConv6, k3 × 3	112 × 112	24	2
4	MBConv6, k5 × 5	56 × 56	40	2
5	MBConv6, k3 × 3	28 × 28	80	3
6	MBConv6, k5 × 5	28 × 28	112	3
7	MBConv6, k5 × 5	14 × 14	192	4
8	MBConv6, k3 × 3	7 × 7	320	1
9	Conv1 × 1&Pooling&FC	7 × 7	1280	1

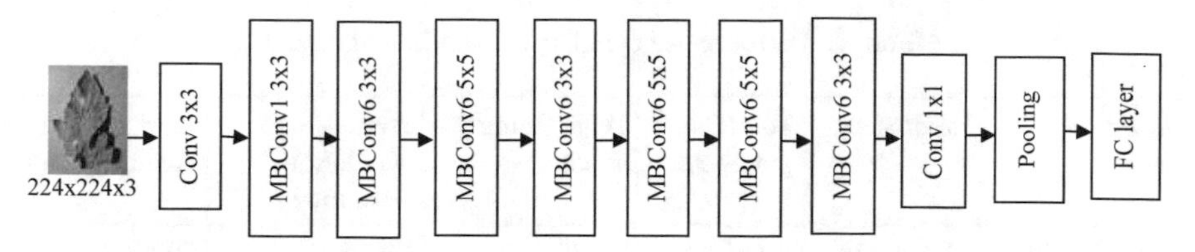

Fig. 3. EfficientNet B0 architecture for Plant Disease Classification

4 Results and Discussions

One popular deep learning model EfficientNet B0 has been proposed in research work to detect and classify the 12 diseases of pepper, potato, and tomato plants along with the healthy images of these plants. We have used the transfer learning approach in the EfficientNet B0 model, trained on ImageNet data and now are trained on a new dataset for our research work. This model has been trained with the learning rate of 0.001 and batch size of 32. Adam optimizer has been used with a momentum of 0.9 and 30% dropout has been used along with L1 regularizer in the model. Rectified linear unit (ReLU) function has been used as an activation function in the architecture. Adam is a combination of Stochastic Gradient Descent and RMSprop with momentum. It scales the learning rate using squared gradients and uses the momentum to move the average of gradients rather than gradients.

The proposed work has been implemented using Google Colaboratory Pro which offers GPU & TPU support. A total of 80%, 10%, and 10% data were used for training, validation, and test data respectively. We have used the early stopping with 3 as a patience value and defined the maximum epoch as 50. Table 3 summarizes the various hyperparameters used in the proposed research work.

Table 3. Hyperparameters used for the proposed model

Model	Input Size	Optimizer	Epoch	Learning Rate	Batch Size	Momentum	Dropout	Activation function
EfficientNet B0	224 × 224	Adam	50	0.001	32	0.9	0.30	ReLU

Table 4 summarizes the performance comparison of different models with EfficientNet B0 on the dataset. As seen from the results, the EfficientNet B0 model has both training and testing accuracy of 99.92% and 99.79% which are higher than that of other models.

Figure 4 displays the accuracy and loss curve of the EfficientNet B0 model for validation and training data. It is evident from the figures that EfficientNet B0 gives approx. 99.92 and 99.81 percent accuracy on training and validation data respectively for 15 disease types. Initially, both training loss and validation loss were large but these losses decreased with an increase in the epoch. The minimum validation loss occurs at epoch = 30. The result of the proposed model has been analyzed using a confusion matrix

Table 4. Performance comparison of different models

Model	Input size	Total parameter	Avg. Training Accuracy (%)	Avg. Validation Accuracy (%)	Avg. Testing Accuracy (%)
DenseNet121	224 × 224	8.1M	99.70	99.75	99.59
DenseNet169	224 × 224	14.3M	99.82	99.67	99.63
DenseNet201	224 × 224	20.2M	99.84	99.83	99.75
EfficientNet B0 (Proposed Model)	224 × 224	5.3M	99.92	99.81	99.79

and shown in Fig. 5. One image in Pepper_H, one in Tomato_EB, one in Tomato_SLS, and two images in Tomato_TS have been misclassified as Pepper_BS, Tomato_SLS, Tomato_MV, and Tomato_TSSM respectively. Out of 15 classes, four classes have a total of 5 samples that have been misclassified. For the remaining classes, the model classifies all images correctly.

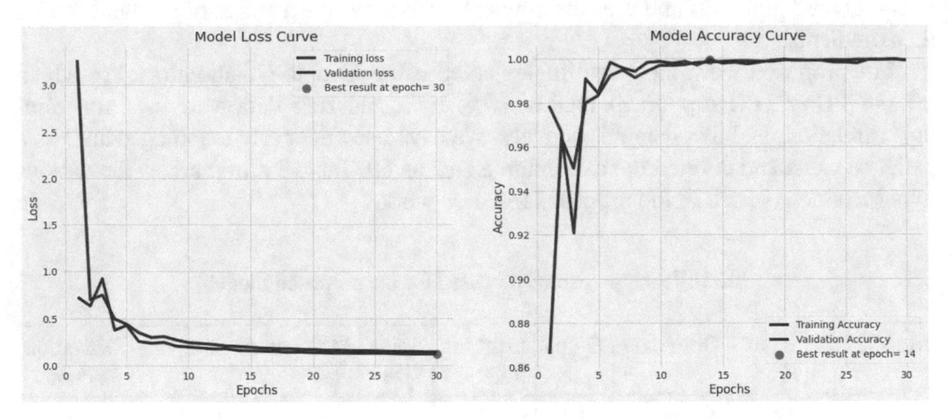

Fig. 4. Model loss and accuracy curve for EfficientNet B0 model

For object detection and classification tasks, many performance metrics are used. Some of them are precision, recall, F_1 score, and classification accuracy and calculated using Eqs. (2–5). Four important terminologies like true positive (TP), false positive (FP), true negative (TN), and false negative (FN) are used for the calculation of these performance metrics.

A true positive gives the number of correctly classified positive examples. True negative counts the total number of correctly classified negative samples. False positive refers to the total number of misclassified positive examples. False negative gives the total number of misclassified negative examples.

Precision gives the ratio of correctly classified positive examples to total positive examples. The recall is the measure of actual positives that have been correctly classified.

F_1 score is the function of precision and recall. It is the harmonic mean of precision and recall. A confusion matrix is a NxN matrix that summarises the all predicted results for a classification model where total number of classes is denoted by N. The confusion matrix for the presented model has been displayed in Fig. 5. Table 5 summarizes the recall, precision and F_1 score of the proposed model.

$$\text{Precision(i)} = \frac{\#TP(i)}{\#TP(i) + \#FP(i)} \tag{2}$$

$$\text{Recall(i)} = \frac{\#TP(i)}{\#TP(i) + \#FN(i)} \tag{3}$$

$$\text{F1 score} = \frac{2 \times Precision \times Recall}{Precision + Recall} \tag{4}$$

$$\text{Classification accuracy} = \frac{\#TP(i) + TN(i)}{\#TP(i) + \#FP(i) + TN(i) + FN(i)} \tag{5}$$

where i is the number of classes.

From Table 5, it can be observed that EfficientNet B0 has a precision value of 0.99 for Pepper_BS, Tomato_MV, Tomato_SLS, and Tomato_TSSM. Similarly, a recall value of 0.99 is there for Pepper_H, Tomato_EB, Tomato_SLS, and Tomato_TS. As seen in Table 4, through comparative analysis with DenseNet121, DenseNet169, and DenseNet201, it is observed that EfficientNet B0 has better classification accuracy for these 15 disease types which proves the utility of the EfficientNet B0 model.

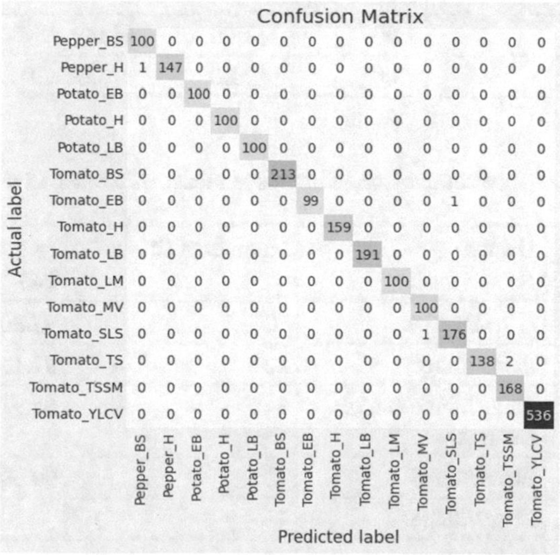

Confusion Matrix

Actual label \ Predicted label	Pepper_BS	Pepper_H	Potato_EB	Potato_H	Potato_LB	Tomato_BS	Tomato_EB	Tomato_H	Tomato_LB	Tomato_LM	Tomato_MV	Tomato_SLS	Tomato_TS	Tomato_TSSM	Tomato_YLCV
Pepper_BS	100	0	0	0	0	0	0	0	0	0	0	0	0	0	0
Pepper_H	1	147	0	0	0	0	0	0	0	0	0	0	0	0	0
Potato_EB	0	0	100	0	0	0	0	0	0	0	0	0	0	0	0
Potato_H	0	0	0	100	0	0	0	0	0	0	0	0	0	0	0
Potato_LB	0	0	0	0	100	0	0	0	0	0	0	0	0	0	0
Tomato_BS	0	0	0	0	0	213	0	0	0	0	0	0	0	0	0
Tomato_EB	0	0	0	0	0	0	99	0	0	0	0	1	0	0	0
Tomato_H	0	0	0	0	0	0	0	159	0	0	0	0	0	0	0
Tomato_LB	0	0	0	0	0	0	0	0	191	0	0	0	0	0	0
Tomato_LM	0	0	0	0	0	0	0	0	0	100	0	0	0	0	0
Tomato_MV	0	0	0	0	0	0	0	0	0	0	100	0	0	0	0
Tomato_SLS	0	0	0	0	0	0	0	0	0	0	1	176	0	0	0
Tomato_TS	0	0	0	0	0	0	0	0	0	0	0	0	138	2	0
Tomato_TSSM	0	0	0	0	0	0	0	0	0	0	0	0	0	168	0
Tomato_YLCV	0	0	0	0	0	0	0	0	0	0	0	0	0	0	536

Fig. 5. Confusion matrix for EfficientNet B0 model

Performance comparison of the presented model has been done with different CNN based models and shown in Table 6. With more classes, the proposed model has better testing accuracy than these CNN based models.

Table 5. Precision, recall, and F_1 score of EfficientNet B0 for 15 disease types.

	Disease Type	EfficientNet B0		
		Precision	Recall	F_1 score
1	Pepper_BS	0.99	1.00	1.00
2	Pepper_H	1.00	0.99	1.00
3	Potato_EB	1.00	1.00	1.00
4	Potato_H	1.00	1.00	1.00
5	Potato_LB	1.00	1.00	1.00
6	Tomato_BS	1.00	1.00	1.00
7	Tomato_EB	1.00	0.99	0.99
8	Tomato_H	1.00	1.00	1.00
9	Tomato_LB	1.00	1.00	1.00
10	Tomato_LM	1.00	1.00	1.00
11	Tomato_MV	0.99	1.00	1.00
12	Tomato_SLS	0.99	0.99	0.99
13	Tomato_TS	1.00	0.99	0.99
14	Tomato_TSSM	0.99	1.00	0.99
15	Tomato_YLCV	1.00	1.00	1.00
Accuracy		1.00		
Macro Avg.		1.00	1.00	1.00
Weighted Avg.		1.00	1.00	1.00

Table 6. Performance comparison of proposed model with other CNN based models

Model	Dataset	Number of Classes	Avg. Testing Accuracy (%)
Custom Model [16]	Own Dataset	2	97.30
DenseNet121 [18]	PlantVillage	10	97.11
VGG with Xgboost [19]	PlantVillage & Corn Dataset	11	95.71
DenseNet121	PlantVillage	15	99.59
DenseNet169	PlantVillage	15	
DenseNet201	PlantVillage	15	
EfficientNet B0 (Proposed Model)	PlantVillage	15	99.79

5 Conclusions

The study incorporates the novel deep learning architecture for automated early diagnosis and an accurate image-based classification approach. The proposed work has been carried out for identification and classification of 12 diseases of pepper, potato, and tomato plants along with healthy images of these plants, and these images used have been taken from the PlantVillage dataset. The proposed EfficientNetB0 models were implemented using the transfer learning approach and achieved an accuracy of 99.79% without data augmentation. The authors believe that the proposed EfficientNet B0 model can be utilized as a tool for identifying plant diseases and assisting farmers in the field. In the future, presented research work can be extended to enhance the robustness of the model by incorporating more disease-infected images for various plants at different disease severity levels.

References

1. Singh, V., Misra, A.K.: Detection of plant leaf diseases using image segmentation and soft computing techniques. Inf. Process. Agric. **4**(1), 41–49 (2017)
2. Khirade, S.D., Patil, A.B.: Plant disease detection using image processing. In: 2015 International Conference on Computing Communication Control and Automation, pp. 768–771. IEEE (2015)
3. Ale, L., Zhang, N., Wu, H., Chen, D., Han, T.: Online proactive caching in mobile edge computing using bidirectional deep recurrent neural network. IEEE Internet Things J. **6**(3), 5520–5530 (2019)
4. Ale, L., Fang, X., Chen, D., Wang, Y., Zhang, N.: The lightweight deep learning model for facial expression recognition. In: 2019 18th IEEE International Conference On Trust, Security And Privacy In Computing And Communications/13th IEEE International Conference On Big Data Science And Engineering (TrustCom/BigDataSE), pp. 707–712. IEEE (2019)
5. Mohanty, S.P., Hughes, D.P., Salathé, M.: Using deep learning for image-based plant disease detection. Front. Plant Sci. **7**, 1419 (2016)
6. Krizhevsky, A., Sutskever, I., Hinton, G.E.: Imagenet classification with deep convolutional neural networks. Adv. Neural Inf. Process. Syst. **25** (2012)
7. Simonyan, K., Zisserman, A.: Very deep convolutional networks for large-scale image recognition. arXiv preprint arXiv:1409.1556 (2014)
8. Szegedy, C., et al.: Going deeper with convolutions. In: Proceedings of the IEEE Conference on Computer Vision and Pattern Recognition, pp. 1–9 (2015)
9. He, K., Zhang, X., Ren, S., Sun, J.: Deep residual learning for image recognition. In: Proceedings of the IEEE Conference on Computer Vision and Pattern Recognition, pp. 770–778 (2016)
10. Huang, G., Liu, Z., Van Der Maaten, L., Weinberger, K.Q.: Densely connected convolutional networks. In: Proceedings of the IEEE Conference on Computer Vision and Pattern Recognition, pp. 4700–4708 (2017)
11. Sladojevic, S., Arsenovic, M., Anderla, A., Culibrk, D., Stefanovic, D.: Deep neural networks-based recognition of plant diseases by leaf image classification. Comput. Intell. Neurosci. **2016** (2016)
12. Lu, Y., Yi, S., Zeng, N., Liu, Y., Zhang, Y.: Identification of rice diseases using deep convolutional neural networks. Neurocomputing **267**, 378–384 (2017)

13. Brahimi, M., Boukhalfa, K., Moussaoui, A.: Deep learning for tomato diseases: classification and symptoms visualization. Appl. Artif. Intell. **31**(4), 299–315 (2017)
14. Wang, G., Sun, Y., Wang, J.: Automatic image-based plant disease severity estimation using deep learning. Comput. Intell. Neurosci. **2017** (2017)
15. Rangarajan, A.K., Purushothaman, R., Ramesh, A.: Tomato crop disease classification using a pre-trained deep learning algorithm. Procedia Comput. Sci. **133**, 1040–1047 (2018)
16. Nigam, S., et al.: Automating yellow rust disease identification in wheat using artificial intelligence (2021)
17. Nigam, S., Jain, R.: Plant disease identification using deep learning: a review. Ind. J. Agric. Sci. (2019)
18. Abbas, A., Jain, S., Gour, M., Vankudothu, S.: Tomato plant disease detection using transfer learning with C-GAN synthetic images. Comput. Electron. Agric. **187**, 106279 (2021)
19. Hassan, S.M., Jasinski, M., Leonowicz, Z., Jasinska, E., Maji, A.K.: Plant disease identification using shallow convolutional neural network. Agronomy **11**(12), 2388 (2021)
20. Hughes, D., Salathé, M.: An open access repository of images on plant health to enable the development of mobile disease diagnostics. arXiv preprint arXiv:1511.08060 (2015)
21. Tan, M., Le, Q.: Efficientnet: Rethinking model scaling for convolutional neural networks. In: International Conference on Machine Learning, pp. 6105–6114. PMLR (2019)
22. Russakovsky, O., et al.: Imagenet large scale visual recognition challenge. Int. J. Comput. Vision **115**(3), 211–252 (2015)
23. Sandler, M., Howard, A., Zhu, M., Zhmoginov, A., Chen, L.C.: Mobilenetv2: inverted residuals and linear bottlenecks. In: Proceedings of the IEEE Conference on Computer Vision and Pattern Recognition, pp. 4510–4520 (2018)

Opinion-Based Machine Learning Approach for Fake News Classification

Poonam Narang[1](✉), Ajay Vikram Singh[1], and Himanshu Monga[2]

[1] AIIT, Amity University, Noida, India
hipoonam@gmail.com

[2] Jawaharlal Nehru Government Engineering College, Sundernagar, India

Abstract. Fake news detection is intended to prevent malicious news from spreading and keep panic from people, which is generated by fake news. So, the importance of detecting fake news can be appreciated undoubtedly. Researchers have tried many ways to classify fake news, like machine learning and deep learning algorithms. In this research, authors have experimented with machine learning algorithms for the binary classification of fake news based on a novel opinion-mining approach. Authors have tried Decision Tree, Support Vector Machine, and K-Nearest Neighbour and have shown that sentiment analysis and these algorithms generate better accuracies than state-of-the-art methods. Authors have demonstrated that the Decision tree causes the highest accuracy on the LIAR dataset, which is 14% more than the approaches mentioned in the existing literature.

Keywords: Fake news · machine learning · opinion mining · decision tree · K-Nearest Neighbour · Support Vector Machine · Social Media

1 Introduction

Social media is a vital aspect of people's daily life, allowing them to express themselves, get information, and communicate. The advancement of technology has permitted quicker and more convenient access to accurate and falsified data, posing a complex problem. Most people feel that the knowledge they acquire from social networking sites is valid and correct, i.e., people are intrinsically biased toward the truth. Additionally, people readily believe and trust what they imagine in their brains. Disseminating such false information has a profoundly damaging effect on the targeted individuals and larger society. In the spring of 2018, an article accompanied by a video claiming that "Cadbury chocolate is tainted with HIV-Positive Blood." This message garnered popularity on Facebook, particularly in South Asian nations like India. The spread of rumors harmed Cadbury's reputation, and even individuals who independently verified the veracity of the stories were afraid to purchase the chocolates [1]. In this regard, researchers are keen to find the truth of social networking information using an automated method. Previous research has demonstrated that authors' writing styles and linguistics can sometimes deduce their personalities and detect dishonesty and misleading statements. Considering this, Zhou et al. [18] have shown some patterns of writing style which can be significantly tilted towards fake news.

H. Sharma et al. (Eds.): ICIVC 2022, PALO 17, pp. 33–42, 2023.
https://doi.org/10.1007/978-3-031-31164-2_4

The nature of fake news which can be interpreted as real time social media news makes it much more challenging to detect fake online news [2]. Most misleading stories are designed to induce confusion and mistrust among readers. Such issues prompted academics to investigate automated methods for determining the ground truth values of bogus text based on article text uploaded on social sites. For this reason, academia and industry persons are collaboratively trying to find out various ways to identify the problem automatically. One solution to this problem is fact-checking conducted by experts. Researchers have also determined that an automated system can identify false news stories more accurately than humans. Automated systems can be valuable for recognize fake news blogs, stories and articles that sway public perception on social networking sites [3].

The main motive of automatic fake news detection is to save humans effort and time in detecting deceptive news and assisting us in preventing its spread. With the advancement of subfields of Computer Science such as Natural Language Processing, Machine Learning and Data Mining the challenge of detecting false news has been investigated from numerous perspectives. In this research, authors have proposed an automated detection technique of fake news from the view point of Natural Language Processing. Opinion mining is a part of Natural Language Processing (NLP) that designs and implements methods, procedures and models to evaluate if a content contains subjective or objective information and, if subjective, if it is positive, neutral, negative, or weak. Opinion Mining is a synonym for Sentiment Analysis. Dickerson et al. [4] proved that human account was sufficient for distinguishing sentiment-related behavior by incorporating multiple sentiment indicators. According to Nejad et al. [5], Sentiment analysis is a crucial way to detect spam and fabricated news.

In this research, authors have proposed a fake news classification method based on sentiment score, a binary classification issue where news is either fake or reliable. This paper's contents are structured as follows: Sect. 2 discusses pertinent work on Fake News and opinion mining for fake news detection. Section 3 describes the description of dataset used in proposed work. Section 4 discusses the supervised machine learning used along with evaluation measures adopted in the research. Section 5 discusses the proposed method and Sect. 6 discusses the results and followed by the conclusion and future work.

The primary contributions of the research can be summed up as follows:

1. This research proposes an approach based on the Sentiment score based fake news classification using three supervised ML (machine learning) algorithms: K- nearest neighbor, Support Vector Machine and Decision tree (DT). The proposed method achieves state-of-the-art performance for the task of fake news identification.
2. The proposed approach has been evaluated on the LIAR dataset and comparison of the results with the existing work has been mentioned in the literature.
3. Feature Selection is a tedious task and takes most of the classification time. Authors have reduced this effort as the proposed method does not need a feature selection task for fake news detection.

2 Related Work

There are numerous methods for detecting fake news. Several people have performed extensive research to clarify this topic to avoid confusion caused by false information. In this part, we discuss prior work on fake news identification techniques, such as rumor detection in news articles and fake news identification.

2.1 Fake News Detection

From an artificial intelligence-based perspective, researchers have studied numerous aspects of the reliability of internet information. Oshikawa et al. [6] investigated the scientific hurdles in identifying fabricated news and machine learning algorithms applied by reseachers can solve multiple tasks. They emphasized the alignment of false news detection task using natural language processing. Faustini et al. [7] trained five datasets in three languages using Nave Bayes, Random Forest, Support Vector Machine(SVM), and K-Nearest Neighbors. Regarding accuracy and F1-Score, they compared the results acquired by a customized collection of features, Word2Vec, bag-of-words and Document-class Distance. Ultimately, they determined that bag-of-words produced the most significant results and SVM and Random Forest performed better than other algorithms.

Jain et al. [8] use Nave Bayes to distinguish between real and fake news. By merging news content and social environment variables. Vedova et al. [9] introduced a unique machine learning false news detection algorithm that improves accuracy by as much as 4.8%. Rasool et al. [10] suggested a unique method for multiclass, multilevel identification of fake news by relabeling of dataset and iterative learning. The strategy is evaluated by applying various supervised machine learning algorithms, such as decision tree and SVM. Their technique exceeded the accuracy in stat of the art with an increase of 39.5% on the LIAR dataset.

2.2 Opinion Mining for Fake News Detection

Researchers have proposed various techniques for opinion mining-based fake news detection. De Seouza et al. [11] analyzed the many features linked to detection of fake news and concluded that sentiment analysis was a beneficial feature for rapidly verifying the veracity of social network content. Sharma et al. [12] addressed false news detection and mitigation solutions emphasizing different computational algorithms to solve these tasks, assembled available data sets related to deceptive news identification, and offered open topics and challenges related to it.

Turney et al. [13] introduced lexicon-based approaches which followed the strategy of mapping subjective scores to words by generating semantic orientation dictionaries.The total word scores might then be summed together to estimate the opinion of a given information. Qurishi et al. [14] employed sentiment analysis to identify doubtful content in Arabic Twitter fraudulent or malicious information. They believed that sentiment measured user behavior, leading to high precision in credibility studies. Bhutani et al. [15] proposed fake news detector based on sentiment analysis, assuming an article's sentiment would determine its authenticity. They utilized Naive-Bayes classification to assess text opinion and used it in Random Forest and Multinomial NaiveBayes classifier and to detect fake news.

3 Dataset Used for Fake News Detection

In this article, the LIAR dataset [16] has been used in which 12.8K human-labeled PolitiFact.com short statements were evaluated for honesty using six labels: "pants-fire," "false," "barely-true," "half-true," "mostly-true," and "true." Each statement's author's meta-data is also provided. TV advertising,News releases, campaign speeches, television and radio interviews, Facebook posts, debates and, Twitter posts are used. The immigration, state budget, economy, taxation, healthcare, education, jobs, federal budget, elections, and candidate biographies are frequently mentioned [17]. Figure 1 shows the visualization of the LIAR dataset with the help of word cloud representation which has been created using python language.

Fig. 1. Word Cloud for LIAR dataset

4 Methods Used for Detection of Fake News

Many ML techniques exist for false news identification. An opinion-based machine learning approach based on the K nearest neighbor, Support Vector Machine and Decision tree has been proposed in this paper for fake news classification. These supervised ML algorithms have been discussed in the current section.

5 Machine Learning Algorithms Used

Decision Tree: Decision Tree classifiers are applicable for both classification and regression. DT has a top-down structure which resembles a tree in which nodes can only be external nodes bound to decision nodes or a label class that is responsible for making decisions. The classifier forecasts the variable by studying the featured data and subdividing the domain into subregions. Multiple characteristics are split based on two criteria: one measure of error and the other information gain. Assume that discrete attribute M have distinct values and L_i is the set containing training dataset L with a value of i.

Gain ratio and information gain can be computed as:

$$Information\,gain = Entropy(L) - \sum\nolimits_{i=1}^{n} \frac{|L_i|}{|L|} Entropy(L_i) \quad (1)$$

$$Gain\,Ratio = \frac{Infromation\,Gain}{-\sum_{i=1}^{n} \frac{|L_i|}{|L|} log \frac{|L_i|}{|L|}} \quad (2)$$

A decision tree is a technique for Supervised Machine Learning in which the data is continually separated based on a particular parameter. It is employed when the result is a discrete variable, such as "true" or "false." Decision trees utilize the Divide-and-Conquer Algorithm. Decision Trees are quick at classifying records, remove unrelated characteristics, and offer a comparable level of accuracy to other classification algorithms across a variety of data sets.

Support Vector Machine: The SVM uses a principle of Structural Risk Minimization. The supporrt vector machine model represents feature vectors extracted from documents of news stories as points in feature space. Instances are separated using a hyperplane in a binary classification problem, so that

$$s^T x + b = 0 \quad (3)$$

here n represents support vectors, s is a coefficient weight dimension vector that is normal to the hyperplane, and x represents the biased term b that is the offset values from data points and origin.

The Lagrangian function can be used to solve w in linear situations. But In the case of nonlinearity, the decision function and kernel trick of w, b and n are stated as follows in Eq. 4:

$$f(x) = \left\{ \sum\nolimits_{j=1}^{n} \alpha_j y_j k(x_i, x) + b \right\} \quad (4)$$

It is characterised by a "best" separating hyperplane and is also known as a discriminative classifier. For binary classification, SVM is utilised. It is used in situations when the output is either true or false. In our research, we employ a linear kernel. Kernels are useful for fitting instances of data that are difficult to separate and/or multidimensional.

K Nearest Neighbour: The k-nearest neighbour (KNN) is one of the most straightforward classification techniques. It first identifies this sample's k nearest neighbours based on a distance measure. A distance metric is crucial to the k-nearest neighbour classifier. The classification will be superior the better this metric reflects label similarity. The most prevalent option is the Minkowski distance. The formula for calculating the Minkowski distance of order p (where p is an integer) between two points is:

$$Dis(A, B) = \left[\sum_{a=1}^{d} |x_a - y_a|^p\right]^{\frac{1}{p}} \tag{5}$$

Here A = (x1, x2, x3,..........xn) and A = (y1, y2, y3,..........yn) belongs to set $\mathbb{R}^n$

Classification using KNN can be utilized well as a fault detection method. This technique is also capable of detecting deceptive news on social media. K-Nearest Neighbor is a simplistic strategy for constructing a classification model that assigns the class label to instances with a problem. KNN is utilized in recommendation systems, semantic search, detection, and anomaly detection.

5.1 Evaluation Measure

Four well-known evaluation metrics, recall, precision, F-measure, and Accuracy, were employed to assess the achievement of the experimental approach.

Precision: Regarding positive observations, precision is the proportion of accurately anticipated words to all predicted positive statements.

$$Precision(PR) = trueP/(trueP + falseP) \tag{6}$$

Accuracy: The most straightforward performance metric is just the proportion of adequately predicted observations to total observations.

$$Accuracy = trueP + trueN / (trueP + falseP + falseN + trueN) \tag{7}$$

Recall: The ratio of accurately predictable positive observations to total actual class observations is known as recall (RC).

$$RC = trueP/(trueP + falseN) \tag{8}$$

F1 Score: The F1 Score is determined by the weighted average of RC and PR.

$$F1\,Score = 2 * (RC * PR)/(RC + PR) \tag{9}$$

Here trueP represents true positive, trueN shows true negative, falseP represents false positive, and falseN represents false negative.

6 Proposed Model for Fake News Classification

In this paper, Fake News detection on the LIAR dataset is classified using Sentiment Score and ML algorithms to organize Fake News labeled with True and False. This research is a novel approach applied to various machine learning algorithms on sentiment scores. In this paper, researchers tried to compare different Machine Learning Algorithms applied to the Sentiment score of Fake news texts.

The proposed fake news classification method is applied to the LIAR Dataset. In this dataset, 10269 Training samples and 1283 testing samples are given. In this research, all training and testing samples are taken for the training and testing purpose of Machine Learning Algorithms.

The Classification here is binary. The True, half-True, and mostly-true are taken as class 1(True), and False, barely-true and pants-fire are taken as class 0 (False). The selected ML algorithms are DT, SVM, and KNN.

Figure 2 describes the algorithm used in this research. It shows the flow diagram where the text from the selected dataset has been taken and processed using the JAVA program to tokenize the words of the text. Then, the text is classified using sentiment analysis and machine learning algorithms.

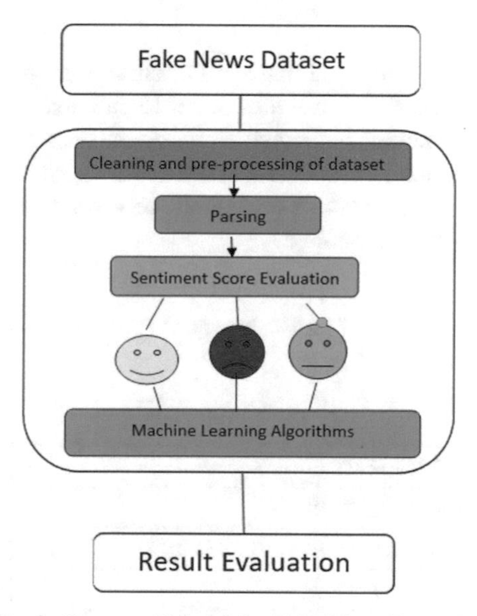

Fig. 2. Proposed Model for Fake News Detection

Decision tree is chosen as it is simple to understand and interpret. It requires little data preparation and can be applied to numeric data, which is the requirement of this research. Decision tree has a logarithmic time complexity with the training data, which is helpful for the large-sized training dataset.

Support Vector machines work well for high dimensional data. Here in the proposed method, three-dimensional data are classified in two-way classification.

Another machine learning algorithm is tested on the sentiment score of the LIAR dataset, which is the K-nearest neighbor, which also has its advantages. It is simple, intuitive, and valuable for non-linear decision boundaries.

These three algorithms are chosen by keeping in mind the usefulness of the machine learning algorithms.

7 Results and Discussions

The results of the applied algorithms are described in Table 1, which shows the performance measures of used machine learning approaches in terms of accuracy, precision, recall and F1- score.

Table 1. Performance measures of Supervised machine learning algorithms

Algorithm	Accuracy	Precision	Recall	F1-Score
Decision Tree	0.5211	0.4142	0.5409	0.4692
Support Vector Machine	**0.4806**	**0.4140**	0.7888	0.5430
KNN	**0.5059**	0.4129	0.6228	0.4965

From Table 1, it can be observed that the Decision Tree gives more accuracy than SVM and KNN. In contrast, SVM provides better Recall and F1-Score.

The literature review noted that previous work on fake news classification using sentiment score is unavailable. In this view, it is a novel approach to the simulated news classification task. But there are some works public on sentiment analysis for Fake news. A comparison table is shown in Table 2.

Table 2. Comparison of Accuracy of LIAR dataset of Proposed Methods with state-of-the-art methods

Algorithm	Accuracy
Hybrid CNN (Wang et al., 2017 [17])	0.27
Multi-Layered SVM (Rasool et al., 2019 [10])	0.39
Multi-Layered Decision Tree (Rasool et al., 2019 [10])	0.39
Random Forest (Bhutani et al., 2019 [15])	0.37
M-Naïve Bayes (Bhutani et al., 2019 [15])	0.25
Proposed SVM	0.48
Proposed Decision Tree	0.52
Proposed KNN	0.51

From the comparison in Table 2, it can be inferred that the proposed approaches have better accuracy than state-of-the-art methods. The earlier strategies mainly work on texts, whereas the proposed method works on sentiment scores. The proposed approach provides greater precision than earlier approaches of machine learning techniques. Bhutani et al., 2019 [15] have taken cosine similarity, tf-IDF, and sentiment score together for classification purposes. The proposed system has selected positive, negative, and neutral sentiment scores for classification purposes. The methods described in this paper have higher accuracy than similar state-of-the-art methods.

8 Conclusion

Fake news identification is the most significant topics in digital age. Deceptive news detection using opinion mining has not been explored much. In this paper, the authors have compared different ML algorithms on the sentiment of fake news. Authors have compared their methods with state-of-the-art methods and shown the advantages. There are ample algorithms available in machine learning. The authors suggest future work for various ML algorithms applied to fake news datasets. In this research, Liar datasets have been tested. The same or extended version of algorithms can be used for different datasets available for Fake News.

References

1. TruthorFiction.. Cadbury Chocolate Eggs Are Infected With HIV-Positive Blood-Fiction (As per Springer style, both city and country names must be present in the affiliations. Accordingly, we have inserted the country name in the affiliation. Please check and confirm if the inserted country name is correct. If not, please provide us with the correct country name.). https://www.truthorfiction.com/cadbury-products-infected-hiv-blood/
2. Zhang, X., Ghorbani, A.A.: An overview of online fake news: Characterization, detection, and discussion. Inf. Process. Manag. **57**(2), 102025 (2020)
3. Jain, P., Kumaraguru, P.: In: Proceedings of the 3rd IKDD Conference on Data Science, 2016, p. 6. ACM (2016)
4. Dickerson, J.P., Kagan, V., Subrahmanian, V.S.: Using sentiment to detect bots on twitter: are humans more opinionated than bots? In: Wu, X., Ester, M., Xu, G. (eds.) Proceedings of the 2014 IEEE/ACM International Conference on Advances in Social Networks Analysis and Mining, ASONAM 2014, Beijing, China, 17–20 August 2014, pp. 620–627. IEEE Computer Society. Washington, DC (2014)
5. Nejad, S.J., Ahmadi-Abkenari, F., Bayat, P.: Opinion spam detection based on supervised sentiment analysis approach. In: 2020 10th International Conference on Computer and Knowledge Engineering (ICCKE), 29 October 2020, pp. 209–214. IEEE (2020)
6. Oshikawa, R., Qian, J., Wang, W.Y.: A survey on natural language processing for fake news detection. In: Calzolari, N., et al. (eds.) Proceedings of the 12th Language Resources and Evaluation Conference, LREC 2020, Marseille, France, 11–16 May 2020, pp. 6086–6093. European Language Resources Association, Paris (2020)
7. Faustini, P.H.A., Covões, T.F.: Fake news detection in multiple platforms and languages. Expert Syst. Appl. **158**, 113503 (2020)
8. Jain, A., Kasbe, A.: Fake news detection. In: 2018 IEEE International Students' Conference on Electrical, Electronics and Computer Science (SCEECS), Bhopal, pp. 1–5 (2018)

9. Della Vedova, M.L., Tacchini, E., Moret, S., Ballarin, G., DiPierro, M., de Alfaro, L.: Automatic online fake news detection combining content and social signals. In: Proceedings of 22nd Conference on Open Innovation Association (FRUCT), pp. 272–279 (2018)
10. Rasool, T., Butt, W.H., Shaukat, A., Akram, M.U.: Multi-label fake news detection using multi-layered supervised learning. In: Proceedings of 11th International Conference on Computation and Automation Engineering (ICCAE), pp. 73–77 (2019)
11. de Souza, J.V., Gomes Jr., J., Souza Filho, F., Oliveira Julio, A., de Souza, J.F.: A systematic mapping on automatic classification of fake news in social media. Soc. Netw. Anal. Min. **10**(1), 1–21 (2020). https://doi.org/10.1007/s13278-020-00659-2
12. Sharma, K., Qian, F., Jiang, H., Ruchansky, N., Zhang, M., Liu, Y.: Combating fake news: a survey on identification and mitigation techniques. ACM Trans. Intell. Syst. Technol. **10**, 21:1-21:42 (2019)
13. Turney, P.D.: Thumbs up or thumbs down? semantic orientation applied to unsupervised classification of reviews. In: Proceedings of the 40th Annual Meeting of the Association for Computational Linguistics, Philadelphia, PA, USA, 6–12 July 2002, pp. 417–424. ACL, Stroudsburg (2002)
14. AlRubaian, M.A., Al-Qurishi, M., Al-Rakhami, M., Rahman, S.M.M., Alamri, A.A.: Multistage credibility analysis model for microblogs. In: Pei, J., Silvestri, F., Tang, J. (eds.) Proceedings of the 2015 IEEE/ACM International Conference on Advances in Social Networks Analysis and Mining, ASONAM 2015, Paris, France, 25–28 August 2015, pp. 1434–1440. ACM, New York (2015)
15. Bhutani, B., Rastogi, N., Sehgal, P., Purwar, A.: fake news detection using sentiment analysis. In: Proceedings of the 2019 Twelfth International Conference on Contemporary Computing, IC3 2019, Noida, India, 8–10 August 2019, pp. 1–5. IEEE, Piscataway (2019)
16. https://www.cs.ucsb.edu/~william/data/liar_dataset.zip
17. Wang, W.Y.: Liar, Liar Pants on Fire": a new benchmark dataset for fake news detection. In: Barzilay, R., Kan, M. (eds.) Proceedings of the 55th Annual Meeting of the Association for Computational Linguistics, ACL 2017, Vancouver, BC, Canada, 30 July–4 August 2017, vol. 2: Short Papers, pp. 422–426. Association for Computational Linguistics, Stroudsburg (2017)
18. Zhou, X., Zafarani, R.: A survey of fake news: Fundamental theories, detection methods, and opportunities. ACM Comput. Surv. (CSUR) **53**(5), 1–40 (2020)

Survey on IoMT: Amalgamation of Technologies-Wearable Body Sensor Network, Wearable Biosensors, ML and DL in IoMT

S. V. K. R. Rajeswari and P. Vijayakumar(✉)

SRM Institute of Science and Technology, Kattankulathur, Chennai 603203, Tamil Nadu, India
vijayakp@srmist.edu.in

Abstract. Boundless reports on the advancement in Internet of Things (IoT) and its applications are presented in recent times. Healthcare has become crucial in today's era. Healthcare is a novel approach to IoT i.e., the Internet of Medical Things (IoMT).There is a need for smart healthcare for better treatment. IoMT involves medical devices that are connected through the internet for health monitoring, therapy, emergency alerts and different applications. The proposed paper throws light on the technologies that are involved in an IoMT system right from the type of sensors, wearable biosensors, wearable body sensor network and the implementation of Machine learning and Deep learning in IoMT applications. By the end of the survey paper, a researcher would have an idea about the type of sensor system and prediction algorithms that has to be chosen to successfully implement in an IoMT system.

Keywords: Artificial Intelligence · Deep learning · IoMT · Machine learning · Wearable body sensor network · Wearable biosensors

1 Introduction

With the emerging technologies and recent trends, Internet Of Things (IoT) has taken a step ahead to enhance the lives of human beings. IoT is the seamless integration of intelligent interfaces of "things" that have virtual personalities, aspects and identities. The "things" in IoT can be gadgets, objects, animals or people. Depending on the everyday changing needs of people, IoT technologies incline towards evolving and updating the current system [1]. According to Gartner's chart, IoT is among the top 5 technologies [2]. IoMT is a healthcare extension of IoT where the things in IoMT are the human body. The human body is the source of medical information that can provide physical and psychological parameters. From the current health market statistics, "Internet of Medical Things (IoMT) market size" is estimated to be United States Dollars (USD) 142.45 billion by 2026, with a Compound Annual Growth Rate (CAGR) of 28.9% on account of the prevalence of chronic diseases [3]. IoMT is built on Wearable Body Sensor Network (WBSN).It is regarded as the key component of IoT in WBSN [4]. WBSN is the distributed system where the network is formed with sensor nodes connected to

H. Sharma et al. (Eds.): ICIVC 2022, PALO 17, pp. 43–55, 2023.
https://doi.org/10.1007/978-3-031-31164-2_5

each other through nodes on a human body. The nodes are connected following a bus, mesh or star topology [5] Wearable biosensors are the heart of any wearable body sensor network. A lot of data is generated in the process of collecting the sensor information while monitoring the health conditions. It has become vital to select a system with great computation power. Artificial Intelligence (AI) transpires as a solution. Implementing IoT and AI has become the breakthrough in the current technology. The main tool of AI is Machine Learning(ML). ML helps to make better decisions by machines in general which can imitate the natural language processing(NLP) of humans [6].

1.1 Contribution

In smart healthcare, for patients suffering from chronic diseases, patients in Intensive Care Unit (ICU), parameter health monitoring has become vital, especially if the disease turns out to be an emergency [7]. This survey paper aims to provide a deep analysis on state of the art and comparison with related works that would thrive in providing solution for future technologies when blending with IoMT. There is a need for survey work on AI and ML as both technologies have more to offer in the field of IoMT. In this paper,

- The structure of IoMT is described with the requirements and methodologies with the current state of the art and the comparison of different works.
- This survey would enlighten the readers about BSN (Body Sensor Network), which is an extension of IoT and Wearable Biosensor technology.
- The main contribution of this research paper lies in investigating, analyzing and concluding a specific methodology. This would be helpful for a researcher to choose a particular technology for their research.
- A brief study on the AI and ML, along with Deep Learning(DL) contributing to different applications is described. The advantages of different algorithms and the application of a particular algorithm for a specific challenge are presented.

This survey paper is structured as follows. IoMT is briefly discussed with literature review, ML and DL in the field of IoMT, algorithms implemented for a particular disease is presented in the Sect. 2. WBSN is thoroughly discussed with current state of the art literature in Sect. 3. Working model of WBSN, wearable biosensors and their importance is briefed in the section. Types of sensors implemented in different research is also presented in the section. The paper ends with conclusion in Sect. 4.

2 Internet of Medical Things

IoMT finds its application without human intervention in monitoring biomedical signals and diagnosing diseases, the interconnection of medical things. Sensors, smart devices, advanced lightweight communication protocols are the things in IoMT [8]. The potential of IoMT allows for more accurate diagnosis, less mistakes and lower costs of care through the assistance of technology, allowing patients to send health information data to doctors [9].

2.1 Illustration of Real-Time Health Monitoring in IoMT

Real-time health monitoring is one of the most beneficial fields of research in IoMT. Continuous monitoring of health is possible by generating alerts to the users by monitoring real-time results. IoMT architecture is depicted in Fig. 1. IoMT architecture is divided into three layers consisting of data acquisition layer, data aggression/pre-processing layer, data analytics. Sensors and electronic circuits comprised together acquire biomedical signals from a patient in an IoMT, which is known as a data acquisition layer. The data is sent to the communication unit to process the biomedical signals through a Wi-Fi or BLE module. A visualization platform i.e., AI takes decisions when sending data through a cloud or antenna while transmitting biomedical data over a network, be it in a temporary or permanent storage unit. Alerts are sent to the external environment like mobile apps for remote monitoring of health in hospitals or to the patients undergoing surgeries. Insight on different diseases and the corresponding learning methods in IoMT are detailed in the next sub section.

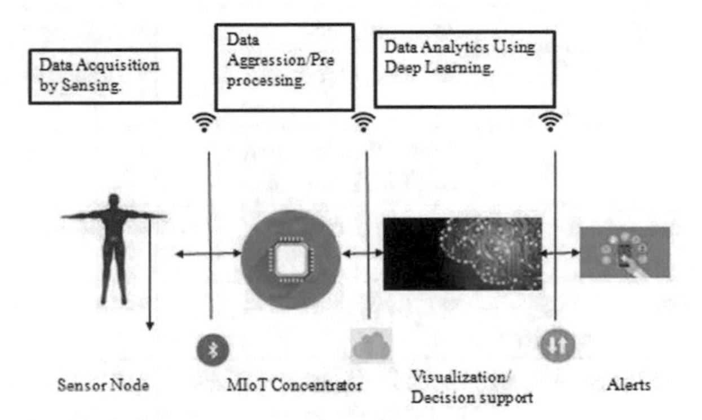

Fig. 1. Block diagram of an IoMT architecture.

2.2 Machine Learning in IoMT

Diagnostics, assistive healthcare and health monitoring are the three beneficial fields of ML applications [10, 11] as discussed in Table 1. The table presents use cases of different disease detection in ML. Assistive healthcare in [12–15] and health monitoring is presented in [16–19]. Chronic disease detection is proposed where the performance of ML algorithms are compared. Logistic regression, random forest, SVM, KNN, naive Bayes classifier, feed-forward neural network have been implemented. Though the integrated model with logistic regression and random forest achieved the best accuracy of 99.83%,the model implemented would work mostly on the data imputation and sample diagnostics [10].

Accuracy can be improved by considering more number of samples. On good amount of samples collected, SVM and DT algorithms are applied on the dataset where excellent accuracy of 99% of is achieved with the KNN algorithm [11]. Automatic classification of knee osteoarthritis severity using gait data is proposed. AUC grades (0–4) of 0.93,

Table 1. Literature review on Machine learning application

Ref	Proposed Work	Algorithm Implemented	Outcome
[10]	Chronic Disease Detection	Combining Logistic regression and Random forest	• Accuracy = 99.83%
[11]	Soft Tissues Tumors	K-Nearest Neighbour	• Accuracy = 99%
[12]	Knee Osteoarthritis	Support Vector Machine	• AUC grades (0–4) = 0.93,0.82,0.83,0.88,0.97 • Sensitivity = 0.70 • Precision = 0.76 and F1-score = 0.71
[13]	Alzheimer's disease	Stacked Encoders and Multikernel Support Vector Machine	• Accuracy = 91.4% • Specificity = 91.67%
[14]	Tongue cancer	Decision Tree	• Accuracy = 88.7%
[15]	Patient fall detection	Decision Tree	• Accuracy = 90%
[16]	Heart disease detection	Hybrid Random Forest with Linear model	• Accuracy = 88.7%
[17]	Heart disease detection	Random classifier	• AUC = 0.79
[18]	Cuffless blood pressure monitoring	AdaBoost regression	• Accuracy = 98%
[19]	Cuffless blood pressure and heart rate monitoring	K-medoids clustering and histogram triangle	• Blood pressure – ME ± STD = −1.4 ± 5.2 – MAE = 4.2 – RMSE = 5.4 • Heart rate – ME ± STD = 0.8 ± 2.7 – MAE = 1.8 – RMSE = 2.8

0.82, 0.83, 0.88, 0.97, Sensitivity of 0.70, Precision of 0.76 and F1-score of 0.71 is achieved [38]. The model has not been implemented on real-time data, which is vital for real-time analysis and results. Diagnosis using multimodal neuroimaging features is discussed for Alzheimer's disease. 91.4% accuracy and 91.67% specificity is achieved with multimodal neuroimaging features with Stacked Encoders and Multikernel Support Vector Machine [13]. The model was implemented in the smaller available dataset and therefore model should be implemented on larger dataset.Under predictive healthcare, tongue cancer prediction is proposed where accuracy of 88.7% is achieved using Decision Tree [14]. Fall detection is discussed with 38 sample collected using Morse fall detection technique which is uncommon in many hospitals. This study has not focused on first-time fallers. It only describes methods for inpatients who are at risk of falling. 90% accuracy is achieved with Decision tree [15]. A prediction model for heart disease detection is developed. Hybrid random forest with a linear model (HRFLM) is implemented where an accuracy of 88.7% is achieved [16]. The accuracy achieved can be improved by introducing other algorithms like nomogram [14]. Non invasive health monitoring in ischemic patients using an activity tracker is proposed. AUC of 0.79 is achieved with Random classifer [17]. Accuracy can be improved by implementing hybrid algorithms .EEG can be implemented in the proposed model for tracking brain activity in the patient. Health monitoring for cuffless measuring blood pressure is proposed where ECG and PPG signals are taken as input for estimating blood pressure. AdaBoost regression is developed on dataset extracted from ICU patients. The data may show variations in blood pressure due to the drugs that are given to the patients during treatment. 98% accuracy is achieved [18]. A wearable for monitoring blood pressure and heart rate using EEG/PPG signal is proposed. K-medoids clustering which is a variant of K NN and histogram triangle .Motion stability is also implemented in the proposed system. For blood pressure monitoring, ME $\pm$ STD $= -1.4 \pm 5.2$, MAE $= 4.2$, RMSE $= 5.4$ and for heart rate, ME $\pm$ STD $= 0.8 \pm 2.7$, MAE $= 1.8$ and RMSE $= 2.8$ is achieved [19]. The proposed system is not designed for continuous monitoring. It can be inferred that ML algorithms has produced good accuracy in every field of healthcare. Performance of each model can be justified by applying models on real time datasets. Real time wearable monitoring devices can be designed with the improved models.

2.3 Deep Learning in IoMT

Subset of ML is Deep Learning. To achieve better accuracy and further development of the proposed model, DL algorithms are implemented. Table 2 represents literature review on DL in healthcare.

Algorithms are implemented depending on the type of disease. ECG signal is reconstructed using a doppler sensor. Correlation coefficient of 0.86 is achieved on implementing hybrid deep learning model using CNN and LSTM [20]. One setback of this approach is that the ECG signal is detected after taking the reading of the heartbeat. First heartbeat should be detected and then an ECG signal must be recorded so that there is no large distortion in the wave signal. An end-to-end classification of raw ECG signals was proposed. DNN algorithm is used for capturing important information from ECG signals which produced an AUC of 0.999 [21]. ECG signals are time series dependent data signals. Long-term dependencies are the feature of LSTM. Implementing LSTM with the same dataset may increase the accuracy. Chronic disease prediction is described in [22]. DNN is implemented to predict chronic kidney disease which produced 99.6% accuracy. Spatial Transformer Network is derived to automatically analyze the Lung Ultrasonography (LUS) images. Accuracy of 96% and dice score of 0.75 is achieved [23]. The model must be tested on larger and heterogeneous dataset such that prediction becomes much accurate and outliers can be identified. Automatic detection of central sleep apnea events is proposed. Where bidirectional long short-term memory (Bi-LSTM) is applied on the dataset. The highest accuracy of 95.1% is achieved using Bi-LSTM [24]. The system is designed on a mat. The readings may turn false positive if position of the mat slides which is a major drawback of a mat-based sleep apnea detector. A double layered fully

Table 2. Literature review on Deep learning application

Ref	Proposed Work	Algorithm Implemented	Outcome
[20]	Reconstructing ECG signal	Hybrid deep learning model with Convolution Neural Network and Long Short Term Memory	• Correlation coefficient = 0.86
[21]	Classifying ECG signals	Deep Neural Network	• AUC = 0.999
[22]	Chronic kidney disease prediction	Deep Neural Network	• Accuracy = 99.6%
[23]	Covid-19 classification	Spatial Transformer Network	• Accuracy = 96% • Dice score of 0.75
[24]	Central sleep apnea detection	Bidirectional long short term memory	• Accuracy = 95.1%
[25]	Regular health factor analysis in IoMT	A double layered fully connected Convolution Neural Network layer	• Accuracy = 70.97% • AUC = 74.93%
[26]	Covid-19 detection	ResNet +	• Precision 86.9% • F1 score = 84.4%
[27]	Human daily activity	Convolution Neural Network	• Accuracy = 93.77%
[28]	Diagnosing dementia	Deep Neural Network	• Accuracy = 73.75%
[29]	Detecting prostate cancer	ResNet-101	• Accuracy = 100% • AUC = 1

connected CNN layer is implemented which produced an accuracy of 70.97% and AUC of 74.93% [25]. A weakly supervised algorithm is proposed for detecting COVID-19 chest CT scans. Where LSTM is implemented which obtained a precision of 81.9% and F1 score of 81.4% [26]. Accuracy can be increased in the proposed model by adding more layers.

Human activity recognition is proposed where CNN is implemented. 93.77% accuracy is achieved [27]. Dementia diagnosis is proposed using neuroimaging where DNN produced an accuracy of 73.75% [28]. Accuracy can be improved by combining two or more algorithms. ResNet-101 algorithm is implemented in detecting prostrate cancer which provides 100% accuracy [29].

3 Wireless Body Sensor Network

In a world when everything is powered by AI, modifying the devices as smart is one of the promising and potential fields in health care. Applications of AI can be found in autonomous driving ex: Tesla, smart home, smart and connected health care (SCH) [30]. WBSN (Wearable Body Sensor Networks) is a distributed system which has nodes in a body wearable sensor network where data is acquired by each node from the body.

For better strategies and to manage WBAN resources, AI is used as a powerful tool for enhancing technology [4]. WBAN consists of three modules. In the first module, wearable feature design, diagnosing fault processing, designing low power sensor and deployment is worked with the sensor being the heart of the first module.

3.1 Working Model of a Wearable Body Sensor Network

Denoising technique, HAR, Context-aware sensing techniques, feature extraction, data classification, data compression are worked in the Data fusion module, which is the second module of the body sensor network. Design of communication protocol (lightweight), Routing algorithm, channel access control, network topology with channel characteristics is worked on in the third module.BSN's are deployed in the human body invasively or non-invasively to monitor the status of a particular organ (invasively) and to monitor physiological parameters on the body, for example EEG, ECG, Heart Rate Monitoring,Blood Pressure monitoring etc. [31]. BSN communicates over radio frequency channels between 850 MHz–2.4 GHz, where Bluetooth Low-End Extension (LEE) is used for short, medium and long-range communications [32]. For communication between sensors and central nodes, a Short-range communication method is required, among which Bluetooth, Zigbee are the most used technologies [33]. Long-range communication refers to communication transmitted over long distances. or Long-range communication, Sig-Fox and LoRaWAN are used widely.

3.2 Literature Review in WBSN

WBSN model consists of three different modules, i.e., Sensor, Data fusion and Network communication where use cases are divided with respect to the different aspects in each module as depicted in Fig. 2. Different aspects of WBSN, i.e., disease detection, energy optimization, privacy, system performance, decision making, risk assessment and congestion control is discussed in this section.

Fig. 2. Model of the body sensor network

IoMT application in detecting Atrial Fibrillation (AF) cardiovascular disease is proposed [34]. AF is detected using ECG signals. While transmitting the raw ECG signals, battery lifetime gets exhausted, which is also one of the concern in a WBSN. For this, an extraction method with low complexity and time-domain features that combines the RR interval and p-wave absence in the ECG, providing high accuracy, is designed with 100% sensitivity, 96% specificity, and 92% energy saving capacity is achieved [35]. Bellman's dynamic programming with distributed optimization framework that helps in mitigating electromagnetic interference is proposed in [36]. Energy and bandwidth consumption is a challenge in a WBSN network. Energy-efficient hardware is developed in [37]. Sparse encoding technique is used to compare real-time ECG signals under the time-frequency domain using wavelet coefficients. Geometry Based Method (GBM) in time domain and Wavelet Transform based Iterative Threshold(WTIT) in time-frequency domain is used to perform wavelet coefficient approximation. With a signal deterioration of 2%, energy consumption has been reduced by 96% [38]. System Performance is one of the aspects of WBSN module. Model-driven process has been derived in [39]. Among different components of a WBSN, coordination is very vital, which is achieved by coordinating messages. By increasing the abstraction level and increasing the reuse of the derived behavior on multiple platforms, system performance efficiency is achieved. Risk assessment and decision support have been discussed for the acute illness management system in [40]. A multi-sensor data fusion model has been proposed. This helps in gathering multi-data irrespective of any number of vital signs that are to be monitored. Ambient Assisted Living (AAL) is proposed in [41]. To recognize motion provided by body with higher scalability and better performance, a Multinomial Naive Bayes is proposed. Overall accuracy of 97.1% is achieved. Other aspects of a WBSN, i.e., congestion control, have been discussed in [42]. Three indicators, i.e., external network environment, current average queue length at an interactive node and traffic priority, are considered where the packets are dropped. WBSN deals with the sensors and data sensing and transmitting from and to various nodes, lack of privacy is a matter of concern. Without altering the delay constraints and by maintaining the original packets, time-dependent priority queue is proposed in [42, 43]. QoS of data remains unaffected in their approach while maintaining privacy. This approach also prioritizes messages for on-time delivery.

3.3 Wearable Biosensors

Biosensor is an analytical device consisting of three components where; the first component, the Bioreactor system is responsible for sensing an analyte which can be biological, chemical etc. It sends data to the transducer, which converts the sensed signal into an electrical signal and then sends out the electrical signal to external output devices. This is the basic working principle of a sensor system.

Importance of Wearable Biosensors
Biosensors, as the names suggest, sense the biological elements which help in real-time detection and diagnosing many diseases, for example monitoring heart rate, ECG, PPG, Body Temperature [18] and EEG, Sweat rate [31, 44].

Sensors implemented in WBSN are presented in Table 3. A respiratory sensor is used for detecting asthma attacks, hyperventilation due to panic attacks and sleep apnea attacks. Lung cancer and obstructions of the airway, tuberculosis. One of the drawbacks of the respiratory sensor used in [5] is that the sensor is susceptible to temperature fluctuations. A pulse sensor is a very common sensor used to detect the pulse rate. It is used to detect cardiac arrest, pulmonary embolisms, and vasovagal syncope in [5]. Carbon dioxide sensor is used for checking gas composition due to gas ranges used in the house, chemicals in pollution etc. as proposed in [31]. It is helpful for individuals who are prone to dust and weather-dependent allergies. Another sensor i.e., the humidity sensor is introduced in [31]. It is used to determine the skin temperature and sweat rate of an

Table 3. Sensors implemented in WBSN

Ref	Sensor	Data Of Interest	Challenge
[5]	Respiratory Sensor	Asthma attacks, Hyperventilation due to panic attacks and Sleep apnea attacks. Lung cancer and obstructions of the airway, Tuberculosis	Limited wearability, accuracy may be compromised due to temperature fluctuations of other sources
[5]	Pulse Sensor	Cardiac arrest, Pulmonary embolisms, and Vasovagal syncope	Research has to be done practically based on the assumptions of the author to detect the diseases if they are accurate
[31]	Carbon dioxide sensor	Indoor air quality for people sensitive to dust and weather dependent allergies	This sensor can be embed and can work in WBAN network
[31]	Humidity sensor	Skin temperature, sweat rate	This sensor can be embed and can work in WBAN network
[44]	Multifunction sensor	EEG signal and Sweat rate	As the method is an indirect way of measuring sweat, there is a delay during changes in the sweat rate measurements
[45]	Accelerometer Sensor	HAR in fall detection for elderly people	Temperature sensitivity. It operates only in particular temperature range
[46]	ECG,PPG	Heart rate variability and stress	This approach calls for a number of sensors making the whole process very expensive
[47]	ECG, PPG, Heart Rate, B.P, Body Temperature	Photoplethysmography, Electrocardiography Heart Rate and Blood Pressure	Scalability &Reliability is an drawback. Secure transmission of data is also another challenge

individual. A device that consists of a humidity sensor helps the caretakers to determine the condition of patients in the ICU or after surgery hours. To detect fall activity in the elderly, the model is tested on already available datasets. Sweat rate and EEG are used as a multifunction sensor and are measured for finding the correlation between sweat rate and EEG [44]. An accelerometer sensor is used for human activity recognition (HAR).HAR is implemented using an accelerometer sensor in [45]. Sweat is a hub of analytes and finding the analytes that get excited during high blood sugar/high heart rate may give more insight into an individual's health condition. Heart rate variability and detecting stress are proposed in [46]. To monitor heart rate and notice variability, many market-based heart rate monitoring devices are implemented. An integrated sensor system that consists of ECG, PPG, Heart Rate, B.P, Body Temperature is proposed in [47]. Bluetooth module is implemented for the data to be sent from the sensor system to IoT gateway. Drawback is due to bluetooth, speed of data transfer reduces with respect to the battery drain. A Wi-Fi module is a solution for better data transmission with high speed than Bluetooth and location detection. Similarly, PPG sensors are also bought from the market. These devices are connected to an individual and data is collected. The whole proposed system is extremely expensive. Designing a sensor system consisting of EEG, ECG, PPG sensors along with a sensor to detect blood glucose level has a novel advantage on an individual's healthcare. Hence a wearable IoT system that can continuously monitor heart rate, heart rhythm, pulse rate (oxygen in hemoglobin) and blood glucose non-invasively will have an immense impact on healthcare.

4 Conclusion

The proposed paper has surveyed different use cases and state of art in a Medical IoT, i.e., IoMT technologies. IoMT is the amalgamation of biosensors that send and share vital data through a secure network to the user. Data acquisition and processing takes place in the things layer that consists of sensors. In the communication layer, data is received where data aggregation takes place. Detection of abnormalities, prediction of diseases is implemented using ML and DL algorithms is thoroughly discussed in the study. This forms data analytics layer. WBSN which is a pillar of an IoMT is presented with literature. Wearable biosensors that contribute in designing smart healthcare system is presented in this paper. The algorithms that perform better with respect to a certain disease is also filtered out and is presented. With respect to the research papers and the current market statistics that have been discussed in this survey paper,it is evident that the blend of different technologies in IoMT has the potential to enhance healthcare and serve humanity in a better way.

References

1. Aman, A.H.E.M., Yadegaridehkordi, E., Attarbashi, Z.S., Hassan, R., Park, Y.: A survey on trend and classification of internet of things reviews. IEEE Access **8**, 111763–111782 (2020). https://doi.org/10.1109/ACCESS.2020.3002932
2. Yadav, E.P., Mittal, E.A., Yadav, H.: IoT: challenges and issues in indian perspective. In: 2018 3rd International Conference On Internet of Things: Smart Innovation and Usages (IoT-SIU), Bhimtal, pp. 1–5 (2018). https://doi.org/10.1109/IoT-SIU.2018.8519869
3. IoMT (Internet of Medical Things) Market Size, Share | Trends (2026). https://www.fortunebusinessinsights.com/
4. Miyandoab, F.D., Ferreira, J.C., Grade, V.M., Tavares, J.M., Silva, F.J., Velez, F.J.: A multifunctional integrated circuit router for body area network wearable systems. IEEE/ACM Trans. Netw. **28**(5), 1981–1994 (2020). https://doi.org/10.1109/TNET.2020.3004550

5. Baker, S.B., Xiang, W., Atkinson, I.: Internet of Things for smart healthcare: technologies, challenges, and opportunities. IEEE Access **5**, 26521–26544 (2017). https://doi.org/10.1109/ACCESS.2017.2775180
6. Ghosh, A., et al.: Artificial Intelligence in Internet of Things. CAAI Trans. Intell. Technol. **3**(4), 208–218 (2018). https://doi.org/10.1049/trit.2018.1008. Accessed 30 July 2019
7. World Health Organization. WHO | Integrated Chronic Disease Prevention and Control. Who.int (2010). www.who.int/chp/about/integrated_cd/en/entity/chp/about/integrated_cd/en/index.html
8. Vishnu, S., Ramson, S.R.J., Jegan, R.: Internet of Medical Things (IoMT) - an overview. In: 2020 5th International Conference on Devices, Circuits and Systems (ICDCS), Coimbatore, India, pp. 101–104 (2020). https://doi.org/10.1109/ICDCS48716.2020.243558
9. "The Importance of IoMT in Healthcare 2020 | Digital Healthcare | Healthcare Global.https://healthcare-digital.com/, www.healthcareglobal.com/digital-healthcare/importance-iomt-healthcare-2020. Accessed 19 Dec 2020
10. Qin, J., Chen, L., Liu, Y., Liu, C., Feng, C., Chen, B.: A machine learning methodology for diagnosing chronic kidney disease. IEEE Access **8**, 20991–21002 (2020). https://doi.org/10.1109/ACCESS.2019.2963053
11. Alaoui, E.A.A., Tekouabou, S.C.K., Hartini, S., Rustam, Z., Silkan, H., Agoujil, S.: Improvement in automated diagnosis of soft tissues tumors using machine learning. Big Data Min. Anal. **4**(1), 33–46 (2021). https://doi.org/10.26599/BDMA.2020.9020023
12. Kwon, S.B., Han, H., Lee, M.C., Kim, H.C., Ku, Y., Ro, D.H.: Machine learning-based automatic classification of knee osteoarthritis severity using gait data and radiographic images. IEEE Access **8**, 120597–120603 (2020). https://doi.org/10.1109/ACCESS.2020.3006335
13. Liu, S., et al.: Multimodal neuroimaging feature learning for multiclass diagnosis of alzheimer's disease. IEEE Trans. Biomed. Eng. **62**(4), 1132–1140 (2015). https://doi.org/10.1109/TBME.2014.2372011
14. Alabi, R.O., et al.: Comparison of nomogram with machine learning techniques for prediction of overall survival in patients with tongue cancer. Int. J. Med. Inf. **145**, 104313 (2021). www.sciencedirect.com/science/article/abs/pii/S1386505620310133, https://doi.org/10.1016/j.ijmedinf.2020.104313. Accessed 29 Mar. 2021
15. Lindberg, D.S., et al.: Identification of important factors in an inpatient fall risk prediction model to improve the quality of care using EHR and electronic administrative data: a machine-learning approach. Int. J. Med. Inf. 143, 104272 (2020)/ https://doi.org/10.1016/j.ijmedinf.2020.104272. Accessed 1 Nov 2020
16. Mohan, S., Thirumalai, C., Srivastava, G.: Effective heart disease prediction using hybrid machine learning techniques. IEEE Access **7**, 81542–81554 (2019). https://doi.org/10.1109/ACCESS.2019.2923707
17. Meng, Y., et al.: A machine learning approach to classifying self-reported health status in a cohort of patients with heart disease using activity tracker data. IEEE J. Biomed. Health Inform. **24**(3), 878–884 (2020). https://doi.org/10.1109/JBHI.2019.2922178
18. Kachuee, M., Kiani, M.M., Mohammadzade, H., Shabany, M.: Cuffless blood pressure estimation algorithms for continuous health-care monitoring. IEEE Trans. Biomed. Eng. **64**(4), 859–869 (2017). https://doi.org/10.1109/TBME.2016.2580904
19. Zhang, Q., Zeng, X., Hu, W., Zhou, D.: A machine learning-empowered system for long-term motion-tolerant wearable monitoring of blood pressure and heart rate with Ear-ECG/PPG. IEEE Access **5**, 10547–10561 (2017). https://doi.org/10.1109/ACCESS.2017.2707472
20. Yamamoto, K., Hiromatsu, R., Ohtsuki, T.: ECG signal reconstruction via doppler sensor by hybrid deep learning model with CNN and LSTM. IEEE Access **8**, 130551–130560 (2020). https://doi.org/10.1109/ACCESS.2020.3009266

21. Xu, S.S., Mak, M.-W., Cheung, C.-C.: Towards end-to-end ECG classification with raw signal extraction and deep neural networks. IEEE J. Biomed. Health Inform. **23**(4), 1574–1584 (2019). https://doi.org/10.1109/JBHI.2018.2871510
22. Chittora, P., et al.: Prediction of chronic kidney disease - a machine learning perspective. IEEE Access **9**, 17312–17334 (2021). https://doi.org/10.1109/ACCESS.2021.3053763
23. Roy, S., et al.: Deep learning for classification and localization of COVID-19 markers in point-of-care lung ultrasound. IEEE Trans. Med. Imaging **39**(8), 2676–2687 (2020). https://doi.org/10.1109/TMI.2020.2994459
24. Azimi, H., Xi, P., Bouchard, M., Goubran, R., Knoefel, F.: Machine learning-based automatic detection of central sleep apnea events from a pressure sensitive mat. IEEE Access **8**, 173428–173439 (2020). https://doi.org/10.1109/ACCESS.2020.3025808
25. Ismail, W.N., Hassan, M.M., Alsalamah, H.A., Fortino, G.: CNN-based health model for regular health factors analysis in internet-of-medical things environment. IEEE Access **8**, 52541–52549 (2020). https://doi.org/10.1109/ACCESS.2020.2980938
26. Mohammed, A., et al.: Weakly-supervised network for detection of COVID-19 in chest CT scans. IEEE Access **8**, 155987–156000 (2020). https://doi.org/10.1109/ACCESS.2020.3018498
27. Yen, C.-T., Liao, J.-X., Huang, Y.-K.: Human daily activity recognition performed using wearable inertial sensors combined with deep learning algorithms. IEEE Access **8**, 174105–174114 (2020). https://doi.org/10.1109/ACCESS.2020.3025938
28. Ahmed, M.R., Zhang, Y., Feng, Z., Lo, B., Inan, O.T., Liao, H.: Neuroimaging and machine learning for dementia diagnosis: recent advancements and future prospects. IEEE Rev. Biomed. Eng. **12**, 19–33 (2019). https://doi.org/10.1109/RBME.2018.2886237
29. Iqbal, S., et al.: Prostate cancer detection using deep learning and traditional techniques. IEEE Access **9**, 27085–27100 (2021). https://doi.org/10.1109/ACCESS.2021.3057654
30. Wang, H., Daneshmand, M., Fang, H.: Artificial Intelligence (AI) driven wireless body area networks: challenges and directions. In: 2019 IEEE International Conference on Industrial Internet (ICII), Orlando, FL, USA, pp. 428–429 (2019). https://doi.org/10.1109/ICII.2019.00079
31. Lai, X., Liu, Q., Wei, X., Wang, W., Zhou, G., Han, G.: A survey of body sensor networks. Sensors **13**(5), 5406–5447 (2013). https://doi.org/10.3390/s130505406
32. Nanjappan, V., et al.: Body sensor networks: Overview of hardware framework and design challenges. In: 2015 International SoC Design Conference (ISOCC), Gyungju, pp. 175–176 (2015). https://doi.org/10.1109/ISOCC.2015.7401775
33. Gholamhosseini, L., Sadoughi, F., Ahmadi, H., Safaei, A.: Health Internet of Things: strengths, weakness, opportunity, and threats. In: 2019 5th International Conference on Web Research (ICWR), Tehran, Iran, pp. 287–296 (2019). https://doi.org/10.1109/ICWR.2019.8765286
34. Almusallam, M., Soudani, A.: Embedded solution for atrial fibrillation detection using smart wireless body sensors. IEEE Sens. J. **19**(14), 5740–5750 (2019). https://doi.org/10.1109/JSEN.2019.2906238
35. Laurijssen, D., Saeys, W., Truijen, S., Daems, W., Steckel, J.: Synchronous wireless body sensor network enabling human body pose estimation. IEEE Access **7**, 49341–49351 (2019). https://doi.org/10.1109/ACCESS.2019.2910636
36. Zhang, L., Hu, J., Guo, C., Xu, H.: Dynamic power optimization for secondary wearable biosensors in e-healthcare leveraging cognitive WBSNs with imperfect spectrum sensing. Fut. Gener. Comput. Syst. **112**, 67–92 (2020). ISSN 0167–739X. https://doi.org/10.1016/j.future.2020.05.013
37. Ghosh, A., et al.: Energy-efficient IoT-health monitoring system using approximate computing. Internet of Things, 100166 (2020). https://doi.org/10.1016/j.iot.2020.100166. Accessed 4 Feb 2020

38. Subramanian, A.K., et al.: PrEEMAC: priority based energy efficient MAC protocol for wireless body sensor networks. Sustain. Comput. Inf. Syst. **30**, 100510 (2021). www.sciencedirect.com/science/article/pii/S2210537921000032, https://doi.org/10.1016/j.suscom.2021.100510. Accessed 16 Mar. 2021
39. Harbouche, A., et al.: Model driven flexible design of a wireless body sensor network for health monitoring. Comput. Netw. **129**, 548–571 (2017). https://doi.org/10.1016/j.comnet.2017.06.014. Accessed 24 Apr 2019
40. Habib, C., et al.: Health risk assessment and decision-making for patient monitoring and decision-support using wireless body sensor networks. Inf. Fusion **47**, 10–22 (2019). https://doi.org/10.1016/j.inffus.2018.06.008. Accessed 1 Sept 2019
41. Syed, L., et al.: Smart healthcare framework for ambient assisted living using IoMT and big data analytics techniques. Fut. Gener. Comput. Syst. **101**, 136–151 (2019). https://doi.org/10.1016/j.future.2019.06.004. Accessed 9 Oct 2020
42. Xu, Q.Y., Li, S.P., Mansour, H.: An intelligent packet drop mechanism in wireless body sensor network for multiple class services based on congestion control. Procedia Comput. Sci. **154**, 453–459 (2019). ISSN 1877–0509, https://doi.org/10.1016/j.procs.2019.06.064
43. Diyanat, A., Khonsari, A., Shafiei, H.: Preservation of temporal privacy in body sensor networks. J. Netw. Comput. Appl. **96**, 62–71 (2017). ISSN 1084–8045, https://doi.org/10.1016/j.jnca.2017.07.015
44. Gao, K.-P., Shen, G.-C., Zhao, N., Jiang, C.-P., Yang, B., Liu, J.-Q.: Wearable multifunction sensor for the detection of forehead EEG signal and sweat rate on skin simultaneously. IEEE Sens. J. **20**(18), 10393–10404 (2020). https://doi.org/10.1109/JSEN.2020.2987969
45. Lu, W., Fan, F., Chu, J., Jing, P., Yuting, S.: Wearable Computing for internet of things: a discriminant approach for human activity recognition. IEEE Internet Things J. **6**(2), 2749–2759 (2019). https://doi.org/10.1109/JIOT.2018.2873594
46. Umair, M., Chalabianloo, N., Sas, C., Ersoy, C.: HRV and stress: a mixed-methods approach for comparison of wearable heart rate sensors for biofeedback. IEEE Access **9**, 14005–14024 (2021). https://doi.org/10.1109/ACCESS.2021.3052131
47. Wu, T., Wu, F., Qiu, C., Redouté, J.-M., Yuce, M.R.: A rigid-flex wearable health monitoring sensor patch for iot-connected healthcare applications. IEEE Internet Things J. **7**(8), 6932–6945 (2020). https://doi.org/10.1109/JIOT.2020.2977164

Machine Learning and Digital Image Processing in Lung Cancer Detection

Saroja S. Bhusare, Veeramma Yatnalli(✉), Sanket G. Patil, Shailesh Kumar, Niranjan Ramanna Havinal, and C. S. Subhash

ECE Department, JSS Academy of Technical Education, Bangalore, Karnataka, India
{sarojasbhusare,veerammayatnalli}@jssateb.ac.in

Abstract. The ability of lung cancer to spread to other body tissues and inflict damage makes it one of the most hazardous cancers in existence. Lung cancer is one of the main reasons for cancer deaths worldwide. Early detection of cancer increases the patient's probability of survival and is the best defense against cancer. Machine learning and deep learning are now the most advanced techniques for solving many categorization problems. The LIDC-IDRI dataset is used in this investigation. Deep CNN, ResNetv101, VGG-16 are three machine learning models which take pre-processed, edge detected, and segmented pictures as the input. Malignancy classifications from various models are compared. The proposed method for detecting lung cancer in this study makes use of machine learning algorithms and image processing, appears to have immense potential. The Deep CNN model outperformed ResNet101, VGG-16 in terms of results. The Deep CNN model achieved an accuracy rate of 99.10%. The ResNet101v2 model achieved an accuracy percentage of 97.1%. VGG-16 model achieved an accuracy rate of 81.90%.

Keywords: Lung Cancer · Machine Learning · Image Processing · Deep CNN · ResNetv101 · VGG-16

1 Introduction

Depending on the cell of origin, location, and hereditary changes, lung cancer can take on several forms. The development of aberrant cells in the lungs causes lung cancer. These cells obstruct the lung's ability to function normally. These unnatural cells proliferate and develop into a tumor. These tumors may be benign or malignant. Patients with lung cancer have the lowest post-diagnosis survival rates, and the number of fatalities is steadily rising. Ten million deaths, or nearly one in six deaths, were caused by cancer in 2020, making it the top cause of death globally. In 2020, there were about 2.2 million new cases of lung cancer. The use of tobacco, drinking alcohol, eating few fruits, having a high body mass index, vegetables, and not exercising accounts for about one third of cancer-related fatalities. According to World Cancer Research Fund International [1], Hungary had the highest overall rate of lung cancer at the end of 2020, with 10,274 cases and 2,206,771 people around the world were counted, of which 1,435,943 were

H. Sharma et al. (Eds.): ICIVC 2022, PALO 17, pp. 56–67, 2023.
https://doi.org/10.1007/978-3-031-31164-2_6

men and 770,828 were women. There was 1,796,144 lung cancer deaths worldwide by the year 2020, with 1,188,679 men and 607,465 women. With 8,920 deaths, Hungary had the highest overall mortality rate [2] as shown in Fig. 1.

Although for all stages, lung cancer has been treated with surgery, radiation therapy, and chemotherapy, the five-year survival rate combined is only 14%. Early cancer identification has a critical role in preventing the growth and spread of cancer cells.

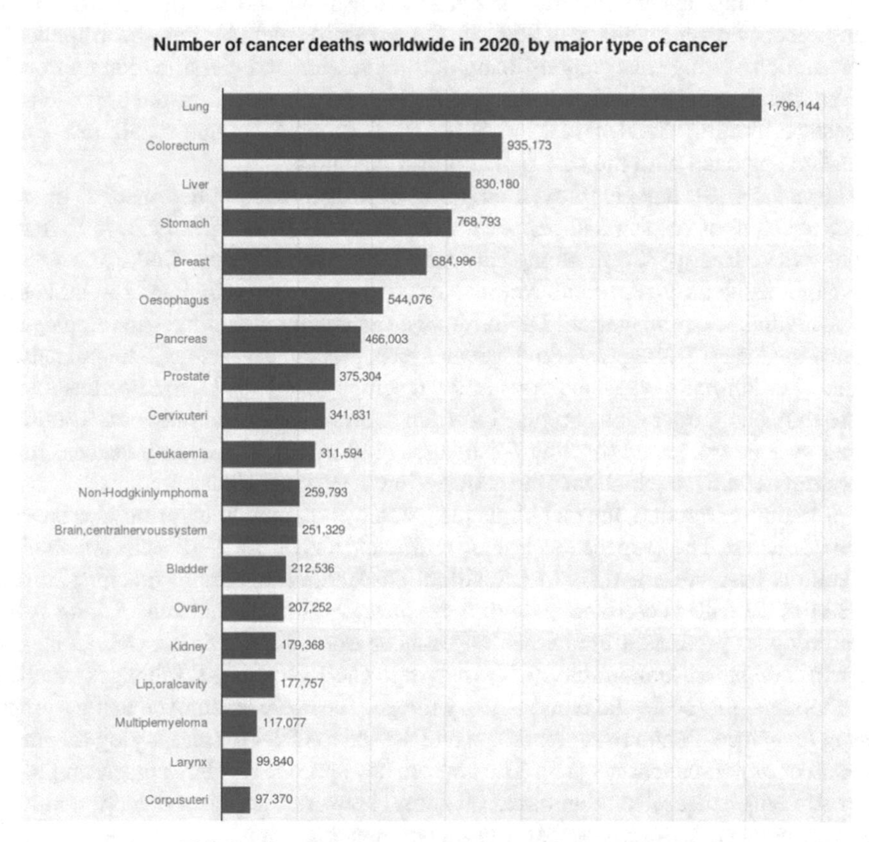

Fig. 1. Number of cancer deaths worldwide 2020 [2]

Currently, a typical process for making a lung cancer diagnosis is low-dose CT scan (LDCT) which is performed on the patient. However, a large or expanding mass of cells indicate malignancy in the patient. There are two issues here. A doctor cannot predict whether a group of cells will expand or not. A patient must wait at least six months before taking another CT scan. This may give a chance for the cells to grow. The distinction between a little and a large cell clump is the second issue.

2 Literature Survey

The authors have used the DICOM (Digital Imaging and Communications in Medicine) format images to implement the probabilistic neural network and compared the results with JPEG format images [3]. The overall accuracy of the system was found to be 86.25%. In the diagnosis of lung diseases, Computed Tomography (CT) scan provides significant information. A novel FPSOCNN [4] reduces the computational complexity of CNN and outperforms compared to other techniques. The authors have implemented watershed segmentation [5] using CT scan images and accurate results are obtained in the detection of cancer. An Artificial Neural Network (ANN) based technique [6] detects the presence or absence of lung cancer in a human body. The ANN model detects presence or absence of lung cancer with an accuracy of 96.67%. Deep learning methods have been applied for the prediction of lung cancer by utilizing both regression and classification approaches [7]. The results obtained with accuracy of 71.18% for the classification approach with the best-performing ANN model, 13.50% (RMSE) and R2 value of 50.66% for the regression approach with the best-performing CNN model.

The authors [8], have employed cropping of images and applied median filters on image, which removes salt and pepper noise present in the images. Feature Extraction is done where Energy, Correlation, Variance, Difference Entropy, Contrast are calculated. For classification, Artificial Neural Networks were used with Feed forward neural network with back-propagation. Digital image processing techniques are employed to enhance the lung CT images. A feed forward back propagation artificial neural network is trained to bifurcate non-cancerous and cancerous images [9]. Binary thresholding, feature extraction, operations are used to train and test the neural network. The implemented system was tested for lung CT images (150 types) with overall success rate of the system as 96.67% which met the expectation of system [10].

For feature extraction, the technique [11] employs segmentation principles based on region of interest. The proposed technique is efficient and promising results are obtained. The authors have presented EDM (Electrical Discharge Machining) machine learning algorithm [12] with vectorized histogram features. A clinical decision-making system for radiologists predicts a malicious lung cancer from SCLC (Small Cell Lung Cancer) with computed tomography (CT) imaging. The RGB-lung CT images were first converted into gray scale and then to binary image. The noise is removed using a preprocessing technique. Features are extracted and are fed to CNN for classifying the images into cancer or non-cancerous [13]. The authors have presented PNN algorithm [14], an automatic lung cancer detection based on neural network for cancer detection in initial stages. A good amount of accuracy and low computation time are obtained and highly suitable for cancer identification. The authors [15] have proposed a technique with preprocessing approaches such as gray scale conversion, noise reduction and binarization. Median filter and segmentation techniques are used to get accurate results. Support Vector Machine (SVM) classifier classifies the positive and negative samples of lung cancer images in the system. In this paper [16], median filter is used for noise removal. K-Mean unsupervised learning algorithm is used for clustering or segmentation which groups the pixel dataset according to certain characteristics. The system has an accuracy of about 90.7%.

The authors [17] have studied the classification of the benign and malignant tumor which was not predicted accurately. Majorly, authors have identified three main gaps identification with a thorough literature survey [18]. Authors [19] have applied various machine learning methods to detect lung cancer from chest radiographs. The authors also used PCA to reduce the dimension of chest radiographs in the ratio of 1/8. Reducing dimension causes feature loss. They used KNN, SVM, DT, ANN models for the classification. The accuracy obtained for KNN, SVM, DT, ANN after application of PCA was found to be 75.68%, 55.41%, 79.97%, 82.43% respectively. A literature survey is made on the classification and prediction of the benign and malignant tumor accurately. In paper [20], CT images are used for Lung cancer detection and features are extracted using ML algorithms. The paper [21] reviews on different methodologies for preprocessing, nodule segmentation, and classification. In paper [22], the technique identifies the early stage of lung cancer and determines the accuracy levels in machine learning algorithms.

3 Materials and Methods

3.1 Dataset

The methodology shown in Fig. 2 makes use of the LIDC-IDRI (Lung Image Data Consortium) dataset which consists of the Computed Tomography (CT) scans with marked-up annotations. The dataset has lesion annotations from thoracic radiologists. There are around 1,018 low-dose lung CTs from 1010 lung patients. Each patient has an annotation file in XML format. The XML file consists of information about the nodules computed by four radiologists. Each patient's name is disguised and has a unique patient ID. Each patient has multiple image slices which vary in numbers for each patient. Each slice is in DICOM format with shape 512 $\times$ 512. The Python library, Pylidc, retrieves the information about the patient's lung nodules which is written in an XML file and is converted into a CSV file. The data set includes the scores given for the malignancy. If the malignancy is above 3, then it is classified as cancerous. If the malignancy is less than 3, then it is classified as non-cancerous and if the malignancy is three, then it is classified as ambiguous. For improved outcomes, more focus is given on the patients who are cancerous and non-cancerous while neglecting patients who are unclear. With the help of the file, DICOM images of patients are stacked and converted into NumPy array for further processing.

3.2 Data Cleaning

The input at this stage is the processed data from the previous step which is in the form of NumPy array of shape 512 * 512. The CSV file is examined and selected only those patients whose slices were cancerous or non-cancerous. This procedure is conducted to improve the ability to distinguish between malignant and healthy lungs. This procedure was conducted using the CSV Python package.

Input CT DICOM Image: DICOM is used to exchange, store, and transmit medical images. Modern radiological imaging has evolved significantly due to DICOM. Radiation therapy, computed tomography, ultrasonography, magnetic resonance imaging, and radiography are just a few of the imaging modalities that use DICOM standards.

Pre-processing: For data preparation, superior quality data is required to create better models. The real data that is available has issues like, noise, incomplete and unreliable. Preprocessing is crucial to enhance the fineness of the data. The anomalies are removed to standardize the data for comparison and enhance the accuracy of the results.

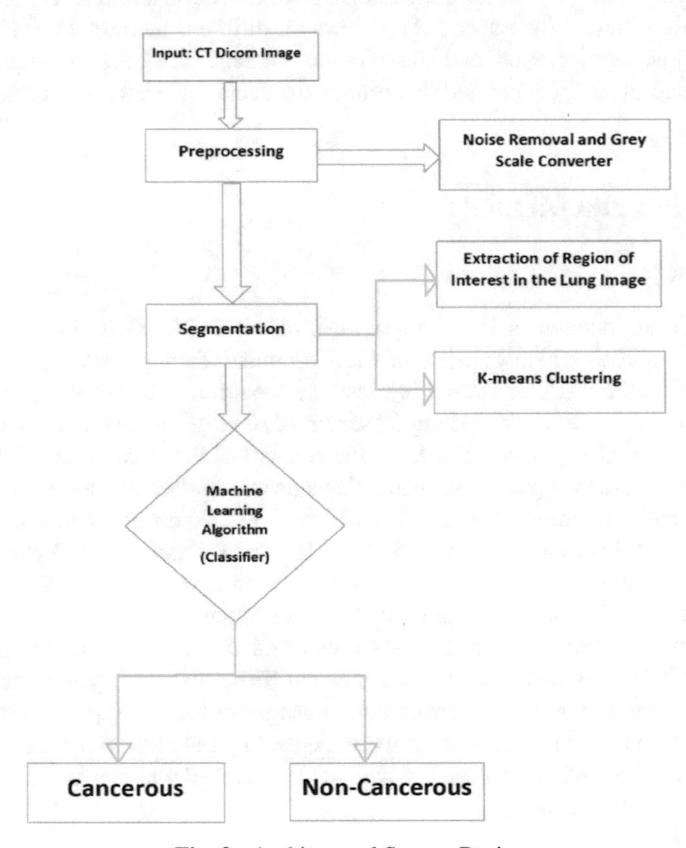

Fig. 2. Architectural System Design

The preprocessing method consists of applying two filters to reduce noise and converting the captured image into grey scale image.

(1) *Median filter.* Median filter is applied on the image followed by anisotropic diffusion filter. The median filter is a straightforward yet effective nonlinear filter. The intensity variance between two pixels is lessened using this filter. The median value is used in place of the centered pixel value in this filter.
(2) *Anisotropic diffusion filter.* Anisotropic diffusion is a nonlinear filter which is popularly used for the restoration of MR image intensities. The edge-stopping function (ESF) allows it to smooth noisy pixels while maintaining the intensity of most of the edges.

Segmentation: Image segmentation is the decomposition of an image into various sub-group of pixels identified as Image Objects. Segmentation can minimize the complexity of the image, making image analysis easier. Segmentation is done for edge detection by applying threshold, dilation, erosion for the image. Thresholding is the simplest technique that segments binary pictures from a gray scale image. Erosion causes the image pixels to shrink. Pixels on object borders are removed through erosion. The minimum value of all the neighborhood pixels is chosen as value of the output pixel. If any of the nearby pixel's value is zero, that pixel is set to 0. Dilation causes the image pixels to expand. Pixels on the object borders are added through dilation. The maximum value of all the neighborhood pixels makes up the value of the output pixel. If any of the nearby pixels contain the value 1, that pixel is set to 1. The input image's sections are labelled based on how connected the pixels are to one another. A single region will be identified when adjacent pixels have the same value. All related regions are given the same integer value in the labelled array. Several properties such as area, minor axis lengths, major axis length, etc., can be measured by region props function.

Machine Learning Models: The methodology focuses on gaining good accuracy for the classification of Lung Cancer. It uses three models of Convolution Neural Networks which are Deep CNN, Resnet101v2 and VGG-16. All the models employ a convolution layer, max-pooling layer, and dense layer. In this work, three 2D convolutions, ReLu activation function, and a flattened pooling layer of varied sizes are employed. The flattened layer's output passes through three dense layers. To get the output classes, SoftMax is fed to the final dense layer.

(1) *Deep CNN.* The Deep CNN Architecture consists of nine layers. It uses max pool. The dropout used was 0.25. Adam optimizer was used. It had 11,925,858 trainable parameters. Model undergoes three 2D-Convolution followed by three Activation function ReLU and MaxPooling2D. This model uses 32 filters and the kernel (5, 5) for the first Conv2D layer. The second Conv2D layer uses thirty-two filters and kernel size (3, 3).
(2) *ResNet101v2.* The ResNet101v2 architecture consists of 101 layers. It is built on ImageNet weights with 1000 classes. ResNet101v2 is treated as the base model. The output of the base model is flattened to convert 2-D data to 1-D vector. Following this layer are a Dense layer, a Dropout layer, and another Dense layer.
(3) *VGG-16.* The VGG-16 architecture consists of 16 layers. It uses max pooling. The output of VGG-16 model is merged with another simple model. Adam is used as an optimizer. It has 119,554,688 trainable parameters.

Neural Networks: Neural networks are one of the most widely utilized algorithms in supervised machine learning. The analogy of the human brain used in neural networks is that neurons activate when sufficiently stimulated. Layers, a collection of nodes and weights between the layers connecting the nodes together make up neural networks. The neural network components are shown in Fig. 3.

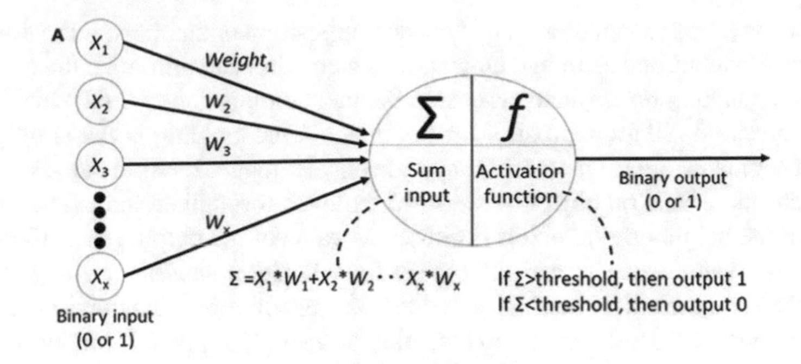

Fig. 3. Neural Network

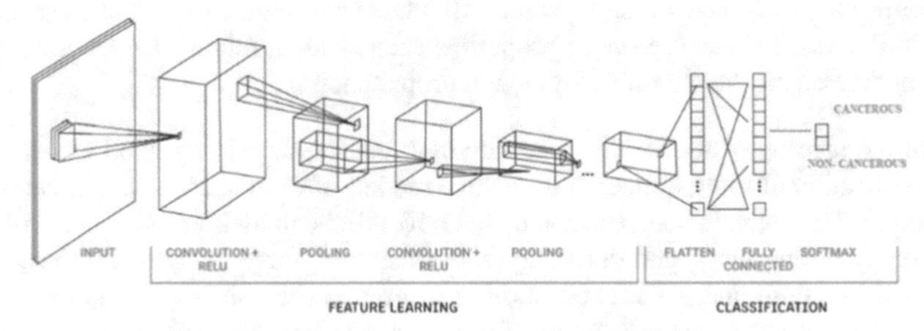

Fig. 4. Basic Convolutional Neural Network Architecture

A machine learning method called deep learning is used to create artificial intelligence (AI) systems. The most popular type of deep convolutional neural network (CNN or DCNN) for pattern recognition in pictures and videos. Deep convolutional neural networks are primarily employed for tasks like object detection, image classification, and recommendation systems. The Basic Convolutional Neural Network Architecture is shown in Fig. 4.

4 Results

The dataset used was the Lung Image Data Consortium (LIDC-IDRI) Lung CT Image Dataset. The images undergo various processes to get segmented images and the image is resized according to the requirement of models. Three different CNN architectures like Deep CNN, ResNet101v2, and VGG-16 are used. The three machine learning models DCNN, ResNet101v2, and VGG-16 are evaluated using the metrics listed in the following sections.

Accuracy: Accuracy is the measure of the count of correctly classified data instances over the total count of data instances. Accuracy determines the best model by identifying relationships and patterns between variables in a dataset based on the input data or training data.

$$\text{Accuracy} = \frac{TN + TP}{TN + FP + TP + FN}$$

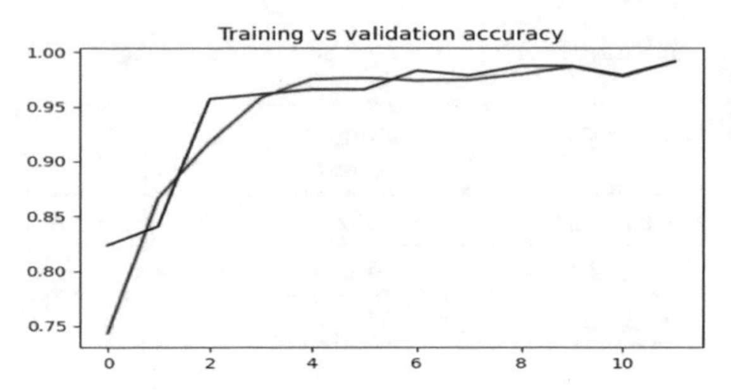

Fig. 5. Deep CNN Accuracy

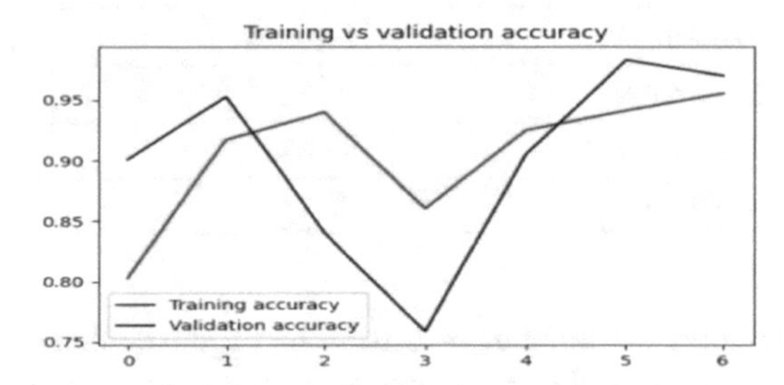

Fig. 6. ResNet101v2 Accuracy

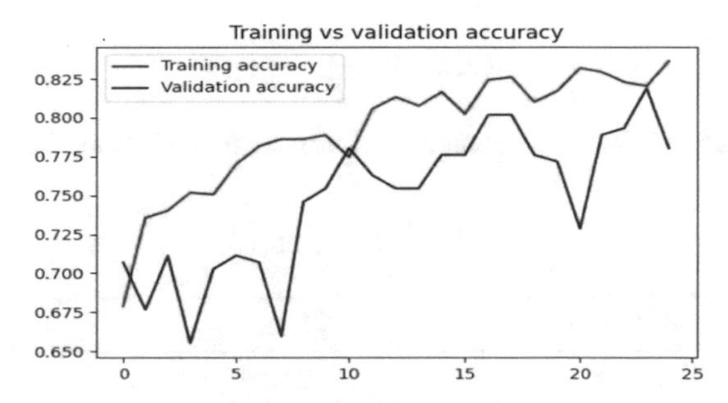

Fig. 7. VGG-16 Accuracy

The training accuracy and validation accuracy results are depicted in Fig. 5, Fig. 6, Fig. 7 and in Table 1.

Precision: Precision is the percentage of correctly classified positive samples (True Positives) to all positive samples that have been classified (either correctly or incorrectly). Precision gauges the percentage of accurate positive forecasts.

$$\text{Precision} = \frac{TP}{FP + TP}$$

The precision obtained from three machine learning models is as shown in Table 2.

Table 1. Training Accuracy and Validation Accuracy.

ML Models	Training Accuracy	Validation Accuracy
Deep CNN	99.29%	99.10%
ResNet101v2	95.54%	97.10%
VGG-16	83.65%	81.95%

Table 2. Precision.

ML Models	Precision
Deep CNN	99.3%
ResNet101v2	98.6%
VGG-16	78%

Recall: Recall examines the number of erroneous negatives that are included in the prediction process rather than the number of false positives the model predicted. It is referred to as the percentage of total positive classes that our model accurately predicted. There must be a strong recall.

$$\text{Recall} = \frac{Tp}{TP + FN}$$

The recall obtained from three machine learning models is as shown in Table 3.

Table 3. Recall

ML Models	Recall
Deep CNN	99.3%
ResNet101v2	96.6%
VGG-16	96.6%

F-Measure: The evaluation metric for a classification is F-measure. F-measure is computed through harmonic means of precision and recall. This statistical parameter measures the accuracy of a test or model. F-measure is used to express the performance of the machine learning model.

$$\text{F} - \text{Measure} = 2 * \frac{Precisin * Recall}{Precioon + Recall}$$

The F-Measure Recall obtained from three machine learning models is as shown in Table 4.

Table 4. F-Measure

ML Models	F-Measure
Deep CNN	99.3%
ResNet101v2	97.6%
VGG-16	86.3%

Loss: Loss function measures the difference between predicted output and the actual output from the model for the single training example. The parameters learnt by the model are determined by minimizing a chosen loss function. Loss functions define what a good prediction is and is not. The training loss and validation loss results are depicted in Fig. 8, Fig. 9, Fig. 10 and in Table 5.

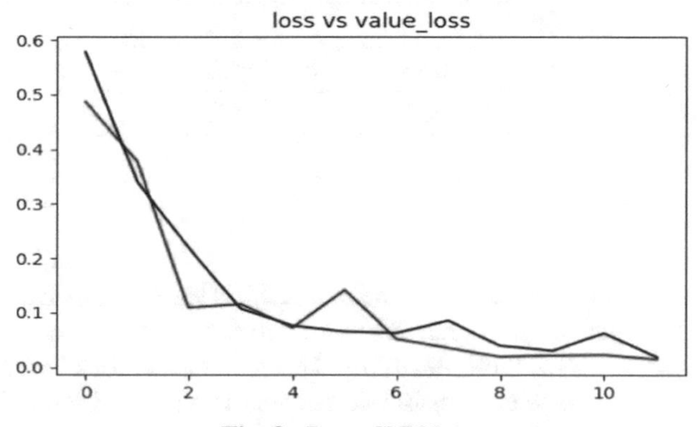

Fig. 8. Deep CNN Loss

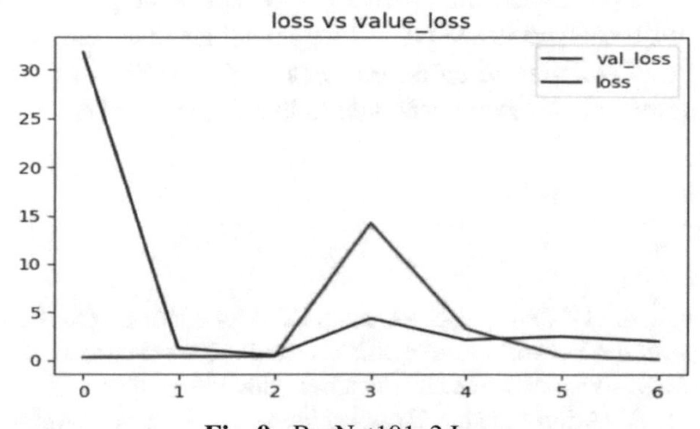

Fig. 9. ResNet101v2 Loss

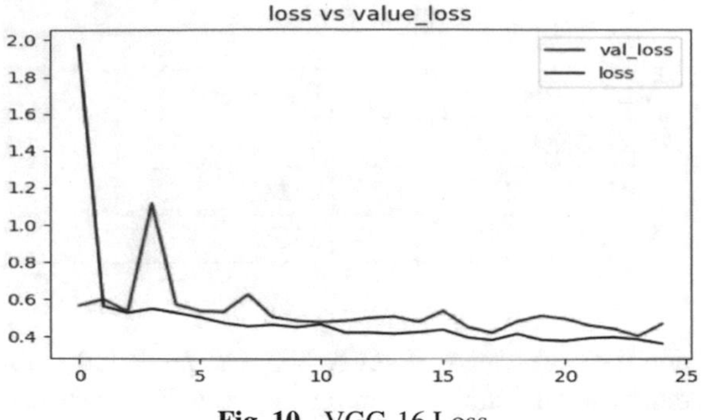

Fig. 10. VGG-16 Loss

Table 5. Training Loss and Validation Loss.

ML Models	Training Loss	Validation Loss
Deep CNN	0.48	0.58
ResNet101v2	0.54	14.27
VGG-16	0.37	1.12

5 Conclusion

The work carried out applies many innovative machine learning algorithms to a system for classifying lung cancer. It is known that preprocessing of the image dataset leads to better results as it allows more flexibility. Using K-means for clustering and edge-detection of the image showed remarkable results. The result from different models seems promising. In terms of results, the Deep CNN model performed better than the ResNet101, VGG-16 models. ResNet101v2 and Deep CNN models showed satisfactory results with low false positive located between 1–5. The Deep CNN model achieved an accuracy rate of 99.10%. The ResNet101v2 model achieved an accuracy percentage of 97.1%. VGG-16 model achieved an accuracy rate of 81.90%. To help the radiologist and for future improvement, the study sought to identify the best accuracy of the cancer result.

References

1. Lung cancer statistics I World Cancer Research Fund International (wcrf.org)
2. Number of cancer deaths worldwide in 2020, by major type of cancer. https://www.statista.com/statistics/288580/number-of-cancer-deathsworldwide-by-type
3. Chethan Dev, K.K., Arjun Palathil, T.A., Panicker, V.: Machine learning based approach for detection of lung cancer in DICOM CT image. In: [Methods in Molecular Biology] Microcalorimetry of Biological Molecules, vol. 1964 (Methods and Protocols) (2019)
4. Asuntha, A., Srinivasan, A.: Deep learning for lung Cancer detection and classification. Multimedia Tools Appl. **79**, 7731–7762 (2020)

5. Abdullah, D.M., Ahmed, N.S.: A review of most recent lung cancer detection techniques using machine learning. Int. J. Sci. Bus. IJSAB Int. **5**(3), 159–173 (2021)
6. Nasser, I.M., Abu-Naser, S.S.: Lung cancer detection using artificial neural network. Int. J. Eng. Inf. Syst. (IJEAIS) **3**(3) (2019)
7. Doppalapudi, S., Qiu, R.G., Badr, Y.: Lung cancer survival period prediction and understanding: deep learning approaches. Int. J. Med. Inf. **148** (2021)
8. Vas, M., Dessai, A.: Lung cancer detection system using lung CT image processing. In: International Conference on Computing, Communication, Control and Automation (ICCUBEA). IEEE Xplore, 13 September 2017
9. Evangeline, P.P., Batri, K.: Detection of lung cancer by machine learning. Int. J. Eng. Res. Technol. (IJERT) **08**(09) (2019)
10. Miah, M.B.A., Yousuf, M.A.: Detection of lung cancer from CT image using image processing and neural network. In: International Conference on Electrical Engineering and Information Communication Technology (ICEEICT). IEEE Xplore, 29 October 2015
11. Al-Tarawneh, M.S.: Lung cancer detection using image processing techniques. Leonardo Electron. J. Pract. Technol. **11**(21), 147–158 (2012)
12. Wu, Q., Zhao, W.: Small-cell lung cancer detection using a supervised machine learning algorithm. In: International Symposium on Computer Science and Intelligent Controls (ISCSIC). IEEE Xplore, 19 February 2018
13. Bhalerao, R.Y., Jani, H.P., Gaitonde, R.K., Raut, V.: A novel approach for detection of lung cancer using digital image processing and convolution neural networks. In: 2019 5th International Conference on Advanced Computing & Communication Systems (ICACCS). IEEE Xplore, 6 June 2019
14. Vaishnavi, D., Arya, K.S., Devi Abirami T., Kavitha, M.N.: Lung cancer detection using machine learning. Int. J. Eng. Res. Technol. (IJERT) **7**(1) (2019)
15. Pratap, G.P., Chauhan, R.P.: Detection of Lung cancer cells using image processing techniques. In: 2016 IEEE 1st International Conference on Power Electronics, Intelligent Control and Energy Systems (ICPEICES). IEEE Xplore, 16 February 2017
16. Rahane, W., Dalvi, H., Magar, Y., Kalane, A., Jondhale, S.: Lung cancer detection using image processing and machine learning healthcare. In: Proceeding of 2018 IEEE International Conference on Current Trends toward Converging Technologies, Coimbatore, India (2018)
17. Sangamithraa, P.B., Govindaraju, S.: Lung tumour detection and classification using EK-mean clustering. In: IEEE WISPNET 2016 Conference (2016)
18. Joshua, E.S.N., Chakkravarthy, M., Bhattacharyya, D.: An extensive review on lung cancer detection using machine learning techniques: a systematic study. In: International Information and Engineering Technology Association, vol. 34, no. 3, pp. 351–359 (2022)
19. Günaydin, Ö., Günay, M., Şengel, Ö.: Comparison of lung cancer detection algorithms. In: 2019 Scientific Meeting on Electrical-Electronics & Biomedical Engineering and Computer Science (EBBT), pp. 1–4 (2019)
20. Karthick, K., Rajkumar, S., Selvanathan, N., Saravanan, U.B., Murali, M., Dhiyanesh, B.: Analysis of lung cancer detection based on the machine learning algorithm and IOT. In: 2021 6th International Conference on Communication and Electronics Systems (ICCES) 2021, pp. 1–8 (2021)
21. Rajalaxmi, R.R., Kavithra, S., Gothai, E., Natesan, P., Thamilselvan, R.: A systematic review of lung cancer prediction using machine learning algorithm. In: 2022 International Conference on Computer Communication and Informatics (ICCCI), 25 January 2022, pp. 1–7 (2022)
22. Ismail, M.B.: Lung Cancer Detection and Classification using Machine Learning Algorithm. Turkish J. Comput. Math. Educ. (TURCOMAT). **12**(13), 7048–7054 (2021)

Video Captioning Using Deep Learning Approach-A Comprehensive Survey

Jaimon Jacob[1(✉)] and V. P. Devassia[2]

[1] Model Engineering College, Ernakulam, Kerala, India
jaimon@mec.ac.in

[2] St. Joseph's College of Engineering and Technology, Palai, Kerala, India

Abstract. Video captioning is a technique used to represent videos using phrases of natural language. This technique automatically generates phrases to explain the contents of input videos. Deep learning-based video captioning methodologies are used for generating descriptions in natural language for a given video or an image. The significance of video captioning in the present era has advanced research in this field during the last few decades. A methodical review of recent works, various benchmarks used for performance evaluation, and various data sets for training are explained in this review paper for the benefit of new researchers. The most widely used neural network variants for visual and spatiotemporal feature extraction, as well as an approach for language generation, are included in this survey paper. The results show that ResNet and VGG are the most widely used visual feature extractors, and 3D convolutional neural networks are the most widely used Spatio-temporal feature extractors. Aside from that, the language model has primarily been used as Long Short-Term Memory (LSTM). The most widely used metrics for video description evaluation are BLEU, ROUGE, METEOR, CIDEr, SPICE, and WMD. This article is intended to be an extensive overview of deep learning-based captioning approaches for a given image and video.

Keywords: Video captioning · Image captioning · Deep learning · Long short term memory

1 Overview

Video is a highly effective form of communication medium, and its usage has grown with the extensive usage of social media and business marketing. Based on the forecast by Global DataSphere of International Data Corporation (IDC), the cumulative amount of data on the planet was estimated to be 44 zettabytes at the beginning of 2020. By 2025, global day-wise data is expected to surpass 463 exabytes. This shows the significance of video processing in the coming years.

Humans can easily describe the visual content from a video clip or an image in natural language by their natural ability. Computing machines can do the same tasks using various video captioning algorithms on available and data sets. This is extremely difficult and time-consuming. It is therefore a challenge in terms of accuracy and processing time

H. Sharma et al. (Eds.): ICIVC 2022, PALO 17, pp. 68–87, 2023.
https://doi.org/10.1007/978-3-031-31164-2_7

to generate a caption automatically from video clips or images. In an image or a video clip, lots of data is present which includes objects, backgrounds, motions, and language. The deep learning algorithm can convey all this information in natural language using the phrases generated by the algorithm used. In a Video captioning system, a list of sentences is generated in natural language that effectively describes the contents of video frames.

When a user searches for a particular topic in a video, say news video, the video clip significant to the topic in the query only needs to be extracted from a long news video and sent to the user. This is the major advantage of the video captioning system. Similarly in the video surveillance system, when a suspicious event occurs, messages can be sent to the persons concerned as text messages for quick response and actions.

An overview of deep learning-based video captioning and image captioning approaches is presented in this survey paper. Image captioning articles are classified in Fig. 1.

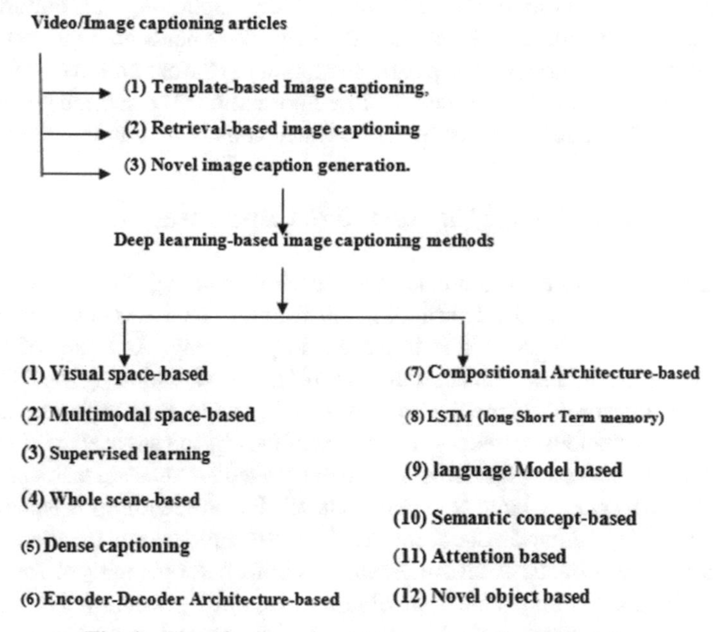

Fig. 1. Classification of current image captioning articles

2 Video Captioning Architecture

Image captioning and video captioning are the two sorts of visual content in natural language. Image captioning is the process of creating captions for an image. Video captioning is like image captioning, but the process is repeated on a series of images derived from video frames. Video captioning entails a sequence of tasks that are divided into two stages. The first is to visually comprehend the video content and grammatically describe the video content is the second.

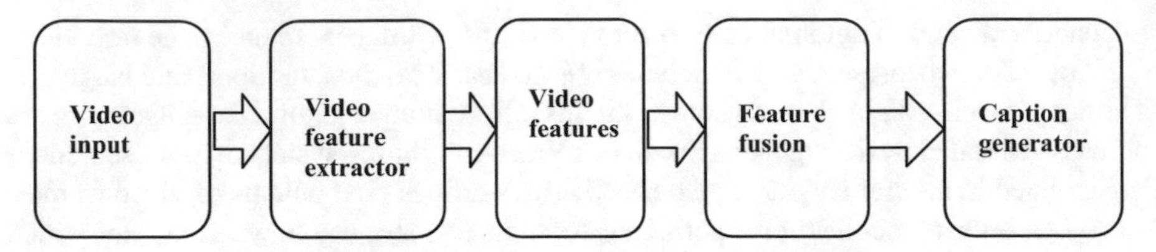

Fig. 2. The architecture of a standard video captioning system.

Videos contain much more information than still images because the scene changes dramatically. Because video captions require more features for text descriptions, video captioning is more difficult than image captioning. As shown in the figure, there are two stages. One is abstraction of video features, while the other is interpretation. Figure 2 illustrates the standard architecture for video captioning systems. A video consists of a lot of spatial and temporal information, which makes video captioning very important. With the video feature extractor, the key objects and their movements are extracted from the video. A feature fusion phase will be performed following the extraction of features from key objects to generate more prominent feature information. The caption generator uses this prominent information to generate captions that contain natural language phrases.

3 Deep Learning-Based Video/Image Captioning Models

It is important to identify the prominent features within an image before creating a caption. There are two methods for determining this feature. The first is based on traditional machine learning, and the second is based on deep learning. As a part of traditional machine learning techniques, handcrafted features are used, including local binary patterns (LBP), histograms of oriented gradients (HOG), scale-invariant feature transforms (SIFTs), and combinations thereof. The features of the objects are used to classify them using SVMs (Support Vector Machines). As hand crafted features are task-specific, they cannot be extracted from a large or diverse dataset. Feature learning is achieved using machine learning techniques in deep learning. These techniques can handle a vast range of video and image formats. A classifier, such as SoftMax, is employed for classification and for feature learning with Convolutional Neural Networks (CNN). The most used captioning algorithms are CNN followed by Recurrent Neural Networks (RNN) (Fig. 3).

Deep machine learning and visual space techniques are used in novel caption generation methods. It is also possible to generate captions using multimodal space. Supervised learning, reinforcement learning, or unsupervised learning are three types of deep learning-based image captioning techniques. Using encoder-decoder architectures or compositional architectures, image captions can be constructed. As opposed to captioning the entire image, dense captioning produces captions for different regions. In addition to attention mechanisms, semantic concepts, and styles, other methods can be used to create images descriptions. LSTMs are most used as language models for captioning images, while CNNs or RNNs are also popular options.

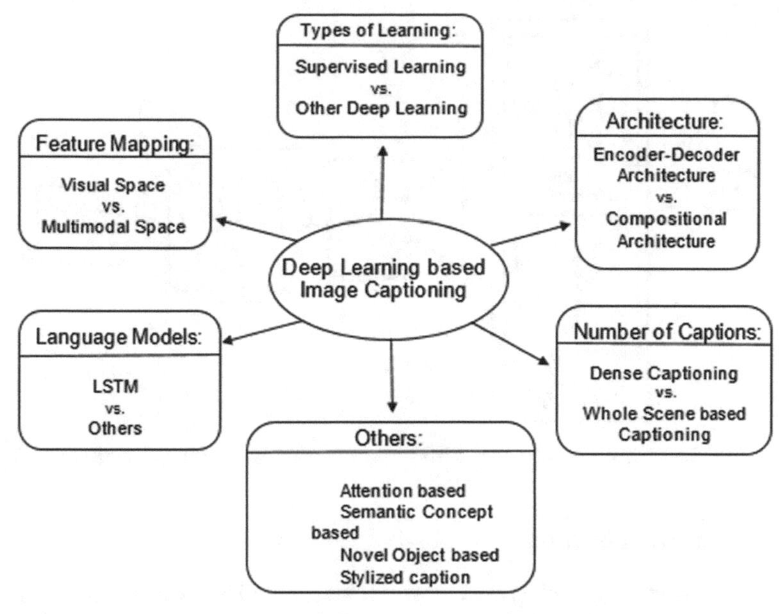

Fig. 3. Deep learning-based image captioning approaches - An overall taxonomy

To generate captions using neural networks, several model architectures are used currently. By analyzing and categorizing data, neural networks are always able to produce trained predictions. In other words, deep learning is the process of learning from data. Deep learning can have a good prediction system when multiple processing layers are used in the model. Data sets with complex structures can be recognized by the back propagation algorithm. From large and diverse collections of images and videos, deep learning networks can automatically generate captions.

Image captions can be generated from visual and multimodal spaces using deep learning-based methods. The language decoder receives both the image features and the corresponding captions independently using the visual space-based method. The multimodal space case, on the other hand, relies on learning the shared space from the images and corresponding captions. Once this representation is created, it is sent to the language decoder. Visual space models allow for a wide variety of caption generation techniques for image captioning.

A language encoder, a vision encoder, a multimodal space encoder, and a language decoder comprise the main components of the block diagram in Fig 4 of multimodal space-based captioning methods. Deep convolutional neural networks are used as feature extractors in the visual part of the algorithm for extracting image features. Language encoders extract the word features and learn how to embed dense feature embeddings into each word. Afterward, it passes the semantic temporal context to the recurrent layers.

In the multimodal space part, the features from the images are mapped into a common space with the features from the words. Once a map has been created, it is given to the language decoder, which generates captions based on it.

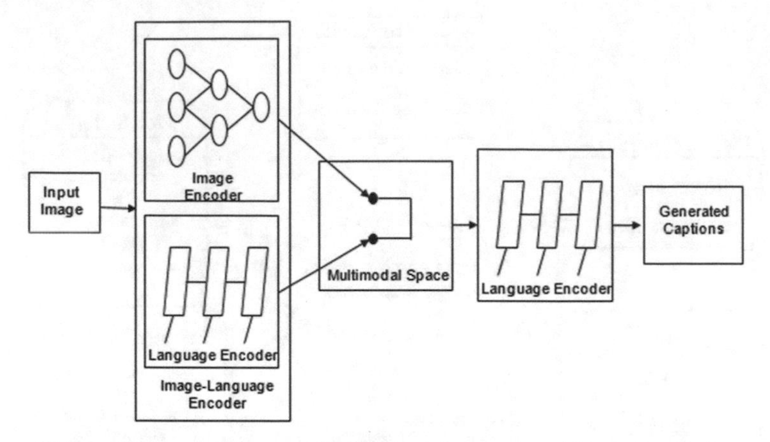

Fig. 4. Schematic Diagram of Multimodal space-based image captioning

3.1 Recurrent Architectures

Most of the deep learning models use Convolutional Neural Networks (CNN) and Recurrent Neural Networks (RNN) or Long Short-Term Memory (LSTM) networks. CNN is used for extracting image features while RNN or LSTM for caption generation. The disadvantages of LSTMs or RNNs as opposed to CNNs are substantial. Because of the Vanishing or Exploding Gradient issue [1], RNN only has memory for a few steps in the past. As a result, it could not model dependencies across a wide range of languages. Parallel processing is also a strength of CNN, whereas RNNs are limited in their use since their training process is sequential from left to right. This means that every state depends upon the previous one. RNNs do not consider a sentence's underlying hierarchical structure, which is useful for caption generation. An accurate representation of the hierarchical structure of a sentence can be achieved with a CNN by layering words that represent a sentence's hierarchy. LSTM (Long Short-Term Memory) [2] networks overcome the limitations of vanishing gradients in RNNs and allow long-term temporal structure to be learned via gradient descent. LSTM techniques are like the standard RNN architecture with one hidden layer, but the nodes of the hidden layer now include a recurrent edge with self-connection and fixed weight that stores information for long periods of time.

The topological inner structure of a recurrent neural network cell is approximated in evolutionary recurrent neural networks [3] by the intermediary associative connections, further evolved by an evolutionary algorithm on the proxy of captioning images. In terms of finding novel neural networks, automatic architecture search is widely used in vision or natural language tasks. On cross-modality tasks, however, a strong emphasis is placed on the associative mechanisms between visual and language models rather than focusing on the best performing convolutional neural network (CNN) or recurrent neural network (RNN).

The diagram in Fig. 5 illustrates various building blocks for Video to Textual Description. The first stage of the process is to extract the image features from the video frames and create a feature vector. Spatial and temporal features are gathered with the help of

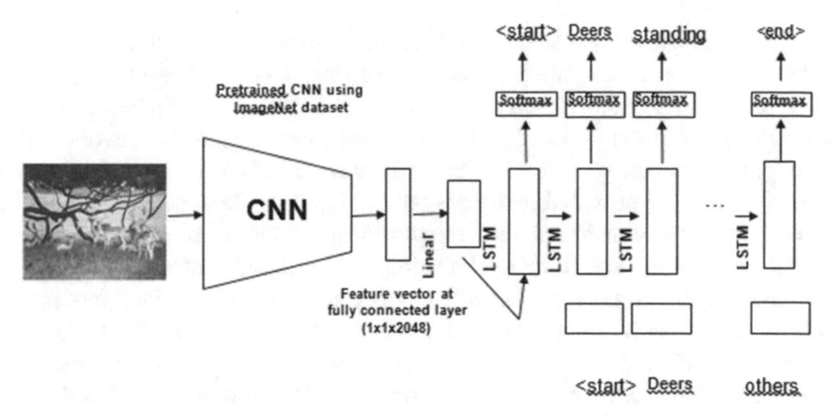

Fig. 5. Building blocks for Video to Textual Description

three-dimensional convolutional neural networks-CNNs. A caption generator uses the features extracted from the video content to create meaningful sentences, as illustrated in Fig. 5. A combination of LSTM and RNN is then used to encode the video using the frame's image features.

To describe the contents of an image automatically, an encoder-decoder framework [4] is used in recent machine translation techniques. As an encoder, we use convolutional neural networs, and as decoders we use recurrent neural networks.

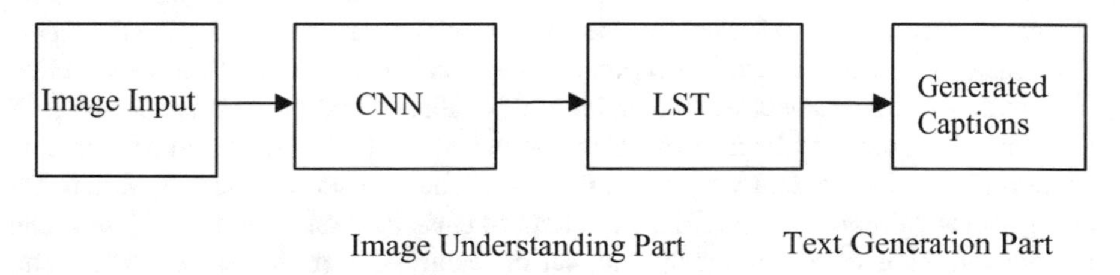

Fig. 6. Simple captioning architecture using encoders and decoders.

The Encoder-Decoder framework is shown in Fig. 6, in which vectors are represented by encoders based on an input image. Decoders produce captions in natural language, corresponding to the vectors they receive. Although the decoder assembles all semantics [5] and feature information about objects, it is unlikely to compile all the structural information needed for future phases of decoding. This network has as a major constraint that the decoder must produce sentences that describe the contents of the image and must also fit the trained model. By doing so, the generating process is balanced and prevents the generated sentences from being unrelated to the image. Several solutions are likely to be found to this problem: (1) Develop attention mechanisms [6] for capturing significant information from images that will generate captions, (2) inject selected relevant information into the decoding process [7] so that each step can access related images. All these models do not adjust or balance themselves according to the input information. One innovative solution is presented in [8] with a model that can learn an additional guiding vector for generating sentences. It is proposed to use a guiding neural network to model the vector and to connect it to the decoder. A decoder used error signals as feedback to adaptively adjust itself in generating captions at each time step. In order to compose the input of the decoder with the current word, a guiding network is used to learn the guiding vector v. Those features are extracted from the input image and the annotation vectors are provided by the encoder as inputs to the guiding network. In [9], we inject guiding information about the input image as part of the decoder step of the traditional encoder-decoder framework for improving performance.

The task of detecting complex events in video captioning is challenging. It is made difficult by video events' complexity and diversity. Many methods using Convolutional Neural Networks (CNNs) have shown promise in the detection of complex events. Several novel deep CNN technologies are presented in [10], including Deep Event Network (DevNet), which identifies events at the video level by aggregating CNN features from key frames. As part of the detection process, evidence pieces related to the results of the detection are also automatically relocatable, both temporally and spatially. Considering the variability of video and moving objects, the models rely primarily on detection and tracking to improve the efficiency of event analysis. Detection and tracking models are more complex, especially when the number of video sequences is large. The problem of complex event detection was addressed by an innovative algorithm in [11]. Initially the input video are passed through various stages of histogram equalization for contrast improvement and smoothing based on the histograms, hybrid feature extraction using Gradient Local Ternary Pattern (GLTP), Histogram of Oriented Gradients (HOG) and Tamura feature (coarseness, contrast and directionality) to extract the feature vectors. The extracted feature vectors are highly dimensional and lead to the 'curse of dimensionality', so a feature selection technique called Ensemble Feature Selection (EFS) is employed to select optimal subsets of features. This paper proposes an EFS algorithm which uses Pearson Correlation coefficient (PCC), correlation-based feature selection (CFS), f-score and variance to determine the best set of multiple features for using geometric mean, thus resulting in enhanced classification accuracy with minimal components. After obtaining an optimal set of feature subsets for each action/event, our algorithm uses Multi Support Vector Machine (MSVM) to classify together the actions/events, and then identifies the relevant ones using Euclidean distance.

Captioning video requires extracting information concerning objects and their properties and behavior that exists in multiple frames. Most of the previous methods have concentrated on methods of extracting representations, including the characteristics of entities and motions in video. Yet it is difficult to determine whether certain words represent video information or are required for grammatical purposes. It is necessary for the deep learning model to have the structure that can learn these two types of words separately to identify the relationship between the video and the description without using natural language processing methods, such as part-of-speech tagging. A multi-representation switching method is used in [12] to enable a model to learn visual characteristics as well as textual characteristics. The method is based on an encoder-decoder architecture. In the proposed method, only two-dimensional (2D) CNNs and RNNs are used to extract video features. In this architecture, the main objective is to construct a model with high generality because a verb can be adequately modeled using a single frame as input during image captioning. In 2D CNNs, entities are extracted from each frame, and in RNNs, motion features are extracted based on the sequence of these entity features. A RNN is used to extract a textual feature from the word sequence information. To generate the word sequence, entity and motion features are extracted from the encoder and textual features from the decoder. As a result, the attention mechanism selects the entity and motion information that is required to produce each word. Selecting entities, motions, and textural features from the extracted entity, a switcher generates the next word.

3.2 Sequence-To-Sequence Architectures

The recurrent models presented in Sect. 3.1 present good performance when inputs and targets are encoded as fixed-dimensional vectors, but they cannot be used when inputs and targets have variable dimensions. When it comes to recurrent neural networks, sequences present a challenge since they require determination and fixing of input dimensionality ahead of time. When the length of a sequence is unknown a priori, it must be mapped into a fixed dimensional vector representation. It is proposed in this regard that sequences are transformed into sequences by means of a mapping model. Using this approach, models learn to map input sequences into vectors of fixed dimensions, then map those vectors back into output sequences based only on minimal assumptions about sequence structure and interpret target sequences based on encoded vector representations of the source sequences. Since Sequence-to-Sequence models [13] are conditional language models, their inputs and outputs are variable in length. Aside from being used as encoder-decoders, they have also been successfully applied in a variety of other fields, including machine translation [14, 15], text summarization [16], dialogue systems [17], as well as image and video captioning [18]. Seq2seq models entail of two matching recurrent neural networks: the first network (encoder) condenses a variable-length source sequence of symbols $x = (x1, x2,\ldots, xT)$ into a rich fixed-length vector depiction v, while the subsequent network (decoder) procedures the vector representation v as its opening hidden state and decodes it into another variable-length objective order of symbols $y = (y1, y2,\ldots, yT')$ by computing the conditional probability of each word yt in the target sequence given the earlier words $y1, y2,\ldots, yt - 1$ and the input x, where length T of input may differ from length T' of output.

In open domain videos, the problem of generating descriptions is complicated not only by the variety of objects, scenes, and actions, but also by the difficulty of determining the salient content while accurately describing the event. Video clips and paired sentences that describe the events depicted in video clips teach the model what is worth describing. In recurrent neural networks (RNNs) [19], Long Short-Term Memory (LSTM) networks have achieved great success in sequence-to-sequence tasks, such as speech recognition and translation [20]. LSTMs are well-suited for describing events in videos due to the inherent sequential nature of videos and language.

In a stacked LSTM, each frame is encoded one by one, using an output from CNNs fitted to the intensity values of each frame as input. A sentence is then generated word by word once all frames have been read. A parallel corpus is used to learn both the encoded and decoded frame and word representations.

Captioning video using sequence-to-sequence models uses encoder-decoder architecture. There are two main parts of the process: the first extracts the features from the video, while the second creates a natural language phrase accompanying it as a caption on the video. Sequential models such as LSTM are employed in this respect.

Inputs and outputs for video captioning are variable in length. Through preprocessing, this problem can be avoided by making the input videos the same size. But in real-life situations, variable-length inputs will need to be handled. A variety of approaches have been used to address this issue; for example, researchers have generated fixed-length video representations [21], combined clips into one video stream [20], designated a fixed quantity of input frames from input videos, and dealt with variable-length video input. Video to text sequence is a sequence-to-sequence technique that is commonly used. Sequence to sequence techniques include LSTM [22], LSTM-E [23], Sequence to Sequence-Video to Text (S2VT), Multi-scale Multi-instance Video Description Networks (MM-VDNs) [24], LK [25], Long-term Recurrent Convolutional Networks (LRCNs) and Video Response Maps (VRM).

3.3 Attention-Based Architectures

Machine translation mainly uses neural encoder-decoder-based approaches [28]. CNN acts as the encoder in the image captions to excerpt the graphic structures of the annotated image, and the RNN acts as the decoder to translate this image illustration into natural language. The problem with these methods is that they cannot analyze the image for a period while they generate the description for the image. In addition, both methods do not consider the spatial size of the image in relation to the caption. Therefore, they create captions that cover the entire image. Since attention-based mechanisms can address these limitations, they are becoming increasingly popular in deep learning. Their output sequences can be generated while dynamically focusing on dissimilar portions of the input image. Methods of this type generally follow the following steps:

(1) CNN provides image information based on the entire scene.
(2) Language generation is the process by which words and phrases are derived from the output of Step 1.
(3) The generated language generation model focuses on the salient portions of a given image each time step.
(4) As part of the language generation model, subtitles are restructured animatedly until the finale state.

The featured content of the image is automatically determined by this method. The attention-based approach has the advantage of focusing on the salient features of the image while generating words that correspond to them. Random hard attention techniques and deterministic soft attention techniques are used to generate attention by this method. In CNN-based approaches, the top layer of ConvNet is used to extract feature information from the image. There is a downside to these techniques that they can lose important information that can be useful in closed captioning. Attention preserves information from the bottom layer rather than the wholly linked layer. Jin et al. have suggested an attention-based captioning technique [29]. Semantic relationships between visual and textual information can be used to derive abstract meanings. In addition, higher-level semantic information can be obtained using context-specific contexts. One of the key differences between this method and other attention-based methods is that it introduces multiple visual regions of the image at multiple scales. A consistent visual description of an object in an image can be extracted, allowing it to perform a variety of operations. To extract scene-specific context, the Latent Dirichlet Allocation (LDA) method is first used to generate a dictionary from all the legends in the dataset. Then, the topic vector is predicted using a multi-layer perceptron. The image context is described using a two-layer LSTM arranged on top of each supplementary.

The use of hidden states in CNN to perform multiple caption checks is that of Wu et al. proposed method [31]. The image is processed by CNN to create a set of fact vectors containing global facts. CNN produces a fact vector containing the global details of the image as an output. Vectors are subject to attention mechanisms like LSTMs. When asked what the objects in the image are, the image inspection engine first asks: What are they? You can extract information about the overall context of the image. This information is passed to the decoder to generate captions for the image. Pedro Soli et al. [32] introduced a methodology to create captions based on region-based attention. The only language models mapped to the image domain in previous attention-based methods were RNN models. In this methodology, image areas are associated with subtitle words using RNN states. As the RNN progresses, it becomes able to predict the next subtitle word and the region of the image that corresponds to that subtitle word. The system generates captions by predicting the next word and the corresponding region of the image at each time step of the RNN. An attention-based method for captioning images was developed by Lu et al. [33].

This method uses adaptive attention in combination with visual sentinels. Current captioning methods focus the image at each time step to determine captions. For example, words and phrases don't need to be represented visually (eg for and a). Subtitle generation can also be affected by these unwanted visual cues. By deciding when to emphasis on image regions and when to emphasis only on speech, the proposed method can decide which type of focus is appropriate (Fig. 7).

3.4 Sequence to Sequence Models with Attention Mechanism

For variable length video, the previous method with fixed length input works better. However, it doesn't work well with variable length inputs. Related to the unpretentious encoder-decoder-based model [34] with variable-length inputs, attention-based encoder-decoder algorithms reach incredibly high levels of performance. Therefore, LSTMs embedded in sequential models cannot describe long-term dependencies. This is addressed in [35] by developing a Temporal Deformable Convolutional Encoder-Decoder Network (TDConvED). It consists of both an encoder network and a decoder network using convolution. When multiple layers are stacked, feedforward convolution can easily capture long-term dependencies and produce a fixed-length output representation. CNN is used to extract features from the sampled video frames and fed into stacked convolution blocks. The encoder then applies a temporal deformable convolution. In addition, sentence generation integrates a time-based attention mechanism.

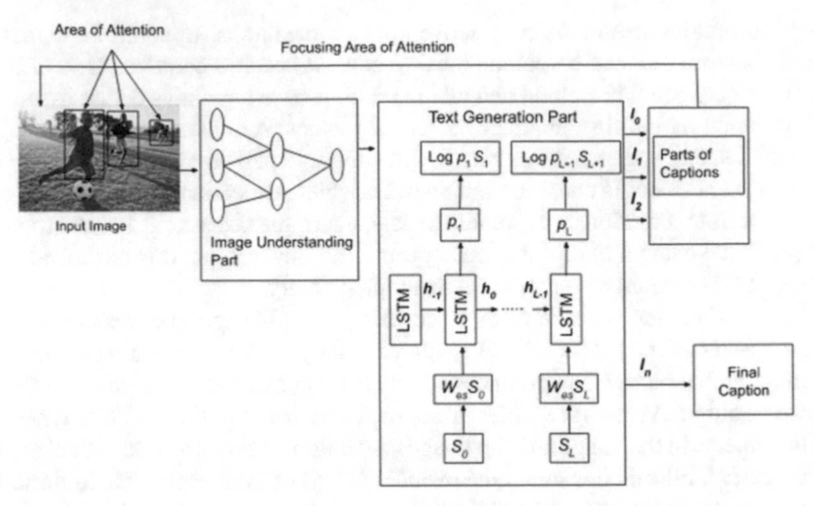

Fig. 7. Attention-based image captioning method – Schematic diagram.

Methods that have achieved good results include Temporal Attention (TA) [36], Generational Adversarial Network (GAN) [37], Bidirectional LSTM (biLSTM) [38] and Attention-based LSTM (aLSTM) [39] as two multifaceted layers of attention (MFATT) [40]. Recent studies have endeavored to insert characteristics into structure learning of video captions. These are characteristics perceived in pictorial content that enable better visual perception by representing semantic information. [41] Proposes a long short-term memory (LSTM) with transmitted semantic characteristics. The semantic attributes of frames within and between videos are identified with this method. Video frames have semantic properties that visually represent static entities and sights, while semantic properties extracted from the video convey the dynamic temporal dynamics of actions and tasks. Transfer units are used to combine video frames and video attributes. To improve the power of video captioning, semantic attributes can be integrated into encoder/decoder frameworks.

3.5 Transformer-Based Methods

Transformers are increasingly used to overcome the limitations of LSTM. Transformers are essentially encoding and decoding layers. In their study, Chen et al. [42] used several transformer-based encoders to encode video features and transformer-based decoders to generate descriptions. (SBAT) is a sparse boundary sensitive transformer presented in [43] and the reserve attention is built into the encoder. We perform boundary-aware aggregation to solve redundancy in videos and to select different video features. In addition, the decoder uses a boundary-aware strategy and multiple attention headers. In addition, a binding intermodal coding scheme is proposed to promote multimodal interactions.

3.6 Hierarchical Methods

Video information can be retrieved from video using hierarchical methods at different levels, in time or space. Since video deals with time series data, the temporal structure and spatial incidents of entities can be processed using temporal hierarchical approaches, spatial hierarchical methods, or a grouping of both as proposed by Pan et al. [44]. It uses encoder-decoder based architecture to retrieve timing information from video. Time transitions between frames can be detected using the HRNE model at different levels of detail. Yu et al. [45] developed a hierarchical RNN (h-RNN) system for extracting video subtitles. The system can create a sentence or a subtitle paragraph. When creating sentences, it uses a spatial-temporal attention mechanism. The hierarchical multimodal video captioning approach is presented in [46] using RNNs with closed repetition units (GRUs) as decoders. Text and image methods are used in this work.

3.7 Deep Reinforcement Learning

Video with a single event performs well with previous methods. A new method of generating video descriptions of long videos with multiple events using hierarchical reinforcement learning was presented in [47]. Textual content and visual context are reinforcement learning environments. The high-level work packages in the methods-based reinforce the design of the sub-goals at every interval step, and the lower-level work packages document the actions needed to achieve the sub-goals. The end-to-end video subtitle model proposed in the work of Li and Gong [48] is trained from video input to subtitle output using multi-task reinforcement learning. Advanced caption-based techniques use sentence-matching measures such as the Bilingual Evaluation Study (BLEU) [49] and the Consensus-Based Image Descriptive Evaluation (CIDER) [50] as their function reward ability. Unfortunately, these techniques do not consider the logical accuracy of the produced sentences. CIDent (Input Enhancement Reward) is an improved technique proposed in [51], where a low participation score penalizes the CIDer reward. Partial matches are only allowed if they are reasonably implied, and inconsistencies are avoided.

3.8 Memory-Based Architectures

Even though RNNs and LSTMs can predict the next token in a sequence, their memory is small and is mainly used to store context information locally, which prevents them from remembering the key details in the past. The recurrent memory of a neural network degrades with time [52]. However, parametric neural networks cannot generalize to complex language tasks because they implicitly incorporate memory into their weights [53 54]. To achieve this goal, non-local dependencies are incorporated into language models to allow them to flexibly adapt to changing environments and update word probabilities based on long-term context. Neurolinguistic models will be augmented with external memory by introducing soft attention mechanisms [55], allowing them to focus on specific parts of the input, an explicit memory that allows prediction words and cache models [56] can be added on top of the pre-trained language model. By incorporating external memory units into the integrated LSTM network, the actual technical requirements for the LSTM are somewhat reduced. To extend the memory resources of RNNs, a neural Turing machine was developed by Graves et al. in [57]. These machines combine a bank of addressable external memory (i.e., random access memory with read and write operations) with an RNN. The CNN and LSTM networks are combined in the C-LSTM network [58] to enable learning of local and global features of sentences as well as temporal and global semantics of sentences. A memory network acting as a dynamic knowledge base designed to serve as a long-term memory for answering questions [59] was developed to facilitate question answering. However, the discrete model is difficult to train as a back propagation model and requires monitoring at every network layer. Memory networks are further expanded so that they can operate in a contiguous space without supervision [60]. In sentence transmission tasks such as machine translation, single-layer LSTM networks are enhanced with the same unlimited discriminant memory rendering performance as deep RNNs [61]. It has been suggested [62] that stacked memory layers can be used in sequence learning to store and access intermediate representations. [63] Developed hierarchical repeat sequence models to argue for batch memories in order to generate relevant responses to question-answering tasks.

3.9 Dense Video Captioning

The approaches described in the preceding divisions are designed to provide captions for tiny videos featuring single events and phrases. As the span of the video growths, the likelihood of additional events occurring in the video increases. For example, answering a question about a video involving multiple events would require taking those events into account. In addition, events may overlap, and some events may start before the previous event ends. To solve this problem, dense captions can be used. Dense captions describe events in a video by precisely timing them. As video captioning develops, a new emphasis will be put on localizing multiple events as well as creating captions for each event that take overlap into account.

Each area of the scene is given a dense caption while other methods annotate the entire scene. A method of creating image captions called DenseCap was proposed by Johnson et al. [64]. It locates all the prominent areas of the image and then generates descriptive text for each of these areas. This type of method typically includes the following steps:

(1) For each region of the given image, proposals are created.
(2) The regions of the image are determined by CNN.
(3) A language model generates captions for each region based on the results from the previous stage.

Figure 8 illustrates a block diagram of a typical dense captioning system. An LSTM language model [65] is combined with a composite network, a dense localization layer, and a dense subtitle network applied in dense captioning. In dense localization, an image is transmitted with a single transition, implicitly predicting regions of interest in the image. Compared to Faster R-CNN or Faster R-CNN's full network [66], this localization layer does not require external regional recommendations. It also follows the same operating principle as Faster R-CNN [67]. Instead of compounding return on investment, [68] uses differential spatial attention mechanisms [69] as well as bilinear interpolation [70]. This change allows the method of copying through the network and selecting areas to work smoothly. The Visual Genome dataset was used to generate regional lore in this research work.

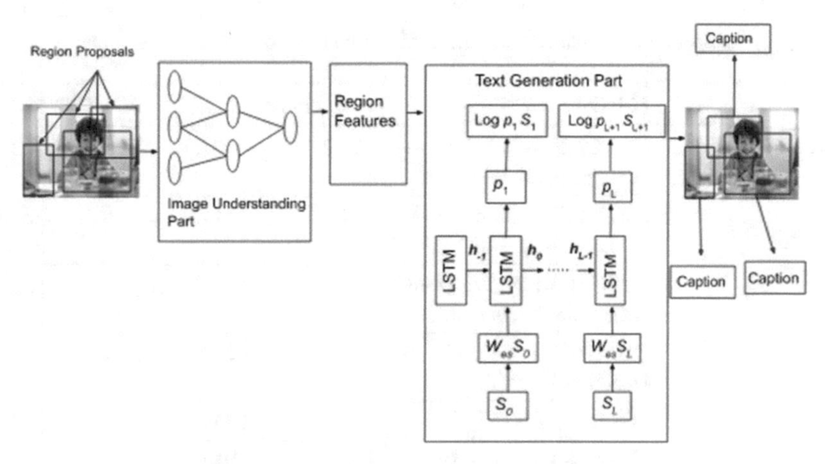

Fig. 8. Dense captioning-Schematic diagram

Work performed by Krishna et al. [71], described the process of detecting and describing all events in the video simultaneously. To accomplish the task, they created actionable recommendations. A set of C3D features [72] are extracted from the video images and passed to the recommendation module. DAP implementation [73] was used in the proposed module. The proposal includes a unique start and end time and a hidden representation. A major contribution of the paper is a large-scale database for dense caption events called ActivityNet Captions. By combining event localization and per-event sentence generation, Li et al. [74] modeled dense video subtitles instead of detecting events before modeling sentence generation. As part of the sentence generation phase, a new descriptive regression design was used in conjunction with a single-shot detection method to locate and time each event described. Sentence Generation (SG) and Timed Event Clauses (TEP) are two key components of this approach. The TEP module generates descriptive propositions and regressions to determine the descriptive complexity of an event. Inferences about an event's descriptive complexity are made using reinforcement learning and semantic properties.

3.10 Performance Analysis Using Evaluation Metrices

All types of models can be evaluated against ground truth using the evaluation metric. The evaluation metrics used in video captioning contribute greatly to assessing how close the model's captions are to human annotation. The five most important metrics in this section are METEOR [75] (Metric for Evaluation of Translation with Explicit ORdering), CIDEr[50](Consensus-based Image Description Evaluation, BLEU[49] (Bilingual

Evaluation Understudy), SPICE[76] (Semantic Propositional Image Captioning Evaluation)., and ROUGE[77] (Recall-Oriented Understudy for Gisting Evaluation). The benchmark results of the studied video captioning approaches [78] discussed in this paper are presented in Table 1.

Table 1. Performance evaluation of various video captioning models

Data Set	Model	METEOR	CIDEr	ROUGE	BLEU
MSVD	2D-CNN + 3D-CNN + compositional decoder	39.2	107.7	78.3	62.3
	2D-CNN + 3D-CNN + compositional LSTM	41.9	111.5	79.5	64.4
	CNN + RNN (with soft attention mechanism)	34.8	88.2	71.7	54.2
	2D CNN + R-CNN + 3D CNN + LSTM	36.5	94.4	73.4	54.6
	LSTM + LSTM	34.3	75	9	52.7
	CNN + LSTM + Two Layer LSTM	34.9	91.0	72.1	53.9
	CNN + temporal deformable convolutional encoder + convolutional decoder + temporal attention mechanism	33.8	76.4	71	53.3
	2D CNN + 3D CNN + LSTM	33.31			52.8
	3D CNN + I3D + Bi-directional LSTM + attention-based LSTM	29.4	48.9	62.0	42.2
MSR-VTT	2D-CNN + 3D-CNN + compositional decoder	31.4	50.6	64.3	45.5
	2D-CNN + 3D-CNN + compositional LSTM	30.1	48.0	63.1	45.6
	CNN + RNN (with soft attention mechanism)	27.5	47.5	60.2	40.9
ActivityNet Captions	LSTM	10.58	39.73		
	3D CNN + LSTM	9.57	28.03		4.37
	C3D + LSTM	10.33	12.93	21.21	2.09
	3D CNN + I3D + Bi-directional LSTM + attention-based LSTM	9.96	28.23	21.17	3.68

4 Conclusion

We include all relevant data in the field of video captions in this survey. The basic architecture of the video captioning system, including visual feature extraction and information management, was presented. This research covers nearly all of the major visual mining methods, from machine learning to more recent deep learning research. We've collected and compared all the current best tactics, along with a statistical analysis of the results. There is also a list of other deep learning-based video captioning approaches currently in development. We have attempted to demonstrate the analysis and comparison between several approaches from different time periods. We found that the performance of strategies improved as technology and research developed. In addition, new methodologies have revealed previously unknown features that are useful in creating more accurate video captions. However, the most challenging aspect of creating video captions is successfully creating subtitles from a variety of videos. When videos contain a lot of activities as well as many objects, the problem becomes more serious. As a result, the accuracy of the video subtitles of the model is degraded. Long video clips with lots of different activities are another challenge, as the models are intended only to record and encode brief actions from video clips. Either way, the newer approaches are attention-based and attention-based modeling to highlight the spatially and temporally relevant elements of a video.

References

1. Kalra, S., Leekha, A.: Survey of convolutional neural networks for image captioning. J. Inf. Optim. Sci. **41**(1), 239–260 (2020)
2. Garbacea, C., Mei, Q.: Neural language generation: Formulation, methods, and evaluation (2020). *arXiv preprint* arXiv:2007.15780
3. Wang, H., Wang, H., Xu, K.: Evolutionary recurrent neural network for image captioning. Neurocomputing **401**, 249–256 (2020)
4. Xu, K., et al.: Show, attend and tell: Neural image caption generation with visual attention. In: International Conference on Machine Learning, pp. 2048–2057. PMLR, June 2015
5. You, Q., Jin, H., Wang, Z., Fang, C., Luo, J.: Image captioning with semantic attention. In: Proceedings of the IEEE Conference on Computer Vision and Pattern Recognition, pp. 4651–4659 (2016)
6. Zhou, L., Xu, C., Koch, P., Corso, J.J.: Watch what you just said: Image captioning with text-conditional attention. In: Proceedings of the on Thematic Workshops of ACM Multimedia 2017, pp. 305–313, October 2017
7. Yao, T., Pan, Y., Li, Y., Qiu, Z., Mei, T.: Boosting image captioning with attributes. In: Proceedings of the IEEE International Conference on Computer Vision, pp. 4894–4902 (2017)
8. Jiang, W., Ma, L., Chen, X., Zhang, H., Liu, W.: Learning to guide decoding for image captioning. In: The Thirty-Second AAAI Conference on Artificial Intelligence, April 2018
9. Jia, X., Gavves, E., Fernando, B., Tuytelaars, T.: Guiding the long-short term memory model for image caption generation. In: Proceedings of the IEEE International Conference on Computer Vision, pp. 2407–2415 (2015)
10. Gan, C., Wang, N., Yang, Y., Yeung, D.Y., Hauptmann, A.G.: DevNet: a deep event network for multimedia event detection and evidence recounting. In: Proceedings of the IEEE Conference on Computer Vision and Pattern Recognition, pp. 2568–2577 (2015)

11. Alamuru, S., Jain, S.: Video event detection, classification and retrieval using ensemble feature selection. Clust. Comput. **24**(4), 2995–3010 (2021). https://doi.org/10.1007/s10586-021-03308-1
12. Kim, H., Lee, S.: A video captioning method based on multi-representation switching for sustainable computing. Sustainability **13**(4), 2250 (2021)
13. Venugopalan, S., Rohrbach, M., Donahue, J., Mooney, R., Darrell, T., Saenko, K.: Sequence to sequence-video to text. In: Proceedings of the IEEE International Conference on Computer Vision, pp. 4534–4542 (2015)
14. Cho, K., et al.: Learning phrase representations using RNN encoder-decoder for statistical machine translation (2014). arXiv preprint arXiv:1406.1078
15. Luong, M.T., Pham, H., Manning, C.D.: Effective approaches to attention-based neural machine translation (2015). *arXiv preprint* arXiv:1508.04025
16. Nallapati, R., Zhou, B., Gulcehre, C., Xiang, B.: Abstractive text summarization using sequence-to-sequence RNNS and beyond (2016). *arXiv preprint* arXiv:1602.06023
17. Vinyals, O., Le, Q.: A neural conversational model (2015). *arXiv preprint* arXiv:1506.05869
18. Vinyals, O., Toshev, A., Bengio, S., Erhan, D.: Show and tell: a neural image caption generator. In: Proceedings of the IEEE Conference on Computer Vision and Pattern Recognition, pp. 3156–3164 (2015)
19. Graves, A., Jaitly, N.: Towards end-to-end speech recognition with recurrent neural networks. In: International Conference on Machine Learning, pp. 1764–1772. PMLR, June 2014
20. Rohrbach, M., Qiu, W., Titov, I., Thater, S., Pinkal, M., Schiele, B.: Translating video content to natural language descriptions. In: Proceedings of the IEEE International Conference on Computer Vision, pp. 433–440 (2013)
21. Guadarrama, S., et al.: YouTube2Text: recognizing and describing arbitrary activities using semantic hierarchies and zero-shot recognition. In: Proceedings of the IEEE International Conference on Computer Vision, pp. 2712–2719 (2013)
22. Venugopalan, S., Xu, H., Donahue, J., Rohrbach, M., Mooney, R., Saenko, K.: Translating videos to natural language using deep recurrent neural networks (2014). *arXiv preprint* arXiv:1412.4729
23. Pan, Y., Mei, T., Yao, T., Li, H., Rui, Y.: Jointly modeling embedding and translation to bridge video and language. In: Proceedings of the IEEE Conference on Computer Vision and Pattern Recognition, pp. 4594–4602 (2016)
24. Xu, H., Venugopalan, S., Ramanishka, V., Rohrbach, M., Saenko, K.: A multi-scale multiple instance video description network (2015). *arXiv preprint* arXiv:1505.05914
25. Venugopalan, S., Hendricks, L.A., Mooney, R., Saenko, K.: Improving LSTM-based video description with linguistic knowledge mined from text (2016). *arXiv preprint* arXiv:1604.01729
26. Donahue, J., et al.: Long-term recurrent convolutional networks for visual recognition and description. In: Proceedings of the IEEE Conference on Computer Vision and Pattern Recognition, pp. 2625–2634 (2015)
27. Nian, F., Li, T., Wang, Y., Wu, X., Ni, B., Xu, C.: Learning explicit video attributes from mid-level representation for video captioning. Comput. Vis. Image Underst. **163**, 126–138 (2017)
28. Sutskever, I., Vinyals, O., Le, Q.V.: Sequence to sequence learning with neural networks. In: Advances in Neural Information Processing Systems, pp. 3104–3112 (2014)
29. Jin, J., Fu, K., Cui, R., Sha, F., Zhang, C.: Aligning where to see and what to tell: image caption with region-based attention and scene factorization (2015). *arXiv preprint* arXiv:1506.06272
30. Blei, D.M., Ng, A.Y., Jordan, M.I.: Latent Dirichlet allocation. J. Mach. Learn. Res. **3**, 993–1022 (2003)
31. Wu, Z.Y.Y.Y.Y., Cohen, R.S.W.W.: Encode, review, and decode: reviewer module for caption generation (2016). *arXiv preprint* arXiv:1605.07912, *3*

32. Pedersoli, M., Lucas, T., Schmid, C., Verbeek, J.: Areas of attention for image captioning. In: Proceedings of the IEEE International Conference on Computer Vision, pp. 1242–1250 (2017)
33. Lu, J., Xiong, C., Parikh, D., Socher, R.: Knowing when to look: adaptive attention via a visual sentinel for image captioning. In: Proceedings of the IEEE Conference on Computer Vision and Pattern Recognition, pp. 375–383 (2017)
34. Cho, K., Courville, A., Bengio, Y.: Describing multimedia content using attention-based encoder-decoder networks. IEEE Trans. Multimedia **17**(11), 1875–1886 (2015)
35. Chen, J., Pan, Y., Li, Y., Yao, T., Chao, H., Mei, T.: Temporal deformable convolutional encoder-decoder networks for video captioning. In: Proceedings of the AAAI Conference on Artificial Intelligence, vol. 33, no. 01, pp. 8167–8174, July 2019
36. Yao, L., et al.: Describing videos by exploiting temporal structure. In: Proceedings of the IEEE International Conference on Computer Vision, pp. 4507–4515 (2015)
37. Yang, Y., et al.: Video captioning by adversarial LSTM. IEEE Trans. Image Process. **27**(11), 5600–5611 (2018)
38. Bin, Y., Yang, Y., Shen, F., Xie, N., Shen, H.T., Li, X.: Describing video with attention-based bidirectional LSTM. IEEE Trans. Cybern. **49**(7), 2631–2641 (2018)
39. Gao, L., Guo, Z., Zhang, H., Xu, X., Shen, H.T.: Video captioning with attention-based LSTM and semantic consistency. IEEE Trans. Multimedia **19**(9), 2045–2055 (2017)
40. Long, X., Gan, C., De Melo, G.: Video captioning with multi-faceted attention. Trans. Assoc. Comput. Linguist. **6**, 173–184 (2018)
41. Pan, Y., Yao, T., Li, H., Mei, T.: Video captioning with transferred semantic attributes. In: Proceedings of the IEEE Conference on Computer Vision and Pattern Recognition, pp. 6504–6512 (2017)
42. Chen, M., Li, Y., Zhang, Z., Huang, S.: TVT: two-view transformer network for video captioning. In: Asian Conference on Machine Learning, pp. 847–862. PMLR, November 2018
43. Jin, T., Huang, S., Chen, M., Li, Y., Zhang, Z.: SBAT: video captioning with sparse boundary-aware transformer (2020). *arXiv preprint* arXiv:2007.11888
44. Pan, P., Xu, Z., Yang, Y., Wu, F., Zhuang, Y.: Hierarchical recurrent neural encoder for video representation with application to captioning. In: Proceedings of the IEEE Conference on Computer Vision and Pattern Recognition, pp. 1029–1038 (2016)
45. Yu, H., Wang, J., Huang, Z., Yang, Y., Xu, W.: Video paragraph captioning using hierarchical recurrent neural networks. In: Proceedings of the IEEE Conference on Computer Vision and Pattern Recognition, pp. 4584–4593 (2016)
46. Liu, A.A., Xu, N., Wong, Y., Li, J., Su, Y.T., Kankanhalli, M.: Hierarchical & multimodal video captioning: discovering and transferring multimodal knowledge for vision to language. Comput. Vis. Image Underst. **163**, 113–125 (2017)
47. Wang, X., Chen, W., Wu, J., Wang, Y.F., Wang, W.Y.: Video captioning via hierarchical reinforcement learning. In: Proceedings of the IEEE Conference on Computer Vision and Pattern Recognition, pp. 4213–4222 (2018)
48. Li, L., Gong, B.: End-to-end video captioning with multitask reinforcement learning. In: 2019 IEEE Winter Conference on Applications of Computer Vision (WACV), pp. 339–348. IEEE, January 2019
49. Papineni, K., Roukos, S., Ward, T., Zhu, W.J.: BLEU: a method for automatic evaluation of machine translation. In: Proceedings of the 40th Annual Meeting of the Association for Computational Linguistics, pp. 311–318, 2002, July
50. Vedantam, R., Lawrence Zitnick, C., Parikh, D.: CIDEr: consensus-based image description evaluation. In: Proceedings of the IEEE Conference on Computer Vision and Pattern Recognition, pp. 4566–4575 (2015)

51. Pasunuru, R., Bansal, M.: Reinforced video captioning with entailment rewards (2017). *arXiv preprint* arXiv:1708.02300
52. Khandelwal, U., He, H., Qi, P., Jurafsky, D.: Sharp nearby, fuzzy far away: how neural language models use context (2018). *arXiv preprint* arXiv:1805.04623
53. Nematzadeh, A., Ruder, S., Yogatama, D.: On memory in human and artificial language processing systems. In: Proceedings of ICLR Workshop on Bridging AI and Cognitive Science (2020)
54. Daniluk, M., Rocktäschel, T., Welbl, J., Riedel, S.: Frustratingly short attention spans in neural language modeling (2017). *arXiv preprint* arXiv:1702.04521
55. Tran, K., Bisazza, A., Monz, C.: Recurrent memory networks for language modeling (2016). *arXiv preprint* arXiv:1601.01272
56. Grave, E., Joulin, A., Usunier, N.: Improving neural language models with a continuous cache (2016). *arXiv preprint* arXiv:1612.04426
57. Graves, A., Wayne, G., Danihelka, I.: Neural turing machines (2014). *arXiv preprint* arXiv: 1410.5401
58. Zhou, C., Sun, C., Liu, Z., Lau, F.: A C-LSTM neural network for text classification (2015). *arXiv preprint* arXiv:1511.08630
59. Weston, J., Chopra, S., Bordes, A.: Memory networks (2014). *arXiv preprint* arXiv:1410. 3916
60. Sukhbaatar, S., Szlam, A., Weston, J., Fergus, R.: End-to-end memory networks (2015). *arXiv preprint* arXiv:1503.08895
61. Grefenstette, E., Hermann, K.M., Suleyman, M., Blunsom, P.: Learning to transduce with unbounded memory. Adv. Neural. Inf. Process. Syst. **28**, 1828–1836 (2015)
62. Meng, F., Lu, Z., Tu, Z., Li, H., Liu, Q.: A deep memory-based architecture for sequence-to-sequence learning (2015). *arXiv preprint* arXiv:1506.06442
63. Kumar, A., et al.: Ask me anything: dynamic memory networks for natural language processing. In: International Conference on Machine Learning, pp. 1378–1387. PMLR, June 2016
64. Johnson, J., Karpathy, A., Fei-Fei, L.: DenseCap: fully convolutional localization networks for dense captioning. In: Proceedings of the IEEE Conference on Computer Vision and Pattern Recognition, pp. 4565–4574 (2016)
65. Han, Y., Li, G.: Describing images with hierarchical concepts and object class localization. In: Proceedings of the 5th ACM on International Conference on Multimedia Retrieval, pp. 251–258, June 2015
66. Girshick, R.: Fast R-CNN. In: Proceedings of the IEEE International Conference on Computer Vision, pp. 1440–1448 (2015)
67. Ren, S., He, K., Girshick, R., Sun, J.: Faster R-CNN: towards real-time object detection with region proposal networks. Adv. Neural. Inf. Process. Syst. **28**, 91–99 (2015)
68. Gregor, K., Danihelka, I., Graves, A., Rezende, D., Wierstra, D.: DRAW: a recurrent neural network for image generation. In: International Conference on Machine Learning, pp. 1462–1471. PMLR, June 2015
69. Jaderberg, M., Simonyan, K., Zisserman, A.: Spatial transformer networks. Adv. Neural. Inf. Process. Syst. **28**, 2017–2025 (2015)
70. Krishna, R., et al.: Visual genome: connecting language and vision using crowdsourced dense image annotations. Int. J. Comput. Vision **123**(1), 32–73 (2017)
71. Krishna, R., Hata, K., Ren, F., Fei-Fei, L., Carlos Niebles, J.: Dense-captioning events in videos. In: Proceedings of the IEEE International Conference on Computer Vision, pp. 706–715 (2017)
72. Ji, S., Xu, W., Yang, M., Yu, K.: 3D convolutional neural networks for human action recognition. IEEE Trans. Pattern Anal. Mach. Intell. **35**(1), 221–231 (2012)

73. Escorcia, V., Caba Heilbron, F., Niebles, J. C., Ghanem, B.: Daps: deep action proposals for action understanding. In: Leibe, B., Matas, J., Sebe, N., Welling, M. (eds.) ECCV 2016. LNCS, vol. 9907, pp. 768–784. Springer, Cham (2016). https://doi.org/10.1007/978-3-319-46487-9_47
74. Li, Y., Yao, T., Pan, Y., Chao, H., Mei, T.: Jointly localizing and describing events for dense video captioning. In: Proceedings of the IEEE Conference on Computer Vision and Pattern Recognition, pp. 7492–7500 (2018)
75. Banerjee, S., Lavie, A.: METEOR: an automatic metric for MT evaluation with improved correlation with human judgments. In: Proceedings of the ACL Workshop on Intrinsic and Extrinsic Evaluation Measures for Machine Translation and/or Summarization, pp. 65–72, June 2005
76. Anderson, P., Fernando, B., Johnson, M., Gould, S.: Spice: semantic propositional image caption evaluation. In: Leibe, B., Matas, J., Sebe, N., Welling, M. (eds.) ECCV 2016. LNCS, vol. 9909, pp. 382–398. Springer, Cham (2016). https://doi.org/10.1007/978-3-319-46454-1_24
77. Lin, C.Y.: ROUGE: a package for automatic evaluation of summaries. In: Text Summarization Branches Out, pp. 74–81, July 2004
78. Islam, S., Dash, A., Seum, A., Raj, A.H., Hossain, T., Shah, F.M.: Exploring video captioning techniques: a comprehensive survey on deep learning methods. SN Comput. Sci. **2**(2), 1–28 (2021)
79. Perez-Martin, J., Bustos, B., Pérez, J.: Attentive visual semantic specialized network for video captioning. In: 2020 25th International Conference on Pattern Recognition (ICPR), pp. 5767–5774. IEEE, January 2021
80. Perez-Martin, J., Bustos, B., Pérez, J.: Improving video captioning with temporal composition of a visual-syntactic embedding. In: Proceedings of the IEEE/CVF Winter Conference on Applications of Computer Vision, pp. 3039–3049 (2021)
81. Liu, S., Ren, Z., Yuan, J.: SibNet: sibling convolutional encoder for video captioning. In: Proceedings of the 26th ACM International Conference on Multimedia, pp. 1425–1434, October 2018
82. Tan, G., Liu, D., Wang, M., Zha, Z.J.: Learning to discretely compose reasoning module networks for video captioning (2020). arXiv preprint arXiv:2007.09049
83. Guo, Y., Zhang, J., Gao, L.: Exploiting long-term temporal dynamics for video captioning. World Wide Web **22**(2), 735–749 (2018). https://doi.org/10.1007/s11280-018-0530-0
84. Wang, B., Ma, L., Zhang, W., Jiang, W., Wang, J., Liu, W.: Controllable video captioning with POS sequence guidance based on gated fusion network. In: Proceedings of the IEEE/CVF International Conference on Computer Vision, pp. 2641–2650 (2019)
85. Wang, J., Wang, W., Huang, Y., Wang, L., Tan, T.: M3: multimodal memory modelling for video captioning. In: Proceedings of the IEEE Conference on Computer Vision and Pattern Recognition, pp. 7512–7520 (2018)
86. Wang, X., Wu, J., Zhang, D., Su, Y., Wang, W.Y.: Learning to compose topic-aware mixture of experts for zero-shot video captioning. In: Proceedings of the AAAI Conference on Artificial Intelligence, vol. 33, no. 01, pp. 8965–8972, July 2019
87. Wang, T., Zheng, H., Yu, M., Tian, Q., Hu, H.: Event-centric hierarchical representation for dense video captioning. IEEE Trans. Circuits Syst. Video Technol. **31**(5), 1890–1900 (2020)
88. Kim, J., Choi, I., Lee, M.: Context aware video caption generation with consecutive differentiable neural computer. Electronics **9**(7), 1162 (2020)
89. Zhang, Z., Xu, D., Ouyang, W., Tan, C.: Show, tell and summarize: dense video captioning using visual cue aided sentence summarization. IEEE Trans. Circuits Syst. Video Technol. **30**(9), 3130–3139 (2019)

A Novel Hybrid Compression Algorithm for Remote Sensing Imagery

Swetha Vura[1,2](✉) and C. R. Yamuna Devi[3]

[1] Department of ECE, SoET, CMR University, Bengaluru, India
vuraswetha@gmail.com
[2] VTU, Belagavi, India
[3] Department of TCE, Dr. Ambedkar Institute of Technology, Bengaluru, India

Abstract. The remote sensing images that are captured by satellites are very high in resolution and involve an enormous amount of memory for storage. These images must be compressed so that more images can be transmitted in the same bandwidth. Extensive study has taken place in the field of compression for remote sensing imagery. In this paper, a novel hybrid algorithm is proposed that increases the compression ratio while maintaining the quality of the image. This hybrid methodology comprises wavelet transform, Run-length, and Arithmetic coding to achieve higher PSNR with minimal MSE. The experimental results up to three levels of wavelet decomposition indicate the range of the obtained parameter values like compression ratio, PSNR, and MSE. The compression is implemented by MATLAB software tool.

Keywords: Remote Sensing · Satellite · Compression · Wavelet Transform · Run-length Encoding · Arithmetic Coding

1 Introduction

Remote sensing is the field where information about an area is obtained without any physical connection with it. Satellite communication integrates remote sensing using radio waves for transmitting and receiving data. The onboard sensors discern and record the emitted or reflected energy. These sensors have the capacity to obtain information at a very large scale which is cost-effective when compared to other acquisitions [1]. The imaging systems capture an area across different wavelength bands present in the electromagnetic spectrum which results in a huge amount of onboard data [2]. Their technology is advanced and incorporates highly capable processing methodologies which account for superior resolutions [3]. The resolutions may be classified as spatial, spectral, temporal, and radiometric resolutions [4]. Some of the Earth Observation satellites launched by the Indian Space Research Organisation (ISRO) are IRS, CARTOSAT, RISAT, RESOURCESAT, EOS, etc. The images captured by satellites fall into the categories of panchromatic, multispectral, and hyperspectral imagery [5, 6]. They consist of several bands which makes it difficult to transmit more images in a single bandwidth. This situation paves the way for the process of compression so that the irrelevancy and

H. Sharma et al. (Eds.): ICIVC 2022, PALO 17, pp. 88–100, 2023.
https://doi.org/10.1007/978-3-031-31164-2_8

redundancy are reduced, and more images can be transmitted. Image compression may be broadly categorized [7] as lossy or lossless. Depending on the application and requirement, any one of the above compression methodologies is chosen for a specific or desired outcome. Innumerous techniques have been proposed in the field of image compression that comprise their own advantages and disadvantages. Compression pertaining to the remote sensing imagery has the onus of reducing the repetitive information in the vast available amount of data. Apart from achieving a good compression ratio, image quality ought to be maintained [8].

This paper proposes a novel hybrid algorithm that has the advantages of wavelet transform and arithmetic coding along with the simplicity of run-length encoding. The rest of the paper is organized as follows. Image Acquisition System is provided in Sect. 2. Wavelet transform is explained in Sect. 3. The coding techniques are discussed in Sect. 4. The proposed methodology is detailed in Sect. 5. Experimental results are presented in Sect. 6 and the conclusion is given in Sect. 7.

2 Image Acquisition System

Image Acquisition is the procedure of sensing and acquiring an image from a particular source for processing or analyzing it. It can be obtained using a single sensor, sensor strip, or array of sensors.

Bhuvan which means Earth in Sanskrit is the official Geo-platform of the Indian Space Research Organisation (ISRO). It has several options to view and analyze images in 2D and 3D formats. It was launched on 12th August 2009 as a web-based GIS tool. This website contains geo-spatial data that includes numerous tools and services for public usage. The imagery of a vast range of satellites obtained from sensors can be viewed, evaluated, and downloaded at no cost. Bhuvan's applications are from various sectors like Water, Forestry, Agriculture, E-governance, Tourism, Urban and Rural, State and Central government, etc. The steps required to extract the imagery are shown in the following flowchart in Fig. 1.

The procedure to select category, subcategory and products are shown in Fig. 2.

The options under Category are

- Satellite/Sensor
- Theme/Products
- Program/Products

If the Satellite/Sensor is chosen, the available subcategories are

- Oceansat-2: OCM
- Resourcesat-1/Resourcesat-2: LISS-III
- IMS-1: Hyperspectral Imager (HySi)
- Cartosat-1
- Resourcesat-1/Resourcesat-2: AWiFS
- SCATSAT-1: Scatterometer

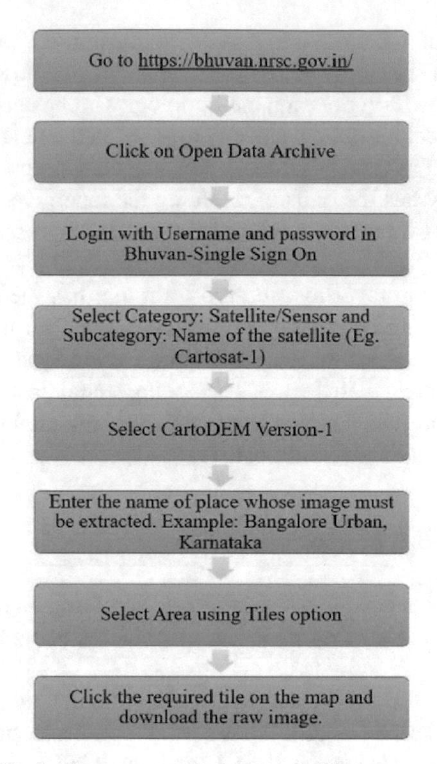

Fig. 1. Steps to extract images in Bhuvan portal

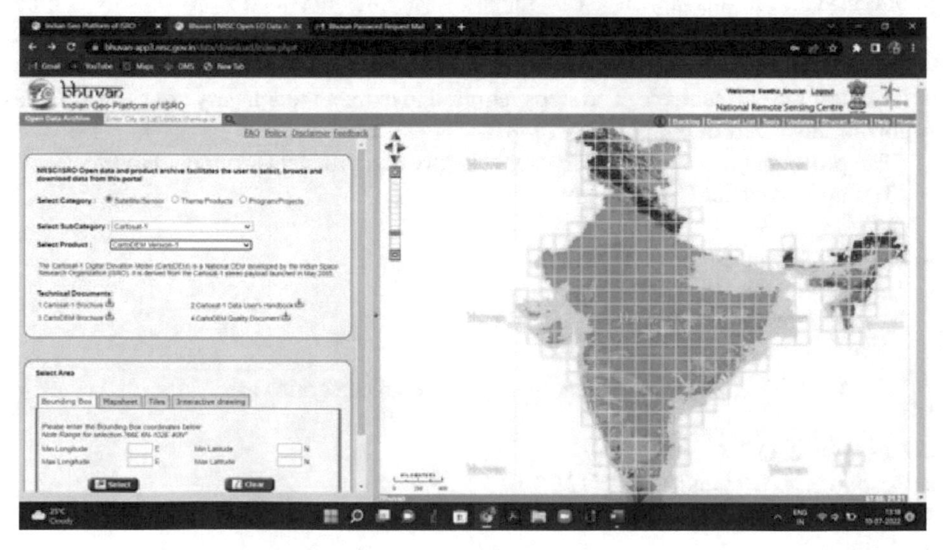

Fig. 2. Selection of Categories

After the selection of the subcategory, various versions are available under Products. The area can be extracted by Bounding Box, Mapsheet, Tiles or Interactive drawing as depicted in Fig. 3.

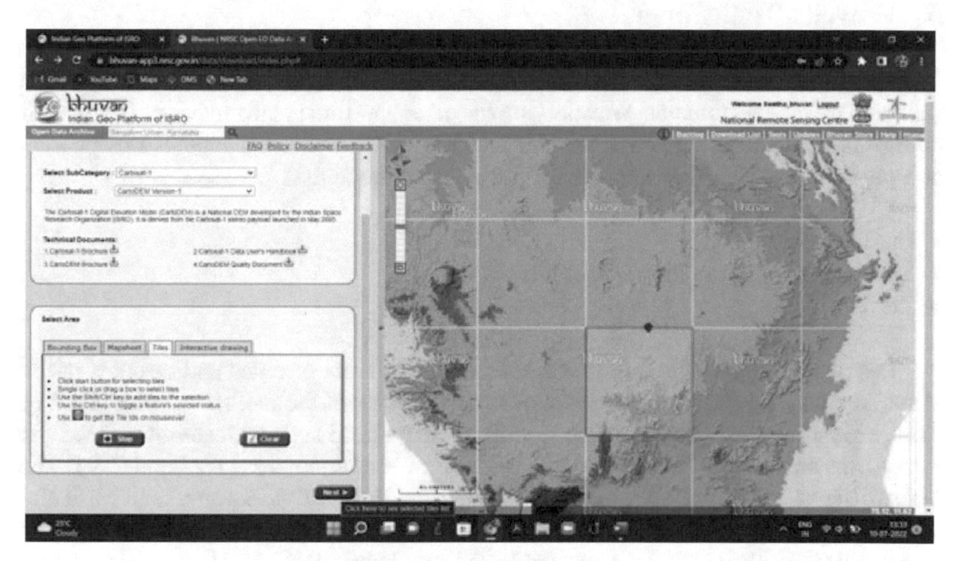

Fig. 3. Tile Selection

The raw image can be viewed and downloaded as in Fig. 4. The Toposheet number, Bounding Box latitude and longitude and Metadata can be understood for each tile.

Fig. 4. Raw Image Download. Note: Image Source for Fig. 2 to Fig. 4 are from Bhuvan Portal.

The images will be downloaded in the form of bands as GeoTIFF files which later have to be merged for further processing.

3 Wavelet Transform

Waves can be understood as oscillations that are periodic in nature and can be mapped on time and frequency domains. Wavelets, on the other hand are localized waves where their energy is centralized in time or frequency. They are time-domain signals that possess compact support [9]. They possess the following characteristics.

- Finite Energy
- Sharp localization in time and frequency
- Zero average

The signals in Fourier analysis are depicted in terms of sines and cosines, but in Wavelet Analysis, they are in translations and dilations of the mother wavelet [10, 11]. This analysis divides the signal into several fragments and later evaluates them. Wavelet transforms are broadly categorized into discrete and continuous as DWT and CWT. The main difference is that CWT works on each scale while DWT works only for a particular subset [12]. The advantage of wavelets is that they can segregate the coarse details from fine details [13]. JPEG 2000 is a wavelet-based compression standard that provides better compression ratios as compared to JPEG. It has the advantage of compressing the entire image rather than block by block as performed in DCT [14]. Some of the most used wavelets are Haar, Daubechies, Mexican Hat, Symlet, Coiflet Meyr, etc.

The Haar wavelet function $\Psi(t)$ shown in Fig. 5 is the simplest one [15] which can be defined by

$$\begin{aligned} \Psi(t) &= 1,\ 0 \le \mathrm{t} < 0.5 \\ &-1,\ 0.5 \le \mathrm{t} < 1 \\ &0,\ \text{otherwise} \end{aligned} \tag{1}$$

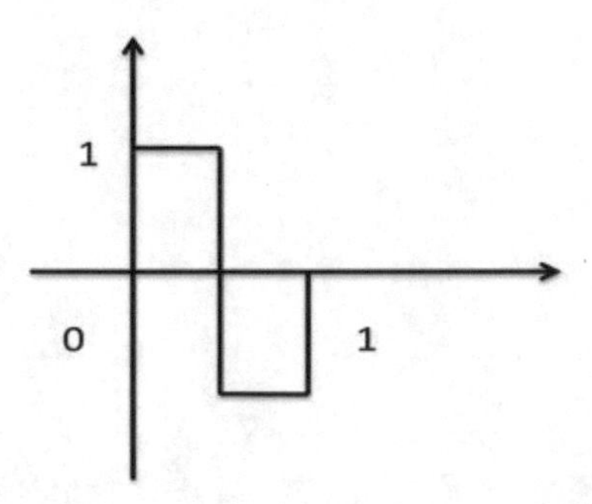

Fig. 5. Haar wavelet function Ψ(t)

The Haar wavelet impulse response can be depicted as $\{1/\sqrt{2}, -1/\sqrt{2}\}$. It can be represented as $1/\sqrt{2}\{1, -1\}$. This can be analyzed as a High Pass Filter (HPF). It computes the finite differences between the consecutive pixels of an image and keeps track of the details.

The Haar scaling function $\Phi(t)$ as shown in Fig. 6 can be defined as

$$\begin{aligned} \Phi(t) &= 1,\ 0 \leq t < 1 \\ &= 0,\ \text{otherwise} \end{aligned} \tag{2}$$

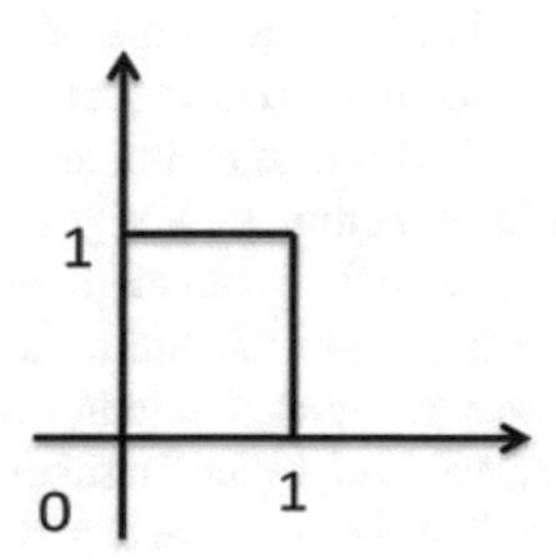

Fig. 6. Haar scaling function $\Phi(t)$

The Haar wavelet impulse response can be depicted as $\{1/\sqrt{2}, 1/\sqrt{2}\}$. It can be represented as $1/\sqrt{2}\{1, 1\}$. This can be analyzed as a Low Pass Filter (LPF). It computes the average between consecutive pixels and averages/smoothes the image.

The wavelet-based compression decomposes the image into many subbands and then performs transformation for each subband individually. The Pyramidal decomposition algorithm calculates the averages and differences in the first iteration. The same is performed on columns for the second iteration. This gives rise to four subbands as depicted in Fig. 7. The top-left subbands comprise the average (LPF) and the remaining subbands

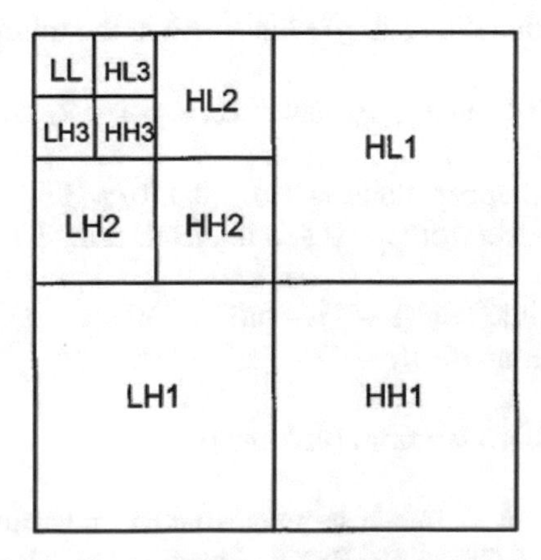

Fig. 7. Image Pyramidal Decomposition

comprise the differences (HPF) [16]. The average is called the coarsest representation and the remaining are the details.

After the image is wavelet transformed, it is quantized and then Run-length and Arithmetic coding is implemented for further compression.

4 Coding Techniques

Various entropy encoding algorithms have been developed to compress images. It is a technique that is implemented after the quantization stage. It assigns a distinctive prefix-free code to each distinct symbol in the input. Some of the image compression entropy algorithms are Huffman coding, Arithmetic coding, etc.

Run-length encoding is a lossless image compression technique where it encodes the image line by line with tracks of a similar color [17]. It works best for images with repetitive data. The sequence of consecutive data elements is replaced by two segments, one represents the count of repetition, and the other, the value in the stream. It means that if a data value 'd' is repeating 'n' times, the long run is replaced by a single value 'nd'. In remote sensing imagery, we find a huge number of similar pixels, particularly in the background of the image. RLE is therefore a perfect compression technique for such satellite images.

Arithmetic coding is a form of entropy encoding which is implemented for lossless image compression. It is executed to overcome the disadvantages of Huffman coding. The main drawback of the Huffman scheme is that it does not work well with symbols of high probability. Arithmetic coding entirely circumvents the idea of replacing input symbols with particular codes. It represents the entire stream of input symbols with a single floating-point number between [0, 1). As and when each symbol is encoded, the range is reduced depending on its probability of occurrence. When the range is suitably low, the partitioning is stopped and the codeword is assigned to a binary fraction that exists in the range [18]. The regularly used characters are stored with more bits and not so regularly used with few bits [19]. The compression ratio obtained is more and provides an optimum result.

The algorithm of arithmetic coding is illustrated in the following steps.

- Consider a sequence of random variables $X = \{x1, x2, x3\ldots\}$ and a sequence of symbols $S = \{s1, s2, s3, \ldots\}$
- Initialize the lower and upper limits as $l(0) = 0$, $u(0) = 1$
- Repeat the interval update from $i = 0$ to n in increments of 1

  o Lower interval limit $l(i) = l(I - 1) - (u(i - 1))Fx(x - 1)$
  o Upper interval limit $u(i) = l(i - 1) - (u(i - 1))Fx(x)$

- Output the code word as a tag from $[l(n), u(n))$

The resultant compressed image is reconstructed and analyzed with the help of evaluation parameters like CR, MSE, PSNR, encoding and decoding times.

Compression ratio (CR) can be defined as the ratio of the size of input image to the compressed image.

$$\text{CR} = \text{Input Image size} / \text{Compressed Image size} \tag{3}$$

Mean Square Error (MSE) compares the input image with the reconstructed image to evaluate the amount of error [20]. It can be depicted by the following formula.

$$\text{MSE} = 1/\text{MN}\left\{\sum\nolimits_{i=1}^{M} (I(i,j)\sum\nolimits_{j=1}^{N} R(i,j))^2\right\} \tag{4}$$

where M and N are rows and columns of the image, I(i, j) is the original input image and R(i, j) is the reconstructed image.

Peak Signal to Noise Ratio (PSNR) is the ratio of maximum signal power to the reconstruction noise power [21]. It is shown as

$$\text{PSNR} = 10\log_{10}\left(\text{MAX}^2/\text{MSE}\right) \tag{5}$$

When the image is represented using 8 bits/sample, MAX value becomes 255 and the Eq. (2) transforms as

$$\text{PSNR} = 10\log_{10}\left(255^2/\text{MSE}\right) \tag{6}$$

The values of MSE and PSNR are inversely proportional to each other.

5 Proposed Methodology

The proposed methodology in this paper presents a hybrid compression scheme that combines DWT, Run-length encoding, and Arithmetic coding techniques. Three decomposition levels are administered in DWT to analyze the evaluation parameters like CR, MSE, and PSNR. The input remote sensing imagery used for the proposed compression technique is downloaded from the Bhuvan Geoportal and ISRO websites.

The block diagram of the proposed method is given in Fig. 8.

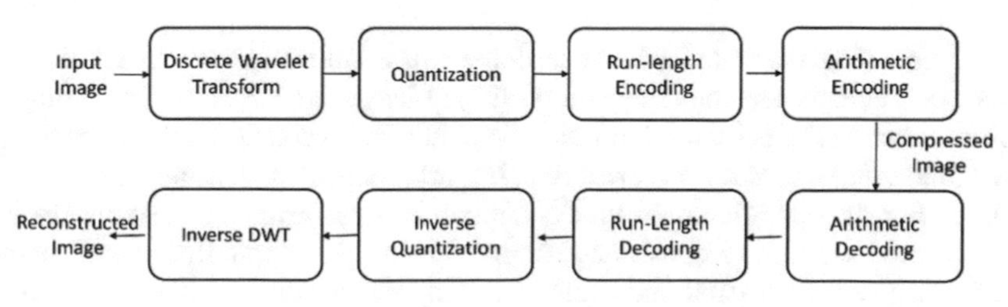

Fig. 8. Block diagram of proposed methodology

The input images are extracted from Bhuvan and ISRO portals. The images are downloaded as bands and the number of bands depends on the remote sensing satellite. They are merged using QGIS software with the required specifications. The images are then subjected to wavelet transform-based decomposition. Three levels of decomposition are implemented and analyzed. The input image x[n] is decomposed using LPF and HPF to obtain approximate and detail coefficients as depicted in 3D wavelet tree structure (Fig. 9). The subband LL3 gives the coarsest representation for third level of decomposition.

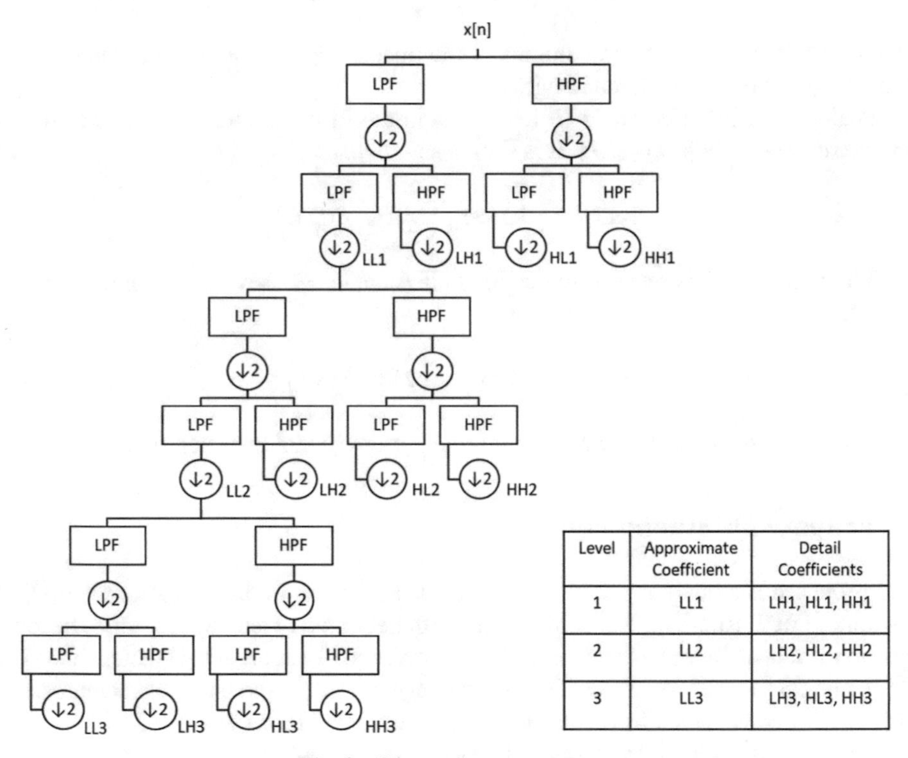

Level	Approximate Coefficient	Detail Coefficients
1	LL1	LH1, HL1, HH1
2	LL2	LH2, HL2, HH2
3	LL3	LH3, HL3, HH3

Fig. 9. 3D wavelet tree structure

After obtaining the transformed coefficients, it is quantized and thresholded. The resultant coefficients are implemented with Run-length and Arithmetic coding algorithms to enhance the compression ratio. To obtain the reconstructed image, the compressed image undergoes reverse operation. It is arithmetic decoded and then Run-length decoded. Then the operation of inverse quantization is performed. At last, inverse DWT is implemented to get the reconstructed image. The original input image and the reconstructed image are compared and evaluated using the parameters like CR, MSE, and PSNR. Encoding and decoding times are also calculated and examined.

6 Experimental Results

In the proposed methodology, various grayscale and color images with a diverse range of resolutions and details are taken into consideration. Wavelet decomposition for three levels is considered using the Haar wavelet. The evaluation metrics like CR, MSE, and PSNR are calculated to examine the proposed hybrid compression methodology. The obtained values from the MATLAB code are tabulated and plotted as graphs for better visualization and understanding.

The image of an area in Bengaluru taken by the remote sensing CARTOSAT-1 satellite is downloaded from Bhuvan Geoportal as a Geo-TIFF file with the metadata as listed in Table 1.

The Geo-TIFF image cannot be viewed directly as it is obtained in the form of bands and must undergo compression and reconstruction to be visualized. The Cartosat-1 image

Table 1. Satellite metadata of Cartosat-1 imagery.

Satellite Metadata	Specifications
Name of the satellite	Cartosat-1
Sensor	PAN (2.5 m) Stereo Data
File Format	GeoTIFF
Bits per Pixel	16 bit
Spatial Resolution	1 arc sec
Spatial Resolution Unit	m

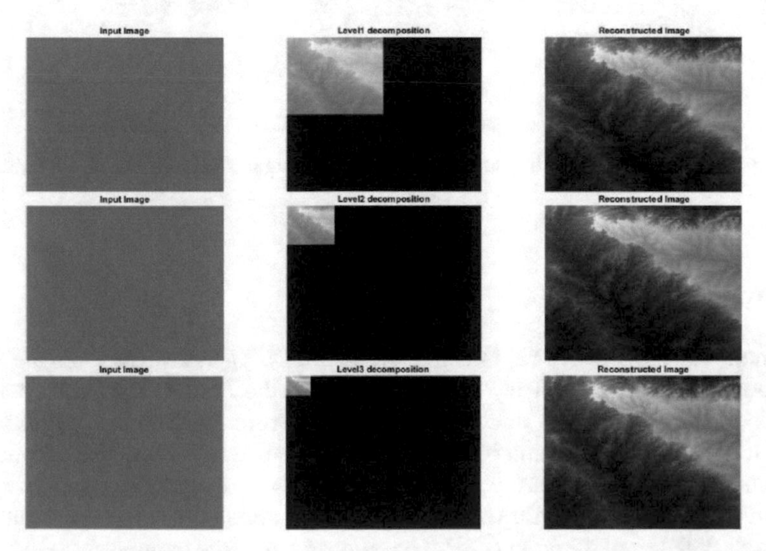

Fig. 10. Bengaluru CARTOSAT-1 input and reconstructed images for three levels of decomposition

with three levels of decomposition can be viewed in Fig. 10. The obtained evaluation parametric values of CR, MSE and PSNR for the image are shown in Table 2.

The comparison of the parametric values of CARTOSAT-1 imagery can be visualized as a plot in Fig. 11.

Table 2. Evaluation parametric values of Bengaluru CARTOSAT-1 imagery.

Levels of decomposition	CR	MSE	PSNR
L1	4.1941	1.3478	46.8686
L2	4.4991	1.3366	46.9047
L3	4.7003	1.3371	46.9031

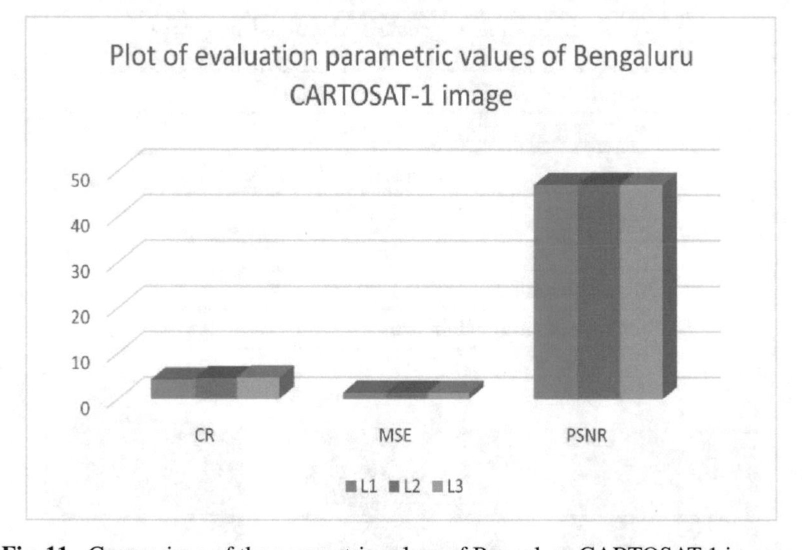

Fig. 11. Comparison of the parametric values of Bengaluru CARTOSAT-1 imagery

7 Conclusion

The remote sensing images captured by the CARTOSAT series satellites are used in the proposed methodology. They are extracted from the Bhuvan Geoportal and ISRO websites. As the images are downloaded as bands from the Bhuvan portal, they are merged into a single multi-band image by the QGIS software. The obtained imagery is wavelet transformed by a Haar wavelet in three levels of decomposition. The resultant transformed coefficients are run-length and arithmetic coded to obtain high compression ratios, high PSNR, and minimal MSE. The comparison of the parametric values clearly indicates that the proposed hybrid compression methodology results in better CR and PSNR compared to other compression techniques for remote sensing imagery.

References

1. Chawla, I., Karthikeyan, L., Mishra, A.K.: A review of remote sensing applications for water security: quantity, quality, and extremes. J. Hydrol. **585**, 124826 (2020). ISSN 0022-1694. https://doi.org/10.1016/j.jhydrol.2020.124826
2. Gunasheela, K.S., Prasantha, H.S.: Satellite image compression-detailed survey of the algorithms. In: Guru, D.S., Vasudev, T., Chethan, H.K., Sharath Kumar, Y.H. (eds.) Proceedings of International Conference on Cognition and Recognition. LNNS, vol. 14, pp. 187–198. Springer, Singapore (2018). https://doi.org/10.1007/978-981-10-5146-3_18
3. Toth, C., Jóźków, G.: Remote sensing platforms and sensors: a survey. ISPRS J. Photogrammetry Remote Sens. **115**, 22–36 (2016). ISSN 0924-2716. https://doi.org/10.1016/j.isprsjprs.2015.10.004
4. Belgiu, M., Stein, A.: Spatiotemporal image fusion in remote sensing. Remote Sens. **11**(7), 818 (2019). https://doi.org/10.3390/rs11070818
5. Sara, D., Mandava, A.K., Kumar, A., Duela, S., Jude, A.: Hyperspectral and multispectral image fusion techniques for high resolution applications: a review. Earth Sci. Inf. **14**(4), 1685–1705 (2021). https://doi.org/10.1007/s12145-021-00621-6
6. Vivone, G.: Multispectral and hyperspectral image fusion in remote sensing: a survey. Inf. Fusion **89**, 405–417 (2023). ISSN 1566-2535. https://doi.org/10.1016/j.inffus.2022.08.032
7. Afjal, M.I., Mamun, M.A., Uddin, M.P.: Band reordering heuristic for lossless satellite image compression with CCSDS. In: 2018 International Conference on Computer, Communication, Chemical, Material and Electronic Engineering (IC4ME2), pp. 1–4 (2018). https://doi.org/10.1109/IC4ME2.2018.8465493
8. Ma, S., Zhang, X., Jia, C., Zhao, Z., Wang, S., Wang, S.: Image and video compression with neural networks: a review. IEEE Trans. Circuits Syst. Video Technol. **30**(6), 1683–1698 (2020). https://doi.org/10.1109/TCSVT.2019.2910119
9. Polikar, R.: The story of wavelets. Phys. Mod. Top. Mech. Electric. Eng., 192–197 (1999)
10. Sifuzzaman, M., Rafiq Islam, M., Ali, M.Z.: Application of wavelet transform and its advantages compared to Fourier transform (2009)
11. Abramovich, F., Bailey, T.C., Sapatinas, T.: Wavelet analysis and its statistical applications. J. Roy. Stat. Soc. Ser. D (Stat.) **49**(1), 1–29 (2000)
12. Alessio, S.M.: Discrete wavelet transform (DWT). In: Digital Signal Processing and Spectral Analysis for Scientists. SCT, pp. 645–714. Springer, Cham (2016). https://doi.org/10.1007/978-3-319-25468-5_14
13. Chowdhury, M., Hoque, M., Khatun, A.: Image compression using discrete wavelet transform. Int. J. Comput. Sci. Issues (IJCSI) **9**(4), 327 (2012)
14. Sridhar, S., Rajesh Kumar, P., Ramanaiah, K.V.: Wavelet transform techniques for image compression-an evaluation. Int. J. Image Graph. Sig. Process. **6**(2), 54 (2014)
15. Xiao, P.: Image compression by wavelet transform. East Tennessee State University (2001)
16. Thakral, S., Manhas, P.: Image processing by using different types of discrete wavelet transform. In: Luhach, A.K., Singh, D., Hsiung, P.-A., Hawari, K.B.G., Lingras, P., Singh, P.K. (eds.) ICAICR 2018. CCIS, vol. 955, pp. 499–507. Springer, Singapore (2019). https://doi.org/10.1007/978-981-13-3140-4_45
17. Abdmouleh, M., Masmoudi, A., Bouhlel, M.: A new method which combines arithmetic coding with RLE for lossless image compression. J. Softw. Eng. Appl. **5**(1), 41–44 (2012). https://doi.org/10.4236/jsea.2012.51007
18. Yang, M., Bourbakis, N.: An overview of lossless digital image compression techniques. In: 48th Midwest Symposium on Circuits and Systems, vol. 2, pp. 1099–1102 (2005). https://doi.org/10.1109/MWSCAS.2005.1594297

19. Boopathiraja, S., Kalavathi, P., Chokkalingam, S.: A hybrid lossless encoding method for compressing multispectral images using LZW and arithmetic coding. Int. J. Comput. Sci. Eng. **6**, 313–318 (2018)
20. Bindu, K., Ganpati, A., Sharma, A.K.: A comparative study of image compression algorithms. Int. J. Res. Comput. Sci. **2**(5), 37 (2012)
21. Kumar, G., et al.: A review: DWT-DCT technique and arithmetic-Huffman coding based image compression. Int. J. Eng. Manuf. **5**(3), 20 (2015)

Optimization of Localization in UAV-Assisted Emergency Communication in Heterogeneous IoT Networks

Vikas Kumar Vaidya(✉) and Vineeta Saxena Nigam

Rajiv Gandhi Proudyogiki Vishwavidyalaya, Bhopal, M.P., India
vikas.vaidya0107@gmail.com

Abstract. The growth of next-generation communication technologies is enabled by UAV-assisted emergency communication systems' high mobility and dynamic topology. However, resource optimization and localization of aerial nodes are significant challenges for wireless communication. This paper proposes improved particle swarm optimization-based aerial node grouping and multi-hop communication with a single ground base station. The proposed algorithms encapsulate genetic algorithms to select fitness constraints on resources. An emergency communication system employing UAVs as flying beacons is proposed to assist the IoT in overcoming communication constraints caused by a natural disaster. Communication resources may be appropriately distributed and increase the quality of the user experience by utilizing the resource allocation technique of UAV-assisted IoT systems. To begin, a model of an IoT system with BS selection is created in order to maximize system throughput. After that, the system model is converted into a stochastic optimization problem using improved particle swarm optimization. The suggested approach simulates and measures standard parameters of UAV networks using MATLAB software. GA, ACO, and PSO are all compared in the suggested algorithm. The results indicate that the suggested approach is much more efficient than existing algorithms.

Keywords: UAV · IoT · PSO · PDR · Next-Generation Communication

1 Introduction

The IoT's diversity brings innovation and technical development to wireless communication systems. The collaboration of UAVs and IoT is a veritable emergency communication system in natural disaster management [1, 2]. The service of UAV-assisted wireless communication outperforms various advantages such as good line of sight (LOS), dynamic movement, and the probability of events. Designing an emergency communication system based on a UAV-assisted model deals with two different areas: the design of the device and the communication model [3, 4]. The method of the device deals with mechanical structure, weight, and whether the device is metallic or non-metallic. The communication system plays a major role in an emergency communication system's

H. Sharma et al. (Eds.): ICIVC 2022, PALO 17, pp. 101–112, 2023.
https://doi.org/10.1007/978-3-031-31164-2_9

battery-operated, unmanned aerial vehicle. The limited resources of a wireless communication system, such as energy and bandwidth, are major challenges in handling the process of unnamed aerial vehicles efficiently. Resource optimization and allocation play a vital role in heterogeneous IoT networks and UAV-assisted systems. The optimization of resources improves the performance of dedicated systems [5–8]. The reported survey suggests that various authors proposed algorithms based on swarm and artificial intelligence. Swarm intelligence algorithms are used to reduce localization errors and increase system throughput. Despite several challenges, the UAV-assisted emergency communication system incorporates various protocols and networks to increase the strength of the communication system. In recent decades, ADHOC networks and cognitive radio networks have provided multi-hop emergency communication. The small segment of UAVs, also known as drones, is famous among emergency response teams for installing the emergency communication system. The drone system collects the data and transmits it to the base station. Data transmission classified the UAV-assisted communication devices into two categories: single-link communication and multiple-link communication. In single-link communication, the device is connected to the base station. Instead, the numerous link devices are associated with different network topologies and transmit data to the base station. The distance of the UAV-assisted system is a significant factor for localization issues in GPS and faces a 10–30 m location error [9–12]. Several authors proposed reducing localization errors. However, most algorithms apply the distance measurement approach to handle this problem. This paper proposes an efficient algorithm based on PSO for efficient localization and data transfer to the base station. Our objective is to design an efficient protocol for UAV networks. The main contribution of the paper is mentioned here. First, the proposed algorithm reduces the convergence problem and detects the actual target of UAV networks. The second objective of this paper is to enhance the performance of data delivery over UAV networks. The rest of the article is organized as follows: Section 2, related work; Sect. 3, the proposed methodology; Sect. 4, experimental analysis; and Sect. 5, conclusion and future work.

2 Related Work

In recent years, the benefits of UAVs for aided communication, such as high mobility and rapid deployment, have piqued the interest of business and academia. UAVs have improved line-of-sight (LOS) pathways due to their aerial superiority and approach to IoT nodes to send or receive data. As a result, UAVs can respond quickly to a wide range of emergency communication needs, including UAV-assisted cellular communication and the essential public network in hotspots, disasters, and isolated highlands. In this [1] article, the author proposes an energy-efficient UAV-aided delay-tolerant network with a socially oriented message splitting approach. The suggested SOCS algorithm using the enhanced social relationship-based forwarding mechanism has been shown to be optimal in extensive simulations. The authors [2] proposed a particle swarm optimization-based channel sharing method for UAV communication. The proposed algorithm reduces the impact of interference and controls power allocation. In this [3] chapter, the author provides an optimization standpoint and a detailed investigation of resource management in UAV-assisted wireless networks. Various restrictions, optimization types, and approaches tailored to UAV-assisted wireless networks are also described. In order to correctly anticipate and explain the composites fading channel in 5G-UAV-based EWC networks, the author of this [4] paper proposes that reflectors, path-loss exponents, fading, and shadowing factors are employed in a novel heterogeneous composites Rayleigh fading model for EWC networks. According to the numbers, this will lead to better communication services for the public after a disaster. Authors [5] describe the iterative algorithms for power allocation and fixed trajectory of localization. The results of methods increase the performance of UAV communication systems. In this [6] article, the author provides the use of a testbed to evaluate the success of UAVs as edge cloud nodes, with an ACS function for controlling and orchestrating a UAV fleet and a focus on the features of UAVs as core networking gear with radio and backhaul capabilities. The author establishes in this [7] paper using a UAV-enabled SWIPT-based emergency communications architecture for IoT networks, disaster scenarios are classified into three categories: dense areas, vast regions, and emergency spots. The effectiveness of the aforementioned strategies is demonstrated by simulation results. Finally, the author highlights future research directions as well as concerns. The author of this [8] created a brand-new UAV-enabled WPT system in which UAVs travel among Internet of Things devices and act as data aggregators and network power providers. By deciding on the optimal course of action, examining the system's effects, and keeping an eye on its behavior, it conducts a functional analysis. In this [9] article, the author proposes that they successfully aid in improving packet reception rate with nominal buffer delays when used in a UAV-enabled LoRa network, according to the research. In terms of packet reception rate and average delay quality of service (QoS) measurements, the simulation results suggest that the proposed technique is practicable. Authors [10] focus on the minimization of response time for UAV based fire monitoring. The proposed algorithm deals with the Markov random field and particle swarm optimization algorithm. The reduced response time increases the utilization of resources.Authors [11] proposed the method of location optimization of UAV emergence communication systems. The localization algorithm applies particle swarm optimization algorithm and incorporates dual fitness function for pre-coded network The author of this [12] article proposes that UAVs function as data collectors, allowing a large number of IoT and cellular UEs to share spectrum in a TDD protocol. According to simulation results, the proposed protocol enhances the EE, and hence the IoT device's lifetime, due to the drone's proximity to the IoT device, while also protecting the UEs using the optimum IoT transmit power. Authors [13] describe the challenges of resource allocation and optimization in different levels of UAV communication. The authors describe the limitation and NP-complete problem of limited constraints in manners of resources utilization in this [14] article, the author develops a method for configuring the HetNet to achieve a decent coverage-rate trade-off while maintaining proper QoS metrics. Their findings indicate that in a multitier Het Net, sub-6GHz and mm-wave UAVs can complement each other, with the sub-6GHz layer providing higher coverage and the mm-wave tier providing superior data throughput. In order to obtain the lowest practical uplink rate across all IoT nodes, the author of this [15] optimizes UAV trajectory, subcarrier, power, and sub-slot distribution while taking the entire downlink rate into consideration. Simulated findings indicate that the suggested approach may optimize the UAV trajectory while also correcting for node mobility. The author of this article proposes maximizing the use of IoT network resources. The proposed approach is compared and contrasted with similar publications based on a variety of scenarios. Their proposed technique appears to be promising based on the findings of the research. Furthermore, the results show speedy convergence, suitability for heterogeneous IoT networks, and minimal assessment task complexity. In this chapter [17], the author considers Unmanned aerial vehicles (UAVs) could be employed in two scenarios to protect the transfer of classified information (a UAV as a flying base station and a UAV as an aerial node). The author next suggests physical layer security solutions and numerically exhibits greater performance advantages using two case studies. Finally, the author highlighted new possibilities for network design change that can be used to inform future research. In this article [18], the author presents a Deep Q-Network (DQN)-based resource allocation strategy to maximize the system's energy efficiency. Simulation results reveal that the proposed DQN-based resource allocation method can greatly improve system efficiency when compared to earlier Q-Learning, random, and maximal resource allocation strategies. In this [19] article, the author considers a CNC that learns observational data before broadcasting it to a network of inter-cluster nodes. According to simulated data, one's technique effectively employs DDPG to develop the ideal plan for municipal arithmetic in stochastic and complicated circumstances, surpassing A3C, DQN, and greedy predicated offloading. In this [20] article, the author introduces the EE at specified target SINR values, which are doubled when UAV base stations are deployed. The authors further demonstrate that exceeding a particular EE point causes the EE to decrease. Their research shows that their proposed strategy beats previous strategies for increasing system total rate or lowering system power usage. Haoran Zhu et al. [21] propose a cooperative algorithm that uses the spider monkey optimization (SMO) algorithm to solve the problem of local minima and speed up the convergence rate of path planning in unmanned combat aerial vehicles.

3 Proposed Methodology

Resource and localization optimization play a vital role in UAV-assisted emergency communication systems. The limited resources are optimized to enhance the performance and network reliability of the communication system. The allocation of resources in wireless communication is a non-convex problem, so swarm intelligence and hybrid optimization methods play an important role in reducing localization errors and increasing the efficiency of network throughput. This paper proposes an improved particle swarm optimization algorithm using a genetic algorithm for the UAV-assisted emergency communication system [13, 14].

Genetic Algorithm

The genetic algorithm plays a vital role in the search and mapping of resource allocation in wireless networks. The main objective of genetic algorithms is the optimization of resources available for a mode of communication. The processing of genetic algorithms is inspired by the process of evaluation and bio-inspired genetics. The resources of the network are represented as genes, and optimal parameters are set as chromosomes. The main function of a genetic algorithm is the fitness function, which can take many forms, including the wheel method, tournament method, and probability-based method. A set of population, crossover, mutation, and fitness constraints [16] are used to explain how genetic algorithms work.

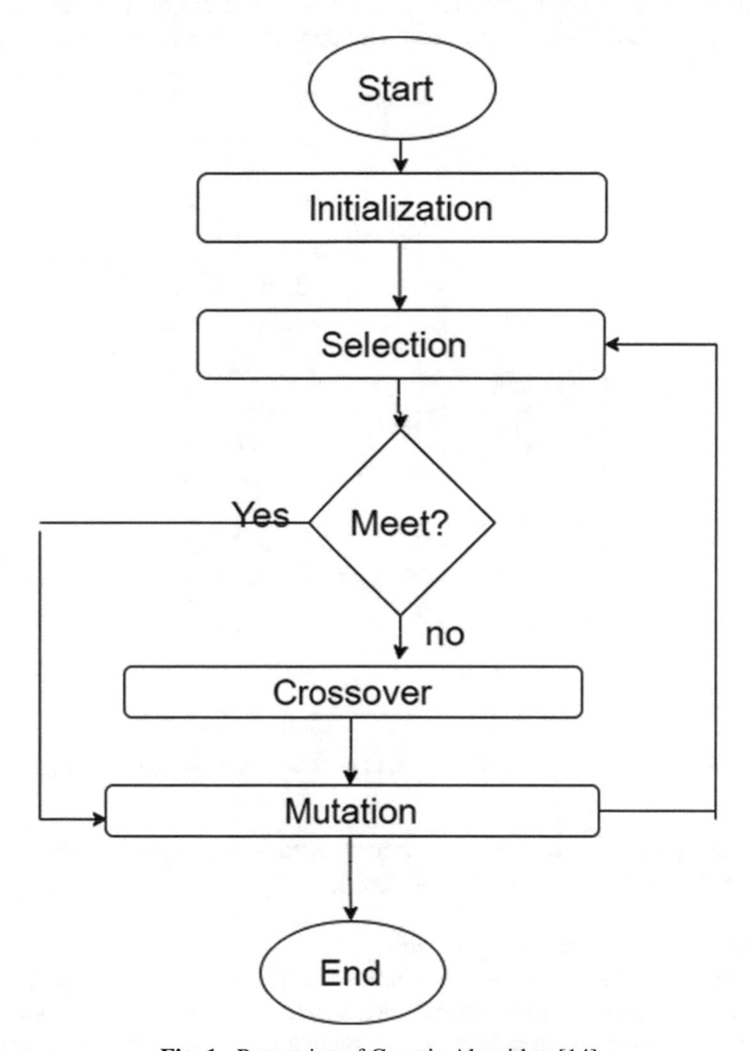

Fig. 1. Processing of Genetic Algorithm [14]

Figure 1 shows the processing steps of a genetic algorithm, which consist of all phases from population to optimal results.

Particle Swarm Optimization (PSO)

Particle swarm optimization is a dynamic population-based heuristic approach. The particle swarm optimization algorithm's working behaviors are inspired by the forks of birds. The velocity and position of particles to satisfy the selection condition of population data are the main processing components of the particle swarm optimization algorithm. An algorithm's processing uses the previous value of the iteration to store the next value and move the particle. The process of iteration maintains the updated position of the particle. The constant acceleration function accelerates the particles for the last iteration of population [12]. The processing of particle swarm optimization is shown in Fig. 2.

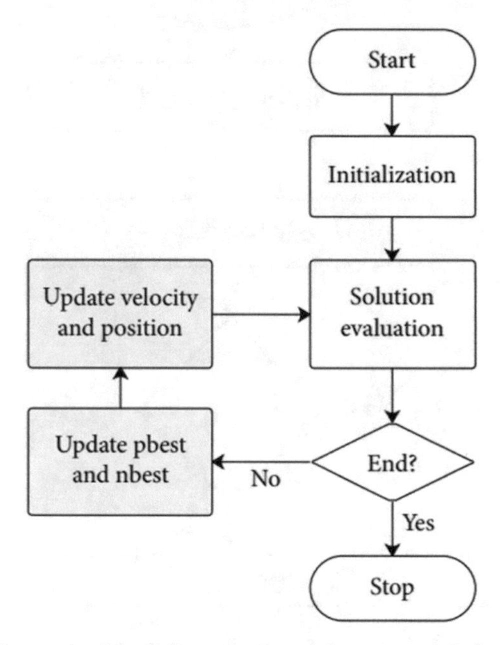

Fig. 2. Processing block diagram of particle swarm optimization [12]

Figure 2 explores the working process of the particle swarm optimization algorithm in all stages, from population to optimal results.

Improved Particle Swarm Optimization

The primary goal of the UAV-assisted network is to improve data reliability and accelerate data delivery to response teams. We proposed an improved particle swarm optimization algorithm for heterogeneous networks. The algorithm's main goal is to calculate the network's optimal weight vector. Now the fitness function is formulated of the objective function that is $f(x) = \frac{1}{u}$. The processing of generation function defined as

$$\emptyset i + 1 = \mu \phi i(1 - \phi i) \quad (1)$$

Here Ø is a random number between intervals [0, 1] and $\mu \in [0, 4]$ is constant.

The processing of the particle swarm optimization (PSO) algorithm faces a problem of global optimization in UAV networks, so apply a genetic algorithm to control the iteration and minimize the deficiency of particle swarm optimization. The positions of particles can be thought of as genes, and the production of genes as fitness constraints. The hybrid optimization selection process is now described as follows:

$$\{Pr\,Pr(Xi) = \frac{g(xi)X\ exp(-h(xi))}{\sum_{j=1}^{N}(g(xj)X\ exp(-h(xj)}g(Xi)$$
$$= \frac{1}{exp(-SNRi)}h(Xi) = \frac{1}{m}\sum_{j=1}^{N}|g(xi) - g(xj)| \tag{2}$$

Here g(Xi) and h(Xi) represent the degree of odder level

The subgroup of particle colon convergence population defined as

$$Var\{Ui\} = \frac{1}{N}\sum_{i=1}^{N}\left(fi(t) - f'(t)\right) \tag{3}$$

Now define the mutation operator for average fitness constraints function of particles

$$X_i^* = xi + cE(0, 1) \tag{4}$$

where c denotes the length of mutation and E(0, 1) is mutation operator

After the mutation estimates the similarity of particles

$$sim(Xi, Xj) = 1 - \frac{||Xi - Xj||}{MaxXi, Xj \in Xc\{|Xi - xj|\}} \tag{5}$$

After the similarity estimation of particle finally assign the weight to available resources.

4 Experimental Results

To evaluate the performance of the improved particle swarm optimization algorithm for UAV-assisted communication systems, use MATLAB software. The simulation was carried over into the Windows operating system with a configuration of Windows 10, a processor I7, and 16 GB of RAM. The analysis of the results focuses on the localization error rate, routing overhead, and packet delivery ratio (PDR). The proposed algorithm compares ACO, PSO, and GA. The simulation parameters of the network are given in Table 1 [10, 12, 17, 20].

Figure 3 depicts the accuracy of localization in relation to the number of UAV nodes. The modified fitness constraint's function reduces the error of localization and enhances the performance of the algorithm instead of GA, ACO, and PSO.

In Fig. 4, we show the mean localization error with respect to the number of UAV nodes. The selection of resources for the UAV network is enhanced by the proposed algorithm. The fitness function of particle swarm optimization fixed the location of UAV nodes.

Table 1. Simulation parameters of UAV-assisted communication network

Simulation Parameters	Value
Network area	1000 m × 1000 m
UAV transmission range	250–300 m
Number of UAVs	10 = 150
Number of base stations	1
Traffic type CBR	CBR
CBR rate 2 Mbps	2 Mbps
Speed	10–30 m/s
Number of rounds	2000
Traffic load	5 messages/s
Message size	250 kB
Initial energy	2 J
UAV transmission power	5 W
Maximum number of iterations	50
Transmission range	Dynamic

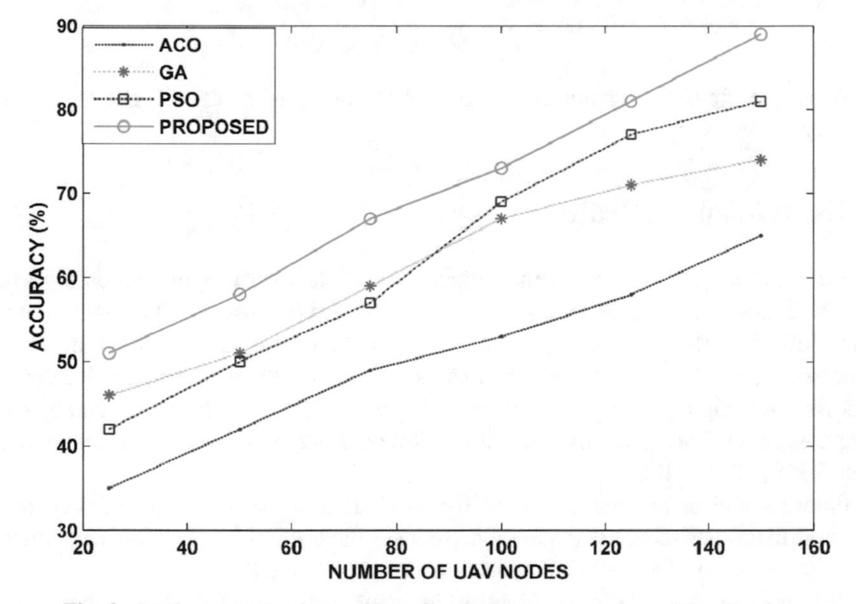

Fig. 3. Performance analysis of Localization accuracy and number of nodes.

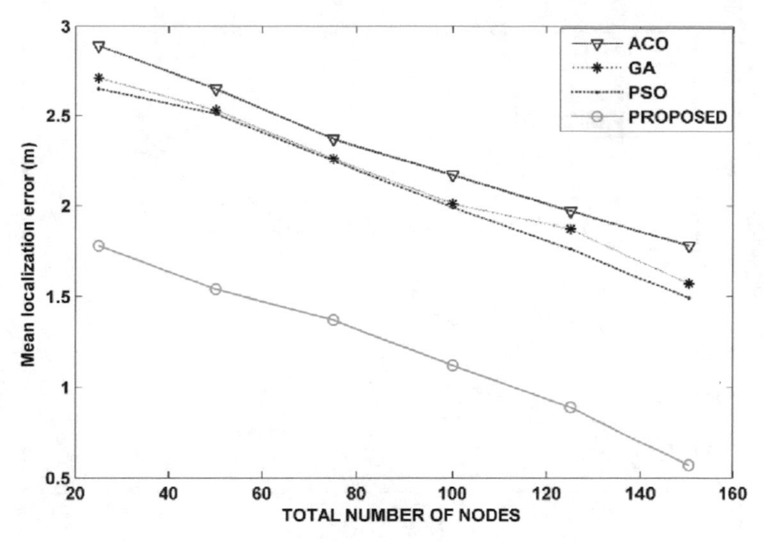

Fig. 4. Performance analysis of Mean localization error and number of nodes.

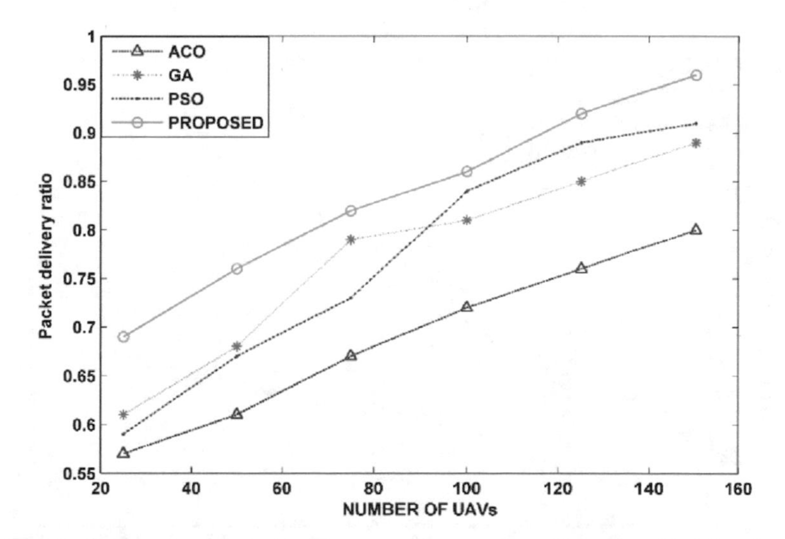

Fig. 5. Performance analysis of Packet delivery ratio and number of UAV nodes.

In Fig. 5, we show the packet delivery ratio of UAV networks. The optimization of resource allocation enhances the packet delivery ratio. The analysis of results using three algorithms (ACO, GA, and PSO) Instead of existing algorithms, the proposed algorithm raises the PDR value.

The overhead routing of UAV networks is depicted in Fig. 6. The overhead of the routing protocol increases the network's response time. The proposed algorithm reduces the network overhead due to optimized resource allocation instead of GA, ACO, and PSO.

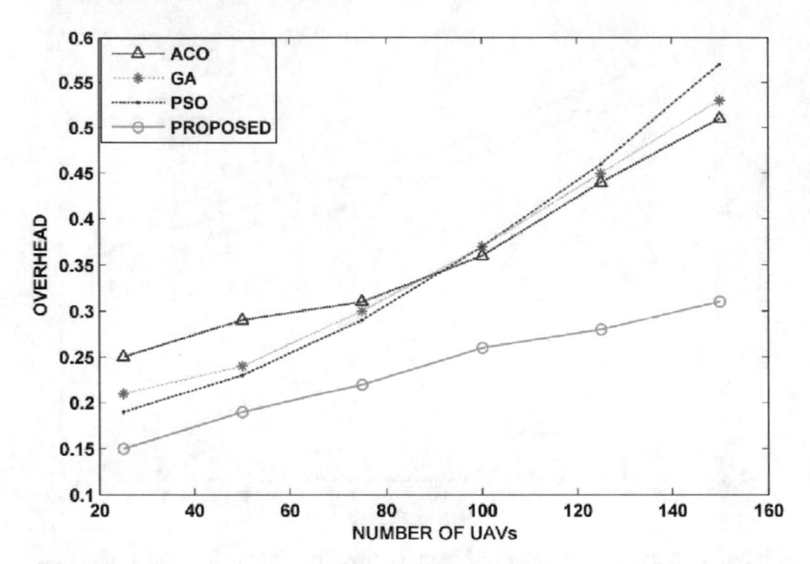

Fig. 6. Performance analysis of routing overhead and number of UAV nodes.

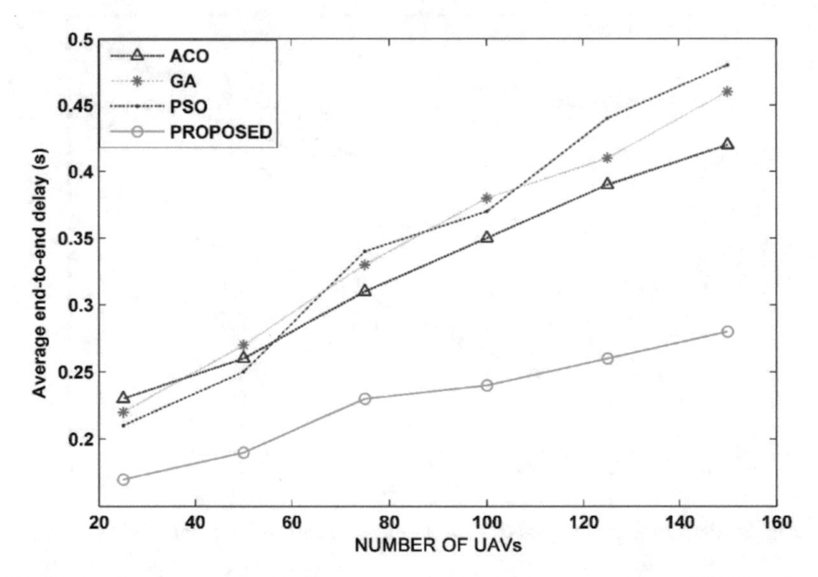

Fig. 7. Performance analysis of average end-to-end delay vs. number of UAV nodes.

In Fig. 7, we show the average end-to-end delivery of packets in UAV networks. The optimization and allocation of resources reduces localization errors and the delivery response time. The analysis of the results suggests that the proposed algorithm is very efficient compared to GA, ACO, and PSO.

5 Conclusion and Future Scope

The high-speed mobility and dynamic topology of UAV networks, resource optimization, and minimization of localization errors are critical tasks. The proposed algorithm efficiently minimizes the localization error and optimizes resources in UAV networks. The proposed algorithm mapped the 3D space of UAV nodes using particle swarm optimization and selected fitness constraints under a single base station. As a result, the proposed algorithm maximizes the system's data transmission rate and enhances the system's throughput and total packet delivery ratio. The algorithm also addresses the resource allocation problem using a genetic algorithm in UAV-assisted systems. We present this challenge to you: you'll need an efficient resource allocation strategy. The simulation results show that the proposed resource allocation technique outperforms previous strategies in terms of system PDR and that the algorithm converges quickly. According to the findings, optimizing limited resources necessitated multi-dimensional and multi-objective functions for the improvement of UAV-based emergency services. The investigation of resource optimization algorithms opens up the possibility of dual and multiple constraints functions for the processing of UAV networks. In the future, more research will be done on swarm intelligence and the use of UAVs as auxiliary equipment, such as UAVs used as relays and UAVs that charge other devices wirelessly.

References

1. Zhang, C., Dong, M., Ota, K.: Heterogeneous mobile networking for lightweight UAV assisted emergency communication. IEEE Trans. Green Commun. Network. **5**(3), 1345–1356 (2021)
2. Pan, Y., et al.: Joint optimization of trajectory and resource allocation for time-constrained UAV-enabled cognitive radio networks. IEEE Trans. Veh. Technol. **71**(5), 5576–5580 (2022)
3. Masroor, R., Naeem, M., Ejaz, W.: Resource management in UAV-assisted wireless networks: an optimization perspective. Ad Hoc Netw. **121**, 102596 (2021)
4. Yao, Z., Cheng, W., Zhang, W., Zhang, H.: Resource allocation for 5G-UAV-based emergency wireless communications. IEEE J. Sel. Areas Commun. **39**(11), 3395–3410 (2021)
5. Nasrollahi, S., Mirrezaei, S.M.: Toward UAV-based communication: improving throughput by optimum trajectory and power allocation. EURASIP J. Wirel. Commun. Network. **2022**(1), 1–16 (2022)
6. Bekkouche, O., Samdanis, K., Bagaa, M., Taleb, T.: A service-based architecture for enabling UAV enhanced network services. IEEE Netw. **34**(4), 328–335 (2020)
7. Feng, W., et al.: UAV-enabled SWIPT in IoT networks for emergency communications. IEEE Wirel. Commun. **27**(5), 140–147 (2020)
8. Lhazmir, S., AitOualhaj, O., Kobbane, A., Mokdad, L.: A decision-making analysis in UAV-enabled wireless power transfer for IoT networks. Simul. Model. Pract. Theory **103**, 102102 (2020)
9. Saraereh, O.A., Alsaraira, A., Khan, I., Uthansakul, P.: Performance evaluation of UAV-enabled LoRa networks for disaster management applications. Sensors **20**(8), 2396 (2020)
10. Sun, L., Wan, L., Wang, X.: Learning-based resource allocation strategy for industrial IoT in UAV-enabled MEC systems. IEEE Trans. Industr. Inf. **17**(7), 5031–5040 (2020)
11. Eltokhey, M.W., Khalighi, M.-A., Ghassemlooy, Z.: UAV location optimization in MISO ZF pre-coded VLC networks. IEEE Wirel. Commun. Lett. **11**(1), 28–32 (2021)
12. Hattab, G., Cabric, D.: Energy-efficient massive IoT shared spectrum access over UAV-enabled cellular networks. IEEE Trans. Commun. **68**(9), 5633–5648 (2020)

13. Xiao, Z., et al.: Resource management in UAV-assisted MEC: state-of-the-art and open challenges. Wirel. Netw. **28**(7), 3305–3322 (2022)
14. Jan, M.A., Hassan, S.A., Jung, H.: QoS-based performance analysis of mmWave UAV-assisted 5G hybrid heterogeneous network. In: 2019 IEEE Global Communications Conference (GLOBECOM), pp. 1–6. IEEE (2019)
15. Na, Z., Zhang, M., Wang, J., Gao, Z.: UAV-assisted wireless powered internet of things: joint trajectory optimization and resource allocation. Ad Hoc Netw. **98**, 102052 (2020)
16. Munaye, Y.Y., Juang, R.-T., Lin, H.-P., Tarekegn, G.B., Lin, D.-B.: Deep reinforcement learning based resource management in UAV-assisted IoT networks. Appl. Sci. **11**(5), 2163 (2021)
17. Li, B., Fei, Z., Zhang, Y., Guizani, M.: Secure UAV communication networks over 5G. IEEE Wirel. Commun. **26**(5), 114–120 (2019)
18. Chen, X., Liu, X., Chen, Y., Jiao, L., Min, G.: Deep Q-network based resource allocation for UAV-assisted ultra-dense networks. Comput. Netw. **196**, 108249 (2021)
19. Seid, A.M., Boateng, G.O., Anokye, S., Kwantwi, T., Sun, G., Liu, G.: Collaborative computation offloading and resource allocation in multi-UAV-assisted IoT networks: a deep reinforcement learning approach. IEEE Internet Things J. **8**(15), 12203–12218 (2021)
20. Chakareski, J., Naqvi, S., Mastronarde, N., Jie, X., Afghah, F., Razi, A.: An energy efficient framework for UAV-assisted millimeter wave 5G heterogeneous cellular networks. IEEE Trans. Green Commun. Network. **3**(1), 37–44 (2019)
21. Zhu, H., Wang, Y., Li, X.: UCAV path planning for avoiding obstacles using cooperative co-evolution spider monkey optimization. Knowl. Based Syst. **246**, 108713 (2022)

Time Domain and Envelope Fault Diagnosis of Rolling Element Bearing

Arvind Singh[1], Arvind Singh Tomar[1(✉)], Pavan Agrawal[2], and Pratesh Jayaswal[1]

[1] Department of Mechanical Engineering, Madhav Institute of Technology and Science, Gwalior, MP 474005, India
arvind_tomar25@rediffmail.com

[2] Department of Mechanical Engineering, Shriram Institute of Information Technology, Banmore, MP 476444, India

Abstract. Rolling bearing vibrations (REB Vibration) are complex. This complex vibration occurs due to geometrical imperfection, a surface defect of the rolling contact bearing, and geometrical errors in nearby or attached components. This research paper focuses on the Time domain analysis of the theoretical mathematical model of an REB and fault detection methods for fault identification of REB (rolling element bearing), different sources of vibration, and vibration measurement techniques like overall vibration level, frequency spectrum, envelope spectrum, etc. Also, characteristic defect frequency may be presented in this paper. A laboratory rolling bearing test has been conducted for various rotation frequencies and compares their deviation from the theoretical ones. In this paper for bearing diagnostics, envelope analysis is used as a valuable vibration analysis approach. The frequency-domain properties enable a quick evaluation of a machine's health without the need for extensive diagnostics. The intermittent effects of a defective bearing were captured using envelope analysis on the modulated signal. It is still reasonable when the vibration signal is low in energy and 'embedded' within extra vibration from the appropriate component.

Keywords: Rolling element bearing · Vibration Analysis · Condition-based maintenance · Fault Detection

Nomenclature:

REB	Rolling element bearing
CWRU	Case Western Reserve University
EDM	Electro-discharge machine
IR	Inner Race
OR	Outer Race
BPFI	Ball pass frequency of the inner race
BPFO	Ball pass frequency of the outer race
BSF	Ball spin frequency
DAQ	Digital Data Acquisition

H. Sharma et al. (Eds.): ICIVC 2022, PALO 17, pp. 113–127, 2023.
https://doi.org/10.1007/978-3-031-31164-2_10

SK	Spectral Kurtosis
FFT	Fast Fourier Transform
ESA	Envelop Spectrum Analysis
HFRT	High-frequency resonance technique

1 Introduction

Roller contact bearing is extensively used in rotating machinery; their successful and safe performance depends on the bearing type chosen and the correctness of all associated elements, like a shaft, gear, nuts, etc. [1]. Generally, REB is failed by fatigue, on the hypothesis that the rolling contact bearing is correctly maintained, run, and fitted. Presently, advanced technology is used in the manufacturing industry and materials [2]. It is generally the fact that the fatigue cycle of bearing, which is associated with subsurface stresses in the outer race of the bearing and the Inner race (IR) of the bearing, etc., is not the limiting part and probably values for smaller than three percent of failures in service [3]. Early bearing failure occurs in some cases because of improper or poor lubrication, temperature limits, contamination, poor attachment/fits, misalignment, and unbalance [4]. For the quiet running of rolling element bearing, the bearing manufacturer has developed vibration analysis as an adequate way of estimating the good condition. A fundamental way is to mount the rolling element bearing on a smooth running shaft and calculate the radial velocity in frequency bands in 50–300 Hz, 300–1800, and 1800–10000 Hz. The rolling element bearing matched three RMS velocity limits in all the above frequency bands [5].

Time-domain is utilized to analyze the vibration data as a function of time. In time-domain investigation statistical characteristic are calculated for vibration signals. By associating the statistical characteristic, the fault of the bearing may be revel [6]. Statistical features of the time domain are employed as input for the training of ANN, and the system is able to classify the rolling element-bearing faults effectively [6].

B. Sreejith et al. [7] include estimated time indices and demographic moments namely RMS value, peak value, shape factor, etc., Kurtosis was suggested as a useful failure indication for bearing signals. To transform time-domain vibration data into a frequency domain, [8, 9] used the Fast Fourier Transformation technique. The peaks of this frequency domain were repeatedly discovered to be multiples of a certain fault frequency estimated using the REB geometry. The major benefit of this technique is that little or no data is lost before the inspection. It helps in detailed analysis and direct fault diagnosis. The disadvantage of this technique is that there is often too much data.

2 Time Domain Analysis and Theoretical Mathematical Models

In the time domain, vibration data collected by vibration transducers are used for rotating machines. In practice, vibration signals which recorded using transducer include various responses from different sources with background noises, are a set of time-indexed data points. The generated vibration signals are difficult to use directly for machine flaw detection; an additional strategy has to figure out particular characteristics of this basic signal that can categorize the signal in the core. These features are sometimes known as characteristics, signatures, or features.

Several categories of methodologies are used for integrated fault identification, raw vibration datasets at the beginning and mature sets of conclusions at the end. There are 3 types of methods to analyze Vibration analysis which are frequency domain methods, time domain methods and time-frequency domain methods.

There are two forms of manual examination of vibration signals used in time domain defect diagnosis: (I) feature-based detection & (II) visual inspection.

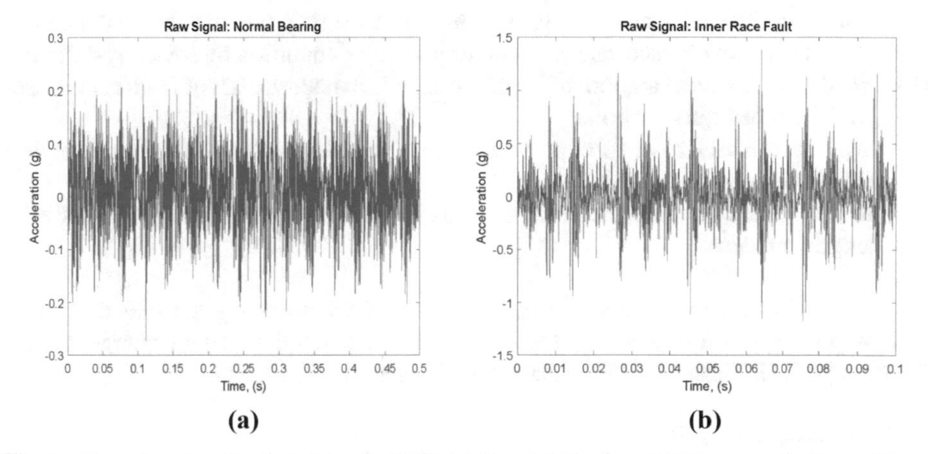

Fig. 1. Time domain vibration data of a REB: (a) Normal Bearing; (b) inner race fault condition.

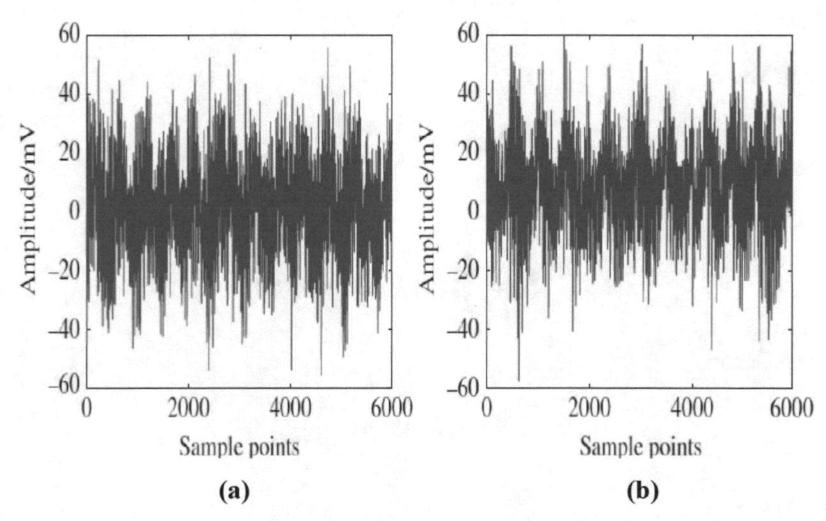

Fig. 2. Time domain vibration data of a REB: (a) worn but unharmed situation; (b) OR fault condition.

By contrasting a recorded vibration signal with one that was already registered from a machine that was in regular functioning, the state of a Visual Inspection machine may be established. Figure 1(a) displays vibration signal in time domain for healthy roller bearings, whereas Fig. 1(b) depicts a roller bearing inner race (IR) problem scenario. For four reasons, visual examination is unreliable in the field for condition monitoring of rotating equipment:

(a) Not all rotating equipment time waveform signals have clear visual distinctions [10] (Guo et al. 2005). e.g., Fig. 2 depicts two types of vibration signals from the source 'roller bearings' in a scratched but defect-free state (Fig. 2(a)) and an OR defect state (Fig. 2(b)). In such a case, relying on inspecting visually the evaluation of time waveform characteristics & determination of condition of device is difficult.
(b) We deal with a huge amount of vibration signals in reality, each of which comprises additional background noise.
(c) We occasionally have to cope with signals of low-amplitude evaluated against chattering backgrounds.
(d) It is important to detect fault as soon as possible so manually checking is a bit lengthy process.

A time-domain included statistical attributes of vibration signals like root mean square, peak level, crest factor, kurtosis and skewness as a trait to detect emergent faults in bearings [11]. These statistical characteristics are listed below [6, 12]:

i. Root mean square (RMS):

$$RMS = \sqrt{\frac{1}{N}\sum_{n=1}^{N} f_n^2} \tag{1}$$

ii. Mean Value:

$$Mean = \frac{1}{N}\sum_{i=1}^{N} f_n \tag{2}$$

iii. Peak value:

$$P_v = \frac{1}{2}\left[max(f_n) - min(f_n)\right] \tag{3}$$

iv. Crest factor:

$$Crest\,factor = \frac{P_v}{RMS} \tag{4}$$

v. Skewness:

$$Skewness = \frac{\frac{1}{N}\sum_{n=1}^{N}\left(f_n - \underline{f}\right)^3}{RMS^3} \tag{5}$$

vi. Kurtosis:

$$Kurtosis = \frac{\frac{1}{N}\sum_{n=1}^{N}\left(f_n - \underline{f}\right)^4}{RMS^4} \tag{6}$$

vii. Variance:

$$variance = \sigma^2 = \frac{\sum_{n=1}^{N}\left(f_n - \underline{f}\right)^2}{N} \tag{7}$$

viii. Standard Deviation:

$$SD = \left(\frac{1}{N-1}\sum_{n=1}^{N}\left(f_n - \underline{f}\right)^2\right)^{\frac{1}{2}} \tag{8}$$

where, $\underline{f}$ is the mean, f_n ($n = 1, 2\ldots\ldots\ldots, N$) is the amplitude at sampling point n and N is the number of sampling points.

To be more specific, Lu et al. [13] and Cui et al. [14] used time-domain signal conditioning approaches to assess bearing condition. The time-domain study provided only a restricted understanding of the issue. Furthermore, vibration signals with a lot of noise might occasionally gives misleading result for the REB failure identification. Furthermore, because to the existence of non-stationary in the signal, such approaches require a correction factor, and as a result, findings may differ. Aside from time-domain analysis, several studies have been discovered that utilize statistical factors to emphasize the impulses induced by the defect in the complicated vibration signal [15]. When evaluated to other statistical properties, Sreejith et al. [7] proposed kurtosis as an acceptable defect indicator.

A defected rolling element bearing (balls or rollers), generally, contact with localized defect during running in the bearing and produces impacts. If inner race velocity is constant, the bearing geometry calculates the repetition rate of these impacts. The bearing equation assumes no slippage between the roller and the raceway surface. But this is inaccurate overall because various factors like sliding and rolling occur simultaneously. Defected rolling bearing frequency spectrum comprises inner race, outer race rotation frequency, harmonics relating to the defect frequency, and sideband due to amplitude modulation. These sidebands correlate with cage frequency rotation values or inner or outer race frequency rotation [16]. In the above mathematical equation, both the IR and OR are rotating. But in the maximum working condition, the outer race is fixed [17]. The following are the main fault frequency of rolling-element bearing:

$$FTF = f_g = \frac{1}{2}\left[f_i\left(1 - \frac{d\cos\theta}{D_p}\right)\right] \tag{9}$$

$$BPFO = \left\{\frac{N}{2}\left[f_i\left(1 - \frac{d\cos\theta}{D_p}\right)\right]\right\} \tag{10}$$

$$BPFI = \left\{\frac{N}{2}\left[f_i\left(1 + \frac{d\cos\theta}{D_p}\right)\right]\right\} \tag{11}$$

$$BSF = f_r = \frac{D_p}{2d}(f_i)\left[1 - \left(\frac{d\ cos\theta}{D_p}\right)^2\right] \quad (12)$$

The following relationship is derived from Eqs. (9), (10), (11), and (12).

$$\text{BPFI} = \text{N}\,(FTF) \quad (13)$$

$$\text{BPFI} = \text{N}\,(f_i - FTF) \quad (14)$$

As a result, the geometry of the bearing must be known to calculate the defect frequencies. If the geometry of bearing is unknown, the following approximate equation called empirical mathematical model (9), (10), (11), and (12) are possible for determining the fault frequencies if the contact angle is unknown. All other parameter is known, then put the value of contact angle = 0 in the mathematical model. If, in some conditions, the number of parameters identified does not allows the use of the mathematical model (9), (10), (11), and (12), alternatively, the empirical relation can be used to determine approximated values for the frequencies generated by a faulty rolling element bearing.

$$\left.\begin{array}{l} \text{FTF} = 0.5(f_i) \\ \text{BPFO} = 0.4\,\text{N}\,(f_i) \\ \text{BPFI} = 0.6\,\text{N}\,(f_i) \end{array}\right\} \quad (15)$$

In accordance with the empirical analytical model Eq. (15), about 40% of the roller contacts a flaw in the outer race during a full rotation of the IR, whereas 60% of the roller contacts a fault in the bearing inner race.

3 Signal Processing Fault Detection Techniques

Condition-based REB diagnostic maintenance employs vibration signals. However, while the time domain displays statistical aspects of vibration signals such as RMS value, kurtosis, and so on, the frequency domain (spectrum analysis) is the most commonly used.

3.1 Frequency Domain

Frequency investigation, often recognized as spectrum investigation, is more prominent and commonly used vibration investigation methods for observe machine conditions [18]. In actuality, frequency domain investigation approaches can disclose data based on frequency aspects challenging to distinguish in the time domain [8]. Numerous rotating machine components, such as bearings, shafts, and fans, commonly generate the observed time domain vibration signals [19]. A single component motion generates a sinusoidal curve with a single frequency and amplitude whereas the multiple component motion generates additional frequencies [20]. To put it another way, each rotary apparatus element creates its individual frequency [21]. However, we occasionally see these formed frequencies in the recorded signal separately but we obtain a total collection of the signals detected by the sensor [22].

The frequency domain is concerned with investigating analytical functions or signals related to frequency reasonably than time. The frequency domain reveals how greatly of a signal resides inside any particular frequency band spanning a range of frequencies. In contrast, the time domain tells how a signal evolves [23]. Furthermore, the time domain aids with REB fault detection, while the frequency domain aids in detecting, isolating, and identifying faults, a process known as the diagnosis. Frequency domain analysis frequently provides earlier detection of a problem's development and the issue's source.

Because defect detection is complicated, signal processing is necessary [24]. FFT (Fast Fourier Transform) is normally utilized in the frequency domain. The shifting defect position modulates the series of impulses. The rolling element experiences rolling and slippage in the load zone. Because of the combined action of rolling and slipping, the impact cannot be seen in the spectrum at the same time.

3.2 Envelope Analysis

Signals from malfunctioning equipment masked with noise are modulated by defect feature, which necessitates demodulation investigation [25]. Envelope analysis, for example, can extract diagnostic data from acquired signals [26]. To detect difficulties in rolling element bearings, a signal-processing approach known as envelope investigation, also known as high-frequency resonance technique (HFRT), is widely utilized [27]. It usually consists of the three steps listed below: Before being treated using FFT, the original vibration data was being filtered by using band-pass filter process, after that it has been corrected or enveloped by folding the lower component of the time waveform over the upper part, typically using the Hilbert-Huang transform HHT (Hilbert-Huang transform) [28]. The envelope spectrum is then produced, revealing significant amplitude peaks, which are unattainable in the FFT domain [29]. In their early study from 1984 (McFadden and Smith 1984) [30], McFadden and Smith assessed the vibration monitoring of REB using the high-frequency resonance approach to investigate the applicability of envelope analysis in diagnosing machine problems.

4 Experimental Investigation

For bearing data at normal conditions and faulty conditions for this study we refer to Case Western Reserve University (CWRU) [16]. For the experiment, a 6205-2RS JEM deep groove ball bearing manufactured by SKF was used. A 2 hp Reliance electric motor was used with load variations from 0 to 3 hp and shaft speed from 1772 to 1730 rpm accordingly. For fault 12,000 Hz sample frequency was taken. For the experiment purpose intended faults were imparted on different locations. Since the study focuses on the IR and OR fault on bearing the defect were also produced on the IR and OR of the REB. And the defects were created by using Electrical discharge machining. Figure 3 shows the used experimental setup [31]. Also, Table 1 shows bearing test conditions, and their bearing characteristic frequency formulae are represented in Table 2. Table 3 represents the numerical value of the bearing characteristic frequency obtained for the

Fig. 3. Experimental set up [31]

Table 1. Bearing functioning condition test

	Type of REB	Motor load	Defect information	Motor Speed (rpm)	REB Condition
Case Study 1	6205-2RS JEM	0 HP	Diameter: 1 mm	1797	Ball fault
			Depth: 0.2 mm		IR Fault
			(IR & OR)		OR Fault
			Depth: 0.1 mm (Ball Fault)		(Centered @ 6:00

rolling element bearing parameter, and Table 4 represents the REB parameter for the test bearing.

Table 2. Bearing Characteristic frequencies.

Fault Location	Failure Frequency
Defect on cage (FTF)	$FTF = f_g = \frac{1}{2}\left[f_i\left(1 - \frac{d\,cos\theta}{D_p}\right)\right]$
Defect on Ball (BSF)	$BSF = f_r = \frac{D_p}{2d}(f_i)\left[1 - \left(\frac{d\,cos\theta}{D_p}\right)^2\right]$
Defect on inner race (BPFI)	$BPFI = \left\{\frac{N}{2}\left[f_i\left(1 + \frac{d\,cos\theta}{D_p}\right)\right]\right\}$
Defect on outer race (BPFO)	$BPFO = \left\{\frac{N}{2}\left[f_i\left(1 - \frac{d\,cos\theta}{D_p}\right)\right]\right\}$

At 1797 rpm, time series vibration signals are collected, identifying both healthy and defective bearing states. At 1797 rpm the vibration signal is recorded and used as sample

Table 3. Characteristic frequencies of REB

Characteristic frequencies (Hz)	Shaft Frequency	FTF	BSF	BPFI	BPFO
Case Study 1	29.95	11.228	141.16	162.18	107.36

Table 4. Rolling element bearing parameters.

	Ball number N	Contact angle α	Ball diameter d	Outside Diameter	Inside Diameter	Pitch Diameter D_p
Case Study 1	9	0	0.00794 m	0.0519 m	0.025 m	0.039 m

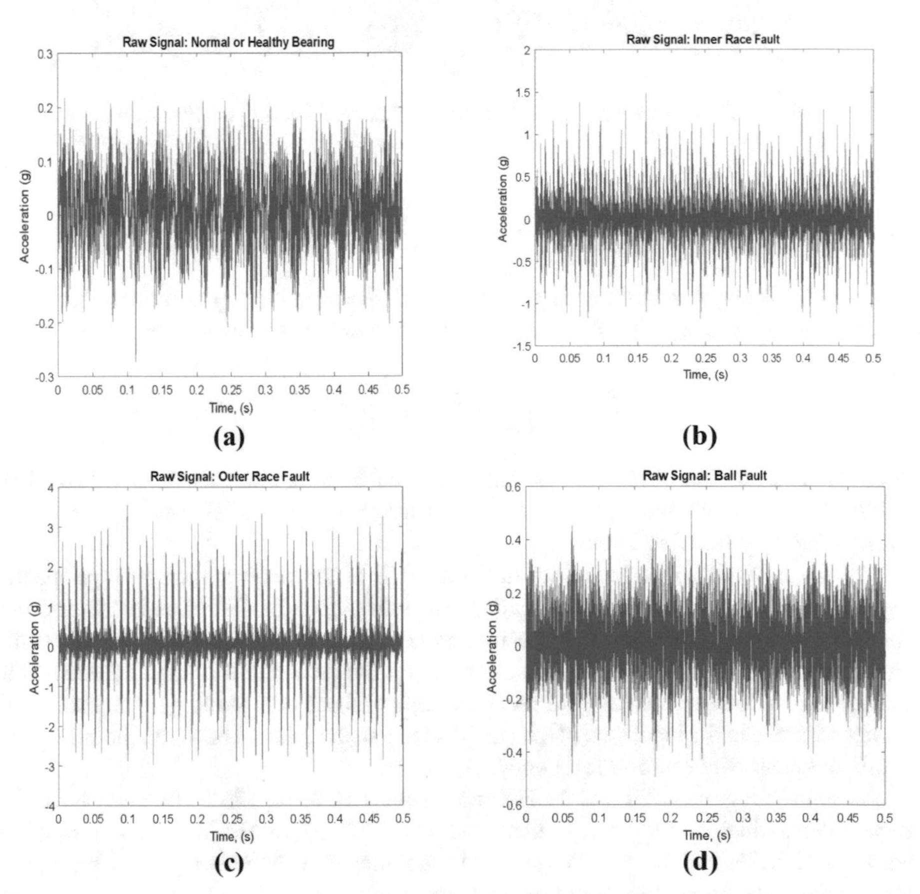

Fig. 4. The time-domain waveform of (a) healthy signal, (b) bearing with IR defect, (c) Bearing with OR defect, and (d) REB with ball defect.

data for investigation. The time domain vibration signal characteristics are presented in Fig. 4 whereas the FFT spectra for this sampled signal are presented in Fig. 5.

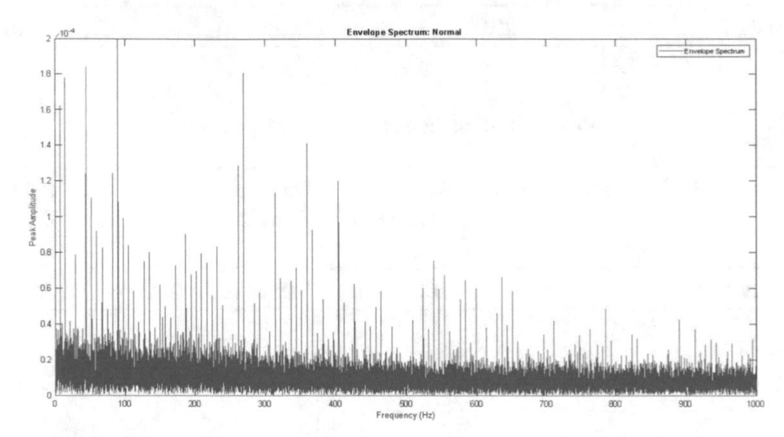

Fig. 5. The frequency-domain waveform of normal bearing healthy signal.

4.1 Case 1 IR Fault Diagnosis

Case 1 is a self-aligning ball bearing with an IR defect, and the suggested approach was used to find the bearing. The BPFI of given bearing can be calculated as follows

$$BPFI = \left\{ \frac{N}{2} \left[f_i \left(1 + \frac{d\ cos\theta}{D_p} \right) \right] \right\}$$

where N is the number of rolling elements, D_p is the bearing pitch diameter, d is the ball diameter, θ is the contact angle and f_i is shaft frequency. The related parameter of the testing bearing is listed in Table 4.

From a previously inserted minor fault on the IR of the bearing, the envelope spectra displayed in Fig. 6 were acquired, revealing the fault frequency of BPFI and its harmonics as other harmonic sidebands may be recognized. These sidebands correspond to the signal that the inner race is spinning. As a result, the defect will touch and depart the load zone, resulting in a change in raceway contact force and the largest amplitude of impact recorded in the load zone [4]. It can be seen that the interval between the sidebands equals the inner race rotation frequency.

As seen in Fig. 6, BPFI and its Harmonics are 161.7 Hz, 323.4 Hz, and 485.1 Hz, respectively, which is the nearly calculated value of 162.18 Hz. It indicated that the bearing has an IR defect as its frequency is modulated at the characteristic frequency derived above in the mathematical expression.

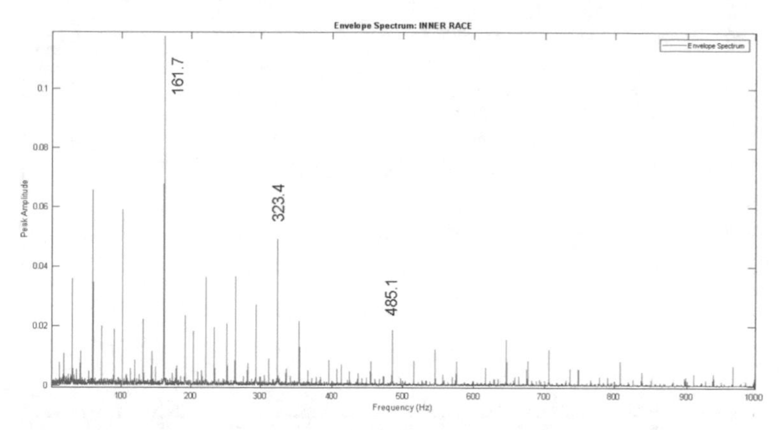

Fig. 6. The envelope spectrum (BPFI)

Figure 6 shows a sample vibration signal with its FFT spectrum at 1797 rpm with IR defect. In some circumstances, the envelope spectrum modulates the OR of the bearing spinning and the defect frequency. Amplitude modulation shows the sideband in the envelope spectra at cage rotation frequency. Another type of defect is the cage defect which is not mentioned in this paper.

4.2 Case 2 OR Fault Diagnosis

In Case 2: The BPFO of bearing can be measured as follows: Table 4 lists the corresponding parameters of the test bearing.

$$BPFO = \left\{ \frac{N}{2} \left[fi \left(1 - \frac{d\, cos\theta}{D_p} \right) \right] \right\}$$

In the second experimental research instance, the shaft speed and frequency are set to 1797 rpm and 29.95 Hz. For the sake of this empirical inquiry, 12000 Hz was chosen as the sampling frequency. Figure 7 demonstrates the envelope spectrum analysis with the maximum frequency of fault characteristics. The most visible component is 107.6 Hz, which matches the projected BPFO (107.36 Hz). The feature frequency (107.36 Hz) and the harmonics of its fault frequency are 215.3 Hz and 322.9 Hz, respectively. As a consequence, the corresponding BPFO, according to the bearing's specifications, is 107.6 Hz.

It stands to reason that the problem characteristic can be found when testing a REB with an OR defect. The outer race defect frequency (BPFO) and its harmonics, as illustrated in Fig. 7, have been categorized. There appear to be no fault sidebands in this specific occurrence. This conclusion arises from the test equipment's fixed bearing and, as a result, no amplitude modulation.

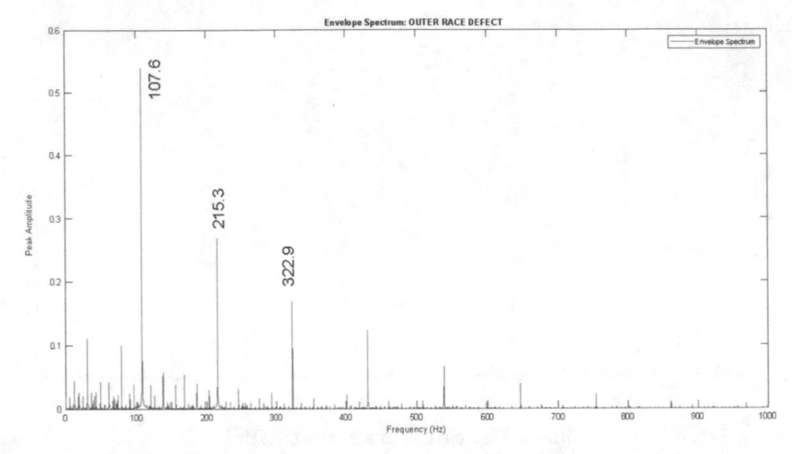

Fig. 7. The envelope spectrum (BPFO)

4.3 Case 3 Ball Fault

The BSF of bearing can be calculated as follows: Table 4 lists the corresponding parameters of the test bearing.

$$BSF = f_r = \frac{D_P}{2d}(f_i)\left[1 - \left(\frac{d\ cos\theta}{D_p}\right)^2\right]$$

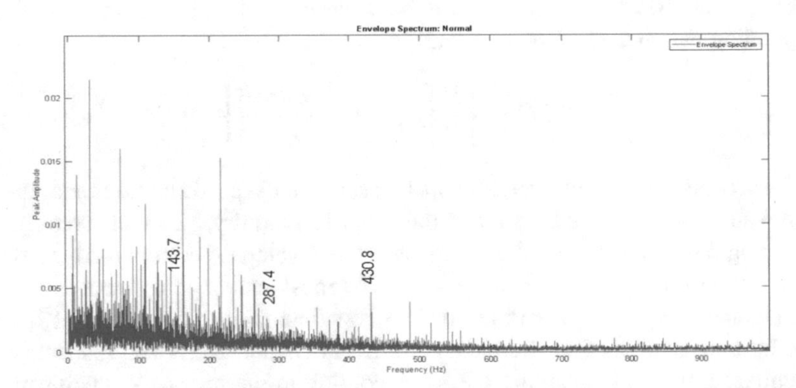

Fig. 8. The envelope spectrum (BSF)

In this experimental research instance, the shaft speed and frequency are set to 1797 rpm and 29.95 Hz. For the sake of this empirical inquiry, 12000 Hz was selected as the sampling frequency. As seen in Fig. 7, the Ball fault and its Harmonics are 143.7 Hz, 287.4 Hz, and 430.8 Hz, respectively, which is the nearly calculated value of 141.16 Hz. It indicated that the bearing has an IR defect as its frequency is modulated at the characteristic frequency derived above in the mathematical expression.

It stands to reason that the fault feature can be found when testing a bearing with an outside race problem. The Ball fault and its harmonics have been categorized in Fig. 8.

5 Conclusions

Experiments were performed while the existing study explains the various fault frequency and sidebands in the frequency spectrum. Analyzing the mathematical framework employed in the derivation of the cited defect frequencies does not reveal any impact. To replicate variations in cage speed, a didactic technique was used to provide an intentionally exaggerated sliding to explain its influence in the spectrum analysis. We highlight the relevance of the related variations within the calculated and include data as an efficient judgment maintenance tool. The application of rolling bearings in experiments carried out while the current investigations reject the influence of other geometric parameters, like the contact angle. The importance of that geometric parameter in the fault diagnosis of defected REB is under ongoing research. In addition to envelope analysis, several advanced techniques such as empirical mode decomposition, which is adaptive in nature, are proposed for future study or research on nonlinear dynamic characteristics and nonstationary vibration signals.

References

1. Al-Naggar, Y.M., Jamil, N., Hassan, M.F., Yusoff, A.R.: Condition monitoring based on IoT for predictive maintenance of CNC machines. Procedia CIRP **102**, 314–318 (2021). ISSN 2212-8271. https://doi.org/10.1016/j.procir.2021.09.054
2. Gupta, V.V.K., Kankar, P.K.: Bearing fault diagnosis using feature ranking methods and fault identification algorithms. Procedia Eng. **144**, 343–350 (2016). ISSN 1877-7058. https://doi.org/10.1016/j.proeng.2016.05.142
3. Wescoat, E., Mears, L., Goodnough, J., Sims, J.: Frequency energy analysis in detecting rolling bearing faults. Procedia Manuf. **48**, 980–991 (2020). ISSN 2351-9789. https://doi.org/10.1016/j.promfg.2020.05.137
4. Jayaswal, P., Wadhwani, A.K., Mulchandani, K.B.: Machine fault signature analysis. Int. J. Rotating Mach. **2008** (2008)
5. Khadersab, A., Shivakumar, S.: Vibration analysis techniques for rotating machinery and its effect on bearing faults. Procedia Manuf. **20**, 247–252 (2018). ISSN 2351-9789. https://doi.org/10.1016/j.promfg.2018.02.036
6. Pratyusha, L.P., Priya, S.V., Naidu, V.P.S.: Bearing health condition monitoring, time domain analysis. Int. J. Adv. Res. Electr. Electron. Instrum. Eng. **3**, 75–82 (2014)
7. Sreejith, B., Verma, A.K., Srividya, A.: Fault diagnosis of rolling element bearing using time-domain features and neural networks. In: 2008 IEEE Region 10 Colloquium and the Third International Conference on Industrial and Information Systems, pp. 226–234 (2008)
8. Saucedo-Dorantes, J.J., Delgado-Prieto, M., Ortega-Redondo, J.A., Osornio-Rios, R.A., Romero-Troncoso, R.D.J.: Multiple-fault detection methodology based on vibration and current analysis applied to bearings in induction motors and gearboxes on the kinematic chain. Shock Vib. **2016**, 1–13 (2016)
9. Marichal, G., Artes, M., Garcia-Prada, J.: An intelligent system for faulty-bearing detection based on vibration spectra. J. Vib. Control **17**(6), 931–942 (2011)
10. Guo, H., Jack, L.B., Nandi, A.K.: Feature generation using genetic programming with application to fault classification. IEEE Trans. Syst. Man Cybern. Part B (Cybern.) **35**(1), 89–99 (2005)
11. Shakya, P., Darpe, A.K., Kulkarni, M.S.: Vibration-based fault diagnosis in rolling element bearings : ranking of various time, frequency and time-frequency domain data-based damage identification parameters. Int. J. Cond. Monit. **3**(2), 53–62 (2013)

12. Bafroui, H.H., Ohadi, A.: Application of wavelet energy and Shannon entropy for feature extraction in gearbox fault detection under varying speed conditions. Neurocomputing **133**, 437–445 (2014)
13. Lu, S., Guo, J., He, Q., Liu, F., Liu, Y., Zhao, J.: A Novel contactless angular resampling method for motor bearing fault diagnosis under variable speed. IEEE Trans. Instrum. Meas. **65**(11), 2538–2550 (2016)
14. Cui, L., Zhang, Y., Zhang, F., Zhang, J., Lee, S.: Vibration response mechanism of faulty outer race rolling element bearings for quantitative analysis. J. Sound Vib. **364**, 67–76 (2016)
15. Khanam, S., Tandon, N., Dutt, J.K.: Fault size estimation in the outer race of ball bearing using discrete wavelet transform of the vibration signal. Procedia Technol. **14**, 12–19 (2014). ISSN 2212-0173. https://doi.org/10.1016/j.protcy.2014.08.003
16. Zhang, Y., Lv, Y., Ge, M.: Time–frequency analysis via complementary ensemble adaptive local iterative filtering and enhanced maximum correlation kurtosis deconvolution for wind turbine fault diagnosis. Energy Rep. **7**, 2418–2435 (2021). ISSN 2352-4847. https://doi.org/10.1016/j.egyr.2021.04.045
17. Udmale, S.S., Patil, S.S., Phalle, V.M., Singh, S.K.: A bearing vibration data analysis based on spectral kurtosis and ConvNet. Soft. Comput. **23**(19), 9341–9359 (2018). https://doi.org/10.1007/s00500-018-3644-5
18. Dwyer, R.: Detection of non-Gaussian signals by frequency domain Kurtosis estimation. In: ICASSP 1983. In: IEEE International Conference on Acoustics, Speech, and Signal Processing, pp. 607–610 (1983). https://doi.org/10.1109/ICASSP.1983.1172264
19. Huang, H., Baddour, N., Liang, M.: Short-time Kurtogram for bearing fault feature extraction under time-varying speed conditions. In: V008T10A035 (2018). https://doi.org/10.1115/DETC2018-85165
20. Antoni, J.: The spectral kurtosis: a useful tool for characterising non-stationary signals. Mech. Syst. Sig. Process. **20**(2), 282–307 (2006). ISSN 0888-3270. https://doi.org/10.1016/j.ymssp.2004.09.001
21. Tomar, A.S., Jayaswal, P.: Envelope spectrum analysis of noisy signal with spectral kurtosis to diagnose bearing defect. In: Singh, M.K., Gautam, R.K. (eds.) Recent Trends in Design, Materials and Manufacturing. LNME. Springer, Singapore (2022). https://doi.org/10.1007/978-981-16-4083-4_23
22. Antoni, J., Randall, R.B.: The spectral kurtosis: application to the vibratory surveillance and diagnostics of rotating machines. Mech. Syst. Sig. Process. **20**(2) 308–331 (2006). ISSN 0888-3270. https://doi.org/10.1016/j.ymssp.2004.09.002
23. Udmale, S.S., Singh, S.K.: Application of spectral kurtosis and improved extreme learning machine for bearing fault classification. IEEE Trans. Instrum. Meas. **68**(11), 4222–4233 (2019). https://doi.org/10.1109/TIM.2018.2890329
24. Mertins, A.: Signal Analysis: Wavelets, Filter Banks, Time-Frequency Transforms and Applications (2001). https://doi.org/10.1002/0470841834.ch2
25. Li, B., Goddu, G., Chow, M.Y.: Detection of common motor bearing faults using frequency-domain vibration signals and a neural network based approach. In: Proceedings of the 1998 American Control Conference, vol. 4, pp. 2032–2036. IEEE (1998)
26. McCormick, A.C., Nandi, A.K.: Real-time classification of rotating shaft loading conditions using artificial neural networks. IEEE Trans. Neural Netw. **8**(3), 748–757 (1997)
27. McCormick, A.C., Nandi, A.K.: Bispectral and trispectral features for machine condition diagnosis. IEE Proc. Vis. Image Sig. Process. **146**(5), 229–234 (1999)
28. McFadden, P.D., Smith, J.D.: Vibration monitoring of rolling element bearings by the high-frequency resonance technique – a review. Tribol. Int. **17**(1), 3–10 (1984)
29. Nandi, A.K., Liu, C., Wong, M.D.: Intelligent vibration signal processing for condition monitoring. In: Proceedings of the International Conference Surveillance, vol. 7, pp. 1–15 (2013). https://surveillance7.sciencesconf.org/resource/page/id/20

30. Randall, R.B., Antoni, J., Chobsaard, S.: A comparison of cyclostationary and envelope analysis in the diagnostics of rolling element bearings. In: 2000 IEEE International Conference on Acoustics, Speech, and Signal Processing. ICASSP 2000. Proceedings, vol. 6, pp. 3882–3885. IEEE (2000)
31. Bearing Data Center, Case Western Reserve University. http://csegroups.case.edu/bearingdatacenter/pages/download-data-file

Predictive Analytics for Fake Currency Detection

P. Antony Seba(✉), R. Selvakumaran, and Dharan Raj

Kumaraguru College of Technology, Coimbatore, Tamil Nadu, India
Sebaantony97@gmail.com

Abstract. Fake currency is the imitation of the real currency without the legal sanction of the government. Producing or using fake currency is a form of forgery and it is illegal. The circulation of fake currency increases inflation and raises demands for goods which also affects the producers and consumers thereby leads to currency devaluation. Classifying fake currency on the fly is a major challenge. The aim of this work is to detect fake currencies using predictive analytic models by considering the statistical characteristics of the fake currency note as features. The US dollar images are considered for classification of fake currency notes. The statistical characteristics variance, skewness, kurtosis and entropy of the fake currency images are extracted from images and are treated as the independent variables for model building. The dataset is splitted into 80:20 as training and test datasets. The classifiers Naïve Bayes, Random Forest, AdaBoost, Logistic Regression, Support Vector Machine, K- Nearest Neighbor and Multi-layer Perceptron are built. The performance metrices accuracy, precision, recall and F1 score of the built models are compared. The predictive analytics models SVM, KNN and MLP yield the best accuracy 100% and these models are not suffering from drift between the train and test dataset.

Keywords: Currency note · variance · skewness · kurtosis · entropy · multilayer perceptron

1 Introduction

The most used methods for transactions are paper money even today. The currency notes that are imitated with illegal sanction of the government is counterfeit currency. Every country contributes a number of security features for securing its currency. Currency counterfeiting is a challenging term for any country. The Currency counterfeiting problem affects the economic and financial growth of a country. The long-term issue in Currency counterfeiting is the identification of notes. The large volume of counterfeit currency increases more circulation of money, leading to a flow in demand for products and services that cause currency depreciation.

The authentic currency notes are made of cotton or cellulose natural material and the cotton is mixed with other textile fibres, linen etc., to provide additional strength against the counterfeiting. The security loop is woven into the currency note during the manufacturing process. The banknote is infused with alcohol, gelatine, and polyvinyl

H. Sharma et al. (Eds.): ICIVC 2022, PALO 17, pp. 128–137, 2023.
https://doi.org/10.1007/978-3-031-31164-2_11

for robustness against normal printing and writing papers. Other unique features are also incorporated in currency notes viz. For instance, Indian currency has anti photocopying feature, European Euro has perforation and US dollar has 3D security ribbon and bell in ink.

The security thread is represented in counterfeit notes by drawing a line with a pencil, printing a line with grey ink, or using aluminum thread while sticking two thin sheets of paper together, which are becoming more similar to originals, making it harder for non-experts to detect them [1]. Machines can identify counterfeit money but are very expensive. Artificial Intelligence techniques have the potential to solve this problem significantly [2]. Machine learning is the part of artificial intelligence that helps in predicting fake currency notes using supervised learning algorithms [3]. The algorithms are first learned by training the model with the training dataset and evaluated using the test dataset. In this work, the US dollar images are considered for prediction of fake currency notes. The image data is converted into structured data by extracting the variance, kurtosis, skewness and entropy as independent features. State of the art predictive analytics models are built in classifying fake currency notes and the models are evaluated. Section 2 of this article elaborates on literature reviews, Sect. 3 elaborates the methodology adopted in classifying the fake currency and Sect. 4 discusses on results yielded by the classifiers.

2 Literature Review

Twana et al. in their review article identified that fake notes are detected and verified using a human-visual system [4]. Most of the research works used Convolutional Neural Networks since it can identify crucial features, Generative Adversarial Networks for recognizing fake notes with better accuracy. Santhi et al. in their proposed work used reinforcement learning and recurrent convolutional neural network with image processing technique to determine the note's transparency [5]. The authors proposed a reliable approach such that the methods are applied one after the other and hence If any segment is missed through the first phase, that might be covered in the second phase. Sasikumar et al. proposed a method to detect fake currency notes by counting the number of interruptions in the thread line and also the authors have calculated the entropy of the currency notes for building a better prediction model using MATLAB [6]. Aman et al. in their work proposed a fake currency detection using K-Nearest Neighbors classifiers and then applied image processing techniques [7]. They have used the banknote authentication dataset created with the high computational and mathematical strategies to give the correct information regarding the entities and features related to the currency notes. Ali et al. proposed a dubbed deep money machine assisted system to detect fake currency using GANs and achieved 80% accuracy [8]. Cesar et al. employed transfer learning strategy in CNN and an AlexNet sequential CNN has been used in detecting fake currency notes by training both models [9]. They have used Columbian banknote dataset and achieved better accuracy. Dereje et al. collected various Ethiopian currencies of different ages and used different optimization techniques for CNN architects to predict the fake currency notes [10, 11]. The authors employed various CNN architectures such as InceptionV3, MobileNetV2, XceptionNet, and ResNet50. MobileNetV2 with RMSProp optimization technique is found to be a robust and reliable model with an accuracy of 96.4%. In this proposed work, structured dataset is used in making classifications rather than images. And it is evident from the results that the machine learning models yield the best accuracy.

3 Methodology

3.1 Dataset

The dataset used in this work is collected from Kaggle repository [12]. The dataset consists of the statistical measures of 1372 US currency note images. The textural features extracted from the currency images by transforming the images using wavelet [13, 14] are the second moment variance, third moment skewness, fourth moment kurtosis and entropy of each image as shown in Table 1.

Table 1. Statistical measures of image dataset

S. No	Features	Type
1	*Variance*	Numeric
2	*Skewness*	Numeric
3	*Entropy*	Numeric
4	*kurtosis*	Numeric
5	*Target*	Categorical

Variance describes the spread of the data i.e., how far the pixels of the image lie away from its mean and is calculated as given in Eq. 1.

$$\sigma^2 = \frac{1}{N}\sum\nolimits_{i=1}^{N} \left(x_i - \underline{x}\right)^2 \tag{1}$$

where N is the total number of pixels, $\underline{x}$ is the mean of the pixels in an image and x_i is the value of each pixel. Skewness measures the degree of asymmetry exhibited by the image. If skewness equals zero, the histogram is symmetric about the mean and the skewness of an image is calculated as given in Eq. 2.

$$\gamma_1 = \frac{1}{N}\sum_{i=1}^{N}\left[\frac{x_i-\mu}{\sigma}\right]^3, \ \mu \text{ is the mean} \tag{2}$$

Kurtosis measures the peakedness of an image. The normal distribution of kurtosis is 0 and is calculated as given in Eq. 3.

$$\gamma_2 = \frac{1}{N}\sum\nolimits_{i=1}^{N}\left[\frac{x_i-\mu}{\sigma}\right]^4 - 3 \tag{3}$$

Entropy of an image is a measure of uncertainty of image information and is defined as shown in Eq. 4. The target attributes values are (fake, not fake)

$$H = -\sum_{k} p_k log_2(p_k) \tag{4}$$

3.2 Data Pre-processing

The dataset does not have outliers and there are no missing values. The 80% of the dataset is considered as a training dataset and the remaining 20% of the dataset is considered as a test dataset. The split up 80:20 is done in a stratified manner i.e., distributed based on target labels to prevent imbalance issues.

3.3 Model Building

Machine learning models Naïve Bayes (NB), Random Forest (RF), AdaBoost (AB), Logistic Regression (LR), Support Vector Machine (SVM), K- Nearest Neighbors (KNN) and Multi-layer perceptron (MLP) are built using the training dataset and are estimated using the test dataset. NB classifier uses a probabilistic approach in classifying the fake currency as stated in Eq. 5.

$$P(A|B) = \frac{P(B|A)XP(A)}{P(B)} \tag{5}$$

where P(A|B) is the posterior probability, P(B|A) is the likelihood, P(A) is the prior probability and P(B) is the marginal probability. Ensemble techniques (bagging and boosting) are gaining popularity in prediction and classification. RF uses bagging technique and reduces variance and thereby prevents overfitting. It consists of many decision trees and the classification is based on majority votes as shown in Fig. 1. The Gini index is used to identify the feature relevance as given in Eq. 6.

$$Gini = 1 - \sum_{i=1}^{n} (p_i)^2 \tag{6}$$

where p_i is the probability of the target. Boosting is an ensemble technique to build a strong classifier from weak classifiers. Initial model is built from the training data and the next model is built by correcting the errors present in the initial model. This iteration is continued and models are added until the maximum as shown in Fig. 2.

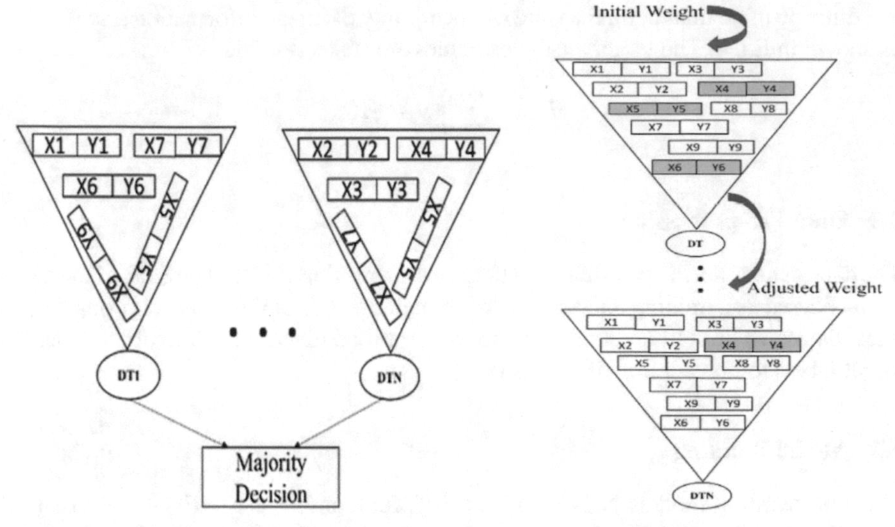

Fig. 1. Random Forest **Fig. 2.** AdaBoost

LR use statistical approach and classifies the dependent variable by analyzing the relationship between independent variables as given in Eq. 7.

$$p(x) = \frac{1}{1 + e^{-(\beta_0 + \beta_1 * variance + \beta_2 * skewness + \beta_3 * kurtosis + \beta_4 * entropy}} \quad (7)$$

MLP is based on neurons and it consists of three layers: input layer, hidden layer, and output layer. Input features can be quantitative or qualitative and each feature are associated with weights and bias. Neurons computational units have weighted inputs and produce an output using an activation function where weighted inputs are summed and passed through transfer function to make it non-linear. The activation function controls the threshold of the neurons and thereby the strength of the output. The entire process of this proposed work is shown in Fig. 3.

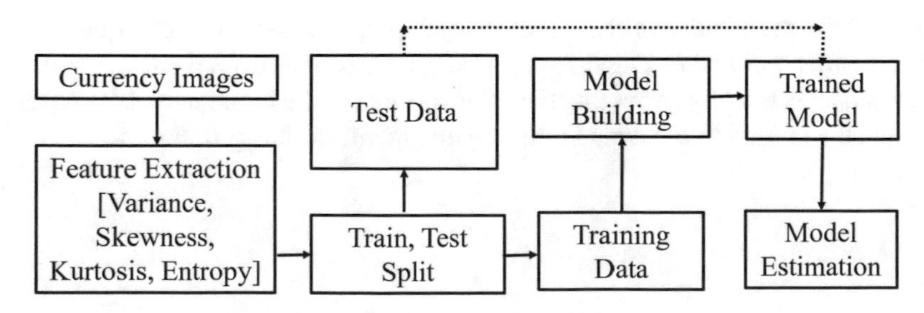

Fig. 3. Fake currency detection

In this work, the MLP architecture is built with hidden layers and neurons per layer. The SVM algorithm is to find a hyperplane in an N-dimensional space that classifies the data points. The hyperplane is just a line when there are two independent variables. The hyperplane becomes a 2-D plane when the number of independent variables is three. The maximum margin hyperplane is identified and the support vectors are used on classification of fake currency detection. KNN is a lazy learning classifier, the majority of its k neighbors is the outcome of the classifier. And the MLP, SVM and KNN yields better accuracy in classifying fake currency.

4 Results and Discussion

The independent variables used in this work and their distribution are given in Fig. 4 (a) to Fig. 4 (d). All the independent variables used in this work are slightly skewed.

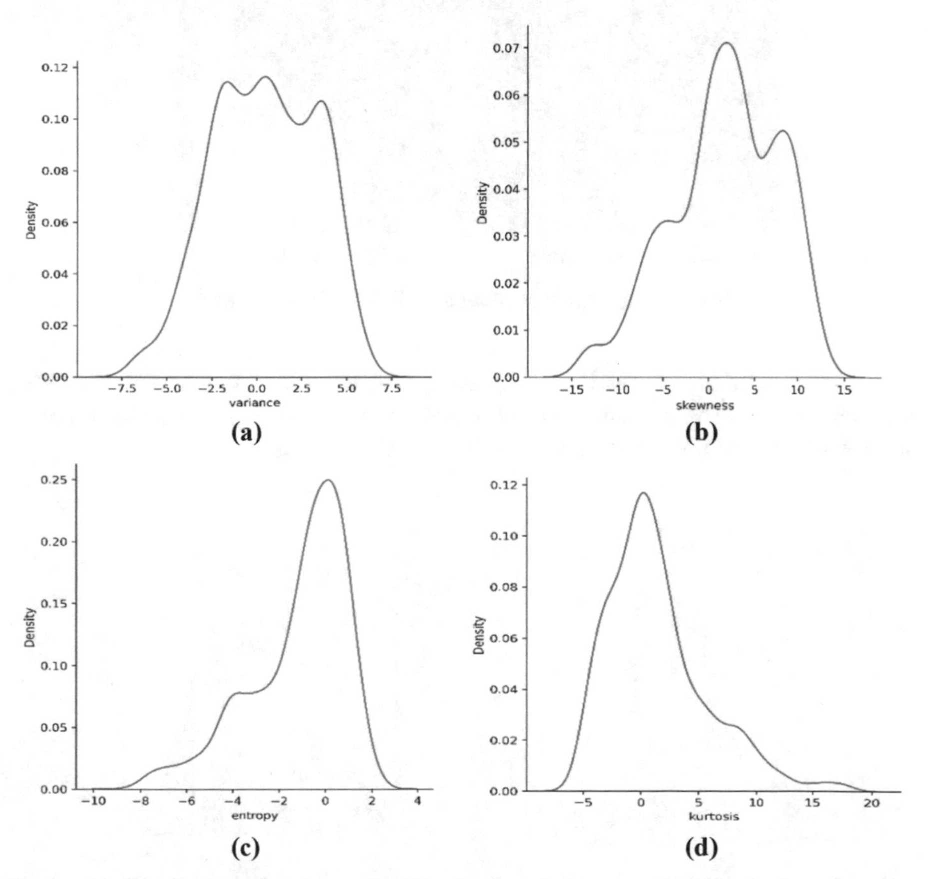

Fig. 4. (a). Distribution of variance. (b). Distribution of skewness. (c). Distribution of entropy. (d). Distribution of kurtosis

The dataset is tested for multicollinearity using correlation coefficient and the heatmap is visualized as in Fig. 5. It is evident from the heatmap that there is no correlation among the independent variables and all the variables are considered for model building.

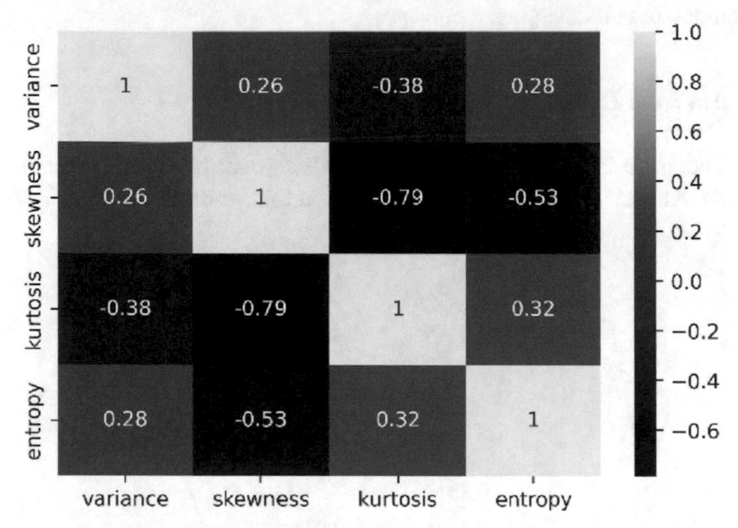

Fig. 5. Correlation coefficient of independent variables

The 80% training dataset is trained using NB, RF, AB, LR, SVM, KNN, MLP classifiers and the 20% test data is used for evaluation. The statistical distribution of the training test and the test set is visualized in Fig. 6 (a) and Fig. 6 (b).

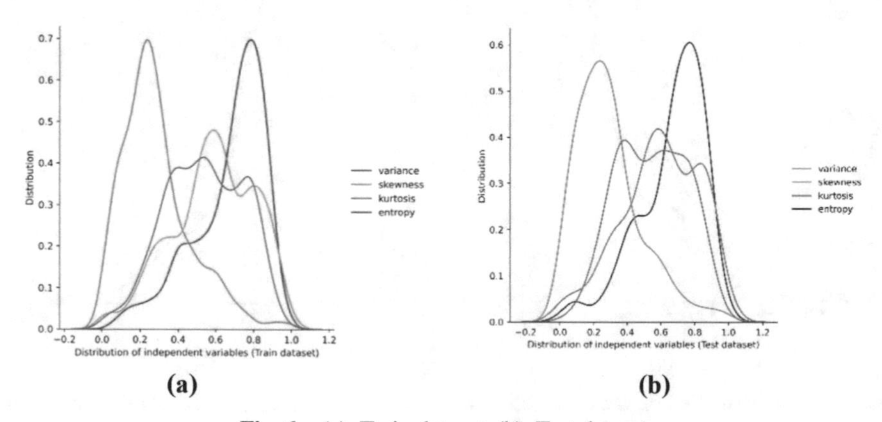

Fig. 6. (a). Train dataset. (b). Test dataset

The performance metrices accuracy, precision, recall and f1 score of the built model is shown in Table 2. The RF classifier is built with 100 estimators and the measure of purity considered in this work is Gini. The AdaBoost uses 50 estimators and the number of stumps is 1. The KNN classifier is built with K as 7. The MLP classifier is built with four hidden layers and four neurons per layer. First three hidden layers use the relu activation function and the fourth layer uses sigmoid activation function that outputs a value between 0 and 1 with an s-shaped distribution. Each layer is connected with bias and weights. MLP is built and estimated for 50, 100 and 150 epochs.

Table 2. Performance metrics

Models	Accuracy	Precision	Recall	F1 score
Naïve Bayes	83.27	82.9	82.9	82.9
Random Forest	98.90	98.83	98.93	98.88
AdaBoost	98.90	98.83	98.93	98.88
Logistic Regression	98.90	98.83	98.93	98.88
Support Vector Machine	100	100	100	100
K-Nearest Neighbours	100	100	100	100
Multilayer Perceptron	100	100	100	100

The Calibration plots compares the probabilistic predictions of a binary classifier and plots the true frequency of the positive label against its predicted probability, for binned predictions and the calibration plots along with brier score for the training and test data of the models NB and MLP are shown in Fig. 7 (a) to Fig. 7 (d).

The built MLP model yields 96.72% accuracy with MSE 2.82 for 50 epochs, 99.27% accuracy with MSE 1.13 for 100 epochs and 100% accuracy with MSE 1.37 for 150 epochs. SVM and KNN also yields 100% accuracy with brier score 0 and the categorical drift score between training and test dataset is less than 0.2 and numerical drift score is less than 0.1.

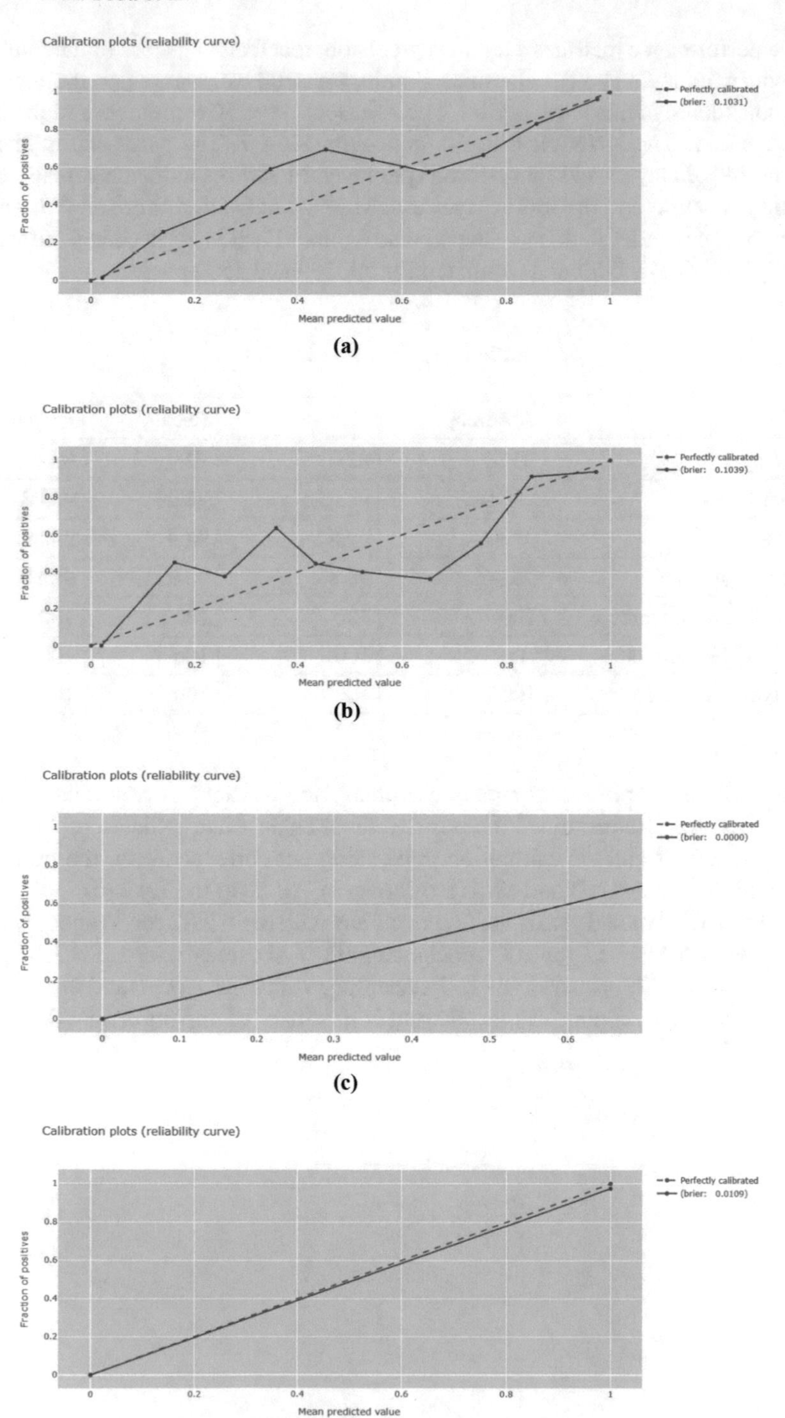

Fig. 7. **(a)** Calibration plot of Naïve Bayes (Train). **(b)** Calibration plot of Naïve Byes (Test). **(c)** Calibration plot of MLP (Train). **(d)** Calibration plot of MLP (Test)

5 Conclusion

The statistical features from US dollar currencies are extracted and a multilayer perceptron is designed in this work and the model is trained with 80% training data and the model has been validated with 20% test data. The MLP, SVM and KNN classifiers outperforms and yields 100% accuracy. Machine learning classifiers yields the best results in classifying fake currencies. The same can also be experimented with different variants of neural networks using the currency images instead using structured data by extracting features from the currency images.

References

1. Dahanukar, R.: MONEY TALKS: "back to the future"—challenges in banknote design. In: Chakrabarti, A., Poovaiah, R., Bokil, P., Kant, V. (eds.) ICoRD 2021. SIST, vol. 221, pp. 191–201. Springer, Singapore (2021). https://doi.org/10.1007/978-981-16-0041-8_17
2. Frosini, F., Angelo, S., Marco Gori, T., Paolo Priami, F.: A neural network-based model for paper currency recognition and verification. IEEE Trans. Neural Netw. **7**(6), 1482–1490 (1996)
3. Roy, A., Halder, B., Garain, U. and Doermann, D.S.: Machine-assisted authentication of paper currency: an experiment on Indian banknotes. Int. J. Doc. Anal. Recogn. **18**(3), 271–285 (2015)
4. Twana Nasih Ahmed, F., Sabat Salih Muhamad, S.: Counterfeit currency recognition using deep learning: a review. Technium **3**(7), 14–26 (2021)
5. Santhi M.V.B.T, F., Hrushikesava Raju, S.S., Adinarayna, S.T., Lokanadham Naidu, V.F., Waris, F.: A hybrid framework for efficient detection of fake currency notes. In: Saini, H.S.F., Singh, R.K.S., Tariq Beg, M.T., Mulaveesala R.F., Mahmood, M.R. (eds.) Innovations in Electronics and Communication Engineering. LNNS, vol. 355. Springer, Singapore (2022). https://doi.org/10.1007/978-981-16-8512-5_24
6. Arya, S., Sasikumar, M.: Fake currency detection. In: International Conference on Recent Advances in Energy-efficient Computing and Communication, pp. 1–4. IEEE (2019)
7. Bhatia, A., Kedia, V., Shroff, A., Kumar, M., Shah, B.K.: Fake currency detection with machine learning algorithm and image processing. In: 5th International Conference on Intelligent Computing and Control Systems (ICICCS), pp. 755–760. IEEE (2021)
8. .Ali, T., Jan, S., Alkhodre, A., Nauman, M., Amin, M., Siddiqui, M.S.: DeepMoney: counterfeit money detection using generative adversarial networks. PeerJ Comput. Sci. **5**, e216 (2019)
9. Pachón, C.G., Ballesteros, D.M., Renza, D.: Fake banknote recognition using deep learning. Appl. Sci. **11**(3) 1281 (2021)
10. Aseffa, D.T., Kalla, H., Mishra, S.: Ethiopian banknote recognition using convolutional neural network and its prototype development using embedded platform. J. Sens. **2022** (2022)
11. Viraktamath, S.V., Tallur, K., Bhadavankar, R.: Review on detection of fake currency using image processing techniques. In: 5th International Conference on Intelligent Computing and Control Systems, pp. 865–870. IEEE Xplore Part Number: CFP21K74-ART (2021)
12. https://www.kaggle.com/datasets/zohaib30/banknotes. Accessed 16 June 2022
13. Gai, F., Shan, S., Guowei Yang, T., Minghua Wan, F.: Employing quaternion wavelet transform for banknote classification. Neurocomputing **118**, 171–178 (2013)
14. Choi, F., Euisun, S., Jongseok Lee, T., Joonhyun Yoon, F.: Feature extraction for bank note classification using wavelet transform. In: 18th International Conference on Pattern Recognition, vol. 2, pp. 934–937. IEEE (2006)

Detecting Spam Comments on YouTube by Combining Multiple Machine Learning Models

B. Aravind(✉) and Anil Kumar Mishra

Amity University, Gurugram, Haryana 122412, India
aravind1027@gmail.com, akmishra2@ggn.amity.edu

Abstract. In recent times, online media has seen an influx with billions of concurrent users. YouTube is one such platform for streaming media, allowing its users to post their ideas in the form of comments. Although it is one of the biggest platforms, it suffers from spam comments from adversaries and bots. To tackle such comments, this paper examines different machine learning models (Multinomial Naive Bayes, Support Vector Machine, Logistic Regression, Decision Tree, and Random Forest) on the comments of five popular videos on the platform. It was found that the Ensemble model with soft voting outperformed the other classifiers when these techniques were coupled with hard and soft voting.

Keywords: Machine learning · Ensemble model · Spam detection

1 Introduction

YouTube is a video-sharing platform, founded in 2005 and was bought out by Google in 2006. Being one of Google's subsidiaries, it has risen to prominence as a major player in the media streaming market. YouTube users can upload, like, share, favorite, report videos, comment their thoughts on videos, and subscribe to other content creators and YouTube channels. As of late 2017, users have been watching more than 1 billion hours of content per day [1]. Gaming is one of the majorly growing industries on the internet and for the last 12 months, users on YouTube have consumed 50 billion minutes of gaming content which shows the scale of the users utilizing the platform.

One of the reasons the site is popular among content creators is that they can monetize on the videos provided they have a collective watch time of over 4000 h on their videos in the last 12 months and have 1000 subscribers to their channel. With more activity on their videos, it will more likely be promoted. Hence, spam comments act as catalysts to some of the creators while lowering the user experience when it comes to major content creators because of abusive or degrading comments irrelevant to the video [2].

Though YouTube has spam filtering as one of its features, adversaries are still able to post such comments and go undetected. We analyzed relevant studies on spam detections and examine ensemble machine learning model with hard voting and soft voting to

H. Sharma et al. (Eds.): ICIVC 2022, PALO 17, pp. 138–149, 2023.
https://doi.org/10.1007/978-3-031-31164-2_12

enhance spam comment detection using a combination of the following machine learning models - Multinomial Naive Bayes, Support Vector Machine, Logistic Regression, Decision Tree, and Random Forest.

The following is a breakdown of the paper's structure: Related work on spam comment detection is reviewed in Sect. 2. Section 3 describes the proposed method for spam comment detection. Applied machine learning models are described in Sect. 4. Evaluation metrics used to analyze the performance are described in Sect. 5. Section 6 contains the results of experiments and conclusion is provided in Sect. 7.

2 Related Work

Detecting spam has been the subject of numerous studies in a variety of domains. Various studies have used various strategies to approach the problem at hand.

Hayoung Oh [2] used Decision Tree Classifier, Logistic regression, Bernoulli Naïve Bayes Classifier, Random Forest Classifier, Support vector machine with linear kernel and Gaussian kernel for detecting spam comments on YouTube. The author used Bag-of-Words for preprocessing the data and compared the individual scores of the above-mentioned algorithms with ensemble hard voting and soft voting models. Bernoulli Naïve Bayes was excluded for both the ensemble algorithms as it had the lowest performance and to have an odd number of input classifiers for the ensemble models. It was observed that ensemble soft voting had the highest performance.

To identify spammers, M. McCord and M. Chuah [3] used machine learning models that were fitted with user and content-centered features. They used twitter data to test their algorithms and determined that the Random Forest model was performing the best among other models.

To detect spam comments on Instagram posts from Indonesian users, Ali Akbar Septiandri and Okiriza Wibisono [4] used length of the comment, frequency of capitalized letters, and usage of emojis in a comment, advertisement and product-oriented words, and textual features, namely BoW, TF-IFD and fastText embeddings, all of which was combined with latent semantic analysis as the main features for the classifiers – Naïve Bayes, SVM and SVM, XGBoost. Combining all three features produced the F1-score of 0.960.

Sohom Bhattacharya et al. [5] used Count-Vectorizer to create a lattice in which each individual word is handled by the framework's segment, and the network's cell worth addresses the coordinate with word check. They utilized the Naive Bayes classifier.

Hammad Afzal et al. [6] compared Naïve Bayes Multinomial, DMNBText, SVM, J48 and Liblinear classifiers to detect spam on SMS and tweets. Their study suggests that both Naïve Bayes Multinomial and DMNBText perform well but Naïve Bayes Multinomial classifier was comparatively better in case of false positives.

To detect spam reviews, the authors, Q. Peng and M. Zhong. [7], used sentiment analysis concepts. They proposed several modifications to the calculation of sentiment score which resulted in performing better than the currently existing models.

Sadia Sharmin et al. [8] collected comments from the videos on YouTube with the help of the platform's API. Each video's comments were manually marked either ham or spam. 10 classifiers were used, which were k-NN, LR, CART, NB (Bernoulli, Gaussian and Multinomial), RF, SVM-L, SVM-P, and SVM-R. It was observed that Naïve Bayes with Bagging out-performed other algorithms.

3 Proposed Method

3.1 Machine Learning Algorithms and Environment

To detect YouTube spam comments, we used five machine learning techniques: Multinomial Naive Bayes classifier, Support vector machine with linear kernel, logistic regression, decision tree, and random forest. We will examine the performance of each algorithm as well as different combinations of these algorithms to determine the optimal combination. All the experiments were performed on the Google Colab.

3.2 Overview

A dataset containing 1956 YouTube comments was used during the experiment. The data was split into training data with 1467 comments (70% of the total comments) and testing data with 489 comments (25% of the total comments) as shown Fig. 1. Then, TF-IDF (Term Frequency—Inverse Document Frequency) which is a technique for calculating significance of a word in the document, is used for preprocessing the data [2].

Each model is individually trained and tested. Further, a combination of 3 out of the 5 models is selected and is trained and tested using the ESM-S (Ensemble with soft-voting) & ESM-H (Ensemble with hard-voting). Finally, a combination of all the models is trained and tested using ESM-S and ESM-H.

3.3 Dataset

The dataset was obtained from UCI repository [9]. These contain comments from popular YouTube videos at the time of collection. All the datasets contain 5 attributes, which are YouTube ID, Author, Date, Content and Class. Class attribute contains two values, 0 and 1 which represent Ham and Spam respectively. We combined all the dataset and dropped all the attributes except for Comment and Class. From the total 1956 comments, 75% (1476 comments) were used to train the models and 25% (489) were used to test the models. The total number of ham and spam in the dataset is provided in Table 1.

3.4 Dataset Processing

Since the input data is textual comment data, it has to be processed in-order for the machine learning algorithms to understand it. TF-IDF is used to process data as it determines how important a word is across a set of documents by assigning values to terms. Term frequency is the count of occurrences of a word with respect to all the words in a document and is a component of the feature vector. A word becomes a feature when it appears in more than 3 documents [10]. Inverse document frequency on the other hand

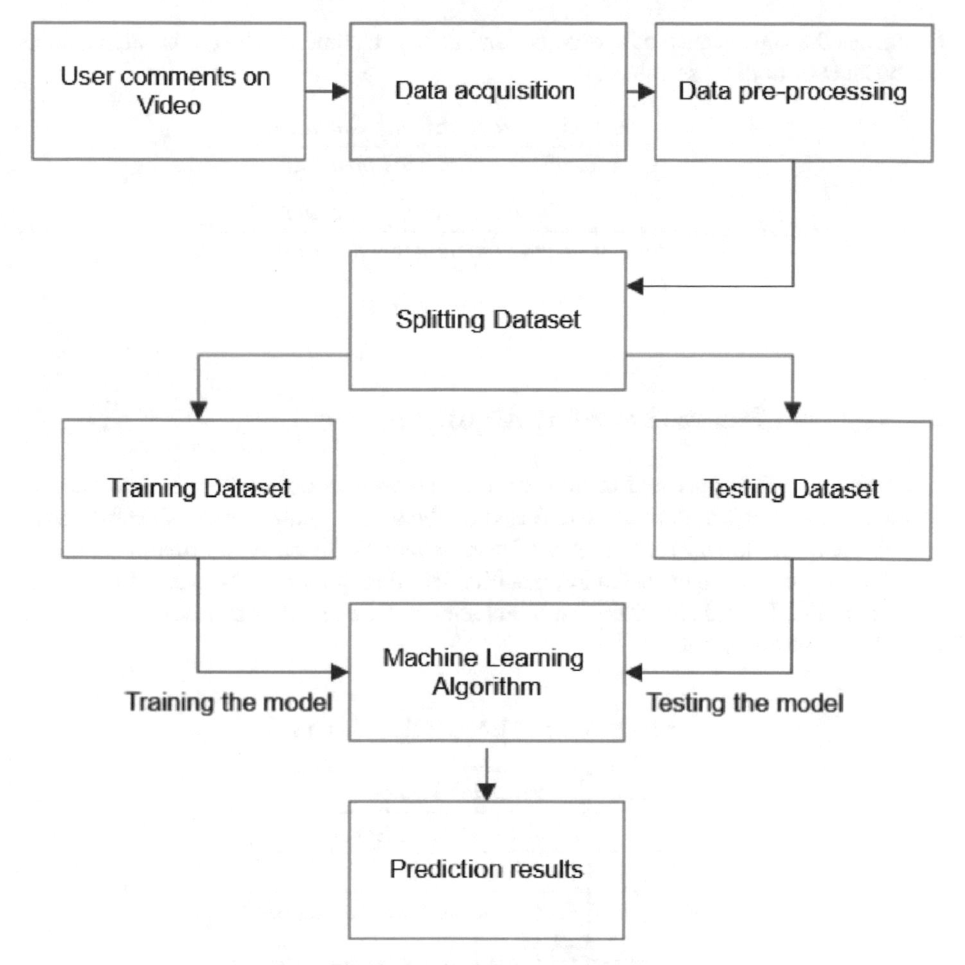

Fig. 1. Methodology overview

Table 1. Dataset overview

Dataset	Count of spam	Count of ham
Psy	175	175
KatyPerry	175	175
LMFAO	236	202
Eminem	245	203
Shakira	196	174
Total	1005	951

quantifies the significance of a word occurring in a document. TF-IDF is calculated as the product of both these values.

$$TF = \frac{Count\ of\ the\ term\ in\ the\ document}{Number\ of\ words\ in\ document} \tag{1}$$

$$IDF = log \frac{Total\ number\ of\ documents}{Number\ of\ documents\ in\ which\ the\ term\ appears} \tag{2}$$

$$TFIDF = TF \times IDF \tag{3}$$

4 Applied Machine Learning Algorithms

The main goal for applying machine learning algorithms on the dataset is to classify whether a comment is spam or ham. Based on the training data, the classification algorithm is used to identify the category of testing dataset. To study information, validate data, and understand their behavior, machine learning is more effective and is usually preferred [11]. Table 2 shows the list of machine learning models that were used for the classifying the comments.

Table 2. Machine learning algorithms used

Abbreviation	Classification Method
NB	Multinomial Naïve Bayes
LR	Logistic Regression
SVM	Support Vector Machine with linear kernel
DT	Decision Tree
RF	Random Forest
ESM-H	Ensemble with hard voting
ESM-S	Ensemble with soft voting

4.1 Multinomial Naïve Bayes

It is a probabilistic model which is popular in NLP and is based on the Bayesian theorem, while assuming Naïve independence between the predictors [11]. It works on the basis of term frequency. The classifier signifies the occurrence and the frequency of a word in the document. Multinomial Naïve Bayes perform better with less number of records [12].

4.2 Logistic Regression

LR is a linear classifier with a curve that can map any real-valued integer to a value that is between 1 and 0, excluding the boundary values [13]. When the dataset can be linearly separated, it works well and has good accuracy for many simple data sets.

4.3 Support Vector Machine

Support vector machine uses support vectors to try to discover the optimal categorization and the division of data. Only the data contained in the support vectors is used to form the model in this method, and the process is unaffected by additional data. SVM looks for the best line to use to separate data to keep it as far apart from the other categories as feasible [14].

4.4 Decision Tree

A decision tree takes a set of input values, runs a series of tests, and returns a single output value as a decision result. The decision tree's input and output values might be continuous or discrete [10]. A decision tree is trained by dividing the source set into subsets based on an attribute value.

4.5 Random Forest

This model is made up of a large number of discrete decision trees that combine together. Each of the decision trees will generate a classification prediction, and the class having the highest votes is chosen as the prediction of the classifier [15]. The model's accuracy increases with increase in the number of decision trees in the random forest classifier which also lowers the chances of over-fitting.

4.6 Ensemble with Hard Voting

Ensemble with hard voting is simply a majority voting where the class predicted is based on the most number of votes.

$$\hat{y} = mode\{C\_1(x), C\_2(x)..C\_m(x)\} \quad (4)$$

4.7 Ensemble with Soft Voting

Soft voting is based on the probabilities of all the input classifiers. The prediction of this ensemble model will be determined by the average of the probabilities of each classifier.

$$\hat{y} = argargmax_i \sum_{j=1}^{m} w_j p_{ij} \quad (5)$$

5 Evaluation Metrics

To comprehend the effectiveness of the outcomes obtained through the machine learning models, various performance evaluation methodologies are proposed. When evaluating a model, it's critical to apply a variety of evaluation indicators because a model may perform well when one measurement from one evaluation metric is used to evaluate the model, but it could perform poorly when another measurement from a different evaluation metric is utilized.

5.1 Confusion Matrix

True positive (TP), true negative (TN), false positive (FP), and false negative (FN) predictions are commonly presented using a confusion matrix. The values are represented in a matrix form while the predicted classes on the X-axis and the true classes on the Y-axis [16]. Figure 2 indicates a confusion matrix.

	Predicted Class (False)	*Predicted Class (True)*
Actual Class (False)	*True Negative (TN)*	*False Positive (FP)*
Actual Class (True)	*False Negative (FN)*	*True Positive (TP)*

Fig. 2. Confusion matrix

5.2 Accuracy

Accuracy is the ratio between correctly classified cases (TP & TN) to the total number of instances [17].

$$ACC = \frac{TP + TN}{TP + TN + FP + FN} \tag{6}$$

5.3 Precision

It denotes the ratio between correctly classified positive instances (TP) out of all the instances classified as positive (TP and FP).

$$P = \frac{TP}{TP + FP} \tag{7}$$

5.4 Recall

It is the ratio between correctly classified positive instances (TP) and all the actual positive instances (TP and FN).

$$R = \frac{TP}{TP + FN} \tag{8}$$

5.5 F1-Score

The F1-score combines precision and recall of a classifier into a single metric by taking the average (harmonic mean) of their values. This score helps in assessing a model's performance with imbalanced dataset.

$$F1{-}Score = \frac{2 \times (P \times R)}{P + R} \tag{9}$$

5.6 ROC Curve

Receiver Operating Characteristic (ROC) curve is a graph that depicts the tradeoff between the true positive rate (TPR) and the false positive rate (FPR) [16]. The area under this curve represents the ability of a classifier to distinguish between the different classes.

6 Results

The dataset was split into 75% training data and 25% testing data. All the 5 machine learning algorithms are then applied. Table 3 contains the list of algorithms along with the accuracy, precision, recall and the f1-score of detected spam comments for the algorithms.

Table 3. Individual experiment results of algorithms

Classifier	Accuracy (%)	Precision (%)	Recall (%)	F1-Score
NB	89.57	87.59	93.92	0.9064
SVM	93.45	95.29	92.40	0.9382
LR	93.45	96.02	91.63	0.9377
DT	92.84	94.88	91.63	0.9323
RF	**94.06**	97.18	91.63	0.9432

It was observed that RF had the highest accuracy of 94.06%, compared to the other algorithms. Further, combinations of 3 of these algorithms were generated and each of the combination was tested with both ESM-S and ESM-H. And finally, a combination of all the 5 algorithms was tested with ESM-S and ESM-H. Table 4 contains the results of this experiment with the values – Acc (Accuracy), P (Precision), R (Recall) and F1-Score.

Table 4. Results of ensemble model

Combinations	ESM-H				ESM-S			
	Acc (%)	P (%)	R (%)	F1-Score	Acc (%)	P (%)	R (%)	F1-Score
(NB, SVM, LR)	93.66	95.67	92.40	0.9400	93.45	94.59	93.16	0.9387
(NB, SVM, DT)	93.45	95.29	92.40	0.9382	94.68	96.11	93.92	0.9500
(NB, SVM, RF)	93.66	95.67	92.40	0.9400	93.86	94.98	93.54	0.9425
(NB, LR, DT)	93.66	96.03	92.02	0.9398	94.88	96.85	93.54	0.9516
(NB, LR, RF)	93.66	96.03	92.02	0.9398	93.45	95.65	92.02	0.9380
(NB, DT, RF)	**94.47**	**96.46**	**93.16**	**0.9478**	94.47	96.46	93.16	0.9478
(SVM, LR, DT)	93.86	95.69	92.78	0.9421	94.68	96.47	93.54	0.9498
(SVM, LR, RF)	93.66	96.03	92.02	0.9398	93.66	95.31	92.78	0.9403
(SVM, DT, RF)	93.86	96.05	92.40	0.9419	**95.09**	**97.23**	**93.54**	**0.9535**
(LR, DT, RF)	93.66	96.40	91.63	0.9396	94.68	96.84	93.16	0.9496
All algorithms	93.45	96.02	91.63	0.9377	93.86	95.69	92.78	0.9421

A combination of SVM, DT and RF yielded the highest accuracy of 95.09% when summed with soft voting. Figure 3 depicts a holistic view of the performance of this combination in the form of confusion matrix.

The evaluation was further extended to ROC curve. This graph contains ROC curves for all the 5 machine learning models along with the highest performing ensemble model, ESM-S combination (SVM, DT, RF). Area under the curve (AUC) was observed to be maximum for ESM-S with AUC = 0.99, which indicates a 99% chance that the model can distinguish between positive class and negative class as shown in Fig. 4.

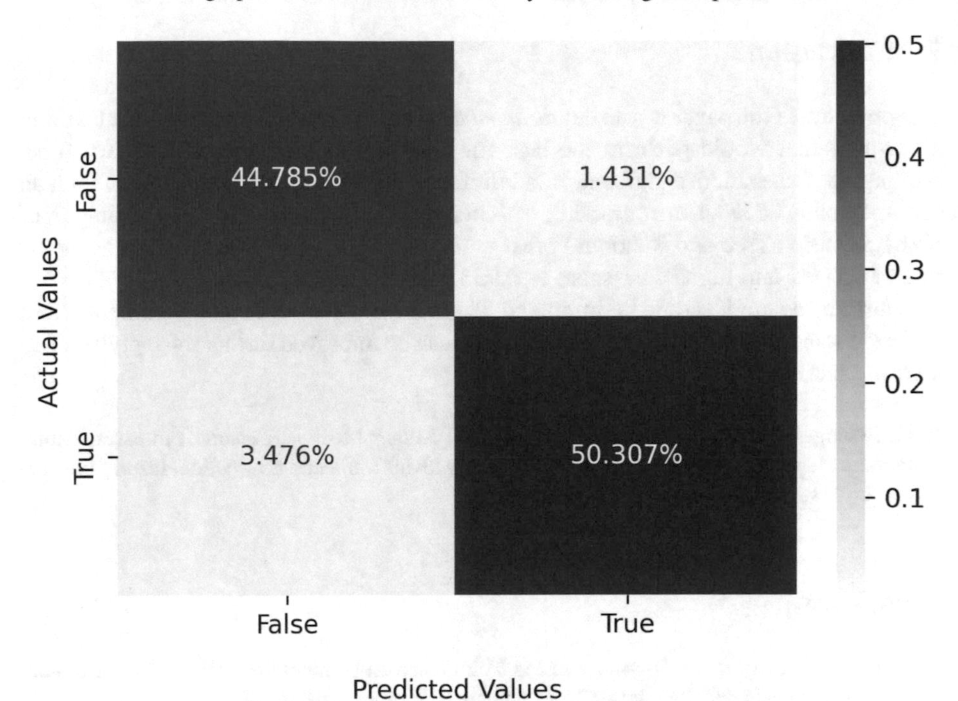

Fig. 3. Confusion matrix for best performing combination of EMS-S model

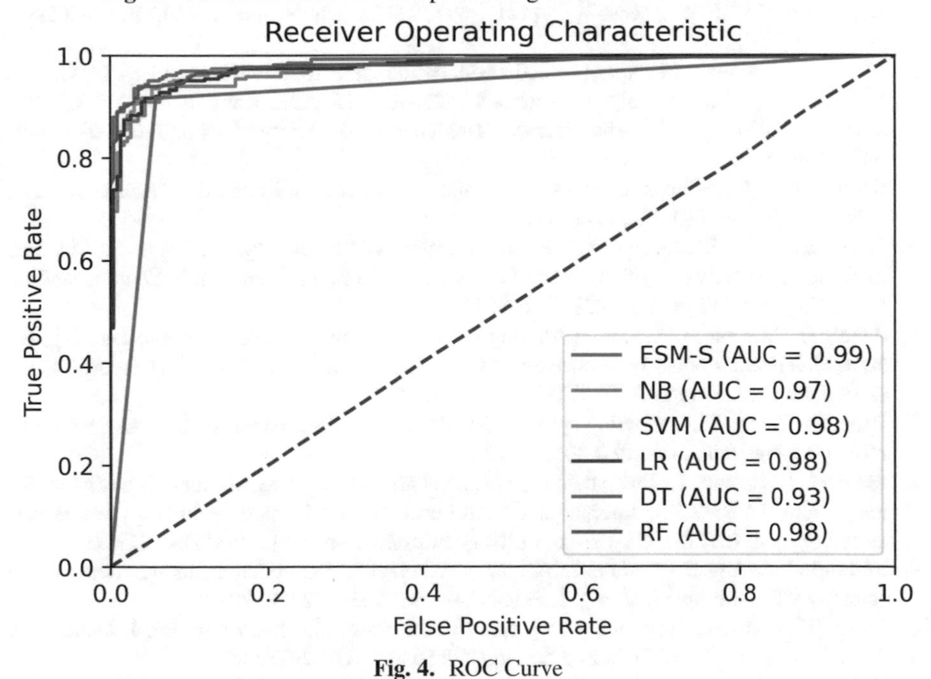

Fig. 4. ROC Curve

7 Conclusion

The objective of the paper was to determine the optimal combination of machine learning algorithms that would perform the best for detecting spam comments on YouTube. Among the 5 classifiers examined, Random forest classifier performed the best with an accuracy of 94.06% when comparing with individual classifiers while a combination of SVM, Decision Tree and Random Forest with soft voting yielded the highest accuracy rate of 95.09% and highest F1-score of 0.9535 for detecting spam.

Further, the model could be improved with a larger dataset of spam comments from different social media platforms. The accuracy can be improved further by applying deep learning techniques.

Acknowledgement. I'm very grateful to Dr. Anil Kumar Mishra, Assistant Professor, Amity University Haryana, who has been a major support with his tremendous direction, inspiration, and constructive suggestions.

References

1. Aiyar, S., Shetty, N.P.: N-gram assisted Youtube spam comment detection. Proc. Comput. Sci. **132**, 174–182 (2018). https://doi.org/10.1016/j.procs.2018.05.181
2. Oh, H.: A YouTube spam comments detection scheme using cascaded ensemble machine learning model. IEEE Access **9**, 144121–144128 (2021). https://doi.org/10.1109/ACCESS.2021.3121508
3. McCord, M., Chuah, M.: Spam detection on Twitter using traditional classifiers. In: Calero, J.M.A., Yang, L.T., Mármol, F.G., García Villalba, L.J., Li, A.X., Wang, Y. (eds.) ATC 2011. LNCS, vol. 6906, pp. 175–186. Springer, Heidelberg (2011). https://doi.org/10.1007/978-3-642-23496-5_13
4. Septiandri, A.A., Wibisono, O.: Detecting spam comments on Indonesia's Instagram posts. J. Phys. Conf. Ser. **801**, 012069 (2017)
5. Bhattacharya, S., Bhattacharjee, S., Das, A., Mitra, A., Bhattacharya, I., Gupta, S.: Machine learning-based Naive Bayes approach for divulgence of Spam Comment in Youtube station. Int. J. Eng. Appl. Phys. **1**(3), 278–284 (2021)
6. Afzal, H., Mehmood, K.: Spam filtering of bi-lingual tweets using machine learning. In: 2016 18th International Conference on Advanced Communication Technology (ICACT), p. 1 (2016). https://doi.org/10.1109/ICACT.2016.7423529
7. Peng, Q., Zhong, M.: Detecting spam review through sentiment analysis. J. Softw. **9** (2014). https://doi.org/10.4304/jsw.9.8.2065-2072
8. Sharmin, S., Zaman, Z.: Spam detection in social media employing machine learning tool for text mining. In: 2017 13th International Conference on Signal-Image Technology & Internet-Based Systems (SITIS), pp. 137–142 (2017). https://doi.org/10.1109/SITIS.2017.32
9. Mehmood, A., On, B.-W., Lee, I., Ashraf, I., Choi, G.S.: Spam comments prediction using stacking with ensemble learning. J. Phys. Conf. Ser. **933**, 012012 (2017)
10. Yang, W., Kwok, L.: Improving blog spam filters via machine learning. Int. J. Data Anal. Tech. Strateg. **9**, 99 (2017). https://doi.org/10.1504/IJDATS.2017.085901
11. Ul Hassan, C.A., Khan, M.S., Shah, M.A.: Comparison of machine learning algorithms in data classification. In: 2018 24th International Conference on Automation and Computing (ICAC), pp. 1–6 (2018). https://doi.org/10.23919/IConAC.2018.8748995

12. Singh, G., Kumar, B., Gaur, L., Tyagi, A.: Comparison between multinomial and Bernoulli Naïve Bayes for text classification. In: 2019 International Conference on Automation, Computational and Technology Management (ICACTM), pp. 593–596 (2019). https://doi.org/10.1109/ICACTM.2019.8776800
13. Bassiouni, M., Shafaey, M., El-Dahshan, E.-S.: Ham and Spam E-mails classification using machine learning techniques. J. Appl. Secur. Res. **13**, 315–331 (2018). https://doi.org/10.1080/19361610.2018.1463136
14. Sohrabi, M.K., Karimi, F.: A feature selection approach to detect spam in the Facebook social network. Arab. J. Sci. Eng. **43**(2), 949–958 (2017). https://doi.org/10.1007/s13369-017-2855-x
15. Yiu, T.: Understanding random forest. Medium, 29 September 2021. https://towardsdatascience.com/understanding-random-forest-58381e0602d2. Retrieved 8 June 2022
16. Czakon, A.J.: 24 evaluation metrics for binary classification (and when to use them). neptune.ai, 5 January 2022. https://neptune.ai/blog/evaluation-metrics-binary-classification. Retrieved 8 June 2022
17. Baccouche, A., Sierra-Sosa, D., Elmaghraby, A., Ahmed, S.: Malicious text identification: deep learning from public comments and emails. Information **11**, 312 (2020). https://doi.org/10.3390/info11060312

A Review of Different Aspects of Human Robot Interaction

A. U. Manas(✉), Sunil Sikka, Manoj Kumar Pandey, and Anil Kumar Mishra

Department of Computer Science and Engineering, Amity University Haryana, Gurugram, India
manasunnikrishnan123@gmail.com, {ssikka,mkpandey, akmishra2}@ggn.amity.edu

Abstract. Human Robot Interaction (HRI) is essential for the robotic research works to grow. In the absence of HRI, capabilities of Artificial Intelligence (AI) might go unexplored. The goal of this paper is to draw attention to the Computer Vision and Natural Language Processing aspects of HRI. This paper reviews Face Recognition, Emotion Recognition and Object Detection as Computer Vision aspects of HRI and Natural Language Processing as a backbone of interaction. Based on the analysis of multiple studies the review provides a simmered study of state-of-the-art works happening in the related area.

Keywords: Human Robot Interaction · Computer Vision · Face Recognition · Emotion Recognition · Object Detection · Natural Language Processing

1 Introduction

A humanoid robot can be defined as the programmable machine which can imitate the tasks of humans as well as their appearance [1]. Humanoid robotics is the field where it assists and entertains humans with the similar appearance as humans. Humanoid robotics is one of the research fields that is new and demanding, that has achieved increasing attention in recent years, and it will continue to play an important role in robotics research [1]. A lot of the 21st century's industrial applications such as assembling of parts, dispensing, picking and placing, welding, soldering, painting, etc. are based on robotics. HRI is the study of understanding, designing, and assessing robotic systems for use by or with humans [2]. Human-Computer Interaction (HCI), Artificial Intelligence (AI), Robotics, Natural Language Understanding (NLU), Natural Language Generation (NLG), Design, and Psychology disciplines play a role in HRI. One of the most important applications of HRI is in Industrial Robots. Industrial robots have already been integrated into manufacturing lines and are working alongside humans. Robots are being used in vital fields such as search and rescue, military battle, mine and bomb detection, scientific research, law enforcement, entertainment, and hospital care in the most technologically sophisticated nations. These new application domains suggest a closer relationship with the user. Humans and robots occupy the same workspace, yet they have different goals in terms of task completion. Researchers are working hard to design new models for better HRI

H. Sharma et al. (Eds.): ICIVC 2022, PALO 17, pp. 150–164, 2023.
https://doi.org/10.1007/978-3-031-31164-2_13

and with less risk factors. In the past, industrial robots were utilized to perform repetitive activities that simply replaced people. In this new era, collaborative robots (cobots) are widely deployed in different work environments to split work with humans [3]. With the advance in AI, it is now possible to design better cobots with physically safe, socially correct and culturally dependent interaction. HRI is the building block which makes this kind of advanced interaction with robots possible. HRI's mission is to create a simple and straightforward way for humans to communicate with robots via speech, gestures, and facial expressions. Dautenhahn K. [4] describes friendly HRI as "Robotiquette" which says "social rules for robot behavior that is comfortable and acceptable to humans". The robot must adapt to the human style of expressing demands and orders, not the other way around. However, everyday situations like houses have far more intricate social regulations than factories or even military environments. As a result, the robot needs perception and comprehension abilities in order to construct dynamic models of its surroundings. It must classify things, identify and detect people, and understand their emotions. The current study majorly reviews the Computer Vision and Natural Language Processing (NLP) aspects of HRI. Face recognition, emotion recognition and object detection are reviewed as 'Vision' aspects of HRI in this study. Nowadays, Machine Learning is used to solve a wide range of Computer Vision tasks such as feature identification, picture segmentation, object recognition, and tracking. Robots are equipped with visual sensors in a variety of applications, from which they learn about their surroundings by completing computer vision challenges. In humanoid robots, Computer Vision is employed for localization, object detection, face detection and recognition. Also, NLP aspects of HRI are reviewed in this study. NLP is a multidisciplinary branch which includes linguistics, computer science, and artificial intelligence concerned with HCI. NLP focuses on teaching computers how to interpret and analyze massive volumes of natural language data [5]. NLP aims to build a computer that can understand the contents of documents. The system can then extract accurate insights from the contents, as well as classify and arrange them.

The rest of the paper is laid out as follows. Section 2 briefly reviews various computer vision research which further covers Face Recognition, Emotion Recognition, Object Detection and NLP approaches. Section 3 describes the findings from the whole review. Finally, Sect. 4 concludes the study and also provides the future directions for research in the area of HRI.

2 Literature Review of Different aspects of HRI

The review of the literature is divided in 2 parts: 1) Computer Vision and 2) NLP in which Computer Vision is further divided in 3 parts: 1) Face Recognition, 2) Emotion Recognition and 3) Object detection.

2.1 Computer Vision

Computer Vision is a branch of AI through which machines extract useful data from photos, videos and other visual aids. Based on this extracted knowledge, the system can conduct actions or make suggestions. When AI helps computers to think, Computer

Vision helps to see, watch and comprehend. Computer Vision is achieved through Pattern Recognition. The underlying technique of Computer Vision is feeding a large number of labeled images and using these to train the model to identify the patterns in images and assign them to the labels of corresponding images. Even though Computer Vision has a lot of applications including self-driving cars to Augmented Reality (AR), this study focuses on the HRI applications of Computer Vision. In HRI systems Computer Vision is majorly used to recognize objects, people and their emotions.

Face Recognition. A Facial Recognition (FR) system compares a human face from a digital image or video frame to a database of faces. FR systems are most commonly used to identify customers via ID verification services, and they work by pinpointing and measuring facial features from a supplied image. Although FR systems as biometric technology are less accurate than iris and fingerprint recognition, they are commonly used due to their contact-less approach [6]. Despite the fact that people can recognize faces without much effort, face recognition is a difficult pattern recognition problem in computing. The goal of fr systems is to recognize a human face from a two-dimensional image that is three-dimensional with variations in appearance, lighting, and facial expression. This computing operation is completed in four phases by FR systems. Face detection is employed first to isolate the face from the background image. In the second phase, the segmented face image is aligned to extract face pose, image size, and photographic properties such as lighting and grayscale. The image is prepared for the third stage, which is facial feature extraction, which precisely locates the face features. Features such as the eyes, nose, and mouth are found and measured in the image to show the face. In the fourth stage, the Face's established feature vector is matched against a database of faces [7].

In the field of face recognition many works have been carried out. In this review some recent advancements of the field are discussed. Ding et al. [8] presented a CNN framework for overcoming video-based facial recognition problems. In surveillance films, human faces are frequently subjected to significant image blur, dramatic position fluctuations, and occlusion. Due to a shortfall in real-world video training data, the authors intentionally blurred training data to create a blur resilient face representation model. CNN trained Blur - insensitive features automatically by supplying training data that included both still pictures and artificially blurred data. The authors introduced a Trunk Branch Ensemble (TBE) CNN to improve the model's robustness to posture variations and occlusions. The triplet loss function was proposed to improve the discriminative power of the TBE CNN representation. Experiments were conducted on popular databases including, PaSC, COX Face and YouTube Faces and state-of-the-art performance was achieved by the proposed model. Liu et al. [9] offers a face recognition model based on Principle Component Analysis (PCA), Genetic Algorithm (GA), and Support Vector Machine (SVM). PCA is used to minimize feature dimension, GA is used to improve search methods, and SVM is used to classify. Population size for GA was taken as 20 and the Roulette approach was utilized with a Mutation probability of 0.02. Fitness function for the model was taken as

$$\phi(t) = lg\frac{P_i}{(1 - P_i)} + \frac{H_t}{P_i} + \frac{H_L - H_t}{1 - P_i} \tag{1}$$

In which, Φ = selection operator, P_i = Probability of selection, $H_t = -\sum_{i=0}^{t} P_i lg P_i$, and $H_L = -\sum_{i=0}^{L-1} P_i lg P_i$.

Authors analyzed the relationship between accuracy and number of features in this research and it shows that after a threshold the increase in number of features has a negative impact on accuracy. The research shows that the number of iterations of GA is directly proportional to the accuracy but inversely affects the efficiency. The best number of iterations is found out to be 10 considering efficiency and accuracy. Grayscale or RGB colour images are commonly used as input formats in face recognition frameworks. Choi et al. [10] proposed a DCNN face recognition framework using Gabor face representations to improve performance. This study is an expansion of the work carried out by Choi et al. [11] and Liu et al. [12], in which Gabor features were extracted using only a single "fixed" Gabor filter applied to the first or second convolutional layer. The authors [10] attempted a new strategy by using various and varied Gabor face representations as input. By changing parameters like sizes and orientations, many types of Gabor filters can be produced. 40 distinct Gabor representations are created using these filters. The model is created in two stages. Models are built in the first section of the ensemble of GDCNN members or base. Each of these models is trained using a specific Gabor face representation. In the second part individual GDCNN members are combined. A confidence-based majority voting (CMV) strategy is created for this purpose, which takes into account the number of votes for an identity label (acquired from the ensemble of GDCNN) as well as the confidence of these votes. The identification of the face photos is determined by combining the number of votes and the accompanying confidence values. The performance of the proposed model is evaluated using four public face databases: FERET, CAS-PEAL-R1, LFW, and MegaFace. Following these tests, it is discovered that the suggested model outperforms models that use grayscale or RGB color face photographs as inputs. The suggested GDCNN method is designed using VGG Face and Lightened CNN. It's the first time an ensemble of DCNNs has been employed to improve Face Recognition performance using distinct Gabor face representations. Although the CNN model reduces computing time when compared to full connection networks, some relevant information is lost during the propagation and calculation process. Lou et al. [13] provides a multi-level information fusion model for the convolution calculation approach, which recovers the discarded feature information and increases the picture recognition rate. ORL Face Database, BioID Face Database, and CASIA Face Image Database all verify the model's accuracy. The results of the proposed model are compared with Eigenface, FisherFace, ICA, 2DPCA and traditional CNN and outstanding performance was observed over the 3 datasets. Serengil et al. [14] described the development of a lightweight face recognition framework. The framework uses OpenCV library to handle detection of frontal faces and eyes. To properly align the face, some trigonometric methods are used. At the location of the eyes, a right-angle triangle is drawn similar to Fig. 1. The lengths of the three sides are then determined by considering one side of the triangle to be horizontal.

$$cosA = \frac{b^2 + c^2 - a^2}{2bc} \tag{2}$$

Then the base image will be rotated A degrees.

Fig. 1. Face alignment [14]

VGG-Facial, FaceNet, OpenFace, DeepFace, DeepId, and Dlib are among the most popular face recognition models included in the framework. It is the first model to include all of these cutting-edge facial recognition models. The composite approach's development has made it easier to transition between different models. FaceNet is the most performant model in this framework, followed by VGG-Face. It makes the face recognition task very easy and quick. The proposed framework is published as an open-source package under MIT license. Face recognition models with a large number of parameters are limited by deep neural networks. Boutros et al. [15] proposed 'Pocket-Net', an incredibly lightweight and accurate FR model, as a solution to this challenge. Using neural architecture search, a new family of lightweight face specific architectures was built in this research. In the training stages, a multi-step knowledge distillation was provided, in which knowledge is distilled from the instructor model to the student model. The smallest proposed lightweight network, 'PocketNetS-128,' with 0.92 million parameters, outperformed state-of-the-art models with up to 4 million parameters (Table 1).

Emotion Recognition. One of the most fundamental aspects of human emotion recognition is facial expression. One can get a sense of someone's feelings by looking at their face. Computer vision, nonverbal human behavior, and HCI all benefit from facial expression recognition. Because of many concerns, such as duplication of moves, huge head postures, and so on, it is a difficult task [17]. The research works in the area of facial emotion recognition is based on image or video networks and neural networks. According to Guo et al. [18] dominant and complementary emotions (e.g. Happily-disgusted, sadly-fearful) are more detailed than the seven classical facial emotions (e.g. happy, disgust, etc.). Very limited number of datasets are available for studies in compound emotions (dominant and complementary emotions). Furthermore, the datasets have fewer categories and the data distributions are imbalanced. Machine learning algorithms are

Table 1. Face Recognition Summary

Approach used	Results	Year	Reference
Single Stage Headless Face Detector combined with CNN model	Effectively handles occlusion in face detection and recognition	2020	[16]
Multi-level information fusion model for the convolution calculation approach	Model recovers rejected feature information and enhances the picture recognition rate Outperformed Eigenface, FisherFace, ICA and 2DPCA algorithms	2020	[13]
A FR framework including VGG-Facial, FaceNet, OpenFace, DeepFace, DeepId, and Dlib	Allows users to switch between multiple FR models making face recognition an easy task	2020	[14]
Neural architecture search is used to build lightweight face specific architectures trained with multi-step knowledge distillation	The smallest proposed lightweight network, 'PocketNetS-128,' with 0.92 million parameters, outperformed state-of-the-art models with up to 4 million parameters	2022	[15]

used to assign labels to data, resulting in mistakes. To address these concerns, Guo et al. [18] provided the iCV-MEFED dataset, which contains 50 types of compound emotions and over 30,000 photos, with labels assigned by psychologists. A competition was organized based on the dataset to have detailed experiments and the top 3 winner methods are discussed in this study. The first ranked method approached the program using the combination of texture and geometrical information. Texture features using AlexNet and landmark displacement as geometric representations of emotions are collectively used. The second ranked method used a combination of unsupervised learning and multiple SVM classifiers. The model was trained using K-means clustering. The third ranked method was implemented by using CNN inception-v3 with central loss function. The experiments show that pairs of compound emotions like surprisingly-happy vs happily-surprised are difficult to identify, and the top 3 methods are also varied in performance for each category. So, a combination of these 3 methods might lead to a more accurate solution. Jain et al. [17] proposed a model to classify a face image into 6 emotional classes using Deep Convolutional Neural Networks. The model trained on Extended CohnKanade (CK+) and Japanese Female Facial Expression (JAFFE) dataset. Gaussian normalization and standard deviation have been used for normalizing the images. The proposed model handles different image sizes without human intervention. Images were stripped of all non-expression elements such as background and hair. The model contains 6 convolutional layers and 2 residual learning blocks. A max-pooling layer follows each convolution layer. After the second and fourth convolution layers, the two-deep residual blocks were implemented. Each residual block contains 4 convolutional layers, two short connections and one skip connection. Softmax activation was used for classification. With 95% accuracy, the suggested model achieved state-of-the-art performance.

The brightness of visible facial images is affected by lighting conditions. Instead of geometric and appearance patterns, thermal facial pictures identify facial temperature distributions. So, it is robust to light conditions. So, to increase performance, Wang et al. [19] used fusion of visible and thermal images. Deep learning methods were used to find the image representations from thermal and visible images. These feature representations were used to train SVM classifiers. In this way, the pair of IR and visible images are only required in the training phase and not in the testing phase. The experimental results using the MAHNOB laughter database shows that when thermal IR images are used in place of visible facial images in training, improved facial representations and a stronger emotion classifier are obtained. The accuracy of the proposed model is 13.14%, 23.03%, 20.47%, and 17.25% higher than SVM2K, MMDBM, DCCA and DCCAE respectively. Motivated by the belief that emotion intensity maps provide further specific information than other methods, Zhang et al. [20] proposed a deep neural network for emotion recognition from images based on emotion intensity learning. This research was the stepping stone to using emotion intensity learning to identify emotions. This network contains 2 classification streams with an intensity prediction stream in between them as shown in Fig. 2. Intensity maps are created using ResNet, specifically ResNet-50 and ResNet-101 structures were used. Emotion-6, the FI-8 dataset, and WEBEmo were used in the trials. The proposed network was also evaluated for image sentiment classification. The experimental results reveal that the suggested network outperformed the current state-of-the-art approaches (Table 2).

Object Detection. In order to react effectively to environmental factors, autonomous assistive robots ought to be able to handle visual information in real time. The ability to consistently identify and distinguish objects is frequently a prerequisite for achieving this goal. Model-based techniques are the most typical method to distinguish geometric forms. These methods begin with capturing a huge number of photographs in various positions as well as from various perspectives. A framework is designed and trained ahead of them. The retrieved characteristics from objects in the environment are then compared to characteristics from already stored frameworks. When real-world scenarios are taken into account, real-time accurate object identification and recognition remains a difficult task. The existence of complex, changeable surroundings with probable occlusions, interactions, and extra optical and structural fluctuations contributes to the issue. Martinez et al. [24] presented a system that computes color, movement, geometrical data and uses these combinations in a statistical way to detect and recognize objects. Color space was used for color computation in this approach due to its robustness to illumination changes and the variations in planar direction of the object. This work proposed a hybrid technique for motion cue data that combines frame segmenting and noise removal with a single - Guassian reference framework. The suggested model used a structural feature in the form of Gabor filters based on front, lateral, upper, and perspective representations. Authors utilized Gabor filters to extract object shapes, SVM as an object structure predictor, and principal component analysis (PCA) for recognition. Multiple experiments were conducted on the developed model. The model performance is evaluated using 3 types of input scenes. First a semi structured scene was considered. The test scene consisted of two genuine, congested scenes wherein the goal object was to be located between a range of commonplace things. Third, a visual data set was employed to

Table 2. Emotion Recognition Summary

Approach used	Results	Year	Reference
Emotion intensity learning network contains 2 classification streams with an intensity prediction stream in between them used to identify emotions	Suggested network outperformed the current state-of-the-art approaches in sentiment classification	2020	[20]
Proposed a Convolutional Neural Networks model incorporated with facial landmark and HOG preprocessing methods trained on FER2013 and Japanese Female Facial Expression (JAFFE) dataset for Emotion Recognition	The accuracy of the proposed model is 91.2% and 74.4% on JAFFE and FER2013 databases respectively	2020	[21]
Two stage recognition networks combined with coarse recognition and fine recognition for compound emotion recognition	Controlled the false classification by utilizing the context information Enhanced the classification of symmetrical emotion labels	2020	[23]
Released a new dataset built on Balanced Twitter for Sentiment Analysis dataset (B-T4SA) containing an image, text and labels. Intermediate fusion on image and text inputs and late fusion on image, text and intermediate fusion's output	Handled the unavailability of labeled multimodal emotional data. The proposed approach achieves the state-of-the-art performance with an accuracy of 90.2%	2021	[22]

assess the suggested precision. In plausible situations, the vision system was also implemented in a humanoid. Liu et al. [25] Introduced two large datasets of object recognition relying on videos. The authors also offered Latent Bi-constraint SVM (LBSVM) for video-based object detection. LBSVM was composed of two driving factors. One of the factors stretched learning films and demanded all sub-sequences to be precisely identified and it was in charge of teaching the predictor to classify evaluation clips of diverse viewpoints. While another factor was that the evaluating criterion is monotonic in terms of the inclusion relationship between video sub-sequences. Office objects and museum sculptures were recognized using the planned LBSVM. Most of the Object recognition models were surpassed by the suggested method. Kasaei et al. [26] described a 3D object identification system that simultaneously identifies item classes as well as analytical features or encodes things. In particular an extension of Latent Dirichlet Allocation (LDA) was proposed in this work. For each separate category, it was useful to study deep geometrical features using easy and accessible attribute information. Clustering was used to identify each category and was reviewed continuously by means of the latest feature information. This approach enabled merging the benefits of local characteristics and deep geometrical features in an effective manner. A large number of tests were run to evaluate the suggested Local-LDA with respect to adaptability, and performance. The performance of Local-LDA was superior to state-of-the-art techniques. Furthermore, the proposed Local-LDA strategy outperformed the Bag-of-Words (BoW) approach in terms of adaptability. Even in case of time complexity, the Local-LDA approach kept the 2nd position right after BoW. For night scene-based object recognition, Wang et al. [27] combined the Region based Convolutional Neural Networks (R-CNN) object identification framework with DCNN and Region Of Interest (ROI) pooling. Figure 2 shows the diagrammatic representation of the proposed network. Night photographs were used as input in the suggested method. To acquire highly accurate classification results, the authors also created simulated object images that are identical to daytime images. The proposed method identified its own night scene data set with an accuracy score of 82.6%, which is significantly greater than the original R-CNN score of 80.4%. As a result, the approach fits the real-world requirements for object identification at night time (Table 3).

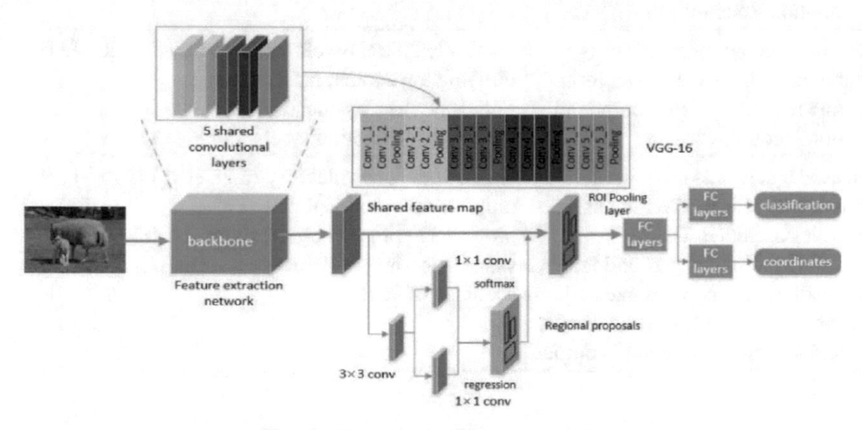

Fig. 2. Faster R-CNN network [27]

2.2 NLP

The use of different algorithms for the study and creation of human vocabulary is known as Natural Language Processing (NLP). For a healthy human-robot interaction, the capacity to express machine knowledge in normal words is critical [30]. NLP is an area of linguistics, computer science, and AI dealing with HCI. NLP is concerned with teaching machines how to interpret and evaluate vast volumes of natural language data. The aim is to develop a machine that can "understand" textual information, especially complexities of linguistic knowledge. The system may then use the text files to retrieve accurate knowledge and ideas, as well as classify and manage them. Speech recognition, NLU and NLG are all major difficulties in NLP. Tse et al. [31] put forward a framework for communication in various functions between humans and machines. This approach collects all of the data from a Human-Computer group and combines it into a fused probability density function (pdf). Semantic accuracy and information preservation were used to construct a system that describes its belief pdfs in English terms.

Table 3. Object Detection Summary

Approach used	Results	Year	Reference
Proposed a 3D object identification system that simultaneously identifies item classes as well as analytical features or encodes things	The performance of Local-LDA was superior to state-of-the-art techniques. Furthermore, the proposed Local-LDA strategy outperformed the Bag-of-Words (BoW) approach in terms of adaptability. Even in case of time complexity, the Local-LDA approach kept the 2nd position right after BoW	2019	[26]
The proposed approach combined R-CNN object identification framework with DCNN and ROI pooling for night scene-based object recognition	The proposed method identified its own night scene data set with an accuracy score of 82.6%, which is significantly greater than the original R-CNN score of 80.4%	2020	[27]
The Proposed object detection algorithm employs the average disparities and position adjustments of the bounding boxes to merge the bounding boxes of each object and combines the connected component technique with disparities values to connect pixels with comparable disparities	Resolved the problem of object detection in overlapping objects	2020	[28]
The suggested Automatic Synthetic Merged Object (ASMO) approach merges several object photos without backdrop images to automatically produce trainset. For merging several objects, the distance and angle are parameterized	Deep learning is used to automatically create a trainset to distinguish obscured items. As a mean average precision, the suggested method has a correct detection rate of 88.43%	2020	[29]

A collection of sentence frameworks was presented for defining complex, multi-model belief pdfs, which enables the generation of optimal expressions through compositions of many statements. The non-parametric Dirichlet process mixture of statement generation improves the model allowing it to automatically calculate the best amount of sentences necessary to explain a particular pdf. Hinaut et al. [32] presented a Recurrent Neural Network (RNN) instructed to point statements to useful information. The proposed model's skill to adjust to different Asian and European languages were investigated. Authors introduced a new learning principle which is not limited to a single language but is more generic in nature. It allows the model to work with a lot of representations. The suggested neural parser attained efficiency in 15 languages from all over the world and from various linguistic families. Additionally, the system can also work with statements with unfamiliar terms. The majority of NLP research has focused on template-based verbalization on symbolic episode data. In contrast to previous approaches, Bärmann et al. [30] investigated a fresh manner of creating an episodic memory using deep learning algorithms. It was based on experiences and the verbalization of those experiences in natural language. As part of the technique, a large data set was compiled that included over a thousand multi-modal robot events captured by both virtual environments and real robot executions. Within the dataset representative natural language inquiries and replies concerning the robot's previous experience was recorded. An episodic memory verbalization model was also shown, which consists of a voice encoder and decoder paired with an LSTM-based auto encoder. Due to the challenge of how to extract emotion-oriented information, efforts in speech emotion recognition are still unable to enhance performance. Zhang et al. [33] presented a deep emotion feature extraction auto encoder with emotion embedding. Instead of batch normalization, instance normalization was used for the proposed approach. The given model combined the description learned by the autoencoder with the data obtained from the openSMILE toolkit, and the resulting feature vector was used to classify emotions (Table 4).

3 Challenges and Discussions

From the study carried out on the recent advancements in Computer Vision and NLP aspects of HRI, current challenges in the relative fields and the scope of improvements in relative areas are understood. Works in Face Recognition, Emotion Recognition, Object Detection and NLP are reviewed in this study. Image blur, occlusions and pose variations are the major difficulties in Face Recognition and huge amounts of work are happening in the field recently. Traditionally, CNN models are used for Face Recognition purposes. In recent works, artificially blurred training data is used to overcome the challenges of image blurs, ensemble of CNN models are used to enhance the accuracy of the current models, different filters are applied on the input images and the model behavior is observed by researchers, user friendly framework consisting of top state of the art models are built to make the Face Recognition task easier and accessible. FR Databases are refurbished with new images and information. Research is being held on how to make FR systems lightweight and more accurate. The review of these works clearly shows that the researches are moving towards a progressive direction with rapid positive results. Emotion Recognition research works are trending in recent times with the advancement in hardware support and with advanced brain study findings. In the past years, Emotion Recognition was focused on 7 fundamental human emotion predictions. Nowadays research work in the field is shifting towards more complex and compound emotions which are really hard to recognize. New databases with labels provided by psychologists are created and tested using deep learning algorithms. Algorithms capable of automatically stripping of the unwanted features from images are fastening the growth of the field. Thermal images are being used in the place of visible ones with the goal of making the model robust to light conditions. These studies with experimental results show that the overall performance of the whole field has increased drastically. Object Detection is an all-time trending and important area of research in HRI point of view. Even though Object Detection is a very old area of research and many studies have already been done

Table 4. NLP Summary

Approach used	Results	Year	Reference
Presented a RNN with more generalized learning principle instructed to point statements to useful information	The proposed neural parser attained efficiency in 15 languages from all over the world and from various linguistic families and is able to work with unfamiliar terms as well	2020	[32]
The proposed CNN model combined self-attention mechanism with gating mechanism for Aspect Based Sentiment Analysis (ABSA)	Performance of both Aspect Category Sentiment Analysis (ACSA) and Aspect Term Sentiment Analysis (ATSA) have been improved	2020	[34]
A large data set was compiled that included over a thousand multi-modal robot events captured by both virtual environments and real robot executions. An episodic memory verbalization model was also shown, which consists of a voice encoder and decoder paired with an LSTM-based auto encoder	Experimental outcomes are great considering the work as a first approach based on episodic-memory-based verbalization. The model can be further improved by feeding more realistic conversational data	2021	[30]
The given model combined the description learned by the autoencoder with the data obtained from the openSMILE toolkit, and the resulting feature vector was used to classify emotions	The model was able to recognize anger and sadness with more accuracy than other emotions. Happiness and neutrality were really difficult for the model to identify. Using a prepossessed BERT model for Speech Emotion Recognition can be done as a future improvement to the model	2021	[33]

associated with the field, various new and exciting works are happening in the field at this point of time. LBSVM object recognition models, Local-LDA strategies and R-CNN with the combination of DCNN have surpassed the older models with significant increase in accuracy. NLP is considered as the backbone in HRI. Without NLP research, HRI study won't be completed. Neural Network advancements and linguistic researches made NLP more and more advanced. Major issues in NLP research are Loss of information during the communication process, models built for a single language might not be compatible for other languages, the shortfall in real world data to training and Speech Emotion Recognition. These issues are being solved by the recent research works in the area. Some of the recent advancements in NLP are data collected from Human-Robot communication are recorded and been used for training, Mixture of Statement model enhanced the performance, generalized learning principle which can be used for multiple languages are being designed, a new concept of episodic-memory-based verbalization is introduced and Speech Emotion Recognition is made easier with pretrained models like BERT. These advancements in each single area like Face Recognition, Emotion Recognition, Object Detection and NLP are contributing to a large growth factor towards the HRI field.

4 Conclusion

This study examined prior research on various elements of HRI in order to assess the present state of research and to encourage additional studies. The study draws attention to the Computer Vision and NLP aspects of HRI. This paper reviewed existing research on Face Recognition, Emotion Recognition, Object Detection and NLP. The paper also summarized each research work on the relative field. This type of collective review from different areas of HRI is rarely done, but it is useful for HRI designing and development fields. A review of different aspects of HRI summarized the issues happening in each relative field and new fascinating ways of resolving them. FR systems are becoming more and more user friendly and easily accessible with the developments of frameworks consisting of top state-of-the-art models, Emotion Recognition researches are transforming from 7 basic human emotion predictions to more complex and compound emotion identifications, new Object Recognition models surpass the performance of the past ones like never before and new NLP systems with multiple language compatibility are trending with high performance. These areas of research are difficult but appealing for future studies.

References

1. Kajita, S., Hirukawa, H., Harada, K., Yokoi, K.: Introduction. In: Introduction to Humanoid Robotics. Springer Tracts in Advanced Robotics, vol. 101. Springer, Cham (2014). https://doi.org/10.1007/978-3-642-54536-8_1
2. Goodrich, M.A., Schultz, A.C.: Human–robot interaction: a survey. Found. Trends® Hum. Comput. Interaction **1**(3), 203–275 (2008). https://doi.org/10.1561/1100000005
3. Lee, J., Park, G., Ahn, S.: A performance evaluation of the collaborative robot system. In: 2021 21st International Conference on Control, Automation and Systems (ICCAS), pp. 1643–1648 (2021). https://doi.org/10.23919/ICCAS52745.2021.9649859
4. Dautenhahn, K.: Socially intelligent robots: dimensions of human-robot interaction. Philos. Trans. R. Soc. London B Biol. Sci. **362**(1480), 679–704 (2007). https://doi.org/10.1098/rstb.2006.2004
5. Zhang, G., Huang, X., Li, S.Z., Wang, Y., Wu, X.: Boosting local binary pattern (LBP)-based face recognition. In: Chinese Conference on Biometric Recognition, pp. 179–186 (2004)
6. Thorat, S.B., Nayak, S.K., Dandale, J.P.: Facial recognition technology: an analysis with scope in India. arXiv preprint
7. Chen, S.K., Chang, Y.H.: 2014 International Conference on Artificial Intelligence and Software Engineering (AISE2014), p. 21. DEStech Publications, Inc. (2014). ISBN 9781605951508
8. Ding, C., Tao, D.: Trunk-branch ensemble convolutional neural networks for video-based face recognition. IEEE Trans. Pattern Anal. Mach. Intell. **40**(4), 1002–1014 (2018). https://doi.org/10.1109/TPAMI.2017.2700390

9. Zhi, H., Liu, S.: Face recognition based on genetic algorithm. J. Vis. Commun. Image R. (2018). https://doi.org/10.1016/j.jvcir.2018.12.012
10. Choi, J.Y., Lee, B.: Ensemble of deep convolutional neural networks with gabor face representations for face recognition. IEEE Trans. Image Process. **29**, 3270–3281 (2020). https://doi.org/10.1109/TIP.2019.2958404
11. Choi, J.Y., Ro, Y.M., Plataniotis, K.N.: Color local texture features for color face recognition. IEEE Trans. Image Process. **21**(3), 1366–1380 (2012)
12. Liu, C., Wechsler, H.: Gabor feature based classification using the enhanced fisher liner discriminant model for face recognition. IEEE Trans. Image Process. **11**(4), 467–476 (2002)
13. Lou, G., Shi, H.: Face image recognition based on convolutional neural network. China Commun. **17**(2), 117–124 (2020). https://doi.org/10.23919/JCC.2020.02.010
14. Serengil, S.I., Ozpinar, A.: LightFace: a hybrid deep face recognition framework. In: 2020 Innovations in Intelligent Systems and Applications Conference (ASYU), pp. 1–5 (2020). https://doi.org/10.1109/ASYU50717.2020.9259802
15. Boutros, F., Siebke, P., Klemt, M., Damer, N., Kirchbuchner, F., Kuijper, A.: PocketNet: extreme lightweight face recognition network using neural architecture search and multistep knowledge distillation. IEEE Access **10**, 46823–46833 (2022). https://doi.org/10.1109/ACCESS.2022.3170561
16. Tsai, A.-C., Ou, Y.-Y., Wu, W.-C., Wang, J.-F.: Occlusion resistant face detection and recognition system. In: 2020 8th International Conference on Orange Technology (ICOT), pp. 1–4 (2020). https://doi.org/10.1109/ICOT51877.2020.9468767
17. Jain, D.K., Shamsolmoali, P., Sehdev, P.: Extended deep neural network for facial emotion recognition. Pattern Recogn. Lett. **120**, 69–74 (2019). ISSN 0167-8655, https://doi.org/10.1016/j.patrec.2019.01.008. https://www.sciencedirect.com/science/article/pii/S016786551930008X
18. Guo, J., et al.: Dominant and complementary emotion recognition from still images of faces. IEEE Access **6**, 26391–26403 (2018). https://doi.org/10.1109/ACCESS.2018.2831927
19. Wang, S., Pan, B., Chen, H., Ji, Q.: Thermal augmented expression recognition. IEEE Trans. Cybern. **48**(7), 2203–2214 (2018). https://doi.org/10.1109/TCYB.2017.2786309
20. Zhang, H., Xu, M.: Weakly supervised emotion intensity prediction for recognition of emotions in images. IEEE Trans. Multimedia **23**, 2033–2044 (2021). https://doi.org/10.1109/TMM.2020.3007352
21. John, A., Abhishek, M.C., Ajayan, A.S., Sanoop, S., Kumar, V.R.: Real-time facial emotion recognition system with improved preprocessing and feature extraction. In: 2020 Third International Conference on Smart Systems and Inventive Technology (ICSSIT), pp. 1328–1333 (2020). https://doi.org/10.1109/ICSSIT48917.2020.9214207
22. Kumar, P., Khokher, V., Gupta, Y., Raman, B.: Hybrid fusion based approach for multimodal emotion recognition with insufficient labeled data. In: 2021 IEEE International Conference on Image Processing (ICIP), pp. 314–318 (2021). https://doi.org/10.1109/ICIP42928.2021.9506714
23. Zhang, Z., Yi, M., Xu, J., Zhang, R., Shen, J.: Two-stage recognition and beyond for compound facial emotion recognition. In: 2020 15th IEEE International Conference on Automatic Face and Gesture Recognition (FG 2020), pp. 900–904 (2020). https://doi.org/10.1109/FG47880.2020.00144
24. Martinez-Martin, E., del Pobil, A.P.: Object detection and recognition for assistive robots: experimentation and implementation. IEEE Robot. Autom. Mag. **24**(3), 123–138 (2017). https://doi.org/10.1109/MRA.2016.2615329
25. Liu, Y., Hoai, M., Shao, M., Kim, T.-K.: Latent bi-constraint SVM for video-based object recognition. IEEE Trans. Circuits Syst. Video Technol. **28**(10), 3044–3052 (2018). https://doi.org/10.1109/TCSVT.2017.2713409

26. Kasaei, S.H., Lopes, L.S., Tomé, A.M.: Local-LDA: open-ended learning of latent topics for 3D object recognition. IEEE Trans. Pattern Anal. Mach. Intell. **42**(10), 2567–2580 (2020). https://doi.org/10.1109/TPAMI.2019.2926459
27. Wang, K., Liu, M.Z.: Object recognition at night scene based on DCGAN and Faster R-CNN. IEEE Access **8**, 193168–193182 (2020). https://doi.org/10.1109/ACCESS.2020.3032981
28. Chiu, C.-C., Lo, W.-C.: An object detection algorithm with disparity values. In: 2020 4th International Conference on Imaging, Signal Processing and Communications (ICISPC), pp. 20–23 (2020). https://doi.org/10.1109/ICISPC51671.2020.00011
29. Lee, T.-H., Kang, Y.-G., Ryu, S., Lee, H.-J.: An ASMO method for CNN-based occluded object detection. In: 2020 IEEE International Conference on Consumer Electronics - Asia (ICCE-Asia), pp. 1–2 (2020). https://doi.org/10.1109/ICCE-Asia49877.2020.9277112
30. Bärmann, L., Peller-Konrad, F., Constantin, S., Asfour, T., Waibel, A.: Deep episodic memory for verbalization of robot experience. IEEE Robot. Autom. Lett. **6**(3), 5808–5815 (2021). https://doi.org/10.1109/LRA.2021.3085166
31. Tse, R., Campbell, M.: Human–robot communications of probabilistic beliefs via a Dirichlet process mixture of statements. IEEE Trans. Rob. **34**(5), 1280–1298 (2018). https://doi.org/10.1109/TRO.2018.2830360
32. Hinaut, X., Twiefel, J.: Teach Your Robot Your Language! Trainable neural parser for modeling human sentence processing: examples for 15 languages. IEEE Trans. Cogn. Dev. Syst. **12**, 179–188 (2019). https://doi.org/10.1109/TCDS.2019.2957006
33. Zhang, C., Xue, L.: Autoencoder with emotion embedding for speech emotion recognition. IEEE Access **9**, 51231–51241 (2021). https://doi.org/10.1109/ACCESS.2021.3069818
34. Yang, J., Yang, J.: Aspect based sentiment analysis with self-attention and gated convolutional networks. In: 2020 IEEE 11th International Conference on Software Engineering and Service Science (ICSESS), pp. 146–149 (2020). https://doi.org/10.1109/ICSESS49938.2020.9237640

Optimized Static and Dynamic Android Malware Analysis Using Ensemble Learning

Samyak Jain(✉), Adya Agrawal, Swapna Sambhav Nayak, and Anil Kumar Kakelli

SCOPE, Vellore Institute of Technology, Vellore 632014, Tamil Nadu, India
jainsamyak2300@gmail.com

Abstract. As the dominant operating system for mobile devices, Android is the prime target of malicious attackers. Installed Android applications provide an opportunity for attackers to bypass the system's security. Therefore, it is vital to study and evaluate Android applications to effectively identify harmful applications. Android applications are analyzed by conventional methods using signature hash-based algorithms or static features-based machine learning approaches. This research proposes optimized ensemble classification models for Android applications. Ensemble models have been trained for both static and dynamic analysis using seven and eight distinct classifiers respectively. These models have been optimized by tuning their hyper-parameters and evaluated using K-fold cross-validation. We were able to acquire an F1 score of 99.27% and an accuracy of 99.47% for static analysis and our dynamic analysis model yielded an F1 score of 96.96% and an accuracy of 96.66%. Our proposed approach overcomes conventional solutions by taking into account both static and dynamic analysis and attaining high accuracy with the help of ensemble models.

Keywords: Android malware classification · Static analysis · Dynamic analysis · Ensemble learning

1 Introduction

Smartphones have grown in popularity as a result of their ability to provide a wide range of services over current networks. Because of its openness and accessibility, the Android platform is currently used by the majority of smartphones, and many developers are drawn to creating new Android applications. However, Android's popularity has also made it a perfect target for hackers to break security and access sensitive user data. Many commercial Android malware detection technologies rely on fixed signatures or identifiers but fail to detect unknown malware. Machine learning approaches do not require malware to be analyzed prior to the definition of fixed signatures and identifiers. Numerous existing research works [3,10,13] approaches incorporate common machine learning algorithms like K-nearest neighbours, Decision Tree, Naive Bayes, and

H. Sharma et al. (Eds.): ICIVC 2022, PALO 17, pp. 165–179, 2023.
https://doi.org/10.1007/978-3-031-31164-2_14

Random Forest. These studies also include feature selection techniques such as Information Gain [2], and Principal Component Analysis[6].

Majority of these studies [3,5,6,10,13] identify malware using static Android features. Static analysis entails evaluating an executable without executing it. Android static analysis includes obtaining attributes from the Android app package (apk). The AndroidManifest.xml file contains the majority of the static features, including needed permissions, activities, intent actions, intent filters, and package names. Other elements, including API calls, are retrieved from the classes.dex file.

Static analysis alone is insufficient to detect unknown malware with accuracy. In Dynamic Analysis, we run the application on an actual device or an emulator to trace all system/external calls made by the application in real time. There is often a pattern of system calls in malware files that may be detected in the majority of APKs. Both forms of analysis have benefits and drawbacks, and they complement one another. There is more research on static analysis but there is scant existing research [9] on dynamic analysis which produces results with very low levels of accuracy. The datasets used are significantly insufficient, and the complete methodology for tracking dynamic system calls has not been documented [15].

Keeping all this in mind, we created a robust method of malware detection in Android by carrying out both static and dynamic analysis and obtained significantly higher levels of accuracy. We implemented various machine learning and deep learning models and then created ensemble learners with the best-performing models thereby achieving high accuracy.

The rest of the paper is structured as follows: Sect. 2 discusses the related work and identifies the limitations of the existing methodologies for android malware detection. Section 3 presents the methodology proposed in this paper by discussing both static and dynamic analysis in detail. The section extensively explains all the approaches used in the paper hence, making them reproducible. The results and discussions are provided in Sect. 4, followed by the conclusion and future work in Sect. 5.

2 Previous Work

The Bayes' algorithm and permission-based features approach for malware identification has been presented by Sharfah and Mohd Faizalon [2]. 10000 samples from 20 malware families were studied, which was critical in the study's performance analysis. They were able to achieve an accuracy of 83%. An Android malware classification framework (ANDMFC) has been developed by Sercan and Ahmet [4] to identify 71 Android malware classes and 10 harmful application categories. It uses ML algorithms, including ensemble and neural network algorithms, to classify malware. The model's minimal accuracy values varied between datasets due to the datasets' imbalanced structure. They were able to achieve 85.14% accuracy.

Jungsoo and Hojin et al. [5] suggested categorizing applications into benign, harmful, and suspicious categories using the YARA rule. Performance measures are absent from the report. In Ahmed et al. 's study [6], they examined seven classification models: Decision Tree, SVM, KNN, Random Forest, Nave Bayes, and Multi-Layer Perceptron. Each classifier was evaluated using a 10-fold cross-validation technique. The article did not account for data imbalance, which could affect model performance. Feature extraction techniques such as PCA and Information Gain were used.

The framework Apk2Img4AndMal by Oğuz Emre Kural et al. [1] transforms binary APK files to grayscale pictures. Convolutional neural networks classify grayscale pictures as benign or malevolent. It achieved 94% accuracy. This approach does not require a feature extraction phase but will fail in obfuscated code situations. Talal A. A Abdullah et al. [3] compared six supervised ML models: KNN, Decision tree, SVM, Random Forest, Naive Bayes, and Logical Regression. Malgenome, including 215 features from 3799 application samples, was used for this investigation. Random Forest achieved 99.37% 10-fold cross-validation accuracy. This paper lacks dynamic analysis, making the technique less credible.

Ferdous Zeaul Islam et al. [10] tested machine-learning algorithms using static Android application characteristics using Drebin dataset. KNN, Decision tree, linear SVM, and random forest are machine learning classifiers. Model accuracy is 96.3 percent. Since dynamic analysis is not conducted, the model may fail in circumstances where the Android application code is obfuscated. Anam Fatima et al. [11] suggested a machine learning Android malware detection technique using evolutionary algorithms for feature selection. Genetic algorithms select features for machine learning classifiers. The SVM model obtained 95% accuracy, and the neural networks reached 94% accuracy. The 40,000-application dataset was provided by IIT Kanpur's C3I center.

As proposed by Aayasha Palikhe et al. [12] Malduonet utilizes a dual-neuralnet framework to examine API call features. Word embedding features are extracted by the first neural network and given to the second auxiliary classifier, which classifies the application as benign or malicious. Accuracy and F1-score were 96.91% and 98.53% respectively. An Android malware detection method based on static analysis is presented by Eslam Amer et al. [13]. Several models of machine learning have been trained, including Decision Tree, Random Forest, SVM, KNN, LDA, MLP, and AdaBoost. Using the ensemble model, an accuracy of 99.3% was achieved.

Santosh K. Smmarwar et al. [14] proposed an ensemble model consisting of weak learners, such as kernel naive bayes, weighted k closest neighbors, cubic support vector machine, fine tree, and meta-learners. Using binary grey wolf, the selection of features was optimized. This approach achieved a maximum accuracy of 96.958%. These meta-learners are used to aggregate results to forecast output.

DATDroid [16] is a dynamic analysis-based Android malware detection tool introduced by Rajan Thangaveloo et al. Weka Tool is utilized throughout the step of feature selection and classification. The accuracy of 91.7% achieved for dynamic analysis is good however, dynamic analysis can only reveal the run time information while static analysis can reveal other information that cannot be extracted through the dynamic process. The benign programs were downloaded from the APKPure market, whereas the malicious applications were downloaded via the Android Malware Genome Project. U.S. Jannat et al. [15] have undertaken both static and dynamic analysis for Android malware applications, with a focus on finding malware in popular Bangladeshi applications. For static analysis, features such as permissions, intents, and API calls have been utilized. The APK was analyzed dynamically by executing it in a sandboxed environment on an emulator, and 15 features were extracted. Static analysis produced an accuracy of 81%, while dynamic analysis produced 93%. For static analysis, the malgenome dataset, a Kaggle dataset, and the Android Wake Lock Research dataset were utilized, however for dynamic analysis, only the malgenome and Android Wake Lock Research datasets were utilized.

ShadowDroid is a dynamic Android malware analysis solution presented by Che-Chun hu et al. [9]. It detects privacy leaks using string comparison techniques. 193 examples of programs, comprising 62 benign applications, 11 botware applications, and 74 ransomware applications, were collected. In the absence of privacy information, this approach can achieve 90% accuracy. The dataset used is quite inadequate. S. Roopak et al. [17] proposed a hybrid-based android malware detection technique by combining the estimated malicious probability values of three distinct naive Bayes classifiers based on API calls, permissions, and system calls using the Bayesian model averaging approach. The proposed methodology has a 94% accuracy rate. It uses three naive Bayes classifiers based on API calls, permissions, and system calls. The training dataset contains samples of malware and 25 samples of goodware.

Following a review of the relevant literature, we found some shortcomings. Most studies rely solely on static analysis for malware detection. The few studies that did dynamic analysis employed extremely limited and inadequate data sets. None of the works on dynamic analysis reached high accuracy. Having discovered the aforementioned research gaps, we focused our efforts on offering both static and dynamic analysis of Android malware and training our ensemble models with higher accuracy. Our proposed approach tries to classify malware effectively by utilizing both static and dynamic information, such as permissions, intents, and API call logs, as well as system call sequences. Ensemble models have been trained with hyper-parameter tuning and stratified k-fold validation for both static and dynamic analysis. Finally, the experimental outcomes are assessed using several performance metrics.

3 Methodology

3.1 Static Analysis

Dataset. We have used the Drebin [7] dataset for static analysis. In this dataset, there are 5560 malware applications and 9476 benign applications. This is the most highly used, extensive and cited dataset for android malware analysis. We have used this dataset to test various machine learning and deep learning models. Figure 1 depicts an example of how our dataset is structured.

	transact	onServiceConnected	bindService	attachInterface	ServiceConnection	android.os.Binder	SEND_SMS	Ljava.lang.Class.getCanonicalName	Lj
0	0	0	0	0	0	0	1	0	
1	0	0	0	0	0	0	1	0	
2	0	0	0	0	0	0	1	0	
3	0	0	0	0	0	0	0	0	
4	0	0	0	0	0	0	0	0	
...	...	...	...	...	...	...	...	...	
15031	1	1	1	1	1	1	0	1	
15032	0	0	0	0	0	0	0	0	
15033	0	0	0	0	0	0	0	0	
15034	1	1	1	1	1	1	0	1	
15035	1	1	1	1	1	1	0	1	

15036 rows × 215 columns

Fig. 1. Drebin Dataset

Data Cleaning

- Removing unwanted data: After examining the Drebin dataset for duplicate and irrelevant observations, we determined that none are present.
- Handling missing data: Since one of the features in the dataset contained null values, that feature was removed.
- Managing Outliers: As our dataset is categorical, every feature has either a value of 0 or 1, hence there are no outliers.
- Imbalanced Data: Since our data set is imbalanced, we have incorporated Stratified K-Fold Cross Validation to address concerns associated with imbalanced data sets.

Feature Selection. Feature selection aims to rank the importance of the existing features in the dataset and discard less important ones. We select the features which are highly dependent on the target variable. We have used the Chi-Square test for feature selection. Chi-square test is used for categorical features in a dataset. We compute Chi-square between each feature and the target and then select the number of features with the highest Chi-square scores. Figure 2. shows the F1 squares of Multi-layer Perceptron model after selecting the corresponding number of features. It was found that the highest F1 score was obtained at 211 features.

Models. To analyse and get best results of malware analysis we have used a variety of models: Logistic regression, KNN, Naive bayes, Decision tree, Random forest, SVM, neural network and Light-GBM model. We have also made an ensemble learning model by combining these models to test for higher accuracy. Logistic regression takes the input variables and models the probability of the binary outcome like true/false, yes/no or in our case benign and malware. A sigmoid function is a mathematical formula used for prediction of the respective classes by setting a threshold value. The SVM (Support vector machine) model creates a hyperplane from the features to classify the data points on the basis of support vectors of the hyperplane. The hyperplane can be of many dimensions

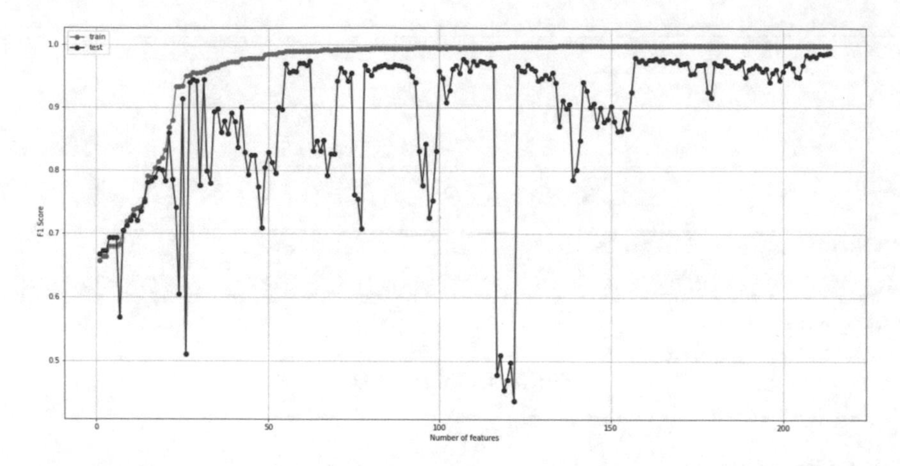

Fig. 2. Number of Features vs F1 Score after Chi-squared test

based on the number of features(line for 2 features, plane for 3 features etc.). We have performed hyperparameter tuning using GridSearchCV in SVM with 'rbf' kernel on two parameters, gamma value, which is kernel coefficient value and c value which is a penalty parameter of the error term. This technique provided us with an optimal value of gamma as 0.5 and c as 10. Random forest uses subsets of the dataset to create multiple decision trees and uses the predictions of all of them in a voting classifier to determine the outcome. With the help of GridSearchCV, we were able to get the optimal value of n-estimators as 282 which inturn yielded a F1 Score of 97.97%.

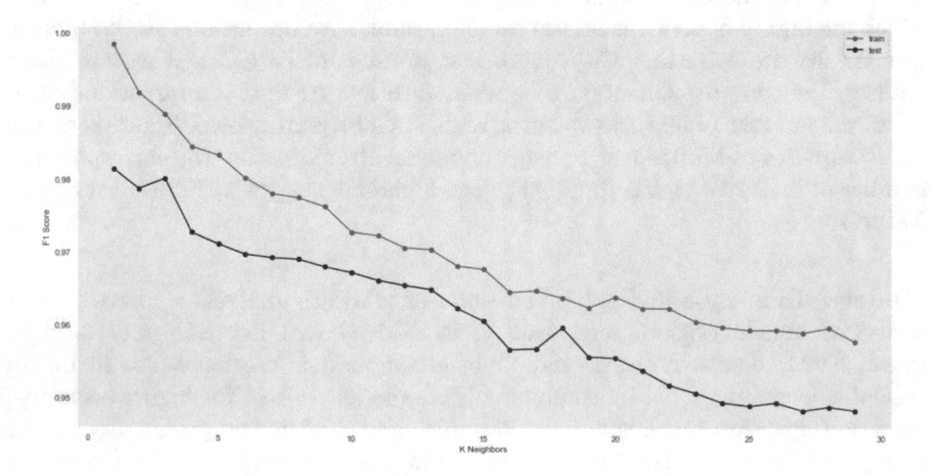

Fig. 3. F1 Score vs K Neighbors Selected

MLP is a feedforward artificial neural network that produces a set of outputs from a set of inputs. Multiple layers of input nodes connected as a directed graph between the input and output layers constitute an MLP. MLP utilises backpropagation for training of the network. GridSearchCV was used to perform hyperparameter tuning which gave us an optimal value of alpha as 0.01. KNN (K- nearest neighbours) uses euclidean distance to classify a specific data-point on the basis of its nearest neighbours. The distance from the data point to "k" nearest data points is taken in consideration and the mode of class value for the neighbours is the class of the data-point. We tested for different k values ranging from 1 to 100 and found k = 3 to be the optimal neighbour value which yields highest Accuracy and F1 Score. In Fig. 3, the red line depicts testset F1-score for k-values between 1–30 and we can see that k = 3 produced the maximum F1-score.

Decision Tree: Our study utilized a decision tree based on CART (classification and regression trees algorithm). This method employs the Gini index as the splitting factor. GridSearchCV was used to determine optimal max-depth values between 1 and 100 based on accuracy, f1-score, and AUC. GridSearchCV found 43 as the ideal max-depth hyperparameter value. Figure 4 depicts the Accuracy, F1-Score, and Area under the Curve for Depth ranging from 1 to 60. Light GBM is a gradient boosting framework employing a tree-based learning technique. We set the boosting type to GOSS: gradient based one side sampling. In Gradient Boosted Decision Trees, GOSS uses data instances without native weight. Data instances with larger gradients contribute more to information gain. GOSS retains instances with larger gradients and conducts random sampling of data.

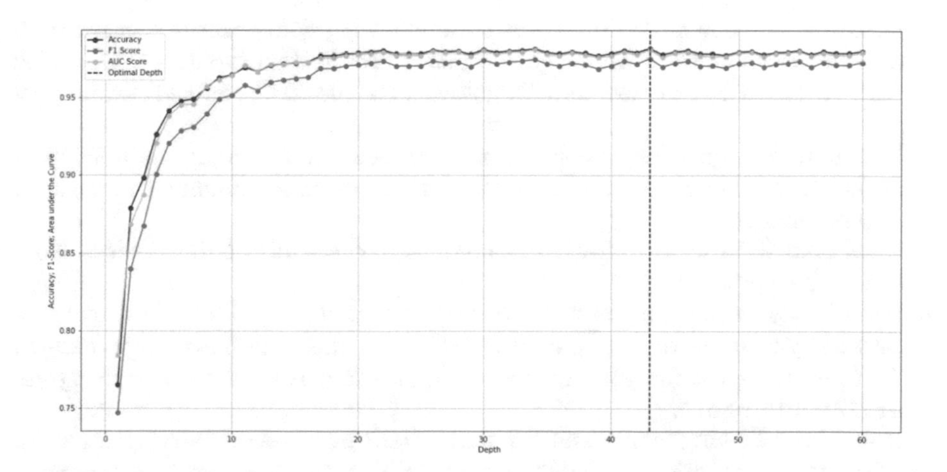

Fig. 4. Decision Tree Depth Hyperparameter Tuning

Ensemble Model: We have developed an ensemble learning model with a hard voting classifier that combines Multilayer Perceptron, Support Vector Machine, LightGBM, K-Nearest Neighbors, Random Forest, Decision Tree, and Logical

Regression. Then, hyper-parameter tuning was conducted, and we determined that the best weights for the aforementioned models are 2, 2, 1, 1, 1, 1, 1 respectively. Since we are working with unbalanced data, we evaluated the model using stratified 10-fold cross validation. The complete proposed methodology for static analysis has been depicted in a flowchart as shown in Fig. 5.

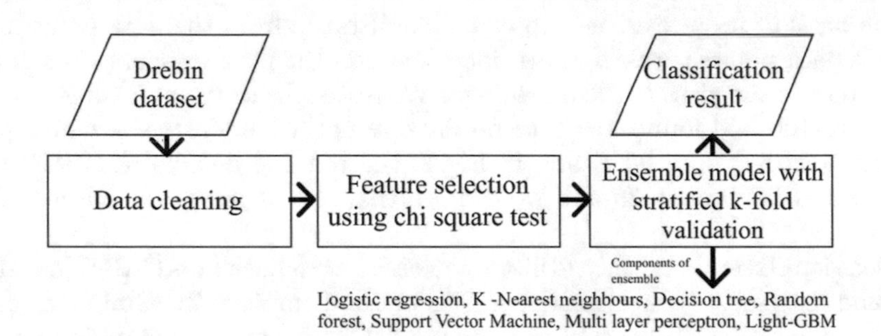

Fig. 5. Static Analysis Workflow

3.2 Dynamic Analysis

Getting Application System Calls Log File: For fetching the system calls dynamically from android apps, we have used ADB(Android Debug Bridge) terminal and strace. Following are the steps to be followed to track the calls:

1. Create a new emulator device with the following configurations: Hardware: a device or emulator without google play store; System image: a 86X image with API level 28 or below and Target with Google APIs enabled and no play store.
2. Navigate to the platform-tools folder inside android sdk folder in your operating system. Open terminal and type.\\`adb.exe help` command to verify if adb is running.
3. Now, execute.\\`adb.exe devices` command to check all running devices. Note the id of your device.
4. Install and run the app you want to test in the emulator. Check the app's process id by running this command : .\\`adb.exe shell ps` <`package name`>. To open the shell for your emulator, Type.\\`adb.exe -s emulator-`<`your device id`> `shell`.
5. Run the following command to start tracking system calls: `strace -p` <process id of app> `-s 1000 -S calls -f -C -o` <`destination path for storing the system call file`>.
6. To create uniform events and run the app externally we shall use the MonkeyRunner tool. While running the monkey tool, we will use strace and track the system calls for all the generated events. To start the monkey tool type the following command in a separate terminal in the same platform-tools folder.\\`adb shell -p -v 500 - s 42`.

7. Once the monkey tool has finished execution, navigate back to the strace terminal and Press ctrl + C to stop tracking system calls.
8. Now we will pull the tracked system calls by this command. \`adb.exe pull <original stored path> <destination path>`.

The proposed workflow for dynamic analysis is shown in Fig. 6.

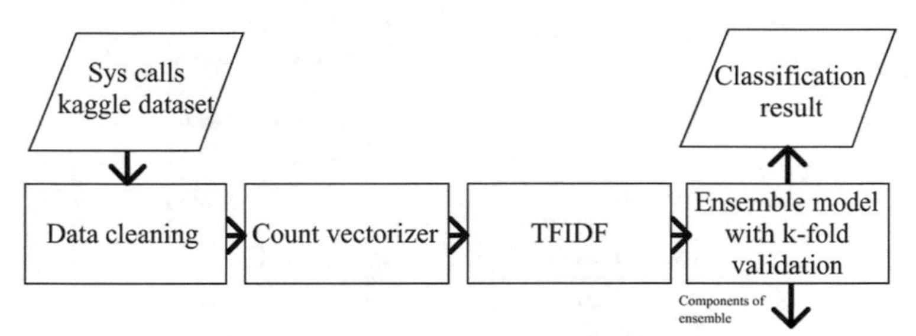

Fig. 6. Dynamic Analysis Workflow

Dataset. The dataset [8] we used comprises 400 system call traces, identical to those we retrieved before, for 200 benign and 200 malicious Android applications. Each line of the trace file provides the timestamp and system call data, including the opcode, for each system call. Some examples of system calls are read(), write(), futex(), epoch_pwait() etc.

	System calls	Target
0	epoll_pwait read getuid32 fstatat64 fstatat64 ...	0
1	fstatat64 faccessat openat getuid32 writev get...	0
2	write futex write write futex getuid32 epoll_p...	0
3	recvfrom ioctl pread64 pread64 mmap2 munmap pr...	0
4	epoll_pwait read getuid32 fstatat64 fstatat64 ...	0
...	...	...
395	clock_gettime epoll_wait read clock_gettime st...	1
396	clock_gettime clock_gettime recvfrom recvfrom ...	1
397	clock_gettime epoll_wait read clock_gettime st...	1
398	clock_gettime epoll_wait read clock_gettime st...	1
399	clock_gettime epoll_wait read clock_gettime st...	1

400 rows × 2 columns

Fig. 7. Post Data Cleaning Data Format

```
array([   0,    0,    0,    0,    0,    0,   31,    6,   40,    0,   11,
          0,   10,  615,    0,    0,    7,    0,    0,    0,   11,    0,
         81,   67,    7,    0,    0, 1485,    0,    0,    0,    0,    0,
          3,    0,    7,    0,    0,  605,  740,    0,    0,    0,   48,
          0,    0,  168,  484,  147,    0,    0,   28,    0,    0,   55,
        243,    0,    0,    0,  216,    0,    7,  862,    0,    0,    0,
         59,    0,    0,    0,    3,    0,    0,    6,    0,    0,    0,
          0,    0,    0,    0,    0,    0,    0,    0,    0,    0,    0,
          0,    0,  896,   66], dtype=int64)
```

Fig. 8. CountVectorizer Output

Data Cleaning. For each trace file we have iterated through all the lines and removed the timestamp and extracted the system call names. Then we combined all those system call names of a particular application and made a sequence. Following data cleaning, the data is in the format depicted in Fig. 7.

```
array([0.        , 0.        , 0.        , 0.        , 0.        ,
       0.        , 0.0143196 , 0.00311924, 0.01577645, 0.        ,
       0.00434936, 0.        , 0.00409454, 0.43303868, 0.        ,
       0.        , 0.00628789, 0.        , 0.        , 0.        ,
       0.00441494, 0.        , 0.03685202, 0.05269481, 0.00892225,
       0.        , 0.        , 0.59750661, 0.        , 0.        ,
       0.        , 0.        , 0.        , 0.00307414, 0.        ,
       0.00663672, 0.        , 0.        , 0.2362502 , 0.28896719,
       0.        , 0.        , 0.        , 0.01995061, 0.        ,
       0.        , 0.06759671, 0.20678284, 0.06186751, 0.        ,
       0.        , 0.02129127, 0.        , 0.        , 0.04196136,
       0.10726207, 0.        , 0.        , 0.        , 0.08434718,
       0.        , 0.0088567 , 0.33576935, 0.        , 0.        ,
       0.        , 0.04892462, 0.        , 0.        , 0.        ,
       0.00298776, 0.        , 0.        , 0.00234298, 0.        ,
       0.        , 0.        , 0.        , 0.        , 0.        ,
       0.        , 0.        , 0.        , 0.        , 0.        ,
       0.        , 0.        , 0.        , 0.        , 0.        ,
       0.35075823, 0.02596629])
```

Fig. 9. TF-IDF Output

Feature Extraction. The frequency representation of system calls carries information about the behaviour of the application. A particular system call may be utilised more in a malicious application than in a benign application, and the system call frequency representation is intended to capture such information. We have analysed the frequency of system calls by using the CountVectorizer function. CountVectorizer created a corpus of 92 distinct system calls. Figure 8 depicts an example of a system call vector taken from a single instance of one of the apps. Next, we have passed the vector through the TF-IDF function. Term frequency-inverse document frequency, is a numerical statistic that reflects how important a system call is to an application trace file. After applying the TF-IDF statistic, Fig. 9 shows how the vector has been transformed.

Models. We have trained different models for dynamic analysis as well, including Decision tree, Random forest, Stochastic Gradient Descent Classifier, Logistic regression, LightGBM, Support Vector Machine and Multilayer Perceptron. SGD Classifier: The SGDClassifier implements a simple stochastic gradient descent learning algorithm that supports various classification loss functions and penalties. SGD is merely an optimization strategy. This classifier was trained with the hinge loss, which is roughly comparable to a linear SVM. This model has also undergone hyperparameter tuning using GridSearchCV, with alpha=1e-3 and max iterations set to 5 as the optimal values. Table 1 lists the tuned hyperparameters for the respective classifiers for dynamic analysis.

Table 1. Hyperparameters for Dynamic analysis classifiers

Models	Hyperparameters
Support Vector Machine	Kernel:rbf
Random Forest	-
Decision Tree	max-depth:7
Multilayer Perceptron	batch-size:220,learning-rate:0.01
K-Nearest Neighbors	n-neighbors:7
Logistic Regression	n-jobs:1, c:1e5
LigttGBM	-
Stochastic Gradient Descent	loss:hinge, alpha:1e-3, $max_{iterations:5}$,

4 Experimental Results

4.1 Static Analysis

We trained multiple classifiers using the Drebin dataset for the static analysis portion of our research, as outlined in Sect. 3. Following that, they were used to train an ensemble model. Table 2 shows the results of the seven classifiers along with our proposed ensemble model. When the cost of false positives is high, precision is often preferred as a metric. It is the ratio of accurately predicted positive occurrences to the total number of predicted positive occurrences. If a malicious program is predicted as benign during malware detection, the user's device is prone to attack. When the cost of false negativity is substantial, recall or sensitivity is given priority. In the situation of unbalanced class distribution, F1-score is a better metric over accuracy. It is the weighted average of precision and recall when false positives and false negatives are considered.

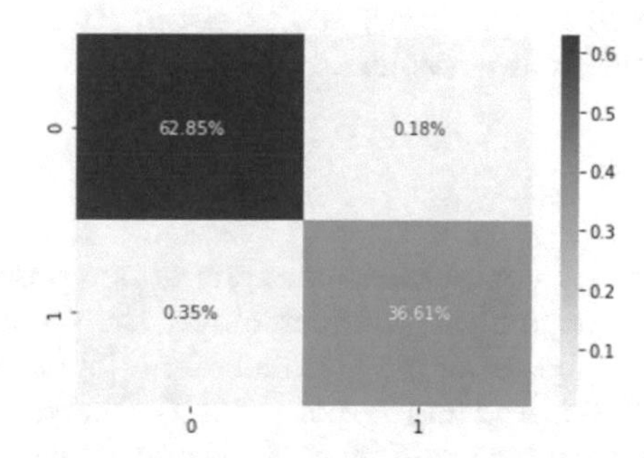

Fig. 10. Static Analysis Ensemble Confusion Matrix

Among the classifiers in Table 2, MLP performed the best with an F1-score of 98.61% and an accuracy of 98.98%. After applying stratified 10-fold cross-validation, our proposed ensemble model attained a high F1-score of 99.27% and accuracy of 99.4677% and outperformed all the other classifiers. The confusion matrix aggregates and depicts the performance of the models. Figure 10 presents the True Positive Rate (TPR), False Negative Rate (FNR), True Negative Rate (TNR), and False Positive Rate (FPR) for our Ensemble model.

Table 2. Static Analysis

Models	Accuracy	F1 Score
Logistic Regression	97.70	96.84
Decision Tree	98.27	97.65
Random Forest	98.49	97.94
KNN	98.54	98.01
LightGBM	98.63	98.13
SVM	98.98	98.60
Multilayer Perceptron	98.98	98.61
Ensemble Model	99.47	99.27

Table 3. Dynamic Analysis

Models	Accuracy	F1 Score
Logistic Regression	87	86.59
Decision Tree	90	90.19
LightGBM	93	93.07
SGD Classifier	93	93.07
SVM	93	93.2
KNN	94	93.75
Multilayer Perceptron	95	94.85
Random Forest	95	95.05
Ensemble Model	96.66	96.96

4.2 Dynamic Analysis

For the dynamic analysis portion of our research, we utilized the Kaggle dataset [8] and eight classifiers were trained, the results of which are shown in Table 3. Random forest outperformed the others, with an F1-score of 95.05%. Multilayer Perceptron came extremely near to that result, with an F1-score of 94.845%. On the other hand, accuracy has dropped significantly for a rudimentary model like the Logistic Regression model with an F1-score of 86.59%. Our proposed ensemble model achieved a 10-fold cross-validation F1 Score of 96.96%. Figure 12 illustrates the F1 scores and accuracy scores of our Proposed Ensemble model along with the eight different classifiers.

5 Discussions

Table 4 gives us a detailed comparative analysis of our proposed model against the related works in static analysis conducted over the years. Of all the research works conducted so far, the paper by Abdullah et al. [3] presented the best work with an accuracy of 99.37% and an F1-score of 99%; whereas our proposed model outperformed it with an accuracy of 99.47% and an F1 score of 99.27%.

In Dynamic Analysis, as shown in Table 5, our results exceeded the results of previous works with an accuracy of 96.66% and an F1 score of 96.96%. For instance, the best work in dynamic analysis was done by Rajan et al. with an accuracy of 91.7%.

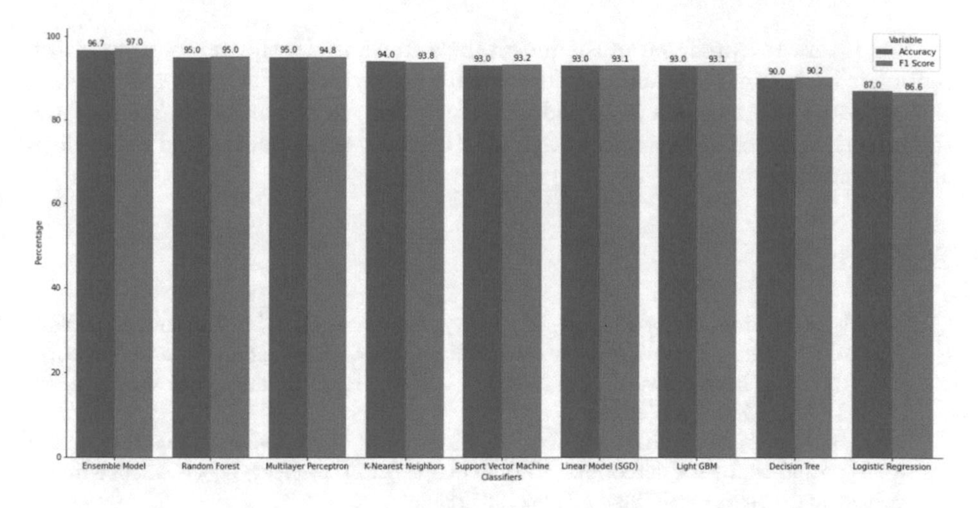

Fig. 11. Dynamic Analysis Classifier Comparision: F1 Score and Accuracy

Table 4. Static Analysis

	Accuracy	F1 Score
Our Proposed Work	99.47	99.27
Palikhe et al. [12]	96.91	98.53
Jannat et al. [15]	81	–
Abdullah et al. [3]	99.37	99
Smmarwar et al. [14]	96.958	95.989

Table 5. Dynamic Analysis

	Accuracy	F1 Score
Proposed Work	96.66	96.96
Rajan et al. [16]	91.7	91.52
Jannat et al. [15]	–	93
Hu et al. [9]	90	–

6 Conclusion

The primary objective of our work was to effectively detect malware behaviour in android apps using static and dynamic analysis. Static analysis deals with the static features like Permissions and API calls of the application, whereas dynamic analysis deals with live tracking of system calls of the application. One must conduct both static and dynamic analysis to create an efficient malware detection system. In our study, we employed seven distinct classifiers and developed an ensemble learning model for static analysis. Similarly, for dynamic analysis, we trained eight distinct classifiers and developed an ensemble model. We collected two datasets and systematically conducted experimentation by pre-processing the dataset, performing feature selection and extraction, optimizing the classifier by hyperparameter tuning and then evaluating it with stratified k-fold cross validation. We discovered that our ensemble model surpassed all other works, with an accuracy of 99.47% in static and 96.66% dynamic analysis.

For future research, we must undertake a more in-depth examination of the sequence of system calls and consider them while training the model. By taking the order of system calls into consideration, we believe we will be able to increase the effectiveness of our system in dynamic analysis using LSTM and other deep learning models.

References

1. Kural, O., Şahin, D., Akleylek, S., Kılıç, E., Ömüral, M.: Apk2Img4AndMal: Android Malware Detection Framework Based on Convolutional Neural Network. In: 2021 6th International Conference on Computer Science and Engineering (UBMK), pp. 731–734 (2021)
2. Mat, S., Razak, M., Kahar, M., Arif, J., Firdaus, A.: A Bayesian probability model for Android malware detection. ICT Express. (2021), https://www.sciencedirect.com/science/article/pii/S2405959521001235
3. Abdullah, T., Ali, W., Abdulghafor, R.: Empirical study on intelligent android malware detection based on supervised machine learning. Int. J. Adv. Comput. Sci. Appl. **11** (2020), https://dx.doi.org/10.14569/IJACSA.2020.0110429
4. Türker, S., Can, A.: AndMFC: Android Malware Family Classification Framework. In: 2019 IEEE 30th International Symposium on Personal, Indoor And Mobile Radio Communications (PIMRC Workshops). 1, 1–6 (2019)
5. Park, J., Chun, H., Jung, S.: API and permission-based classification system for Android malware analysis. In: 2018 International Conference on Information Networking (ICOIN), pp. 930–935 (2018)
6. Fiky, A., Elshenawy, A., Madkour, M. Detection of android malware using machine learning. In: 2021 International Mobile, Intelligent, and Ubiquitous Computing Conference (MIUCC), pp. 9–16 (2021)
7. Yerima, S.: Android malware dataset for machine learning 2. (figshare 2018,2), https://tinyurl.com/drebinData
8. Razgallah, A.: Android Apps system calls traces. (kaggle 2021,3), https://www.kaggle.com/razgallah/android-apps-system-calls-traces
9. Hu, C., Jeng, T., Chen, Y.: Dynamic Android Malware Analysis with De-Identification of Personal Identifiable Information. (Association for Computing Machinery, 2020), https://doi.org/10.1145/3418688.3418694
10. Islam, F., Jamil, A., Momen, S.: Evaluation of Machine Learning Methods for Android Malware Detection using Static Features. In: 2021 IEEE International Conference On Artificial Intelligence in Engineering and Technology (IICAIET), pp. 1–6 (2021)
11. Fatima, A., Maurya, R., Dutta, M., Burget, R., Masek, J.: Android Malware Detection Using Genetic Algorithm based Optimized Feature Selection and Machine Learning. In: 2019 42nd International Conference On Telecommunications and Signal Processing (TSP), pp. 220–223 (2019)
12. Palikhe, A., Li, L., Tian, F., Kar, D., Zhang, N., Zhang, W.: MalDuoNet: A Dual-Net Framework to Detect Android Malware. In: 2021 RIVF International Conference on Computing and Communication Technologies (RIVF), pp. 1–6 (2021)
13. Amer, E.: Permission-based approach for android malware analysis through ensemble-based voting model. In: 2021 International Mobile, Intelligent, And Ubiquitous Computing Conference (MIUCC), pp. 135–139 (2021)

14. Smmarwar, S., Gupta, G., Kumar, S., Kumar, P.: An optimized and efficient android malware detection framework for future sustainable computing. Sustain. Energy Technol. Assessments. **54**, 102852 (2022). https://www.sciencedirect.com/science/article/pii/S2213138822009006
15. Jannat, U., Hasnayeen, S., Bashar Shuhan, M., Ferdous, M.: Analysis and detection of malware in android applications using machine learning. In: 2019 International Conference on Electrical, Computer and Communication Engineering (ECCE), pp. 1–7 (2019)
16. Thangaveloo, R., Jing, W., Chiew, K., Abdullah, J.: DATDroid: Dynamic analysis technique in android malware detection. Int. J. Adv. Sci. Eng. Inform. Technol. **10**, 536 (2020,3)
17. Roopak, S., Thomas, T., Emmanuel, S.: Android malware detection mechanism based on bayesian model averaging. Recent Findings In Intelligent Computing Techniques, pp. 87–96 (2019)
18. Han, H., Lim, S., Suh, K., Park, S., Cho, S., Park, M.: Enhanced android malware detection: an svm-based machine learning approach. In: 2020 IEEE International Conference on Big Data and Smart Computing (BigComp), pp. 75–81 (2020)
19. Islam, F., Jamil, A., Momen, S.: Evaluation of machine learning methods for android malware detection using static features. In: 2021 IEEE International Conference on Artificial Intelligence in Engineering and Technology (IICAIET), pp. 1–6 (2021)
20. Mohamed, S., Ashaf, M., Ehab, A., Abdalla, O., Metwaie, H., Amer, E.: Detecting Malicious Android Applications Based on API calls and Permissions Using Machine learning Algorithms (2021)
21. Arslan, R.: Identify Type of Android Malware with Machine Learning Based Ensemble Model. In: 2021 5th International Symposium on Multidisciplinary Studies and Innovative Technologies (ISMSIT), pp. 628–632 (2021)

Geometry Enhancements from Visual Content: Going Beyond Ground Truth

Liran Azaria(✉) and Dan Raviv

Tel Aviv University, Tel Aviv, Israel
liranazaria@mail.tau.ac.il, darav@tauex.tau.ac.il

Abstract. This work presents a new self-supervised cyclic architecture that extracts high-frequency patterns from images and re-insert them as geometric features. This procedure allows us to enhance the resolution and bittage of low-cost depth sensors capturing fine details on the one hand and being loyal to the scanned scene on the other. We present state-of-the-art results for depth super-resolution tasks as well as visually attractive, enhanced generated 3D models. Our main focus is on situations in which there is no large-scale accurate ground-truth depth map to train on, since the depth maps are captured by a low-cost scanner, making it either noisy or quantized.

Keywords: Computational Photography · Deep Learning · 3D geometry · Super resolution · Unsupervised Learning

1 Introduction

RGBD cameras are important in numerous applications spread across multiple fields such as robotics [2,6,14,33], virtual reality [2,5,6,24,31], augmented reality [2,5,14, 24] and human-computer interaction [2,5,14]. They provide fast and reliable depth maps that allow us to distinguish between objects and background, and to have a better understanding of our environment. Unfortunately, due to physical limitations of the sensors, the resulted depth maps are usually at a lower resolution than the corresponding RGB image [12,13,23] and are often interpolated in a simplistic manner to provide an output that seems to have a higher resolution. In addition, when using a low-cost sensor the depth pixels contain missing values and are either highly quantized or noisy. Several approaches are commonly used in order to improve depth maps. 1) super-resolution (SR) of the depth map without additional information [18,31,37,40]; These methods are usually divided into two subcategories: the first is filter-based methods [15,35], that are very fast and simple, however, generate blurring and artifacts. The second is optimization-based methods [19,20], which design complex regularization or explore prior to constrain the reconstruction of important high-resolution (HR) depth map. These methods require solving the global energy minimization problem which is

Supplementary Information The online version contains supplementary material available at https://doi.org/10.1007/978-3-031-31164-2_15.

H. Sharma et al. (Eds.): ICIVC 2022, PALO 17, pp. 180–194, 2023.
https://doi.org/10.1007/978-3-031-31164-2_15

time-consuming [11,36] 2) depth prediction from a single RGB image; These methods face a different problem since there are infinite distinct 3D scenes that can produce the same 2D RGB image [8,9,16], meaning that reconstructing the 3D scene from the RGB image requires finding the inverse of a non-injective function. 3) RGB information fused with the quantized low-resolution (LR) depth in order to produce a finer, more detailed depth map. The RGB image is usually available and containing HR information [18,31,37,40].

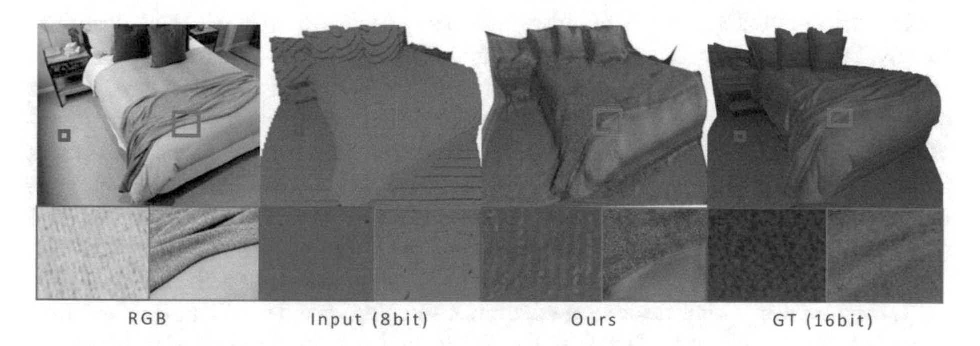

Fig. 1. Refined depth for an image taken from the Matterport3D data-set [3] displayed as a mesh in 3D space. From left to right: RGB image, depth input at 8-bit (simulating a low-cost scanner), our geometrically enhanced depth, ground-truth at 16-bit. We were able to retrieve missing geometric features not visible to the low-cost depth scanner yet captured in a regular camera. Furthermore, we were able to reconstruct details that are hidden by the noise of the 16-bit ground truth such as the texture of the carpet.

In this work, we tackle depth enhancement task using new architecture for image-depth fusion. We report state-of-the-art results on known benchmarks but even more appealing is our ability to transfer fine details from images to depth, generating 3D models that better explain the geometry going beyond the physical limitation of the depth sensor. To use known benchmarks that have been captured using expensive scanners and simulate a low-cost sensor, we quantized the depth map into 8-bits and used it as our only depth map.

1.1 Contributions

Our contributions are three-fold:

- Introduce the first self-supervised architecture for geometric enhancements guided by visual content.
- Propose new loss functions for soft data-fusion.
- We present a zero-shot depth super-resolution method based on a DNN that requires no prior training and can be optimized from a single RGBD image.

2 Related Work

We identify three different research lines of work for depth multi-resolution: Super-resolution from depth; Image to depth, usually from a single image; and Color guided depth super-resolution. We focus on the latter.

2.1 Depth Super Resolution

Under this umbrella, we distinguish between axiomatic methods and learning-based [11,36], usually deep learning techniques [31]. Image super-resolution [10], which is an extremely popular technique, inspired many depth SR algorithms, where the network remains almost identical but the trained data is replaced with relevant sampling. [34] created a method that specifically fits depth maps and used for simultaneous SR and denoising. They use a coupled dictionary learning method with locality coordinate constraints to reconstruct the high-resolution depth map. [21] proposed a method that combines SR with point-cloud completion for upsampling of both uniform and nonuniform grids. [7] combines recent learning-based up-sampling with variational SR. First, they trained a high-resolution edge prior to using an external data-set, and then they used this as guidance in the pipeline. Due to the lack of additional information that exists in the RGB image, none of these methods can reconstruct small structures that were lost during the acquisition process. The distinction between real-world edges and edges that results from artifacts is also more difficult and may cause errors.

2.2 Image to Depth Estimation

The physical limitations of depth sensors raised an important question - can we predict depth from a single image given enough data just from viewing an image. Surprisingly, when the networks are rich enough and trained on enough data the answer is yes in many scenarios [4]. [32] created a pair-wise ranking loss and a sampling strategy. Instead of randomly sampling point pairs, they guide the sampling to better characterize structure of important regions based on the low-level edge maps and high-level object instance masks. [16] uses densely encoded features, given dilated convolutions, in order to achieve more effective guidance. To that end, they utilize novel local planar guidance layers located at multiple stages in the decoding phase. [8] uses dilated convolutions and avoids down-sampling via spatial pooling. This allows them to obtain HR feature information. In addition, they trained their network using an ordinary regression loss, which achieved higher accuracy and faster convergence. Unlike these methods, we base our prediction on a lower resolution depth map that provides us with a decent estimation of depth. This solves one of the greatest issues of depth estimation - perspective.

2.3 Color Guided Depth Super-Resolution

Images provide important details for inferring the geometry. The direction of the normal, for example, is correlated to color changes in the image [31]. The challenge here is that the modality is different, and similar depth might be represented in an image with different color values, and on the other hand, different depth might have similar color. Among the popular solutions we find, [18] which created an optimization framework for color guided depth map restoration, by adopting a robust penalty function to model the smoothness term of their model. [37] formulate a separate guidance branch as prior knowledge to recover the depth details in multiple scales. [13] combine heterogeneous depth and color data to tackle the depth super-resolution and shape-from-shading problems. [17] created a correlation-controlled color guidance block, using this block, they designed a network that consists of two multi-scale sub-networks that provide guidance and estimate depth. Last, in order to create photorealistic depth maps that fit the real world, [31] created a loss function that measures the quality of depth map upsampling using renderings of resulting 3D surfaces.

2.4 Deep Zero-Shot Models

Lack of training data raised new interesting problems referred to as few-shot-learning and zero-shot-learning. in layman's terms not only we do not have labels, but we have not seen any sampled data from the distribution examined. One popular way to achieve that is to allow the network to train directly in test time, optimizing for a specific end case. [26] created an unconditional image generator that can be used for a variety of tasks, such as paint-to-image, editing of images, and super-resolution. [29] is used for super-resolution of natural images by exploiting the internal recurrence of information inside a single image. [38] created a non-stationary texture synthesis method based on a generative adversarial network (GAN). [30] exploit both external and internal information for zeros-shot super-resolution, where one single gradient update can yield considerable results. In our work, we show that our loss functions are robust enough to train on the provided image during test - and provide highly appealing results.

3 Proposed Method

Our method focuses on super-resolution and reconstruction of fine details that were lost due to the imperfect acquisition process of the depth map. Simultaneously, we fix the depth artifacts created over smooth areas, which we refer to as stairs artifacts which is a byproduct of quantization, displayed in Figs. 1 and 4. To that end, we designed a depth super-resolution (DSR) module that takes as input an $H \times W$ depth map with the corresponding $2H \times 2W$ gray-scale image and output a refined $2H \times 2W$ depth map, shown in Fig. 3. We use this module sequentially three times, to get an $\times 8$ resolution gain, shown in Fig. 2. Hence, the network can be formulated as:

$$D_{HR} = f(I_{HR}, f(I_{DS2}, f(I_{DS4}, D_{LR}))) \quad (1)$$

where D_{HR} is the predicted HR depth, D_{LR} is the low-resolution depth, and I_{DSx} is the RGB image down-sampled x times.

3.1 Architecture

Our architecture uses three consecutive upsampling modules called Depth Super-Resolution, DSR in short, each one up-sample the resolution by a factor of $\times 2$ and refines the depth. Then, a refinement module, identical to the one inside each DSR module, is used for fine-tuning before returning the final output, as shown in Fig. 2. Our modular design allows up-sampling by different factors - stacking n DSR modules resulting in 2^n resolution increase.

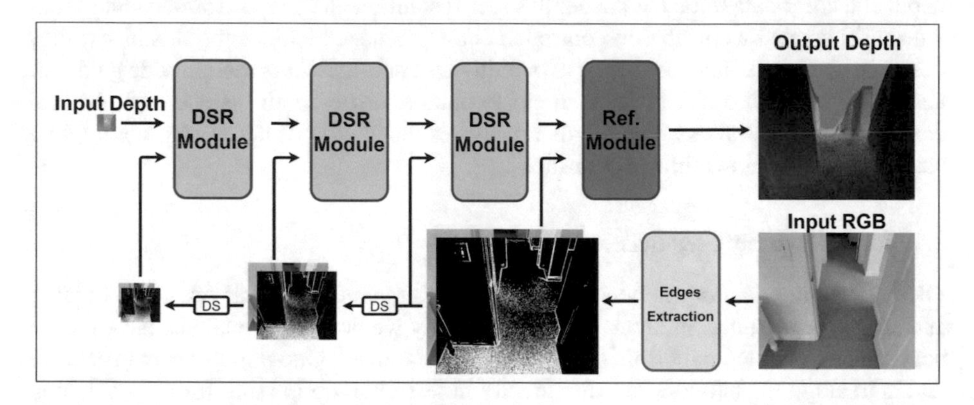

Fig. 2. Full Network Architecture: The network takes as input a high-resolution RGB image a well as a low-resolution depth map,returning an unquantized high-resolution refined depth-map. After passing the RGB image to the edge extraction module, it is downsampled to fit each DSR module's input, as shown in Fig. 3.

DSR Module. Each DSR module, displayed in Fig. 3, receives two inputs. The first is the higher ($\times 2$) resolution grayscale and image with concatenated edges for enhancing fine details, and the second is the LR depth map. The depth input passes through two 3×3 convolutional layers, a 3×3 deconvolution layer, that upsamples the depth to the height and width of the RGB image, and another three 3×3 convolutional layers. At this point, when both inputs are at the same resolution, we concatenate the two channels.

The last unit in the pipeline is a refinement module. It starts with a Dilated Inception Module, inspired by [27] and [28] with dilation rates of [1, 2, 4, 8]. Then, ten convolutional layers, where the last one is a 1×1 convolution, that return a single channel depth map, used as input for the next block and for loss calculation for this block.

Cycle Module. One of the main challenges we face here is the lack of reliable labels. As the title implies, we do not have a reliable ground truth. To overcome this limitation, we suggest transferring information from the geometry space back to the image space. But, unlike other dmain-transfer networks, we focus on other the high frequencies or edges in this case. Specifically, this module starts by applying edge extraction to the predicted depth map. The edges are concatenated to the predicted depth and together, they pass our geometry-to-image transformer. This module, applied for each DSR output, is constructed from 15 consecutive 3×3 convolutions and ends with a single 1×1 convolution layer.

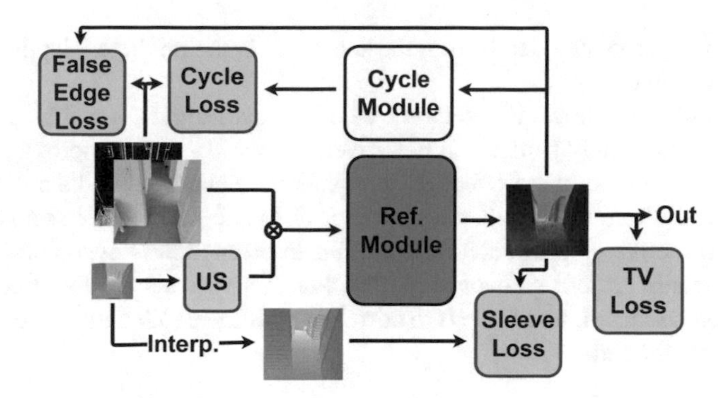

Fig. 3. Depth Super-Resolution (DSR) Module: This module take a lower-resolution ($H \times W$) depth map and a high-resolution ($2H \times 2W$) RGB edges to return an higher-resolution ($2H \times 2W$), refined depth map. the lower-resolution depth passes through an up-sample (US) module that generates a higher-resolution feature map. Concatenated to the edges the feature map passes the refinement module to create the high-resolution depth output.

Edge Extraction. The edges extracted from the color image contain important information that is required for quality refinement and SR of depth maps. However, since the RGB image and the depth map are two different modalities, the magnitude of the derivatives can be washed out. To face this issue we start by transforming the RGB image into a grayscale image. Then, we used the Sobel operator for edge extraction. Next, by thresholding the magnitude of the operator, using the p percentile (p is set to 90 by default) in that image as a threshold, we created a binary map that shows "edge" vs "no edges" according to the RGB image.

3.2 Losses

The loss functions used during training were crafted especially to fit the task of self-supervised depth refinement, under the premise that although the ground truth is absent, it can be approximated with a certain error factor by a simple interpolation of the input. Our network calculates loss for each scale and penalizes known artifacts that exist in quantized depth maps, such as "Stairs Artifact". Specifically, we introduce four different losses. Sleeve loss 3.2, which provides some level of tolerance, not punishing at all if the solution close enough to the interpolated quantized depth. Cycle loss 3.2, making sure fine details appear in the geometry. False edge loss 3.2, which removes geometric artifacts that appear in the interpolated quantized depth. Last, Total Variation loss 3.2, for a smooth solution. Let us elaborate on each loss function.

Sleeve Loss. After approximating the ground truth depth data with the up-sampled lower resolution depth map, we train the network in a manner that allows changes and corrections of the input channel even in locations we received data from the depth sensor. Here we propose to use a sleeve loss, which is defined by

$$L_{sleeve} = max(|Y - \hat{Y}| - s, 0) \tag{2}$$

where Y is an interpolated depth from the sensor, $\hat{Y}$ is the predicted depth and s is the allowed error sleeve.

This formulation allows the network to detach its output from the approximated ground truth without any penalty up to a given threshold - s, the approximation error. Due to the native smoothness effect of convolutional neural networks not only do we get a smooth solution, but it falls within a pipe shape along with the scanned, binned, depth data. The obvious drawback of using this loss function is over-smoothness, and indeed fine details are not recovered by this one. Here we set the low frequencies of the domain on one hand, and force it to correlate with the depth data we received as an input on the other hand.

Cycle Loss. Due to the limitations of current low-cost scanners, and the difficulty in acquiring large-scale data-sets for supervised training we chose to focus on unsupervised solutions - learnable and zero/few-shot. To achieve this goal, motivated by the cyclic loss presented in [39], we have built a transformer network 3.1, that transfers the edges of the predicted depth into the domain of the edges extracted from a grayscale image. These edges are compared to the genuine edges of the grayscale image, extracted using the above-mentioned Edge extraction module 3.1. We chose to cycle back and predict only the edges and not the image since it is a simpler task, and focus just on the necessity required to guide the depth reconstruction unit. The comparison is done as a simple L_1 norm.

False-Edge Loss. Due to our choice of approximating the ground truth by interpolating the LR quantized depth map, our approximation contains many false edges. Any smooth surface at an inclined slope contains the above-mentioned Stairs Artifact. To deal with this issue, we created the False Edge loss that relies on the premise that depth edges that are reliable to the scanned scene can be found in both the depth map and the grayscale image. We use our edge extraction module to find the edges in the grayscale image. Then, we sum the edges of the predicted depth that correspond to smooth areas on the RGB image, meaning edges that are found in the depth map but not in the grayscale image. Calculated as follows:

$$L_{fe}(d, gs) = \sum E_d(1 - E_{gs}) \tag{3}$$

where d and gs are the depth map and grayscale image, E_d and E_{gs} are the edges extracted from the depth map and grayscale image accordingly.

Figure 4 shows a visual representation of this phenomena, the green edges are edges that fit the RGB image (around the lights and the rectangle created by the drop ceiling), and as a result will not yield loss. The red edges are the result of the stairs-artifact and do not appear in the RGB image nor the real world. The summation of these edges is the definition of the False-Edge loss. Differently from the Cycle-loss, which compares edges in the image domain, this loss is applied at the geometry domain and only punishes false edges (edges that exist where edges should not be found and are marked in red. This loss does not punish the absence of true edges (edges that exists where edges should be found and are marked in green).

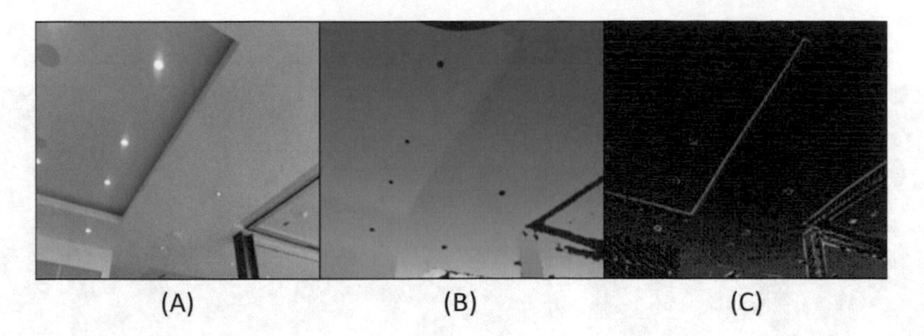

Fig. 4. Example of false and true depth edges. A) is an RGB image of a ceiling. B) Is the corresponding depth map, and C) are the quantized depth map edges in which green marks true edges and red are false edges created by the stairs artifact.

Total-Variation Regularization. In order to force smoother predictions, we added a total variation (TV) regularization. Similarly to [1], formulated as:

$$R_{tv}(d) = \sum_{i,j}(\sqrt{|d_{i,j} - d_{i+1,j}|^2 + |d_{i,j} - d_{j+1,i}|^2}) \quad (4)$$

where $R_{tv}(d)$ is the TV regularization value for the depth map d and $d_{i,j}$ is the depth pixel at the (i, j) location.

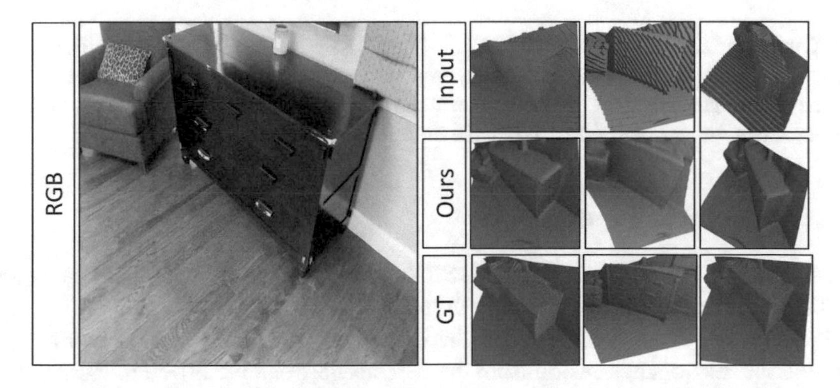

Fig. 5. Results for multi-view test: presenting our results in a natural view for AR/VR environment. Not only we remove the quantization artifacts, but we also keep edges and corners of the furniture.

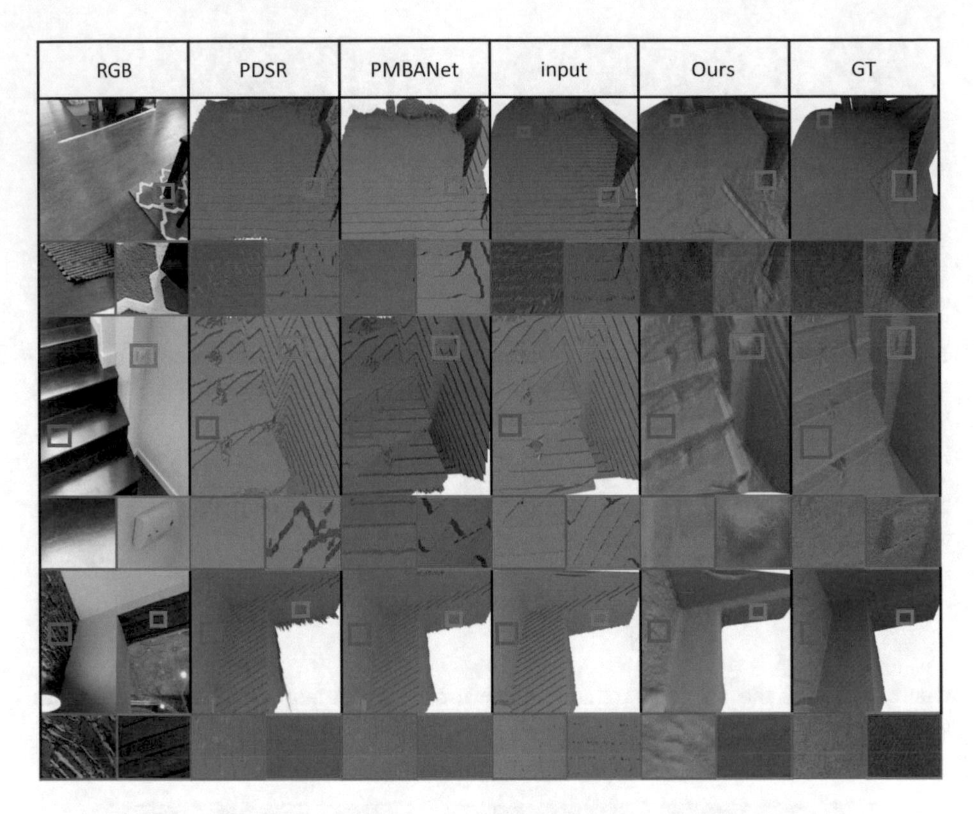

Fig. 6. Qualitative results for Matterport3D [3]: The proposed method was able to retrieve curved shapes, edges, and corners while smoothing only relevant parts in the provided depth, restoring even fine details that were lost in the noise of the ground truth such as the carpet on the top row or the texture of the wall and the wood on the bottom row.

3.3 Implementation Details

Our architecture is built out of three DSR blocks followed by a refinement module. Each DSR block contains one cycle module and one Refinement module.

The zero-shot training is accomplished simply by optimizing the network on one image without any changes aside from the weights between the different losses.

We chose to construct each module in the following way: All layers (both convolutional and deconvolutional) have 8 output channels, with exception of the final two layers in each block/module. The second to last layer outputs $k \geq 8$ channels (k is a

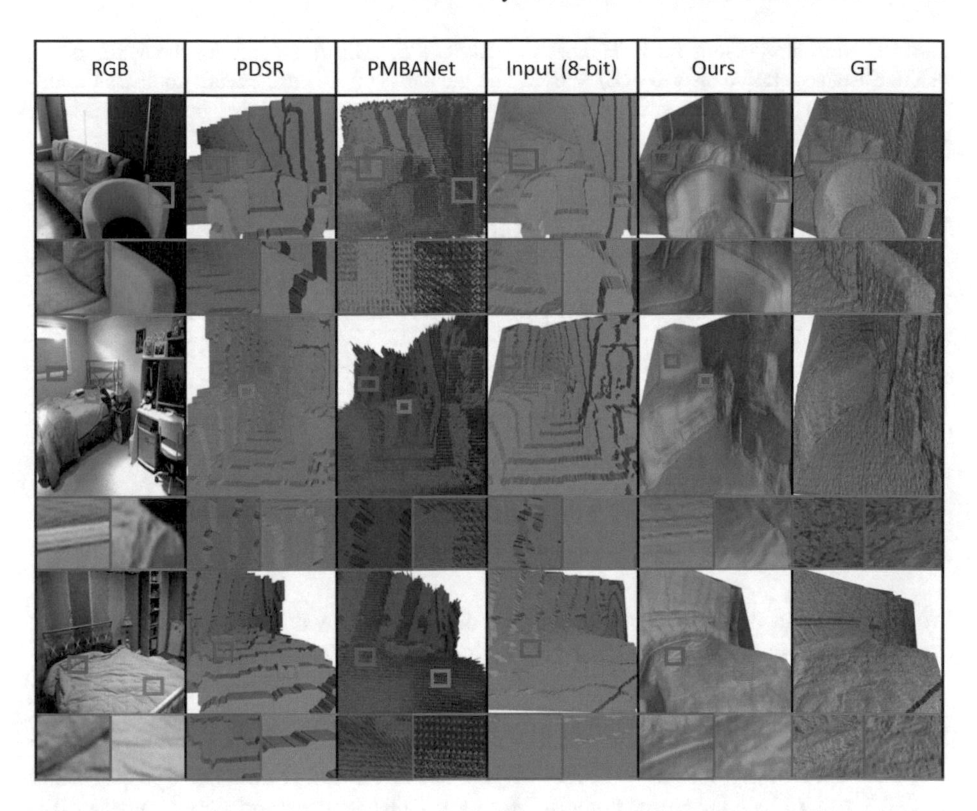

Fig. 7. Qualitative results for NYUv2 [22]: Reconstructed depth information for in-doors scenes. The proposed method was able to retrieve curved shapes, edges, and corners while removing artifacts and smoothing only relevant parts. It can be seen that although the ground truth contain noise, especially visible on smooth surfaces, the predicted depth is smoother and cleaner.

hyperparameter, set to 16 by default) and the last layer outputs 1 channel (meant to fit the dimensions of the output to the dimensions of the depth map). Hence, for each given module the number of filters for each layer will be as follows: [8, 8, 8, ..., 8, k, 1].

For optimization, we chose Adam optimizer with $\mu = 0.9$, $\beta_1 = 0.9$, $\beta_2 = 0.99$ and $\epsilon = 10^{-8}$. The initial learning rate is set to 10^{-4} and decreased by 0.5 every 10 epochs. Zero-shot training was done with the same parameters. We trained the network using a single NVIDIA RTX 2080ti.

For each visual test, we divided each data-set for 70% train and 30% test. We used the HR depth maps as ground truth. For the input of the networks, we downsampled the HR depth maps $\times 8$ using Nearest-Neighbor (NN) interpolation, then converted the result to 8 bit. Similarly to other works, we used several common metrics to evaluate our performance. Specifically, we used Mean Square Error (MSE), Root Mean Square Error (rMSE), Mean Absolute Error (MAE), Absolute Relative Error (ARE), Peak Signal-to-Noise Ratio (PSNR), and Median Error (ME).

Table 1. Numerical results for ETH3D stereo dataset [25]: This table presents the MSE, rMSE, MAE, absolute relative error (ARE), PSNR, and median error of our network compared to different methods (by running their published code and using pre-trained weights when those are given). We outperform all methods by a large margin.

Method	MSE	rMSE	MAE	ARE	PSNR	ME
earest	0.904	0.819	0.763	0.299	26.953	0.504
Bilinear	0.895	0.814	0.762	0.298	27.013	0.503
RGDR [18]	0.767	0.78	0.728	0.277	26.962	0.702
PMBANet [37]	0.806	0.799	0.734	0.281	26.883	0.737
PDSR [31]	0.635	0.69	0.639	0.250	28.286	0.631
Ours	0.205	**0.278**	0.173	**0.053**	39.5	**0.046**
Ours (zero-shot)	**0.196**	0.280	**0.172**	0.054	**39.903**	0.05

4 Results

We evaluated our method quantitatively and qualitatively by refining and enhancing the depth resolution of several common RGBD indoor data-sets.

4.1 Data-sets

- **Matterport3D** [3]**:** This large and diverse indoors RGB-D data-set was created in 2017 using the Matterport camera. Both the RGB images and depth maps in this data-set are at a resolution of 1280×1024.
- **NYUv2** [22]**:** This RGB-D data-set was created in 2012 using a Kinect depth sensor, originally for semantic segmentation tasks. Both the RGB images and depth maps in this data-set are at a resolution of 640×480. The Train/Test split was done using the provided official split.
- **ETH3D** [25]**:** The creation of this data-set was motivated by the limitations of existing multi-view stereo benchmarks. Since many of these limitations are similar to the limitations in many RGB-D data-sets it fits our needs almost perfectly. Both the RGB images and depth maps in this data-set are at a resolution of 6048×4032, the depth is sparse but not quantized. This data-sets depth is very clean, containing nose levels that are barley visible.

Our tests focused on two different scenarios. The first, qualitative, comparing our results to different methods on datasets for which ground-truth exists, but is either quantized or noisy and does not represent the real world accurately enough. This tests the visual appeal of our results. For that test we chose Matterport3D [3] and NYUv2 [22]. The second, quantitive, we compared our results to other methods numerically using ETH3D [25] and simulated a scenario in which we cannot capture a reliable dataset and need to either train self-supervised methods or train on different datasets and count on a small enough domain gap. The former test whether the model represents the human expectation of the real-world depth, while the latter, which requires nearly perfect depth maps, allows us to measure whether the model represents the real-world depth accurately.

4.2 Qualitative Results

For this evaluation, we used Matterport3D [3] and NYUv2 [22]. These data sets allow us to demonstrate our results in a variety of scenes, taken from widely known sources. As shown in Figs. 6 and 7 Our results are more photo-realistic than those of other methods and are often better than the ground truth. We restore many of the fine details that were lost during quantization and did not exist in the networks' input depth. Furthermore, our restoration does not contain the noise that is prominent in the ground truth. Additionally, it is visible that the supervised methods recreated the artifacts that existed in the input and by that, generated a less realistic depth map that represents the real world unreliably. For the second qualitative evaluation, we rotated the mesh created by the projected depth map. This allows us to demonstrate how the results can reliably represent the real world, and that details are restored in a reliable manner that fits the human expectation of depth, see Fig. 5. A comparison between zero-shot and unsupervised training can be seen in Fig. 1 and the supplements.

4.3 Quantitive Results

These tests were done on indoor scans from the ETH3D data-set [25]. This data set's sparse, yet clean and unquantized depth allowed us to create a low-resolution quantized depth input and test the model's performance using reliable ground truth. The depth maps in this data set contain fine details and minimal levels of noise, making it as close as possible to the scanned scene. This allows us to compare our method's results to other top-rated super-resolution methods, under a scaling factor of $\times 8$, as shown in Table 1. To compare our results to supervised methods under the scenario in which no reliable ground truth exist for training, we used pre-trained models that were trained on other data sets. In this tested scenario, our model reaches superior results compared to all others. We also show comparable results between our trained and zero-shot methods.

4.4 Ablation Study

In order to show the importance of each element of the network, we conducted an ablation study. When possible, we removed the tested element completely, for example, the cycle loss. The sleeve loss was not completely removed, since its absence did not allow the network to converge to any sort of displayable results, so its ablation study was by replacing it with $L1$ (absolute difference) loss. The results in Table 2 show superior results for the complete model.

4.5 Failed Cases

Our network often fails over paintings and writings that cause string edges in the RGB images and does not effect depth. For dealing with these failed cases we can research the following approach: First, training a different network that segments painting, text, and other strong misleading gradients. Then, we can block them in the RGB image, preventing the gradients from reaching the network.

Table 2. Numerical results for ablation study. Each line represents the results of the network with a different part missing

	MSE	rMSE	MAE	ABS REL	PSNR	Median Err
Sleeve loss	0.364	0.376	0.250	0.078	35.80	0.254
False Edge loss	**0.193**	0.293	0.228	0.075	37.985	0.239
Total Variation	0.379	0.353	0.246	0.065	38.985	0.065
Cycle loss	6.248	1.797	1.556	0.572	4.351	4.645
Full Architecture	0.196	**0.280**	**0.172**	**0.054**	**39.903**	**0.05**

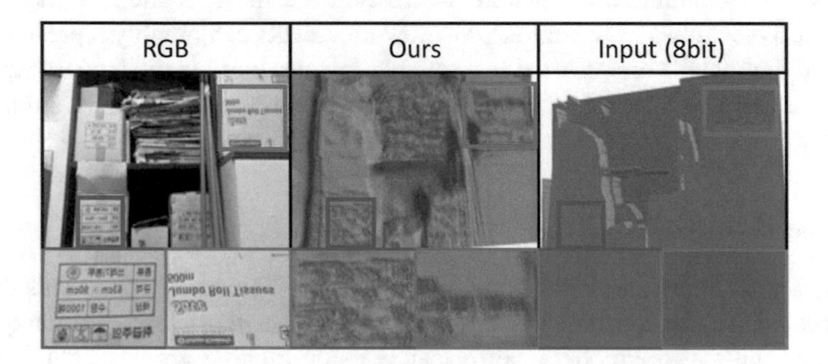

Fig. 8. Failure example: Image taken from [4], the text appears in the predicted depth as causing depth changes. This is caused by the strong gradients in the RGB image that the letters create.

5 Conclusion

We presented the first unsupervised zero-shot neural network architecture for geometric depth refinement and super-resolution. Our method is based on a novel network architecture and loss functions, that not only outperforms known methods on given benchmarks but can retrieve fine details of the geometry that can not be captured in low-cost depth scanners. Due to the lack of other self-supervised depth super-resolution methods, we compared our results to pre-trained supervised methods, trained on different data sets. Under this scenario, the other methods deal with a domain gap that our self-supervised method can avoid. This allowed us to achieve superior results in a variety of tests.

References

1. Aly, H.A., Dubois, E.: Image up-sampling using total-variation regularization with a new observation model. IEEE Trans. Image Process **14**(10), 1647–1659 (2005)
2. Canessa, A., Chessa, M., Gibaldi, A., Sabatini, S.P., Solari, F.: Calibrated depth and color cameras for accurate 3d interaction in a stereoscopic augmented reality environment. J. Visual Commun. Image Representation **25**(1), 227–237 (2014)
3. Chang, A., et al.: Matterport3d: Learning from rgb-d data in indoor environments. In: International Conference on 3D Vision (3DV) (2017)

4. Cho, J., Min, D., Kim, Y., Sohn, K.: A large rgb-d dataset for semi-supervised monocular depth estimation. arXiv preprint arXiv:1904.10230 (2019)
5. Du, H., et al.: Interactive 3d modeling of indoor environments with a consumer depth camera. In: Proceedings of the 13th International Conference on Ubiquitous Computing, pp. 75–84 (2011)
6. El Jamiy, F., Marsh, R.: Survey on depth perception in head mounted displays: distance estimation in virtual reality, augmented reality, and mixed reality. IET Image Proc. **13**(5), 707–712 (2019)
7. Ferstl, D., Ruther, M., Bischof, H.: Variational depth superresolution using example-based edge representations. In: Proceedings of the IEEE International Conference on Computer Vision, pp. 513–521 (2015)
8. Fu, H., Gong, M., Wang, C., Batmanghelich, K., Tao, D.: Deep ordinal regression network for monocular depth estimation. In: Proceedings of the IEEE Conference on Computer Vision and Pattern Recognition (CVPR) (2018)
9. Godard, C., Mac Aodha, O., Firman, M., Brostow, G.J.: Digging into self-supervised monocular depth estimation. In: Proceedings of the IEEE International Conference on Computer Vision, pp. 3828–3838 (2019)
10. Gu, J., Lu, H., Zuo, W., Dong, C.: Blind super-resolution with iterative kernel correction. In: 2019 IEEE/CVF Conference on Computer Vision and Pattern Recognition (CVPR), pp. 1604–1613 (2019)
11. Guo, C., Li, C., Guo, J., Cong, R., Huazhu, F., Han, P.: Hierarchical features driven residual learning for depth map super-resolution. IEEE Trans. Image Process. **28**(5), 2545–2557 (2018)
12. Haefner, B., Peng, S., Verma, A., Quéau, Y., Cremers, D.: Photometric depth superresolution. IEEE Trans. Pattern Anal. Mach. Intell. **42**(10), 2453–2464 (2020)
13. Haefner, B., Quéau, Y., Möllenhoff, T., Cremers, D.: Fight ill-posedness with ill-posedness: Single-shot variational depth super-resolution from shading. In: Proceedings of the IEEE Conference on Computer Vision and Pattern Recognition, pp. 164–174 (2018)
14. Izadi, S., et al.: Kinectfusion: real-time 3d reconstruction and interaction using a moving depth camera. In: Proceedings of the 24th Annual ACM Symposium on User Interface Software and Technology, pp. 559–568 (2011)
15. Kim, C., Yu, H., Yang, G.: Depth super resolution using bilateral filter. In: 2011 4th International Congress on Image and Signal Processing, vol. 2, pp. 1067–1071 (2011)
16. Lee, J.H., Han, M.K., Ko, D.W., Suh, I.H.: From big to small: Multi-scale local planar guidance for monocular depth estimation. arXiv preprint arXiv:1907.10326 (2019)
17. Li, T., Lin, H., Dong, X., Zhang, X.: Depth image super-resolution using correlation-controlled color guidance and multi-scale symmetric network. Pattern Recogn. **107**,107513 (2020)
18. Liu, W., Chen, X., Yang, J., Qiang, W.: Robust color guided depth map restoration. IEEE Trans. Image Process. **26**(1), 315–327 (2016)
19. Liu, X., Zhai, D., Chen, R., Ji, X., Zhao, D., Gao, W.: Depth super-resolution via joint color-guided internal and external regularizations. IEEE Trans. Image Process. **28**(4), 1636–1645 (2019)
20. Lu, J., Min, D., Pahwa, R.S., Do, M.N., et al.: A revisit to mrf-based depth map super-resolution and enhancement. In: 2011 IEEE International Conference on Acoustics, Speech and Signal Processing (ICASSP), pp. 985–988 (2011)
21. Mandal, S., Bhavsar, A. and Sao, A.K.: Depth map restoration from undersampled data. IEEE Trans. Image Process. **26**(1), 119–134 (2016)
22. Silberman, P.K.N.: Derek Hoiem and Rob Fergus. Indoor segmentation and support inference from rgbd images, In ECCV (2012)

23. Or-El, R., Hershkovitz, R., Wetzler, A., Rosman, G., Bruckstein, A.M., Kimmel, R.: Real-time depth refinement for specular objects. In: 2016 IEEE Conference on Computer Vision and Pattern Recognition (CVPR) (2016)
24. Peng, S., Haefner, B., Quéau, Y., Cremers, D.: Depth super-resolution meets uncalibrated photometric stereo. In: Proceedings of the IEEE International Conference on Computer Vision Workshops, pp. 2961–2968 (2017)
25. Schops, T., et al.: A multi-view stereo benchmark with high-resolution images and multi-camera videos. In: Proceedings of the IEEE Conference on Computer Vision and Pattern Recognition (CVPR), pp. 3260–3269 (2017)
26. Shaham, T.R., Dekel, T., Michaeli, T.: Learning a generative model from a single natural image. In: Proceedings of the IEEE International Conference on Computer Vision, pp. 4570–4580 (2019)
27. Shi, W., Jiang, F., Zhao, D.: Single image super-resolution with dilated convolution based multi-scale information learning inception module. In: 2017 IEEE International Conference on Image Processing (ICIP), pp. 977–981 (2017)
28. Shi, W., Jiang, F., Zhao, D.: Single image super-resolution with dilated convolution based multi-scale information learning inception module. In: 2017 IEEE International Conference on Image Processing (ICIP), pp. 977–981. IEEE, (2017)
29. Shocher, A., Cohen, N., Irani, M.: "zero-shot" super-resolution using deep internal learning. In: Proceedings of the IEEE Conference on Computer Vision and Pattern Recognition (CVPR) (2018)
30. Soh, J.W., Cho, S., Cho, N.I.: Meta-transfer learning for zero-shot super-resolution. In: Proceedings of the IEEE/CVF Conference on Computer Vision and Pattern Recognition, pp. 3516–3525 (2020)
31. Voynov, O., et al.: Perceptual deep depth super-resolution. In: Proceedings of the IEEE International Conference on Computer Vision, pp. 5653–5663 (2019)
32. Xian, K., Zhang, J., Wang, O., Mai, L., Lin, Z., Cao, Z.: Structure-guided ranking loss for single image depth prediction. In: Proceedings of the IEEE/CVF Conference on Computer Vision and Pattern Recognition, pp. 611–620 (2020)
33. Xiao, Y., Cao, X., Zhu, X., Yang, R., Zheng, Y.: Joint convolutional neural pyramid for depth map super-resolution (2018)
34. Xie, J., Feris, R.S., Yu, S.S., Sun, M.T.: Joint super resolution and denoising from a single depth image. IEEE Trans. Multimed. **17**(9), 1525–1537 (2015)
35. Yang, Q., Yang, R., Davis, J., Nister, D.: Spatial-depth super resolution for range images. In: 2007 IEEE Conference on Computer Vision and Pattern Recognition, pp. 1–8 (2007)
36. Ye, X., Duan, X., Li, H.: Depth super-resolution with deep edge-inference network and edge-guided depth filling. In: 2018 IEEE International Conference on Acoustics, Speech and Signal Processing (ICASSP), pp. 1398–1402. IEEE (2018)
37. Ye, X., et al.: Pmbanet: progressive multi-branch aggregation network for scene depth super-resolution. IEEE Trans. Image Process. **29**, 7427–7442 (2020)
38. Zhou, Y., Zhu, Z., Bai, X., Lischinski, D., Cohen-Or, D., Huang, H.: Non-stationary texture synthesis by adversarial expansion. arXiv preprint arXiv:1805.04487 (2018)
39. Zhu, J.Y., Park, T., Isola, P., Efros, A.A.: Unpaired image-to-image translation using cycle-consistent adversarial networks. In: The IEEE International Conference on Computer Vision (ICCV), Oct (2017)
40. Zuo, Y., Qiang, W., Fang, Y., An, P., Huang, L., Chen, Z.: Multi-scale frequency reconstruction for guided depth map super-resolution via deep residual network. IEEE Trans. Circuits Syst. Video Technol. **30**(2), 297–306 (2019)

Comparing Neural Architectures to Find the Best Model Suited for Edge Devices

Bhavneet Singh[1] and Jai Mansukhani[2](✉)

[1] Computer Science, Guru Gobind Singh Indraprastha University, Dwarka, India
[2] Electrical and Communication Engineering, Guru Gobind Singh Indraprastha University, Dwarka, India
jaimansukhani267@gmail.com

Abstract. Training large-scale neural network models is computationally expensive and demands a great deal of resources. It is an important area of study with a lot of potential for the future of the AI industry. In recent years, the power of computer hardware has significantly improved and we have new breakthroughs in deep learning. With these innovations, the computational cost of training large neural network models has declined by at least 10 folds in high- and average-performance machines. In this research, we explore NAL, AutoML, and other frameworks to determine the best suitable model for edge devices. The biggest improvements compared to reference models can be acquired if the NAS algorithm is co-designed with the corresponding inference engine.

1 Introduction

The training of large machine learning models is typically done on specific high-performance computers, but may still require a lot of time. This makes it hard to tune hyper-parameters or the model architecture in an intensive way. Therefore a way to reduce the design time of human interaction and know-how is needed. This is not the case for smaller models targeting embedded systems as their architecture is often much simpler resulting in training times of just a few minutes for mid-size applications. Finding the best suitable model architecture to run on edge devices, which are known to have several resource constraints like computational power and memory size with typical values in the range of several KB to very few MB, is currently a highly relevant topic being explored by many researchers worldwide. After a brief introduction of some core concepts of Machine Learning like NAS, AutoML and available Frameworks, a survey on papers on the aforementioned subject follows in Sect. 2. The benchmark results for these approaches are compared as far as possible in Sect. 3 before the core results and ideas for further work on the topics are summarized in Sect. 4.

1.1 Neural Architecture Search and AutoML

Short training times enable the application of optimization techniques to either improve the accuracy or the size of the resulting models in a meaningful way. Neural Architecture

H. Sharma et al. (Eds.): ICIVC 2022, PALO 17, pp. 195–203, 2023.
https://doi.org/10.1007/978-3-031-31164-2_16

Search methods especially benefit from such targets as they allow many design iterations and therefore more promising results in a shorter time. AutoML workflows, which combine NAS with further automated process steps as shown in Fig. 1, are already being used in research groups and will become more relevant for the future AI industry [9]. While Data Preparation and Feature Engineering are prerequisite steps to apply to NAS, automated machine learning may make full use of the potential of Neural Architecture Search during the generation and estimation of models. Search space exploration and optimization methods like heuristics or evolutionary algorithms may be used to gain architectures fulfilling the given constraints. There are many methods which can be used for model compression in the final estimation step, where weight-sharing is one of the most relevant ones. In the method of weight-sharing a larger super-network is typically trained to provide a higher number of possible sub-networks for evaluation. This reduces the search cost as the parameters of overlapping sub-networks are only computed once [4]. Other approaches to reduce the model complexity are the quantization of floating point model state values and parameters to smaller fixed-point data types or pruning which removes a certain amount of parameters determined by a given pruning method [16].

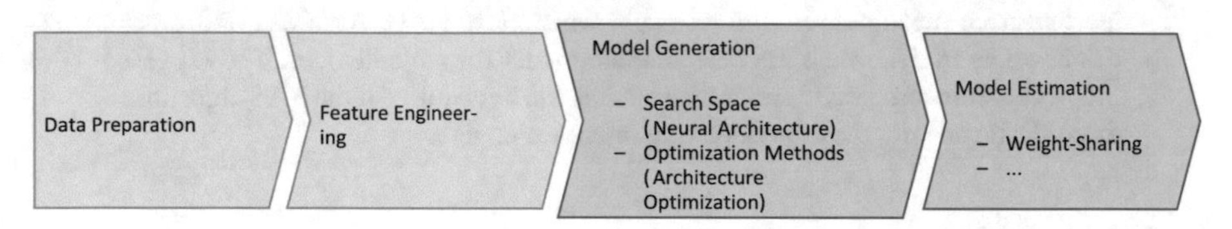

Fig. 1. Typical AutoML Flow [9] (Highlighted Steps are relevant for NAS)

1.2 Frameworks

To design machine learning models and deploy them on a target device some framework or library is typically used. As later shown in this paper, this choice can have a large impact on the performance and size of the resulting design. Especially the "engine" used for running the inference on a microcontroller device may be a bottleneck or provide some room for improvements. While Tensorflow Lite (TFLite) mostly targets mobile devices, there is a relatively novel but popular library called Tensorflow Lite Micro (TFLM) with kernels, e.g. methods to execute operations of the model graph, specialized to resource constrained architectures [5]. The term TinyML is often used in the context of TFLM but may also refer to machine learning models running on edge devices in general.

Custom kernels often referred to as CMSIS-NN exist for some targets with hardware accelerators for typical Digital Signal Processing (DSP) instructions using neural network calculations like Multiply and Accumulate [11]. The data in Table 1 also presents the relevance of both frameworks compared to less widely used MicroTVM and Pytorch, which are not going to be discussed here. Each completely custom framework proposed in a paper has to be compared to the aforementioned state of the art approaches.

Table 1. Frameworks used in the discussed papers (°: comparisons only)

Frameworks	TFLite (Micro)	CMSIS-NN	MicroTVM	Pytorch	Custom
MCUNet [14]	◦	◦	◦		•
μNAS [12]	•	•			
SpArSe [7]	N/A	N/A	N/A	N/A	N/A
MicroNets [1]	•	•			
Once-for-All [2]				•	

2 Literature Survey

This section presents a literature review of neural architecture search (NAS). This review was conducted to shed light on the various types of NAS techniques and how they work to produce novel neural architectures. We define NAS broadly to refer to a family of methods that automatically explore configurable space and help facilitate the desirable goal of generating novel neural architectures automatically. The space that can be explored includes but is not limited to: (a) number of layers, (b) number or type of neurons within each layer, and (c) types of connection between neurons between different layers.

Although the following brief overview on research papers focuses on the proposed NAS methods instead of the inference engine or model types, the framework used for the design and deployment sometimes also need consideration due to specific model-optimizations, which may only have certain effects when applied together with the suitable counterparts.

Beside frameworks there exist many research results on NAS performance for popular reference models like ImageNet [6] and CIFAR-10 [10]. Unfortunately they are initially targeted at devices offering more computing power. Due to their large size or operation count (computational demand) not all of them are suitable for comparing NAS results on microcontrollers. To demonstrate the possibilities of machine learning on edge devices some popular application spaces like Image Recognition (ImageNet), Visual Wake Words and Speech Commands (Keyword Spotting, KWS), can be used. There also exist many datasets for the classification of handwritten characters or digits (MNIST). Applications like Anomaly Detection or DSP Classification have higher relevance for the real world but often lack representative datasets, hence why they are only used in some of the research papers.

2.1 MCUNet

The MCUNet system-model co-design framework consists of two main components. The first one is a Neural Architecture Search algorithm suitable for MCUs and a lightweight inference engine based on code generation. The latter is called TinyEngine and profits from memory scheduling, which takes the overall network topology into account instead of only considering individual layers for optimization. This leads to smaller models in terms of memory requirements and operation count and therefore to an increase in the size of the NAS search space. The two-stage algorithm of the so-called TinyNAS component optimizes the aforementioned search space according to given resource constraints before specializing its properties to handle various constraints at once [14]. Inference results in comparison with the state of the art Tensorflow Lite Micro library and CMSIS-NN [11] are presented in Sect. 3.

TinyNAS at first searches for the best search space configuration $S*$ in the set of possible configuration S which is a non-trivial task. By using the cumulative distribution-function (CDF) of FLOPS instead of the CDF of model accuracy the high cost of training each network can be avoided. One-shot architecture search making use of weight sharing in combination with evolutionary search determines the resulting model suitable for the given set of constraints.

By performing in-place depth-wise convolutions using a single temporal buffer, TinyEngine can reduce the peak memory for N channels from $2N$ to $N + 1$. Additionally, methods such as adaptive scheduling or loop unrolling are applied on parts of the model architecture during code generation to reduce the number of required operations for running a neural network model [14].

2.2 μNAS

Similar to TinyNAS, μNAS generates mid-tier MCU-level networks with sizes from 0.5 KB to 64 KB, which are primarily intended for image classification tasks. Multi-objective optimization takes RAM-size, persistent storage and processor speed into account [12]. The experiment results in the recently published paper already featuring comparisons with for example the MCUNet framework.

Different to traditional approaches for GPU- or mobile-level NAS design, μNAS was designed following two special design requirements: A highly granular search space and the accurate computation of used resources. The resulting fourstep search procedure is based on the concepts of aging evolution and pruning and is briefly shown in Fig. 2.

2.3 SpArSe

To demonstrate the possibility of designing Convolutional Neural Networks (CNNs) which generalize well via AutoML, while still being small enough to fit onto memory-constrained MCUs, [7] proposes an architecture search method combining neural architecture search with pruning. The resulting applications for the IoT field definitely display a success of the aforementioned challenge.

Using a balanced multi-objective optimization problem operating on the problem space $\Omega = \{\alpha, \theta, \omega, \Theta\}$ composed of the network connectivity graph α, network weights

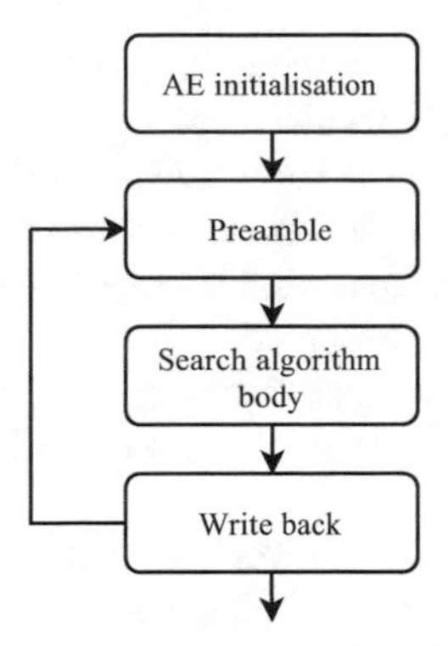

Fig. 2. General search procedure for μNAS [12]

ω, operations θ and training hyper-parameters Θ small but performant CNNs can be generated. The three objective functions are used to design the SpArSe architectures with L being the number of network layers. After a model size compression via pruning the best suitable subspaces are determined using Bayesian optimization. Instead of weight-sharing, Network morphism is applied in combination with random scalarizations [7] to evaluate each configuration Ω^n at low computational costs.

- $f_1(\Omega) = 1 - \text{ValidationAccuracy}(\Omega)$
- $f_2(\Omega) = \text{ModelSize}(\Omega)$
- $f_3(\Omega) = \max_{l \in 1,\ldots,L} \text{WorkingMemory}_l(\Omega)$

2.4 MicroNets

To employ a new differentiable NAS (DNAS) realization for edge devices with tight memory and performance constraints in [1] the properties of NAS search spaces for MCU model design were studied first. The most important result of this step is the correlation between the model latency and the model operation count under a uniform prior over models in the search space. Therefore the op count is used as viable proxy to latency when searching for the optimal network architecture. Designed models can target three different size-classes (S/M/L). Another aspect of NAS is the underlying latency/energy model which is also mentioned in [1]. TinyML performance for industry standard benchmarks is shortly mentioned in 3.

After deriving the correlation between the layer operation count and the layer latency and the energy model the paper proposes MircoNet Models following a DNAS approach. Instead of using model constraints to restrict the feasible space, regularization terms are incorporated in the DNAS objective function. Compression techniques such as 'Sub-Byte' quantization follow [1].

2.5 Once-for-All (OFA)

While the generation and inference of models especially designed for microcontrollers was the main objective in the previously mentioned works, the Once-for-All approach [2] is built to design optimal networks suitable for a wide range of devices. Usually this is a costly goal as training one model for each architectural setting is required. By decoupling the training and search, it is possible to get a sub network which can be then optimized in multiple dimensions like depth, width and kernel size. As a novel progressive shrinking algorithm, a generalized pruning algorithm is proposed and state-of-the-art performance especially for the mobile setting presented.

The following special CNN techniques are applied in the OFA search space generation process to create a single large Once-for-All network.

- Elastic depth: arbitrary numbers of layers per unit
- Elastic width: arbitrary numbers of channels per layer
- Elastic kernel size: arbitrary kernel sizes
- Elastic resolution: arbitrary input image sizes

Smaller sub-networks can make use of this trained OFA network by exploiting weight sharing and progressive shrinking. In the deployment stage no additional expensive training time is required due to the decoupling [2].

Considering a lot of other relevant research results on lightweight machine learning unrelated to NAS which use optimization techniques such as mixed low bitwidth quantization [17]. Implementing CNNs on MCUs with standard convolutions instead of depth-wise convolutions by trading accuracy with computational cost also offers appealing results as proposed in [18]. Another model compression method is proposed in [13] and uses a reordering of the model moderation to reduce its memory footprint.

3 Methodology

This study offers a rundown of the latest advancement in machine learning frameworks and compares the different approaches used for use cases in mobile phones. The comparison is detailed in the following paragraphs.

There are a lot of metrics to evaluate how the proposed NAS methods improve or degrade the performance of the resulting models. Often the model accuracy after deployment and the size of the generated program are primarily considered because of the given tight memory constraints in terms of program memory and SRAM usage on edge devices. If inference latency or frequency is relevant, the execution time, count of instructions or number of Multiply or Accumulate operations will be taken into account. This is a more accurate approach than counting Floating Point Operations (FLOPS) on embedded systems. It may also be helpful to evaluate metrics like training time, search cost or even the resulting energy consumption as done in paper [2].

ProxylessNAS is proposed in [3] and mainly focuses on mobile target devices. It is not included in the previous section but shortly explained here as three of the MCU-class NAS approaches mention it as a reference design. In [15] a survey on deep learning techniques, which are critical for edge intelligence implementation, statements on the performance of MCUNet, Sparse and OFA can be found. Unfortunately the table is missing important information such as the search cost.

By using the MCUNet framework instead of Tensorflow Lite Micro and CMSIS-NN the required amount of MCU-memory drops by the large factor of 3.4 while the inference time is 1.7- to 3.3-times shorter [14]. This achievement is likely due to the interpreter based inference engine used in the TFLite Micro Library which introduces a large memory overhead compared to TinyEngine's codegeneration approach. In addition the layer-level optimization strategies which are used in the MicroTVM framework as well are sub-optimal, which lead to further advantages for MCUNet [14].

Model accuracy was first evaluated by comparing top1 ImageNet results which achieved >70% on common microcontrollers. Meanwhile the required SRAM memory and flash compared to quantized MobileNetV2 and ResNet18 is 3.5 or 5.7 times smaller and especially for visual and audio wake words, which are popular reference implementations, the MCUNet framework is faster and also smaller than MobileNetV2 and ProxylessNAS-based solutions [14].

Experiments with μNAS have been conducted on image classification data sets. Either the top-1 classification accuracy can be increased by up to 4.8%, the memory footprint can be lowered by 4–13 times or alternatively a reduction of the number of multiply-accumulate operations by $\approx 900\times$ is possible according to [12]. Further interesting numbers can be found in a table of pareto-optimal architectures as it can be seen that μNAS outperforms SpArSe in multiple character classification.

SpArSe achieves to provide a more accurate and up to $4.35\times$ smaller models compared to previous approaches on IoT data sets. Unfortunately no comparisons with either of the other papers are available, hence why performance on standard datasets and the Bonsai-Net paper [8] are the only references in this case. Character classification with SpArSe turned out to have similar higher accuracy or drastically lower memory requirements and network sizes.

MicroNet models have been deployed on MCUs to compete with Tensorflow Lite Micro in experiments. Visual wake words, audio keyword spotting and anomaly detection have been compared with MobileNetV2, ProxylessNAS, Tensorflow Lite Micro and MSNET [1].

The paper also states the pareto optimality of MicroNet models against MCUNet for KWS tasks.

Using the Once-for-All tools over 1000 sub-networks suitable to run on several constrained target hardware platforms could be obtained. Regardless of the actual latency constraints it was possible to maintain the level of accuracy after the initial training. Actual numbers are provided for ImageNet top-1 accuracy under the mobile setting with sub-600 million Multiply-Accumulate operations for running the network while achieving an accuracy of 80% [2].

Even on edge devices the results are promising as either an increase of ImageNet top1 accuracy by 4.0% or a 1.5−2.6 times lower latency compared with MobileNetV3 and EfficientNet at constant accuracy level is possible. It is emphasized that the amount of GPU hours for training also drops by many orders of magnitude resulting in less CO2 emission [2].

4 Conclusion

The proposed study centres around finding the best suitable model architecture to run on edge devices with the comparisons of some core concepts of Machine Learning like NAS,

AutoML, and available Frameworks. The study leads to the application of optimization techniques to either improve the accuracy or the size of the resulting model focusing on short training times.

To summarize the results of the recent work on Neural Architecture Search for edge devices, it is easy to see that it is a highly relevant research topic which offers a lot of innovation for the future AI industry.

The biggest improvements compared to reference models can be acquired if the NAS algorithm is co-designed with the corresponding inference engine. The reason for this is that a larger search space provides more opportunities to find optimal designs, due to faster model invocation.

All approaches are published under an open-source license which should allow other researchers to build up on these work to obtain even better results. For example Tensorflow Lite Micro, which is the most relevant microcontroller inference framework, could profit a lot if NAS would be integrated in the design workflow. Beside the reusability improvement this could also make it easier to gather data for comparisons with other approaches.

Recognizing upcoming work with the employment of the dataset of some classification problems a test case can be presented and comparing them on different performance measures.

References

1. Banbury, C., et al.: MicroNets: neural network architectures for deploying TinyML applications on commodity microcontrollers (2020)
2. Cai, H., Gan, C., Han, S.: Once for all: train one network and specialize it for efficient deployment. arXiv, pp. 1–15 (2019)
3. Cai, H., Zhu, L., Han, S.: ProxylessNAS: direct neural architecture search on target task and hardware. arXiv, pp. 1–13 (2018)
4. Chu, X., Zhang, B., Xu, R., Li, J.: FairNAS: rethinking evaluation fairness of weight sharing neural architecture search (2019)
5. David, R., et al.: TensorFlow Lite Micro: embedded machine learning on TinyML systems (2020)
6. Deng, J., Dong, W., Socher, R., Li, L.-J., Li, K., Fei-Fei, L.: ImageNet: a LargeScale hierarchical image database. In: CVPR 2009 (2009)
7. Fedorov, I., Adams, R.P., Mattina, M., Whatmough, P.N.: SpArSe: sparse architecture search for CNNs on resource-constrained microcontrollers. arXiv, pp. 1–26 (2019)
8. Geada, R., Prangle, D., McGough, A.S.: Bonsai-Net: Oneshot Neural Architecture Search via differentiable pruners (2020)
9. He, X., Zhao, K., Chu, X.: AutoML: a survey of the state-of-the Art. arXiv, (Dl) (2019)
10. Krizhevsky, A., Nair, V., Hinton, G.: Cifar-10 (Canadian Institute for Advanced Research)
11. Lai, L., Suda, N., Chandra, V.: CMSIS-NN: efficient neural network Kernels for arm cortex-M CPUs. arXiv, pp. 1–10 (2018)
12. Liberis, E., Dudziak, L., Lane, N.D.: μNAS: constrained neural architecture search for microcontrollers (2020)
13. Liberis, E., Lane, N.D.: Neural networks on microcontrollers: saving memory at inference via operator reordering. arXiv, pp. 1–8 (2019)
14. Lin, J., Chen, W.-M., Lin, Y., Cohn, J., Gan, C., Han, S.: MCUNet: tiny deep learning on IoT devices. (NeurIPS), pp. 1–12 (2020)

15. Liu, D., Kong, H., Luo, X., Liu, W., Subramaniam, R.: Bringing AI to edge: from deep learning's perspective, pp. 1–23 (2020)
16. Molchanov, P., Mallya, A., Tyree, S., Frosio, I., Kautz, J.: Importance estimation for neural network pruning. In: Proceedings of the IEEE Computer Society Conference on Computer Vision and Pattern Recognition (2019)
17. Rusci, M., Capotondi, A., Benini, L.: Memory-driven mixed low precision quantization for enabling deep network inference on microcontrollers. arXiv (2019)
18. Rusci, M., Fariselli, M., Capotondi, A., Benini, L.: Leveraging automated mixed-low-precision quantization for tiny edge microcontrollers. arXiv, pp. 1–12 (2020)

Hand Anatomy and Neural Network Based Recognition of Isolated and Real-Life Words of Indian Sign Language

Akansha Tyagi(✉) and Sandhya Bansal

Department of Computer Science and Engineering, Maharishi Markandeshwar (Deemed to be University), Mullana, Ambala, Haryana 133207, India
akan7sha@gmail.com

Abstract. The primary source of communication used by deaf-mute people is sign language. The task of automatically recognizing sign language motions is difficult. For the recognition of Indian sign language, a model based on hand anatomy and neural networks is proposed in this research. Features are extracted through Features from the Accelerated Segment Test (FAST) and Scale Invariant Feature Transform (SIFT) techniques, and hand anatomy is utilized for the selection of hand landmarks out of these detected points. Finally, training and testing of the model have been done by the neural network. Experimentation was performed on challenging and publicly available datasets for isolated single-letter gestures (alphabets and numbers) and word gestures (various real-life gestures) with different backgrounds and lighting conditions. Experimental results prove that fast and highly accurate results are achieved by the proposed model when compared with other available models. An accuracy of 98.90% has been achieved for isolated gestures and 97.58% for word gestures.

Keywords: Human-Computer Interaction · Indian Sign Language · Hand Geometry · Neural Network · FAST · SIFT

1 Introduction

Hand Gesture Recognition (HGR) serves as an interface for deaf-mute people. The use of HGR in vision-based systems motivates the study of Automatic Sign Language Recognition systems (ASLR). ASLR applies to many of the domains featured for the deaf-mute community. Even though various device-based recognition systems like sensors and gloves have been recently used, However, these types of systems require the utilization of movement sensors or hand gloves to precisely distinguish the places of the various fingers. Although these are successful and can represent pretty much every sign, they require the utilization of some exceptionally delicate equipment that everybody can't utilize and is uncomfortable to use. Vision-based recognition could be the alternative. Vision-based recognition is becoming widespread due to the significant scope of application areas found in the literature [1]. ASLR using machine learning and soft computing has been a field of interest for a long time. Scientists have utilized a few

H. Sharma et al. (Eds.): ICIVC 2022, PALO 17, pp. 204–219, 2023.
https://doi.org/10.1007/978-3-031-31164-2_17

methodologies and have made a ton of progress in preparing distinctive machine and profound learning models that can perceive signs compared to different words. However, most of the studies done are for American Sign Language (ASL). ASLR for Indian Sign Language (ISL) is still in the development phase. Moreover, ISL comprises 6,000 words commonly used in the Indian subcontinent and has no standardized ISL. ASLR of static gestures is a difficult task since it involves several processes, including the acquisition of images, image pre-processing, feature extraction, and gesture recognition. The extraction of features, in particular, has a vital function to play in the accurate recognition of gestures since it influences the performance of a classifier.

In the literature, various feature extraction techniques are proposed. In [2, 3], the authors provide a review of these algorithms and their implementation. YCbCr values are utilized to extract the hand in [4]. The Discrete Wavelet Transform (DWT) was utilized to extract features, which were then passed through the Hidden Markov Model (HMM). In addition, K-nearest Neighbor (K-NN) is employed to classify the data. In [5], the user's hand is retrieved using contour representation, though with a glove on. The classification of ASL digits 0–9 was done using ANN. Experimental results show an accuracy of 90% for gesture recognition. Furthermore, as in [6] a hierarchical centroid method is used to extract features from isolated character-level ISL gestures, and k-nearest neighbour and Naive Bayes classifiers are used for classification. An accuracy of 97% was achieved. In [7], a co-articulation method is proposed to classify the British Sign Language (BSL) signs. A dataset of 1000 keywords in 1000 h of video is also created to localize the sign-instances keywords automatically. Further, [8] proposed a vision-based SLR system for 2000 words/glosses using two deep learning models. One is based on visual appearance, and the other is based on a 2D human pose. At the top-10 words, the proposed model had a precision of 62.63%.

The author used a hybridized Scale Invariant Feature Transform (SIFT) with adaptive thresholding and Gaussian blur for feature extraction [9]. To recognize ISL motions, a Convolutional Neural Network (CNN) was deployed. A deep learning-based hand pose, an aware model for RGB and depth SLR videos has also been demonstrated, in addition to [10]. The Histogram of Oriented Gradients (HOG) is used as a feature extractor in [11], while K-NN is used as a classifier. The power of CNN was used [12] to recognize hand motions in RGB images. On the NUS-II complex dataset and the ASL fingerspelling dataset, the proposed architecture is evaluated. A uniform architecture, StaDNet, is proposed in [12]. Two novel approaches to ASL recognition were proposed in [13]. The first approach uses a hybrid of Speed Up Robust Features (SURF) and Hu moments invariant methods for accurate and fast recognition of gestures. The second approach uses the hidden Markov model for the recognition of fingerspelled words. The problem with the above-proposed techniques is that they produce a large feature vector length. To reduce this length and to keep only relevant and reduced feature vectors, hand kinematics may play an important role. In [15], a novel method for hand landmark detection and localization in RGB images is proposed. Hand transformation and template matching are used for the analysis of skin-masked directional images. Both the contour and the parameters of the hand mask detect landmarks. The localization error of the landmark is used to recognize it. Furthermore, independent of the context, [16] provides a system based on landmark detection in RGB images. For hand gesture recognition, point-based features, distance features, angle features, geometric features, and whole-hand-based features are used. These parameters are optimized using the Grey Wolf metaheuristic. For classifying and recognizing gestures, a reweighted genetic algorithm is used.

From the mined literature, it is clear that commonly used methods for ASLR are deep learning, computer vision techniques, and hand landmark detection. However, deep learning methods are computationally expensive and require a lot of training datasets, while computer vision techniques generate a huge feature vector that is used as a feature extractor [17] and cannot work for real-life gesture recognition. Moreover, SIFT and SURF methods are stable most of the time, but they are slow [18]. Features from the Accelerated Segment Test (FAST) are fast, real-time feature descriptors [19]. Although the hybridization of FAST and SIFT (F-SIFT) for gesture recognition generates features fast, it sometimes produces useless features [20]. Based on these observations, in this paper, a novel approach that combines hand kinematics, SIFT, FAST, and neural networks (NN) is proposed for the recognition of ISL. The main contributions of this paper are:

(1) The paper proposes a fast and accurate hybrid recognizer that combines the advantages of FAST, SIFT, NN, and hand kinematics for the recognition of ISL sign and word datasets.
(2) FAST-SIFT was used to detect keypoints, which decreases time complexity and improves the efficiency of state-of-the-art approaches.
(3) These key points are then localized using the geometrical methods, making the model effective by removing the redundant features.
(4) The NN with 3 hidden layers is implemented for the classification of gestures.
(5) The proposed method is tested for challenging and publicly available datasets for isolated single-letter gestures (alphabets and numbers) and ISL word gestures (various real-life gestures) with different backgrounds and lighting conditions for the ISL and Arabic languages.

Based on the hand anatomy, the proposed model overcomes the limitation of gesture recognition in complex backgrounds. Hence, it will also be helpful in recognition of various regional sign languages such as Bengali Sign language, Bangalore sign language, Hyderabad sign language, Delhi sign language, Punjab sign language and Chennai sign language [41].

The remaining sections of the paper are organized as follows: The proposed methodology and algorithm are described in depth in Sect. 2. The dataset utilized throughout the experiment is described in Sect. 3. The experimental results achieved using the proposed approach are presented in Sect. 4. Section 5 concludes with a conclusion and suggestions for future development.

2 FiST_HGNN

The proposed model is categorized into four stages. In the first phase, the palm is detected from the RGB image. After palm detection, key points are detected using the F-SIFT [17] approach in the second phase. Hand anatomy is utilized for the localization of 21 key points out of the features generated after phase 2 because the proposed model is based on the observation that hand gestures can be recognized by locating hand joints and angles. Finally, each gesture is classified using a trained NN in the classification phase (phase 4). The model's overall architecture is depicted in Fig. 1.

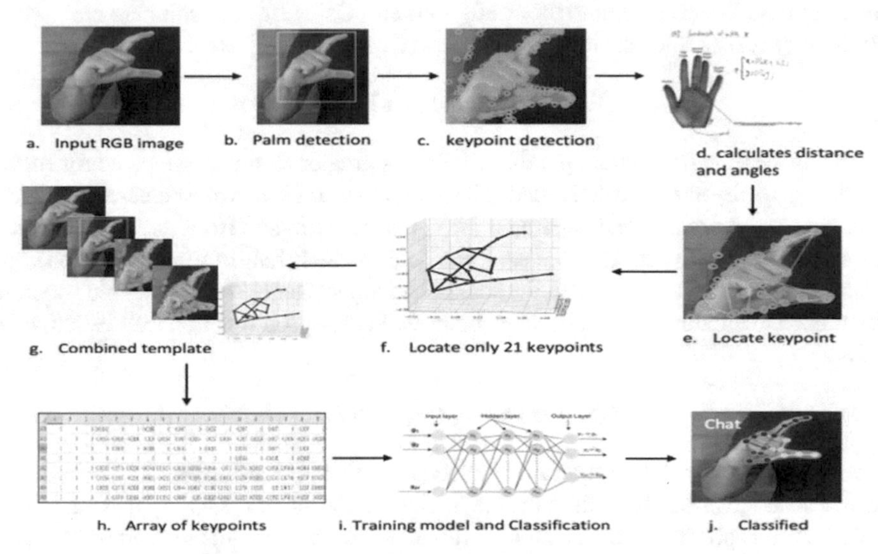

Fig. 1. Overall architecture of the proposed model.

The detailed description of the proposed model is as follows:

2.1 Palm Detection

In this phase, the input is the RGB image of a hand, and the output is the palm from the image. To extract the hand from the image, an ordinary rectangular frame is used. The basic principle is that an object in an image has the same intensity of pixels in a particular region. The non-max suppression concept is used to create the rectangular frame around the hand region of interest. Using grid intersection and union, it analyzes neighboring rectangular areas at a particular position in a detection window, adds up the pixel intensities of each region, and determines the difference between these sums [40]. An image is divided into $x * x$ grids; the value of x is calculated based on the image size. This process is repeated several times till the hand part is extracted for all the images in the dataset. This is the first stage that takes all the image's pixels as input. The equation can be formulated as in Eq. (1):

$$I_m = (P_x, f(x), x \in \{1, \ldots\ldots\ldots, x - 1\}) \tag{1}$$

where I_m is the image output of the phase, P_x denotes the pixel value of the image, and $f(x)$ are the features extracted from the image x. This phase enables us to infer the part, making it easier to detect the key points.

2.2 Keypoints Detection

The next step is to detect keypoints and remove the redundant pixels. For each detected point, a scoring function v is computed, which is the sum of the absolute difference between pixels in the center and the contiguous arc. Using Eq. 2 to compare the v values of two neighboring interest sites [38], the lower values are rejected.

$$I_m(x, y) = \{0, \quad v(x, y) \leq \tau 1, \quad v(x, y) > \tau \tag{2}$$

where τ is the threshold value, pixels with values greater than τ are selected for further processing, while others are discarded. The next step is to localize the detected keypoints. Scale spaces are created and separated into octaves produced from the convolution of the gaussian kernel at different scales. Each octave size is half of the previous one, and the number of octaves depends on the scale of an input image. A Gaussian blur operator is then used to create the blurred image octaves. Scale-space function [39] is defined as Eq. (3)-

$$L(x, y, \sigma) = G(x, y, \sigma) * I_m(x, y) \tag{3}$$

where * is the convolution operator and $G(x, y, \sigma)$ is a variable-scale Gaussian. To find the scale-invariant keypoints the Laplacian of Gaussian (LOG) approximations is applied. At this phase, the final output is the image with keypoints located on the hand. The output of this stage can be formulated as in Eq. (4):

$$I_m = \sum_{k_p=1}^{k_p+1} \left\| P_x^{k_p} - L^{k_p} \right\|_2 \tag{4}$$

where k_p denotes the keypoints, $P_x^{k_p}$ denotes the pixel with higher threshold values, whereas L^{k_p} denotes the pixels with lower threshold values. $\left\| P_x^{k_p} - L^{k_p} \right\|_2$ denotes the normalization of the pixels, with a lower bound of 2. The output of this stage is the image with keypoints located on the higher intensity pixels.

2.3 Keypoint Localization

After detecting keypoints, a two-dimensional array of coordinates, where each coordinate is produced corresponds to one of the keypoints in hand. The location of the k_p keypoint is denoted by $P_{kp} = \left(x_{kp}, y_{kp}\right)$, and the location of the wrist is denoted as $P_1 = (x_1, y_1)$. To locate the wrist coordinate, the distance between two extreme points (w_x and w_y) is calculated, as shown in the below Eq. (5).

$$w_{xy} = \frac{\theta\left(w_x + w_y\right)}{2} \tag{5}$$

where w_{xy} is the wrist center point, θ is the angle associated with the wrist. w_x and w_y are the wrist coordinates at x and y coordinates, respectively. Further, a template is created for each joint using the method mentioned by the author in [21]. The distance of each joint is calculated using the formulated Eq. (6).

$$D_j = \{\frac{x_{kp} + y_{kp} + \sqrt{\emptyset}}{2(x_{kp}, y_{kp}) + w \quad otherwise} \text{ when } \emptyset \geq 0 \tag{6}$$

$$\emptyset = 2w^2 - (x_{kp} - y_{kp})^2 \tag{7}$$

where, x_{kp} and y_{kp} is the distance in x and y coordinates, respectively, $x_{kp} = (D_{x+1,m}, D_{x-1,m})$, and $y_{kp} = (D_{y,n+1}, D_{y,n-1})$, $\emptyset$ denotes the angle for the wrist corresponding to x and y axis.

All coordinates of these keypoints are subtracted by the coordinates of the hand wrist to verify that features are invariant to shifting. It indicates that rather than using absolute positions, relative positions are used. The transformation of the coordinates can be extracted by Eq. (8).

$$P_{kp} = P_{kp} - P_1, k_p \in \{2, \ldots\ldots.., 21\} \tag{8}$$

Here P_{kp} is a 2D vector. After this modification, the wrist coordinate is aligned to the coordinates' origin. To verify that the features are invariant to scaling, the coordinates of these keypoints are all scaled by the maximum norm value using the Eq. (9) below.

$$I_m = ||P_{kp}||_2 \tag{9}$$

$$P_{kp} = \frac{P_{kp}}{||P_{kp}||_2}, k_p \in \{2, \ldots\ldots.., 21\} \tag{10}$$

where $||P_{kp}||_2$ implies the norm distance between them P_{kp} and the origin with the lower bound value of 2. The norms of these coordinates are all smaller than or equal to 1 after scaling, which aids in the normalization of features. Then, ignoring the wrist coordinates, these coordinates are concatenated one by one to form a single feature vector with a length of $20 * x$. The classifier is then fed the processed feature vector for training and testing.

2.4 Classification and Prediction

A collection of training samples is provided with labels, assuming that in the feature space, these features will cluster around multiple centers. The proposed model is a feedforward with four hidden layers. It has sixty-three neurons in the input layer, 128, 64, 32, 16, for the first, second, third and fourth hidden layer respectively. Twenty-five neurons in the output layer, as shown in Fig. 2. Sixty-three neurons in the input layer correspond to the vector, k_p for each gesture, g_s category.

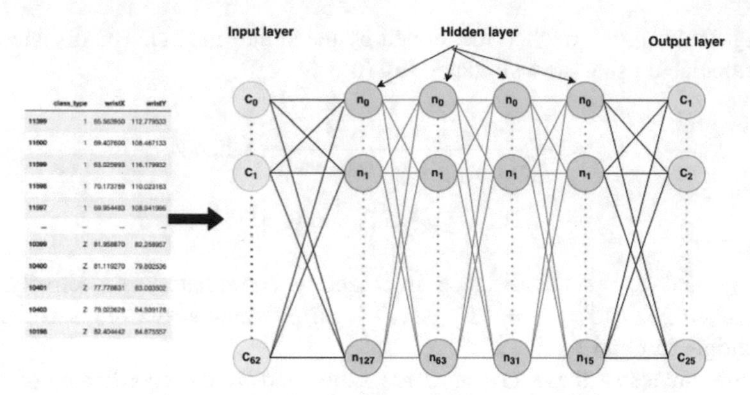

Fig. 2. Architecture of the Neural Network

Since the approach, NN is a multi-layer network with n layers ($n \geq 3$) and the model has x^n neurons at the last layer. Let g^1 the neuron of the input layer and the g^n is the neuron of the nth layer. Considering also that there are n neurons at each layer and the desired output is Y. The activation function σ_g for each layer can be different. The difference between the current node output and the desired output is the error function to minimize, which can be formulated as:

$$e = \frac{1}{4}\sum_{n=1}^{n^m} (Y_n - g_n^m)^2 \tag{11}$$

where n^m denotes the number of neurons, n in a layer m. To find the error function e, the backpropagation algorithm is applied. The weights in the model are initialized randomly. The gradient error function is computed concerning the weights (w) and bias z, to modify the initial weights. The output, Y_n^m of neuron n in a layer m is formulated as:

$$g_n^m = \sigma_g\left(y_n^m\right) = \sigma_g\left[\sum_{i=1}^{n^{i-1}} w_{n,i}^m g_i^{m-1} + z_n^m\right] \tag{12}$$

where $\sigma_g\left(y_n^m\right)$ is the activation function of a neuron n for layer m.

The weights for the model can be modified according to the formulated equation:

$$w_{n,i}^m(m+1) = w_{n,i}^m(n) + \Delta w_{n,i}^m \tag{13}$$

$$\Delta w_{n,i}^m = -\alpha \frac{\partial e}{\partial w_{n,i}^m} = \alpha \delta_n^m g_n^{m-1} \tag{14}$$

where α is the learning rate $l_{r[0,1]}$, the bias is given as:

$$\Delta z_n^m = \alpha \delta_n^m \tag{15}$$

The δ is formulated as:

$$\delta_n^m = \begin{cases} (Y_n - g_n^m)\sigma_g(y_n^m) \ \ if\ \ m = n \sum_{i=1}^{n^{i+1}} (\delta_n^{i+1} w_{n,i}^{i+1})\sigma_g(y_n^m) \quad if\, 4 \le i \le n \end{cases} \tag{16}$$

The activation function used is sigmoid. The equation for the activation function is:

$$\sigma(y) = \frac{1}{1 + e^{-y}} \tag{17}$$

and its derivatives can be given as:

$$\sigma(y) = \sigma(y)\left[1 - \sigma(y)\right] \tag{18}$$

The $l_{r[0,1]}$ is used to reduce the uncertainty of the neurons and increase the speed of the training. Let's assume that R denotes the iteration in the training, then the updated weight value for any node mapping can be formulated as:

$$\omega_R = \omega_{R-1} + \Delta\omega + l_r(\omega_{R-1} - \omega_{R-2}) \tag{19}$$

With a batch size of 128, $\alpha = 0.1$ and $l_r = 0.1$ the model was trained for 20 epochs. All parameters gained during the training phase are designated as Q. The probability can be calculated as $p(Q)$ giving a testing sample x_t. And further, the approval or rejection of the testing image as an actual gesture image is determined by using the formulated equation:

$$p(Q) > \tau \tag{20}$$

This gesture is classified as a non-gesture pattern if the prospect is less than a certain threshold value τ, otherwise; it is classified as an aiming gesture. The probability for every single image can be calculated as $p(x_t)$. An algorithm for detecting and locating keypoints is explained in Algorithm 1. The algorithm for training and testing is given in Algorithm 2.

Algorithm 1: Detecting Keypoints and Locating Keypoints

Parameters for the Detection Algorithm: I_m = image, P_{kp} : pixel value in the image, D_j: distance between the center of two adjacent features, τ: threshold.

1. $Load\ dataset;$
2. $An\ RGB\ image\ I_m = I^{h*w*3}$, h *is the height*, w *is the width and* 3 *is the RGB channel value.*
3. $Function\ DETECT_{KEYPOINTS}(Min\_X, Min\_Y, Max\_X, Max\_Y):$
4. $I_m \leftarrow 1;$
5. $for\ I_m := 1\ to\ x\ do$
6. $P_{kp} \leftarrow locate(Min\_X, Min\_Y, Max\_X, Max\_Y);$
7. $if\ keypoints\_Locate\ (Min\_X, Min\_Y, Max\_X, Max\_Y)\ \ then$
8. $k \leftarrow Determine\_Region$
9. $Add\_Element(idx\_X, idx\_Y)$
10. $idx\_X \in [Min\_X, Max\_X] \forall P_{kp};$
11. $idx\_Y \in [Min\_Y, Max\_Y] \forall P_{kp};$
12. $Locate\_Keypoints\ kp$
13. $if\ ||kp|| > \tau; \qquad where\ \tau = 0.5$
14. $then$
15. $getangle \leftarrow kp\ in\ range(idx\_X, idx\_Y);$
16. $else$
17. $getdistt \leftarrow locate(D_j);$
18. $endif$
19. $if\ P_{kp} \neq P_{kp+1}\ then$
20. $Add\ elements(P_{kp}, P_{kp+1});$
21. $kp \leftarrow kp + 1;$
22. $end\ for$
23. $return\ keypoints\ P_{kp};$
24. $end\ function;$

Algorithm 2: Training and Testing of model:
Parameters for training and Testing model-, k_i: kernel size, s_i: stride, p_i: padding, n_i: number of features, r_i: receptive field of a feature in the layer.

1. *Load the dataset* Q_l *and divide it in ratio* 70:30 *i.e.* Q_{tr} *and* Q_{te};
2. *Function GESTURE_PREDICTION*(G_s)
3. *for layer* (l_m)*with* n_0 = *image size*; $\alpha = 0.1$; $l_r = 0.1$;
4. *layer.append(mappingLayer)*;
5. *return mlmodel.save('Keypoint_detector.mlmodel')*;
6. *for each* $Q_{tr} \in Q$;
7. *m = Keypoint_detector.mlmode*(W_i, N_d);
8. *m.load (Keypoint_detector.mlmode)*;
9. *kp* (x, y, z) = *(kp Array*[x], *kp Array*[y], *kp Array*[z], 0.0);
10. *Find Error*;
11. $e = \frac{1}{4}\sum_{n=1}^{n^m} (Y_n - g_n^m)^2$;
12. $M = arg|(x_k | y_k = g, x_k \in X_k)|$
13. *Evaluate Accuracy*$(Q_{tr}$ *and* $Q_{te})$;
14. *predicted gesture* = $G_s(Q_{tr} : Q_{te})$;
15. *else*;
16. *end for*;
17. *return predicted gesture*;
18. *end function*;

3 Dataset Description

To validate the performance of the FiST_HGNN, two types of datasets are considered. The first category consists of isolated single-letter gestures, and in the second category, we considered ISL-two hand gestures and words under uniform and complex backgrounds. Isolated hand-letter signs, gestures representing alphabets and digits from the ISL, Arabic language (ArSL) [23], and NUS-II [24] datasets are considered in isolated gestures. Further, two hand ISL alphabets and numbers [27], sign word [25] and Wadhawan, A. & Kumar P [13] are considered in ISL word category. However, the ArSL [23] and NUS-II [24] are the benchmark dataset, which are commonly used for SLR systems. A detailed description of the used dataset is given in Table 1.

4 Experimental Results

The neural network is used to train the model. The system features an Intel® core i5 processor running at 1.8 GHz, 8 GB of RAM, and 256 caches per core for a total of 3MB cache. The dataset is shuffled using the random-seed method and divided into the ratio of 70:30 as training and testing, respectively. The dataset is shuffled to prevent biases of the model towards certain features. The 21-keypoints are chosen with the help of mediapipe tool present in python.

4.1 Performance Metrics

The performance of the FiST_HGNN is evaluated based on accuracy and time taken by the model. Accuracy gives us the percentage of all correct predictions out of total

Table 1. Description of Dataset

Category	Dataset	Background	Total Images	Gesture	Classes	Samples
Isolated single letter gestures	ISL alphabets [21] and digits [22]	Uniform	6,988	Alphabets except for J and Z, digits	34	A 5
	ArSL [23]	Uniform	54094	Arabic alphabets	32	
	NUS-II [24]	Complex	2750	Different Gestures	10	
ISL Word gestures	ISL alphabets and numbers[27]	Uniform	42000	All alphabets and numbers(1 to 9)	35	B C
	Sign-Word [26]	Complex	18,000	Real-life gestures like call, cold, yes, wash, etc.	20	CALL CLOSE
	Wadhawan, A. & Kumar P. [13]	Complex	1,400	Afraid, Doctor, Pray, Sick.	4	

predictions made as follows.

$$Accuracy = \frac{TP + TN}{TP + FP + TN + FN} \tag{21}$$

Time taken by the model is the total time taken for training and testing the gestures.

4.2 Accuracy Comparison

Figure 3 shows the accurate comparison on isolated single letter gestures. From Fig. 3 it is clear that the proposed model FiST_HGNN achieves 4.63% and 3.62% better results than FiST_CNN [17] and Pointbased+fullhand [15], respectively, on ISL-alphabet & digits. Similarly, for the NUS-II dataset, an improvement of 9.25% and 7.18% is achieved over the other two approaches. For the ArSL dataset, 5.63% and 4.57% improved results are obtained by the FiST_HGNN. It is also clear that the proposed approach works well for both uniform and complex gestures. Figure 3 shows the accuracy achieved for ISL Words and gestures that are used in real life. Figure 4 demonstrates that the results achieved by the FiST_HGNN are better on the sign word dataset [25] than both the compared approaches, but it achieves relatively fewer results on the Ankita Wadhawan [13] dataset, this is because the compared approach uses a 2 CNN model for gesture recognition and hence increases the complexity of the model when compared with the FiST_HGNN, which classifies the gestures with 3 hidden layers. Moreover, it also works with just relevant features (21) because of the hand anatomy technique.

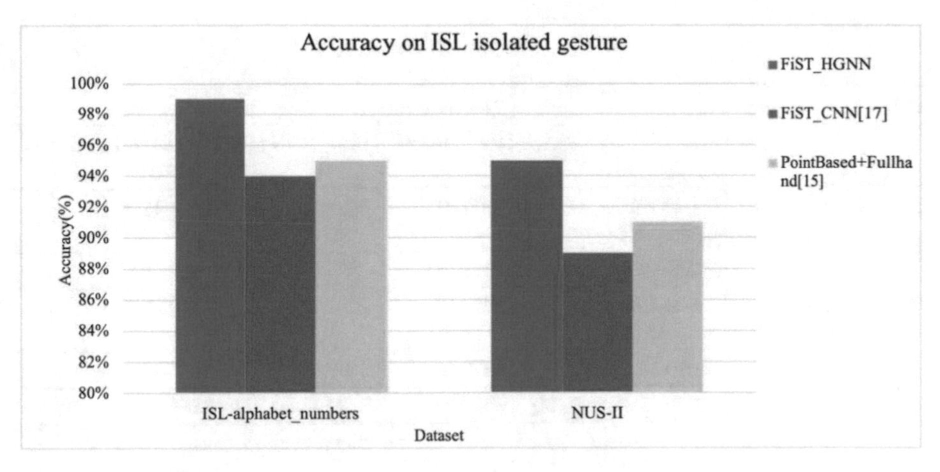

Fig. 3. Accuracy Comparison of Isolated single letter gestures

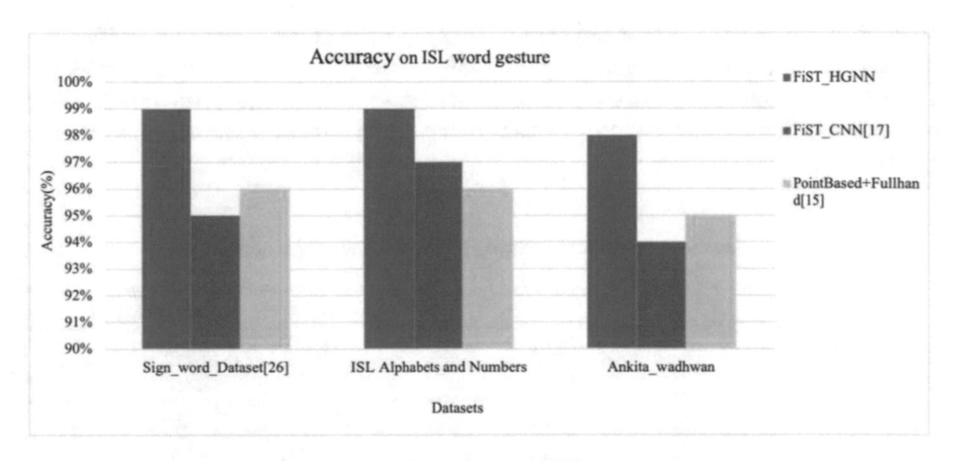

Fig. 4. Accuracy Comparison of ISL word gestures

4.3 Time Comparison

The time taken by each model for training and testing on the different datasets is presented in Table 2. The proposed method recognizes gestures relatively quickly, as shown in the table, attributable to the FAST technique's fast detection and localization of keypoints. Locating the key points in images improves the classification accuracy, as the model is trained only on the necessary coordinates, making the system more efficient and effective. Moreover, another reason for this is that reducing the training parameters also reduces the number of layers and mapping nodes, reducing the time complexity of the model. All these factors make the methods quite effective.

Table 2. Comparison based on time

Dataset		Time (second)		
		FiST_HGNN	FiST_CNN [17]	PointBased+Full Hand [15]
Isolated Single Letter	ISL alphabets [22] and digits [23]	3217.43	4628.34	3924.45
	NUS-II [24]	778.54	889.56	990.76
ISL Word gestures	ISL alphabets and numbers [27]	2287.67	2987.90	3130.87
	Sign-Word [26]	3278.65	5678.89	3657.34
	Ankita Wadhwan [14]	1543.22	3189.43	2164.59

4.4 Comparison to the Other State of Art Algorithms

Table 3 shows the accuracy compared with other state of art algorithms in literature for isolated single letter gestures. From Table 3, the NUS-II [24] dataset achieves a better result on the FiST_HGNN. For the ArSL dataset, it shows relatively lower results but with less complexity. For ISL alphabets and digits dataset it outperforms all other approaches.

Table 3. Accuracy comparison with other approaches

Dataset	Author	Classifier	Accuracy (%)
NUS-II [24]	[27]	SVM	92.50
	[28]	SVM	92.5
	[29]	SVM	94.36
	[30]	SVM	94.6
	FiST_HGNN	NN	95.78
ArSL [23]	[31]	CNN	98.06
	[32]	CNN	99.57
	[33]	CNN	97.6
	FiST_HGNN	NN	95.65
ISL alphabets [21] and digits [22]	[34]	NN	63.78
	[27]	SVM	90
	[35]	SVM	93.4
	[36]	SVM	91.3
	[37]	ANN	90
	FiST_HGNN	NN	97.58

5 Conclusion

A model for the recognition of ASLR has been established in this research. The proposed approach uses the hybridization of computer vision techniques FiST_CNN, hand anatomy, and NN for the recognition of ISL gestures. The FAST technique provides the rapid detection of keypoints. Relevant keypoints are selected using hand anatomy. As a classifier, NN is used. The proposed method was challenged on two types of ISL datasets (single letters and words). The proposed approach achieves an accuracy of 98.90% for isolated gestures and 97.58% for word gestures and outperforms the other approaches in the literature. In future work can be extended to improve the recognition of gestures where the hand overlaps with each other. The model will be extended to dynamic gestures and daily-life gestures.

References

1. Tyagi, A., Bansal, S.: Feature extraction technique for vision-based indian sign language recognition system: a review. In: Computational Methods and Data Engineering, pp. 39–53 (2021)
2. Adeyanju, I.A., Bello, O.O., Adegboye, M.A.: Machine learning methods for sign language recognition: a critical review and analysis. Intell. Syst. Appl. **12**, 200056 (2021)
3. Tyagi, A., Bansal, S., Kashyap, A.: Comparative analysis of feature detection and extraction techniques for vision-based ISLR system. In: 2020 Sixth International Conference on Parallel, Distributed and Grid Computing (PDGC), pp. 515–520. IEEE, November 2020
4. Candrasari, E.B., Novamizanti, L., Aulia, S.: Discrete Wavelet Transform on static hand gesture recognition. J. Phys. Conf. Ser. **1367**(1), 012022 (2019). IOP Publishing
5. Jalal, A., Khalid, N., Kim, K.: Automatic recognition of human interaction via hybrid descriptors and maximum entropy markov model using depth sensors. Entropy **22**(8), 817 (2020)
6. Sahoo, A.K., Sarangi, P.K., Gupta, R.: Indian sign language recognition using a novel feature extraction technique. In: Sharma, T.K., Ahn, C.W., Verma, O.P., Panigrahi, B.K. (eds.) Soft Computing: Theories and Applications. AISC, vol. 1380, pp. 299–310. Springer, Singapore (2022). https://doi.org/10.1007/978-981-16-1740-9_25
7. Albanie, S., et al.: BSL-1K: scaling up co-articulated sign language recognition using mouthing cues. In: Vedaldi, A., Bischof, H., Brox, T., Frahm, J.-M. (eds.) ECCV 2020. LNCS, vol. 12356, pp. 35–53. Springer, Cham (2020). https://doi.org/10.1007/978-3-030-58621-8_3
8. Li, D., Rodriguez, C., Yu, X., Li, H.: Word-level deep sign language recognition from the video: A new large-scale dataset and methods comparison. In: Proceedings of the IEEE/CVF Winter Conference on Applications of Computer Vision, pp. 1459–1469 (2020)
9. Dudhal, A., Mathkar, H., Jain, A., Kadam, O., Shirole, M.: Hybrid SIFT feature extraction approach for Indian sign language recognition system based on CNN. In: Pandian, D., Fernando, X., Baig, Z., Shi, F. (eds.) ISMAC 2018. LNCVB, vol. 30, pp. 727–738. Springer, Cham (2019). https://doi.org/10.1007/978-3-030-00665-5_72
10. Mahmud, I., Tabassum, T., Uddin, M. P., Ali, E., Nitu, A.M., Afjal, M.I.: Efficient noise reduction and HOG feature extraction for sign language recognition. In: 2018 International Conference on Advancement in Electrical and Electronic Engineering (ICAEEE), pp. 1–4. IEEE, November 2018

11. Adithya, V., Rajesh, R.: A deep convolutional neural network approach for static hand gesture recognition. Procedia Comput. Sci. **171**, 2353–2361 (2020)
12. Mazhar, O., Ramdani, S., Cherubini, A.: A deep learning framework for recognizing both static and dynamic gestures. Sensors **21**(6), 2227 (2021)
13. Wadhawan, A., Kumar, P.: Deep learning-based sign language recognition system for static signs. Neural Comput. Appl. **32**(12), 7957–7968 (2020). https://doi.org/10.1007/s00521-019-04691-y
14. Rekha, J., Bhattacharya, J., Majumder, S.: Hand gesture recognition for sign language: A new hybrid approach. In: Proceedings of the International Conference on Image Processing, Computer Vision, and Pattern Recognition (IPCV), p. 1. The Steering Committee of the World Congress in Computer Science, Computer Engineering and Applied Computing (WorldComp) (2011)
15. Grzejszczak, T., Kawulok, M., Galuszka, A.: Hand landmarks detection and localization in color images. Multimedia Tools Appl. **75**(23), 16363–16387 (2015). https://doi.org/10.1007/s11042-015-2934-5
16. Ansar, H., Jalal, A., Gochoo, M., Kim, K.: Hand gesture recognition based on auto-landmark localization and reweighted genetic algorithm for healthcare muscle activities. Sustainability **13**(5), 2961 (2021)
17. Tyagi, A., Bansal, S.: Hybrid FiST_CNN approach for feature extraction for vision-based indian sign language recognition. Int. Arab J. Inf. Technol. **19**(3), 403–411 (2022)
18. O'Mahony, N., et al.: Deep learning vs. traditional computer vision. In: Arai, K., Kapoor, S. (eds.) Advances in Computer Vision, CVC 2019. AISC, vol. 943, pp. 128–144. Springer, Cham (2020). https://doi.org/10.1007/978-3-030-17795-9_10
19. El-Gayar, M.M., Soliman, H.: A comparative study of image low level feature extraction algorithms. Egypt. Inform. J. **14**(2), 175–181 (2013)
20. Csóka, F., Polec, J., Csóka, T., Kačur, J.: Recognition of sign language from high resolution images using adaptive feature extraction and classification. Int. J. Electron. Telecommun. **65** (2019)
21. Chen Chen, F., Appendino, S., Battezzato, A., Favetto, A., Mousavi, M., Pescarmona, F.: Constraint study for a hand exoskeleton: human hand kinematics and dynamics. J. Robot. **2013** (2013)
22. https://github.com/imRishabhGupta/Indian-Sign-Language-Recognition
23. https://www.kaggle.com/ardamavi/sign-language-digits-dataset
24. Latif, G., Mohammad, N., Alghazo, J., AlKhalaf, R., AlKhalaf, R.: ArASL: Arabic alphabets sign language dataset. Data Brief **23**, 103777 (2019)
25. www.ece.nus.edu.sg/stfpage/elepv/NUS-HandSet/
26. https://www.u-aizu.ac.jp/labs/is-pp/pplab/swr/sign_word_dataset.zip.
27. https://drive.google.com/drive/folders/1keWr7-X8aR4YMotY2m8SlEHlyruDDdVi
28. Kaur, B., Joshi, G., Vig, R.: Indian sign language recognition using Krawtchouk moment-based local features. Imaging Sci. J. **65**(3), 171–179 (2017). https://doi.org/10.1080/13682199.2017.1311524
29. Adithya, V., Rajesh, R.: An efficient method for hand posture recognition using spatial histogram coding of NCT coefficients. In: 2018 IEEE Recent Advances in Intelligent Computational Systems, RAICS 2018, pp. 16–20 (2019). https://doi.org/10.1109/RAICS.2018.8635066
30. Pisharady, P.K., Vadakkepat, P., Loh, A.P.: Attention based detection and recognition of hand postures against complex backgrounds. Int. J. Comput. Vision **101**(3), 403–419 (2013). https://doi.org/10.1007/s11263-012-0560-5
31. Vishwakarma, D.K.: Hand gesture recognition using shape and texture evidences in complex background. In: 2017 International Conference on Inventive Computing and Informatics (ICICI), pp. 278–283. IEEE, November 2017

32. Elsayed, E.K., Fathy, D.R.: Sign language semantic translation system using ontology and deep learning. Int. J. Adv. Comput. Sci. Appl. **11**(1), 141–147 (2020). https://doi.org/10.14569/ijacsa.2020.0110118
33. Saleh, Y., Issa, G.F.: Arabic sign language recognition through deep neural networks fine-tuning. Int. J. Online Biomed. Eng. **16**(5), 71–83 (2020). https://doi.org/10.3991/IJOE.V16I05.13087
34. Ansari, Z., Harit, G.: Nearest neighbour classification of Indian sign language gestures using kinect camera. Sadhana **41**(2), 161–182 (2016). https://doi.org/10.1007/s12046-015-0405-3
35. Joshi, G., Singh, S., Vig, R.: Taguchi-TOPSIS based HOG parameter selection for complex background sign language recognition. J. Vis. Commun. Image Represent. **71**, 102834 (2020). https://doi.org/10.1016/j.jvcir.2020.102834
36. Anand, M.S., Kumar, N.M., Kumaresan, A.: An efficient framework for Indian sign language recognition using wavelet transform. Circ. Syst. **07**(08), 1874–1883 (2016). https://doi.org/10.4236/cs.2016.78162
37. Rao, G.A., Kishore, P.V.V.: Selfie video based continuous Indian sign language recognition system. Ain Shams Eng. J. **9**(4), 1929–1939 (2018). https://doi.org/10.1016/j.asej.2016.10.013
38. Lowe, D.G.: Object Recognition from Local Scale-Invariant Features (1999)
39. Lowe, D.G.: Distinctive image features from scale-invariant keypoints. Int. J. Comput. Vis. **60**(2), 91–110 (2004)
40. Rosten, E., Porter, R., Drummond, T.: Faster and better: a machine learning approach to corner detection. IEEE Trans. Pattern Anal. Mach. Intell. **32**(1), 105–119 (2010)
41. Wadhawan, A., Kumar, P.: Sign language recognition systems: a decade systematic literature review. Arch. Comput. Methods Eng. **28**(3), 785–813 (2019). https://doi.org/10.1007/s11831-019-09384-2

Drones: Architecture, Vulnerabilities, Attacks and Countermeasures

Noshin A. Sabuwala(✉) and Rohin D. Daruwala

Veermata Jijabai Technological Institute, Mumbai, India
{nasabuwala_p21,rddaruwala}@el.vjti.ac.in

Abstract. There is significant surge in the popularity of Unmanned Aerial Vehicles (UAVs), often called drones, with an ongoing increase in demand due to their multi-functional uses. The capacity of the drones to respond to people's wants accounts for their pervasiveness. Drones with extended functions and capabilities when supplied with communication equipment can be deployed to appropriate places in the field to supplement the public networks operate more efficiently and in vital missions such as infrastructure monitoring operations. To be effective, an unmanned aerial system (UAS) must interact securely with its network's entities, such as ground control stations, other aircraft and UAS, air traffic control systems, and navigation satellite systems. UAVs are exposed to a dangerous and costly world of cyber dangers as a result of cyber technology and connections. Information interchange between the Ground Control Station (GCS) and the UAV is dependent on communication lines, which are vulnerable to cyber attacks. As a result, detective, defensive, and preventative countermeasures are critical. This study gives a brief summary of exploitation of drones vulnerabilities and cyber-attacks divided into three groups, data interception, manipulation, and interruption. It also analyzes the countermeasures proposed by various researchers for security of UAS networks against cyber-attacks. This review would be beneficial to the readers to understand and evaluate the issues in existing strategies for cyber security of UAVs.

Keywords: Operational Technology (OT) · Unmanned Aerial Vehicle (UAV) · Drones · Security · Vulnerabilities · Cyber Attacks

1 Introduction

The reason why drones are so prevalent is because they are able respond to people's needs. Drones provide users a bird's eye view which may be triggered and utilized practically anywhere and at any time. To be effective, an unmanned aerial system (UAS) must interact securely with its network's entities, such as ground control stations, other aircraft and UAS, air traffic control systems, and navigation satellite systems. Information interchange between the GCS and the UAV is dependent on communication lines, which are vulnerable to cyber-attacks. A UAV relies on many sensors, for example, to find and compute its GPS coordinates, making sensor attacks lethal. Furthermore, these systems are

H. Sharma et al. (Eds.): ICIVC 2022, PALO 17, pp. 220–232, 2023.
https://doi.org/10.1007/978-3-031-31164-2_18

susceptible to a variety of cyber threats including message manipulation, injection, GPS spoofing, and jamming. These attacks can confuse pilots and air traffic controllers along with UAS and aircraft, resulting in crashes and/or hijacking. Security is of the utmost significance in such networks as critical information may be transferred between various network entities. As a result, detective, defensive, and preventative countermeasures are critical. In this study, discussion about drone architecture, communication architectures for networking UAVs, and examination of the cyber security issues of drone networks is done. To comprehend the possible risk and to protect Operational Technology (OT) systems, this study highlights their difference from traditional Information Technology (IT) systems. Furthermore, focus is on methodically recognising and categorising the cyber-attacks according to their types (data interception, manipulation, and interruption) and defense strategies, thus highlighting similarities and key characteristics of attacks that highlight the specific problems provided in protecting OT systems.

2 Drone Architecture

A drone architecture is made up of three components.[1]:

1. The Flight Controller is its primary unit of processing. [2].
2. GCS gives the operators on land the required functionalities to monitor and/or control UAVs from a distance throughout their operations. GCSs varies based on the kind, size, and mission of the drone.
3. The wireless links known as Data Links govern the information flow connecting the GCS and the drone which is determined by the UAV's operational range. Based on the distance from the GCS[2], drones' control can be categorized as Visual Line-of-Sight (VLOS), controlled using radio waves and Distance Beyond Visual Line-of-Sight (BVLOS), controlled through satellite communications.

3 Communication Architectures

Four communication architectures that define how data travels between the UAV and the ground station or other UAV are discussed in this section[3].

1. Centralized Network: This architecture defines GCS as the central node which connects all UAVs. GCS transmits and receives the control and command data to the UAVs as they are not directly connected to one other. The GCS acts as a relay for communication between two UAVs. This type is represented in Fig 1.A.
2. UAV Ad Hoc Network: As can be seen in Fig 1.B, in this network, a backbone UAV acts as the ad hoc network's gateway, passing data between the GCS and the other UAVs. The gateway UAV requires two radios to communicate with the GCS and other UAVs.

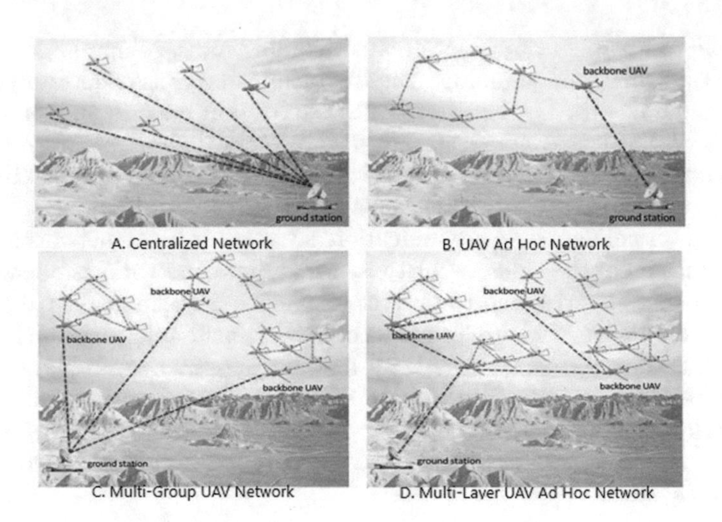

Fig. 1. Communication Architectures [3]

3. Multi-Group UAV Network: In this network, UAVs form an ad hoc network within a group connecting its backbone UAV to the GCS. Intra-group communications are conducted in a UAV ad hoc network, while inter-group communications are performed via respective backbone UAVs and the GCS. This type is illustrated in Fig 1.C.
4. Multi-Layer UAV Ad Hoc Network: Individual groups of UAVs create a UAV ad hoc network, which is the lower layer of this network. The backbone UAVs of all groups constitutes the upper-layer. Any communication between two UAV groups need not go through the GCS. The GCS simply handles information data that is sent to it as shown in Fig 1.D.

4 Drones Communications Types

The division of the drone communications, as illustrated in Fig 2, is as follows:

1. Drone-To-Drone (D2D): Peer-to-Peer (P2P) communication can be used to emulate D2D communication. As a result, it would be subject to many sorts of P2P attacks like D-DoS and sybil attacks. Such communication is not yet standardized.
2. Drone-To-Ground Control Station (D2GCS): This communication relies on wireless technologies such as Bluetooth and Wi-Fi. Most D2GCS communications are open and insecure, relying on a single factor authentication that can be readily cracked, rendering them subject to active (man-in-the-middle) and passive (eavesdropping) attacks.
3. Drone-To-Network (D2N): D2N enables network selection based on the desired level of security. It may incorporate 3 GHz, 4 GHz, 4G+ (LTE), and 5 GHz cellular communications. Such wireless communications need security.

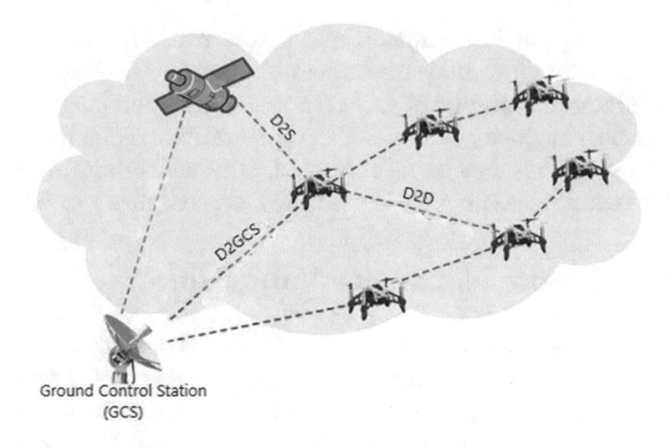

Fig. 2. Drones Communications Types [4]

4. Drone-To-Satellite (D2S): This sort of communication is required for sending real-time coordinates via the Global Positioning System (GPS). This permits any drone to be summoned back to its original station if it wanders outside the line of sight or control. Satellite communications are thought to be safe and secure. They do, however, have a high cost and demand a lot of upkeep[1].

5 IT (Information Technology) vs OT (Operational Technology) Cyber Security Differences

Every logic executed in a control system having straight influence on the physical environment necessitates that security be prioritised. Off-the-rack commercial solutions for cyber security in IT may not work well with specific applications and operating systems on OT systems, to which IT professionals are not familiar. Factors such as time-criticality, continuous availability, large physical base, restricted computation resources, social acceptance, wide interface between analog and digital signals, legacy issues, user aversion to change, and cost effectiveness make system security a unique challenge. Since completing a task after the deadline is deemed pointless and might generate a cascade impact in the actual world, OT systems are frequently hard real-time systems. Latency is particularly detrimental to the performance of an OT system and the operational time frames between an event and response from system impose strict restrictions i.e. failing to meet the deadline results in the system failing completely. Hard real-time systems differ from soft real-time systems by whether or not they meet the deadline and not by the duration of time period. On the other hand, soft real-time systems, like live audio-video systems, may tolerate some delay and respond by lowering service quality, such as losing frames when presenting a video. In soft real-time systems, minor violations of time limits result in diminished quality as opposed to system crash. Additionally, the physical nature of the activities performed by the OT system, as well as the processes inside each job, necessitates frequent interruption and restarting. The usage of traditional encryption block techniques may be impossible due to the time aspect and job interruptions. The vulnerability of OT as a real-time operating system (RTOS) is due to its crucial. Majority of the OT equipments are embedded systems that work for several years without being rebooted, thus collecting fragmentation. As a result, buffer overflow is a bigger issue in OT as compared to typical IT.[5]

6 How Are Drone Networks Vulnerable?

As UAS networks have distinct security needs than traditional networks, using protection tactics designed for other wireless networks is ineffective in UAS networks. These difficulties open fresh research opportunities in this area, necessitating the development of security approaches that take UAS-specific security requirements into account which are summarised below:

- Low latency is frequently required for data transmission between UAVs and between the GCS and the UAVs. The resource limits of UAVs and the latency constraint provide security difficulties for drone networks.
- Attackers take advantage of UAV sensor inputs and jeopardize mission's efficacy by changing settings to mislead the sensors. This might lead to authorities reacting incorrectly to false circumstances.
- Most UAVs have global positioning system (GPS) module, for example, to record its starting position before taking off to autonomously return home. Additionally, a UAV must use GPS to detect whether it is in a no-fly zone. Civil GPS has two threats: jamming and spoofing, as opposed to military encrypted GPS [6,7]. The UAV system's dependability would be severely harmed if an opponent was able to feed the module incorrect GPS signals.
- Another possible flaw arises in the communication module of UAVs. Wi-Fi or Zigbee technologies are commonly used to communicate with GCS, posing a danger of opponents intercepting or interfering with wireless transmissions. An opponent, for example, might block essential information, preventing authorities from escaping a hazardous zone. A denial-of-service attack is another conceivable hazard. A malicious attacker may keep a server occupied by flooding it with useless packets, preventing it from receiving any useful data from UAVs. The authorities may lose control of the UAVs because of this [8]. It is demonstrated in [9] that by rerouting the traffic of the telemetry channel, a Man-in-the-Middle attack may be carried out efficiently from kilometres away. The paper has demonstrated that by reverse-engineering the software used in the UAV system's communication, they can insert packets and control the UAV.
- Attacks can also wreak havoc on system recovery systems. A faulty error-handling mechanism, for example, might cause serious system failures by putting the system in an inconsistent state. Similarly, attackers can tamper with diagnostic to declare a faulty unit non-faulty and a non-defective unit faulty, isolate non-faulty units instead of faulty ones, tamper with reconfiguration to turn off non-faulty units and turn on faulty ones, and soon. [10]

7 Drones Vulnerabilities, Attacks and Their Countermeasures

The literature review of how drones can be easily intercepted, manipulated and interrupted and their countermeasures provided by various researchers is summarised in Table 1.

Table 1. Drones Vulnerabilities, Attacks and Their Countermeasures

Attack Type	Vulnerability	Attacks	Countermeasures
Interception (Attack on Confidentiality)	Keylogging	Keylogging in the form of keystroke monitoring systems, recorders, loggers, tracking software, sniffers, snoopware, and computer activity monitoring software[11]	- In [12], Aviv et al. proposed to examine every software before making it available in the market. - In [13], Simon et al. proposed the idea of whitelist. - In [14], Cai et al. suggested to warn the user when sensitive information is used by an application
Interception (Attack on Confidentiality)	Eavesdropping	In an UAV network, when there is a lack of or weak encryption, attackers can eavesdrop communications between UAV and the network entities[15–17]	- In [8], a Homomorphic Encryption (HE) method is presented by He et al., and based on the additive HE, data aggregation schemes presented in [18, 19]. - In [20], Cui et al. proposed an iterative sub-optimal algorithm and in [21], by generating jamming signals, a caching UAV assisted secure transmission scheme is proposed by Zhao et al. - In [22], Lee et al. presented an iterative algorithm by optimizing the user scheduling variables, the transmit power, and the UAVs trajectory. A scheme based on P-CCCP for secure communication in a dual-UAV enabled system was presented in [23]. - In [24], Sharma et al. proposed a scheme to protect UAV Heterogeneous Network as Functional Encryption (FE)
Manipulation (Attack on Integrity and Authenticity)	GPS Spoofing	- In [25,26] authors showed that the information can be easily recorded, and/or altered, giving hackers complete control of the drone. - In [8], SkyJet enables turns drones in its vicinity them into "zombie drones". - In [27,28], a GPS signal simulator was presented using Universal Software Radio Peripheral (USRP). - In [16], Hornet Mini-rotorcraft UAV was hijacked by transmitting weak counterfeit GPS signals and increased the counterfeit signal power. - Sedjelmaci et al. in [29] generated fake signals and then broadcast with USRP (Universal Software Radio Peripheral	- In [30], Mitchell et al. proposed a specification-based behaviour-rule adaptive IDS detecting malign UAVs. - In [29], a hierarchical intrusion detection & response mechanism was given by Sedjelmaci et al. - In [31], Hooper et al. presented a multi-layer security framework as a defence-in-depth mechanism to guard UAVs against zero-day vulnerabilities. - Schemes integrating code timing authentication in cryptographic code origin to authenticate GPS schemes [32] and using reserved and available bits for periodic signature [33]. - A scheme for identifying the relationships between two narrow-band receivers' unknown encrypted GPS signals. [34]. - The error distribution between the inertial navigation system of the UAV and the GPS is used in the ML detection approach in [35]. - Allocating extra channels to each satellite to track the greatest and weakest environmental signals. [36]. - In [37] detecting spoofing attacks with the use of wireless aerial surveillance like ADS-B and classic localization techniques like TDoA

(continued)

Table 1. (*continued*)

Attack Type	Vulnerability	Attacks	Countermeasures
Manipulation (Attack on Integrity and Authenticity)	ADS-B Attacks (Message Injection / Deletion / Modification): Telemetry feeds use non-secure wireless transmission [38]	- Hackers install malware on GCS system by injecting a reverse-shell TCP payload the drone's memory[1,39,40] - Falliere et. al at Symantec [41] states that Stuxnet reprograms Programmable Logic Controllers and the incident is hidden from the network operations centre [5,42]. - In [43], Damien et al. presented Attack Injection into Avionic Systems through Application Code Mutation. - Telemetry feeds allow infected files to be sent from the drone to the GCS and vice versa[15,44–46]. - Modifying messages by bit flipping, overshadowing, and/or combination of message deletion and injection[27,47]	- In [48], A rule-based intrusion detection scheme was proposed by Strohmeier et al. to detect false data injection attacks in 40 s. - [46] proposed a lightweight PKI-based mechanism in which a node delivers its digital signature (DS) such that neighbouring nodes receive it every m messages. - [49] presented a scheme based on cryptographic primitives in depth of the format-preserving encryption (FPE) and the timed efficient stream loss-tolerant authentication protocol (TESLA). - [50] The encryption process is divided into "key encryption" and "data encryption" and when a new stage is started, key encryption is performed. - [51,52] proposed Fusing ADS-B data with data from other surveillance systems. - [53,54] presented a scheme which is based on multilateration and uses ADS-B signals received in localization by FDoA, ToA, and ADoA. - Detecting attacks by message injection using the transponder type to classify the aircraft [55] and by modeling the flight paths [56]
Interruption (Attack on Availability)	Deauthentication Attacks	To discontinue their communication, both nodes get deauthentication packets. The attacker then uses other tactics like replay attack or message injection to impersonate a genuine ground controller and gain control of the UAV. [57]	- For protecting communication over Wi-Fi, encrypting with proper length of encryption key and bonus bits is done[58]. - Disable the broadcast of UAS's SSID after establishing a connection with trustworthy people, conceal the access point. - Limit access to pre-registered media access control (MAC) addresses only
Interruption (Attack on Availability)	Wi-Fi Jamming, Buffer Overflow, Replay, ARP Cache Poison	- By installing a Raspberry Pi, the hacker was able to jam the drone's frequency and entice it to reconnect to his Wi-Fi. [59]. - Buffer Overflow and Replay occurs by attacker becoming the controller and capturing network traffic via Wireshark [60], and from the /proc/stats directory embedded statistics[61–63]. - ARP Cache Poison by continuously executing a malign Python script called "Scapy" till the device disconnects from the drone[64,65]	- In [66,67], anomaly-based learning algorithms were presented. - In [68], Chen et al. used Elliptic Curve Cryptography, hashing, and digital signature to present a Traceable and Privacy-Preserving Authentication scheme. - [69,70] The system learns from the signals it receives, and controls the transceiver by making decisions. - [71] To determine the best option, a game is played between the worst-case attack and security approaches. - [72] Reinforcement learning is used to find a flight path to protect from the jamming area. [73] Antenna array is used to find the direction of the incoming signal to suppress the signals in that direction. - [74] detected the jamming attack using signal decomposition and nullified updating the array antenna weights

8 Discussion and Areas Open for Research

This section summarises drawbacks in the existing detection techniques and the areas where research is needed:

One issue is that most of the approaches need changes to the UAS network architecture and communication standards, making these solutions impractical. For example, the US-FAA mandates that ADS-B communications be sent in plain text across unencrypted data links to ensure that they are accessible to every recipient in the network, making encryption-based approaches impractical. As a result, developing countermeasures capable of dealing with various attacks with the current infrastructure and protocols would be a big task. A UAS's functioning and navigation are reliant on GPS. In environments when GPS is unavailable, this reliance makes UAS navigation difficult. When a UAS's GPS signal is lost or blocked, it can't fulfil its mission. As a result, effective procedures must be developed to allow the UAS to traverse securely in GPS-denied settings. Other concern that went unnoticed is the UAS's various subsystems' security such as radars, lidars, autopilot, processing units, power subsystems and drive units. Attacking them may lead to UAS operation failure. For example, an attack might cause the UAS processing units to do extra processing cycles, causing the battery to deplete quicker. Drones are vulnerable to jamming and de-authentication attempts due to their lack of frequency hopping[1]. To prevent such attacks, the optimal approach incorporates the flexibility to switch between frequencies. Another issue is that the majority of present UAS networks are self-organizing networks. Planning, setup, optimization, administration, and network restoration are all automated in these networks making them vulnerable to security risks like unauthorised links, unlawful access, malign control, and assaults. Hence, additional research may be focused on creating strategies for providing unified network protection. The UAS may transfer and broadcast private audio, video, and picture data along with command/control data, ADS-B, and GPS making them the subject of security attacks aimed at disrupting the UAS network's functionality and compromising the privacy of information and persons.

9 Conclusion

Increasing applications of drones has ushered in a new epoch of unmanned aerial vehicles in civilian as well as military sectors, with numerous advantages like economic and industrial benefits, owing to their easy-to-use, flexible, and autonomous nature, as well as energy and cost efficiency. Their usage, however, resulted in a slew of security, safety, and privacy concerns, which expressed themselves in the form of a slew of cyber assaults, threats, and difficulties, all of which are mentioned and discussed in this study. Furthermore, successful drone detection, interception, and hijacking experiments using jamming and de-authentication are highlighted, following the systematic hacking cycle, demonstrating how easily drones can be caught, particularly from UAV communication channels. These attacks are categorised into three classes, message interception, manipulation, and interruption. The countermeasures against cyber-attacks proposed to secure UAV networks are also analysed.

References

1. Yaacoub, J.P., Noura, H., Salman, O., Chehab, A.: Security analysis of drones systems: Attacks, limitations, and recommendations. Internet Things **11**, 100218 (2020)
2. Altawy, R., Youssef, A.M.: Security, privacy, and safety aspects of civilian drones: a survey dl.acm.org, **1**(7), 1–25 (2016). https://dl.acm.org/doi/abs/10.1145/3001836

3. Li, J., Zhou, Y., Lamont, L.: Communication architectures and protocols for networking unmanned aerial vehicles. ieeexplore.ieee.org, 2013. https://ieeexplore.ieee.org/abstract/document/6825193/
4. Kumar, K., Kumar, S., Kaiwartya, O., Kashyap, P.K., Lloret, J., Song, H.: Drone assisted flying ad-hoc networks: Mobility and service oriented modeling using neuro-fuzzy. Ad Hoc Netw. **106**, 102242 (2020). https://www.sciencedirect.com/science/article/pii/S1570870520301062
5. Zhu, B., Joseph, A., Sastry, S.: A taxonomy of cyber attacks on scada systems, Proceedings - 2011 IEEE International Conferences on Internet of Things and Cyber. Phys. Social Comput., iThings/CPSCom **2011**, 380–388 (2011)
6. O'Brien, B.J., Baran, D.G., Luu, B.B.: Ad hoc networking for unmanned ground vehicles: Design and evaluation at command, control, communications, computers, intelligence, surveillance and reconnaissance on-the-move (2006)
7. Zhi, Y., Fu, Z., Sun, X., Yu, J.: Security and privacy issues of uav: A survey. Mobile Netw. Appl. **25**, 95–101, 1 (2019). https://link.springer.com/article/10.1007/s11036-018-1193-x
8. He, D., Chan, S., Guizani, M.: Drone-assisted public safety networks: The security aspect, IEEE Communications Magazine (2017). https://ieeexplore.ieee.org/abstract/document/7891797/
9. Rodday, N.M., Schmidt, R.D.O., Pras, A.: Exploring security vulnerabilities of unmanned aerial vehicles. In: Proceedings of the NOMS 2016–2016 IEEE/IFIP Network Operations and Management Symposium, pp. 993–994, 6 (2016)
10. Dessiatnikoff, A., Deswarte, Y., Alata, E., Nicomette, V.: Potential attacks on onboard aerospace systems. IEEE Security Privacy. **10**, 71–74, 7 (2012)
11. Manesh, M.R., Kaabouch, N.: Cyber-attacks on unmanned aerial system networks: Detection, countermeasure, and future research directions. Comput. Secur. **85**, 386–401, 8 (2019)
12. Aviv, A.J., Sapp, B., Blaze, M., Smith, J.M.: Practicality of Accelerometer Side Channels on Smartphones. In: Proceedings of the 28th Annual Computer Security Applications Conference on - ACSAC '12 (2012)
13. Simon, L., Anderson, R.: PIN Skimmer: Inferring PINs Through The Camera and Microphone. In: Proceedings of the Third ACM workshop on Security and privacy in smartphones & mobile devices - SPSM '13, 2013. https://doi.org/10.1145/2516760.2516770
14. Cai, L., Machiraju, S., Chen, H.: Defending against sensor-sniffing attacks on mobile phones. In: SIGCOMM 2009 - Proceedings of the 2009 SIGCOMM Conference and Co-Located Workshops, MobiHeld 2009, pp. 31–36, Aug (2009)
15. McCallie, D., Butts, J., Mills, R.: Security analysis of the ADS-B implementation in the next generation air transportation system. Int. J. Crit. Infrastruct. Prot. **4**(2), 78–87 (2011)
16. Humphreys, T.: Statement on the vulnerability of civil unmanned aerial vehicles and other systems to civil GPS spoofing. rnl.ae.utexas.edu (2012). https://rnl.ae.utexas.edu/images/stories/files/papers/Testimony-Humphreys.pdf
17. Manesh, M.R., Kaabouch, N.: Analysis of vulnerabilities, attacks,countermeasures and overall risk of the Automatic Dependent Surveillance-Broadcast (ADS-B) system. Elsevier. https://www.sciencedirect.com/science/article/pii/S1874548217300446
18. Shim, K.A., Park, C.M.: A secure data aggregation scheme based on appropriate cryptographic primitives in heterogeneous wireless sensor networks. IEEE Trans. Parallel Distrib. Syst. **26**(8), 2128–2139 (2015)

19. Zhang, G., Wu, Q., Cui, M., Zhang, R.: Securing uav communications via joint trajectory and power control. IEEE Trans. Wireless Commun. **18**(2), 1376–1389 (2019)
20. Cui, M., Zhang, G., Wu, Q., Ng, D.W.K.: Robust trajectory and transmit power design for secure uav communications. IEEE Trans. Veh. Technol. **67**(9), 9042–9046 (2018)
21. Zhao, N., et al.: Caching UAV assisted secure transmission in hyper-dense networks based on interference alignment. IEEE Trans. Commun. **66**(5), 2281–2294 (2018)
22. Lee, H., Eom, S., Park, J., Lee, I.: UAV-aided secure communications with cooperative jamming. IEEE Trans. Veh. Technol. **67**(10), 9385–9392 (2018)
23. Cai, Y., Cui, F., Shi, Q., Zhao, M., Li, G.Y.: Dual-UAV-Enabled secure communications: Joint trajectory design and user scheduling. IEEE J. Selected Areas Commun. **36**(9), 1972–1985 (2018)
24. Sharma, D., Rashid, A., Gupta, S., Gupta, S.K.: A functional encryption technique in UAV integrated HetNet: a proposed model, ijssst.info. https://ijssst.info/Vol-20/No-S1/paper7.pdf
25. Kerns, A.J., Shepard, D.P., Bhatti, J.A., Humphreys, T.E.: Unmanned Aircraft Capture and Control Via GPS Spoofing. J. Field Robot. **31**(4), 617–636 (2014) https://onlinelibrary.wiley.com/doi/full/10.1002/rob.21513
26. Seo, S.H., Lee, B.H., Im, S.H., Jee, G.I.: Effect of spoofing on unmanned aerial vehicle using counterfeited GPS signal. J. Positioning Navigation, Timing **4**(2), 57–65 (2015). http://dx.http//www.gnss.or.krPrint
27. Tippenhauer, N.O., Pöpper, C., Rasmussen, K.B., Capkun, S.: On the requirements for successful GPS spoofing attacks. In: Proceedings of the ACM Conference on Computer and Communications Security, pp. 75–85 (2011)
28. Hartmann, K., Giles, K.: UAV exploitation: A new domain for cyber power. In: International Conference on Cyber Conflict, CYCON, vol. 2016-August, pp. 205–221 (2016)
29. Sedjelmaci, H., Senouci, S.M., Ansari, N.: A hierarchical detection and response system to enhance security against lethal cyber-attacks in uav networks. IEEE Trans. Syst. Man, Cybern.: Syst. **48**, 1594–1606 (2018)
30. Mitchell, R., Chen, I.R.: Adaptive intrusion detection of malicious unmanned air vehicles using behavior rule specifications. IEEE Trans. Syst., Man, Cybern. Syst. **44**(5), 593–604 (2014)
31. Hooper, M. et al.: Securing commercial WiFi-based UAVs from common security attacks. In: Proceedings - IEEE Military Communications Conference MILCOM, pp. 1213–1218 (2016)
32. Wesson, K., Rothlisberger, M., Humphreys, T.: Practical cryptographiccivil gps signal authentication. Navigation **59**(3), 177–193 (2012) https://www.ion.org/publications/abstract.cfm?articleID=102576
33. Kerns, A.J., Wesson, K.D., Humphreys, T.E.: A blueprint for civil gps navigation message authentication. In: Record - IEEE PLANS, Position Location and Navigation Symposium, pp. 262–269 (2014)
34. O'Hanlon, B.W., Psiaki, M.L., Bhatti, J.A., Shepard, D.P., Humphreys, T.E.: Real-time gps spoofing detection via correlation of encrypted signals. Navigation **60**, 267–278 (2013). https://onlinelibrary.wiley.com/doi/full/10.1002/navi.44
35. Panice, G., et al.: A svm-based detection approach for gps spoofing attacks to uav. In: ICAC 2017–2017 23rd IEEE International Conference on Automation and Computing: Addressing Global Challenges through Automation and Computing, vol. 10 (2017)

36. Ranganathan, A., Ólafsdóttir, H., Capkun, S.: Spree: A spoofing resistant gps receiver. In: Proceedings of the Annual International Conference on Mobile Computing and Networking, MOBICOM, vol. 0, pp. 348–360 10 (2016)
37. Jansen, K., Schäfer, M., Moser, D., Lenders, V., Päpper, C., Schmitt, J.: Crowd-gps-sec: Leveraging crowdsourcing to detect and localize gps spoofing attacks. ieeexplore.ieee.org. https://ieeexplore.ieee.org/abstract/document/8418651/
38. Lin, X., et al.: Mobile network-connected drones: Field trials, simulations, and design insights. ieeexplore.ieee.org. https://ieeexplore.ieee.org/abstract/document/8758988/
39. Kim, A., Wampler, B., Goppert, J., Hwang, I., Aldridge, H.: Cyber attack vulnerabilities analysis for unmanned aerial vehicles. In: AIAA Infotech at Aerospace Conference and Exhibit 2012 (2012). https://arc.aiaa.org/doi/abs/10.2514/6.2012-2438
40. Shashok, N.: Analysis of vulnerabilities in modern unmanned aircraft systems,. cs.tufts.edu (2017). http://www.cs.tufts.edu/comp/116/archive/fall2017/nshashok.pdf
41. Falliere, N., Murchu, L.O., Chien, E.: W32. stuxnet dossier. Symantec Security Response, 14(February), pp. 1–69, (2011) http://large.stanford.edu/courses/2011/ph241/grayson2/docs/w32_stuxnet_dossier.pdf
42. Matrosov, A., Rodionov, E., Harley, D., Malcho, J.: Stuxnet under the microscope. ESET LLC, pp. 1–85, 2010. http://www.eset.com/us/resources/white-papers/Stuxnet
43. Damien, A., Feyt, N., Nicomette, V., Alata, E., Kaâniche, M.: Attack injection into avionic systems through application code mutation. In: AIAA/IEEE Digital Avionics Systems Conference - Proceedings, vol. 2019-September, 9 (2019)
44. AAbdallah, A., Ali, M.Z., Mišić, J. and Mišić, V.B.: Effcient security scheme for disaster surveillance UAV communication networks. Information **10**(2), 43 (2019https://www.mdpi.com/2078-2489/10/2/43
45. Kovar, D.: Uavs, iot, and cybersecurity (2016). https://www.usenix.org/conference/lisa16/conference-program/presentation/kovar
46. Costin, A., Francillon, A.: Ghost in the air(traffic): On insecurity of ads-b protocol and practical attacks on ads-b devices. 1 (2012)
47. Wilhelm, M., Schmitt, J.B., Lenders, V.: Practical message manipulation attacks in ieee 802.15. 4 wireless networks. lenders.ch. https://www.lenders.ch/publications/conferences/Pilates12.pdf
48. Strohmeier, M., Lenders, V., Martinovic, I.: Intrusion detection for airborne communication using phy-layer information. Lecture Notes in Computer Science (including subseries Lecture Notes in Artificial Intelligence and Lecture Notes in Bioinformatics), vol. 9148, pp. 67–77, 2015. https://link.springer.com/chapter/10.1007/978-3-319-20550-2_4
49. Yang, H., Zhou, Q., Yao, M., Lu, R., Li, H., Zhang, X.: A practical and compatible cryptographic solution to ads-b security. IEEE Internet of Things J. **6**, 3322–3334 (2019)
50. Baek, J., Hableel, E., Byon, Y.J., Wong, D.S., Jang, K., Yeo, H.: How to protect ads-b: Confidentiality framework and efficient realization based on staged identity-based encryption. IEEE Trans. Intell. Transp. Syst. 18 690–700, 3 (2017)
51. Kovell, B., Mellish, B., Newman, T., Kajopaiye, O.: Comparative analysis of ads-b verification techniques (2012)
52. Liu, W., Wei, J., Liang, M., Cao, Y., Hwang, I.: Multi-sensor fusion and fault detection using hybrid estimation for air traffic surveillance. IEEE Trans. Aerosp. Electron. Syst. **49**, 2323–2339 (2013)

53. Kaune, R.: Wide area multilateration using ads-b transponder signals — ieee conference publication — ieee xplore (2012). https://ieeexplore.ieee.org/document/6289874
54. Nijsure, Y.A., Kaddoum, G., Gagnon, G., Gagnon, F., Yuen, C., Mahapatra, R.: Adaptive air-to-ground secure communication system based on ads-b and wide-area multilateration. IEEE Trans. Veh. Technol. **65**, 3150–3165 (2016)
55. Leonardi, M., Piracci, E., Galati, G.: ADS-B jamming mitigation: a solution based on a multichannel receiver. IEEE Aerosp. Electron. Syst. Mag. **32**(11), 44–51 (2017)
56. Habler, E., Shabtai, A.: Using lstm encoder-decoder algorithm for detecting anomalous ads-b messages. Comput. Secur. **78**, 155–173 (2018)
57. Javaid, A.Y., Sun, W., Devabhaktuni, V.K., Alam, M.: Cyber security threat analysis and modeling of an unmanned aerial vehicle system. In: 2012 IEEE International Conference on Technologies for Homeland Security, HST 2012, pp. 585–590, (2012)
58. He, D., Chan, S., Guizani, M.: Communication security of unmanned aerial vehicles. IEEE Wireless Commun. **24** 134–139, 12 (2016)
59. Westerlund, O., Asif, R.: Drone hacking with raspberry-pi 3 and wifi pineapple: Security and privacy threats for the internet-of-things. In: 2019 1st International Conference on Unmanned Vehicle Systems-Oman, UVS 2019, 3 (2019)
60. Shepard, D.P., Bhatti, J.A., Humphreys, T.E., Fansler, A.A.: Evaluation of smart grid and civilian uav vulnerability to gps spoofing attacks — request pdf (2012) https://www.researchgate.net/publication/259183414_Evaluation_of_Smart_Grid_and_Civilian_UAV_Vulnerability_to_GPS_Spoofing_Attacks
61. Bonilla, C.A.T., Parra, O.J.S., Forero, J.H.D,: Common security attacks on drones, pp. 4982–4988 (2018) http://www.ripublication.com
62. Daubert, J., Boopalan, D., Muhlhauser, M., Vasilomanolakis, E.: Honeydrone: A medium-interaction unmanned aerial vehicle honeypot. In: IEEE/IFIP Network Operations and Management Symposium: Cognitive Management in a Cyber World, NOMS 2018, pp. 1–6, 7 (2018)
63. Lindley, J., Coulton, P.: Game of drones. In: CHI PLAY 2015 - Proceedings of the 2015 Annual Symposium on Computer-Human Interaction in Play, pp. 613–618, 10 (2015)
64. French, R., Ranganathan, D.P.: Cyber attacks and defense framework for unmanned aerial systems (uas) environment — journal of unmanned aerial systems. http://www.uasjournal.org/volume-three/article/cyber-attacks-and-defense-framework-unmanned-aerial-systems-uas-environment
65. Booker, B.M., Gupta, A., Guvenc, A.L.: Effects of hacking an unmanned aerial vehicle connected to the cloud (2018). https://kb.osu.edu/handle/1811/84561
66. Rani, C., Modares, H., Sriram, R., Mikulski, D., Lewis, F.L.: Security of unmanned aerial vehicle systems against cyber-physical attacks. https://doi.org/10.1177/1548512915617252,vol. 13, pp. 331-342, 11 (2015). https://journals.sagepub.com/doi/10.1177/1548512915617252
67. Condomines, J.P., Zhang, R., Larrieu, N.: Network intrusion detection system for UAV ad-hoc communication: from methodology design to real test validation. Ad Hoc Netw. **90**, 101759, 7 (2019)
68. Chen, C.L., Deng, Y.Y., Weng, W., Chen, C.H., Chiu, Y.J., Wu, C.M.: A traceable and privacy-preserving authentication for UAV communication control system. Electronics 2020, **9**, 62, 2020. https://www.mdpi.com/2079-9292/9/1/62

69. Chen, C.L., Deng, Y.Y., Weng, W., Chen, C.H., Chiu, Y.J., Wu, C.M.: A traceable and privacy-preserving authentication for UAV communication control system. Electronics 2020, **9**, 62 (2020). https://www.mdpi.com/2079-9292/9/1/62
70. Reyes, H., Gellerman, N., Kaabouch, N.: A cognitive radio system for improving the reliability and security of UAS/UAV networks. IEEE Aerosp. Conf. Proc. **2015** (2015)
71. Slater, D., Tague, P., Poovendran, R., Li, M.: A game-theoretic framework for jamming attacks and mitigation in commercial aircraft wireless networks. In: AIAA Infotech at Aerospace Conference and Exhibit and AIAA Unmanned...Unlimited Conference (2009). https://arc.aiaa.org/doi/abs/10.2514/6.2009-1819
72. Johansson, R., Hammar, P., Thoren, P.: On simulation-based adaptive UAS behavior during jamming. In: 2017 Workshop on Research. Education and Development of Unmanned Aerial Systems, RED-UAS **2017**, 78–83 (2017)
73. Ni, S., Cui, J., Cheng, N., Liao, Y.: Detection and elimination method for deception jamming based on an antenna array. **14**(5), pp. 2018 (2018). https://journals.sagepub.com/doi/full/10.1177/1550147718774466
74. Wu, R., Chen, G., Wang, W., Lu, D., Wang, L.: Jamming suppression for ADS-B based on a cross-antenna array. In: ICNS 2015 - Innovation in Operations, Implementation Benefits and Integration of the CNS Infrastructure, Conference Proceedings, pp. K31–K39, (2015)

Classifications of Real-Time Facial Emotions Using Deep Learning Algorithms with CNN Architecture

Bommisetty Hema Mallika(✉), G. Usha, A. Allirani, and V. G. Rajendran

Department of Electronics and Communication Engineering, SRC, SASTRA Deemed University, Kumbakonam, Tamil Nadu, India
mallika29102002@gmail.com

Abstract. In today's world, it is necessary to detect one's emotions, as emotion can use for public safety by identifying the emotion from facial expressions. In addition to understanding human behavior, human emotion detection also covers identifying mental illnesses and artificially created human emotions. It is still challenging to recognize human emotions accurately, so we use Deep Learning or Machine Learning techniques. Pre-processing, face identification, feature extraction, and emotion classification are typically the four phases of two standard techniques based on geometry and appearance. The existing method suggested a Deep Learning architecture based on Convolutional Neural Networks (CNN) for emotion identification from images. The Dataset used to test is The Facial Emotion Recognition Challenge (FER-2013). For the FER2013, the accuracy rate for different architectures is 53.58%, 60.52%, and 64.4%, respectively. Deep Convolutional Neural Network is used in this work to enhance emotion recognition. The performance of the proposed technique is evaluated with FER2013 and CK+. To get the best accuracy, we experimented with different depths, maximum pooling layers, or alternative topologies and eventually reached 89.42% accuracy. We use this high-accuracy architecture to demonstrate how real-time emotion identification works and gives consistent results using a webcam.

Keywords: Deep Convolutional Neural Network · FER · CK+ · Facial expression · Emotion recognition

1 Introduction

Emotion is a nervous system-related mental state that includes feelings, perceptions, behavioural responses, and a level of pleasure or annoyance [1]. The neural networks are used in Artificial Intelligence (A.I.) networks for different uses to identify faces in photos and videos. Most methods handle visual data and look for recurring patterns in human features in pictures or videos. Face detection is a surveillance tool that may be used in crowd control and law enforcement. In this, we provide a technique for recognizing seven emotions; using facial expressions, one may express disdain, neutral fear, joy, sadness, and surprise. Prior studies employed Deep Learning algorithms to develop

H. Sharma et al. (Eds.): ICIVC 2022, PALO 17, pp. 233–244, 2023.
https://doi.org/10.1007/978-3-031-31164-2_19

emotional-based models of facial expressions to identify emotions [2]. Emotions are critical in forming and maintaining relationships and successful interpersonal communication. Emotions are expressed through words, gestures, and facial expressions. We encounter a wide range of emotions in our daily lives. However, we cannot comprehend most of them since they are not visible to the naked eye, so Machine Learning and Deep Learning are becoming increasingly important.

A lot of information is lost during a regular Human-Computer Interaction (HCI) since it doesn't consider the users' emotional states. In comparison, users are more effective and want emotion-sensitive HCI systems [3]. With the rising need for leisure, business, physical and psychological well-being, and education-related applications, interest in emotional computing has grown recently. Due to this, several emotionally sensitive HCI system products have been created over the past several years. However, a long-term fix for this study area has not yet been proposed. A crucial aspect of human contact is the commonality of facial emotions and body language. In the eighteenth century, Charles Darwin wrote about widely used facial expressions, which are crucial in non-verbal communication [4]. Ekman Friesen stated that some emotions were consistently connected with facial behaviours in 1971 [5]. Both humans and animals exhibit particular muscular movements corresponding to particular mental states. Nicholson et al. [6] are recommended for those interested in studying emotion categorization by voice recognition. Several Deep Learning algorithms are used to recognize the seven human emotions of anger, disgust, neutrality, fear, pleasure, surprise, and sorrow. "Facial emotion recognition" refers to recognizing human emotions from the face. Human facial expressions are frequently researched for several causes and circumstances [12]. The phrase Face Emotion Recognition refers to the process of recognizing human emotions on the face. Human facial expressions are commonly studied for a variety of reasons in a variety of situations. Face identification, facial feature extraction, and emotion recognition are the three critical processes of an autonomous Facial Emotion Detection System [13]. Emotion Recognition (E.R.) systems employ computer-based methods to identify expressions in real-time. The accuracy rate must be high for the computer to identify and classify emotions correctly. CNN is our preferred method for accomplishing this [14]. It is a neural network approach consisting of neurons that receive input and produce output using an activation function. Convolution is the activation function in the CNN model. This model consumes much less time than other models. CNN has created a Real-Time Facial Emotion Detection System using the critical information it can extract and learn from photographs. Most relevant information about facial emotions is conveyed through the lips, eyes, brows, and other facial expressions [15].

2 Literature Survey

An attentional convolutional network-based deep learning technique developed by Noel Jaymon, Sushma Nagdeote, Aayush Yadav, and Ryan Rodrigues can distinguish significant facial features and produce improved results when used to the FER2013 dataset [1]. Jay B. et al. evaluated real-world human talks to identify relevant emotional sentiments of prosody, disfluency, and lexical cues [2]. Schuhler et al. suggested two strategies and compared the results. By using continuous hidden Markovian models that take into account many states and rely on low-level features from samples taken at that specific instant rather than general statistics, the strategy that follows increases temporal complexity [3]. Charles Darwin published a list of widely used facial expressions in the nineteenth century that are crucial in nonverbal communication [4]. In 1971, Ekman Friesen asserted that certain emotions were consistently connected with facial behaviour. W. V. Friesens and P. Ekman looked into the Face and Emotion Constants Across Cultures. Journal of personality and social psychology [5].

Humans and animals produce specific muscle movements that belong to a certain mental state. People interested in research on emotion classification via speech recognition are referred to Nicholson et al. [6]. In their article, Medhat Walaa et al. contrasted and examined the findings of currently established methods for emotion detection (based on revised and lexical knowledge) and a proposed technique based on common sense knowledge stored in the EmotiNet: A Knowledge Base for Emotion Detection. [7] The current techniques are primarily concerned with word-level text analysis and are typically only able to identify a few number of sentimental phrases [8]. Krizhevsky and Hinton provide a seminal paper on universal automatic picture classification. The deep neural network demonstrated in this paper functions similarly to the human visual brain. Using the CIFAR-10 dataset and a self-developed tagged array of 60,000 photos divided into 10 groups, a model for categorising objects from photographs is created. The visualisation of the network's filters is a significant research result that allows for evaluation of the way the model decomposes the images [9].

The yearly Imagenet challenges were first introduced in 2010, and since then, the enormous collection of labelled data has been extensively utilised in papers [10]. The ImageNet LSVRC-2010 contest used 1.2 million high-resolution photos to train a network with five convolutional, three max pooling, and three fully connected layers in a later study by Krizhevsky et al. [11]. Poria Soujanya et al. in their paper, since about 90% of the relevant literature appears to use these three modalities, the authors primarily emphasized how we can use audio, visual, and text information for multimodal affect analysis. Only when there are multiple channels or modes, such as visual, audio, text, gestures, etc., can it be used for multimodal affect analysis [12]. French Hollet et al. demonstrated that the XCeption architecture outperforms the Inception V3 model on the ImageNet dataset and significantly outperforms the Inception V3 model on a larger image classification dataset with 350 million images and 17,000 classes [13].

3 Methodology

3.1 Datasets

Here, the FER2013 [16] and CK+ datasets are used. Grayscale portraits of faces measuring 48 × 48 pixels make up the data. The faces have been automatically registered such that each is roughly in the exact location and takes up a similar amount of space. Each face must be assigned to one of seven groups depending on the emotion shown in the facial expression.0 = furious(angry), 1 = disgust, 2 = fear, 3 = happy, 4 = sad, 5 = surprise, and 6 = neutral. 28,709 photos make up the training set, whereas 3,589 images make up the public test set. CK+ [17] contains a total of 981 images of 48 * 48 pixels.

3.2 Implementation

Data Pre-processing: The preparation stage is relatively straightforward: normalizing data first per image, then per pixel. To normalize data per image, the mean value across each image is deducted, and then the image's standard is set from error to 3.125. To normalize the data per pixel, we must first compute the average picture throughout the training set [3]. The average of all comparable pixels (i.e., pixels with the exact dimensions) in overall training photos is used to calculate each pixel in the mean image. The mean score of each pixel in each training sample is then subtracted, and the standard error of each pixel across all training photos is set to 1. Both training data and test data are subject to the pre-processing stage. This can be seen in Fig. 1.

Data Augmentation: Data augmentation approaches are used to increase the number of training samples due to the limited training data to prevent overfitting and enhance recognition accuracy [4]. The below listed consecutive transformations for each image are carried out: There is a 0.5 possibility of mirroring the image. Rotate the image at a chance angle between −45 and 45° (in degrees). Scale down the image from its original 48 × 48 dimensions in the FER-2013 dataset, CK+, to a random size. Select a random 42 by 42-pixel crop for the rescaled image. Resizing the image is the second step, as in Fig. 1.

Training: The network for 20 epochs is trained to guarantee the ideal level of accuracy converges 20, 000 photos from the FER-2013 [16] dataset, and 981 images from the CK+ [17] dataset can be used for training. The FER-2013 database additionally takes advantage of freshly created sample sets and verification (2000 images) (1000 images). After the training and testing of the model, it displays the number of emotions in the final testing and validation set. As a result, emotion detection using deep convolutional neural networks can enhance the performance of a network with additional data. The accuracy can be more significant on all validation and test sets than in earlier runs.

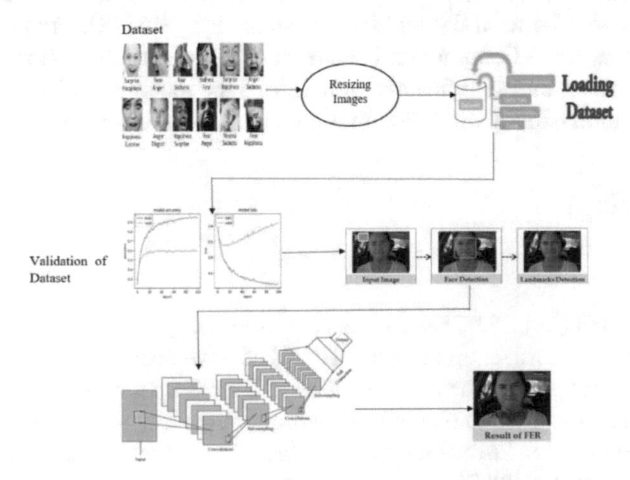

Fig. 1. Proposed Model For Emotion Recognition

Testing: A training set and a validation set from the original Dataset are created. Then, the validation accuracy during training is tracked, and recorded the model state with the most remarkable accuracy on the validation set. The ultimate accuracy is estimated by applying the trained model to the test sets.

3.3 Convolutional Neural Network

CNN architecture is used to identify the correct emotion by comparing it with the given Dataset as in Fig. 1.

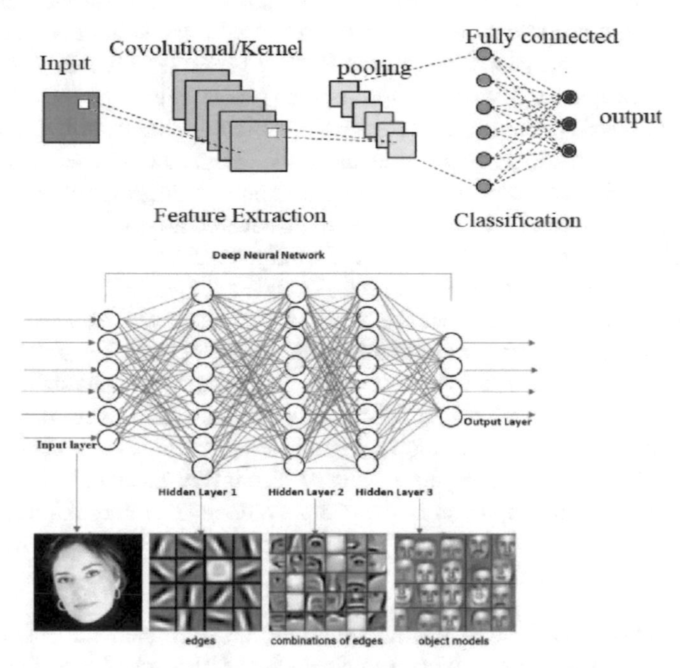

Fig. 2. CNN Architecture

A CNN is made up of a pooling layer (Pool), batch normalization, and a convolution layer (Conv) with rectified linear unit (ReLU) activation function. The fully connected (Fc), dropout, SoftMax, and classification output layers are included in the final layers. The pooling Layer helps reduce the representation's spatial size, and the types are Maximum pooling and Average pooling [2].

As seen in Fig. 2, it contains a fully connected layer that aids in mapping input and output representations. Edges, shapes, textures, and objects are just a few of the patterns that convolutional layer filters may identify in an image. Convolutional layers, which make up the majority of CNNs, apply filters (or kernels) to our input picture to identify fine characteristics like edges, colours, and textures. These traits, such as detecting the nose, mouth, and eyes, become more crucial as we move further into the network (for emotion recognition).

Convolutional neural networks, or ConvNets, are built from three layers. The Convolutional Layer, the foundation of CNN, is in charge of carrying out convolutional processes. The Kernel/Filter is the element in this layer responsible for carrying out the convolution operation. Up until the complete picture is scanned, the kernel makes horizontal and vertical adjustments based on the stride rate. The kernel has more depth while being smaller than an image. The kernel height and width will be insignificant if the image comprises three (RGB) channels, but the depth will be relevant. Convolutional layers also include the Nonlinear activation function, which is important. A nonlinear activation function receives input from linear techniques like convolution. However, mathematical models of real neuron activity have already been constructed using smooth nonlinear functions, such as the sigmoid or hyperbolic tangent (tanh) function. The most used nonlinear activation function today is the rectified linear unit (ReLU). To reduce dimensionality, use the Pooling Layer (POOL). It aids in lowering the amount of computing power required for data processing. Each input is coupled to each neuron in the fully connected layer (F.C.), which functions with flattened inputs. The mathematical functional procedures are often performed at a few additional F.C. levels after the flattened vector have been delivered. The categorizing process starts at this point. The presence of F.C. layers is often discovered after CNN designs.

This work determines the optimal model for the given class issue using six pre-trained CNN models: Alexnet, Vgg16, Squeezenet, Googlenet, Resnet18, and Xception. The input size of ResNet is 224 * 224, and Networks with many layers may be readily trained without increasing the number of train errors. SqueezeNet has a 1-standalone convolution layer, 2–9 fire modules, a final convolution layer, and an input size is 227 * 227. GoogleNet is a deep convolutional neural network with 22 layers. It is a subset of Google researchers' Deep Convolutional Neural Network called the Inception Network. The input Size is 224 * 224. Xception**,** the Input size is 229 * 229 [1]. Depth, wise Separable Convolutions are employed in the Xception deep convolutional neural network design. Xception refers to an "extreme" Inception module. Here we have used a trained network to transfer learning.

AlexNet: Input size is 227 * 227. The network's design resembled that of LeNet by Yann LeCun et al. quite a bit, although it was more complex, with stacked convolutional layers, and more filters per layer. It introduced ReLU activations after each fully connected and convolutional layer. AlexNet's network is split into two streams since it was trained over six days using two Nvidia Geforce GTX 580 GPUs concurrently.

GoogleNet: The network used a CNN that was modeled after LeNet but included a brand-new function called an inception module. RMSprop, batch normalizing, and image distortions were all used. This module substantially reduces the parameters by using small convolutions. Although they used a 22-layer deep CNN, there were just 4 million parameters instead of 60 million (AlexNet) in their architecture.

VGGNet: VGG has 16-layer architecture, has 3×3 convolution layers similar to Alexnet, but has a lot of filters. 2–4 weeks of training on 4 GPUs. The community now views it as the best option for extracting characteristics from photos. However, handling VGGNet's 138 million parameters can be a little difficult. Input size is 224 * 224.

ResNet: ResNet consists of four layers with identical behavior after one phase of convolution and pooling. The same pattern is followed by every layer. The popular disappearing gradient is one of the issues that ResNets addresses. The gradients from which the loss function is produced soon decline to zero after numerous applications of the chain rule when the network is deep enough. As a result, no learning takes place since the weights' values are never changed. ResNets may be immediately traversed by gradients without the need for connections between earlier layers and the starting filters.

SqueezeNet: SqueezeNet is an 18-layer deep convolutional neural network. A pre-trained network version of over a million photos from the ImageNet database may be loaded. Consequently, the network has learned detailed feature representations for a diverse set of pictures.

Xception: Xception is a 71-layer deep convolutional neural network. A pre-trained network version of over a million photos from the ImageNet database may be loaded. The network's picture input size is 299 by 299 pixels.

Transfer Learning Using Pre-trained Network: A pre-trained deep learning network is modified to learn a new task through transfer learning [4]. The following stages are used for transfer learning:

- Choose an already-trained network, i.e., an existing network
- Replace new layers tailored to the updated data set for the final ones
- Indicate in the training graphics the updated number of classes
- Resize the images, and define the network's training parameters (Solver name or training parameters, starting learning rate, number of epochs, mini-batch scale, validation data, and validation frequency).
- Examine a network that has been trained by categorizing test or validation images.

And then extract images using a pretrained network. It is employed as feature extractor.

Facial Emotion Recognition

The three primary functions of an automated facial emotion detection system are face identification, facial feature extraction, and emotion recognition.

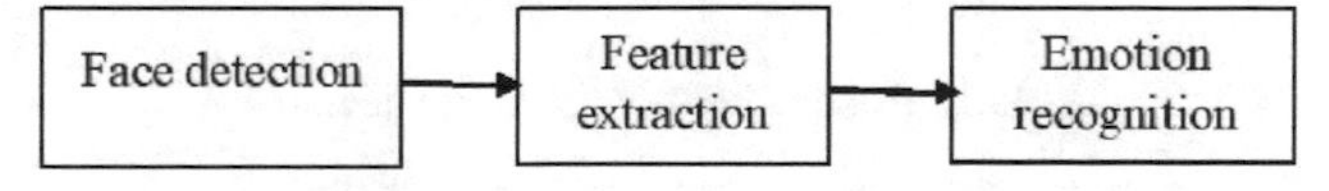

Fig. 3. Facial Emotion Recognition

Face Detection

Face detection is the first stage, which is shown in Fig. 3. Collecting faces from situations is known as "Face detection." Consequently, the system recognizes a specific picture area as a face. Face tracking, position estimation, and compression are just a few of the uses of this approach. The ability to locate a face in any supplied picture or frame. The result is the discovered faces' bounding box coordinates.

Feature Extraction

The following step is "feature extraction," as per Fig. 3, which involves selecting pertinent face characteristics from the data. It is possible to add particular facial features, modifications, perspectives, or measurements that are either relevant to humans or not (such as eye spacing). Two further applications of this phase are emotion recognition and face feature tracking.

The principal use of this approach is to make the system computationally very fast by reducing a big amount of data into a smaller number of feature sets. It is tinkering with different technological issues such as, Pose modifications such as angles, Differences in facial expressions, Poor lighting conditions. The main idea underlying feature extraction is to extract the exceptional (unique) traits that are provided. The rate of facial emotion recognition is determined by the retrieved unique characteristics. Our technological team is working hard again to make the face extraction procedures as impressive as possible. Typically, developers employ static picture information to apply filtering and geometric characteristics (2D & 3D).

Emotion Recognition: Finally, the system recognizes the face. The system would report an identity from a database in an identification task. A classification method, a comparison mechanism, and an accuracy metric are used at this stage. This stage employs techniques in many different areas to carry out a categorization process. Humans are accustomed to interpreting nonverbal clues from facial emotions. Computers are getting better at reading emotions. So, how can we identify emotions in images? Using an open-source data set called Face Emotion Recognition (FER) from Kaggle, we built a CNN to recognize emotions. Emotions are classified into seven groups: happy, sad, afraid, disgusted, furious, neutral, and astonished. We employed photo augmentations and a 6-layered Convolutional Neural Network (CNN) to improve model performance.

4 Results

Here, the highest accuracy architecture is used in determining Real-time emotion recognition. The real-time input is received from the user, and it recognizes emotion by comparing it with the already loaded Dataset. The results are listed in Table 1 and the bar chart in Fig. 4.

Validation accuracy for CK+ Dataset Using Different Architectures

Table 1. Accuracies of various architectures

Architecture	Input size	Accuracy (%)
AlexNet	227 * 227	84.3
GoogleNet	224 * 224	84.6
SqueezeNet	227 * 227	85.6
Xception	229 * 229	86.1
Resnet18	224 * 224	87.71
VGG16	224 * 224	89.42

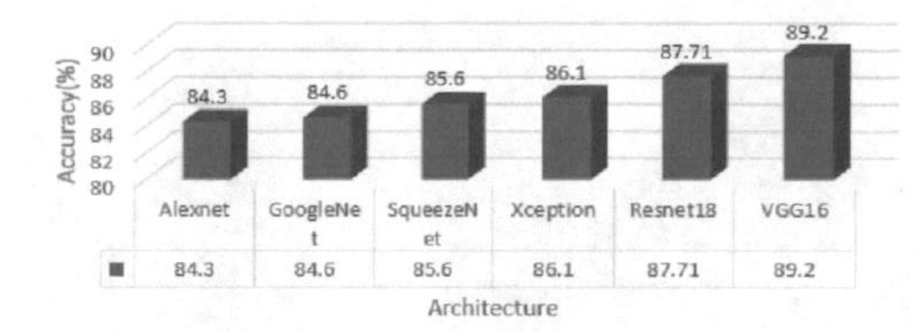

Fig. 4. Accuracy Graph for Various Architecture

As the highest accuracy for VGG16 is witnessed, it is used in Real-Time Emotion Recognition. The Real-time results are shown in Fig. 5.

Real-Time Results

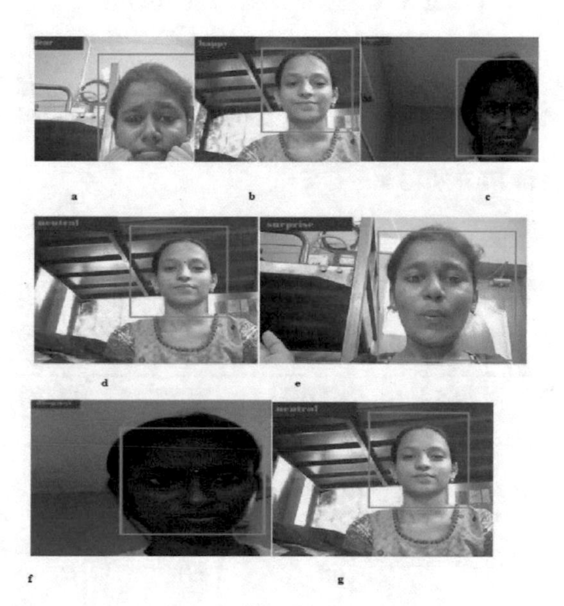

Fig. 5. Real-Time Emotion Detection, 8a-fear, 8b-happy, 8c-anger, 8d-sad, 8e-surprise, 8f-disgust, 8g-neutral

5 Discussion

Sometimes, it may be challenging to differentiate between neutral and anger, surprise and neutral, sadness and surprise, etc. This causes errors in detecting emotion. A confusion matrix represents this (Fig. 6). When FER2013 is used with AlexNet, a Validation

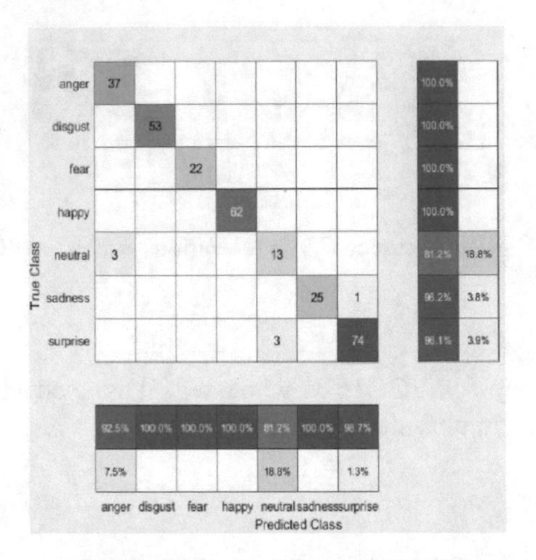

Fig. 6. Confusion Matrix of VGG16

accuracy of 64.3% can be witnessed, which is very low accuracy. If CK+ for the same architecture is used, an accuracy of 84.3% can be witnessed.

Validation Details of FER2013 and CK+

Table 2. Validation Details for Different Architectures using CK+ Dataset

	VGG16	AlexNet	GoogleNet	SqueezeNet	Xception	ResNet18
Validation Accuracy	89.42%	84.3%	84.6%	85.6%	86.1%	87.71%
No. of Epochs	20	20	20	20	20	20
Iterations	1360	1360	1360	1360	1360	1360
Iteration per epoch	68	68	68	68	68	68
Frequency (iterations)	68	68	68	68	68	68
Hardware resource	Single CPU	Single CPU	Single CPU	Single CPU	Single CPU	Single CPU
Learning Rate Schedule	Constant	Constant	Constant	Constant	Constant	Constant
Learning Rate	0.0003	0.0003	0.0003	0.0003	0.0003	0.0003

The accuracies obtained using the existing model are

Table 3. Validation Details Using FER2013 Dataset

	Simple Model	Inception Model	Xception model
Validation Accuracy	53.58%	60.52%	64.4%
No. of Epochs	30	150	150

Comparison of Existing and Proposed Models

Table 2 shows the validation details of the proposed model, and Table 3 shows the validation details of the existing model. From the existing model, only 6 out of 7 emotions are recognized due to less accuracy, and all seven are correctly recognized from the proposed model.

6 Conclusion

The highest accuracy of 89.2% is obtained with VGG16. The model functions effectively in identifying all seven emotions in user-provided images. The model lacks resilience in detecting emotions in real-time. The model can be used in crime identification, public safety, Education, Employment, and Personalized Service Provision. The suggested model is investigated with training and test sample images, and its performance is assessed compared to the pre-existing model. The experimental findings reveal that the suggested model performs better in emotion detection than earlier models mentioned in the literature. The study's results demonstrate that the suggested model provides cutting-edge results on the two datasets.

In Future work, it can be implemented in the remaining models with many images.

References

1. Li, S., Deng, W.: Deep facial expression recognition: a survey. arXiv preprint arXiv:1804.08348 (2018)
2. Correa, E., Jonker, A., Ozo, M., Stolk, R.: Emotion recognition using deep convolutional neural networks. Technical report IN4015 (2016)
3. Tian, Y.I., Kanade, T., Cohn, J.F.: Recognizing action units for facial expression analysis. IEEE Trans. Pattern Anal. Mach. Intell. **23**(2), 97–115 (2001)
4. Darwin, C.R.: The Expression of the Emotions in Man and Animals. John Murray, London (1872)
5. Ekman, P., Friesen, W.V.: Constants across cultures in the face and emotion. J. Pers. Soc. Psychol. 17, 124–129 (1971)

6. Nicholson, J., Takahashi, K., Nakatsu, R.: Emotion recognition in speech using neural networks. Neural Comput. Appl. **9**, 290–296 (2000)
7. Fasel, B., Luettin, J.: Automatic facial expression analysis: a survey. Pattern Recogn. **36**(1), 259–275 (2003)
8. Krizhevsky, A., Hinton, G.: Learning multiple layers of features from tiny images (2009)
9. Deng, J., Dong, W., Socher, R., Li, L.-J., Li, K., Fei-Fei, L.: ImageNet: a large-scale hierarchical image database. In: IEEE Conference on Computer Vision and Pattern Recognition, CVPR 2009, pp. 248–255. IEEE (2009)
10. Krizhevsky, A., Sutskever, I., Hinton, G.E.: ImageNet classification with deep convolutional neural networks. In: Advances in Neural Information Processing Systems, pp. 1097–1105 (2012)
11. Lv, Y., Feng, Z., Xu, C.: Facial expression recognition via deep learning. In: 2014 International Conference on Smart Computing (SMARTCOMP), pp. 303–308. IEEE (2014)
12. Hermitian, F., Sohrabi, M.K.: A survey on classification techniques for opinion mining and sentiment analysis. Artif. Intell. Rev. **52**, 1495–1545 (2019)
13. Sharef, N.M., Zin, H.M., Nadali, S.: Overview and future opportunities of sentiment analysis approaches for big data. J. Comput. Sci. **12**, 153–168 (2016)
14. Minaee, S., Abdolrashidi, A.: Deep-emotion: facial expression recognition using attentional convolutional network. In: Computer Vision and Pattern Recognition (2019)
15. Ajay, B.S., Anirudh, C.R., Karthik Joshi, S., Keshava, B.N., Asha, N.: Emotion detection using machine learning. Int. J. Recent Trends Eng. Res. (IJRTER) (2017)
16. FER2013dataset. https://www.kaggle.com/datasets/msambare/fer2013
17. CK+Dataset. https://www.kaggle.com/datasets/shawon10/ckplus

Healthcare Data Security in Cloud Environment

Garima Chawla and Syed Wajahat Abbas Rizvi(✉)

Department of Computer Science and Engineering, Amity University, Uttar Pradesh, Noida, India

swarizvi@lko.amity.edu

Abstract. As of late, the cloud offers the advantage of putting away the potential documents to a far-off database that can be recovered on-demand. The infrastructure of healthcare cloud is utilized to store the clinical records that thus help in overseeing and following the records of the client. The essential consideration suppliers utilize Electronic Medical Record (EMR) to do the diagnosis and also assess the medical information. Big data analytics is a significant innovation in tremendous business regions that incorporates social media, banking, medical welfare and machine-censored data. Healthcare cloud offers several storage options, for example, one drive, drop box business, Google drive where security requirements and privacy concerns such as data robbery attacks are considered as major breaches of security in the cloud.

In this research paper, I have analyzed the feasibility for applying artificial intelligence techniques to improve the healthcare data security in cloud computing environment. A modified decoy strategy is proposed to guarantee enhanced security for client's Medical Big Data. The system identifies who the attacker is and reverts back the data, for example, access date and time, IP address to the client. The tri-party confirmation key purposes key arrangement convention which is demonstrated by cryptography which is of the form bilinear pairing. It likewise offers the facility to obstruct the account from additional access. The parameters such as computational complexity is diminished and throughput is enhanced when using the bilinear pairing's triple-des algorithm.

Keywords: Cloud Security · Data Privacy · Artificial Intelligence Techniques · Electronic Medical Record (EMR) · Medical Big Data (MBD)

1 Introduction

Cloud computing is an emerging computing system that assimilates and incorporates the advantages of distributed computing, utility computing, grid computing, virtualization technology, parallel computing, and other computer technologies. It provides substantial virtualization, low-cost service, large-scale computation and data storage, high reliability and high expansibility [1]. It is not only needful but also important to understand different cloud computing systems and examine the cloud computing security problems and strategies in light of cloud computing principles and characteristics [2].

Cloud computing's security issue is critical, and it has the potential to stifle the technology's rapid growth. The main security matter in question and issues in cloud

H. Sharma et al. (Eds.): ICIVC 2022, PALO 17, pp. 245–253, 2023.
https://doi.org/10.1007/978-3-031-31164-2_20

computing include data privacy and availability of service. A specific security technique will not address to label the cloud computing security problem [3]. Hence, a combination of old and new approaches in addition to innovative methods will be required to safeguard the entire cloud computing system. Utilizing Artificial Intelligence technologies, the healthcare data security in cloud computing environment can be moved along. Probably the accepted procedures to stay aware of medical care data security incorporate embracing AI-driven advancements [4].

To access the facilities and share the healthcare data, a hybrid cloud is provided. This hybrid cloud stores the clinical records of the patients. Not only this, it additionally helps to manage the client's record and tracks the records for them, nevertheless whether the patient gets across any area of the planet. It has a few security issues, for example, policy issues, licensing, data protection and prevention and transparency. To accomplish a protected and safe healthcare service, cloud is utilized in our proposed technique in the client's Medical Big Data [5].

The rest of the paper is organized as follows; Sect. 1.1 describes the Review of Literature; Sect. 1.2 shows the goal of this paper whereas the Hypothesis is included in Sect. 1.3. Section 2 deals with the methodology adopted including the data collection, analysis and implementation which discusses the various training, testing and validation datasets utilized in this study. Results and discussions are presented in the Sect. 3. Finally, the paper concludes in Sect. 4 with the future scope and limitations of this study.

1.1 Review of Literature

The analysis of medical big data shows that providing protection features for the security and privacy using hybrid cloud technology offers different algorithms as well as mechanisms towards the security in and of the cloud. Dissecting the various technologies, it offered the alternative model for security [6, 7, 8]. It utilizes bilinear pairing convention. The parameters such as computational complexity is viewed as in which in the outcome it diminishes by 20% in this proposed model. The time utilization is the only constraint in the proposed work that is high [9, 10].

The impediments like medical service, privacy and lightweight for healthcare cloud can be defeated by Attribute Based Signature (ABS) technology. By giving the more grounded security model, it centers around the time utilization that diminishes the response time by 0.25 s in the result. The proposed framework has a few restrictions which it is reasonable for limited scope healthcare access [11].

Further developing security and protection characteristic based information partaking in cloud computing innovation has given different algorithms and mechanisms towards the security in the cloud. Dissecting the few strategies, it offered the alternative model for security utilizing the Ciphertext Technology-Attribute based encryption (CP-ABE). The parameters such as computational overhead is viewed as in which it diminishes by 40% in the outcome in this work. The proposed model thus, gives the powerless security model [12, 13].

1.2 Objective

With the multiplication of data in the field of healthcare and medical care frameworks towards cloud, keeping up with the gigantic volume of delicate information becomes compulsory [14]. Utilizing Artificial Intelligence technologies, the healthcare data security in cloud computing environment can be moved along. Probably the accepted procedures to stay aware of medical care data security incorporate embracing AI-driven advancements [15].

The project deals with analyzing the feasibility of using artificial intelligence technologies to improve healthcare data security in cloud computing environment. To distinguish and answer the power and speed of the present cyber-attack techniques, associations need the most recent security strategies in light of the fact that conventional ones will not be having the option to determine the test [16]. With such AI-empowered applications, recognizing oddities and sending ongoing notices about cyber-attacks across the whole organization becomes simpler than previously.

1.3 Hypothesis

A clever brilliant medical services implementation, along with simulation and design uses healthcare using 4.0 techniques. It offers several algorithms and mechanisms towards the security in the cloud [17]. Examining the few strategies, it gave the alternative model for security which utilizes Blockchain, which in turn makes use of the current dataset. The parameters such as time utilization is viewed as in which it decreases by 1s in output in this proposed model. The limitation that is offered by the proposed work lies in adding the extra highlights to the framework [18].

Secure unquestionable database supporting proficient dynamic activities in cloud computing offer different algorithms and mechanisms towards the security of the cloud. Dissecting the few methods, it shows the alternative model of security utilizing BLS signature which utilizes the current dataset [19]. The parameters such as time throughput is viewed as in which the throughput is expanded by 20% in the outcome in this proposed model. The one restriction which is offered by the proposed work is that the pairing is not effective [20].

2 Methodology

2.1 Data Collection

Data security is an overwhelming assignment for infosec experts and IT department. Every year, organizations spare a sizable part of their financial schemes and plans for the security of information technology, safeguarding their associations from programmers' expectation on accessing information through savage power, social designing or taking advantage of weaknesses and vulnerabilities.

Breaches are expensive occasions that outcome for multimillion-dollar legal claims and casualty settlement reserves. On the off-chance that organizations need motivation to put resources into information security, they must just consider the worth put on private information by the courts. Phishing and ransomware additionally are on ascent and are thought about as significant dangers. Organizations should safeguard the information so it can't spill out by means of social designing or malware.

They ought to likewise survey their risk and the assurances their ongoing security ventures offer and pursue choices appropriately. This needs a remarkable degree of perceivability and visibility which most associations lack at the present time.

Numerous associations understand that the worth of information and the expense to safeguard information are expanding at the same time, making it close to difficult to safeguard information simply by layering on greater security. All things being equal, infosec groups and IT teams should think inventively about the data insurance systems and techniques.

2.2 Analysis

Encryption is one of the most fundamental ideas of information security, as essentially encoding delicate information can go far towards accomplishing protection and consistence orders and guarding touchy and delicate data from outsiders. Encryption is definitely not a simple recommendation, because associations should choose the algorithm for encryption that matches with their venture's security prerequisites.

Two of the most widely recognized types of encryptions are symmetric encryption and asymmetric encryption. Symmetric encryption includes changing over plaintext to ciphertext involving a similar key for encoding and decoding. On the other hand, asymmetric encryption utilizes two related keys, one for encoding the information and the other for decoding it. Symmetric encryption has a large number of forms, involving Triple DES and Advanced Encryption Standard. Asymmetric encryption also has a large number of forms, involving RSA and Diffie-Hellman key exchange. Organizations that would rather not encode their data should decide the need of information through grouping.

The attacker is identified using user- profiling and then a message is shipped off the client that his record is being used by some outsider. It consists of data like IP address, access time and date. Presently to transfer the clinical records by the client, the tri-party should be imparted. A secret key is created from the Private Key Generator and it is accessed by the tri-party. Presently each party validates with one more party to covertly speak with others. Now, the clinical records are encoded utilizing the triple-des calculation. This encoded information is put away in the cloud. They are then decoded for reviewing utilizing this calculation.

2.3 Implementation

Understanding the Dataset. For this project, a multimedia dataset which is of 128 megabytes is utilized as the input. It has 150 entries to it. Multimedia dataset of 128 MB with 150 sections is utilized as a contribution to this undertaking. This media dataset is then changed over into figured cipher-text that is the utilized as information for this task. The result of this project undertaken is likewise a code text. Table 1 and Table 2 show the hardware and software requirements specification respectively.

Download Website. https://www.kaggle.com/datasets/kmader/siim-medical-images.

Table 1. Hardware Requirements Specification.

Hardware Requirements	Minimum Requirements
Processor	i5 or above
Hard Disk	10 GB
RAM	8 GB
Monitor	13" Coloured
Mouse	Optical
Keyboard	122 Keys

Table 2. Software Requirements Specification.

Software Requirements	Minimum Requirements
Platform	Windows XP/7, Linux or MacOS
Operating System	Windows XP/7, Linux or MacOS
Technology	Machine Learning - Python
Scripting Language	Python
IDE	Jupyter Notebook

3 Result

3.1 Result Finding

The point at which the client transfers the clinical record, the encryption cycle is finished. The clinical record is encoded by asymmetric strategy - Triple DES. A secret key that is created by the confidential key generator. Presently the first record is changed over into the code picture. Now, this encoded code picture is kept safely in cloud storage with the assistance of a cloud specialist co-op (see Fig. 1).

Decoding is the method involved with changing over encoded information into the plain text. The code picture put away in the cloud specialist co-op is decoded into the

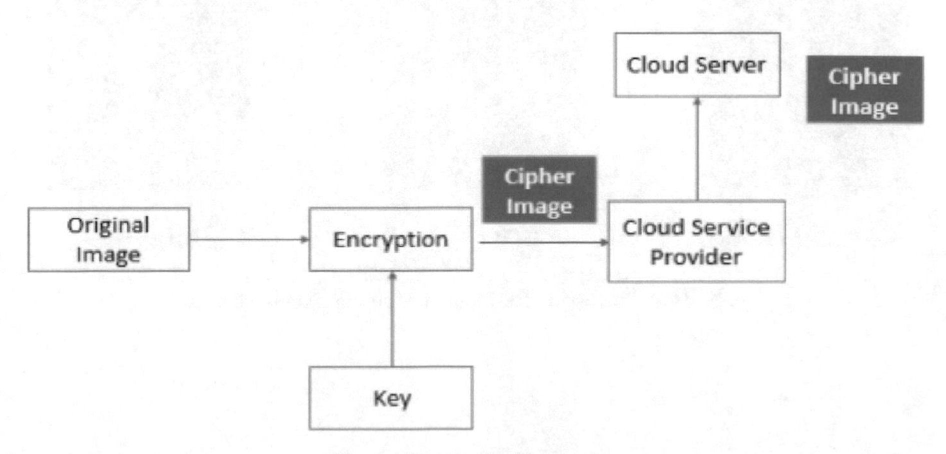

Fig. 1. Process for Encryption

first and foremost information with the help of the secret key which is produced by the PKG and the first information is made accessible to the client (see Fig. 2).

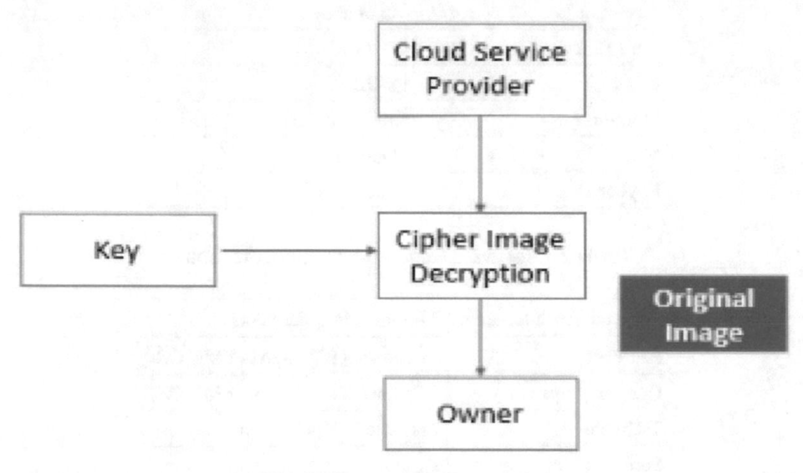

Fig. 2. Process for Decryption

3.2 Result

Input Determination. A plain clinical picture is considered as the input information. Cipher-text is the design for each input information (after encoding).

Output Determination. A plain clinical picture is the output data (after decoding). Cipher-text is the design for each result information. The following Fig. 3 shows the input and output data processing.

Fig. 3. Data Processing for Input and Output Specifications.

3.3 Discussion

Cloud Computing is an arising worldview with the objective to offer efficient resources for computing, gigantic data stockpiling limit and adaptable data sharing administrations. The objective of this project is to improve healthcare data security in cloud environment. The project deals with analyzing the feasibility of using artificial intelligence technologies to improve healthcare data security in cloud computing environment.

The great development of data persuaded business and clients and was driven by the cloud-top elements to outsource their information to the cloud storage frameworks. Nonetheless, the secrecy and honesty of the re-appropriated delicate information in remote cloud administrations are turning into a central issue. Information should be encoded preceding putting away it in the possibly conniving cloud. Existing customary encryption frameworks force a significant weight of overseeing documents and encryption of procedure on information proprietors. They experience the ill effects of serious usability, efficiency and security issues and a few plans are improper for safeguarding the data on the cloud.

4 Conclusion

For protecting and safeguarding the data of a patient, a model for the cloud data security is developed in the Electronic Health Record (EHR) system wherein double level of security has been offered. By utilizing the proposed procedure, the throughput can be expanded by 20% whereas the time consumption and computational complexity can likewise be diminished altogether compared to the existing procedures.

4.1 Inference

Relocation of an association's data to the cloud is a vital and complex choice. Prior to moving information into the cloud, the security difficulties ought to be moderated. A decent cloud specialist co-op ought to screen the safeguarded healthcare data life cycle. Different safety efforts like type of authentication and encryption techniques, detection for intrusion, verification procedures and firewalls ought to be additionally checked.

4.2 Future Scope

Machine Learning and Artificial Intelligence would be the key in consistence endeavors going ahead. Organizations are hoping to robotize few of their regulatory consistence processes, involving extraction and data location. Inventories, like security master Michael Cobb noted, become obsolete except if robotized examining devices are conveyed to support information disclosure catch by recording customary depictions of all repositories and applications where individual data dwells. Automation, as he would see it, is the main way enormous associations can stay consistent with a huge volume of information that is organized and unorganized and put away in the cloud and in the huge data centers.

Cutting edge technology could likewise assist organizations with conforming to other consistence commands, like PCI DSS. Few security specialists recommend considering a zero-trust model as a security methodology for the firms and organizations which have lagged behind on consistence. Owing to zero trust, organizations would take a gander at the full lifecycle of information the executives and widen their concentration past installment card information to different types of individual information, involving monetary information, user data and intellectual property. They would make not a single supposition on where information is normal in sight or the way things are being utilized, just that the risk should be relieved.

Data security would stay a huge test into the future, however imaginative uses of zero-trust models, machine learning and artificial intelligence would help infosec teams and information technology safeguard information and guarantee user privacy.

4.3 Limitation

Security is the principal issue that frustrates the quick reception of the cloud computing technology in the medical services industry. The qualities and advantages of cloud computing far surpass its risks and dangers. Security prerequisites are progressively challenging to meet without a critical interest in manpower and infrastructure. The difficulty is that security is adversely relative to user accommodation. As such, the more modern the safety efforts, the less agreeable the users, and accordingly, they will be less disposed to utilize the cloud administration.

In this paper, I observed that the studied arrangements are not comprehensive in nature, those approaches to some extent settle the challenge for security and privacy. A large portion of those arrangements address a contributor to the issue, and they neglected to adjust all going against security necessities. The issue is that an increase got in one aspect causes a misfortune in another aspect. In future, I will propose an all-encompassing arrangement that endeavors to adjust all going against necessities.

References

1. Joshi, M., Joshi, K., Finin, T.: Attribute based encryption for secure access to cloud based EHR systems. In: 2018 IEEE 11th International Conference on Cloud Computing (CLOUD), San Francisco, CA, 2018, pp. 932–935, (2018) https://doi.org/10.1109/CLOUD.2018.00139
2. Nishad, L.S., Kumar, A.S., Beniwal, S.: Security, privacy issues and challenges in cloud computing: a survey. In: ICTCS '16: Proceedings of the Second International Conference on Information and Communication Technology for Competitive Strategies, March 2019, Article No.: 47, pp. 1–7 https://doi.org/10.1145/2905055.2905253
3. Sabharwal, R., Rizvi, S.W.A.: Impact of Convolutional Neural Networks for Recognizing Facial Expressions: Deep Learning Perspective. In: Sharma, H., Vyas, V.K., Pandey, R.K., Prasad, M. (eds.) Proceedings of the International Conference on Intelligent Vision and Computing (ICIVC 2021), pp. 74–84. Springer International Publishing, Cham (2022). https://doi.org/10.1007/978-3-030-97196-0_6
4. Liu, J., Tang, H., Sun, R., Du, X., Guizani, M.: Lightweight and privacy-preserving medical services access for healthcare cloud. IEEE Trans. Dependable Secure Comput. **7**, 106951–106961 (2019). https://doi.org/10.1109/ACCESS.2019.2931917
5. Mukherjee, M., Ferrag, M.A., Maglaras, L., Derhab, A., Aazam, M.: Security and Privacy Issues and Solutions for Fog. In: Yang, Y., Huang, J., Zhang, T., Weinman, J. (eds.) Fog

and Fogonomics: Challenges and Practices of Fog Computing, Communication, Networking, Strategy, and Economics, pp. 353–374. Wiley (2020). https://doi.org/10.1002/9781119501121.ch14
6. Zhang, L., Cui, Y., Mu, Y.: Improving security and privacy attribute based data sharing incloud computing. IEEE Syst. J. **14**(1), 387–397 (March 2020). https://doi.org/10.1109/JSYST.2019.2911391
7. Kumar, A., Krishnamurthi, R., Nayyar, A., Sharma, K., Grover, V., Hossain, E.: A Novel Smart Healthcare Design, Simulation, and Implementation Using Healthcare 4.0 Processes. IEEE Access **8**, 118433–118471 (2020). https://doi.org/10.1109/ACCESS.2020.3004790
8. Rizvi, S.W.A., Singh, V.K., Khan, R.A.: The state of the art in software reliability prediction: software metrics and fuzzy logic perspective. Adv. Intell. Syst. Comput., Springer **433**, 629–637 (2016)
9. Catherine, M., et al.: Survey on AI-Based Multimodal Models for Emotion Detection. In: High-Perf, pp. 307–324. Springer International Publishing, Modelling Big Data Applications (2019)
10. Shu, L., et al.: A review of emotion recognition using physiological signals. Sensors **18**(7), 2074 (2018). https://doi.org/10.3390/s18072074
11. Kim, D.M., Baddar, W.J., Jang, J., Ro, Y.M.: Multi-objective based spatio-temporal representation learning robast to expression intensity variations for expression recognition. IEEE Trans. Affect. Comput. **10**(2), 223–236 (2019)
12. Rizvi, S.W.A., Khan, R.A.: Improving software requirements through formal methods. Int. J. Inform. Comput. Technol. **3**(11), 1217–1223 (2013)
13. Pantic, M., Valstar, M., Rademaker, R., Maat, L.: Web-based database for facial expression analysis, In: IEEE International Conference on Multimedia and Expo, p. 5, (2005) https://doi.org/10.1109/ICME.2005.1521424
14. Valstar, M.F., Jiang, B., Mehu, M., Pantic, M., Scherer, K.: The first facial expre. recognition and analysis challenge. In: Face and GCesture, pp. 921–926, (2011) https://doi.org/10.1109/FG.2011.5771374
15. Parveen, H., Rizvi, S.W.A., Shukla, P.: Disease Risk Level Prediction using Ensemble Classifiers: An Algorithmic Analysis. IEEE Xplore INSPEC Accession Number: 21662591 In: 12th International Conference on Cloud Computing, Data Science & Engineering (Confluence), https://doi.org/10.1109/Confluence52989.2022.9734121, March 2022, pp. 585–590 (2022)
16. Anagnostopoulos, C.-N., Iliou, T., Giannoukos, I.: Features and classifiers for emotion recognition from speech: a survey from 2000 to 2011. Artif. Intell. Rev. **43**(2), 155–177 (2012). https://doi.org/10.1007/s10462-012-9368-5
17. Sariyanidi, E., Gunes, H., Cavallaro, A.: Automatic analysis for facial: a survey of registration, representation, and recognition. IEEE Trans. Pattern Anal. Mach. Intell. **37**(6), (2014) https://doi.org/10.1109/TPAMI.2014.2366127
18. Parveen, H., Rizvi, S.W.A., Shukla, P.: Parametric analysis on disease risk prediction system using ensemble classifier. In: Lecture Notes on Data Engineering and Communication Technologies, https://doi.org/10.1007/978-981-16-9113-3_53, June 2022, ISSN 2367–4512, ISSN 2367–4520 (electronic), Vol. 111, pp. 719–737
19. Rizvi, S.W.A., Singh, V.K., Khan, R.A.: Software reliability prediction using fuzzy inference system: early-stage perspective. Int. J. Comput. Appl. **145**(10), 16–23 (2016)
20. Lucey, P., Cohn, J.F., Kanade, T., Saragih, J., Ambadar, Z., Matthews, I.: The Extended Cohn-Kanade Dataset: A complete dataset for action and emotion-specified expression. In: 2010 IEEE Computer Society Conference on Computer Vision and Pattern Recognition – Workshops, pp. 94–101 (2010) https://doi.org/10.1109/CVPRW.2010.5543262

Statistical Assessment of Spatial Autocorrelation on Air Quality in Bengaluru, India

Jyothi Gupta(✉)

School of Architecture, CHRIST (Deemed to be University), Bengaluru, India
ms.gupta.18@ucl.ac.uk

Abstract. *Background* The term "air quality" refers to how clean or unhealthy the air is. Polluted air may be harmful to human health as well as the health of the ecosystem, thus it's crucial to monitor air quality. Through previous studies it has explained the significance of outdoor air quality using spatial techniques - Geographical information System (GIS).

Objective Using Spatial autocorrelation techniques - GIS and Moran's I to analyze the spatial patterns for air quality in Bengaluru, Karnataka, India, 2015.

Methods Ten years of air quality data for Bengaluru from 1995 to 2015. The temperature, annual rainfall and CO_2 content in air was considered. Each 198 wards of Bengaluru were geocoded in an R Project tool with spdep library. The hedonic regression model was applied to estimate the GIS based distribution of air contaminants for each ward. To investigate the geographical connection between temperature, rise and yearly rainfall, spatial auto-correlation method (Global Moran's I) was used. The primary reason and the importance of spatial autocorrelation are that statistics rely on observations being independent of one another.

Results At the Central Business District (CBD) of Bengaluru, concentrations of CO2 showed statistical significance spatially clustered. The p-value is 0.74 and Moran I standard deviation is −0.65 whereas in monte-Carlo simulation of Moran I the statistic -s 0.006 with p value 0.77. It examines positive and negative correlation for the normal distribution due to air quality. Finally, autocorrelation is visualized in a GIS map, then the assumptions of observations independent of one another are broken.

Conclusion These results help to inform that the air pollution problem is increasing temperature and annual heavy rainfall in Bengaluru, 1995–2015. Interesting Spatial patterns were observed using spatial autocorrelation modeling and GIS in Bengaluru.

Keywords: Spatial Pattern · Carbon dioxide · Spatial Autocorrelation (SPAC) · Air quality · GIS · Urban Models · Geospatial · Smart cities

1 Introduction

1.1 India Air Quality Index (AQI)

India is a developing country with its GDP growth increasing by 8.2% in 2018, and its economic growth must excel to 7.4% in the upcoming financial year referred to by

H. Sharma et al. (Eds.): ICIVC 2022, PALO 17, pp. 254–265, 2023.
https://doi.org/10.1007/978-3-031-31164-2_21

Wikipedia, [23] With the change in the service sectors, there are many infrastructure activities in progress and various new investments adding to India's development. In addition, with the rise in population, there is an increase in vehicle movements which was not designed presently for the growth rate. Figure 1 represents the increase in CO2 content in time for air quality per million, the statistical data studied for Bengaluru. This impact leads to a concern about the emission of carbon dioxide in the environment affecting air quality in the country. This paper models the rise in environmental factors caused by the increase in traffic concerning spatial patterns. The geospatial patterns of the leading mode of SPAC are similar and positively correlated. Fa-xiu et al. [1] proposed that the air quality is not closely correlated to the North-west region compared to the meridian zone of India with increased traffic congestion and growth in population. Goodchild [2] projected the air impact is a significant concern with the high CO_2 and other pollutant matter coupled with SPAC methods, which measure spatial dependence.

There is an increased air pollutants component due to the spread in road widening in urban areas to ease the traffic flows. Font et al. [3] In this paper, an approach to SPAC model and analyze the data points of rising temperature, carbon dioxide emission, and other pollutants in the environment are presented, caused by traffic. Nandita et al. [4] results show that Bengaluru being the silicon city of India, the city center has critical air impurities causing an impact on the environment and people's health. Air pollution is mostly caused by population increase, industrial activity, transportation patterns, and urbanizations.

1.2 WHO Standards for Air Quality

Major cities' air conditions have deteriorated significantly because of rapid industrialization and urbanization. As per WHO reports, 92% of the public around the planet are exposed toward air pollution at levels over the WHO limit. Tian et al. [5] emphasizes on Human illness and mortality as well as harm to the natural and built environments are linked to pollutants in air, incorporating both nature-occurred and man-made compounds. Hong [6] underlines that urban city core air quality index (AQI) continues to deteriorate, hurting housing sectors surroundings and endangering people's health despite the concerted efforts of both international and national organizations over the past decade to reduce significant pollutant emissions. The use of information and communication technology (ICT) develops to solve the problems of enhancing city air pollution levels. Through and through a network of widely spaced monitoring stations, antenna systems offer a potent instrument for real-time air quality monitoring. To monitor air quality in close to real- time, portable air pollution sensors and GPS technology are added to an existing sensor network. In many nations, the public may access air quality information systems. Hong [6] expressed that with the spread of wireless communication 4G, 5G and mobile technologies, individuals may now readily contact air quality data streaming services without geographical or spatial restrictions thanks to mobile devices like smartphones and tablets. The public's knowledge of urban atmospheric issues has, however, increased, and air pollution is increasingly recognized as a social as well as an environmental concern.

In several European cities, the air quality regulations are exceeded. Given that poor air quality is thought to be responsible for 240 000 premature deaths annually in the UK alone, this poses a danger to public health. Air quality states must get rid of standard exceedances for a variety of pollutants by specified deadlines. Mitchell et al. [7] discussed the European countries who have created a national air quality plan standard (NAQS) that details policies, activities, plus roles for managing air quality to accomplish these goals. In accordance with the NAQS, local governments oversee determining air quality control zones and action plans in places where goals are not anticipated to be attained [7]

1.3 Air Pollutants Impact

The objective of this study is to use Spatial autocorrelation techniques - GIS and Moran's I to analyze the spatial patterns for air quality in Bengaluru, Karnataka, India, from 1995 to 2015.

A dynamic and intricate system, the air environment. Some pollutants, including SO2, NO2, PM10, and O3, have an impact on the quality of the air. Pollutant concentrations are continually changing. But the observed information utilized in studies are often gathered over a time, such as per-sixty minutes, next-days, one-month, and one-year average data. The collection and analysis of direct live data generated every second per minute is challenging. As a result, this method of collecting is viewed as a grey system. In a hoary system, certain pollutants data is known but others are unknown. Ren et al. [8] signified the India's Ambient Air Quality Index Standards (NAAQS) are qualifying pollutants established by the Condensation Particle Counter Battery (CPCB) and are listed in Table 1 netted from Ripinen et al. [9].

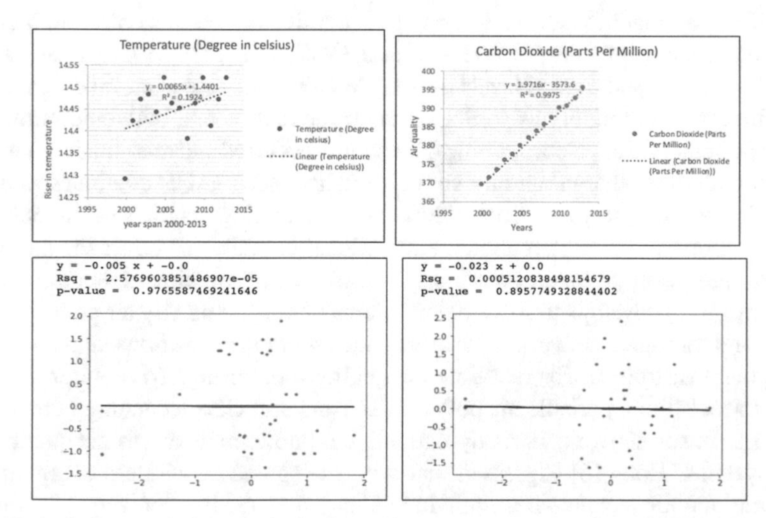

Fig. 1. Hedonic Regression Analysis for Annual temperature and carbon dioxide with statistical data

Nagendra et al. [10] summarized the AQI criteria into a normative scale. Such high levels and frequency of air pollution exacerbate the problem by making it difficult for citizens to assess existing air quality. Kumar [11] perspective, AQIs are commonly used

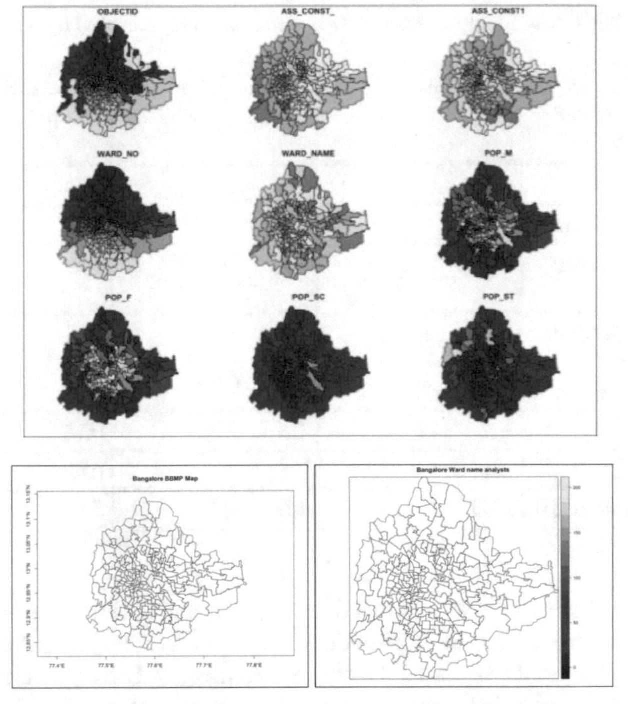

Fig. 2. Bengaluru City Population Census 1995 data with 198 wards governed by the Bruhat Bengaluru Mahanagara Palike (BBMP)

by local environmental authorities to indicate the state of existing air quality. Steven et al. [12] and Murena [13] concluded that air quality index (AQI) is the most accurate indicator of how well the air is for human health.

There are numbers created due to the accumulation of NOx, SO2, O3, CH4, and other impurities in the urban areas of Asia predominately owing to study by Irfan [14] and Paliwal [15]. Barros [16] interesting charts out London dynamics and segregation of areas with low, medium, and high AQI zones.

In the study quoted by Tobler [17] from the University of California, the first rule of GIS is Tobler's First Law of Geography Everything is connected to everything else. However, things that are close together have a stronger connection than those that are far apart.

2 Data and Methods

Bengaluru City Climate Data has been collected to measure the air quality for ten years for a span from 1995 to 2015. Data is available online and can be downloaded for research purposes. The variables were Temperature (high/low) in degree Celsius and rainfall in days/months' timeline as shown in Fig. 2. Bengaluru city is under Tropical zone considering its geographic location.

Table 1. Ambient Air Quality Index (AQI) Indian Standard presented by Riipinen et al. [9]

Pollutant (24/8 hourly value) Time Weighted Average (TWA)	Industries groups	Residential, rural groups	Seniors, Heart Problem groups
SO_2 Sulfur dioxide	120	80	30
NO_2 Oxides of Nitrogen	120	80	30
Suspended Particulate Matter (SPM)	500	200	100
Respirable Suspended Particulate Matter (RSPM)	150	100	75
Lead (Pb)	1.5	1.0	0.75
Ammonia (NH_4)	0.4	0.4	0.4
Carbon Monoxide (CO)	10.0	4.0	2.0

Bengaluru is facing excess unplanned transportation CO2 emission. Wikipedia [26] helped this study to collect Carbon dioxide data from community in different area of the Bengaluru city. Rspatial [27] eased the shape data files for BBMP Bengaluru city with 198 wards were downloaded for research purposes represented in Fig. 2.

The research methods used started with hedonic regression model in Fig. 1 to estimate the GIS based distribution using R '#spdep' library. Further to investigate the statistical assessment between variables, SPAC Global Moran's I and Geary ratio is used. The importance of these spatial coordinates relies on the observation in Table 2 generated by R for autocorrelation between dimensions.

3 Methodology

Globally there is a rising concern raised by Srebotnjak [18] and Woo et al. [19] regarding the impact of pollution in significant parts of the world, especially in the industrial areas, urban areas of the metropolitan countries, which have been developing uncontrolled transportation zones. In Asia, pollution in cities is growing in the industrial zones by about 100–254 ug/m^3 than the most developed nations. Indeed, North America and Europe have a pollution concentration of 30–60 ug/m^3, as described by Cohen et al. [20].

In addition, the air quality data was available for an increase in carbon dioxide content in the air with a steep growth with its coefficient as 0.99 for ten years. However, the annual Temperature and rainfall rate shows a significance score with the line flat on the data points, which is an unusual case in the statistical model for forecasting the accuracy. The list below helps us to apply main elements in urban growth and urbanizations.

- Hasty increase in industrialization
- Dehydrated environment leading to dry and humid weather

- Expansion in regions
- Unforeseen transportation in countries
- Burning of low superiority coal (oil, gas, energy), petroleum (diesel) and other fuels.

Table 2. Statistical matrix of SPAC result created using R programming

Monte-Carlo simulation of Moran's I	Number of simulations +1:600
Statistic	−0.006
Observed Rank	138
p-value	0.77–0.74
Moran I Standard deviation	−0.65
Coefficient	−0.0061
Alternative Hypothesis	Greater

The capital of India, Delhi, is facing challenges around the levels of pollution leading to highly polluted air as per the table below is data for Delhi for PM10. People in other cities like Mumbai, Bengaluru, Chennai, Kolkata are experiencing a significant crisis of pollutants which is an essential factor of health and other correlated factors to diseases in the country leading to a lesser rate of life expectancy, especially in India. The air is arguably unhealthy, caused by the increased population, with added movements and traffic on the road, which was not built to take the load. Urbanization needs to be with widening highways and public transport or shared vehicles to work, reducing the impurities to accumulate in the environment.

3.1 Spatial Analysis

In this study, we explored the spatial pattern of data points in ArcMap GIS software. The visualization on the GIS data and statistics, noted earlier, is based on a buffer zone of 200 m on the major roads in Bengaluru with a significant concentration of traffic pollutants above the threshold levels. The observation was created in Fig. 3(a)(b)(c)(d) of on-air impurities with different buffer zones ranging from 50 m to 600 m to study the relationship between roads and air pollution density, classified and calculated in other areas. This measure is noted and tabulated as a ratio of the proportion of class area to the total buffered size giving the density of pollutants on the road, which could be geo-referenced and digitized. Nevertheless, researchers have calculated and tested the significance coefficient to accumulate impurities on the road with maximum traffic movements. This study cannot conclude their statistical analysis created using the SPAC model.

3.2 Spatial Autocorrelation (SPAC) Techniques

Geo-Spatial autocorrelation was introduced in this paper to analyze the air quality identified as similar objects to other neighbouring things. However, 'Auto' means self, and

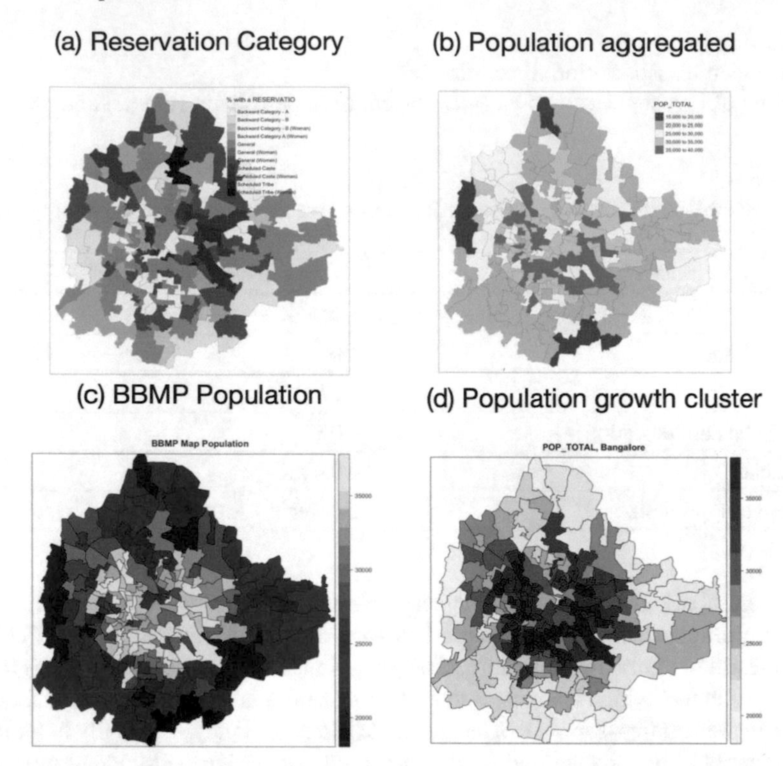

Fig. 3. Spatial Analysis and GIS data visualisation for Bengaluru city

'correlation' implies association. Hence, it measures how close the objects are located and mapped with each other. The reason to use the spatial auto-correlation model is to help analyze the data clusters and dispersion across the country. Research Questions for our paper such as 'Is the air quality in big cities an isolated case' or 'Is the particular air pattern clustered spread across places' can we well understand and answer using autocorrelation analysis.

While we examine Spatial Patterns of Air quality in India, Karnataka and for this article Bengaluru in Fig. 4(a) (b) due to increasing traffic, we use the empirical method. The method explains the traffic congestion grows more than proportionately beside the size of the city. So, the measure of air quality for city size when our statistical parameter could compute a power-law scaling formula for the air quality below.

3.3 Equations

The Eqs. 1–3 are not unique, there is an additional cross-product term in the denominator of Geary equation whereas in Moran it is using deviations from the mean and not directly computed. These are two important pointers for spatial autocorrelation. These global indicators are used in the presence of spatial autocorrelation. In terms of latitude and longitude, India is located around the equator between 6.44 and 35.30° north and 68.7 and 97.25° east. Thus, the coordinate system is set in the R file for GIS mapping analysis.

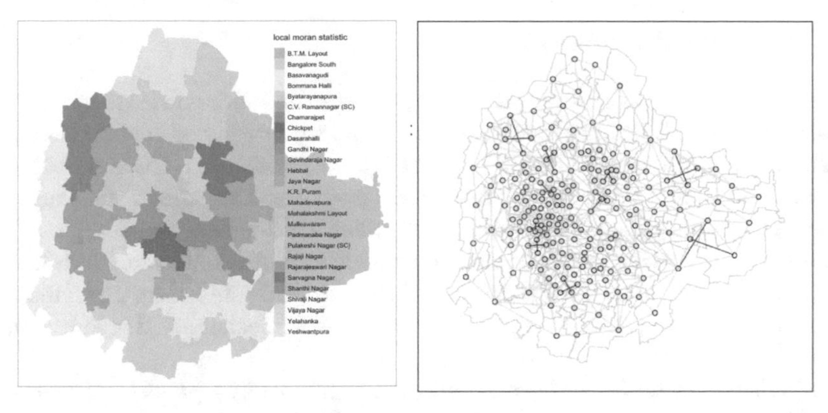

Fig. 4. **(a)** Global Autocorrelation using Moran I technique **(b)** Inverse distance for Moran's I

Result of Moran's I shall be between from − 1 to + 1. Positive spatial autocorrelation is indicated by values extensively above −1/(N−1) while negative spatial autocorrelation is indicated by values drastically below −1/(N − 1). Moran's I value can be converted to z-scores for the purpose of evaluating the statistical hypotheses for this study.

"Moran's I is inversely related to Geary's C, but it is not identical" explained in Wikipedia. However, Geary's C is more sensitive to local spatial autocorrelation than Moran's I, which is a measure of global spatial autocorrelation.

The expected value E(I) for the Maran's I for spatial Autocorrelation could be defined. Big data samples from a large population (where N leads to infinity), the E(I) = 0. The z score for the statistical function is further computed. The relationship between a value and the mean of a group of values is quantified by a Z-score. The Z-score is expressed in units of standard deviations from the mean. Mitchell et al. [7] believed when a data point's Z-score is 0, it means that it has the same score as the mean.

SPAC – Spatial Autocorrelation is represented by Eq. 1.

$$\text{SPAC} = \frac{\sum_{i=1}^{n} \cdot \sum_{j=1}^{n} \cdot c_{ij} w_{ij}}{\sum_{i=1}^{n} \cdot \sum_{j=1}^{n} \cdot w_{ij}} \tag{1}$$

$$c = \frac{(N-1) \sum i \sum j w_{ij} \ \left(x_i - x_j\right)2)}{2\left(\sum i \sum jj w_{ij}\right) \sum i\left(x_i - \underline{x}\right)2} \tag{2}$$

$$I = \frac{N \sum i \sum j w_{ij} \ \left(x_i - \underline{x}\right) \left(x_j - \underline{x}\right)}{\left(\sum i \sum jj w_{ij}\right) \sum i\left(x_i - \underline{x}\right)2} \tag{3}$$

Here in Eq. 1–3: n - number of object variables (pollution, infrastructure)

i,j – object variable of data.

x_i – Value of the object i attribute.

c_{ij} – degree of correlation of i and j attributes.

w_{ij} – degree of correlation of i and j location.

$ifc_{ij} = \left(x_i - x_j\right)2$ Geary c ratio is defined in Eq. 2 of the object i,j attribute.

$ifc_{ij} = \left(x_i - \underline{x}\right)\left(x_j - \underline{x}\right)$ Moran Index I is defined in Eq. 3.

4 Results and Discussion

The spatiotemporal distribution in different properties and variations at the ground level is plotted in Fig. 5. Literature review, Author Singh has researched about transportation and climatic implications. Although there are few traces of gasses (CO and NO2) which are active and significant in the cross-sectional evidence. This study data analysis will be useful in deciding the uncertainties in big cities development and at regional level. The ratio and the distribution of the data points shows the data collections and clusters of climate models and improvised properties of particle size distribution. Thus, the spatial patterns between air quality for India southwest zone especially in Bengaluru city are tabulated.

Referring to the result, we have 11 links in Bengaluru city regions from rook's algorithm. Chowdeswari ward, Kengeri, Yestwanpura, Mallasandra, Sanjaya Nagar, Hebbal Shivaji Nagar, Varthuru, Kadugodi, Singasandra, Marathahalli with 3 links. One connected region: Agaram with 11 links. Now we have queen and weight list objects. There were 7 least connected regions Chowdeswari ward, Kengeri, Mallasandra, Sanjaya Nagar, Hebbal, Kadugodi, Singasandra with 3 links. However, three most connected regions are identified. HMT ward Sampangiram Nagar Agaram with 11 links. The Fig. 5 is the constructing spatial weight result out using R programming.

This short paper has focused on causes of transport in India urban areas which have a vast number of industrial projects and attraction to these pollutants with the case study of Bengaluru on their major roads of pollutant spatial patterns. It is also concluded that tools like GIS are central to creating techniques such as buffering and transforming the data on a visualization front as well as providing statistics via R programming for coefficient values score. The study of air pollution is a critical topic in study for cities throughout the world whether it's Europe, America or Asia countries like India and China.

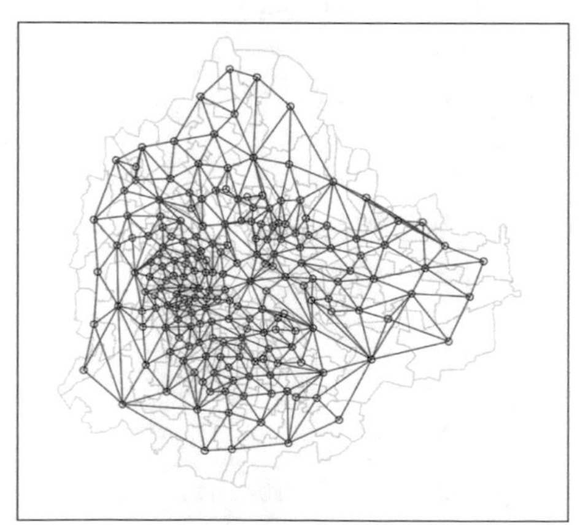

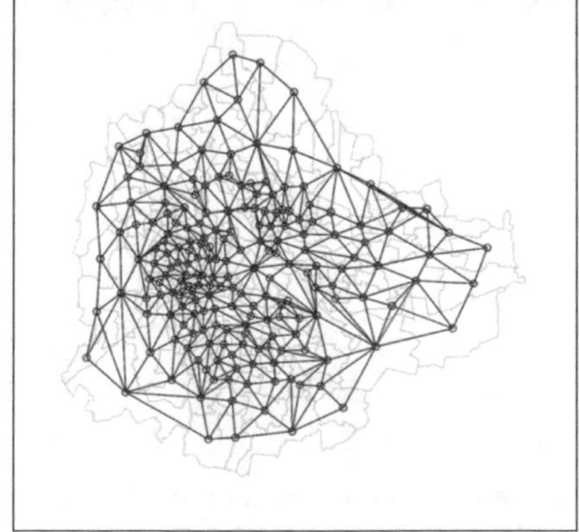

Fig. 5. Constructing spatial weight matrices using the k-nearest neighbour's criteria and 'Queen' and 'Rook' method. (a) Queen (b) KNN 2-Neigh

Breathing such air can affect the heart and cause chronic cardiovascular problems. Air quality has been worsening across India in bigger cities amid falling temperatures.

Although, the only good news with this COVID-19 pandemic is the hazardous air quality has become more breathable as compared during the year 2018–19. We therefore end with a sensible level of assurance that air quality properties must be very seriously considered for climate models, and we need to predict the health interactions across countries like India where the standard of living needs to be concentrated and not to be neglected.

Inadequately maintained mobility for the main cities and in-efficient infrastructure have seen an increase in the amount of pollution affecting people's health and attraction of severe disease and illness specifically for seniors and children which are sensitive to the environment. Further research is planned on the pollutants and quality of air in other states and cities of India to provide a more precise study. Gupta [21] assumed that Environmentalists claim that a significant portion of Bengaluru's air pollution is brought on by vehicle traffic. The city experienced an 80% increase in respiratory ailments brought on by air pollution between 2015 and 2020. Gupta et al. [22] studied air pollution in the construction industry, Bengaluru is far better than Delhi but significantly worse than major US cities.

A Greenpeace Southeast Asia analysis ranked Bengaluru as the third worst Indian city for air pollution in 2020, with 12,000 projected deaths. At 7.30 in the early hours, the city logged an AQI Index of 101, enlisting the city within top ten amongst India's maximum contaminated capitals. Monday morning, Bengaluru AQI index peaked to 101 again, listing it as a polluted city in India. Researcher said, "Vehicular emissions have decreased in the city." Three factors, according to Khaja (Syed Khaja, senior environmentalist, KSPCB.), explain why the AQI has continued to be satisfactory. People were compelled to stay inside during Covid-19, which decreased the city's pollution levels. "After Covid, climatic factors are causing levels to decline," reported by Karnataka State Pollution Control Board (KSPCB).

Cohen et al. [20] works have found that due to construction and real estate growing demand in Bengaluru, there is a crowd accumulating in the central district. Thus, increasing the air pollutant through construction and building industrial challenges.

5 Conclusion

The study emphasized the basic statistical knowledge on "air quality" in metropolitan Bengaluru city in India. The AQI calculation helped us to comprehend the harmful impact of pollutants of air by Central Control record [24]. Alonso et al. [23] used the statistical assessment of the SPAC model to capture the weight matrix in two methods exploring the linkages based on the complexity of networks. Significance score difference in interesting spatial patterns in south India is observed. Times' magazine [25] requested for further research on more cities of India highlighting the concentration of air pollutants and its strategies. We also need to infer the post Covid-19 situation by measuring the air quality for major metropolitan or crowded urban cities in India. It has been stressed at this point that air pollution is measured using GIS mapping techniques. As a conclusion, the statistical results help us to conclude that there is change in excess temperature and heavy rainfall during the 10 years' time span 1995- 2015 of Bengaluru city. The range of these autocorrelated parameters where spatially patterns at central Bengaluru have been indicated. Therefore, further readings and false positive correlation on specific air pollutants are not enough to come to a final recommendation, hence further assessment shall be continued with comparison with other parameters.

References

1. Fa-xiu, Z., Yi, Z., Fei, H., Qi, L.: Spatial pattern of the air-sea interaction near the South China Sea during winter. Chin. J. Oceanol. Limnol. **17**(2), 132–141 (1999). https://doi.org/10.1007/BF02842711
2. Goodchild, M.F.: GIScience, Geography, Form, and Process. Annals of the Association of American Geographers, 2004 - Taylor & Francis, vol. 94, No. 4, 2004, p. 7, 2008, https://www.jstor.org/stable/3694087
3. Font, A., Baker, T., Mudway, I.S., Purdie, E., Dunster, C., Fuller, G.W.: Degradation in urban air quality from construction activity and increased traffic arising from a road widening scheme. Sci. Total Environ. **497–498**, 123–132 (2014). https://doi.org/10.1016/j.scitotenv.2014.07.060
4. Nandita, S., Banerjee, T., Deboudt, K., Sorek-Hamer, M., Singh, R.S., Mall, R.K.: Aerosol chemistry, transport, and climatic implications during extreme biomass burning emissions over the Indo-Gangetic Plain (2018). https://doi.org/10.5194/acp-18-14197-2018
5. Tian, M., et al.: Characteristics of aerosol pollution during heavy haze events in Suzhou. China. Atmos. Chemistry Phys. **16**(11), 7357–7371 (2016). https://doi.org/10.5194/acp-16-7357-2016
6. Hong, S.: Air quality context information model for ubiquitous public access to geographic information. ISPRS Int. J. Geo Inf. **7**(8), 316 (2018). https://doi.org/10.3390/ijgi7080316
7. Mitchell, G., Namdeo, A., Milne, D.: The air quality impact of cordon and distance based road user charging: An empirical study of Leeds, UK. Atmos. Environ. **39**(33), 6231–6242 (2005). https://doi.org/10.1016/j.atmosenv.2005.07.005
8. Ren, X., Luo, Z., Qin, S., Shu, X., Zhang, Y.: A new method for evaluating air quality using an ideal grey close function cluster correlation analysis method. Sci. Rep. **11**(1), 1–7 (2021). https://doi.org/10.1038/s41598-021-02880-1
9. Riipinen, I., et al.: Applying the Condensation Particle Counter Battery (CPCB) to study the water-affinity of freshly-formed 2–9 nm particles in boreal forest. Atmos. Chem. Phys. **9**(10), 3317–3330 (2009). https://doi.org/10.5194/acp-9-3317-2009
10. Shiva Nagendra, S.M., Venugopal, K., Jones, S.L.: Assessment of air quality near traffic intersections in Bangalore city using air quality indices. Transp. Res. Part D: Trans. Environ. **12**(3), 167–176 (2007) https://doi.org/10.1016/j.trd.2007.01.005
11. Kumar, R.G.P.: The Air Quality Assessment of Northern Hilly City in India (2018) Kalpana Corporation. http://ijep.co.in/, https://openresearch.surrey.ac.uk/esploro/outputs/journalArticle/The-Air-Quality-Assessment-of-Northern-Hilly-City-in-India/99515199102346 (Accessed Aug. 22 2022)
12. Bortnick, S.M., Coutant, B.W., Eberly, S.I.: Using Continuous PM2.5 monitoring data to report an air quality index. J. Air Waste Manage. Assoc. **52**(1), 9 (2002) https://doi.org/10.1080/10473289.2002.10470763
13. Murena, F.: Measuring air quality over large urban areas: development and application of an air pollution index at the urban area of Naples. Atmos. Environ. **38**(36), 6195–6202 (2004). https://doi.org/10.1016/j.atmosenv.2004.07.023
14. Irfan, U.: Why India's air pollution is so horrendous. Vox, (2018). https://www.vox.com/2018/5/8/17316978/india-pollution-levels-air-delhi-health (Accessed Aug. 22 2022)

15. Paliwal, A.: Why india struggles to predict the weather over its lands. The Wire. https://thewire.in/environment/imd-weather-prediction-forecast-monsoons-drought-agrometeorology-kharif (Accessed Aug. 22 2022)
16. Barros, M.S.J.: Dynamic segregation in London. Birkbeck, University of London, GISRUK Department of Geography (2018)
17. Tobler, W.R.: A computer movie simulating urban growth in the Detroit region. Econ. Geogr. **46**(sup1), 234–240 (1970). https://doi.org/10.2307/143141
18. Srebotnjak, T.: The role of environmental statisticians in environmental policy: the case of performance measurement. Environ. Sci. Policy **10**(5), 405–418 (2007)
19. Woo, C.-K., Horowitz, I., Moore, J., Pacheco, A.: The impact of wind generation on the electricity spot-market price level and variance: the Texas experience. Energy Policy **39**(7), 3939–3944 (2011)
20. Cohen, A.J., et al.: The global burden of disease due to outdoor air pollution. J. Toxicol. Environ. Health A **68**(13–14), 1301–1307 (2005)
21. Gupta, J.: Challenges of Project Deliverables in Construction Industry in Bangalore (2015). CEAESID- 2014, Print ISSN (2349–8404)
22. Jyothi, G., Subrahmanian, R.R.: Design Challenges of Project Deliverables in Construction Industry in Bangalore (2015), ISSN 2278–0181
23. Alonso, M.J., Wolf, S., Jørgensen, R.B., Madsen, H., Mathisen, H.M.: A methodology for the selection of pollutants for ensuring good indoor air quality using the de-trended cross-correlation function. Build. Environ. **209**, 108668 (2022)
24. "Central Control Room for Air Quality Management," https://app.cpcbccr.com/ccr/#/login. https://app.cpcbccr.com/ccr/#/login (Accessed Aug. 22 2022)
25. http://timesofindia.indiatimes.com/articleshow/80372978.cms?utm_source=contentofinterest&utm_medium=text&utm_campaign=cppst
26. https://en.wikipedia.org/wiki/Economic_development_in_India
27. https://rspatial.org/raster/analysis/3-spauto.html

Apoidolia: A New Psychological Phenomenon Detected by Pattern Creation with Image Processing Software Together with Dirichlet Distributions and Confusion Matrices

Silvia Boschetti[1](✉), Jakub Binter[1,2], and Hermann Prossinger[3]

[1] Faculty of Science, Charles University, Prague, Czech Republic
silvia.boschetti@natur.cuni.cz

[2] Faculty of Social and Economic Studies, University of Jan Evangelista Purkyně, ÚstíNad Labem, Czech Republic

[3] Department of Evolutionary Anthropology, University of Vienna, Vienna, Austria

Abstract. Traditionally, a strict dichotomy between belief and science is commonplace; recently, scientists investigating this complex phenomenon have designed studies to more thoroughly approach this dichotomy. Cognitive and perceptual processes developed under evolutionary pressures, especially stressful and harsh conditions, result in an enhanced perception of patterns. This has led to the emergence of a bias to perceive visual patterns that are manifestly random, a phenomenon called pareidolia ('overperception'), whereas the bias of assigning meaningfulness to general random patterns is apophenia. Methodological constraints in visual perception studies have led researchers to focus only on false perception—pareidolia. We extended the methodology by using image processing software to generate random maps and random maps of varying transparency overlying nonrandom patterns to identify not only false perception but also lack of perception. We used several questionnaires. For every query in every questionnaire, we classified two groups: those participants who never made a mistake versus those who made a mistake (which would be either perceiving a pattern where there is none or not perceiving an existing pattern) at least once. We then estimated the two groups' Dirichlet distributions of the responses, and calculated the confusion matrix to find significant differences. The Rational Experiential Inventory yielded a significant differentiation between the two groups. In addition to perceivers with pareidolia, we found that some perceivers failed to identify an existing pattern. We call this psychological phenomenon apoidolia ('underperception') —seeing no pattern when there is one. To our knowledge this is the first time this psychological phenomenon has been empirically detected.

Keywords: Apoidolia · Apophenia · Pareidolia · Dirichlet Distribution · Confusion Matrix

H. Sharma et al. (Eds.): ICIVC 2022, PALO 17, pp. 266–276, 2023.
https://doi.org/10.1007/978-3-031-31164-2_22

1 Introduction

In recent years, the number of scientific studies of religion and supernatural beliefs has increased, primarily due to novel approaches, novel methodologies and novel experimental procedures. One approach to study such phenomena is the cognitive study of religion, which focuses on those cognitive functions and mechanisms that support religious thinking and beliefs [1]. These mechanisms may have an evolutionary adaptive function in specific contexts and may therefore shape perception and interpretation of the perceived world [2]. In order to ensure survival, humans must rely on the ability to detect patterns and correctly identify their meaning; only then are they able to correctly/optimally respond [3]. This process of pattern identification is, however, subject to errors and mistakes. The error management theory (EMT) postulated that a false positive error (detecting a pattern where there is none) has an evolutionary advantage over a false negative error (not detecting a pattern where there is one), especially during stressful conditions. Indeed, the risk of not perceiving a dangerous animal (typically an almost camouflaged snake) is much higher than the cost of perceiving such an animal when it is actually not there [4]. The perception of patterns and 'meaningful' interconnections among elements—even when they are actually not present (i.e., illusory pattern perception) is called apophenia [5], and is named pareidolia when such phenomena arise in the context of visual stimuli [6]. In studies conducted on a non-clinical population, pareidolia was associated with different types of supernatural beliefs (including religious beliefs, beliefs in coincidence as well as beliefs in conspiracy and also magical beliefs) [7, 8]. Indeed, individual differences in perceptual processes during the elaboration and interpretation of external stimuli could work as substrate for religious and supernatural beliefs [9].

Previous studies mainly focused on false recognition of faces [7, 10]—face pareidolia —, which is seeing the presence of structure in objects with a collection of patterns that resemble the elements of (for example) a face; consequently, participants would therefore identify such patterns as images of faces. Since a face constitutes a very specific type of stimulus, with high evolutionary relevance for humans, a specific brain area is dedicated to processing these stimuli [6, 11]. The results obtained using such stimuli may not be repeatable when studying general illusory pattern perceptions (apophenia), however.

Our study aims to investigate the errors in pattern perception and identification without restricting the results to specific adapted stimuli (i.e. faces) and to clarify the relationship of such perceptual errors with beliefs and thinking styles.

We constructed our own stimuli by using a multidimensional random number generator in MATHEMATICA (from Wolfram Technologies®), to produce random maps (in which, therefore, no discernible patterns were present and the entropy was maximized) and then underlying them with identifiable patterns—concretely with geometric shapes such as octagons and triangles. Such shapes are common in nature [12]. We used these geometric figures as patterns in order to avoid biases involved in the perception of biological objects (such as leaves). We controlled the transparency of the random maps to manipulate the difficulty of perception and thereby the ambiguity of the stimulus.

Using these types of stimuli, we could investigate not only the presence of false positive errors in perception (pareidolic perception) but also for the presence of false negative errors (which we call apoidolic perception). To our knowledge, we are the first to identify and statistically evaluate both, pareidolic and apoidolic perception in a non-clinical population.

2 Materials and Methods

2.1 Participants

Our sample consisted of 105 participants from the Czech Republic, aged between 18 and 50 years (*mean* $\pm$ *SD* $= 33.7 \pm 7.7$ years); 67 were women (*mean* $\pm$ *SD* $= 33.6 \pm 7.5$ years) and 38 men (*mean* $\pm$ *SD* $= 33.8 \pm 8.1$ years). The participants were recruited from a science-oriented web community and the data were collected online.

2.2 Stimuli

We used a multidimensional random number generator to produce random maps of colored squares (henceforth called random patterns) and underlying them with several geometrical objects (Fig. 1). This method ensures a decrease in entropy as transparency is increased. We present three types of stimuli for each series: one in which the geometrical figures were fully covered with the random pattern (No Pattern condition), one in which the transparency of the random map was increased, allowing for the geometrical figures to be partially visible (Partial condition) and a third image in which the transparency had been further decreased and the geometrical figures were well identifiable (Reveal condition) while the (almost transparent) random map remained partially visible. We constructed three series of stimuli; each series contained one stimulus with three conditions (No Pattern, Partial and Reveal) for a total of nine experimental stimuli presented to every participant. To avoid learning bias during the data collection, we mixed the experimental stimuli with filler stimuli.

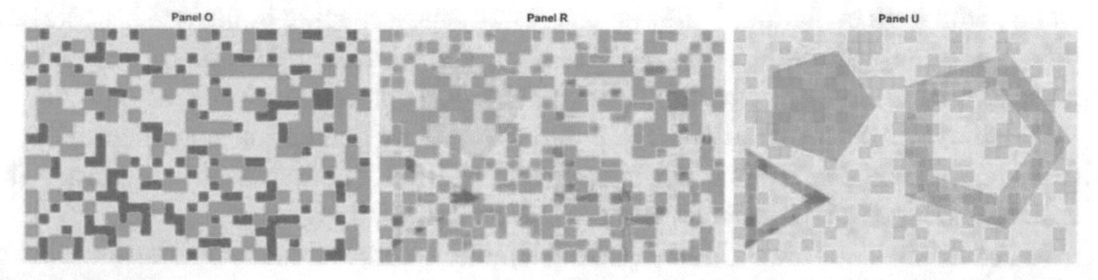

Fig. 1. Three of the patterns displayed to the participants. Panel O: the random pattern (No Pattern condition); Panel R: sample of the random pattern with underlying figures partially revealed (Partial condition), and Panel U: sample of the random pattern with underlying figures almost completely revealed (Reveal condition). The *x*- and *y*-coordinates of the colors (from a sample of a color triple) were randomly generated; the colors of the squares were also randomly chosen. The transparency of the random pattern was increased in seven steps from opacity (Panel O) to almost complete transparency (Panel U). The colors of the geometric objects were not drawn from the color triples, nor were they changed when increasing the transparency of the random overlay pattern.

2.3 Questionnaires

We used a suite of questionnaires (listed below) to determine different types of supernatural and religious beliefs and thinking styles. The questionnaires were translated into Czech by translators with psychological training; their reliability was confirmed by back-translation.

The Questionnaire on Coincidence [13] measures perceived coincidence experiences. This questionnaire has two parts; Part A measures the occurrence of episodes classified as different types of coincidence, while Part B investigates possible explanations for the occurrence of coincidences. Due to the type of study we were conducting and the analysis we chose in this paper, we currently used only Part A.

The Religion Commitment Inventory (RCI-10) [14] tests religious beliefs—especially the commitment to a religious group or community.

The Illusory Beliefs Inventory (IBI) [15] measures magical thinking and has three subscales to identify separate components: (a) magical beliefs (general beliefs in magic), (b) spirituality (beliefs in a higher power) and (c) thought-action-fusion (beliefs that thoughts can shape reality).

The Rational/Experiential Multimodal Inventory (REIm) [16] measures the rational and experiential thinking style as two separate ways to process information. The questionnaire consists of four subscales, one scale for rationality and three separate scales for different aspects of the experiential thinking style: intuition, emotionality and imagination.

2.4 Analytical Approach

We inventoried pattern recognition responses as either correct or erroneous. Depending on the stimulus presented, the error could be a false positive error (a pattern was identified when no pattern was present, as when the No Pattern condition had been presented) or a false negative error (no pattern was identified when a pattern was present, as when the Partial condition and the Reveal condition had been presented). In order to avoid the 'Lady Tasting Tea' fallacy/error [17], several random patterns and several degrees of transparency were used.

For numerous reasons, we decided not to compute a questionnaire index but to use a query-by-query approach to explore the relevance of the actual items (query scores) of the questionnaires.

A Bayesian approach was used. For each query in each questionnaire, two heat maps were constructed: the one with no errors and one with at least one error (Fig. A-1). For each query, the response frequencies are Dirichlet-distributed with five concentration parameters (the questionnaires used 5-option Likert scales). As described in the Appendix: for each query, overlap of the *pdf* s (probability density functions) of the two Dirichlet distributions, one for correct pattern detection versus one for incorrect pattern detection (either pareidolia or apoidolia) occurs. The overlap can be used to construct the confusion matrix; the significance level we adopt is 10% (the level for confusion matrices equivalent to 5% in conventional significance tests [18]).

3 Results

It should be highlighted that—as opposed to traditional classical test theory—we used the query-by-query approach, because all the steps are based on an underlying mathematical theory. We describe the difference (at 10% significance level [18]) between subjects that correctly identified the stimuli and subjects that committed an error separately for each condition: No Pattern condition (no pattern was present) versus Partial condition (the geometrical figures were partially visible) versus Reveal condition (the geometrical figures were well identifiable).

Table 1. Summary of significant queries for each questionnaire. The table lists the number of significant queries per questionnaire and their fraction of the entire questionnaire (in %). * indicates 80% significant queries (or higher).

Questionnaires	**False positive** (No Pattern)	**False negative** (Partial)	**False negative** (Reveal)
Coincidence (7 queries)	5 (71%)	6 (86%)*	4 (57%)
RCI (10 queries)	2 (20%)	2 (20%)	5 (50%)
IBI-Magical Beliefs (10 queries)	3 (30%)	2 (20%)	10 (100%)*
IBI-Spirituality (8 queries)	4 (50%)	8 (100%)*	8 (100%)*
IBI- TAF (5 queries)	3 (60%)	2 (40%)	3 (60%)
REIm-Rationality (12 queries)	12 (100%)*	11 (92%)*	12 (100%)*
REIm-Emotion (10 queries)	8 (80%)*	9 (90%)*	10 (100%)*
REIm-Imagination (10 queries)	8 (80%)*	9 (90%)*	9 (90%)*
REIm-Intuition (10 queries)	9 (90%)*	9 (90%)*	10 (100%)*

For the Coincidence questionnaire (Part A): in the No Pattern condition, five of the seven queries were significant (queries number 1, 2, 3, 6, and 7); in the Partial condition, six queries were significant (queries number 1, 2, 4, 5, 6, and 7), while for the Reveal condition four queries were significant (queries number 1, 2, 4, and 6).

For RCI-10 (10 queries): in the No Pattern condition, two queries were significant (queries number 1 and 6); in the Partial condition, only two queries were significant (queries number 5 and 10), while for the Reveal condition five queries were significant (queries number 2, 4, 5, 6, and 9).

For the subscale Thought-Action-Fusion (5 queries): in the No Pattern condition three queries were significant (queries number 1, 3 and 4); in the Partial condition two queries were significant (queries number 1 and 5), and three in the Reveal condition (queries number 2, 3 and 4).

For the subscale Magical beliefs (10 queries) of the IBI: in the No Pattern condition three queries were significant (queries number 1, 6 and 8); for the Partial condition two queries were significant (queries number 5 and 6), while for the Reveal condition all 10 queries were significant. For the subscale Spirituality (8 queries): in the No Pattern condition four queries were significant (queries number 2, 3, 4, and 8); in both the Partial and Reveal condition all eight queries were significant.

For the subscale Rationality of the REIm: in the condition No Pattern all 12 queries were significant; the same results were obtained for the Reveal condition; for the Partial condition 11 queries were significant (the only exception was query number 12).

For the subscale Emotion (12 queries): in the No Pattern condition eight queries were significant (exceptions were queries number 7 and 10); in the Partial condition nine queries were significant (all except query number 10), and in the Reveal condition all queries were significant.

For the subscale Imagination (10 queries): in the No Pattern condition eight queries were significant (all except queries number 5 and 10); in the Partial condition nine queries were significant (all except query number 9); for the Reveal condition nine queries were significant (all except query number 3).

For the subscale Intuition (10 queries): in both the No Pattern and in the Partial condition nine queries were significant (except for query number 9 and number 1, respectively); for the Reveal condition all queries were significant.

In Table 1 we listed the significances at 10% significance level [18]. For a cut-off, we introduce an 80% threshold for the whole scale or subscale to be considered of impact in pattern perception.

4 Discussion

In this present study we investigated pattern recognition by humans with a focus on the type of error in connection with religious and supernatural beliefs and thinking styles [7, 8].

The stimuli provided the possibility of studying not only false positive errors (pareidolia) but also false negative errors. To our knowledge, this is the first time that the false negative errors have been taken into consideration, together with false positive errors, when studying a non-clinical population in the context of apophenia in visual perception. We referred to this false negative category of errors as apoidolia (as opposed to pareidolia).

We emphasize that the study presented here is an example of a multidisciplinary approach: where psychological investigation techniques were combined with approaches based on technology and on Bayesian statistics in order to provide novel insights. The stimuli used in previous investigations were often limited to face pareidolia, while we studied general pattern perception that is unrelated to faces. In one study [19], in which the authors used the Rorschach's ink blots as stimuli, complexity was analyzed based on fractals. That approach focusses on describing complexity of non-objects (unclear shapes) triggering the cognitive response.

Our approach focusses on providing the raters with stimuli that are specifically geometrically defined shapes (such as triangles, squares, and other regular polygons) that can be observed in nature.

The main problem was to provide randomness in some rigorous way. We used a random number generator that is state-of-the-art (MATHEMATICA v12.2 from WOLFRAM Technologies®), thereby ensuring a high level of control over the stimulus formation. The use of geometric shapes covered with differing levels of transparency of a random pattern provides absolute control over the pattern presence. Geometric shapes are present in nature in crystals, leaves, nests, blooms, bones, ornaments (butterfly wings, zebra or leopard skin, peacock tail, to name a few), shells, etc. Therefore, the choice of the shapes is relevant for perception processes.

In addition, we present, for the first time, a rigorous Bayesian analysis of differences between Likert-type response sequences (vectors); to our knowledge, this approach of calculation of confusion matrices based on Dirichlet distributions has never been done before.

Among the questionnaires used in the study, the RCI-10 was least able to detect differences between the participants that correctly identified the stimuli in contrast to those who made at least one error. This finding could be due to the generally low religiosity in the specific study cohort, since the respondents were recruited from a scientifically oriented community in a country with a very low religiosity (Czech Republic). In contrast, the best results discriminating between the two groups, independent of the conditions, were obtained using the REIm with each of the subscales having a very high fraction of queries (80% or higher) being significant at 10% level for all the conditions. The Coincidence questionnaire was notably good at distinguishing participants with pareidolia; it performed even better for apoidolia—but only in the Partial condition. We found that, for the IBI, the two subscales Magical beliefs and Spirituality were extremely good at distinguishing subjects with apoidolic perception, specifically in the Reveal condition, but not with pareidolic perception.

Interestingly, none of the questionnaires used were better at discriminating between pareidolia and apoidolia. This finding suggests that the presence of an error in the pattern recognition is more important than the specific type of error.

In summary, the results of the current study indicate that, in the context of pattern recognition, the thinking style is more important than the specific belief spectrum of the participants. If the study of visual apophenia involved biological stimuli, such as faces or animal appearances, further cognitive processes may become involved and the role of specific beliefs may become more influential. By using geometrical figures, we avoided such situations and consequently the role of thinking styles emerged in a more identifiable, significant manner.

One limitation of the current study is that our results do not allow us to understand how the overall distribution of responses of the two groups of participants differentiate, even if the results are very reliable in terms of what characteristic is different. To further elucidate this issue, future work is necessary: it will then be necessary to recruit a larger, multicultural sample and incorporate the use of artificial neural networks.

Author contributions. JB and SB designed the research framework. HP produced the stimuli and chose the frame sequences needed for subsequent analyses. HP designed the statistical tests and wrote the computer programs. SB, HP and JB co-authored the manuscript. All authors contributed to revisions of the manuscript. All authors have read and agreed to the published version of the manuscript.

Funding. The research and data collection was supported by Charles University Science Foundation (Grant No. 1404120: "Cognitive Bias, Agency Detection and Pattern Recognition: An Intercultural Quantitative and Experimental Study of Supernatural Belief").

Institutional Review Board Statement. The study was conducted in accordance with the Declaration of Helsinki, and approved by the Institutional Review Board of Faculty of Science, Charles University, Prague, Czech Republic (protocol code 2018/08 approved on April 2, 2018).

Institutional Review Board Statement. All participants were of age, and provided an informed consent (online) prior to the beginning of the research. All participants were informed that they may revoke their consent and/or terminate the participation at any moment without needing to provide a reason.

Conflicts of Interest. The authors declare no conflict of interest.

Appendix

Consider Fig. A-1. We observe two heat maps of a questionnaire and a stimulus (Coincidence questionnaire for the stimulus in Reveal condition): one heat map for the numbers of entries for 'Correctly identified' and one for 'At least once wrong'. Visually, the heat maps appear different. We ask whether the distribution of entries for each query are *significantly* different. The significance can be determined by calculating the confusion matrix—one for each query. In Fig. A-1, the confusion matrices for each query are displayed in the graph.

We simplify the notation: one heat map we label A and the other B; the query we label q. To calculate the confusion matrix with entries $\left(p_{A_{TRUE}}\, p_{A_{FALSE}}\, p_{B_{FALSE}}\, p_{B_{TRUE}}\right)$, where $p_{A_{TRUE}}$, etc., are the probabilities of the respective entries. We cannot calculate these entries using the Frequentist (Laplace) approach for two reasons: (1) the entries are too small to be considered a satisfactory limit of large sample size, and (2) we do not know the (analytic) boundary between *TRUE* and *FALSE*. We therefore use Bayesian statistics. We first describe the method for a one-dimensional problem (Fig. A-2), in which we can define a boundary. Consider a query with only two entries for 'Correctly identified' (A) ($\left\{s_{A_1}, 1 - s_{A_1}\right\} = \{\, 3712 \,\}$) and only two entries for 'At least once wrong' (B) ($\left\{s_{B_1}, 1 - s_{B_1}\right\} = \{\, 69 \,\}$). For each query, the entries are Beta-distributed (Fig. A-2), with likelihood functions ${}_A(s)$ and ${}_B(s)$.

In the one-dimensional problem (Fig. A-2), calculating the confusion matrix is straightforward, because the probabilities of the likelihoods are integrals. Beyond the one-dimensional distributions of entries for a query, however, the problem is very much more difficult, because there is no boundary that can be analytically specified for integration: the likelihood functions are Dirichlet distributions. In these cases, we use Monte Carlo methods to find the probabilities.

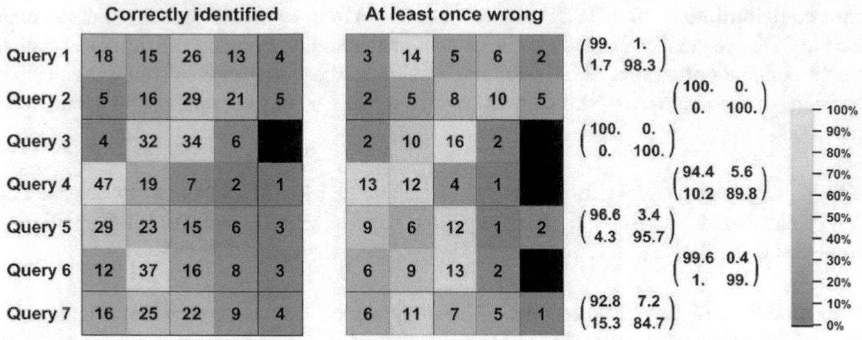

	Correctly identified					At least once wrong					Confusion matrix
Query 1	18	15	26	13	4	3	14	5	6	2	(99. 1. / 1.7 98.3)
Query 2	5	16	29	21	5	2	5	8	10	5	(100. 0. / 0. 100.)
Query 3	4	32	34	6		2	10	16	2		(100. 0. / 0. 100.)
Query 4	47	19	7	2	1	13	12	4	1		(94.4 5.6 / 10.2 89.8)
Query 5	29	23	15	6	3	9	6	12	1	2	(96.6 3.4 / 4.3 95.7)
Query 6	12	37	16	8	3	6	9	13	2		(99.6 0.4 / 1. 99.)
Query 7	16	25	22	9	4	6	11	7	5	1	(92.8 7.2 / 15.3 84.7)

Scale: 100%, 90%, 80%, 70%, 60%, 50%, 40%, 30%, 20%, 10%, 0%

Fig. A-1. The heat map of one questionnaire with the entries and the confusion matrices displayed. Observe that the sum of the number of entries for each query for 'correctly identified' is not the same as for 'at least once wrong'. Black squares occur when there are no entries. The shade of yellow, on the other hand, identifies the fraction of occurrences. Thus, for example, the 3rd entry for Query 5 for 'Correctly identified' is darker than the same entry for 'At least once wrong', although numerically the entries are comparably close.

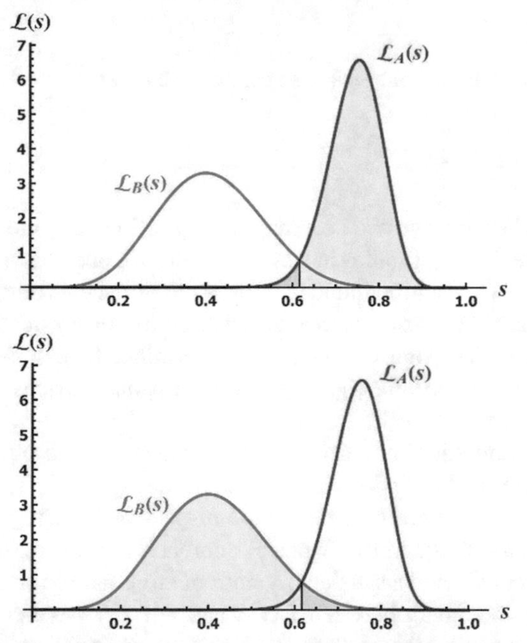

Fig. A-2. The likelihood functions for two samples (A, magenta) ($\{s_{A_1}, 1 - s_{A_1}\} = \{\ 3712\ \}$) and ($B$, green) ($\{s_{B_1}, 1 - s_{B_1}\} = \{69\}$). The likelihood functions are Beta distributions. In the top graph, the probability of $_A(s)$ for TRUE is shaded light magenta, while the probability for FALSE is shaded gray. In the bottom graph, the probability of $_B(s)$ for TRUE is shaded light green, while the probability for FALSE is shaded gray. The vertical black line shows the boundary between TRUE and FALSE. In this (numerical) example, the gray area in the top graph is 0.0235, and in the bottom graph the gray area is 0.0437; the confusion matrix is therefore (0.976 0.0235 0.0437 0.956). The distributions are therefore significantly different at 10% significance level [12].

We note that for a randomly chosen value for s_A in Fig. A-2, the condition TRUE is when $_A(s_A) > _B(s_A)$; likewise for a randomly chosen value for s_B in Fig. A-2, the condition TRUE is when $_B(s_B) > _A(s_B)$. This determination of TRUE and FALSE can be used in higher dimensions as well, because $s_A = \{s_{A_1}, \cdots, s_{A_k}\}$ (for k entries for a query) and the (pseudo) random number generator is defined for the Dirichlet distribution $(s_{A_1} + 1, \cdots, s_{A_k} + 1)$, irrespective of the number k of different entries for a given query. If n_{ran} random numbers are generated (in the paper presented here, $n_{ran} = 175000$), then the confusion matrix is $\left(\frac{n_{A_{TRUE}}}{n_{ran}} \frac{n_{A_{FALSE}}}{n_{ran}} \frac{n_{B_{FALSE}}}{n_{ran}} \frac{n_{B_{TRUE}}}{n_{ran}}\right)$. These ratios in the confusion matrix are not the probabilities defined by the Frequentist (Laplace) limit, but rather the Monte Carlo method of approximating an integral—equivalent to the Bayesian probability as graphed in Fig. A-2 in the one-dimensional case.

References

1. Willard, A.K., Norenzayan, A.: Cognitive biases explain religious be-lief, paranormal belief, and belief in life's purpose. Cognition **129**(2), 379–391 (2013)
2. Kirkpatrick, L.A.: Toward an evolutionary psychology of religion and personality. J. Pers. **67**(6), 921–952 (1999)
3. Jordan, B.: Pattern recognition in human evolution and why it matters for ethnography, anthropology, and society. In: Advancing Ethnography in Corporate Environments, pp. 193–214. Routledge (2016)
4. Haselton, M., Galperin, A.: Error management and the evolution of cognitive bias. Soc. Think. Interpers. Behav. **45**, 63 (2012)
5. Fyfe, S., Williams, C., Mason, O.J., Pickup, G.J.: Apophenia, theory of mind and schizotypy: perceiving meaning and intentionality in randomness. Cortex **44**(10), 1316–1325 (2008)
6. Liu, J., Li, J., Feng, L., Li, L., Tian, J., Lee, K.: Seeing Jesus in toast: neural and behavioral correlates of face pareidolia. Cortex **53**, 60–77 (2014)
7. Zhou, L.F., Meng, M.: Do you see the "face"? Individual differences in face pareidolia. J. Pac. Rim Psychol. **14**, e2 (2020)
8. Van Leeuwen, N., Van Elk, M.: Seeking the supernatural: The interactive religious experience model. Religion, Brain & Behavior **9**(3), 221–251 (2019)
9. Weinberger, A.B., Gallagher, N.M., Warren, Z.J., English, G.A., Moghaddam, F.M., Green, A.E.: Implicit pattern learning predicts individual differences in belief in God in the United States and Afghanistan. Nat. Commun. **11**(1), 1–12 (2020)
10. Wardle, S.G., Taubert, J., Teichmann, L., Baker, C.I.: Rapid and dynamic processing of face pareidolia in the human brain. Nat. Commun. **11**(1), 1–14 (2020)
11. Pascalis, O., Kelly, D.J.: The origins of face processing in humans: Phylogeny and ontogeny. Perspect. Psychol. Sci. **4**(2), 200–209 (2009)
12. Mandelbrot, B.B., Mandelbrot, B.B.: The fractal geometry of nature, vol. 1. WH freeman, New York (1982)
13. Bressan, P.: The connection between random sequences, everyday coincidences, and belief in the paranormal. Appli. Cognitive Psychol: Offi. J. Soc. Appli. Res. Memory Cognit. **16**(1), 17–34 (2002)
14. Worthington Jr., E.L., et al.: The Religious Commitment Inventory-10: Development, refinement, and validation of a brief scale for research and counseling. J. Counseling Psychol. **50**(1), 84 (2003)
15. Kingdon, B.L., Egan, S.J., Rees, C.S.: The Illusory Beliefs Inventory: A new measure of magical thinking and its relationship with obsessive compulsive disorder. Behav. Cognitive Psychoth. **40**(1), 39–53 (2012)

16. Norris, P., Epstein, S.: An experiential thinking style: Its facets and relations with objective and subjective criterion measures. J. Pers. **79**(5), 1043–1080 (2011)
17. Jackson, D.L.: Review of The Lady Tasting Tea: How Statistics Revolutionized Science in the Twentieth Century, by David Salsburg. Struct. Equ. Modeling **10**(4), 651–655 (2003)
18. Caelen, O.: A Bayesian interpretation of the confusion matrix. Ann. Math. Artif. Intell. **81**(3–4), 429–450 (2017). https://doi.org/10.1007/s10472-017-9564-8
19. Taylor, R.P., et al.: Seeing shapes in seemingly random spatial patterns: Fractal analysis of Rorschach inkblots. PloS One **12**(2) (2017)

Understanding Social Media Engagement in Response to Disaster Fundraising Attempts During Australian Bushfires

Yohei Nii[1], Chahat Raj[1], Muhammad Salman Tiwana[1], Mahendra Samarawickrama[2], Simeon Simoff[3], Tony Jan[4], and Mukesh Prasad[1](✉)

[1] School of Computer Science, FEIT, University of Technology Sydney, Sydney, Australia
mukesh.prasad@uts.edu.au

[2] Department of Data Science and Analytics, Australia Red Cross, Sydney, Australia

[3] School of Computer, Data and Mathematical Sciences, Western Sydney University, Sydney, Australia

[4] Faculty of Design and Creative Technology, Torrens University Australia, Sydney, Australia

Abstract. This article studies the impact of social media posts specific to a disaster incident – the Australian bushfires of 2019–2020. We analyse the social media content posted by the Australian Red Cross Organization's Facebook page, and the user generated comments on their posts. We identify user sentiments in response to the natural disaster and towards the organization's fundraising attempts. This study shall enable the stakeholders to understand how the general public reacts to fundraising protocols at the times of unforeseen disasters. It shall also allow policymakers to design sustainable goals to promote healthy donation behaviour through social media platforms. Further, we also analyse how benchmark Natural Language Processing tools, namely, VADER, Afinn, and TextBlob, perform in an unsupervised scenario to perform sentiment classification. Overall VADER results were best among the other algorithms Afinn and TextBlob in the term of accuracy, precision, recall and f1 score performance measure.

Keywords: Sentiment analysis · social media engagement · disaster fundraising · Australian bushfires

1 Introduction

The innovation of internet technology and communication through social media provides a new strategy for the fundraising process that a Non-Profit Organization (NPO) uses to achieve an organization's mission. Social media such as Facebook, Twitter, and Instagram are omnipresent, and their interactive designs have changed the way people communicate. Through social media platforms, users can connect with people across the Globe, create their content, and interactively share their opinion or thoughts. According to the Global NGO Technology Report, 93% of NPOs in Australia & New Zealand regularly utilize social media to engage their supporters and donors [1]. It is essential for

H. Sharma et al. (Eds.): ICIVC 2022, PALO 17, pp. 277–289, 2023.
https://doi.org/10.1007/978-3-031-31164-2_23

NGOs with limited budgets to provide comprehensive communications to various stakeholders quickly and efficiently. The use of social media is an effective tool to connect with supporters and engage with the audience. Regarding fundraising, social networking sites can effectively reach potential donor groups [2]. Social media creates the opportunity to communicate with the potential audience and disseminate information about brands, eventually leading to increased charitable contributions. Researchers have focused on social media usage for building a relationship with the audience in the past decades. [3] developed a conceptual model of driving brand advocacy and reciprocity to improve customer-based brand equity. Brand advocacy refers to positive recommendations from those customers who are strongly connected with brands. Users actively participate on social media platforms that allow advocacy through bidirectional conversations rather than one-way communication. This interactive conversation builds emotional attachment and satisfaction with that brand [4]. The communication conducted on social media also affects the other users and shapes their opinion of the brand. Another critical aspect of the relationship between brands and users is trust. A shared positive brand experience on social media establishes a secure relationship among social media platforms [5]. Social media users are also influenced by each other's experiences, opinions, sentiments, and preferences. Social media engagement is a core factor in mobilizing volunteers, raising awareness, and influencing potential donors.

Reciprocity represents the practice of mutual exchange between people. Many social media users express and share their ideas to find someone who can give them feedback. This reciprocal information sharing facilitates the engagement of social media users. [6] states that reciprocity and brand advocacy are critical mechanisms for effectively improving brand equity. However, the contents of information on social media need to be unique, authentic, and accurate to disseminate the information to a population. As of January 2021, there were 4.66 billion active internet users, 59.5% of the global population in the world [7]. Facebook is one of the biggest social media platforms, with roughly 2.89 billion monthly active users as of April 2021 [8]. There is a tremendous amount of information on the internet, and it would be easy to fail to take advantage of information without any specific strategy. Australian bushfires started in September 2019. The Australian Red Cross organization has solicited bushfire donations from October 2019 to June 2021. The time-series bushfire donation data from the Australian red cross from 2011–2021 was analysed for this study, and we specifically selected the 2019–2021 time period for this research, owing to the massive disastrous bushfires and large donation amounts in its reciprocity. The Guardian reported that Andrew Constance, the Liberal MP for the Bega in New South Wales, criticizes the fundraising decision by the Australian Red Cross [9].

This article provides a brief literature review on the impact of social media, the function of social media as brand advocacy by organizations, and the technological advancements in machine learning and natural language processing for sentiment analysis in Sect. 2. In Sect. 3, we discuss the data collection and annotation strategy, and the unsupervised methods employed for sentiment classification. Section 4 illustrates the results of social media content analysis and sentiment analysis depicting Facebook users' behaviour and emotions in response to the disaster. We also compare the sentiment classification performance of VADER, Afinn, and TextBlob algorithms. Section 5 summarizes the article.

2 Literature Review

Using social media and data extracted from the online platform is critical for disaster management, and it faces many challenges in filtering the correct data. The main focus is text analysis for developing a solution for the authorities to get information about the disaster and respond quickly to emergency services. [10] used supervised and unsupervised machine learning and deep learning techniques based on data collected from real-time online sources. One of the biggest challenges is limited information in the data using informal language. Social media contains actionable content related to advice, precautionary measures, and fundraising after disaster hits; thus, it is imperative to filter the correct data from given information.

To minimize this problem, the blockchain framework improves the accuracy and security of sharing wrong information. The system was trained to use block chaining and machine learning (ML) pipeline-based techniques to map the data automatically due to the crisis caused after the disaster. The data gathered is handled utilizing directed learning and data information mining strategies to deliver related knowledge reports that progressively sum up to significant data about continuous occurrences, hence giving a significant choice to help and rescue the teams. In the current environment around the world, many examinations of English social media take place, while very little work on refining the information about Arabic social media content is produced consistently. Their work centered on the characterization of Arabic feeds data and the utilization of unfavourable weather patterns in the UAE as an experiment. Support Vector Machines (SVM) and Polynomial Networks (PN) classifiers were used for analysis and without the help of stemming, and the outcomes showed a high degree of characterization exactness and a quick reaction time [11].

[12] discussed various Machine Learning Techniques used for analysing and classifying data collected from online sources. Multidimensional models of classification and recognition are used for pandemic diagnoses, monitoring, and prediction. Disaster management is one of the top trending issues on social media; therefore, ML algorithms are used for constructing different models, and these models can be merged with other techniques to improve the results of classification and recognition [13]. They also addressed various phases of ML algorithms used to predict and determine the early sign of disaster. This can minimize the disaster risk, unusual social interaction, and suspicious issues. They also highlighted some of the challenges faced during their survey as specific data is required to analyse for ML algorithms due to which data accuracy is affected.

Twitter data in tweets and comments were used for the sentimental analysis of text, and R language was used to retrieve and analyse the data collected from social media. By applying text mining techniques to find the correlation between the negative public opinion about unemployment after a crisis or disaster occurs. This also helps to find the key features that can affect the employment rate when an incident can cause this situation [14]. Social media data was used to detect the early clue about three different brands' products that causes allergies and adverse event from customers' feedback. [15] try to analyse the data gathered from Facebook and Twitter comments. Two approaches were applied to classify the data, lexicon-based sentiment classification and machine learning-based sentiment classification. Regarding classification, they also compared the results from both techniques and showed results with negative scores from both agree closely while there was a sharp change in the result for positive and neutral.

The internet and smart devices have brought a revolutionary change in this world, and media coverage through social media has become very easy for everyone [16]. Thus, public opinion mining through online feedback is beneficial in targeting the correct related information for users. They showed that Bert (bidirectional encoder representation from transformers) model is improved and fine-tuned to classify the data in the form of comments. The results obtained from the pre-trained Bert model are processed using a constructed neural network model. The accuracy and validity were 85.83%. They also compared the obtained results with other models and found that the Bert model results performed best among all three models used for classification. Religion is significant in accordance with extremism, as stated by [17]. Translation of social media data in two local languages, Sinhala, and Tamil, was initially done, and then tried to predict religious extremism in the context of data extracted from social media in Sri Lanka after a bombing incident occurred. The three well-known algorithms, Naive Bayes, SVM, and Random Forest, for evaluation of text data, were used, in which they have shown Naive Bayes algorithm result accuracy was 81% for Sinhala tweets data, and on the other side, random forest algorithm has shown best results of 73% accuracy rate for the Tamil language tweets.

Disaster affecting in the form of bushfires, earthquakes, floods, cyclones, and heat-waves around the world on social media is discussed every day. It is interesting to study the nature of disasters hitting different places on our globe and through social media the opining of users related to disaster can be studied. Location oriented disasters are a very interesting topic to understand the opinion of how people react differently. Thus, disaster situations are of prime importance for experts and decision makers. [18] presented a new automated approach by using natural language processing and artificial intelligence named entity recognition (NER), irregularity detection, regression, and Getis Ord Gi algorithms to do sentiment analysis of data extracted from social media related to disaster. They proposed a system having sufficient knowledge of social media data including 39 different languages and a data set comprising 67515 tweets. The algorithm extracted 9727 locations around the world with more than 70% reliability through live location feeds collected from twitter of specific regions. The algorithm with the accuracy of 97% is achieved after automatic classification. In the regard they also highlighted the research gaps related to inaccuracy of disaster related tweets classification, limited languages used for analysis and classification, and lack of location oriented live feedback of disaster. Limited online information using mobile apps is another issue. With the help of a proposed approach using mobile devices by implementing AL and NPL, organizations would be able to tackle more productively the crises and emergencies caused by disaster. They also suggested that their data set can be Merage with NASA's global landslide inventory for future prospective [19].

Traditional text mining on social media, i.e., Twitter, suffers from many challenges such as limited characters' restrictions and the noisy nature of tweets posted. Event detection from Twitter is badly affected by the nature of the tweet. Thus, it is crucial to evaluate the segmentation and sentiment of the tweet. SegAnalysis framework was used to handle these challenges faced on social media. [20] in their work, the segmentation was performed using POS (Part of speech) tags to fetch data. Naive Bayes classification and online clustering techniques detect the subject. The Naive Bayes and clustering mechanism help identify the event detection and reduce computational overheads. Segmentation is also very beneficial in preserving the semantic meaning of tweets NER (Named Entity Recognition). In this work, the authors highlighted some challenges related to any disaster, i.e., earthquake, flood, hurricane, or human disaster such as terrorist attacks. They try to analyse the data collected from online sources Facebook, Twitter, and Instagram, among these, most of the time, using API, Twitter is given preference by the researcher in analysing the data using different Machine learning and deep learning techniques. This study collects feedback about research articles for disasters or crises regarding early detection and warning, response after the disaster, and damage caused. This helps other researchers work on three aspects of any disaster before, during, and after user reactions to the crisis. It will support response teams in taking quick action on any disaster or crisis. Their studies showed that researchers face several issues and problems during classification and analysis to improve the precision, recall and accuracy of the detected information gathered from online sources. Their work also provides a source for exploring new ways to understand different techniques used to detect disaster through social media data [21].

It is very important to study the aftereffects of a disaster once it occurs. The reaction of people on social media gives feedback and their intentions to react with the situation. This also enables researchers to study the general as well as individual behavior of people. To find the sentiment in the data, based on the opinion in the response of any action plan tried by organizations to be implemented. A very limited work-related bushfire was addressed previously. Thus, in this research we identified user sentiments in response to the natural disaster and towards the organization's fundraising attempts specific to a disaster incident – the Australian bushfires of 2019–2020.

3 Data Preparation and Methods

The Australian Red Cross organization published 186 posts on its Facebook page from October 2019 to March 2020. We have collected a corpora of all the text content of the related posts and their comments from the Australian Red Cross organization Facebook page using the Facebook Graph API. The data fields of the corpora include text data of the post which create and update the time of the post, post type, post URL, and the social engagement count such as share, comment, likes, and so forth. Some of the comments are filtered out concerning the privacy filtering settings of Facebook users. We also excluded the replies from the comments data. The posts on the Facebook page were categorized into four types. Figure 1 shows the graphical representation of the post count in each category. Among the 186 posts published on the page, 86 posts are photo (image) type. Link posts are posted with a URL link to a donation site or lifehack articles on Red Cross homepages. Album posts often reports the Australian Red Cross activities with multiple photos. It was founded that only 14 videos were posted in the five-month duration as shown in Fig. 1 below.

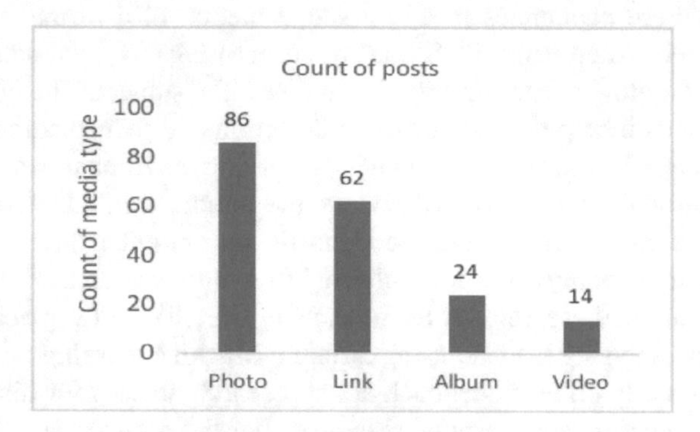

Fig. 1. Count of Media Type

3.1 Sentiment Annotation

To understand the user behaviour closely, we perform sentiment analysis on the Facebook data to categorize each comment as positive, negative, or neutral. Understanding the users' sentiment through their comments on the Australian Red Cross Facebook posts highlights the necessity of social media engagement for disaster fundraising and related events. Since the comments data collected from Facebook Graph by using API does not contain sentiment values, three human annotators manually labelled all the comments, considering the inter-coder reliability.

The process of manual annotation requires a set of specific rules to produce consistent sentiment labels. Due to the nature of the subjectivity of natural language, the lack of specification leaves the annotator in doubt over how to label certain kinds of instances. [22] proposes two annotation schemes: A simple and semantic-role-based sentiment questionnaire. A simple sentiment questionnaire focuses on the language itself. If a text contains negative words such as bad, sad, violent, etc., then the text is labelled as negative even if the speaker's intention is positive for that expression. This annotation scheme is easy to use and cost- effective. However, the simplicity of this questionnaire may not suit the social media text because there are various expressions used in social media, and it is vital to understand the context of the texts. A semantic role-based sentiment questionnaire focuses on not only a word itself but also the target of opinion and the speaker's emotional state. The important aspect of this questionnaire is that it takes account of the entity in a text. For instance, if the text criticizes the Australian Red Cross, Australian Red Cross is the Primary Target of Opinion (PTO), and if the speaker's expression is negative towards the organization, the sentence is likely to be negative. This questionnaire obtains rich information from the social media context and helps label sentiment values. Given such a practical framework, we employed [22] semantic role-based sentiment questionnaire to label the comment data on the Australia Red Cross Facebook posts. To carry out the sentiment annotation task, each human annotator answers the set of questions to categorize each text item into a sentiment category as positive, negative, or neutral as shown in Table 1.

Table 1. Set of questions answered by each annotator during manual sentiment annotation.

Q1	How to read the text, and the speaker's emotional state can best be described as positive, negative or a neutral state?
Q2	How to read the text, and identify the entity towards which opinion is being expressed or the entity towards which the speaker's attitude can be determined?
Q3	What best describes the speaker's attitude, evaluation, or judgment towards the primary target of opinion (PTO)?
Q4	What best describes the sentimental impact of the primary target of opinion (PTO) on most people?

3.2 Methods

Natural language processing (NLP) is a sub-domain of artificial intelligence. It handles human language and assists computer processes, derives, and understands its content and context in the same way human beings can. The information technology industry has accelerated computational performance exponentially in the last couple of decades. Accordingly, researchers are utilizing NLP to deduce more value from a large amount of data. However, the growing amount of social media data is voluminous, complex, and even unmanageable to be processed manually. NLP enables the extraction of useful information from large amounts of textual data and the analysis of content information for collective insight [23]. Sentiment analysis incorporates natural language processing to analyse people's opinions, sentiments, attitudes, and emotions from the text. This technique transforms unstructured text data into structured and qualitative text data labelled with sentiment categories. Conducting sentiment analysis for Facebook comments shall provide an overview of public opinion towards the Australian Red Cross and help them design strategies to communicate with their target audience.

In addition to the human annotation, we utilize Valence Aware Dictionary for Sentiment Reasoning (VADER), Afinn, and Textblob to compute machine-generated sentiments. This additional task for sentiment analysis is performed to measure the capability of unsupervised machine learning techniques. This is helpful in scenarios where user-generated data is in vast volumes and cannot be processed by manual human-based annotations. Evaluation of these machine learning techniques shall aid non-profit and non-government organizations in carrying out independent machine-based tasks for social engagement understanding [23]. Afinn is the simplest, yet popular lexicons used for sentiment analysis. It contains 3300+ words with a polarity score associated with each word. TextBlob is another lexicon-based sentiment analyser which comes with predefined rules (word and weight dictionary), which has scores that help to calculate a sentence's polarity. VADER is a rule-based model for general sentiment analysis and microblog content such as text on Facebook and Twitter [24]. There are several advantages of VADER performing in social media contexts. Firstly, the VADER lexicon performs exceptionally well on social media [25]. Text data in social media contains many punctuation marks and simple words such as emojis, acronyms, and slang that are typically removed in the pre-processing step. However, these algorithms can efficiently produce sentiment analysis scores from those ambiguous words. Computationally, these do not require any training data, i.e., they work in an unsupervised fashion. Training machine learning models can be a time-consuming and complicated process and requires huge volumes of annotated data. The employed algorithms eliminate the necessity of labelled data and makes the opinion analysis process robust.

Text pre-processing is a crucial step for performing data analysis in machine learning. The unstructured or raw data contains noise, which adversely affects the overall performance of a machine learning model [26]. We performed text pre-processing prior to performing the sentiment analysis using VADER, Afinn and Textblob. The implementation is carried out on Google Colab using Python version 3. The raw data was cleaned by removing the hashtags and non-English words from the comments. However, we didn't removed punctuation marks, emojis, or acronyms from the text as they affect the sentiment intensity and score. The 'polarity scores' method in the "Sentiment Intensity Analyzer" object produces a compound score that is the sum of positive, negative, and neutral scores ranging between -1 (negative) and $+1$ (positive) from a given text. A text with more than a 0.05 compound score is classified as positive. A text with less than or equal to a -0.05 compound score is negative. In other cases, a text is classified as neutral.

4 Results

4.1 Content Analysis

This subsection discusses the type of content on the Australian Red Cross Facebook page and analyses its constituency. Table 2 shows the shares, comments, and likes on each type of media content, i.e., albums, photos, videos, and links. It is observed that album posts, on average, get a more significant number of likes and shares, followed by photos, videos, and links, in that order. However, posts with videos engage the viewers more towards leaving a comment followed by the album, photos, and links. It can be inferred that users are less likely to engage with a URL post. The presence of high visual media content, such as multiple photos or videos uploaded together as an album in a post, demonstrate the highest engagement. Multiple media uploads can be used to target and trigger a greater audience. Photo and video posts earn their fair share of engagement through likes, shares, and comments. The table also implies that the visual media types tend to attract more viewers and let them comment more quickly.

4.2 Sentiment Analysis

Figure 2 shows the sentiment distribution of 4761 comments. Out of 4761 comments, 2507 comments are classified as positive by VADER, 1297 comments are classified as

Negative, and 957 comments are classified as Neutral. The number of positive comments significantly differs from the comment distribution by the human rater. Considering the number of negative comments is quite similar, either one might fail to distinguish between positive and neutral comments. VADER classified half of the comments as positive. Most of the classified neutral comments have a sentiment score of zero. This is partly because of the shortness of the comment. Many common words appear both in positive and negative comments.

There are instances where positive comments are classified as negative or neutral as shown in the Tables 3, 4, and 5. This is attributed to the fact that some positive or neutral sentences containing negation words like 'not,' 'never,' or 'no' are labelled with negative sentiments. The presence of negation words is impacting the sentiment score by reducing it. We can observe the limitation that machine-generated sentiment scores are not context-sensitive but only focus on word sentiment individually. In such a case, human-rated annotations are more trustworthy as it seeks human cognition and context awareness to label a particular comment.

Table 6 provides Accuracy (A), Precision (P), Recall (R) and F1 scores (F1) for VADER, Afinn and TextBlob. It was observed that amongst the three algorithms, VADER has the highest scores, followed by Afinn and TextBlob. Figure 3 illustrates the frequency of occurrence of comments based on their sentiment scores. We can affirm the observation derived from figure, as the positive comments have significantly higher occurrence than the negative ones, along with a massive spike around the neutral score of 0, which explains the presence of roughly a fourth of comments being sentiment neutral.

Table 2. Number of social engagement attributes per post

Media Type	# Shares	# Comments	# Likes
Album	216	44	869
Link	89	18	186
Photo	137	42	337
Video	97	73	146

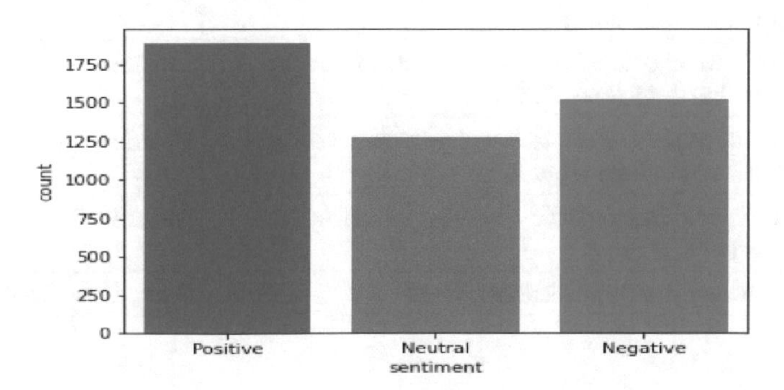

Fig. 2. Sentiment distribution of comments.

Table 3. Example of comments classified as positive.

Row ID	**Message**	**Sentiment**	**Sentiment Score**
79	Done. Happy to help beautiful country that I could hope visit one day❤️	Positive	0.9591
80	Done ✅ keep up your wonderful work❤️	Positive	0.5719
81	Donated. Sending love and praying for all victims of tragedy ❤️ 🙏	Positive	0.6369
82	Donated. Keep up your generous work. 👏	Positive	0.5106
83	Done just now keep up the fantastic work ❤️	Positive	0.5574

Table 4. Example of comments classified as negative.

Row ID	Message	Sentiment	Sentiment Score
138	The people who have been left devastated by the bushfires need that money now dont hold	Negative	−0.7964
156	Stop donating through red cross	Negative	−0.2960
159	Is it true that you are taking a ridiculous of what australias and others have donated to…	Negative	−0.5062
166	Now pass on all the money pffft red cross so disappointed	Negative	−0.6113
221	Donors for the tragic bush fires please do not donate to the red cross the red cross intends	Negative	−0.8555

Table 5. Example of comments classified as neutral.

Row ID	Message	Sentiment	Sentiment Score
15	I have just had the same issue on the 1st attempt but on 2nd it went through	Neutral	0.0000
222	should we register if we have evacuated from our homes but not to one of the eva.	Neutral	0.0000
452	how much per dollar raised will go to the recipients of the bushfires genuine quest…	Neutral	0.0000
1271	who will be independently auditing this financial program	Neutral	0.0000
3804	the money should be distributed now	Neutral	0.0000

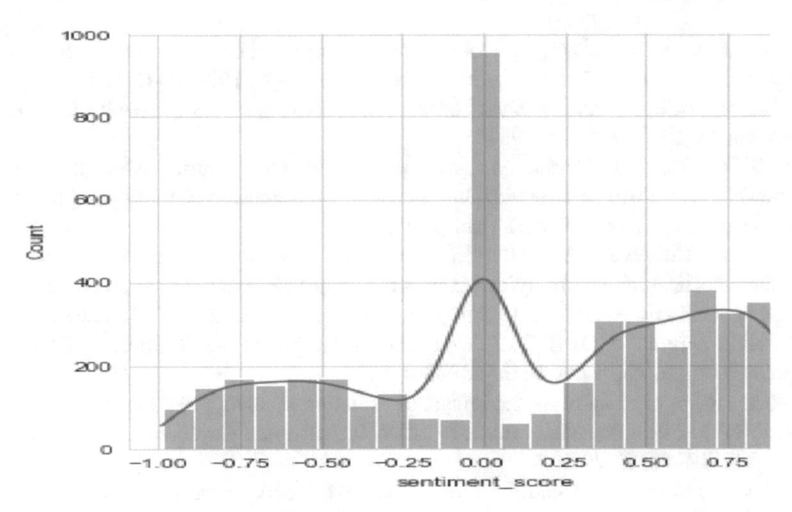

Fig. 3. Distribution of sentiment scores

Table 6. Sentiment classification results.

Algorithm	A (%)	P (%)	R (%)	F1 (%)
VADER	62.21	61.55	62.21	61.06
Afinn	58.16	58.11	58.16	57.26
TextBlob	51.96	55.12	51.96	52.11

5 Conclusion

This study identifies that most of the sentiment of the user-generated comments is positive from Oct. 2019 to Mar. 2020. In accordance, the social engagement attributes such as share, likes, and reach are also high in the same period of time. Although there are strong positive correlations between each social engagement attribute, they do not necessarily correlate with the amount of donation. The results suggest the necessity of developing robust unsupervised algorithms for sentiment classification that can handle the challenges of conflicting text, understand contextual meaning, and provide accurate sentiment annotations in the absence of labelled data resources. Such a development is highly demanded for real world use-cases like natural disasters and emergencies. It shall allow the policymakers and various stakeholders to better understand people's opinion, devise social media-based fundraising strategies, and positively influence their donation behaviour.

References

1. Nonprofit Tech for Good. Trends in Giving Report (2018). https://assets-global.website-files.com/5da60733afec9db1fb998273/5de6d4688ad4f942828cd561_2018-Giving-Report-English.pdf

2. Waters, R.D., et al.: Engaging stakeholders through social networking: How nonprofit organizations are using Facebook. Public Relat. Rev. **35**(2), 102–106 (2009)
3. Mathur, M.: Building brand advocacy on social media to improve brand equity. Int. J. Electron. Mark. Retail. **10**(2), 150–172 (2019)
4. Tsai, T.-Y., Lin, C.-T., Prasad, M.: An intelligent customer churn prediction and response framework. In: 2019 IEEE 14th International Conference on Intelligent Systems and Knowledge Engineering (ISKE). IEEE (2019)
5. Agarwal, A., Chivukula, A.S., Bhuyan, M.H., Jan, T., Narayan, B., Prasad, M.: Identification and classification of cyberbullying posts: a recurrent neural network approach using under-sampling and class weighting. In: Yang, H., Pasupa, K., Leung, A.C.-S., Kwok, J.T., Chan, J.H., King, I. (eds.) ICONIP 2020. CCIS, vol. 1333, pp. 113–120. Springer, Cham (2020). https://doi.org/10.1007/978-3-030-63823-8_14
6. Sawhney, R., et al.: Exploring the impact of evolutionary computing based feature selection in suicidal ideation detection. In: 2019 IEEE International Conference on Fuzzy Systems (FUZZ-IEEE). IEEE (2019)
7. Statista. Number of monthly active Facebook users worldwide as of 2nd quarter 2021 (2021). https://www.statista.com/statistics/264810/number-of-monthly-active-facebook-users-worldwide/
8. Raj, C., et al.: Cyberbullying detection: hybrid models based on machine learning and natural language processing techniques. Electronics **10**(22), 2810 (2021)
9. Gomes, H.L.: Australian Red Cross defends spending 10% of bushfire donations on office costs (2020). https://www.theguardian.com/australia-news/2020/jan/24/australian-red-cross-defends-spending-10-of-bushfire-donations-on-office-costs
10. Shahbazi, Z., Byun, Y.-C.: Blockchain-based event detection and trust verification using natural language processing and machine learning. IEEE Access **10**, 5790–5800 (2021)
11. Alkhatibl, M., El Barachi, M., Shaalan, K.: Using Arabic social media feeds for incident and emergency management in smart cities. In: 2018 3rd International Conference on Smart and Sustainable Technologies (SpliTech), pp. 1–6. IEEE (2018)
12. Chamola, V., et al.: Disaster and pandemic management using machine learning: a survey. IEEE Internet Things J. **8**(21), 16047–16071 (2020)
13. Rajora, S., et al.: A comparative study of machine learning techniques for credit card fraud detection based on time variance. In: 2018 IEEE Symposium Series on Computational Intelligence (SSCI). IEEE (2018)
14. Nirmala, C. R., G. M. Roopa, and KR Naveen Kumar. "Twitter data analysis for unemployment crisis." *2015 International Conference on Applied and Theoretical Computing and Communication Technology (iCATccT)*. IEEE, 2015
15. Isah, H., Trundle, P., Neagu, D.: Social media analysis for product safety using text mining and sentiment analysis. In: 2014 14th UK workshop on computational intelligence (UKCI). IEEE (2014)
16. Liang, H., Tang, B., Cao, S.: Sentiment analysis of comment texts on converged media platforms based on BERT model. In: 2021 International Conference on Culture-oriented Science & Technology (ICCST). IEEE (2021)
17. Fernando, A., Wijayasiriwardhane, T.K.: Identifying religious extremism-based threats in Sri-Lanka using bilingual social media intelligence. In: 2020 International Research Conference on Smart Computing and Systems Engineering (SCSE). IEEE (2020)
18. Sufi, F.K., Khalil, I.: Automated disaster monitoring from social media posts using ai-based location intelligence and sentiment analysis. IEEE Trans. Comput. Soc. Syst. (2022)
19. NASA Global Landslide Catalog Points. The NASA Cooperative Open Online Landslide Repository (COOLR) Points, Downloadable (2021). https://maps.nccs.nasa.gov/arcgis/home/item.html?id=eec7aee8d2e040c7b8d3ee5fd0e0d7b9

20. Patil, M., Chavan, H.K.: Event based sentiment analysis of Twitter data. In: 2018 Second International Conference on Computing Methodologies and Communication (ICCMC). IEEE (2018)
21. Dwarakanath, L., et al.: "Automated machine learning approaches for emergency response and coordination via social media in the aftermath of a disaster: a review. IEEE Access **9**, 68917–68931 (2021)
22. Mohammad, S.: A practical guide to sentiment annotation: challenges and solutions. In: Proceedings of the 7th Workshop on Computational Approaches to Subjectivity, Sentiment and Social Media Analysis (2016)
23. Adhikari, S., et al.: Exploiting linguistic information from Nepali transcripts for early detection of Alzheimer's disease using natural language processing and machine learning techniques. Int. J. Hum.-Comput. Stud. **160**, 102761 (2022)
24. Kabade, V., et al.: Machine learning techniques for differential diagnosis of vertigo and dizziness: a review. Sensors **21**(22), 7565 (2021)
25. Hutto, C., Gilbert, E.: Vader: A parsimonious rule-based model for sentiment analysis of social media text. In: Proceedings of the International AAAI Conference on Web and Social Media, vol. 8, no. 1 (2014)
26. Tanveer, M., et al.: Machine learning techniques for the diagnosis of Alzheimer's disease: a review. ACM Trans. Multimedia Comput. Commun. Appl. (TOMM) **16**(1s): 1–35 (2020)

Evaluating Image Data Augmentation Technique Utilizing Hadamard Walsh Space for Image Classification

Vaishali Suryawanshi(✉), Tanuja Sarode, Nimit Jhunjhunwala, and Hamza Khan

Computer Enginnering Department, Thadomal Shahani Engineering College, Mumbai, India
{vaishali.suryawanshi,tanuja.sarode}@thadomal.org

Abstract. Deep neural networks are complex networks that need millions of parameters to be trained and perform well for huge datasets, however result in poor generalization for small datasets. Data augmentation is an effective method of addressing generalization where an existing dataset is enlarged by generating synthetic images. Commonly used geometric transformation-based augmentation techniques such as rotation, flipping, random cropping etc. cannot be applied generally to all datasets. This paper proposes an effective and yet easy to implement data augmentation technique utilizing frequency domain coefficients obtained using the Hadamard and Walsh transforms which we believe can be applied generally to any dataset. The experiments performed using the VGG16 model on the Cifar-10 dataset show that the Hadamard based technique outperforms the single geometric transformation technique. It also achieves comparable results even with a smaller dataset when compared with multiple geometric transformation techniques. Further, the complexity of proposed techniques is less than that of traditional geometric transformation techniques.

Keywords: Image Data Augmentation · Generalization · Image Classification · Hadamard Transform · Walsh Transform · Deep Neural Network (DNN) · MCC

1 Introduction

Recent trends in the field of deep learning technology have surged interest in research in various fields such as computer vision, natural language processing and time series applications. In computer vision, deep neural networks have been widely used for classical applications such as image classification, object recognition, image segmen-tation etc. However, there are many complex tasks such as medical imaging analysis, precision agriculture, crime detection, artificial intelligence solutions by robotics based on computer vision applications, ecology image processing etc. where the DNNs are not as successful [1]. The primary reason is the unavailability of enough data [1–12] to train the complex DNN. Due to the scarcity of the data, and the com-plexity of the DNN, the DNN remembers the training data too well but performance is degraded for test data resulting in poor generalization.

H. Sharma et al. (Eds.): ICIVC 2022, PALO 17, pp. 290–301, 2023.
https://doi.org/10.1007/978-3-031-31164-2_24

The generalization issue is addressed by various methods such as reducing the capacity of the network using explicit regularization, introducing stochasticity in the input, input transformations and ensembles. Image data augmentation is an effective method for improving the generalization of the network. The dataset is enlarged by creating new instances of the images by introducing variability in the images using the existing data distribution and some transformation method. With the enlarged dataset the DNN fails to remember all the true or accidental features, introducing error in train-ing. When the validation set is presented to the DNN, the validation loss is reduced thereby improving the generalization.

Currently, a lot of research work is being carried out in the field of computer vision-based tasks such as image classification, segmentation, recognition, tracking etc. Most of the researchers working on DNN based models agree that bigger datasets result in better Deep Learning models. However, preparing the bigger datasets is a laborious job due to the manual effort required for collecting and labeling data. Also, the scarcity of availability of data may be attributed to the expensive data collection process or confidentiality issues [1].

The application of image data augmentation techniques is twofold: first it can create anonymized, labeled data of required size in less time and minimum effort, saving the laborious task of manually collecting and labeling the data. Secondly, it can be used for enlarging the dataset to improve the efficiency of the deep learning models. For applications where scarcity of data is common such as autonomous robotic weed management systems in precision farming, identification of plant diseases, medical image analysis [1], ecological imaging, crime detection etc.

2 Related Work

Data augmentation is widely applied in the field of Natural Language Processing (NLP) by random insertion, random swap, text substitution, synonym replacement etc. In computer vision applications image data augmentation is generally applied by color and Geometric Transformations (GT), random erasing, adversarial training, neural style transfer and combinations of these methods. Oversampling methods can be used to create synthetic images by methods of mixing, feature space augmentations etc. [1].

The data augmentation methods utilizing the input space are most common in the literature [2–7]. Though the standard geometric transformations such as random cropping, rotation, random erasing, horizontal/vertical flipping, translation, shear, scaling etc. have been widely used by the researchers for various applications such as detecting Covid-19 [3], melanoma Detection [4] their application to different datasets is still questioned. Taylor and Nitschke [2] used flipping, rotation and cropping for GTs and color jittering, enhancement and PCA method for color transformations for generic data augmentations on Caltech101 dataset using a shallow CNN architecture and found that GTs outperformed color-based transformations with cropping giving better results. Whereas authors [3] found that the IDA based on geometric transformations in X-ray images was not an effective strategy for detecting COVID-19. Authors [4] found improved performance when Color, GT, mixing etc. were applied on Melanoma classification dataset on different models pretrained on ImageNet. Data augmentation based on horizontal flipping, rotation and scaling was carried out in the study [5], on Kaggle Cat vs Dog set using pretrained CNN models ResNet50, VGG16 [12] demonstrating that the results with data augmentation for VGG16 were better than ResNet50.

Inoue in [6] proposed a Sample Pairing technique where a random image is overlayed onto the selected image and simple average of pixels are taken to create synthetic images and observed reduced error rate when the experiments were conducted on GoogleNet for CIFAR-10 [13] and ISLRVC datasets. Summers and Dinneen [7], explored non-linear methods such as horizontal concat, vertical con-cat, random row interval, random square etc. for mixed example data augmentation, where the authors generated the synthetic image by randomly selecting two images from the dataset and applying the non-linear methods. This research claims that improvement is observed on the ResNet18 on CIFAR-10, CIFAR100 and Caltech256 datasets.

Methods have also been proposed to use the deep learning-based complex models such as Generative Adversarial Networks (GANs), Variational Autoencoders (VAE), Neural style transfer etc. to generate synthetic images utilizing the vector codes [8–11]. DeVries and Taylor [8] used LSTM based sequence autoencoder to generate the context vector and applied augmentation techniques such as adding Gausssian noise and interpolation/extrapolation of the context vectors and found that extrapolation performs better while for data augmentation in feature space. Liu et al. [9] used linear interpolation on feature space generated using adversarial autoencoders and found improvement in the performance on Cifar-10 and ISLRVC dataset. In [10], the authors used GTs and GAN to generate synthetic face images and obtained improved results for combination of horizontal reflection, translation and GAN. IDA using neural style transfer was attempted by Perez and Wang [11], where the authors experimented in Dogs vs Cat and Dogs vs Fish dataset obtaining improvement in results.

DCGAN generated liver lesion samples are tested by FridAdar et al. to observe improvement in sensitivity and specificity [14]. Scarcity of Cardiac Magnetic Resonance (CMR) images was addressed using a framework of image segmentation and synthesis. Convolutional VAE were used to synthesize images of fruits and vegetables to address representation learning, where the authors noted significant improvement [15]. Recently color space and GTs based image data augmentation methods are found to be effective as defense strategy against inversion attack in federated learning [16]. Abhishek et al. created Out of Data Distribution (ODD) datasets using Negative Data Augmentation (NDA) and used NDA as additional source of synthetic data for the discriminator while training GAN networks for self-supervised representation learning and observed improved performance [17]. In [18], authors proposed a technique to improve diversity and recognize objects from partial information.

3 Theoretical Background

3.1 Hadamard Transform

Hadamard transform (HT) discovered by James Joseph Sylvester in 1867, is a orthogonal, symmetric and non-sinusoidal transform that decomposes a signal into a set of orthogonal, rectangular waveforms called Walsh functions. The transformation matrix is real, and consists of only two only two possible values $+1$ and -1 [19, 20]. Hadamard transform find its application in a large number of fields such as power spectrum analysis, filtering, image steganography, image compression [21], and processing of speech and medical signals, medical and biological image processing, digital holography, multiplexing and coding in communications, describing nonlinear signals, solving nonlinear differential equations, and logical design and analysis [22].

The Kronecker product [23] can be used to generate Hadamard matrices of any order that is a positive power of 2 (except 1) using the 2×2 base Hadamard matrix H_2 as given below:

$$H_2 = [+1 \quad +1 \quad +1 \quad -1] \tag{1}$$

The Kronecker product of a Hadamard Matrix of order 2 gives a Hadamard Matrix of order 4. Similarly, the Kronecker product of a Hadamard Matrix of order n gives a Hadamard Matrix of order 2n. Hence, by taking the Kronecker product of matrix H_2 with itself n times to give a Hadamard Matrix of order 2n [23].

The Walsh Transform can be obtained from Hadamard Transform by simply rearranging its rows according to the number of times the sign of an element changes when traversing a row horizontally [19].

3.2 Deep Convolutional Neural Network (CNN)

Deep convolutional neural networks have been at the center of current research trend in the field of deep learning, though applications have been developed by using deep NN in early 1990, LeCun et al. [24] developed Optical Character Recognition (OCR), it is only recently it gained popularity when Krizhevskey et al. Applied deep CNN and won the ImageNet Classification Challenge in 2012. In the current research, we have used Deep CNN for our image classification task.

3.2.1 VGG16 Architecture

Simonyan and Zisserman [12] evaluated the convolutional networks of varying depths utilizing convolutional filter size as 3×3 for their Large Scale Image Recognition ImageNet 2014 Challenge. These different CNN configurations with depth 11 to 19 are now popularly known as VGG (Visual Geometry Group) Net.

For experimentation purposes we have used the VGG-16 model as shown in Fig. 1. in total, there are 21 layers (13 convolution layers, 5 max-pooling layers, and 3 dense layers) out of which 16 are weighted layers where the parameters are learnt. In our experiments, the input layer takes an image of dimension (32, 32, 3) as an input where the 3 refers to the three channels (RGB) of an image. The convolution and max-pooling layers are arranged consistently throughout the architecture where the convolution layer has a 3*3 filter with a stride of 1 while the max-pooling layers have a 2*2 filter with a stride of 2. Conv-1, Conv-2, Conv-3 have a total of 64, 128, 256 filters while both Conv-4 and Conv-5 layers have 512 filters. There are 3 fully connected (FCs) layers in which there are 4096 channels each in the case of the first two layers, while the third layer has 1000 channels. In the end, there is a SoftMax layer with 10 nodes corresponding to 10 categories of the dataset.

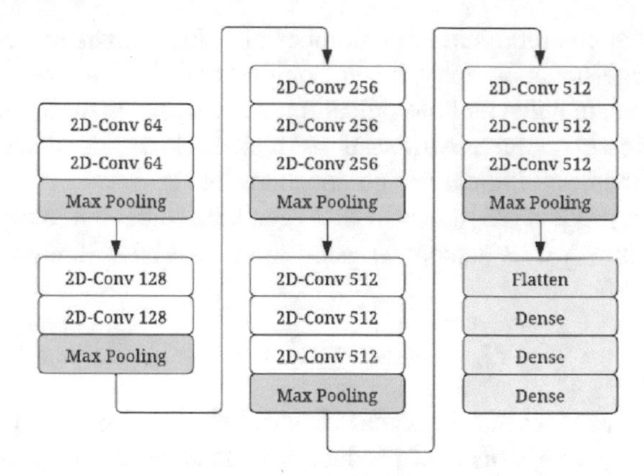

Fig. 1. VGG16 Architecture

4 Methodology

This section describes the methodology followed in this work. In this work, a data augmentation technique based on Hadamard transform is proposed. The technique is applied on the RGB color image.

4.1 Proposed Technique of Image Augmentation utilizing Hadamard Space

In the proposed technique, the image is first split into red, green and blue channels. Then the Hadamard transform is applied on each channel of the image to obtain the frequency domain coefficients of the corresponding channel. The augmented images are reconstructed from these frequency domain coefficients by taking into account only 70% of the coefficients.

The training dataset is augmented with the Hadamard transform reconstructed images. For each image in the training set, one image reconstructed from the Hadamard space is added in the training set, thus increasing the size of the dataset twice as the original. The images in the test dataset are not augmented.

4.2 Image Data Augmentation utilizing the Walsh Space

Here, similar procedure as described in Sect. 4.1 is followed except that the image is transformed using Walsh matrix and 70% of the frequency coefficients from the Walsh space are utilized for constructing the new image.

4.3 Training and Performance Evaluation

The VGG16 model is trained on the original dataset and the benchmark results are obtained using various confusion matrix-based performance evaluation parameters such as accuracy, precision, recall, F1- measure, Normalized MCC and AUC [25, 26].

Further we train the model on the augmented dataset and measure the performance of the model using performance evaluation metrics based on confusion matrix. Results are analyzed to gain insight into the effectiveness of the proposed technique and also. Compared to that of commonly used geometric transformation technique.

5 Experimental Setup

All the experiments were carried out in Google Colab environment with GPU processor using VGG 16 model with SGD optimizer and categorical cross entropy loss function. To examine the effectiveness of the data augmentation technique alone, we trained the model without applying any external regularization techniques such as drop out, batch normalization, pretraining etc. The open-source Octave environment was used for manipulating the Hadamard Walsh space of the images from the dataset.

5.1 Dataset Description

In this set of experiments, we have used CIFAR-10 dataset [14] which is popular for image classification. The CIFAR 10 dataset consists of 60000 images from 10 different classes, each image of size 32×32, with 6000 images per class. There are 50000 training images and 10000 test images. The 50000 training images are divided into 5 batches each batch containing 10000 images. The images from all the classes are chosen randomly for each batch.

5.2 Experiment Details

Following set of experiments are conducted to study image data augmentation.

5.2.1 Establish Benchmark Results on Cifar10 Dataset

Experiments are conducted to establish the benchmark results on the dataset where we used the original training batches (no augmentation applied) and the VGG16 is trained on each of the five training batches containing 10000 original images and tested on 2500 random images from the test batch.

5.2.2 Experiment on Hadamard/Walsh Augmented Dataset

The dataset is enlarged using the proposed technique with 10000 Original Batch images and 10000 images reconstructed using Hadamard(HB)/Walsh (WB) technique creating a training set of total 20000 images. Performance of the model is evaluated on randomly selected 5000 images from the test batch. Results of OBHB and OBWB are summarized in Table 1, Table 2 respectively.

5.2.3 Experiments on Geometric Transformation Technique

Further we conducted following experiments to compare the results of the proposed technique with that of existing Geometric Transformation (GT) techniques.

1. Single Geometric Transformation (SGT): On each batch of Cifar 10 dataset, we applied rotation = 20 and the model is trained using a training batch of 20000 images (10000 original + 10000 Rotated). We used 2500 images for testing. Results of the experiments are summarized in Table 4.
2. Multiple Geometric Transformations (MGT): A set of geometric transformations with Rotation Range = 20, width shift = 0.1, height shift = 0.1, horizontal flip = TRUE is applied on the original batches creating a training set of sizc 50000 image (10000 original + 40000 GT images). For testing, 2500 random images from the test batch are used. The results of the experiments are summarized in Table 5.

6 Results and Discussion

In this section, results of various experiments are presented and analyzed. We used experimental setup as described in Sect. 5.2.

6.1 Benchmark Results

Experiments are conducted with the set-up described Sect. 5.2.1. Table 1 summarizes the results of benchmark results obtained on Cifar-10. It is observed that Batch3 results are better than all other batches.

Table 1. Benchmark results on Cifar-10 dataset

S. No	Evaluation Parameter	Batch1	Batch2	Batch3	Batch4	Batch5
1	Accuracy	0.6092	0.6004	0.6164	0.5844	0.6012
2	Weighted Precision	0.6196	0.6096	0.628	0.59906	0.6089
3	Weighted Recall	0.6092	0.6004	0.6164	0.5844	0.6012
4	Weighted Avg. F1-Measure	0.6102	0.6036	0.6193	0.5863	0.604

6.2 Results of Experiments on Hadamard Augmented Dataset and Walsh Augmented Dataset

Using the methodology described in Sect. 3 and same set-up as described in Sect. 5.2.2, for each Cifar-10 batch a training set of 20000 images (Original Batch10000 + Hadamard Batch 10000 (OBHB)) by utilizing Hadamard space and test set of 5000 images, performance of the proposed technique is evaluated, the results are tabulated in Table 2. Similar results are obtained for Walsh Augmented dataset with Original 10000 + Walsh Batch 10000 (OBWB)), the results are presented in Table 3. The analysis shows OBHB has better average performance than OBWB.

Table 2. Results of OBHB on Cifar-10 dataset

S. No.	Evaluation Parameter	Batch1	Batch2	Batch3	Batch4	Batch5
1	Accuracy	0.6082	0.5994	0.617	0.6094	0.6184
2	Weighted Avg. Precision	0.6127	0.6090	0.6245	0.6107	0.6254
3	Weighted Avg. Recall	0.6082	0.5994	0.617	0.6094	0.6184
4	Weighted Avg. F1-Measure	0.6068	0.6015	0.6182	0.6093	0.6201
5	MCC Measure [+1, −1]	0.5654	0.558	0.5751	0.5661	0.5764
6	Normalized MCC [0, 1]	**0.7827**	**0.779**	**0.7875**	**0.7830**	**0.7882**

Table 3. Results of OBWB on Cifar-10 dataset

S. No.	Evaluation Parameter	Batch1	Batch2	Batch3	Batch4	Batch5
1	Accuracy	0.549	0.6172	0.53	0.6198	0.6084
2	Weighted Precision	0.5517	0.6208	0.5291	0.6218	0.61151
3	Weighted Recall	0.549	0.6172	0.53	0.6198	0.6084
4	Weighted Avg. F1-Measure	0.5479	0.6177	0.5392	0.6197	0.6089
5	MCC Measure [+1, −1]	0.4990	0.5750	0.4778	0.5777	0.5651
6	Normalized MCC [0, 1]	**0.7495**	**0.7875**	**0.7389**	**0.7888**	**0.7825**

6.3 Results of Experiments on SGT and MGT Augmented Datasets

As described in Sect. 5.2.3 experiments are carried out using SGT and MGT techniques and the results are tabulated in Table 4 and Table 5 respectively. Comparative analysis of SGT and MGT shows that the performance of the MGT technique is better than that of the SGT. However, with MGT the model is trained with 30000 more number of images than SGT.

Table 4. Results of SGT on Cifar-10 dataset

S. No.	Evaluation Parameter	Batch1	Batch2	Batch3	Batch4	Batch5
1	Accuracy	0.57	0.578	0.6288	0.6032	0.6612
2	Weighted Precision	0.5712	0.5764	0.6338	0.6010	0.6546
3	Weighted Recall	0.57	0.578	0.6288	0.6032	0.6612
4	Weighted Avg. F1-Measure	0.5667	0.5754	0.6211	0.5976	0.6538
5	MCC Measure [+1, −1]	0.5229	0.5314	0.5893	0.5597	0.6243
6	Normalized MCC [0, 1]	**0.7614**	**0.7657**	**0.7946**	**0.7798**	**0.8121**

Table 5. Results of MGTS on Cifar-10 dataset

S. No.	Evaluation Parameter	Batch1	Batch2	Batch3	Batch4	Batch5
1	Accuracy	0.6863	0.6967	0.7007	0.6958	0.7124
2	Weighted Precision	0.6866	0.6999	0.7022	0.6921	0.7108
3	Weighted Recall	0.6863	0.6967	0.7007	0.6958	0.7124
4	Weighted Avg. F1-Measure	0.6791	0.6881	0.6947	0.6891	0.7077
5	MCC Measure [+1, −1]	0.6530	0.6654	0.6690	0.6629	0.6812
6	Normalized MCC [0, 1]	**0.8265**	**0.8327**	**0.8345**	**0.8314**	**0.8406**

6.4 Category Wise Performance Evaluation using AUC score

The AUC score is also calculated for each batch for all the techniques and the results are presented in Table 6. Analysis of the table shows MGT has better average AUC and OBHB has outperformed SGT in 8 different categories and comparable in 2 categories. OBHB is better in discriminating Cat category and the results of OBWB are approachable to that of SGT.

Table 6. Category wise Average (of Batch1 to Batch5) AUC score for various technique

	SGT	**MGT**	**OBHB**	**OBWB**
airplane	0.8127	0.8503	0.8259	0.8044
automobile	0.8736	0.9139	0.8810	0.8620
bird	0.7116	0.7549	0.7301	0.6908
cat	0.6527	0.6877	**0.6935**	0.6643
deer	0.7382	0.7858	0.7250	0.7040
dog	0.7195	0.7674	0.7356	0.7068
frog	0.8319	0.8905	0.8201	0.8085
horse	0.8220	0.8777	0.8213	0.7939
ship	0.8493	0.9055	0.8570	0.8433
truck	0.8262	0.8902	0.8333	0.8154
Average	**0.78377**	**0.83239**	**0.79228**	**0.76934**

Table 7 shows the comparison of all the techniques we implemented with Light Augmentation (LA: horizontal flip, horizontal and vertical translation) and Heavy Augmentation (HA: scaling, rotation, shear, contrast and brightness adjustment) applied by Alex et al. [27] on All CNN, DenseNET and WRN and ODD technique using GAN [17]. We observe that MGT has highest AUC score but it also uses more training data. OBHB outperforms SGT with respect to AUC score and OBWB outperforms SGT using normalized MCC.

Table 7. Final Comparison table

Technique (Training size/Model)	Accuracy	Avg. Normalized MCC	Avg. AUC
SGT (20000), VGG16	0.6082	0.7827	0.7837
MGT(50000), VGG16	**0.6983**	0.8331	0.8323
OBHB (20000), VGG16	0.6104	0.7694	0.7922
OBWB(20000), VGG16	0.5849	0.7840	0.7693
LA(20000),All CNN, ResNET, WRN [27]	67.55	–	–
HA(30000), All CNN, ResNET, WRN [27]	68.69	–	–
ODD (BigGAN) [17]	–	–	0.56
ODD (JigSaw) [17]	–	–	0.63
ODD (EBM) [17]	–	–	0.60

6.5 Discussion

In this research, we explored Hadamard and Walsh latent space technique, where we found that the proposed techniques are computationally less complex than some of the GT techniques such as rotation and GAN based techniques. The advantages of the proposed techniques are that once a coefficient matrix is obtained, a huge dataset can be created in a short span of time by utilizing variable number of coefficients to construct new images. These reconstructed images will have different levels of visual perception adding some diversity to the image dataset. Our experimental results show that the proposed techniques are effective in discriminating most of the natural image categories of Cifar-10 class categories such as bird, cat, dog etc. compared to SGT technique and approachable to that of MGT technique even with a smaller training set.

7 Conclusion and Future Work

Most of the current research is carried out either using geometric transformations-based image data augmentation techniques or GAN based techniques for generating synthetic images and the latent space is comparatively less explored for generating new images. In this research, we investigated Hadamard/Walsh latent space for its utilization for image data augmentation IDA on Cifar-10 dataset using VGG16 based image classification model. The experiments carried out reveal that the proposed OBHB technique gives better results than SGT technique. The complexity of the proposed technique is less than the traditional geometrical transformation. Further, analysis of the category wise AUC score for all the techniques exposed that the proposed techniques are better in discriminating certain natural image categories than the geometrical transformations techniques.

In future, we would like to investigate the effectiveness of the proposed technique using different models such as VGG19, ResNET, WRN etc. and different datasets.

References

1. Shorten, C., Khoshgoftaar, T.M.: A survey on image data augmentation for deep learning. J. Big Data **6**, 60 (2019)
2. Taylor, L., Nitschke, G.: Improving deep learning with generic data augmentation. In: 2018 IEEE Symposium Series on Computational Intelligence (SSCI). (2018)
3. Elgendi, M., et al.: The effectiveness of image augmentation in deep learning networks for detecting COVID-19: A geometric transformation perspective. Front. Med. **8**, 629134 (2021)
4. Perez, F., Vasconcelos, C., Avila, S., Valle, E.: Data augmentation for skin lesion analysis. In: Stoyanov, D., et al. (ed.) CARE/CLIP/OR 2.0/ISIC -2018. LNCS, vol. 11041, pp. 303–311. Springer, Cham (2018). https://doi.org/10.1007/978-3-030-01201-4_33
5. Poojary, R., Raina, R., Mondal, A.K.: Effect of data-augmentation on fine-tuned CNN model performance. IAES Int. J. Artif. Intell. (IJ-AI) 10, 84 (2021)
6. Inoue, H.: Data Augmentation by Pairing Samples for Images Classification. arXiv:1801.029 29v2 (2018)
7. Summers, C., Dinneen, M.J.: Improved mixed-example data augmentation. In: 2019 IEEE Winter Conference on Applications of Computer Vision (WACV). (2019)
8. DeVries, T., Taylor, G.W.: Dataset Augmentation in Feature Space. arXiv.1702.05538 (2017)
9. Liu, X., et al.: Data augmentation via latent space interpolation for Image Classification. In: 2018 24th International Conference on Pattern Recognition (ICPR) (2018)
10. Porcu, S., Floris, A., Atzori, L.: Evaluation of data augmentation techniques for facial expression recognition systems. Electronics **9**, 1892 (2020)
11. Perez, L., Wang, J.: The effectiveness of data augmentation in image classification using deep learning, pp. 1–8 (2017). arXiv:1712.04621
12. Simonyan, K., Zisserman, A.: Very deep convolutional networks for large-scale image recognition (2014). https://arxiv.org/abs/1409.1556
13. Krizhevsky, A., Nair, V., Hinton, G.: https://www.cs.toronto.edu/~kriz/cifar.html
14. Frid-Adar, M., Diamant, I., Klang, E., Amitai, M., Goldberger, J., Greenspan, H.: Gan-based synthetic medical image augmentation for increased CNN performance in liver lesion classification. Neurocomputing **321**, 321–331 (2018)
15. Amirrajab, S., Al Khalil, Y., Lorenz, C., Weese, J., Pluim, J., Breeuwer, M.: Label-informed cardiac magnetic resonance image synthesis through conditional generative adversarial networks. Comput. Med. Imaging Graph. **101**, 102123 (2022)
16. Shin, S., Boyapati, M., Suo, K., Kang, K., Son, J.: An empirical analysis of image augmentation against model inversion attack in Federated Learning. Cluster Computing (2022)
17. Sinha, A., Ayush, K., Song, J., Uzkent, B., Jin, H., Ermon, S.: Negative data augmentation (2021). arXiv preprint arXiv:2102.05113
18. Han, J., et al.: You Only Cut Once: Boosting Data Augmentation with a Single Cut (2022). arXiv preprint arXiv:2201.12078
19. Jain, A.K.: In: Fundamentals of Digital Image Processing, pp. 155–157. Prentice-Hall of India, New Delhi (2006)
20. Jayathilake, A., Perera, I., Chamikara, M.: Discrete Walsh-Hadamard transform in signal processing. Int. J. Res. Inf. Technol. **1**, 80–89 (2013)
21. Win, K.T., Htwe, N.A.A.: Image compression based on modified Walsh-Hadamard transform (MWHT). Int. J. Adv. Comput. Eng. Netw. (2015)
22. Horadam, K.J.: Hadamard Matrices and Their Applications. Princeton University Press, Princeton (2012)
23. Cook, J.D.: An application of Kronecker product (2020). https://www.johndcook.com/blog/2020/04/13/kronecker-product-hadamard-matrix/

24. Howard, R.E., et al.: Optical character recognition: a technology driver for Neural Networks. In: IEEE International Symposium on Circuits and Systems (1990)
25. Chicco, D., Jurman, G.: The advantages of the Matthews correlation coefficient (MCC) over F1 score and accuracy in binary classification evaluation. BMC Genom. **21**, 1–13 (2020)
26. Chicco, D., Tötsch, N., Jurman, G.: The matthews correlation coefficient (MCC) is more reliable than balanced accuracy, bookmaker informedness, and markedness in two-class confusion matrix evaluation. BioData Mining **14**, 1–22 (2021)
27. Hernández-García, A., König, P.: Further advantages of data augmentation on convolutional neural networks. In: Kůrková, V., Manolopoulos, Y., Hammer, B., Iliadis, L., Maglogiannis, I. (eds.) ICANN 2018. LNCS, vol. 11139, pp. 95–103. Springer, Cham (2018). https://doi.org/10.1007/978-3-030-01418-6_10

A Library-Based Dimensionality Reduction Scheme Using Nonlinear Moment Matching

Aijaz Ahmad Khan(✉), Danish Rafiq, and Mohammad Abid Bazaz

National Institute of Technology Srinagar, Jammu and Kashmir 190006, India
aijazkhan013@gmail.com

Abstract. Numerical simulations are crucial for the study, design, and control of dynamical systems. During the modeling process, models characterized by a large number of differential equations frequently emerge due to the increasing complexity and accuracy demands. Nonlinearities also appear in many practical applications that further complicate the subsequent numerical analysis. Model order reduction (MOR) has been a preferred solution for reducing the computational complexity of large-scale systems. One of the widely employed MOR techniques, i.e., *moment matching*, has been successfully extended to nonlinear systems; however, the choice of an appropriate signal generator remains a fundamental limitation of this method. The current study proposes new criteria to obtain the relevant signal generator dynamics by constructing a signal generator-based library of the nonlinear candidate terms that characterizing the steady-state behavior of the interconnected system. We also present a generalized case scenario wherein we have no knowledge about the system. We have used the Ring-Grid model and the standard IEEE 118-Bus systems to demonstrate the validity of the proposed criteria.

Keywords: Dynamical systems · Model order reduction · Nonlinear moment-matching

1 Introduction

Many physical and engineering problems are modeled as large scale dynamical systems which are generally complex. For instance, weather prediction, power systems, molecular systems, mechanical systems, fluid dynamics, etc., are represented by complex models that involve large degree of freedoms either due to a fine discretization of complex geometries (fluid flow around an aircraft structure) using the finite elements methods (FEM) or involve thousands of individual components, such as in power systems. For such systems, tasks like transient analysis, uncertainty quantification and control becomes computationally demanding due to limitations on speed and memory. Besides, achieving a desired control or design requires repeated system evaluations, which quickly becomes infeasible when large-scale systems are involved. This computational bottleneck can be

H. Sharma et al. (Eds.): ICIVC 2022, PALO 17, pp. 302–313, 2023.
https://doi.org/10.1007/978-3-031-31164-2_25

reduced to a great extent by using the MOR techniques. During MOR, one seeks to replace the given mathematical representation of a system or a process with a reduced order model (ROM) that is computationally inexpensive and retains the essential dynamics of the true model. Thus, MOR is a compromise between model order and how well the model preserves the desired properties of the system. The ROM obtained is then used to replace the full order model (FOM) to obtain faster computations and save memory requirements, hence achieving the goal.

There are well-established techniques and underlying ideas for the reduction of linear models. Some well known techniques are based on modal analysis (MA), singular value decomposition (SVD), moment matching (MM) and SVD-Krylov methods (combination of SVD and Krylov) and balanced truncation. A detailed analysis of these methods is presented in [1–3]. These methods have enjoyed a lot of success for reducing a variety of models raging from circuits [4] to fluid flows [5]. However, many practical systems are nonlinear in nature. Therefore, MOR for nonlinear systems have gained lot of interest in recent years due to its wide applications. For weakly nonlinear systems, a lifting transformation is performed to obtain local models, such as linear time-invariant (LTI), linear time-varying (LTV) or Volterra, then the reduction is carried using standard linear MOR methods. For general nonlinear systems, methods such as proper orthogonal decomposition [6], trajectory piece-wise linear [7,8], missing point estimation [9] are widely used. However, these methods involve significant computational costs for obtaining the reduced subspace by taking expensive measurements of the FOM. Recently, moment-matching techniques have been extended to nonlinear systems [10–13]. This method is based on time-domain notion of moment-matching and nonlinear steady-state interpretations by solving an underlying *Sylvester* partial differential equation (PDE). This techniques has been used to derive families of ROMs achieving moment-matching [15] and has been extended to time-delay systems [16]. However, it involves the expensive simulation of the Sylvester PDE. Towards this direction, Maria et al. have proposed a simplified version of nonlinear moment matching (NLMM) to obtain a practical algorithm that simplifies the underlying nonlinear Sylvester PDE to a system of nonlinear algebraic systems [17,18]. This method has been recently extended to parametric systems in Refs. [19,20]. An updated review of moment-matching methods is also available in Refs. [21,24].

Nonlinear moment matching is superior over POD and TPWL in the sense that it offers a "simulation-free" architecture to obtain the projection basis for the reduced manifold without integrating the expensive FOM [18]. The method involves two stages: a *training* stage and a *testing* stage. During the training stage, NLMM seeks to construct projection basis for the reduced subspace by solving the underlying *Sylvester*-like PDE which can be simplified to a system of nonlinear algebraic system using the step-by-step procedure proposed in Ref. [18]. In the testing stage, the reduced equations are integrated and back projected to obtain the approximation of the original system. The online computation can be made even faster by employing hyper-reduction-based methods as demonstrated in Refs. [22,23]. The training stage of NLMM also involves defining the signal generator dynamics required to characterize the steady-state notion of the moment-matching. As we will show, the choice of the signal generator plays a crucial role in determining the quality of ROMs and imposes a significant challenge in the overall NLMM procedure. In this manuscript, we describe a strategy to select a suitable signal generator by creating a library of candidate nonlinear terms required to achieve moment-matching. Towards this aim, we provide a numerical algorithm that selects the best signal generator based on pre-defined error tolerance and we demonstrate this idea on two benchmark power system models.

The remainder of the manuscript is sectioned as follows. To introduce the problem, we begin with the concept of nonlinear MOR via moment matching in Sect. 2. We describe the moment-matching perspective to nonlinear systems. In Sect. 3, we present the notion of signal generators with respect to nonlinear dynamical systems, and discuss the different types of signal generators and their interpretations. Then, we present a numerical scheme to arrive at a suitable signal generator in an intelligent and efficient fashion. Then, in Sect. 4, we apply this idea to reduce two power system models: a Ring Grid Network and IEEE 118-Bus System using NLMM and demonstrate the applicability of the scheme. Finally, we provide the concluding remarks and some future perspectives in Sect. 5.

2 Dimensionality Reduction via Nonlinear Moment Matching

The idea of moment-matching for nonlinear systems was first presented by Astolfi [10] based on the concepts of output regulation of nonlinear systems [29,30]. To demonstrate the idea, consider a time-invariant, nonlinear, MIMO system represented in state-space model as follows:

$$\begin{aligned} \boldsymbol{E}\dot{\boldsymbol{x}}(t) &= \boldsymbol{f}(\boldsymbol{x}(t), \boldsymbol{u}(t)), \quad \boldsymbol{x}(0) = \boldsymbol{x}_0 \\ \boldsymbol{y}(t) &= \boldsymbol{h}(\boldsymbol{x}(t)) \end{aligned} \tag{1}$$

where, $\boldsymbol{E} \in \mathbb{R}^{n\times n}$, $\boldsymbol{x}(t) \in \mathbb{R}^n$, $\boldsymbol{u}(t) \in \mathbb{R}^m$, $\boldsymbol{y}(t) \in \mathbb{R}^p$. Now, consider a nonlinear signal generator as shown below:

$$\begin{aligned} \dot{\boldsymbol{x}}_{\boldsymbol{r}}^{v}(t) &= \boldsymbol{s}_v \boldsymbol{x}_{\boldsymbol{r}}^{v}(t), \quad \boldsymbol{x}_{\boldsymbol{r}}^{v}(0) = \boldsymbol{x}_{\boldsymbol{r},\boldsymbol{0}}^{v} \neq \boldsymbol{0} \\ \boldsymbol{u}(t) &= \boldsymbol{r}\boldsymbol{x}_{\boldsymbol{r}}^{v}(t), \end{aligned} \tag{2}$$

where $\boldsymbol{s}_v(\boldsymbol{x}_{\boldsymbol{r}}^{v}) : \mathbb{R}^r \to \mathbb{R}^r$, $\boldsymbol{r}(\boldsymbol{x}_{\boldsymbol{r}}^{v}) : \mathbb{R}^r \to \mathbb{R}^m$ are smooth mappings such that $\boldsymbol{s}_v(0) = 0$ and $\boldsymbol{r}(0) = 0$. Note that there are following assumptions:

1. Signal generator $(\boldsymbol{s}_v, \boldsymbol{r}, \boldsymbol{x}_{\boldsymbol{r},\boldsymbol{0}}^{v})$ must be observable i.e., for a pair of initial conditions $\boldsymbol{x}_{\boldsymbol{r},\boldsymbol{a}}^{v}(0) \neq \boldsymbol{x}_{\boldsymbol{r},\boldsymbol{b}}^{v}(0)$, the corresponding trajectories $\boldsymbol{r}(\boldsymbol{x}_{\boldsymbol{r},\boldsymbol{a}}^{v}(t))$ and $\boldsymbol{r}(\boldsymbol{x}_{\boldsymbol{r},\boldsymbol{b}}^{v}(t))$ do not coincide.
2. Signal generator is neutrally stable [30].

Thus the signal generator (2) allows us to captures the relevant dynamics of the FOM (1) required to characterize the steady-state behavior of the system given as:

$$\boldsymbol{y}_{ss}(t) = \boldsymbol{h}(\boldsymbol{x}_{ss}(t)) = \boldsymbol{h}(\boldsymbol{\nu}(\boldsymbol{x}_r^v)(t)) \tag{3}$$

where $\boldsymbol{\nu}$ is the unique solution of following Sylvester PDE:

$$\boldsymbol{E}\frac{\partial \boldsymbol{\nu}(\boldsymbol{x}_r^v)}{\partial \boldsymbol{x}_r^v}\boldsymbol{s}_v(\boldsymbol{x}_r^v) = \boldsymbol{f}(\boldsymbol{\nu}(\boldsymbol{x}_r^v), \boldsymbol{r}(\boldsymbol{x}_r^v)) \tag{4}$$

Lemma 1. *[11] The zeroth nonlinear moments $\boldsymbol{m}_0(\boldsymbol{s}_v(\boldsymbol{x}_r^v(t)), \boldsymbol{r}(\boldsymbol{x}_r^v(t)), \boldsymbol{x}_{r,0}^v)$ of FOM (1) at $\{\boldsymbol{s}_v(\boldsymbol{x}_r^v(t)), \boldsymbol{r}(\boldsymbol{x}_r^v(t)), \boldsymbol{x}_{r,0}^v\}$ are related to the steady-state response*

$$\boldsymbol{y}_{ss}(t) = \boldsymbol{h}(\boldsymbol{\nu}(\boldsymbol{x}_r^v(t))) := \boldsymbol{m}_0(\boldsymbol{s}_v(\boldsymbol{x}_r^v(t)), \boldsymbol{r}(\boldsymbol{x}_r^v(t)), \boldsymbol{x}_{r,0}^v), \tag{5}$$

of the system (1) where $\boldsymbol{\nu}(\boldsymbol{x}_r^v)$ is the unique solution of the Sylvester PDE (4)

This lemma allows us to derive a reduced model that matches the steady-state of the FOM (1) and is given as:

$$\begin{aligned} \boldsymbol{E}_r\dot{\boldsymbol{x}}_r(t) &= \boldsymbol{f}_r(\boldsymbol{x}_r(t), \boldsymbol{u}(t)), \quad \boldsymbol{x}(0) = \boldsymbol{x}_0 \\ \boldsymbol{y}_r(t) &= \boldsymbol{h}_r(\boldsymbol{x}_r(t)) \end{aligned} \tag{6}$$

where $\boldsymbol{E}_r = \tilde{\boldsymbol{W}}_{\boldsymbol{x}_r}^T \boldsymbol{E}\tilde{\boldsymbol{V}}_{\boldsymbol{x}_r}$ and $\boldsymbol{f}_r(\boldsymbol{x}_r, \boldsymbol{u}(t)) = \tilde{\boldsymbol{W}}_{\boldsymbol{x}_r}^T \boldsymbol{f}(\boldsymbol{\nu}(\boldsymbol{x}_r), \boldsymbol{u}(t))$ with $\tilde{\boldsymbol{V}}_{\boldsymbol{x}_r} = \dfrac{\partial \boldsymbol{v}(\boldsymbol{x}_r(t))}{\partial \boldsymbol{x}_r(t)} \in \mathbb{R}^{n\times r}$ and $\tilde{\boldsymbol{W}}_{\boldsymbol{x}_r} = \dfrac{\partial \boldsymbol{\phi}(\boldsymbol{\omega}(\boldsymbol{x}, \boldsymbol{u}))}{\partial \boldsymbol{x}}|_{\boldsymbol{x}=\boldsymbol{v}(\boldsymbol{x}_r)} \in \mathbb{R}^{n\times r}, \boldsymbol{\phi}(\boldsymbol{\omega}(\boldsymbol{x}, \boldsymbol{u})) : \mathbb{R}^n \rightarrow \mathbb{R}^r$.

Theorem 1. *[14] Consider the system (1) is interconnected to signal generator (2) where $(\boldsymbol{s}_v, \boldsymbol{r}, \boldsymbol{x}_{r,0}^v)$ is considered to be observable and neutrally stable. If $\boldsymbol{\nu}(\boldsymbol{x}_r^v)$ is the unique solution of Sylvester PDE (4), then, the exponentially stable ROM (6) exactly matches the (locally-defined) steady-state response of the output of interconnected system (see Fig. 1), i.e.,*

$$\begin{aligned} \boldsymbol{e}(t) &= \boldsymbol{y}(t) - \boldsymbol{y}_r(t) \\ \boldsymbol{e}(t) &= \boldsymbol{h}(\boldsymbol{x}(t)) - \boldsymbol{h}(\boldsymbol{\nu}(\boldsymbol{x}_r(t))) = 0. \end{aligned} \tag{7}$$

In the next section, we will discuss the choice of signal generator used to derive the ROM.

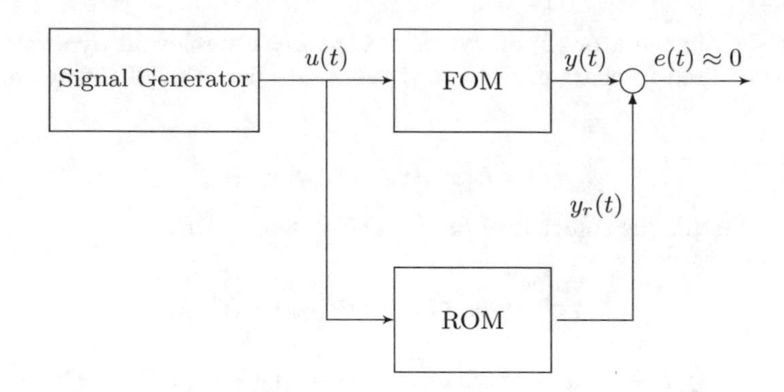

Fig. 1. Interconnection between the nonlinear FOM and ROM and the signal generator

3 Selecting the Right Signal Generator

As discussed in previous section, deriving a reduced model that achieves moment-matching with the FOM corresponds to finding a system that matches the steady-state response of FOM in presence of the inputs generated by an appropriate signal generator. Thus, the choice of signal generator used for moment-matching plays a crucial role for for dictating the quality of approximation. In this following, we present a numerical strategy to obtain an efficient signal generator for NLMM.

3.1 Classification of Signal Generators

A classification of signal generators have been mentioned in Ref. [18] and we reproduce it here for completeness.

Linear Signal Generator: The most basic and obvious choice is to design a linear signal generator given as:

$$\begin{aligned} \dot{\boldsymbol{x}}_r^v(t) &= \boldsymbol{S}_v(\boldsymbol{x}_r^v(t)),\ \boldsymbol{x}_r^v(\boldsymbol{0}) = \boldsymbol{x}_{r,0}^v \neq \boldsymbol{0} \\ \boldsymbol{u}(t) &= \boldsymbol{R}\boldsymbol{x}_r^v(t) \end{aligned} \tag{8}$$

Using a linear signal generator (8) corresponds to exciting the FOM (1) with exponential inputs given as $\boldsymbol{u}(t) = \boldsymbol{R}\boldsymbol{x}_r^v(t) = \boldsymbol{R}e^{\boldsymbol{S}_v t}\boldsymbol{x}_{r,0}^v$ parametrized by the shift matrix $\boldsymbol{S}_v \in \mathbb{R}^{r\times r} = \text{diag}(\sigma_1, \sigma_2, ..., \sigma_r)$ where $\sigma = \delta \pm i\omega \in \mathbb{C}$ and tangential matrix $\boldsymbol{R}$. The shifts σ_i can be chosen as: (i) *pure real*, i.e., $(\delta \neq 0, \omega = 0)$, describing growing or decaying exponentials depending on δ, (ii) *pure imaginary* $(\delta = 0, \omega \neq 0)$ corresponding to permanent oscillations and (iii) *complex* $(\delta \neq 0, \omega \neq 0)$. If $\delta < 0$, then $x_r^v = \exp^{\delta t}(\cos(\omega t) \pm i\sin(\omega t))$ describes a decaying oscillation, whereas for $\delta > 0$ represents a growing oscillation.

Nonlinear Signal Generator: The nonlinear signal generator given as (2) is a more general case whereby the nonlinear mappings $\boldsymbol{s}_v(\boldsymbol{x}_r^v) : \mathbb{R}^r \rightarrow \mathbb{R}^r$, and $\boldsymbol{r}(\boldsymbol{x}_r^v) : \mathbb{R}^r \rightarrow \mathbb{R}^m$ with $\boldsymbol{s}_v(0) = \boldsymbol{0}$ and $\boldsymbol{r}(0) = \boldsymbol{0}$ characterize the moments. Using a nonlinear system corresponding to exciting the system with user defined inputs $\boldsymbol{u}(t) = \boldsymbol{r}(\boldsymbol{x}_r^v(t))$. Maria et al. [18] have provided step-by-step simplifications to avoid a nonlinear projection and to approximate the PDE (4) with nonlinear algebraic system.

Zero Signal Generator: This special case of signal generator can be represented as:

$$\dot{\boldsymbol{x}}_r^v(t) = \boldsymbol{s}_v(\boldsymbol{x}_r^v(t)) = \boldsymbol{0} \tag{9}$$

This means that $\boldsymbol{x}_r^v(t) = \boldsymbol{x}_{r,0}^v =$ constant and $\boldsymbol{u}(t) = \boldsymbol{R}\boldsymbol{x}_r^v(t) = \boldsymbol{R}\boldsymbol{x}_{r,0}^v =$ constant. Thus, using a zero signal generator implies exciting the system with constant input signals.

3.2 A Criteria to Choose Signal Generator

An ideal signal generator is to be chosen in such a way that:

1. $\boldsymbol{x}_r^v(t)$ forms representative eigen functions of the corresponding nonlinear system.
2. $\boldsymbol{u}(t)$ is a *persistently exciting* input that excites the essential dynamics of the system.

To describe the proposed methodology for selecting the right signal generator, we distinguish between the following cases.

Case I: When we have partial knowledge about the system. Let us assume that we have some prior domain knowledge about the system that we can leverage to arrive at the right signal generator and correspondingly the reduced model. For this case, the choice of signal generator can be made by studying the shape of the eigenfunctions of the linearized system and an appropriate signal generator can be designed to match this shape. For instance, exponential functions with desired exponents can be tuned to match the shape of eigenfunctions of the systems. Similarly, sinusoidal functions of specific amplitude and frequency can be generated for models exhibiting periodic oscillations. Hereby the time period T and initial conditions $x_r(0)$ can be selected by pre-analyzing the value range that the signals $x_r^v(t)$ and $\dot{x}_r^v(t)$ cover. Depending upon the underlying nonlinearity of the system the following signal-generator library can be constructed:

$$\Xi = \left[A_1 \sin(\omega_1(t))\; A_2 \sin(\omega_2(t))\; ...\; A_n \sin(\omega_n(t))\right] \tag{10}$$

where A_n and ω_n are the amplitudes and frequencies range of interest. Hereby, it is assumed that the underlying nonlinearity is sinusoidal in nature and we only provide a sin sweep across the frequency. A similar library can be constructed for different underlying nonlinearities such as exponential, polynomial terms, etc.

Case II: Where we have no knowledge about the system For this case, we propose to construct a generalized library consisting of all the representative signal generators such as zero, linear and nonlinear signal generators given as:

$$\Xi = \left[const\ poly(t)\ sin(t)\ exp(t) \right] \tag{11}$$

where *const* contains terms with varying gains for zero signal generator, $poly(t)$ contains terms like $[x_r^v(t),\ x_r^{2v}(t),\ ...,\ x_r^{pv}(t)]$, $sin(t) = \lfloor k\sin(kt)\ k\cos(kt) \rfloor$ for $k = 1, 2, ..., s$ and $exp(t)$ contains terms for exponential functions, i.e., $[ae^{bt}]$ where $(a, b) = 1, 2, ..., e$. Thus, the entire library contains p polynomial terms, s sin and cosine terms, e exponential and a single constant term. The choice of these parameters is problem specific. Since we don't have any knowledge of the system, the hope here is that by selecting the most dominant nonlinearities, we will be able to capture the underlying nonlinear phenomena to achieve moment-matching with the FOM.

Having selected the signal generator library Ξ, depending upon the case discussed above, we then proceed by finding the most suitable signal generator from the respective library. This is achieved using a greedy search algorithm that evaluates the error for each signal generator candidate against the true solution and depending upon the predefined error tolerance, the signal generator with the least error is stored and used for final evaluation of the ROM. The numerical algorithm is given as follows:

Algorithm 1. Signal Generator selection algorithm

Input: $err\_tol, \boldsymbol{\Xi}, \boldsymbol{x}, l =$ total number of terms in Ξ
Output: optimal $\boldsymbol{x}_r^v(t), \boldsymbol{V}_{\text{NLMM}}$
1: **for** $i = 1$ to l **do**
2: Solve the ith signal generator to obtain $\boldsymbol{x}_{r,i}^v$
3: Solve the NLMM algorithm [18] to obtain $\boldsymbol{V}_i$
4: Solve the ROM (24) and store the present error $\boldsymbol{e}_i = |(\boldsymbol{x} - \boldsymbol{V}_i\boldsymbol{x}_{r,i})|_{\mathcal{L}_2}$
5: **if** $err\_tol > \boldsymbol{e}_i$ **then**
6: return $\boldsymbol{x}_{r,i}^v, \boldsymbol{V}_{\text{NLMM}}$
7: break
8: **else**
9: continue
10: **end if**
11: **end for**

In the above algorithm, the outer *for* loop is used to run each signal generator and evaluate the corresponding ROM. The algorithm terminates early if any signal generator yields an error tolerance below the desired accuracy using the *if* condition, otherwise the algorithm continues upto the last term in the library. A signal generator that reaches the set desired accuracy is stored and the corresponding basis $\boldsymbol{V}_{\text{NLMM}}$ is used to obtain the best ROM (6). However, if the algorithm stops without any early termination, then the user can make an informed decision by checking in the error logs stored in $\boldsymbol{e}$ and tweak the signal generator library by adding more terms in the series or introducing new terms. Please note that the numerical scheme described in this paper is just to guide the user for a better implementation of the NLMM methods rather than manually setting the signal generator each time. The overall idea of the scheme is presented as a flowchart in Fig. 2. In the next section, we will demonstrate the implementation of the proposed scheme on two power system models.

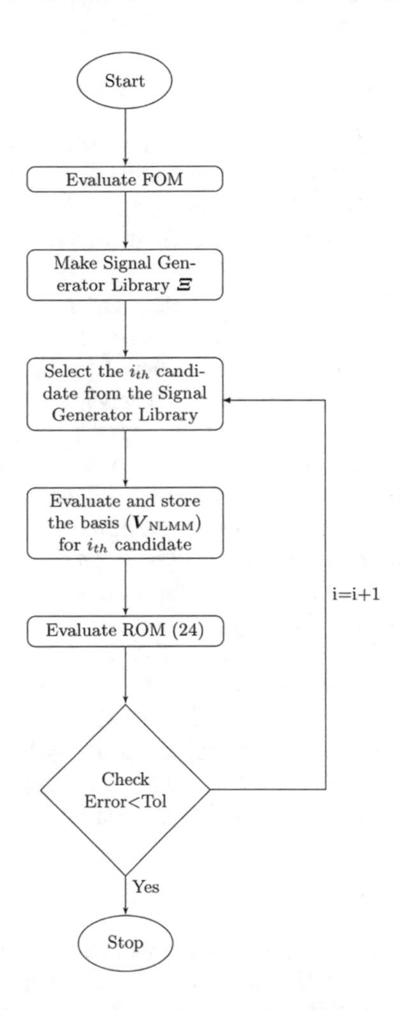

Fig. 2. Flowchart depicting choosing of signal generator from library

4 Numerical Simulation

To demonstrate the application of the proposed scheme, we have simulated two power system models: (i) a Ring-Grid network constituting of 1000 generators, and (ii) the IEEE 118-Bus system. Both the models are simulated in MATLAB environment on i5-4210 CPU, 8 GB RAM and 1.7 GHz processor.

4.1 Ring Grid Model

We first consider the Ring Grid model topology. The system consists of n identical generators arranged in a ring and connected to a common reference bus discussed in [31]. The governing swing equation is given as follows:

$$m_i\ddot{\delta}_i+d_i\dot{\delta}_i = p_i^m - bsin(\delta_i) - b_{int}[sin(\delta_i-\delta_{i+1})+sin(\delta_i-\delta_{i-1})] \quad i = 1, 2, ..., n \quad (12)$$

where, m_i is the mass of ith generator, d_i is the damping of the ith generator, p^m represents the power generated by the generators, b is the susceptance between generator and slack node, b_{int} is the susceptance between consecutive generators and N is the number of generators. The corresponding values were taken from [31].

The full order model was integrated using the implicit *Euler's* scheme with a step size of $dt = 0.01$ and for a total time of $T = 50s$. To proceed with reduction, two reduced models were constructed (i) NLMM_1, obtained by using a sinusoidal signal generator (using prior knowledge about the system) and (ii) NLMM_2, obtained by using a signal generator from the proposed scheme (where we assume no knowledge about the system). After constructing the library as discussed in Sect. 3 and using the Algorithm 1, we obtained a polynomial signal generator given as $x_r^v(t) = x^3$. The output responses of both the ROMs and the corresponding error plots are depicted in Fig. 3. As can be seen from the figures, both the ROMs capture the true output response very well. Since NLMM_1 ROM was obtained using a sinusoidal signal generator $x_r^v(t) = -4800 \cdot cos(2\pi \cdot 9600 \cdot t)$, it was expected to perform better than NLMM_2 ROM as it was designed by leveraging prior knowledge about the system. However, NLMM_2 still performs satisfactorily given the fact it didn't involve any system knowledge and was obtained from a generalized signal generator library. The computational costs and ℓ_2-errors for this case are enlisted in Table 1. It is observed that both ROMs yield ≈ 6 fold reduction in the total CPU time than the FOM.

4.2 IEEE 118-Bus System

The benchmark IEEE 118-Bus system represents an approximation of American Electric Power System in the U.S Midwest (1962). The model features 19 generators, 35 synchronous condensers, 177 lines, 9 transformers and 91 loads [26]. A state-space representation of the model was obtained using the method presented in [27] and was integrated using implicit *Euler's* method with $dt = 0.01$. Similar to previous case, we constructed two ROMs. The first ROM NLMM_1 was obtained by only using sinusoidal terms in the library whereas the second ROM was obtained by taking all nonlinear terms as discussed in Sect. 3. The Algorithm 1 resulted in signal generator $x_r^v(t) = -550 \cdot cos(2\pi \cdot 1100 \cdot t + \pi)$ for sinusoidal and $x_r^v(t) = x^5$ for second case. The output responses (average of rotor angles and frequency) and the corresponding error plots are presented in Fig. 4 respectively while the CPU times are mentioned in Table 1. Again, we can see that the proposed method yielded satisfactorily ROM performance for this case.

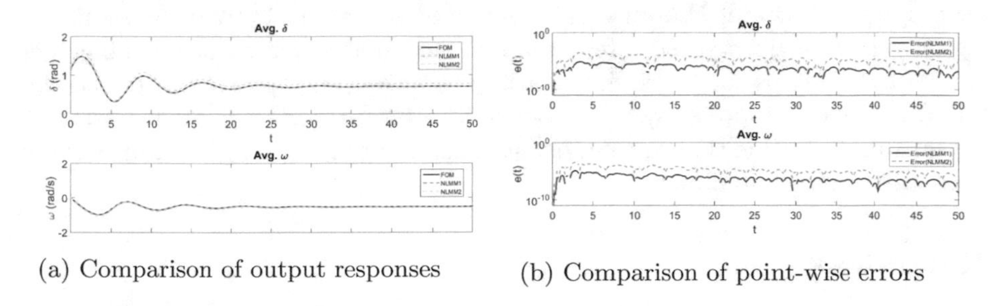

(a) Comparison of output responses (b) Comparison of point-wise errors

Fig. 3. Ring Grid model results

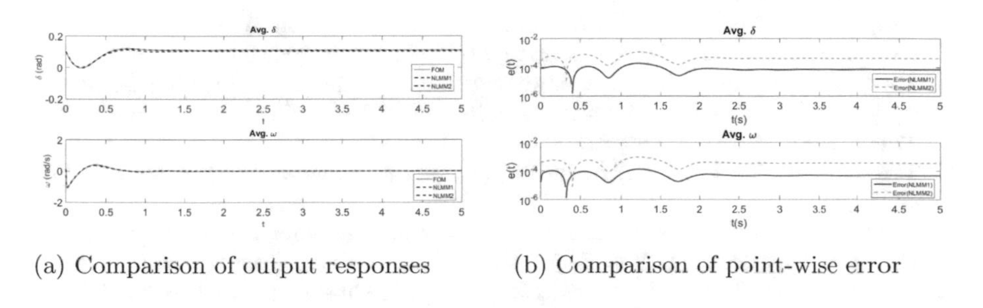

(a) Comparison of output responses (b) Comparison of point-wise error

Fig. 4. IEEE 118 Bus system

Table 1. Quantitative comparison of the CPU times and the relative errors

Technique	Ring Grid model		IEEE 118 Bus model	
	CPU time	ℓ_2 error	CPU time	ℓ_2 error
FOM	258.67	-	243.51	-
$NLMM_1$	44.7	1.6×10^{-6}	62.24	1.6×10^{-5}
$NLMM_2$	28.8	3.5×10^{-4}	42.06	5.3×10^{-3}

5 Conclusion and Future Perspective

In this contribution, we discussed the model reduction problem via moment-matching via frequency and time-domain perspectives. We highlighted the main issue in the numerical computation of NLMM based ROMs. As such, we presented a novel strategy to automate the NLMM method by making a suitable signal generator library-based reduction method that avoids manual tuning. The resulting ROMs perform well for a variety of testing signals. We illustrated this on two benchmark power system models. The future work will involve using an optimization routine that automatically picks the best signal generator without manually tuning the library.

References

1. Antoulas, A.C.: Approximation of Large-scale Dynamical Systems. SIAM, Philadelphia, USA (2005)
2. Benner, P., Mehrmann, V., Sorensen, D.C.: Dimension Reduction of Large-scale Systems. In: Lecture Notes in Computational Science and Engineering, vol. 45, Springer, Berlin, Germany (2005)
3. Gugercin, S., Antoulas, A.C.: A survey of model reduction by balanced truncation and some new results. Int. J. Control **77**(8), 748–766 (2004)
4. Manetti, S., Piccirilli, M.C.: A singular-value decomposition approach for ambiguity group determination in analog circuits. IEEE Trans. Circ. Syst.I: Fundamental Theor. Appl. **50**(4), 477–487 (2003)
5. Taira, K., et al.: Modal analysis of fluid flows: an overview. AIAA J. **55**(12), 4013–4041 (2017)
6. Moore, B.: Principal component analysis in linear systems: controllability, observability and model reduction. IEEE Trans. Autom. Control **26**, 17–32 (1981)
7. Rewienski, M., White, J.: A trajectory piecewise-linear approach to model order reduction and fast simulation of nonlinear circuits and micromachined devices. IEEE Trans. Comput.-Aided Design **22**(2) (2003)
8. Vasilyev, D., Rewienski, M., White, J.: A TBR-based trajectory piecewise-linear algorithm for generating accurate low-order models for nonlinear analog circuits and MEMS. In: Proceedings 2003 Design Automation Conference IEEE, pp. 490–495 (2003)
9. Astrid, P., Weiland, S., Willcox, K., Backx, T.: Missing point estimation in models described by proper orthogonal decomposition. IEEE Trans. Autom. Control **53**(10), 2237–2251 (2008)
10. Astolf, A.: Model reduction by moment matching. In: 7th IFAC symposium on nonlinear control systems. Elsevier, pp. 577–584 (2007)
11. Ionescu, T.C., Astolf, A.: Families of reduced order models that achieve nonlinear moment matching. In: 2013 American Control Conference. IEEE (2013)
12. Ionescu, T.C., Astolf, A.: Nonlinear moment matching- based model order reduction. IEEE Trans. Autom. Control **61**(10), 2837–2847 (2015)
13. Astolf, A.: A new look at model reduction by moment matching for linear systems. In: 2007 46th IEEE Conference on Decision and Control. IEEE, pp. 4361–4366 (2007)
14. Astolfi, A.: Model reduction by moment matching for linear and nonlinear systems. IEEE Trans. Autom. Control **55**(10), 2321–2336 (2010)

15. Ionescu, T.C., Astolf, A., Colaneri, P.: Families of moment matching based, low order approximations for linear systems. Syst. Control Lett. **64**, 47–56 (2014)
16. Scarciotti, G., Astolfi, A.: Data-driven model reduction by moment matching for linear and nonlinear systems. Automatica **79**, 340–351 (2017)
17. Varona, M.C., Nico, S., Lohmann, B.: Nonlinear moment matching for the simulation-free reduction of structural systems. IFAC-Papers OnLine, Vienna, Austria, IFAC **52**, 328–333 (2019)
18. Maria, C.V.: Model Reduction of Nonlinear Dynamical Systems by System-Theoretic Methods, Ph.D. Thesis, Technical University of Munich (2020)
19. Rafiq, D., Bazaz, M.A.: A framework for parametric reduction in large-scale nonlinear dynamical systems. Nonlinear Dyn. **102**(3), 1897–1908 (2020). https://doi.org/10.1007/s11071-020-05970-3
20. Rafiq, D., Bazaz, M.A.: Adaptive parametric sampling scheme for nonlinear model order reduction. Nonlinear Dyn. **107**, 813–828 (2022)
21. Rafiq, D., Bazaz, M.A.: Model Order Reduction via Moment-Matching: A State of the Art Review. In: Archives of Computational Methods in Engineering, vol. 29 (2021). https://doi.org/10.1007/s11831-021-09618-2
22. Rafiq, D., Bazaz, M.A.: A comprehensive scheme for reduction of nonlinear dynamical systems. Int. J. Dyn. Control **8**(2), 361–369 (2020)
23. Rafiq, D., Bazaz, M.A.: Nonlinear model order reduction via nonlinear moment matching with dynamic mode decomposition. Int. J. Nonlinear Mech. **128**, 103625 (2021)
24. Freund, R.W.: Model reduction methods based on Krylov subspaces. Acta Numer **12**, 267–319 (2003)
25. Gallivan, K., Vandendorpe, A., Van Dooren, P.: Sylvester equations and projection-based model reduction. J. Comput. Appl. Math. **162**(1), 213–229 (2004)
26. https://icseg.iti.illinois.edu/ieee-118-bus-system
27. Nishikawa, T., Motter, A.E.: Comparative analysis of existing models for power-grid synchronization. New J. Phys. **17**(1), 015012 (2015)
28. Gunupudi, P.K., Nakhla, M.S.: Model-reduction of nonlinear circuits using Krylov-space techniques. In: Proceedings of the 36th annual ACM/IEEE design automation conference, pp. 13–16 (1999)
29. Isidori, A.: Nonlinear Control Systs., 3rd edn. Springer, Berlin (1995)
30. Isidori, A., Byrnes, C.: Output regulation of nonlinear systems. IEEE Trans. Autom. Control **35**(2), 131–140 (1990)
31. Malik, M.H., Borzacchiello, D., Chinesta, F., Diez, P.: Reduced order modeling for transient simulation of power systems using trajectory piece-wise linear approximation. Adv. Model. Simul. Eng. Sci. **3**(1), 1–18 (2016). https://doi.org/10.1186/s40323-016-0084-6

Performance Evaluation of Machine and Deep Transfer Learning Techniques for the Classification of Alzheimer Disease Using MRI Images

Archana Wamanrao Bhade(✉) and G. R. Bamnote

Department of Computer Science and Engineering, Prof. Ram Meghe Institute of Research and Technology, Badnera, Amravati 444701, India
awbhade@gmail.com

Abstract. Alzheimer is a neurological disorder that causes cell damage and memory loss. The late symptoms of Alzheimer disease necessitate early identification to minimise the progression of disease and mortality rate. Machine learning and deep learning techniques are required to thoroughly read the MRI images and extract the most relevant information about disease progression. Transfer learning approaches assist in the reduction of computational requirements and minimise the overfitting issues. This study identifies the most appropriate technique for the classification of Alzheimer's disease using machine learning, deep learning, and transfer learning techniques. The classification is performed on the ADNI MRI dataset with the minimal pre-processing of data to compare the results on the same scale. Results suggested that transfer learning approaches outperformed the other algorithms and can be used with other feature extraction techniques to further improve model performance.

Keywords: Alzheimer disease · machine learning · transfer learning · MRI images

1 Introduction

Alzheimer's disease (AD) is a leading neurological disorder and a typical form of dementia that causes the loss of an individual's decision-making and judgment power, destroying brain cells, thinking ability, communication skills, cognitive functioning and ultimately lead to memory loss [1]. Early manifestation of Alzheimer is not reported rather the symptoms get notified after 60 years of age, and late detection makes it an irreversible disease [2]. More than 44 million people are affected with Alzheimer, and total patients are expected to get double by 2050 [1]. The process of recording the symptoms of patients, analyzing those symptoms and providing appropriate treatment requires significant efforts and facilities [3, 4]. A detailed analysis of brain functions is required to determine the disorders in brain functioning. The comparison of data concerning

H. Sharma et al. (Eds.): ICIVC 2022, PALO 17, pp. 314–327, 2023.
https://doi.org/10.1007/978-3-031-31164-2_26

abnormal functioning with the brain's normal functioning can provide helpful information and could be used as biomarkers for analysis and for determining the reasons for brain disorders. Therefore, significant efforts are required to develop techniques for early detection of AD symptoms, especially during the pre-symptomatic stage, to prevent the further spreading of the disease [5]. The traditional network-based modelling methods are worked on correlation techniques and are very sensitive to brain function networking [6, 7].

Neuro-imaging data have increasingly been used successfully to classify AD using machine learning (ML) and deep learning (DL) techniques, which provide promising solutions to diagnose and treat the disease [8–12]. The successfully implemented used ML techniques support vector machine (SVM) [13], K nearest neighbour (KNN) [14], Random Forest [15], and decision tree [16]. In contrast, convolution neural network (CNN), VGG16, InceptionV3, ResNet50 provides superior results among DL techniques [10, 17]. The application of ML and DL techniques for the early detection of disease make it a prominent area of research in the computer-aided diagnosis of disease. Alvarez et al. [18] applied ensemble SVM to classify Alzheimer disease using a binary class dataset. The author applied component-based feature extraction and obtained high accuracy. Khedher et al. [19] used a multi-class dataset consisting of 818 images to classify the MRI dataset to classify Alzheimer database images. The algorithm generated using ML techniques provides interactive platform for the visualization, pre-processing, analysis and quantification of results obtained from data analysis [20]. The selection of global and local properties improvises the results significantly but, at the same time, also causes the loss of some data [21]. The complexity of the brain structure cannot be determined using a single feature; therefore, multiple features are required to maximise the model accuracy.

Feature identification by DL approaches involve sequential stratified progression of complex nonlinear transformations, following automated classification of data, majorly image-based data [22, 23]. Involvement of multiple transformations during feature identification allow unsheathing of appropriate information by excluding redundancy and dimensionality along with preventing any significant loss of functional data [24, 25]. The convolution neural network (CNN) models work by allocating similar weights to multiple neurons and vice-versa for feature selection [26]; which evident the major requirements of CNN models like large training sample size, high computational power and pre-processing of raw data for effective working. Liu et al. [27] segregated neuroimaging dataset in variant classes using output generated by auto-encoder and softmax regression techniques for early prognosis of carcinogenic state of cells. Glozman and Liba [28] implemented image pre-processing techniques for designing training sample set and further trained the dataset obtained from ADNI through ImageNet dataset using transfer learning techniques. Korolev et al. [29] applied the transfer learning and neural network model to design the classification model. The review of studies suggested the applicability of DL technique for image data classification over ML techniques.

In this study, various machine learning and deep learning approaches are used to classify the ADNI dataset. The limitation and advantages of each method are discussed to quantify the best approach for developing the futuristic model. The 3D ADNI multi-class dataset is converted to 2D for the normalization and data pre-processing. The skull-stripping is performed to extract the relevant information from the 3D dataset using HD-BET, and all the approaches are compared.

2 Methods

2.1 Database

The image and data archive (IDA) provides extensive neuroimaging data for visualization, analysis, and further exploration. The Alzheimer's disease Neuroimaging Initiative (ADNI) dataset is used from the resource pool of IDA, consisting of significant images to develop an algorithm. ADNI consist of several images collected from 1.5T and 3T scanners of MRI. ADNI1 dataset consisting of multi-class two years' complete dataset from 1.5T scanner is used in this study. There are 2042 images in the ADNI1 dataset, most prominent among another ADNI1 datasets. The sagittal view of MRI images showing the development of Alzheimer from standard control (NC) is shown in Fig. 1.

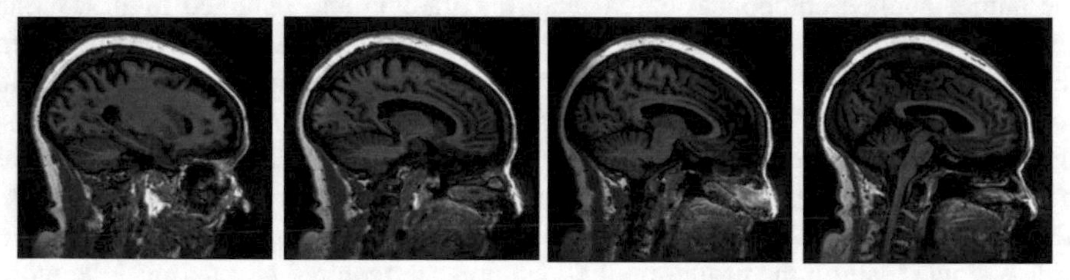

Fig. 1. The sagittal view of MRI from NC to the development of AD (from left to right)

Data Pre-processing. The corrupted files, which are unreadable, are removed from the database. The revised images used for the classification are 1886 after removing 158 corrupted images. All the images are registered and classified in category AD, mild cognitive impairment (MCI), and normal control (NC). The AD, MCI and NC consist of 243, 1117 and 526 images, respectively. The spatial normalisation is performed to ensure the spatial structure of all the dataset images is as similar as possible [30]. The skull stripping is performed using the HD-BET algorithm. HD-BET is a brain extraction tool trained with precontrast T1-w, postcontrast T1-w, T2-w and FLAIR sequences. HD-BET's operation and computational performance were found significantly better than other brain extraction tools such as FSL BET, AFNI 3DSkullStrip, Brainsuite BSE, ROBEX and BeaST [30, 31]. The skull stripping allows reducing the volumetric data generated from the MRI scan by removing other skull parts such as eyes, neck, muscles, nose etc.

Dataset Partitioning. The registered, skull stripped and labelled dataset divided into three sets for modelling. For the model training, testing, and validation, the dataset divided into 70, 15, and 15 ratios for 2D and 3D datasets, as shown in Table 1. After removing corrupted images, the 1886 images classified into three classes. MCI contains a maximum of 1117 images compared to other classes, followed by NC and AD.

Dataset for Models. The slicing of the three-dimensional images is performed manually at every two frames, after some tests. The higher and bottom cuts did not show any significant information. The images are sliced at every second frame between 10 to 40, as the frame above ten and after 40 do not cover essential information for the selection of 16 cuts from an image. However, slices below 40 contain some brainstem. The three-dimensional images are transformed into two-dimensional images by obtaining nine slices and placing them in a 4×4 two-dimensional grid.

Table 1. Number of sample images in each class

Dataset Class	Training	Testing	Validation
AD	170	37	36
MCI	782	168	167
NC	368	79	79

2.2 Methodology

In this study, various machine learning and deep learning techniques are applied to the ADNI1 dataset to determine the applicability of different techniques and measure the performance of techniques. Among machine learning techniques, support vector machine (SVM), random forest and KNN are used. Whereas, among deep learning techniques, 2D-CNN, 3D-CNN, InceptionV3, VGG16 and ResNet50 was applied to classify the dataset in the defined three classes. Transfer learning techniques are also compared to identify the model with high accuracy and the best computational requirement.

Machine Learning Models. SVM is a commonly used technique for data classification studies that works on the maximum margin principle. SVM divides the data into different classes by creating a hyperplane and various kernels for fitting the various data types that handle the data non-linearity and complexity efficiently. The distance from the hyperplane is maximized with the minimization of $\|w\|$ [32]. Structural risk management (SRM) provides stability to the SVM model and improves performance [13]. Random forest (RF) is a computationally efficient classification technique based on the ensemble of trees approach. Each tree represents a molecule of n-dimension having a separate output. The output of all the trees combined to generate the final model output. Classification and regression tree (CART) allow the tree's growth concerning features associated with each tree. The splitting of nodes carried out at every iteration from the subset, and PCA on every subset allows the convergence of similar features to develop a new tree [33]. Naive Bayes (NB) consider all the data points as independent attributes and design the classes using a conditional probability for n number of training data points. The error generated from independent attribute classification handled with optimal bayesian networking by manipulating structure, attributes and instances. The structure manipulation includes the extension of structure to handle the data feature dependencies [34]. The K-nearest neighbour (KNN) is a non-parametric classification based supervised learning technique. The training of samples in KNN carried out by training dataset, which contains both the input, output data, and classifies them based on the nearest neighbour method. The performance of the model validated by comparing the classified data with the part of training data [35].

Deep Transfer Learning Models. Deep neural network (DNN) and convolution neural network (CNN) has become the readily applied technique for various application especially medical imaging due to generality [27, 36]. Convolutional Neural Networks (CNN)'s core is convolutional layers that can extract local features (edges) across an input image through convolution. Each node in a convolutional layer is connected to a small subset of spatially connected neurons. A max-pooling layer follows convolutional layers to reduce computational complexity, which reduces the size of feature maps by selecting the maximum feature response in a local neighborhood. Pairs of convolutional and pooling layers are followed by several fully connected layers, where a neuron in one layer has connections to all activations in the previous layer [25]. Fully connected layers help to learn non-linear relationships among the local features extracted by convolutional layers. Finally, a soft-max layer follows the fully connected layers, which normalizes the outputs to desired levels. CNN's are trained with the back-propagation algorithm, wherein each iteration, weights associated with the neurons in the convolutional layers are updated to minimize a cost function [36]. When training from scratch, the weights are typically initialized randomly, drawing from a normal distribution. The CNN was applied in this study using both the 2D and 3D datasets.

However, to obtain high accuracy in data classification, sufficient training of models is required, and a significant obstacle in obtaining the precise model is the availability of the dataset. The transfer learning approach effectively handles such problems by using the pre-trained network [37, 38]. The transfer learning techniques use the knowledge obtained from the training of large datasets of a similar application area and reduce the requirement of high computational requirement (GPUs) and long processing time of each iteration. In this study, the deep learning algorithm DNN, 2D CNN and 3D CNN are compared with the VGG16, ResNet50, and Inception V3 to classify the ADNI dataset into three classes.

3 Results and Discussion

The dataset is pre-processed to remove the corrupted images that do not deliver any information and spatially normalised to ensure the similar structure of all the images. HD-BET brain extraction tool applied for skull stripping. The HD-BET outperformed other tools such as FSL BET, AFNI 3DSkullStrip, Brainsuite BSE, ROBEX and BEaST and observed better performance by + 1.33 +2.63 units of DICE coefficient. The skull stripping reduces the volumetric data generated from MRI scans by removing other skull parts such as the eyes, neck, muscles, and nose.

The machine learning and deep transfer learning techniques were applied to the pre-processed data after resizing, shaping, slicing and dividing data. VGG16 is being used for feature extraction for machine learning models. Data augmentation is performed to improve the image classification with rescaling, shear and zooming. The zooming performed of 0.2, and the shear range is also 0.2 with the horizontal flip of the dataset. Data augmentation with rescaling carried out at 1/0.255 applied on training, validation and testing dataset.

3.1 Training and Validation of the Model

Training of model acquires the sample size of 16 units, where all the training and validation protocol is described using ratio determination of total image to sample size. The procedure of design involves passage of multiple steps in the back and forth direction of network every time during progression. Evaluation of designed model is carried out through variant measures like confusion matrix, accuracy, precision, recall, and F1-score. Formulation of confusion matrix act as one of the critical measures during evaluation study as it assesses about the pertinence state of model in four combinations extracted out from positive and negative prediction of actual and estimated results. The accuracy, precision, and recall. The linear SVC is used to train the SVM model, Gaussian NB for Naive Bayes. The KNN operated in the range of 0 to 100, where maximum accuracy observed at $k = 12$ while training the model.

Table 2. Layer architecture of CNN 2D and CNN 3D model

Layer (type)	Output Shape	Param	Layer (type)	Output Shape	Param
input_2 (Input Layer)	[(None, 440, 320, 3)]	0	input_1 (InputLayer)	[(None, 80, 110, 80, 1)]	0
conv2d_2 (Conv2D)	(None, 440, 320, 64)	1792	conv3d (Conv3D)	(None, 78, 108, 78, 8)	224
batch_normalization_2 (Batch	(None, 440, 320, 64)	256	conv3d_1 (Conv3D)	(None, 76, 106, 76, 16)	3472
max_pooling2d_2 (MaxPooling2)	(None, 220, 160, 64)	0	max_pooling3d (MaxPooling3D)	(None, 38, 53, 38,	0
conv2d_3 (Conv2D)	(None, 220, 160, 128)	73856	conv3d_2 (Conv3D)	(None, 36, 51, 36,32)	13856
batch_normalization_3 (Batch	(None, 220, 160, 128)	512	conv3d_3 (Conv3D)	(None, 34, 49, 34,64)	55360

(continued)

Table 2. (*continued*)

Layer (type)	Output Shape	Param	Layer (type)	Output Shape	Param
max_pooling2d_3 (MaxPooling2)	(None, 110, 80, 128)	0	max_pooling3d_1 (MaxPooling3	(None, 17, 24, 17,34)	0
activation_3 (Activation)	(None, 110, 80, 128)	0	batch_normalization (BatchNo	(None, 17, 24, 17,64)	256
dropout_3 (Dropout)	(None, 110, 80, 128)	0	flatten (Flatten)	(None, 443904)	0
flatten_1 (Flatten)	(None, 1126400)	0	dense (Dense)	(None, 512)	227279360
dense_1 (Dense)	(None, 3)	3379203	dropout (Dropout)	(None, 512)	0
Total params:	3,455,619		dense_1 (Dense)	(None, 256)	131328
Trainable params:	3,455,235		dropout_1 (Dropout)	(None, 256)	0
Non-trainable params:	384		dense_2 (Dense)	(None, 3)	771
			Total params:	227,484,627	
			Trainable params:	227,484,499	
			Non-trainable params:	128	

The basic structure of the 2D and 3D CNN model is adopted, consisting of multiple convolution operations, batch normalization, max pooling, dense. The convolution is performed using the ReLu activation function, and a dense layer is formed using softmax activation. In 3D CNN, multiple convolution layers are designed with varying filters, and max pooling is carried out to extract the most relevant features using the ReLu activation function. The batch normalization is performed to convolution output before feeding into MLP architecture. The MLP architecture is created with dense layers 512, 256 and 3, performed dropout to avoid overfitting and regularization. The detailed, layered structure of 2D and 3D CNN is shown in Table 2.

Similarly, the transfer learning models modes were developed using the basic architecture and same batch size. VGG16, InceptionV3 and ResNet50 models were developed, and training and validation accuracy are determined. The maximum training accuracy was observed in VGG16 of 0.974, followed by a 2D CNN model of 0.839.

3.2 Testing of Model

The model was tested on 15% of the data, and iterations for each model were determined automatically based on the indexes of loss and accuracy function. Table 3 contains the testing results of various ML and DL approaches adopted in this study. The performance evaluation measures adopted for the study was precision, recall, and F1 score. The values obtained from confusion matrix results are employed to calculate the other evaluation parameters, including accuracy, precision, F1-score, and recall. The evaluation parameters are calculated using the following expressions:

$$Accuracy = \frac{TP + TN}{TP + FP + FN + TN}$$

$$Precision = \frac{TP}{TP + FP}$$

$$Recall = \frac{TP}{TP + FN}$$

$$F1 - Score = 2 * \frac{Precision * Recall}{Precision + Recall}$$

MCI contains the maximum images among the three classes and shows better results. The SVM shows the highest precision for MCI over RF, NB and KNN, whereas RF shows the highest accuracy among machine learning approaches considering all the classes. The SVM shows 58% accuracy of the model, whereas RF shows 59% of accuracy. The ratio of true positive responses from actual samples against the sum total of both true positives and the responses labelled corrected instead of being incorrect is calculated and designated as the recall value. If the responses predicted by model as positive, are actually negative or incorrectly estimated by the model, then the model fails to fit the criteria of applicability for study. The stabilized value of both precision and recall, estimated for the model defines the most significant value of F1 score. While if, the notable difference exists in precision and recall value, the model resulted in the lower value of F1 score. The F1 score observed in the RF model was 74%. Figure 2 shows the confusion matrix from three classes of machine learning techniques.

Table 3. Performance evaluation of testing of machine and deep learning techniques

Technique	Classes	Precision	Recall	F1-Score	Support	Accuracy
SVM	CN	0.40	0.39	0.40	64	0.58
	MCI	0.65	0.77	0.71	166	
	AD	0.42	0.19	0.26	52	
	Macro Average	0.49	0.45	0.46	282	
	Weighted Average	0.55	0.58	0.55	282	
NB	CN	0.37	0.23	0.29	64	0.57
	MCI	0.61	0.87	0.71	166	
	AD	0.25	0.02	0.04	52	

(continued)

Table 3. (*continued*)

Technique	Classes	Precision	Recall	F1-Score	Support	Accuracy
	Macro Average	0.41	0.37	0.35	282	
	Weighted Average	0.49	0.57	0.49	282	
RF	CN	0.38	0.09	0.15	64	0.59
	MCI	0.60	0.96	0.74	166	
	AD	1.0	0.02	0.04	52	
	Macro Average	0.66	0.36	0.31	282	
	Weighted Average	0.62	0.59	0.48	282	
KNN	CN	0.29	0.23	0.26	64	0.54
	MCI	0.59	0.81	0.68	166	
	AD	0.75	0.06	0.11	52	
	Macro Average	0.54	0.37	0.35	282	
	Weighted Average	0.55	0.54	0.48	282	
CNN 2D	CN	0.465	0.422	0.44	64	0.606
	MCI	0.637	0.813	0.71	166	
	AD	0.750	0.173	0.28	52	
	Macro Average	0.617	0.469	0.47	282	
	Weighted Average	0.618	0.606	0.57	282	
CNN 3D	CN	0	0	0	78	0.574
	MCI	0.574	1.0	0.72	162	
	AD	0	0	0	42	
	Macro Average	0.191	0.333	0.24	282	
	Weighted Average	0.330	0.574	0.41	282	
DNN	CN	0.483	0.506	0.49	83	0.613
	MCI	0.666	0.723	0.69	166	
	AD	0.733	0.333	0.45	33	
	Macro Average	0.627	0.521	0.54	282	
	Weighted Average	0.620	0.613	0.60	282	
VGG16	CN	0.552	0.500	0.52	64	0.670
	MCI	0.712	0.819	0.76	166	
	AD	0.636	0.404	0.49	52	
	Macro Average	0.633	0.574	0.59	282	
	Weighted Average	0.662	0.670	0.65	282	
InceptionV3	CN	0.405	0.468	0.43	64	0.589
	MCI	0.642	0.735	0.68	166	
	AD	0.777	0.269	0.39	52	
	Macro Average	0.608	0.491	0.50	282	
	Weighted Average	0.613	0.588	0.57	282	
ResNet50	CN	0	0	0	64	0.592
	MCI	0.592	0.988	0.74	166	
	AD	0.600	0.577	0.10	52	
	Macro Average	0.397	0.348	0.28	282	
	Weighted Average	0.459	0.592	0.45	282	

The deep learning and transfer learning approaches show significant improvement over machine learning approaches. The in-depth analysis of the dataset, data normalization and max-pooling improvise the data classification. The 2D CNN model produces the highest precision for AD class followed by MCI, whereas Recall and F1Score were found highest from MCI and the overall accuracy observed was 0.606. The ROC curve shows the relationship between sensitivity and specificity. The curve near the diagonal line shows low accuracy, and the curve near the left top corner shows improved accuracy. The micro-average ROC curve of the 2D-CNN model shows an accuracy of 75%, as shown in Fig. 3. The 3D-CNN shows low accuracy compared to 2D-CNN. The 3D CNN model receives input from 4 dimensions, where the colour channel form the fourth dimension, whereas 2D CNN receives input from 3 dimensions, including height, width and depth having two kernel sides. The input data for 3D CNN consume significant space; therefore, input images are resampled to reduce the space requirement and pre-processed data consume 25 GB of RAM space. The 2D CNN input layer has the size of (420, 320, 3) against the input layer of 3D CNN (80, 11, 80, 1) due to resource limitation.

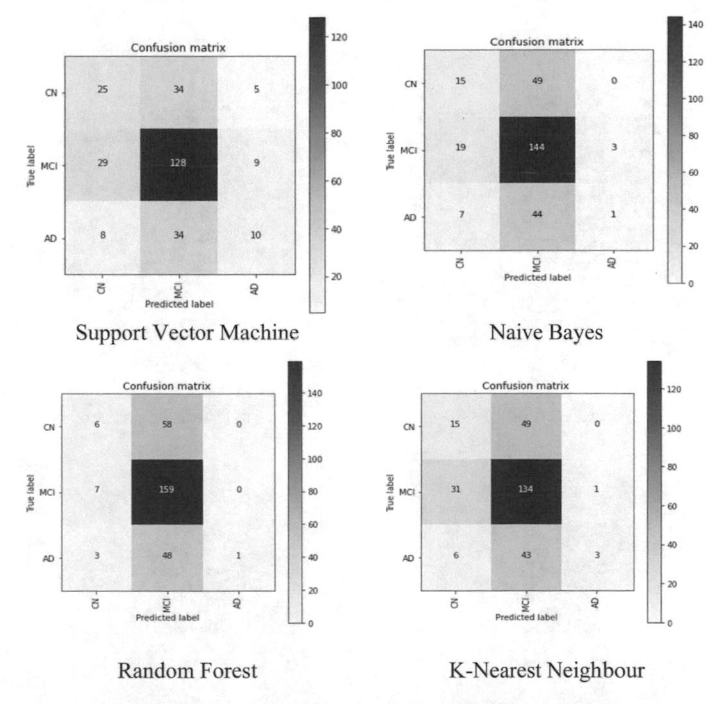

Fig. 2. Confusion matrix of machine learning techniques

The DNN model shows overall accuracy better than the 2D CNN model and observed improved precision for CN and MCI class, whereas 2D CNN shows slightly better precision for AD class. Similarly, DNN also shows better recall and F1-Score values. Among the transfer learning approaches that worked on the pre-trained dataset, VGG16 shows the highest accuracy, recall and F1 score. VGG16 shows the highest precision for MCI, while InceptionV3 shows the highest precision for the AD class. ResNet50 shows the highest Recall value for MCI, while the highest overall accuracy was observed in VGG16 compared to all the machine learning and deep learning models. The training results were found highly accurate compared to testing results, as the number of input images was more during model training and form the base of the model. A significant improvement can also be observed in the results using a large number of images.

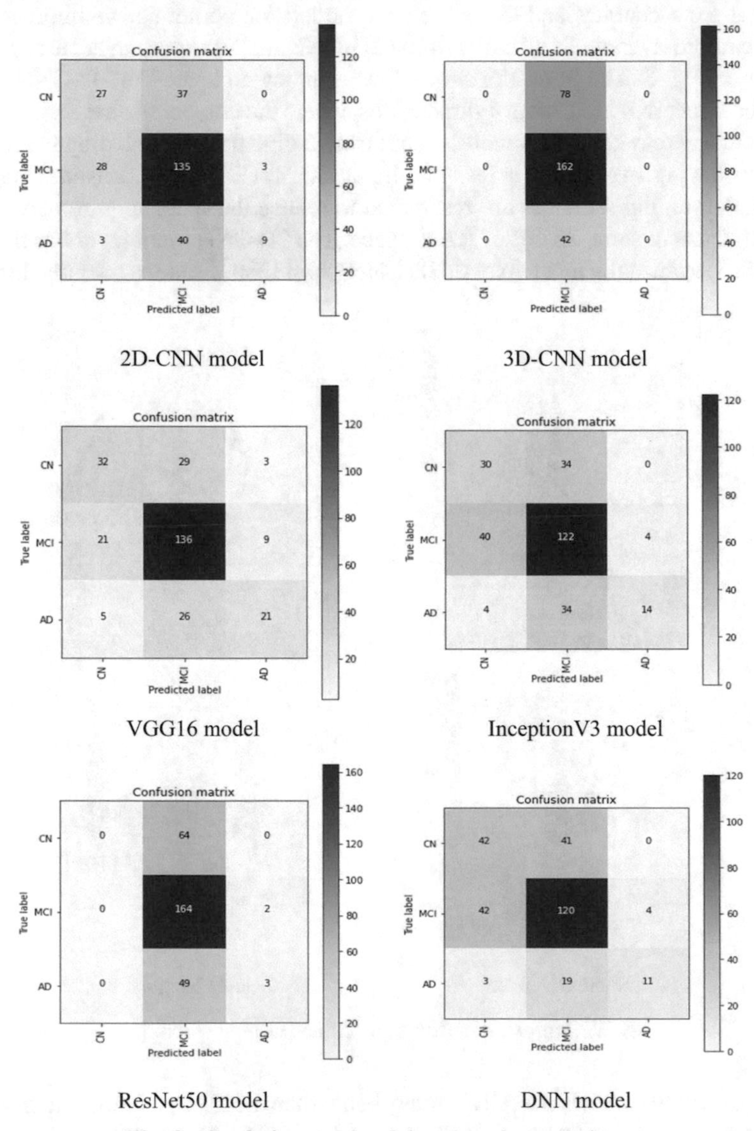

Fig. 3. Confusion matrix of deep learning techniques

The study shows the effectiveness of transfer learning approaches, which shows the highest accuracy with minimum pre-processing of the dataset and low computational space is required as the input data is pre-trained on a larger dataset. The highest overall accuracy obtained from the VGG16 model while testing is 67%, followed by the DNN model of 61.3%. However, the accuracy obtained using the transfer learning model is low compared to studies in literature; the present study involves no major pre-processing and features extraction techniques or combination of multiple techniques [39]. This study represents the preliminary step for assessing the most appropriate model for high-resolution image classification. Several features that cause the development of Alzheimer or dementia can be applied to assess the images statistically and computationally. The images can be improvised using correlation, contrast, homogeneity, entropy and statistical features.

4 Conclusion

Detection of Alzheimer disease at the early stage is a complex problem and but patients detected with MCI at the early stage can recover entirely or fully develop Alzheimer in case of late detection. In this study, the performance of various machine learning, deep learning and transfer learning approaches were evaluated. The techniques are evaluated using accuracy, precision, recall and F1 score. The comparison of testing results of models showed that VGG16 outperformed the other techniques. VGG16 shows better accuracy with minimal pre-processing techniques and without compromising sensitivity and specificity.

The VGG16 model is proposed to couple with other machine learning technique for the data improvisation and extraction of more relevant features for the detection of Alzheimer disease. The deep transfer learning approach can be supported with the relevant feature extraction and data improvisation technique to improve the model accuracy for further studies. Discrepancies in the model's performance also need to be assessed by testing and validating the pre-processing approaches and their related applications, utilized for image feature enhancement in the study. Variant state of the art techniques can be potentially integrated with the designed model for constructing rigid, efficient and robust classifier.

References

1. Association, A.: 2018 Alzheimer's disease facts and figures. Alzheimers Dement. **14**(3), 367–429 (2018)
2. Geerlings, M.I., den Heijer, T., Koudstaal, P.J., Hofman, A., Breteler, M.M.B.: History of depression, depressive symptoms, and medial temporal lobe atrophy and the risk of Alzheimer disease. Neurology **70**(15), 1258–1264 (2008)
3. Guillozet, A.L., Weintraub, S., Mash, D.C., Mesulam, M.M.: Neurofibrillary tangles, amyloid, and memory in aging and mild cognitive impairment. Arch. Neurol. **60**(5), 729–736 (2003)
4. Reitz, C., Mayeux, R.: Alzheimer disease: epidemiology, diagnostic criteria, risk factors and biomarkers. Biochem. Pharmacol. **88**(4), 640–651 (2014)

5. Dickerson, B.C., Sperling, R.A.: Large-scale functional brain network abnormalities in Alzheimer's disease: insights from functional neuroimaging. Behav. Neurol. **21**(12), 63–75 (2009)
6. Simpson, S.L., Bowman, F.D., Laurienti, P.J.: Analyzing complex functional brain networks: fusing statistics and network science to understand the brain. Statist. Surv. **7**, 1 (2013)
7. Braun, U., et al.: Dynamic reconfiguration of frontal brain networks during executive cognition in humans. Proc. Natl. Acad. Sci. **112**(37), 11678–11683 (2015)
8. Marsh, H.W., Morin, A.J., Parker, P.D., Kaur, G.: Exploratory structural equation modeling: an integration of the best features of exploratory and confirmatory factor analysis. Annu. Rev. Clin. Psychol. **10**, 85–110 (2014)
9. Vieira, S., Pinaya, W.H., Mechelli, A.: Using deep learning to investigate the neuroimaging correlates of psychiatric and neurological disorders: Methods and applications. Neurosci. Biobehav. Rev. **74**, 58–75 (2017)
10. Razzak, M.I., Naz, S., Zaib, A.: Deep learning for medical image processing: overview, challenges and the future. In: Dey, N., Ashour, A.S., Borra, S. (eds.) Classification in BioApps. LNCVB, vol. 26, pp. 323–350. Springer, Cham (2018). https://doi.org/10.1007/978-3-319-65981-7_12
11. Rauschert, S., Raubenheimer, K., Melton, P.E., Huang, R.C.: Machine learning and clinical epigenetics: a review of challenges for diagnosis and classification. Clin. Epigenetics **12**(1), 1–11 (2020)
12. Abuhmed, T., El-Sappagh, S., Alonso, J.M.: Robust hybrid deep learning models for Alzheimer's progression detection. Knowl.-Based Syst. **213**, 106688 (2021)
13. Battineni, G., Chintalapudi, N., Amenta, F.: Machine learning in medicine: Performance calculation of dementia prediction by support vector machines (SVM). Inf. Med. Unlocked **16**, 100200 (2019)
14. Jha, D., Kwon, G.R.: Alzheimer disease detection in MRI using curvelet transform with KNN. J. Korean Inst. Inf. Technol. **14**(8), 121 (2016)
15. Alickovic, E., Subasi, A.: Automatic detection of alzheimer disease based on histogram and random forest. In: Badnjevic, A., Škrbić, R., Pokvić, L.G. (eds.) CMBEBIH 2019. IP, vol. 73, pp. 91–96. Springer, Cham (2020). https://doi.org/10.1007/978-3-030-17971-7_14
16. Dana, A.D., Alashqur, A.: Using decision tree classification to assist in the prediction of Alzheimer's disease. In: 2014 6th international conference on computer science and information technology (CSIT), pp. 122–126. IEEE (2014)
17. Mehmood, A., Maqsood, M., Bashir, M., Shuyuan, Y.: A deep siamese convolution neural network for multi-class classification of alzheimer disease. Brain Sci. **10**(2), 84 (2020)
18. Álvarez, I., et al.: Automatic classification system for the diagnosis of alzheimer disease using Component-Based SVM aggregations. In: Köppen, M., Kasabov, N., Coghill, G. (eds.) Advances in Neuro-Information Processing, pp. 402–409. Springer, Heidelberg (2009). https://doi.org/10.1007/978-3-642-03040-6_49
19. Khedher, L., Ramírez, J., Górriz, J.M., Brahim, A., Illán, I.A.: Independent component analysis-based classification of Alzheimer's disease from segmented MRI data. In: Vicente, J.M.F., Álvarez-Sánchez, J.R., de la Paz, F., López, F.J.T.-M., Adeli, H. (eds.) IWINAC 2015. LNCS, vol. 9107, pp. 78–87. Springer, Cham (2015). https://doi.org/10.1007/978-3-319-18914-7_9
20. Sheng, J., et al.: A novel joint HCPMMP method for automatically classifying Alzheimer's and different stage MCI patients. Behav. Brain Res. **365**, 210–221 (2019)
21. Mishra, S., Beheshti, I., Khanna, P., Initiative, A.D.N.: A statistical region selection and randomized volumetric features selection framework for early detection of Alzheimer's disease. Int. J. Imaging Syst. Technol. **28**(4), 302–314 (2018)

22. Islam, J., Zhang, Y.: A novel deep learning based multi-class classification method for Alzheimer's disease detection using brain MRI data. In: Zeng, Y., et al. (eds.) BI 2017. LNCS (LNAI), vol. 10654, pp. 213–222. Springer, Cham (2017). https://doi.org/10.1007/978-3-319-70772-3_20
23. Jain, R., Jain, N., Aggarwal, A., Hemanth, D.J.: Convolutional neural network based Alzheimer's disease classification from magnetic resonance brain images. Cogn. Syst. Res. **57**, 147–159 (2019)
24. Dimiduk, D.M., Holm, E.A., Niezgoda, S.R.: Perspectives on the impact of machine learning, deep learning, and artificial intelligence on materials, processes, and structures engineering. Integr. Mater. Manuf. Innov. **7**(3), 157–172 (2018)
25. Tanveer, M., et al.: Machine learning techniques for the diagnosis of Alzheimer's disease: a review. ACM Trans. Multimedia Comput. Commun. Appl. (TOMM) **16**(1s), 1–35 (2020)
26. Khvostikov, A., Aderghal, K., Benois-Pineau, J., Krylov, A., Catheline, G.: 3D CNN-based classification using sMRI and MD-DTI images for Alzheimer disease studies (2018). arXiv preprint arXiv:1801.05968
27. Liu, W., Wang, Z., Liu, X., Zeng, N., Liu, Y., Alsaadi, F.E.: A survey of deep neural network architectures and their applications. Neurocomputing **234**, 11–26 (2017)
28. Glozman, T., Liba, O.: Hidden cues: Deep learning for Alzheimer's disease classification CS331B project final report (2016)
29. Korolev, S., Safiullin, A., Belyaev, M., Dodonova, Y.: Residual and plain convolutional neural networks for 3D brain MRI classification. In: 2017 IEEE 14th International Symposium on Biomedical Imaging (ISBI 2017), pp. 835–838. IEEE (2017)
30. Park, T., Liu, M.Y., Wang, T.C., Zhu, J.Y.: Semantic image synthesis with spatially-adaptive normalization. In: Proceedings of the IEEE/CVF Conference on Computer Vision and Pattern Recognition, pp. 2337–2346 (2019))
31. Isensee, F., Petersen, J., Kohl, S.A., Jäger, P.F., Maier-Hein, K.H.: nnu-net: Breaking the spell on successful medical image segmentation, vol. 1, pp. 1–8. arXiv preprint arXiv:1904.08128 (2019)
32. Suthaharan, S.: Support vector machine. In: Machine learning models and algorithms for big data classification. ISIS, vol. 36, pp. 207–235. Springer, Boston, MA (2016). https://doi.org/10.1007/978-1-4899-7641-3_9
33. Biau, G., Scornet, E.: A random forest guided tour. TEST **25**(2), 197–227 (2016). https://doi.org/10.1007/s11749-016-0481-7
34. Webb, G.I., Keogh, E., Miikkulainen, R.: Naïve bayes. Encycl. Mach. Learn. **15**, 713–714 (2010)
35. Guo, G., Wang, H., Bell, D., Bi, Y., Greer, K.: KNN model-based approach in classification. In: Meersman, R., Tari, Z., Schmidt, D.C. (eds.) OTM 2003. LNCS, vol. 2888, pp. 986–996. Springer, Heidelberg (2003). https://doi.org/10.1007/978-3-540-39964-3_62
36. Buhrmester, V., Münch, D., Arens, M.: Analysis of explainers of black box deep neural networks for computer vision: a survey (2019). arXiv preprint arXiv:1911.12116
37. Hon, M., Khan, N.M.: Towards Alzheimer's disease classification through transfer learning. In: 2017 IEEE International conference on bioinformatics and biomedicine (BIBM), pp. 1166–1169. IEEE (2017)
38. Maqsood, M., et al.: Transfer learning assisted classification and detection of Alzheimer's disease stages using 3D MRI scans. Sensors **19**(11), 2645 (2019)
39. Oh, K., Chung, Y.C., Kim, K.W., Kim, W.S., Oh, I.S.: Classification and visualization of Alzheimer's disease using volumetric convolutional neural network and transfer learning. Sci. Rep. **9**(1), 1–16 (2019)

Multiobjective Optimization of Friction Welding of 15CDV6 Alloy Steel Rods Using Grey Relational Analysis in the Taguchi Method

P. Anchana[1,2](✉) and P. M. Ajith[2]

[1] APJ Abdul Kalam Technological University, Trivandrum, Kerala 695 016, India
anchanap1990@gmail.com

[2] Department of Mechanical Engineering, College of Engineering Trivandrum, Thiruvananthapuram, Kerala 695016, India

Abstract. Optimization helps to identify the ideal combination of multiple process parameters which is based on the output quality. In this study, a friction welding experiment was designed based on Taguchi's design of experiment methodology and multi objective optimization was done based on Grey Relational Analysis (GRA). Friction welding experiments of 15CDV6 alloy steel rods based on Taguchi's L9 orthogonal array were conducted. The friction pressure (FP), upsetting pressure (UP), speed of rotation (S) and friction time were selected as the input variables and ultimate tensile strength and microhardness of the joint were selected as the output responses. The best combination of process parameters was discovered using grey relational analysis, and ANOVA revealed that friction pressure and upsetting pressure are the most influential. It is concluded that GRA may provide good ability to predict the friction welding process parameters to weld 15CDV6 alloy steel.

Keywords: 15CDV6 alloy steel · Friction welding · Taguchi's design · GRA · ANOVA

1 Introduction

Friction welding is a solid-state joining process in which the joint quality depends upon the proper selection and combination of process parameters such as friction pressure, upsetting pressure, speed of rotation, friction time, upsetting time, and burn off length. It is widely used in many industries as it has many advantages over other fusion welding techniques. The joint quality is defined on the basis of the values of ultimate tensile strength and hardness of the weld. The process consists of two phases- first phase is forging and second phase includes extrusion. The rate of forging and extrusion depends upon the process parameters and thus joint quality.

15CDV6 alloy steel rods are high strength low alloy steel in which all the alloying elements remain less than 5% in proportion by weight. It is widely used in different industries such as defence, aerospace, and power generation sectors because of its high strength, toughness, and ductility. The welding of 15CDV6 alloy steel is possible to all

H. Sharma et al. (Eds.): ICIVC 2022, PALO 17, pp. 328–337, 2023.
https://doi.org/10.1007/978-3-031-31164-2_27

welding processes such as oxyacetylene, electric arc, resistance, electron beam, and laser beam welding. But high heat input causes coarse weld bead formation during welding of 15CDV6 and thus reduces the toughness [1].

More trials should be undertaken for standard experimental design approaches as the number of independent variables increases. As a solution to this problem, the Taguchi method (TM) provides a simple, effective, and systematic tool for optimising designs for quality, performance, and cost. Grey relational analysis (GRA), invented by Deng, is one of the ways used to analyse uncertainties in multi objective choice problems in circumstances of inadequacy and uncertainty, and it provides simpler solutions than quantitative analysis methods. Situations in which there is no information are labelled as black, while those in which there is perfect information are labelled as white. GRA consolidates multi-objective optimization problems into a single approach [2, 3].

Asif et al. [4] used both the Taguchi technique and Grey relational analysis to compare multi response optimization of friction welding process parameters of UNS S31803 duplex stainless steel joints. GRA was found to be a better strategy for optimising multiple solutions than the Taguchi method. The four process parameters were heating pressure, heating time, upsetting pressure, and upsetting time, and the responses were tensile strength, hardness, impact toughness, and corrosion rate. Grey Relational Analysis was used by Sundararajan and Shanmugam [5] to predict the optimum combinations of the friction welding process parameters to obtain good UTS and hardness during welding of aluminium metal matrix composite. Adalarasan et al. [6] conducted friction welding experiments on AA6061-T6 and AA2024-T6 and trials were designed by Taguchi's L9 orthogonal array.

Aydin et al. [7] used the Taguchi-based Grey Relational Analysis to study the multi response optimization of friction stir welding for an ideal parametric combination of tensile strength and elongation. Using analysis of variance, the most significant parameters for tensile strength and elongation were found (ANOVA). RSM coupled with GRA coupled with entropy for obtaining optimal condition for friction stir welding of aluminium alloys AA 2024 and AA 6061 has been attempted by Vijayan and Rao [8]. The main aim of the work is to improve the tensile behaviour characteristics such as ultimate tensile strength and tensile elongation. Jain et al. [9] studied the influence of different friction stir welding parameters on AA6082-T6 and AA5083-O alloy welding quality by using Taguchi, Grey relational analysis and weight method. Shamsudeen and Dhas [10] applied response surface approach and GRA to examine multi-objective optimization of the process parameters of AA 5052 H32 aluminium alloy by FSW process. The influence of each input welding parameter on the output responses was investigated using ANOVA. Wakchaure et al. [11] used a hybrid Taguchi-Grey relational analysis-ANN method to optimise the FSW process parameters and investigated the effects of tool geometry, tool rotational speed, tool tilt angle, and welding speed on mechanical properties. Chanakyan et al. [12] adopted GRA to examine the effect of rotation speed, travel speed, and varied diameter of tool pins of AA2024 and AA6061 alloys during FSW. ANOVA was used to investigate the effects of process parameters. Ghetiya et al. [13] adopted GRA to perform parametric optimization in FSW of AA 8011 for tensile strength, microhardness, and input power. The significance of process variables on grey relational grade was investigated using ANOVA.

Lin et al. optimized the weld bead geometry of Inconel 718 alloy welds in an activated gas tungsten arc welding using Taguchi method, Grey relational analysis, and a neural network. Most of the papers focused on the parametric optimization using GRA in FSW process. There is little studies were conducted on the optimization of process parameters such as friction pressure, upsetting pressure, speed of rotation and friction time on the friction welding joints of 15CDV6 alloy steel rods using GRA. In this paper, work is focused on the multi objective optimization of friction welding of 15CDV6 alloy steel rods using GRA. The ultimate tensile strength and microhardness were selected as the output responses.

2 Experimental Procedure

2.1 Material and Method

In the rotary friction welding experiments, 15CDV6 alloy steel rods were used as the base material, which had a dimension of 16 mm diameter and 70 mm length. Table 1 represents the chemical composition of the material.

Table 1. Chemical compositions of 15CDV6 alloy steel.

Element	C	Si	Mn	Cr	Mo	V	Fe
Min (%wt.)	0.12–0.18	0.13	0.8–1.0	1.25–1.5	0.80–1.0	0.20–0.3	Bal

A computerized FW machine was used for study. Figure 1 depicts the experimental setup.

Fig. 1. Experimental arrangements for rotary friction welding experiments.

In order to acquire the best FW parameters, Taguchi's L9 orthogonal array was used in the experimental design. Friction pressure, upsetting pressure, rotational speed, and friction time were chosen as input parameters, whereas ultimate tensile strength and microhardness were chosen as output parameters. Table 2 shows the various levels of

Table 2. Process parameters for friction welding of 15CDV6 alloy steel rods.

	1	2	3
FP(MPa)	10	50	90
UP(MPa)	10	50	90
Speed(rpm)	1000	1500	2000
FT(s)	6	8	10

control factors. The Taguchi approach was used to create an experimental plan for four FW process parameters with three levels (3^3) (L9 orthogonal array).

Measurements

The ultimate tensile strength and microhardness were measured during FW of 15CDV6 alloy steel. The ultimate tensile strength was measured using Universal tensile testing machine and the microhardness was measured by Vickers microhardness tester. Two process variables, ultimate tensile strength (MPa) and microhardness (HV) were measured. Experimental runs with performance results are shown in Table 3.

Table 3. Taguchi L_9 orthogonal array and performance results.

Experiment number	Friction pressure (MPa)	Upsetting pressure (MPa)	Speed (rpm)	Friction time (s)	Ultimate tensile strength (MPa)	Hardness (HV)
1	10	10	1000	6	1280	392
2	10	50	1500	8	1282	395
3	10	90	2000	10	1285	398
4	50	10	1500	10	1292	401
5	50	50	2000	6	1298	403
6	50	90	1000	8	1300	405
7	90	10	2000	8	1305	403
8	90	50	1000	10	1318	410
9	90	90	1500	6	1320	411

3 Grey Relational Analysis

A grey system is one in which some of the information is known and the rest is unknown. Grey relational analysis can be used to solve complex interrelationships between the specified performance criteria. The grey relational grade (GRG) is described favourably as an indicator of many performance criteria for evaluation through the analysis. It is used to reduce multi-objective optimization problems to single-objective problems[14].

The steps involved in GRA are as follow:

Step 1: Grey relational generation

The first step involves the normalization (In the range between 0 and 1) of experimental results based on the type of performance response. Larger-the-better and lower-the- better criteria are there. In the present study, as ultimate tensile strength and hardness were to be maximized the larger-the-better criteria was selected, which is given in Eqs. 1:

$$Yi = \frac{Y_i^k(k) - \min\left(Y_i^0(k)\right)}{\max\left(Y_i^0(k)\right) - \min(Y_i^0(k)}, i = 1, 2, \ldots, p; k = 1, 2, \ldots, q \quad (1)$$

where p is the number of experiments, q is the number of output variables. In this paper, p = 9 and q = 2 is considered. In Eq. (1), *Yi* is the comparable sequence after data normalization, $Y_i^k(k)$ is the original value of UTS and HV, max $Y_i^0(k)$andmin$Y_i^0(k)$ are the highest and smallest value of $Y_i^0(k)$ for kth response respectively.

Table 4 represents the processed data after grey relational generation. The normalized values lie between zero and one, where one represents the best normalized result.

Table 4. Experimental results after normalization.

Experiment number	UTS (MPa)	Hardness(HV)
Ideal value	1	1
1	0	0
2	0.05	0.1579
3	0.125	0.3158
4	0.3	0.4737
5	0.45	0.5789
6	0.5	0.6842
7	0.625	0.5789
8	0.95	0.9474
9	1	1

Step 2: Grey relational coefficient

The relationship between ideal and the actual experimental results can be obtained from grey relational coefficients. After normalization, the grey relational coefficient $\gamma_i(k)$ can be calculated as:

$$\gamma_i(\mathrm{k}) = \frac{\Delta_{min} + \zeta\,\Delta_{max}}{\Delta_{0i}(k) + \zeta\,\Delta_{max}} \quad (2)$$

$$0 < \gamma_i(k) \leq 1 \quad (3)$$

where $\Delta_{0i}(k)$ is called the quality loss function, which is the absolute difference of reference sequence $Y_0(k)$ and comparability sequence $Y_i(k)$,

$$\Delta_{0i}(k) = |Y_0(k) - Y_i(k)|| \tag{4}$$

$$\Delta_{min} = \min_{\forall j \in i} \min_{\forall k} \left| Y_0(k) - Y_j(k) \right| \tag{5}$$

$$\Delta_{max} = \max_{\forall j \in i} \max_{\forall k} \left| Y_0(k) - Y_j(k) \right| \tag{6}$$

where ζ is the distinguishing coefficient and $\zeta \in [0,1]$. Here it was taken as 0.5 and the grey relational coefficients calculated using Eq. (2) was given in Table 5.

Step 3: Grey relational grade and grey relational ordering

The grey relationship grade indicates the degree of closeness between the comparability sequence and the reference sequence. It is an average of the multi-objective grey relational coefficients.

$$GRG = \frac{1}{n} \sum\nolimits_{(k=1)}^{n} \zeta_i(k) \tag{7}$$

where n is the number of output variables and here it is 2. Higher the value of GRG, more similarity between the comparability sequence and reference sequence. Table 5 shows the grey relational coefficients and grade for 9 experiments. From Table 5, the experiment 9 had the highest grey relational grade and the optimal FW settings for maximum ultimate tensile strength and microhardness.

Table 5. Grey relational coefficients and grade.

Experiment number	UTS (MPa)	Hardness (HV)	Grade	Grey order
1	0.3333	0.3333	0.3333	9
2	0.3448	0.3725	0.3586	8
3	0.3636	0.4222	0.3929	7
4	0.4167	0.4872	0.4519	6
5	0.4762	0.5429	0.5095	5
6	0.5000	0.6129	0.5564	4
7	0.5714	0.5429	0.5571	3
8	0.9091	0.9048	0.9069	2
9	1	1	1	1

Table 6 shows the means of the grey relational grade for each level of controllable parameters obtained from Table 5. The GRG value can be used to analyse multiple performance parameters. The greater the value (bold in Table 6), the better the output performance. The optimal values of process parameters are as follows: friction pressure of 90 MPa (level 3), upsetting pressure of 90 MPa (level 3), speed of 1500 rpm (level 2), and friction time of 6 s(level 1). Figure 2 shows the friction welding parameters in relation to the GRG.

Table 6. Response table for GRG.

Parameters	1	2	3	Rank
FP	0.3616	0.5059	**0.8213**	1(0.4597)
UP	0.4474	0.5917	**0.6497**	2(0.2023)
S	0.5989	**0.6035**	0.4865	4(0.117)
FT	**0.6142**	0.4907	0.5839	3(0.1235)

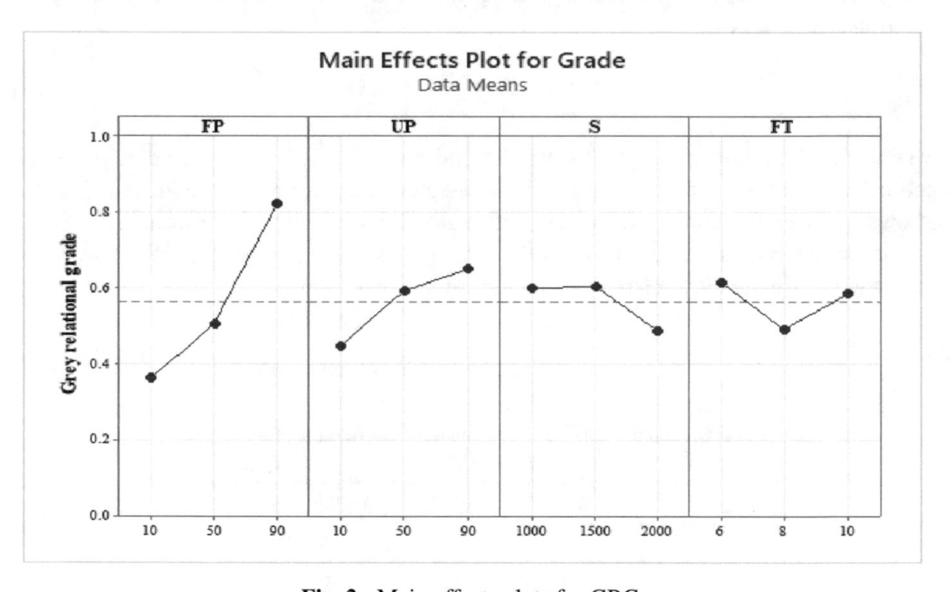

Fig. 2. Main effects plots for GRG.

4 Anova

The Analysis of Variance (ANOVA) is a method used for obtaining the significance of various input parameters on the performance characteristics. The level of confidence 95% were selected for the analysis. The percentage contribution by friction pressure, upsetting pressure, speed and friction time in the total sum of the squared deviation was used to study the effect of input variables on the performance characteristics. Figure 3 represents the percentage of contribution of input parameters on the output variables from the ANOVA results and it is clear that friction pressure has the highest influence on both ultimate tensile strength and hardness. Table 7 represents the results of ANOVA for the GRG values.

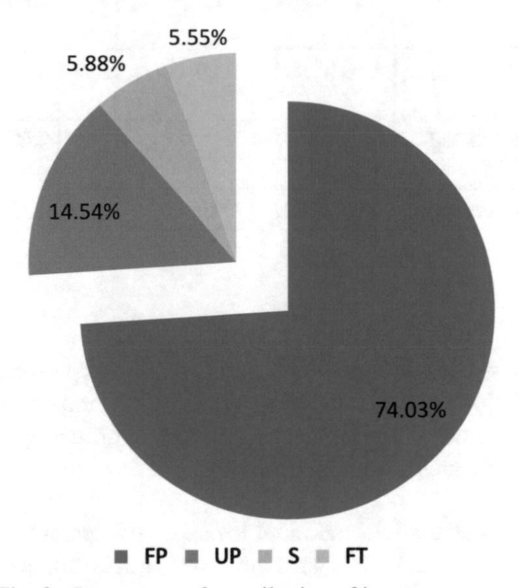

Fig. 3. Percentage of contribution of input parameters.

Table 7. Results of the ANOVA.

Source	DF	Seq SS	Contribution	Adj SS	Adj MS
FP	2	0.33163	74.03%	0.33163	0.16581
UP	2	0.06512	14.54%	0.06512	0.03256
S	2	0.02634	5.88%	0.02634	0.01317
FT	2	0.02486	5.55%	0.02486	0.01243

5 Confirmation Test

The A3B3C2D1 was chosen for the confirmation test because it is an excellent combination of FW process parameters established through GRA. Table 8 shows that the performance parameters of FW were greatly improved by 0.0185.

Table 8. Confirmation test results.

		Setting level	UTS (MPa)	HV	GRG	Improvement in GRG
Initial controllable parameters		A1B1C1D1	1280	392	0.3333	
Optimal parameters	Predicted	A3B3C2D1			0.6721	
	Experiment	A3B3C2D1	1289	390	0.3518	**.0185**

6 Conclusions

In this study, friction welding of 15CDV6 alloy steel rods was performed using Taguchi's L9 method, using friction pressure, upsetting pressure, rotation speed, and friction duration as input parameters and multi-objective optimization using GRA. Based on the findings, the following conclusions can be drawn:

- Based on the GRG calculation, the ideal levels of FW parameters for the required output quality are A3B3C2D1. It is possible to maximise ultimate tensile strength and micro hardness using this combination.
- Based on the ANOVA of the GRG results, it is obvious that friction and upsetting pressure have the greatest influence on output reactions.
- The test experiment was carried out, and this strategy enhanced the grey relational grade of the outputs by.0185.

Acknowledgement. The authors thank Dr. V. Balasubramanian, Professor and Director in CEMAJOR, Annamalai University, for providing the facilities for conducting friction welding experiments. We also thank Deputy Director, Materials and Mechanical Entity. VSSC, Thumba, Trivandrum, for providing 15CDV6 alloy steel rods for conducting the experiments.

References

1. Ramesh, M.V.L., Rao, P.S., Rao, V.V.: Microstructure and mechanical properties of laser beam welds of 15CDV6 steel. Def. Sci. J. **65**, 339–342 (2015)
2. Nayak, S.K., Patro, J.K., Dewangan, S., Gangopadhyay, S.: Multi-objective optimization of machining parameters during dry turning of AISI 304 austenitic stainless steel using grey relational analysis. Procedia Mater. Sci. **6**, 701–708 (2014)
3. Rajendrakumar, S., Sivam, S.P.S.S., Saravanan, K., Moorthy, K.S., Kumaran, D.: Multi-attribute decision making parametric optimization and modelling in friction stir welding through grey relation taguchi analysis and ANOVA: a case study. In: AIP Conference Proceedings, vol. 2034 (2018)
4. Asif, M.M., Shrikrishna, K.A., Sathiya, P.: Optimization of process parameters of friction welding of UNS S31803 duplex stainless steels joints. Adv. Manuf. **4**(1), 55–65 (2015). https://doi.org/10.1007/s40436-015-0130-5

5. Sundararajan, S.K., Shanmugam, S.K.: Multi-objective optimization of friction welding process parameters using grey relational analysis for joining aluminium metal matrix composite. Medziagotyra **24**, 222–229 (2018)
6. Adalarasan, R., Santhanakumar, M., Sundaram, A.S.: Optimization of friction welding parameters for AA6061-T6/AA2024-T6 joints using Taguchi-Simulated Annealing (TSA) algorithm. Appl. Mech. Mater. **592–594**, 595–599 (2014)
7. Aydin, H., Bayram, A., Esme, U., Kazancoglu, Y., Guven, O.: Application of grey relation analysis (GRA) and Taguchi method for the parametric optimization of friction stir welding (FSW) process. Uporaba greyjeve analize (GRA) in Taguchijeve metode za parametrično optimizacijo varjenja z vrtilno-tornim procesom (F. Mater. Tehnol. **44**, 205–211 (2010)
8. Vijayan, D., Rao, V.S.: A parametric optimization of FSW process using RSM based grey relational analysis approach. Int. Rev. Mech. Eng. **8**, 328–337 (2014)
9. Jain, S., Sharm, N., Gupta, R.: Dissimilar alloys (AA6082/AA5083) joining by FSW and parametric optimization using Taguchi, grey relational and weight method. Eng. Solid Mech. **6**, 51–66 (2018)
10. Shamsudeen, S., Dhas, J.E.R.: Optimization of multiple performance characteristics of friction stir welded joint with grey relational analysis. Mater. Res. **21** (2018)
11. Wakchaure, K.N., Thakur, A.G., Gadakh, V., Kumar, A.: Multi-objective optimization of friction stir welding of aluminium alloy 6082–T6 using hybrid taguchi-grey relation analysis-ANN method. Mater. Today Proc. **5**, 7150–7159 (2018)
12. Chanakyan, C., et al.: Parametric optimization for friction stir welding with AA2024 and AA6061 aluminium alloys by ANOVA and GRG. Mater. Today Proc. **27**, 707–711 (2020)
13. Ghetiya, N.D., Patel, K.M., Kavar, A.J.: Multi-objective optimization of FSW process parameters of aluminium alloy using taguchi-based grey relational analysis. Trans. Indian Inst. Met. **69**(4), 917–923 (2015). https://doi.org/10.1007/s12666-015-0581-1
14. Üstüntağ, S., Şenyiğit, E., Mezarcıöz, S., Türksoy, H.G.: Optimization of coating process conditions for denim fabrics by Taguchi method and grey relational analysis. J. Nat. Fibers **19**, 685–699 (2022)

Voting Based Classification System for Malaria Parasite Detection

Pavan Mohan Neelamraju(✉), Bhanu Sankar Penugonda, Anirudh Koganti, Abhiram Unnam, and Kajal Tiwari

Department of Electronics and Communication Engineering, SRM University – AP, Andhra Pradesh, India
npavanmohan3@gmail.com

Abstract. Malaria is a potentially fatal disease caused by plasmodium parasites that infect certain types of mosquitoes that feed on human blood. It usually causes a severe illness with high fever, chills, and flu-like symptoms. Malaria, as an epidemic disease, necessitates prompt and accurate diagnosis to provide appropriate treatment. In practice, the gold standard for malaria diagnosis is the microscopic evaluation of blood smear images, in which the pathologist visually examines the stained slide under a light microscope. This is a subjective, error-prone, and time-consuming visual inspection. In the following work, a voting-based classification method is proposed that can detect the slide with Plasmodium parasites. The proposed modality concentrates on increasing the accuracy of parasite detection by using many of the commonly known image similarity measures. Further, classification metrics like accuracy, precision, recall, specificity, F1 score and negative predictive value are utilized to understand the performance of the currently developed technique. Utilization of the proposed model and its implementations in many other fields are to be more focused.

Keywords: Image binarization · Euler number · Parasitisation · Haematology

1 Introduction

Malaria is caused by parasites that are transmitted through the bites of female Anopheles mosquitoes. In severe cases, parasite-infected red blood cells (RBCs) cause symptoms such as fever, malaise, seizures, and coma. Together with better treatments and mosquito control, fast and reliable malaria diagnosis and treatments are one of the most effective ways of combating the disease. Microscopic evaluation, in which an expert slide reader visually inspects blood slides for parasites, accounts for more than half of all malaria diagnosis worldwide. This is a time-consuming and potentially error-prone process because hundreds of millions of slides are inspected worldwide each year. Accurate parasite identification is required for proper malaria diagnosis and treatment.

Parasite counts are used to assess treatment efficacy, detect drug resistance, and determine disease severity. However, microscopic diagnostics is not standardized and is heavily reliant on the microscopist's experience and expertise. Malaria is an acute

H. Sharma et al. (Eds.): ICIVC 2022, PALO 17, pp. 338–347, 2023.
https://doi.org/10.1007/978-3-031-31164-2_28

disease which might become severe and can grow into a life-threatening disease. A system that can automatically identify and quantify malaria parasites on a blood slide has several advantages. Primarily, it provides a reliable and standardized interpretation of blood films and reduces diagnostic costs by reducing workload. Further, the image analysis on thin blood smears could aid in the identification of Plasmodium parasite life stages such as rings, trophozoites, schizonts, and gametocytes.

Although both thick and thin blood smears are commonly used to quantify malaria parasitaemia, many computer-assisted malaria screening tools currently available rely on thin blood smears. Thick smears are primarily used for rapid initial identification of malaria infection but quantifying parasites and determining species can be difficult in cases where parasitaemia is high. On thin smears, parasite numbers per microscopy field are lower, and individual parasites are more clearly distinguishable from the background, allowing for more precise parasite quantification and differentiation between different species and parasite stages.

In the following work, a novel Voting Based Classification System for Malaria Parasite Detection was developed, which can be utilized as a complementary testing procedure to the current manual procedure of identification. Further, the following paper illustrates some of the previous works pertaining to the problem statement of malaria parasite detection, followed by the steps of data acquisition and model-building procedures in Sect. 3. Finally, the work is concluded by providing a few notable insights and possible steps for future advancement in image-based protozoan detection.

2 Background

Many research studies and investigations were done in the past to identify and evaluate blood specimens and the presence of protozoans. In particular, the past research work in the following field can be broadly classified into two different categories such as designing machine-based visual systems [1, 2] and algorithm-based evaluation. In the domain of algorithm-based evaluation, several studies were done to identify the parasite of the genus Plasmodium. Development of Machine Learning models was done to classify the images based on the presence of parasites. Implementation of such techniques took place to ameliorate the persisting manual procedure of microscopy which is found to be time-consuming [3, 4].

Along similar lines, several image processing techniques were developed that can enhance the quality of blood smear images using segmentation [5], colour spaces [6] and image generators [7]. Consequently, the usage of deep learning models to track and identify the smear sample has seen a drastic increase, particularly due to their levels of accuracy and availability of data. The first smartphone-based deep learning model for malaria parasite detection was developed and the modality was identified to be automated [8]. Models such as stacked Convolutional Neural Networks (CNN) were constructed, where the accuracy was recorded to be 99.98% with good classification metrics [9]. An unsupervised deep learning computer-aided diagnosis scheme was proposed for the identification of parasites in thin blood smear images [10].

To overcome the problem of overfitting, optimization is generally done [11]. Along with the same, models were also constructed to track the stages from the collected image dataset [12]. However, in building confidence in image classification and improving accuracy, ensemble techniques were also utilized in decision-making [13]. Techniques such as cell augmentation [14] and CNN [15] have been widely utilized to understand the trends in blood samples. The recent shift toward automation accelerated the research works in the direction of Machine and Artificial Intelligence [16]. These techniques have been adaptable and are customized according to existing conditions or geographical entities [17]. Two distinct deep learning models based on convolutional neural networks are presented for identifying malaria from blood cell images are proposed and it is more effective than the traditional, time-consuming approach to malaria detection, and assisted physicians in detecting the disease more quickly and easily [18]. The method approach is achieved with the fastest examination at 0.25 s/image, which is 20 times faster than the previous method and this could be a promising alternative to CAD systems for fast parasite detection [19].

The proposed CNN model, implemented using a 5-fold cross-validation approach, yielded the best results to date in deep learning malaria detection [20]. The development of a red blood cell detection and extraction framework to enable the processing and analysis of single cells Preliminary results from 193 patients (148 P. Falciparum infected and 45 uninfected) show that the framework achieved cell count accuracy of 92.2% [21]. A training approach is used in three different colour spaces: RGB, HSV, and GGB. In RGB, HSV, and GGB colour spaces, the proposed method achieved an accuracy of 0.9940, 0.9936, and 0.9947, respectively [22]. A Distributed Bragg Reflector (DBR) with a central microcavity is used. A DBR sensor is made up of multiple layers of two different types of dielectric materials; different layers produce different refractive indices in response to an incident light beam. The parasite is detected by the central cavity based on the RI and wavelength of the RBCs [23]. A CNN model with two fully connected layers and five conventional layers is proposed. In the CNN model, various activation functions such as Sigmoid, Tanh, ReLU, Leaky ReLU, and Swish are used, and each function has a different accuracy, with Swish and ReLU being the most accurate [24].

Deep CNN models are prepared for the prediction of infected and uninfected blood cells after the image pre-processing methods are completed. The average size of parasites and uninfected directories is compared. RBC images are subjected to conditions such as zoom, width, and so on during the data pre-processing phase. The images are then pooled and convolved, and final outputs are obtained using the Sigmoid activation function [25]. From the above-mentioned related works, it could be concluded that most of the research pertaining to the following field included the usage of deep learning techniques, which take a lot of time for model-building. However, the current paper proposes a simple image-processing-based modality that can work on thin blood smear images with less data dependence and a computationally friendly approach.

3 Methodology

3.1 Model Implementation

The data for the following study was collected from an open-source dataset. It contains image-segmented cells which are acquired from thin blood smear slides. The images were segregated into parasitized and uninfected images manually by an expert slide reader at Mahidol-Oxford Tropical Medicine Research Unit in Bangkok, Thailand [26].

The images were found to have varied resolutions. However, the nonuniformity in image resolutions is neglected as the implemented processes involve operations pertaining to the structure and the morphology of the regions present in the image.

The data repository contains 25,000 cell images, out of which, 1500 images were collected for the process of model building. Of the collected 1500 images, 750 images contain samples with parasites, while the remaining images are uninfected. These images were utilized for the purpose of training. These images were selected randomly from the repository and are not pre-processed, as they are already segmented.

For the purpose of the testing phase, 500 images were utilized, out of which, 250 images are parasitized, and the remaining 250 images are non-parasitised. The major goal of the current methodology is to identify and create a smart rule-based voting system with decision-making capability, which works upon traditional image processing tools to classify the images.

3.2 Data Acquisition

The implementation of the current model is based on several commonly used image processing techniques. A congregation of basic image processing tools was developed by identifying patterns in the collected images manually.

The developed voting-based system basically involves three methods, each of them working individually upon a provided input image. The methods are as follows:

- Identification of parasites using Euler's Number.
- Contours and closed boundary-based identification.
- Region-based identification.

The process flow of the current modality involves the binarization of the given input image. The blood smear image with red, green, and blue channels is binarized using the median of digital numbers available.

In the collected images, the range of digital numbers is found to vary from 0 to 255, as the radiometric resolution of the images is identified to be 8 bits per pixel. Hence, the threshold limit for binarization is chosen to be 128, which is the midpoint of the pixel distribution.

Further to the process of binarization, Euler's number of the black and white image is computed, which provides the difference between the number of objects and the number of vacant areas in the object. If the Euler's number was determined to be negative, the image is tagged as a sample with a parasite, otherwise, the image is predicted to be uninfected. An example of the same is shown in Fig. 1.

The second method involves working on the binarized image. The main objective of the second method is to identify the closed contours in the slide image, simultaneously, masking the outliers and redundant information. To accomplish the same, the region representing the parasite needs to be delineated clearly. The morphological operation of erosion was utilized to erode the empty spaces in the image, which in turn emphasizes the regions representing the parasite.

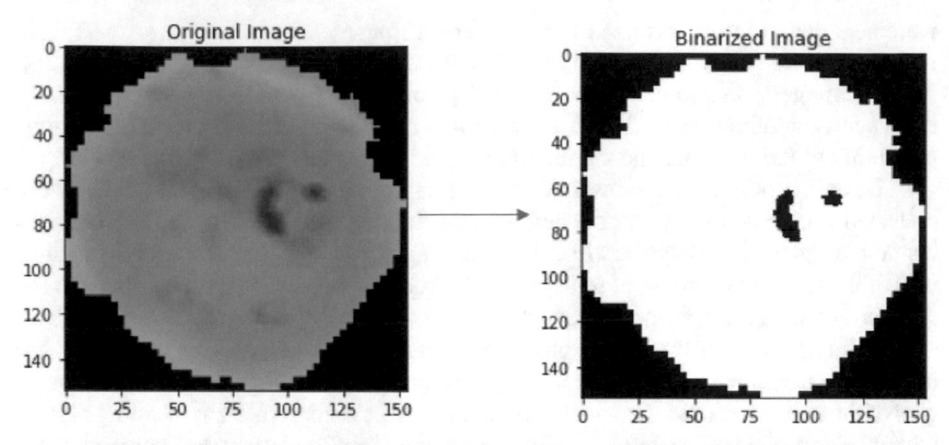

Fig. 1. Conversion of RGB colour image to Binary (Black and White) Image.

Consequently, a flood-filling algorithm is utilized to concentrate on the parasite gulfs, which are recognizable using contour identification techniques. The unnecessary region around the stain is flooded. A stack-based recursive algorithm is developed to work recursively. Thus, the number of closed contours or the closed boundary in the final image is computed. In case of any closed boundaries resembling the parasitized regions, the method tags the image to be infected. In the absence of closed boundaries or contours, the image would be tagged as a non-infected smear image. The procedure followed in the second method is illustrated in Fig. 2.

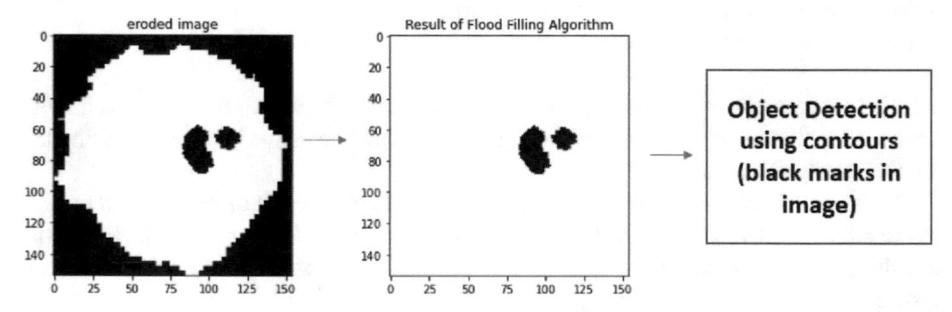

Fig. 2. Illustration of object detection using flood filling algorithm.

The final method of parasite identification involves the implementation of a region-identification method. The flood-filled image is inverted i.e., each black pixel is converted to white and vice versa. Further, the number of regions in the image is computed using the Depth First Search technique in which, each sub-region is visited for every execution time. The total number of executions in each turn provides the total number of regions or closed components in the image. If the total number of regions is greater than one, the image is considered a smear image with parasitisation. Otherwise, the input image is characterized as an uninfected sample. The same procedure can be observed in Fig. 3.

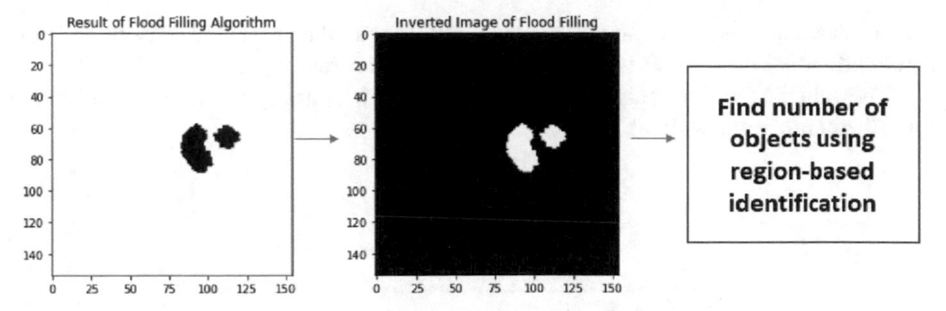

Fig. 3. Flowchart illustrating region-based identification method.

All the above-discussed methods individually produce an outcome based on their respective approaches. Further, the mode or the most repeated outcome out of the mentioned three methods is considered the final output by the voting-based classification system. The complete methodology of the following paper is illustrated in Fig. 4.

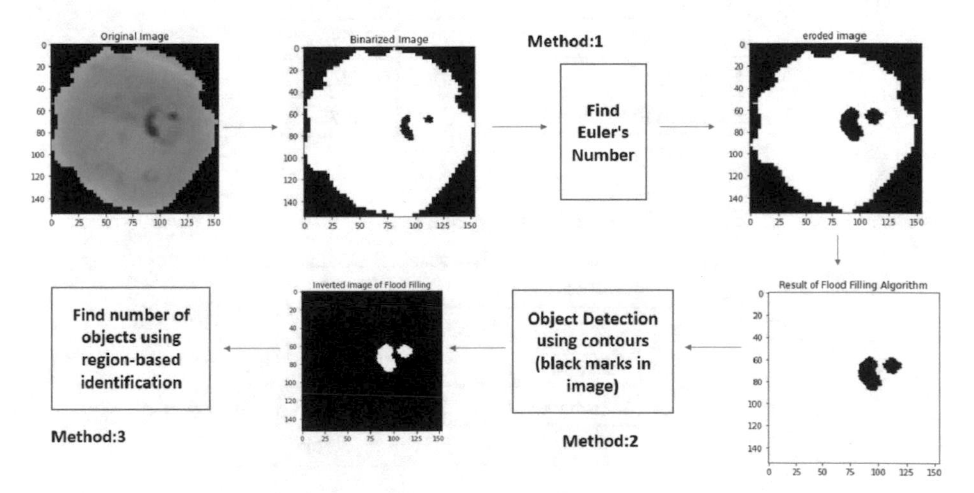

Fig. 4. Mind map showing the complete flow of implementation.

4 Results

From the above context and the explanation of methodology, it can be observed that the decision-making process involves simple and easily implementable tools of image processing. The developed voting-based classification system with three embedded methods was tested with 500 images. Out of these 500 images, 250 images are samples of infected smear while the remaining are uninfected.

The resolution of the testing images is also preserved intact without any further pre-processing and cleaning procedures. In this regard, it is essential to acknowledge that image-cleaning procedures should not be confused with the procedures described in the three individual procedures that were discussed in the previous section.

The values of various classification validation metrics are tabulated in Table 1 along with the confusion matrix in Table 2.

Table 1. Classification Metrics.

Metric	Value
True Positive	228
True Negative	223
False Positive	27
False Negative	22
Accuracy	90.2%
Precision	89.4%
Recall	91.2%
Specificity	89.2%
F1 Score	90.29%
Negative Predictive Value	91.02%

Table 2. Confusion Matrix.

	Actual Values			
Predicted Values		Infected	Uninfected	
	Infected	228	22	250
	Uninfected	27	223	250
		255	245	

Using the methodology described above, the proposed method will be able to identify and classify blood smear images with an accuracy of 90.2%. The malaria-infected Giemsa-stained image is collected for testing. The infected cells are highlighted. Different parameters were used to determine the performance of the proposed model as stated previously. The images used in this work were processed and certain features were extracted. Working on the proposed model, it can be identified that the accuracy of the following model in tagging the blood smear sample is comparatively higher than any of the existing techniques, considering the data utilized for the model-building. Also, it is pivotal to recollect that the process of model-building and runtime are comparatively minuscule in comparison to some of the deep learning methods of greater significance. Along similar lines, it should be noted that the currently developed model is computationally inexpensive, similar to its dependency on huge volumes of data.

Based upon the results of the following paper, the analysis of various classification and disease tagging techniques can be replicated in several other avenues by utilization of preliminary image processing tools and methods. When the individual components or basic tools are combined as a whole, it can be identified that the resulting algorithm set has far better implementations pertaining to confidence in classification.

For developing better results, the following model can be utilized parallelly along with manual testing. Which can still have a greater role to play by taking less data, computational flexibility, and processing time as key advantages.

5 Conclusion

In the realm of haematology, malaria is a terrible illness that kills millions of lives every year. Therefore, the urgent demand is for quick diagnosis and appropriate treatment. A manual microscope takes a lot of time and is prone to human error. In this research, we investigated the parasite by extracting its features from thick blood smear images. This research is the first step in making it easier for medical professionals to diagnose thin blood smears, using computationally flexible low-end techniques. The image processing methodology is used to tag the parasite, in which pictorial data is provided to a simplistic image processing modality to obtain the needed output in a reliable, accurate, and time-saving manner. The algorithm is used to identify the malaria parasite which in turn reduces the need for technicians and saves time. By spotting patterns in the gathered images, classification has been done to aid in the detection of parasite and uninfected malaria images. As possible avenues for the future, the effect of resolution, varying training and testing dataset sizes can be observed.

References

1. Gowda, R.B., Manna, M.S., K, S., Sharan, P.: Detection of plasmodium falciparum parasite intraerythrocytic stages using one dimensional distributed bragg reflector biosensor. In: 2021 IEEE 9th Region 10 Humanitarian Technology Conference (R10-HTC), pp. 1–6 (2021)
2. Wang, W., et al.: Laser-induced surface acoustic wave sensing-based malaria parasite detection and analysis. IEEE Trans. Instrum. Meas. **71**, 1–9 (2022)
3. Telang, H., Sonawane, K.: Effective performance of bins approach for classification of malaria parasite using machine learning. In: 2020 IEEE 5th International Conference on Computing Communication and Automation (ICCCA), pp. 427–432 (2020)
4. Rameen, I., Shahadat, A., Mehreen, M., Razzaq, S., Asghar, M.A., Khan, M.J.: Leveraging supervised machine learning techniques for identification of malaria cells using blood smears. In: 2021 International Conference on Digital Futures and Transformative Technologies (ICoDT2), pp. 1–6 (2021)
5. Aris, T.A., Nasir, A.S.A., Mohamed, Z.: A robust segmentation of malaria parasites detection using fast k-means and enhanced k-means clustering algorithms. In: 2021 IEEE International Conference on Signal and Image Processing Applications (ICSIPA), pp. 128–133 (2021)
6. Setiawan, A.W., Faisal, A., Resfita, N., Rahman, Y.A.: Detection of malaria parasites using thresholding in RGB, YCbCr and lab color spaces. In: 2021 International Seminar on Application for Technology of Information and Communication (iSemantic), pp. 70–75 (2021)

7. Maduri, P.K., Agrawal, S.S, Rai, A., Chaubey, S.: Malaria detection using image processing and machine learning. In: 2021 3rd International Conference on Advances in Computing, Communication Control and Networking (ICAC3N), pp. 1789–1792 (2021)
8. Yang, F., et al.: Deep learning for smartphone-based malaria parasite detection in thick blood smears. IEEE J. Biomed. Health Inform. **24**, 1427–1438 (2020)
9. Umer, M., Sadiq, S., Ahmad, M., Ullah, S., Choi, G.S., Mehmood, A.: A novel stacked cnn for malarial parasite detection in thin blood smear images. IEEE Access. **8**, 93782–93792 (2020)
10. Pattanaik, P.A., Mittal, M., Khan, M.Z.: Unsupervised deep learning CAD scheme for the detection of malaria in blood smear microscopic images. IEEE Access. **8**, 94936–94946 (2020)
11. Prakash, S.S., Kovoor, B.C., Visakha, K.: Convolutional neural network based malaria parasite infection detection using thin microscopic blood smear samples. In: 2020 Second International Conference on Inventive Research in Computing Applications (ICIRCA), pp. 308–313 (2020)
12. Hasan Sifat, M.M., Islam, M.M.: A fully automated system to detect malaria parasites and their stages from the blood smear. In: 2020 IEEE Region 10 Symposium (TENSYMP), pp. 1351–1354 (2020)
13. Ragb, H.K., Dover, I.T., Ali, R.: Deep convolutional neural network ensemble for improved malaria parasite detection. In: 2020 IEEE Applied Imagery Pattern Recognition Workshop (AIPR), pp. 1–10 (2020)
14. Kassim, Y.M., Jaeger, S.: A cell augmentation tool for blood smear analysis. In: 2020 IEEE Applied Imagery Pattern Recognition Workshop (AIPR), pp. 1–6 (2020)
15. Raj, M., Sharma, R., Sain, D.: A deep convolutional neural network for detection of malaria parasite in thin blood smear images. In: 2021 10th IEEE International Conference on Communication Systems and Network Technologies (CSNT), pp. 510–514 (2021)
16. Paul, S., Batra, S.: A review on computational methods based on machine learning and deep learning techniques for malaria detection. In: 2021 9th International Conference on Reliability, Infocom Technologies and Optimization (Trends and Future Directions) (ICRITO), pp. 1–5 (2021)
17. Nakasi, R., Tusubira, J.F., Zawedde, A., Mansourian, A., Mwebaze, E.: A web-based intelligence platform for diagnosis of malaria in thick blood smear images: a case for a developing country. In: 2020 IEEE/CVF Conference on Computer Vision and Pattern Recognition Workshops (CVPRW), pp. 4238–4244 (2020)
18. Taha, B., Liza, F.R.: Automatic identification of malaria-infected cells using deep convolutional neural network. In: 2021 24th International Conference on Computer and Information Technology (ICCIT), pp. 1–5 (2021)
19. Nugroho, H.A., Nurfauzi, R.: Deep learning approach for malaria parasite detection in thick blood smear images. In: 2021 17th International Conference on Quality in Research (QIR): International Symposium on Electrical and Computer Engineering, pp. 114–118 (2021)
20. Joshi, A.M., Das, A.K., Dhal, S.: Deep learning based approach for malaria detection in blood cell images. In: 2020 IEEE Region 10 Conference (TENCON), pp. 241–246 (2020)
21. Ufuktepe, D.K., et al.: Deep learning-based cell detection and extraction in thin blood smears for malaria diagnosis. In: 2021 IEEE Applied Imagery Pattern Recognition Workshop (AIPR), pp. 1–6 (2021)
22. Nautre, A., Nugroho, H.A., Frannita, E.L., Nurfauzi, R.: Detection of malaria parasites in thin red blood smear using a segmentation approach with U-Net. In: 2020 3rd International Conference on Biomedical Engineering (IBIOMED), pp. 55–59 (2020)
23. Khadim, E.U., Shah, S.A., Wagan, R.A.: Evaluation of activation functions in CNN model for detection of malaria parasite using blood smear images. In: 2021 International Conference on Innovative Computing (ICIC), pp. 1–6 (2021)

24. Nugroho, H.A., Nurfauzi, R.: GGB color normalization and faster-RCNN techniques for malaria parasite detection. In: 2021 IEEE International Biomedical Instrumentation and Technology Conference (IBITeC), pp. 109–113 (2021)
25. Paul, A., Bania, R.K.: Malaria parasite classification using deep convolutional neural network. In: 2021 International Conference on Computational Intelligence and Computing Applications (ICCICA), pp. 1–6 (2021)
26. Kassim, Y.M., et al.: Clustering-based dual deep learning architecture for detecting red blood cells in malaria diagnostic smears. IEEE J. Biomed. Health Inform. **25**, 1735–1746 (2021)

Acoustic and Visual Sensor Fusion-Based Rover System for Safe Navigation in Deformed Terrain

Pranav Pothapragada(✉) and Pavan Mohan Neelamraju

Department of Electronics and Communication Engineering, SRM University – AP, Andhra Pradesh, India
p.pranavnarayan@gmail.com

Abstract. With the growing development of autonomous navigation technologies, transportation systems have become safer and hassle-free. However, this decreasing necessity for human intervention has been both a boon and a bane for novel technologies. Utilising autonomous driving systems in highly rugged terrains has become very sophisticated and challenging. Upgradation of persisting navigation technologies in deformed terrains can improve accessibility, economy, security, traffic, and travel time. Therefore, the current preliminary research illustrates an acoustic wave and visual navigation-based rover system that can better navigate by monitoring the terra-formations using the acoustic and visual sensors to receive the reflected signal and image, respectively. The implementation tries to map the sensed world to a digital twin that the rover navigates through. The realisation of the model and the essential metrics are illustrated. The extension of the proposed technology to a grander scale in automobiles must be examined in several other environments and more focused on.

Keywords: Acoustic Waves · transportation systems · Sensor Fusion · Rover

1 Introduction

Autonomous driving and control have become pivotal aspects of automation and transport engineering. The advantages of automation have far-reaching impacts, yielding considerable benefits in terms of safety, accessibility, expenditure, productivity, and gain. The development of intelligent navigation systems aids in a high degree of autonomy, further reducing unsafe driving practices. On the other hand, intelligent decision-making devices have a more significant role in enhancing autonomous vehicles' performance. Mainly, this leads to self-sufficiency providing better chances for quick and safe mobility.

The prevailing model of autonomous vehicles works principally on lane detection and control mechanisms. However, ameliorating the existing modality by considering the characteristics of the terrain has a prominent role in increasing resource efficiency, safety, and control.

Thus, terrain change identification is pivotal in planning and preparing an autonomous navigator. Given the same, the external surface in contact with the vehicle should be represented in a manner much more suitable to the decision-making system

H. Sharma et al. (Eds.): ICIVC 2022, PALO 17, pp. 348–357, 2023.
https://doi.org/10.1007/978-3-031-31164-2_29

whilst attempting to accelerate at high speeds in rugged terrains. In achieving such a sophisticated system, it is equally important to consider both the decision-making system's data-handling capabilities and the computation speed. Similarly, the accuracy in identifying the terrain type and articulating its features will increase the system's reliability.

In the present research work, emphasis is placed on building a preliminary terrain feature characterisation model that can work upon the textural features of the surface. The model utilises the capability of acoustic and visual data and produces a quantified outcome illustrating the characteristics of the varied terrain. The acoustic and visual data collection domains act as self-complementing structures in providing concrete decision support. The results of the preliminary study will help us establish a better understanding of the texture-based classification of surface types and the capabilities of data extraction. In addition, terrain-custom navigation systems lead to more accurate and safer alternatives to persisting mobility opportunities. Terrain surface characterisation and decision-making are the basic foundations for this work.

The paper focuses on providing a summarised review of the works done, thereby enlisting the merits of the developed model in Sect. 3. The modality's data acquisition, implementation and development have been thoroughly discussed in Sect. 3, followed by the results.

2 Background

Many past research studies and investigations were conducted to understand the structural signature of navigation paths. A road structure monitoring system was presented to recognize humps and potholes. An Ultrasonic sensor was utilized to extract signals and aid in the real-time analysis [1]. The aforementioned method expanded the study by utilizing optimization approaches, which led to the development of a more effective road monitoring system. Similar to the previous technique, a Convolutional Neural Network was developed and deployed as an android application [2]. Correspondingly, deep learning techniques were implemented with Global Positioning System (GPS) to communicate the insights to other vehicles [3]. The principal aspect of the prior studies was the use of sensor-based systems to understand the road surface trend and patterns [4].

Along similar lines, studies were done to identify the types of barriers using raw sensor signals [5]. A multitude of sensory data was used to extrapolate the insights from a traffic environment. The data was augmented with popular deep-learning models to classify the road material [6]. Techniques such as Frequency Modulated Continuous Wave radar were employed for continuous real-time salt concentration detection and black ice on winter roads [7]. In identifying road deformations, depth-based models were found to significantly impact computational complexity and processing time [8]. However, supplementing different forms of data and coupled decision-making were some key aspects that can ameliorate conventional data-based models.

Regarding the same, road consistency evaluation was done by statistical models like probabilistic logistic modelling, paving the way for data integrity [9]. Consequently, research works were proposed pertaining to alerting [10], automatic detection [11, 12] of potholes and related irregularities on the surface [13, 14].

Image processing methods were widely utilized in identifying and reporting deformed surfaces [15]. With the growing impact of text data and extractable visual features, there has been a significant improvement in the degradation level estimation of road structures [16]. Tools such as advanced air-coupled 3D ground-penetrating radar were used to help with non-destructive, accurate, and rapid testing of pavement distress [17]. The results showed an accuracy of 98.73%. Albeit, the long-standing image processing models can further be improvised by associating them with pre-trained mathematical methods, which can help in transcending the ill effects of noise and escalate the accuracy in analysing the condition of navigation paths.

In contrast to traditional image processing techniques, computer-vision-based modalities were proposed to enhance the accuracy of distress detection and increase the immunity to image data noise [18]. Enhancement of DeepLabv3+ was done for semantic image segmentation to detect cracks and deformations. Along with the same, models like YOLOv3, in association with techniques such as Bayesian search, were constructed to precisely point out and localize concealed pavement cracks with the help of ground penetrating radar [19]. The above-discussed models can be advanced further by incorporating data acquisition from various sensory approaches [20], resulting in the design of self-complimenting pavement surface assessment techniques.

Most importantly, combining several data acquisition modes significantly impacts the computational complexity and processing speeds. Several studies were carried out to elucidate information-gathering techniques in terrain monitoring and evaluation studies. Technologies such as Ground Penetrating Radar, Visual Sensors and Mobile Laser Scanning devices were employed to construct an automatic survey system to understand path fretting [21].

For autonomous driving, a video-based assistance system was developed to track the unusual changes in a driver's behaviour [22]. Likewise, video processing has also been implemented in the generation of high-definition maps along with a CNN classifier [23]. Continuous changes in video-driven models took place, leading to more adaptable means of monitoring. Research works based on miniature tools like smartphones [24], and change detection methods [25] were proposed to understand and track the conditions of the navigating path [26]. Summarising the literature, it can be observed that the models are becoming more sophisticated and increased accuracy.

In conclusion, we can say that even though much research has been done in the field, most of it focuses on general analysis, and very little has been done to derive conclusions from data collected from numerous sensor systems. The following is a list of our contributions to the existing work:

- Creating a computationally flexible decision-making model.
- Building a self-complimenting model with the capability of considering both visual and acoustic sources of data.

3 Methodology

The model introduced in this paper tries to determine an Acoustic-visual sensor fusion system to gather important information for autonomous vehicles, understand the terrain better, and make appropriate decisions. Firstly, an Acoustic sensor works on acoustic backscatter to understand the surface on which the vehicle is moving. Furthermore, Ultrasound is used explicitly to eliminate interference from external noise.

Fig. 1. Image of path collected for Hough-transform based lane detection model.

For this paper, a modified sensor module based on HC-SR04 was used to pulse 40 kHz sine waves. The signal and threshold outputs from the amplification circuit in the sensor module are taken as inputs. They correspond to pins 9 and 10 of the onboard microcontroller. The rest of the system is left as is for a known transmission pattern. In this model, the ultrasound sensor was positioned in front of the rover, 0.2 m from the surface, at a 20° angle from the ground normal. The receiving ultrasonic speaker then captures this data and sends it to an onboard Raspberry Pi 3 for processing.

Fig. 2. Segmented Image of the collected data sample

The average power of the measured data is then calculated on the Raspberry Pi. K-Nearest Neighbour (KNN) classifier was used to classify these surfaces depending on the average power of the backscattered signal. The classification was made on multiple surfaces the vehicles could be moving on, classifying between concrete, dirt, and grass.

The second part of the methodology deals with the implementation of image processing techniques to identify the change in terrain. In particular, the modality tries to extrapolate the variation in terrain depending upon successive frames collected from the camera, which is mounted on the rover. The collected frame of images is pre-processed using Hough transform. This resulted in the removal of redundant information from the image, like pavement borders and paths sideways. The outcome of the Hough-transform-based pavement detection algorithm can be observed in Fig. 1.

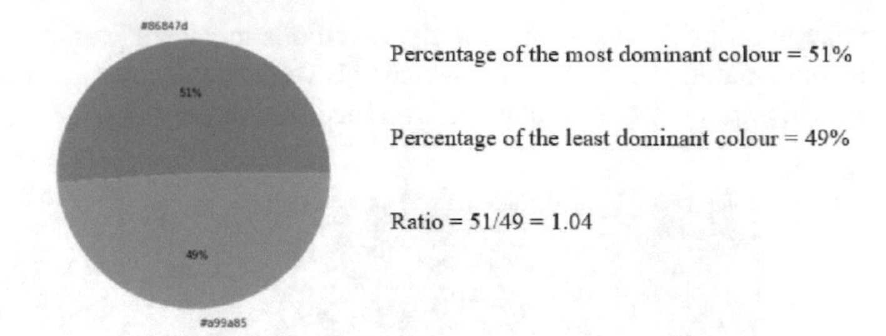

Fig. 3. Percentage composition of most and least dominant colour segments

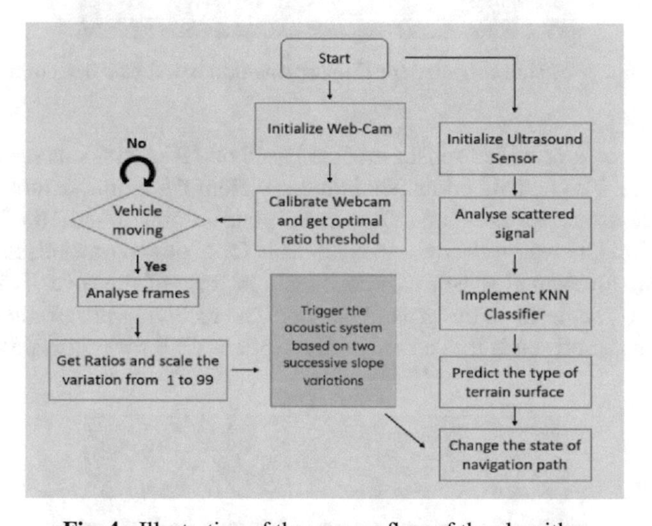

Fig. 4. Illustration of the process flow of the algorithm

The subsequent analysis involves segmentation of the image into two different regions depending upon the colours present in the image. Importantly, the two dominant colours in the image are identified as a primary step, followed by mapping each picture element (pixel) to its nearest dominant colour depending upon the digital numbers in the layers of red, green, and blue channels. To realise this system, the current work employed k-means clustering-based image segmentation. The segmented version of the image is illustrated in Fig. 2.

Finally, the last step of the image processing part deals with counting the total number of pixels mapped to each cluster. The ratio of the number of pixels with the most dominant colour to the number of pixels with the least dominant colour is computed as shown in Fig. 3.

Upon identifying the ratios computed at successive samples or frames, the trend in ratios of two consecutive samples is examined. For example, if a local minimum is attained in the ratio graph, the acoustic-based model is triggered. The advantages of analysing the rate of change of slopes mitigates the pressure on a single model and gives a chance for better decision-making using a multitude of data sources. This further enhances the reliability of navigation path feature characterization.

Sub-figures (a), (b), (c) and (d) with computed ratios of 10.85, 1.38, 1.17 and 99.0 respectively.

Fig. 5. Illustration of a few example images with their respective ratios

For sensor fusion, the sensor data from both the sensors is gathered. Initially, only the visual sensor continually sends data to the Raspberry Pi 3 board. The aforementioned algorithm allows the system to understand whenever there is a change in the type of road, although the specific type of road is left unknown. Whenever a change in the type of road is found, the Raspberry Pi enables the Ultrasound system, which starts pulsing 40 kHz waves and receives the backscattered data.

This data is then processed to extract information about the road surface, which could be used to make better decisions when autonomous driving is used. After this loop is executed, the system moves back to the visual loop and starts rechecking for changes in the type of road. A decision Flow-Chart of the algorithm is depicted in Fig. 4.

4 Results

Many novel modalities have been proposed with the increasing significance of terrain characteristic identification and the subsequent benefits. In particular, tools and techniques for using acoustic and visual sensory systems have played a pivotal role in navigation surface textural analysis. These techniques, combined with the persisting autonomous vehicles, yield valuable results by providing surface-dependent motorable control.

In the current study, the utilisation of image processing techniques to identify terrain change detection provided better results in terms of various classification metrics. The model was validated with 100 different images/frames, out of which 50 images were found to have a change in the navigation surface, while the remaining images were with consistent surface texture. The computed ratio for each image is used to identify the change in the terrain or the navigating path. According to the methodology, the path is expected to have a greater change in surface texture when the computed ratio is around 1. On the other hand, images with a ratio of 99 are found to have consistent surfaces. Also, some of the images utilised in the testing phase are shown in Fig. 5, along with their computed ratios.

Table 1. Performance across different types of surfaces.

Terrain Type	Number of Samples	Number of Correct	Number of Incorrect	Accuracy using KNN
Concrete	31	28	3	90.32%
Dirt	31	25	6	80.64%
Grass	31	30	1	96.77%

Complementing the performance of the visual-sensor-based system, we have the acoustic-based sensory system characterising the path texture depending on the ultrasound reflections. Three different kinds of surfaces were included as part of the study: concrete, dirt, and grass. The rover system is tested on 93 different sample spaces covering all the above-discussed kinds of surfaces, after creating a KNN model with 378 samples covering all the terrain-types equally.

Table 2. Confusion Matrix.

	Actual Values				
Predicted Values		Concrete	Dirt	Grass	
	Concrete	28	2	1	31
	Dirt	4	25	2	31
	Grass	0	1	30	31
		32	28	33	

In each of the surface-type, the accuracy values were computed in Table 1. The confusion matrix for all the terrain types is shown in Table 2. Correspondingly, the overall accuracy of the model, along with the error rate, is shown in Table 3.

Table 3. Derived Accuracy and Error Rate.

Metric	Value
Overall Accuracy	89.25%
Error Rate	10.75%

5 Conclusion

Research on improving autonomous systems contributes to humankind in various paradigms such as safety, intelligent fuel conservation, travel expenses and reduced traffic. Given such advantages, it is the need of the hour to develop smart and efficient automobiles with the capacity of terrain identification, which can aid in speed and acceleration control in hostile environments.

In achieving the goal, the paper extrapolated some of the possible domains of implementation and tools that can help analyse the variation in the navigating surface. Some of the key-note features include the utilisation of simple unsupervised Machine Learning techniques such as k-Means Clustering to identify terrain changes. Implementing the current model helped mitigate the ill effects of supervised model training, which utilises large amounts of data.

Furthermore, implementing an ultrasonic sensor-based system paved the path for real-time terrain analysis with greater accuracy and classification metrics. The outcomes of the study illustrate the need for multi-sensory concatenation of decision-making devices which have a greater role to play in shaping the world of tomorrow. With the growing dependence on data, it is also essential to search for computational models that are more user-friendly and less data-dependent simultaneously. As the world drives and thrives under the shade of automation, the process of real-time assessment needs to be cost-effective and computationally feasible.

Unlike other implementations, the image-processing model utilises the periodic sample of frames, where the whole data is utilised for outcome determination along with the acoustic system. This extensive utilization of image data along with the acoustic model allows the system to work reliably in the darkness, where other visual systems fail. Along with the ultrasonic-powered model, various other information collection sources can be studied that can further enhance the overall accuracy.

Along similar lines, the progress in the domains of sensor systems and Micro-electromechanical devices can help detect various other environmental impacts and changes that significantly influence autonomous driving technologies.

References

1. Chellaswamy, C., Famitha, H., Anusuya, T., Amirthavarshini, S.B.: IoT based humps and pothole detection on roads and information sharing. In: 2018 International Conference on Computation of Power, Energy, Information and Communication (ICCPEIC), pp. 084–090. IEEE (2018)
2. Rahman, A., Mustafa, R., Hossain, M.S.: Real-time pothole detection and localization using convolutional neural network. In: Arefin, M.S., Kaiser, M.S., Bandyopadhyay, A., Ahad, M.A.R., Ray, K. (eds.) Proceedings of the International Conference on Big Data, IoT, and Machine Learning. LNDECT, vol. 95, pp. 579–592. Springer, Singapore (2022). https://doi.org/10.1007/978-981-16-6636-0_44
3. Chellaswamy, C., Saravanan, M., Kanchana, E., Shalini, J., Others: Deep learning based pothole detection and reporting system. In: 2020 7th International Conference on Smart Structures and Systems (ICSSS), pp. 1–6. IEEE (2020)
4. Rateke, T., Justen, K.A., Von Wangenheim, A.: Road surface classification with images captured from low-cost camera-road traversing knowledge (rtk) dataset. Revista de Informática Teórica e Aplicada **26**, 50–64 (2019)
5. Springer, M., Ament, C.: A mobile and modular low-cost sensor system for road surface recognition using a bicycle. In: 2020 IEEE International Conference on Multisensor Fusion and Integration for Intelligent Systems (MFI), pp. 360–366. IEEE (2020)
6. Menegazzo, J., von Wangenheim, A.: Multi-contextual and multi-aspect analysis for road surface type classification through inertial sensors and deep learning. In: 2020 X Brazilian Symposium on Computing Systems Engineering (SBESC), pp. 1–8. IEEE (2020)

7. Blanche, J., Mitchell, D., Flynn, D.: Run-time analysis of road surface conditions using non-contact microwave sensing. In: 2020 IEEE Global Conference on Artificial Intelligence and Internet of Things (GCAIoT), pp. 1–6. IEEE (2020)
8. Mandal, V., Mussah, A.R., Adu-Gyamfi, Y.: Deep learning frameworks for pavement distress classification: a comparative analysis. In: 2020 IEEE International Conference on Big Data (Big Data), pp. 5577–5583. IEEE (2020)
9. Tasmin, T., Wang, J., Dia, H., Richards, D., Tushar, Q.: A probabilistic approach to evaluate the relationship between visual and quantified pavement distress data using logistic regression. In: 2020 IEEE Asia-Pacific Conference on Computer Science and Data Engineering (CSDE), pp. 1–4. IEEE (2020)
10. Reddy, E.J., Reddy, P.N., Maithreyi, G., Balaji, M.B.C., Dash, S.K., Kumari, K.A.: Development and analysis of pothole detection and alert based on NodeMCU. In: 2020 International Conference on Emerging Trends in Information Technology and Engineering (ic-ETITE), pp. 1–5. IEEE (2020)
11. Madli, R., Hebbar, S., Pattar, P., Golla, V.: Automatic detection and notification of potholes and humps on roads to aid drivers. IEEE Sens. J. **15**, 4313–4318 (2015)
12. Gorintla, S., Kumar, B.A., Chanadana, B.S., Sai, N.R., Kumar, G.S.C.: Deep-learning-based intelligent PotholeEye+ detection pavement distress detection system. In: 2022 International Conference on Applied Artificial Intelligence and Computing (ICAAIC), pp. 1864–1869. IEEE (2022)
13. Nguyen, T., Lechner, B., Wong, Y.D.: Response-based methods to measure road surface irregularity: a state-of-the-art review. Eur. Transp. Res. Rev. **11**(1), 1–18 (2019). https://doi.org/10.1186/s12544-019-0380-6
14. Rateke, T., von Wangenheim, A.: Road surface detection and differentiation considering surface damages. Auton. Robot. **45**(2), 299–312 (2021). https://doi.org/10.1007/s10514-020-09964-3
15. Lõuk, R., Tepljakov, A., Riid, A.: A two-stream context-aware ConvNet for pavement distress detection. In: 2020 43rd International Conference on Telecommunications and Signal Processing (TSP), pp. 270–273. IEEE (2020)
16. Ogawa, N., Maeda, K., Ogawa, T., Haseyama, M.: Degradation level estimation of road structures via attention branch network with text data. In: 2021 IEEE International Conference on Consumer Electronics-Taiwan (ICCE-TW), pp. 1–2. IEEE (2021)
17. Liang, X., Yu, X., Chen, C., Jin, Y., Huang, J.: Automatic classification of pavement distress using 3D ground-penetrating radar and deep convolutional neural network. IEEE Trans. Intell. Transp. Syst. **23**(11), 22269–22277 (2022). https://doi.org/10.1109/TITS.2022.3197712
18. Sun, X., Xie, Y., Jiang, L., Cao, Y., Liu, B.: DMA-Net: DeepLab with multi-scale attention for pavement crack segmentation. IEEE Trans. Intell. Transp. Syst. **23**(10), 18392–18403 (2022)
19. Liu, Z., Gu, X., Yang, H., Wang, L., Chen, Y., Wang, D.: Novel YOLOv3 model with structure and hyperparameter optimization for detection of pavement concealed cracks in GPR images. IEEE Trans. Intell. Transp. Syst. **23**(11), 22258–22268 (2022)
20. Christodoulou, S.E., Hadjidemetriou, G.M., Kyriakou, C.: Pavement defects detection and classification using smartphone-based vibration and video signals. In: Workshop of the European Group for Intelligent Computing in Engineering, pp. 125–138 (2018)
21. Li, W., Burrow, M., Metje, N., Ghataora, G.: Automatic road survey by using vehicle mounted laser for road asset management. IEEE Access. **8**, 94643–94653 (2020)
22. Zajic, G., Popovic, K., Gavrovska, A., Reljin, I., Reljin, B.: Video-based assistance for autonomous driving. In: 2020 Zooming Innovation in Consumer Technologies Conference (ZINC), pp. 151–154. IEEE (2020)
23. Golovnin, O.K., Rybnikov, D.V.: Video processing method for high-definition maps generation. In: 2020 International Multi-Conference on Industrial Engineering and Modern Technologies (FarEastCon), pp. 1–5. IEEE (2020)

24. Jingxiang, Z., Jinxi, Z., Dandan, C.: Preliminary study on pavement performance characteristics based on intelligent method. In: 2020 IEEE 9th Joint International Information Technology and Artificial Intelligence Conference (ITAIC), pp. 1484–1490. IEEE (2020)
25. Lembach, J., Stricker, R., Gross, H.-M.: Efficient implementation of regional mutual information for the registration of road images. In: 2020 Tenth International Conference on Image Processing Theory, Tools and Applications (IPTA), pp. 1–6. IEEE (2020)
26. Auriacombe, O., Vassilev, V., Pinel, N.: Dual-polarised radiometer for road surface characterisation. J. Infrared Millimeter Terahertz Waves **43**(1–2), 108–124 (2022). https://doi.org/10.1007/s10762-022-00847-5

Feature Selection with Genetic Algorithm on Healthcare Datasets

Luke Oluwaseye Joel(✉), Wesley Doorsamy, and Babu Sena Paul

Institute for Intelligent Systems, University of Johannesburg, Johannesburg, South Africa
oluwaseyejoel@gmail.com

Abstract. One of the major problems in building machine learning models is identifying relevant features that will positively impact the performance of your model. Removing elements that do not contribute significantly (called irrelevant features) to the target variable is an important task to be carried out by data scientists or researchers to achieve an optimal model. This study tries to improve the performance of the classification of five (5) different machine learning algorithms - Decision tree, K-nearest neighbor, Random forest, Naive Bayes and Adaboost - using the genetic algorithm feature selection method. These classification and feature selection methods were examined using two healthcare - heart disease and breast cancer - datasets. The four metrics employed are precision, recall, f1-score, and accuracy. The results show that the feature selection improves the performance of the algorithms, and that the Random forest algorithm performs the best for both datasets.

Keywords: Genetic Algorithm · Machine learning · Feature Selection · Healthcare Datasets

1 Introduction

The maintenance or improvement of humans' health through prevention, diagnosis and treatment is one of the major priorities today [17]. Hence, the need for more research in the healthcare sector. The choice of these two healthcare datasets is based on how deadly the two diseases under consideration are. Heart disease causes about one in four deaths in the United States of America [25], making it a leading cause of morbidity in the US and various countries of the world. Also, breast cancer comes second in the grounds of death for women [12,28] after lung cancer and its early diagnosis can reduce its mortality rate by 40% [13].

Data collected by different sensors, through the internet of things or other methods, often contain some significant noisy and irrelevant features that could negatively affect the machine learning model and be computationally expensive. Hence, the need to remove features that do not contribute significantly (called irrelevant features) to the target variable. Like imputation of missing values in a given dataset [18], feature selection is a necessary task to be carried out by data scientists or researchers to achieve an optimal machine learning model.

H. Sharma et al. (Eds.): ICIVC 2022, PALO 17, pp. 358–374, 2023.
https://doi.org/10.1007/978-3-031-31164-2_30

Performing this feature selection [1,2,26] process helps develop models that are easy to interpret; achieve shorter training time; obtain an enhanced generalisation of the machine learning models by reducing overfitting; and to minimize variable redundancy. This study uses genetic algorithm to do feature selection with different machine learning classifiers to evaluate the feature selected on heart disease and breast cancer datasets. The performance of these machine learning algorithms was evaluated using precision, recall, f1-score, and accuracy metrics.

The remaining part of this paper is organized as follows. Section 2 explains the different categories of feature selection methods. Section 3 explains the genetic algorithm employed in selecting the features in this study. Section 4 describes the different machine learning algorithms used to evaluate features subsets in the selection process. Section 5 gives an overview of the two healthcare datasets employed in this paper. The results and discussion for this study are provided in Sect. 6. And lastly, Sect. 7 is the conclusion of this study.

2 Feature Selection Methods

Feature selection techniques can be broadly divided into three main categories, namely filter, wrapper and embedded methods [5]. Figure 1 shows the graphical explanation of how these three popular feature selection methods function. In the filter method, the selection of features is independent of the machine learning algorithm [6,8,41]. The selection is made based on the characteristics of the

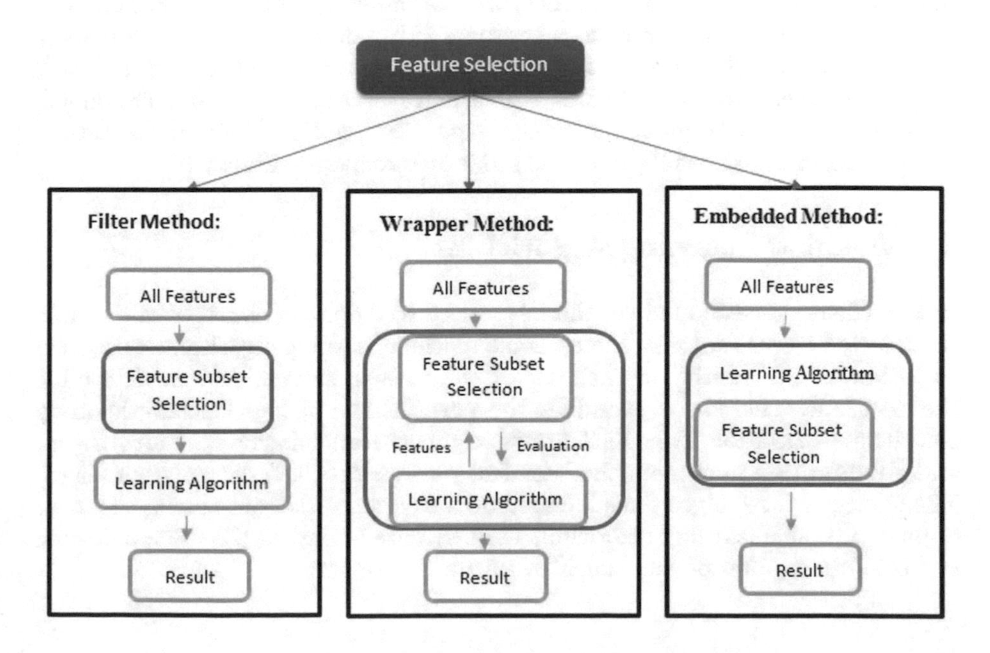

Fig. 1. Types of Feature Selection Methods

features in the dataset [5,6,8,41]. Hence, filter methods are known to be less computationally expensive but are suitable for the quick removal of irrelevant features.

Unlike the filter method, the wrapper method is computationally expensive because the selection process is considered as a search problem where features are selected and evaluated using the machine learning algorithm [8,34]. The final selected features are output based on the evaluation done by the learning algorithm [37]. On the other hand, in the embedded method, feature selection is incorporated as part of the machine learning algorithm [8]. This gives an iterative process such that only features which contribute the most to the training for a particular iteration are selected. Hence, embedded method combines the advantages of both filter and wrapper methods [37].

3 Genetic Algorithms

Genetic Algorithms (GAs) [14,23] are search-based optimization methods that are based on the concepts of natural selection and genetics [15,21]. GA is a subset of evolutionary computation. It is faster and more efficient than traditional methods and does not require any derivative information. With respect to feature selection, GA is an efficient method for selecting relevant features in a dataset which could help to obtain better results [16].

The first step of the GA is to generate a population of possible features. The fitness of each subset is evaluated, from this population, using some criteria. And this will help to identify which subsets will continue to the next generation. The subsets that survive to the next generation will undergo a process of crossover and mutation to reproduce new feature subsets. This process will run for a certain number of generations, and the best subset of features among the population will be selected. This paper uses the genetic algorithm module for feature selection, called "`sklearn-genetic`", from the scikit-learn software library [7].

4 Machine Learning Algorithms

In the sklearn-genetic module, there is a need to employ a classifier to evaluate the selected feature subsets. Hence, some machine learning algorithms were used to estimate the "merit" of each set of the feature subsets and to determine the best one. This study examines the performance of five machine learning algorithms - Decision Tree, K-Nearest Neighbor, Random Forest, Naive Bayes, and Adaboost - to estimate the best feature subset of the datasets employed. The choice of these algorithms is based on a random pick from among the most popular machine learning algorithms used by data scientists [29]. The following is a brief description of each machine learning algorithm.

4.1 Decision Tree

Decision tree (DT) algorithm is a supervised learning algorithm that can be used for both regression and classification problems [9,32]. The algorithm uses tree representation to show the predictions that result from a series of feature-based splits starting with a root node and ending with two or more leaf nodes. The leaf nodes (or terminal nodes) are the output of the decision tree algorithm and cannot be split any further [4,9]. Whereas the nodes obtained after splitting the root node are called decision nodes. These will be divided if a leaf node is not reached. The tree-like structure represented in the DT algorithm makes it easy to understand and interpret. The DT algorithm is capable of handling both numerical and categorical data.

4.2 K-Nearest Neighbor

K-nearest neighbor (KNN), also called lazy learner algorithm, is a supervised learning algorithm used for both classification and regression problems [3]. It works on identifying the nearest neighbor to the tested data by calculating the distances between the tested and the trained data [35]. The tested data will then be assigned to the class of its nearest neighbor based on the value of k (which represents the number of nearest neighbors). The choice of the value k has a significant impact on the performance of KNN algorithm [44]. Hence, this value must be chosen carefully. One of the most used methods to calculate distances in KNN is the Euclidean distance [3,35]. Others are Manhattan, Minkowski, and Chebyshev distances.

4.3 Random Forest

Random forest (RF) is a supervised learning algorithm that can be used for both classification and regression problems [11,24]. RF algorithm uses the concept of ensemble learning in its operation. That is, it combines multiple classifiers to solve any complex problem. The RF algorithm consists of many decision trees, and hence, its outcome is based on the predictions of these decision trees. Thereby, it overcomes the limitations of a decision tree algorithm, increasing the precision of the results [24,27]. RF has an advantage of taking lesser training time compared to other machine learning algorithms and it can produce highly accurate predictions even with missing data [24].

4.4 Naive Bayes

Naive Bayes (NB) is a classification algorithm based on Bayes' theorem. This theorem, which is used to determine the conditional probability of two or more events, states that the probability of an event, A, occurring can be obtained from a given probability of another event, B, that has occurred before. The mathematical expression for this theorem can be found in [10,38,40]. It is also known as a probabilistic classifier because it incorporates conditional independence assumptions in its models [10]. It assumes that each feature in a dataset is independent and contributes equally to the target class. NB is fast and easy to use among the machine learning algorithms [38]. Some popular NB algorithms are Gaussian, multinomial and Bernoulli NB [40]. This study uses the Gaussian NB.

4.5 Adaboost

Adaboost, short form for the adaptive boosting algorithm, is an ensemble method in machine learning. It uses the concept of assigning equal weights to each training sample at the beginning and re-assigning weights to each sample in the subsequent steps, with higher weights given to the incorrect samples and lower weights to the correct samples [39]. This procedure is repeated until the errors are minimized. That is, the Adaboost algorithm tries to build a strong classifier from several weak classifiers [39,43]. Adaboost algorithm has the advantage of avoiding overfitting [43]. The algorithm has attracted some attention among researchers in machine learning and motivated the development of several versions of the algorithm which have proven to be competitive in many applications [33,36,39,42,43].

5 Datasets

The two healthcare datasets to be used in this study are the heart disease and breast cancer datasets. The heart disease data has 14 features including the dependent feature (called "target") and 303 entries. In comparison, the breast cancer dataset has 32 features including the dependent feature (called "diagnosis") and 569 entries. The description of these datasets is given below.

5.1 Heart Disease Dataset

Heart disease, also called cardiovascular disease, describes several conditions that involve blocked or narrowed blood vessels, which can lead to stroke, chest pain, heart failure, or heart attack. This disease causes about one in four deaths in the United States of America [25], making it a leading cause of morbidity in the US. The heart disease dataset is taken from the popular dataset database [20]. The names and the datatype of each of the features is shown in Fig. 2. The dependent feature, called "target" is "0" indicating the presence of heart disease, and "1" indicating the absence of heart disease. The proportion of "0" in the dataset is 137, while the proportion of "1" in the dataset is 165, as shown in Fig. 3.

5.2 Breast Cancer Dataset

Breast cancer is an abnormal growth of cells in a woman's breast. This abnormal growth, also called tumor, can be cancerous (malignant) or non-cancerous

```
<class 'pandas.core.frame.DataFrame'>
RangeIndex: 303 entries, 0 to 302
Data columns (total 14 columns):
 #   Column    Non-Null Count  Dtype
---  ------    --------------  -----
 0   age       303 non-null    int64
 1   sex       303 non-null    int64
 2   cp        303 non-null    int64
 3   trestbps  303 non-null    int64
 4   chol      303 non-null    int64
 5   fbs       303 non-null    int64
 6   restecg   303 non-null    int64
 7   thalach   303 non-null    int64
 8   exang     303 non-null    int64
 9   oldpeak   303 non-null    float64
 10  slope     303 non-null    int64
 11  ca        303 non-null    int64
 12  thal      303 non-null    int64
 13  target    303 non-null    int64
dtypes: float64(1), int64(13)
memory usage: 33.3 KB
```

Fig. 2. The Heart Disease Dataset

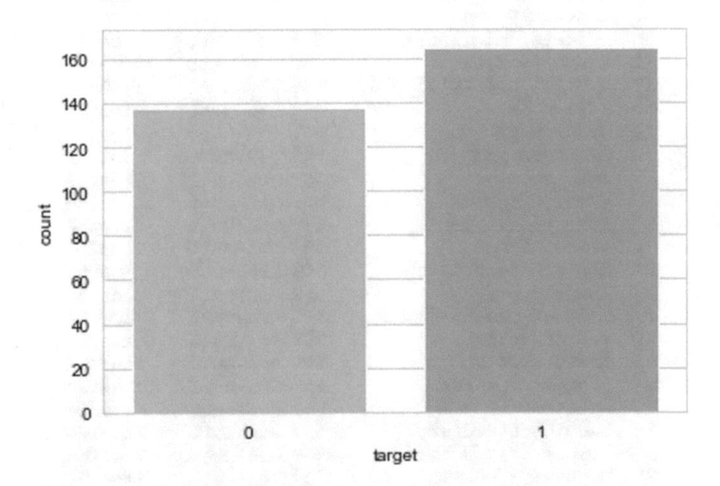

Fig. 3. The Distribution of the 'target' Dependent Feature

(benign). The former grows rapidly, invades, and destroys nearly normal tissues in the breast, while the latter grows slowly and do not spread. Hence, though, sometimes painful and potentially dangerous, the benign tumors do not pose a serious threat that malignant tumor do. Breast cancer comes second in the causes of death for women after lung cancer [12,28], and its early diagnosis can reduce its mortality rate by 40% [13].

The breast cancer dataset, for this study, is also taken from the popular dataset database [19]. The names and data type of each of the features is shown in Fig. 4. The dependent feature, called "diagnosis", is "M" for malignant (cancerous) and "B" for benign (noncancerous). The class distribution of this dependent feature is shown in Fig. 5. The diagnosis for malignant has 212 entries while benign has 357 entries. As observed from Fig. 4, the dependent feature, (called "diagnosis"), has an object as data type, which means it is either a "M" or "B". However, to feed this into our feature selection and classification algorithms, we used the transformer called "`LabelEncoder`" from the"`sklearn.preprocessing`" module to encode it into "1" or "0". Note that

```
<class 'pandas.core.frame.DataFrame'>
RangeIndex: 569 entries, 0 to 568
Data columns (total 32 columns):
 #   Column                   Non-Null Count  Dtype
---  ------                   --------------  -----
 0   id                       569 non-null    int64
 1   diagnosis                569 non-null    object
 2   radius_mean              569 non-null    float64
 3   texture_mean             569 non-null    float64
 4   perimeter_mean           569 non-null    float64
 5   area_mean                569 non-null    float64
 6   smoothness_mean          569 non-null    float64
 7   compactness_mean         569 non-null    float64
 8   concavity_mean           569 non-null    float64
 9   concave points_mean      569 non-null    float64
 10  symmetry_mean            569 non-null    float64
 11  fractal_dimension_mean   569 non-null    float64
 12  radius_se                569 non-null    float64
 13  texture_se               569 non-null    float64
 14  perimeter_se             569 non-null    float64
 15  area_se                  569 non-null    float64
 16  smoothness_se            569 non-null    float64
 17  compactness_se           569 non-null    float64
 18  concavity_se             569 non-null    float64
 19  concave points_se        569 non-null    float64
 20  symmetry_se              569 non-null    float64
 21  fractal_dimension_se     569 non-null    float64
 22  radius_worst             569 non-null    float64
 23  texture_worst            569 non-null    float64
 24  perimeter_worst          569 non-null    float64
 25  area_worst               569 non-null    float64
 26  smoothness_worst         569 non-null    float64
 27  compactness_worst        569 non-null    float64
 28  concavity_worst          569 non-null    float64
 29  concave points_worst     569 non-null    float64
 30  symmetry_worst           569 non-null    float64
 31  fractal_dimension_worst  569 non-null    float64
dtypes: float64(30), int64(1), object(1)
memory usage: 142.4+ KB
```

Fig. 4. The Cancer Dataset

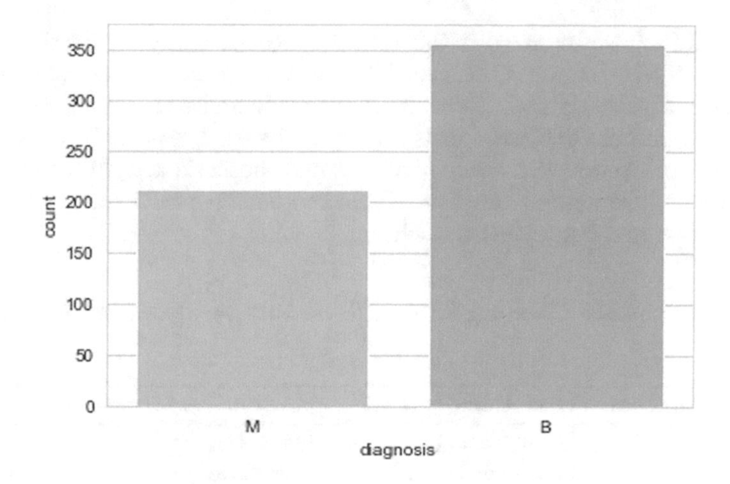

Fig. 5. The Distribution of the 'diagnosis' Dependent Feature

the 'id' feature was not used for analysis because it contains a string of numbers to identify each patient.

6 Results and Discussion

The platform used for both feature selection and prediction on the two healthcare datasets is the scikit-learn (sklearn) package [31], coded on Jupyter notebook. Scikit-learn is a free Python module for machine learning projects. It contains different classification, regression, clustering, and feature selection algorithms. The percentage of the training/testing sets splits for both datasets is 70/30. For the GA used for feature selection, an experiment was conducted to determine the suitable values of the mutation and crossover rates. Table 1 shows the different mutation and crossover rates used.

Table 1. The Different Mutation and Crossover Rates

Mutation Rate	Crossover Rate
0.05	0.4, 0.6, 0.8
0.1	0.4, 0.6, 0.8
0.2	0.4, 0.6, 0.8

The number of the selected features by the different machine learning algorithms on heart disease and breast cancer datasets is given in Table 2 and 3, respectively. The tables also contain the mutation and crossover rates where the best and worst scores are attained. The following algorithms - DT, KNN and RF - achieved their best performances in the heart disease dataset (Table 2) when mutation and crossover rates is 0.2 and 0.4, respectively. While the following algorithms - KNN, RF and Gaussian NB - have their worst performances in the breast cancer disease dataset (Table 3) at the same mutation and crossover rates. The highest number of selected features on the heart disease dataset is 10 out of 13 features. And the lowest number of features chosen is 4. While on the breast cancer dataset, the highest number of selected features is 17 out of 29 features. And the lowest number of features chosen is 4.

Table 2. Mutation and Crossover Rates with the Number of Selected Features (Heart Dataset)

Algorithm	Best Performance	Worst Performance
DT	Mutation = 0.2 Crossover = 0.4 Number of features = 6	Mutation = 0.1 Crossover = 0.4 Number of features = 8
KNN	Mutation = 0.2 Crossover = 0.4 Number of features = 8	Mutation = 0.1 Crossover = 0.6 Number of features = 4
RF	Mutation = 0.2 Crossover = 0.4 Number of features = 8	Mutation = 0.2 Crossover = 0.6 Number of features = 10
NB	Mutation = 0.1 Crossover = 0.6 Number of features = 8	Mutation = 0.1 Crossover = 0.8 Number of features = 10
Adaboost	Mutation = 0.1 Crossover = 0.6 Number of features = 7	Mutation = 0.2 Crossover = 0.6 Number of features = 8

The results of the best performance of the algorithms for the different mutation and crossover rates are given in Figs. 6 and 8. While the results for the worst performance are given in Figs. 7 and 9. The graphs show that RF algorithm performs the best for both datasets. This performance is followed by KNN and NB algorithms for heart disease and breast cancer datasets. Adaboost and NB algorithms perform the same on the heart disease dataset and they are the worst-performing algorithms among them. While in the breast cancer dataset, the worst-performing algorithm is KNN.

Figures 10, 11, 12, 13 and 14 compare the performances of each machine learning algorithm without and with feature selection. The results for the feature selection includes both the best and worst performances. Generally, as observed in the figures, the results of the best performance when feature selection is done are better than those without feature selection for both datasets. Also, the results

Table 3. Mutation and Crossover Rates with the Number of Selected Features (Breast Cancer Dataset)

Algorithm	Best Performance	Worst Performance
DT	Mutation = 0.2 Crossover = 0.4 Number of features = 9	Mutation = 0.05 Crossover = 0.8 Number of features = 8
KNN	Mutation = 0.1 Crossover = 0.8 Number of features = 4	Mutation = 0.2 Crossover = 0.4 Number of features = 6
RF	Mutation = 0.2 Crossover = 0.6 Number of features = 8	Mutation = 0.2 Crossover = 0.4 Number of features = 12
NB	Mutation = 0.1 Crossover = 0.4 Number of features = 9	Mutation = 0.2 Crossover = 0.4 Number of features = 12
Adaboost	Mutation = 0.05 Crossover = 0.6 Number of features = 17	Mutation = 0.05 Crossover = 0.8 Number of features = 17

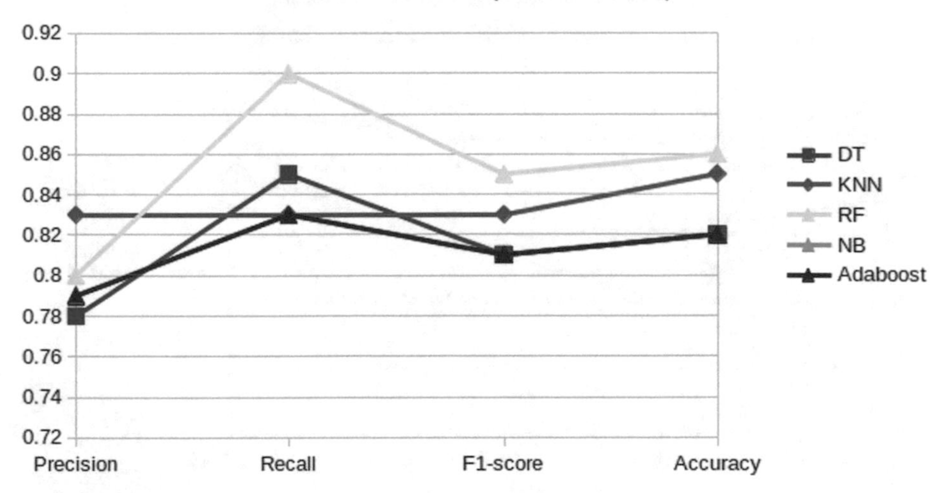

Fig. 6. The Comparison of the Different Evaluation Metrics for the Best Performance (Heart Disease)

without feature selection are better than those for the worst performance when feature selection is performed.

However, there are few exceptions to this as seen in Fig. 10 – 14. For example, in the result of the DT algorithm (Fig. 10), the precision scores for the best performance and the one without feature selection are the same on the heart

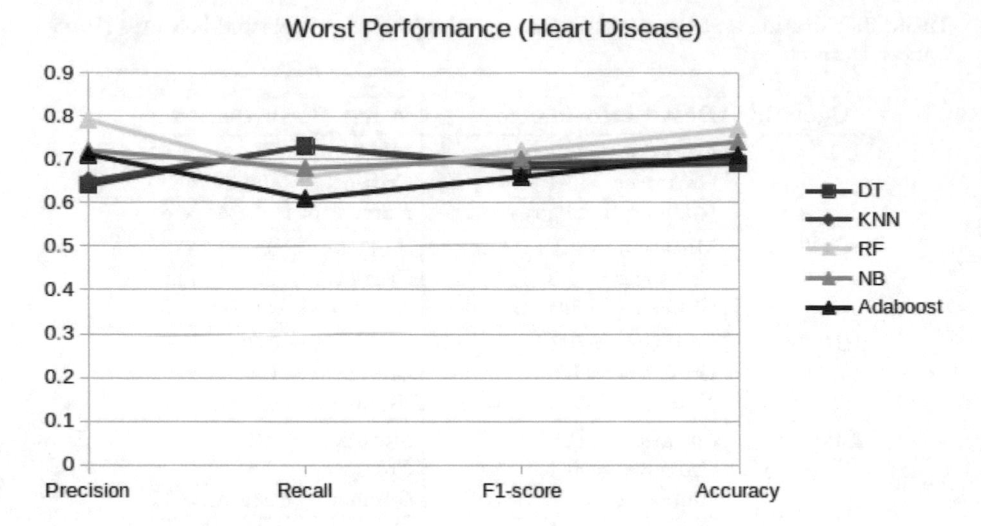

Fig. 7. The Comparison of the Different Evaluation Metrics for the Worst Performance (Heart Disease)

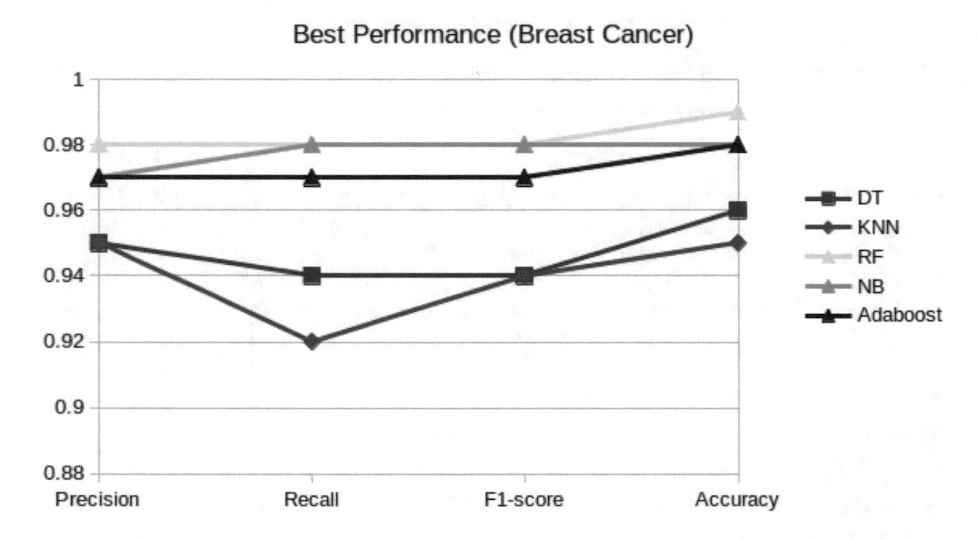

Fig. 8. The Comparison of the Different Evaluation Metrics for the Best Performance (Breast Cancer Disease)

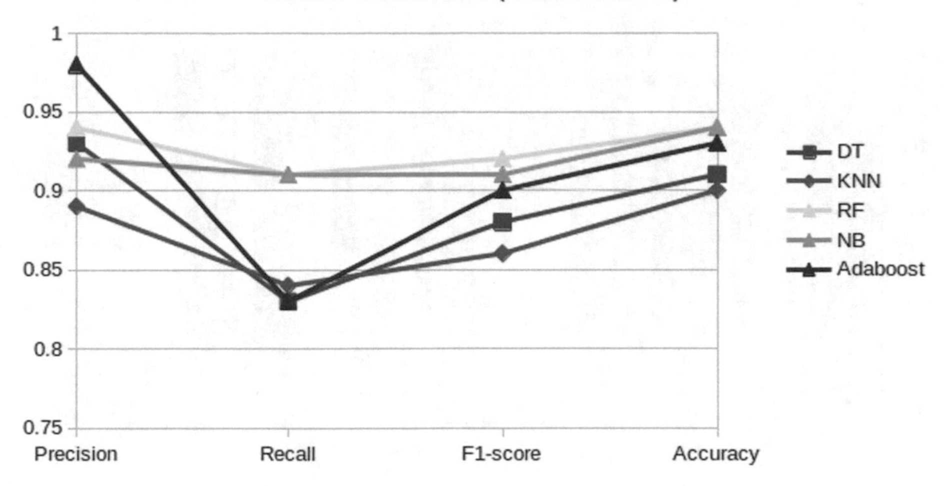

Fig. 9. The Comparison of the Different Evaluation Metrics for the Best Performance (Breast Cancer Disease)

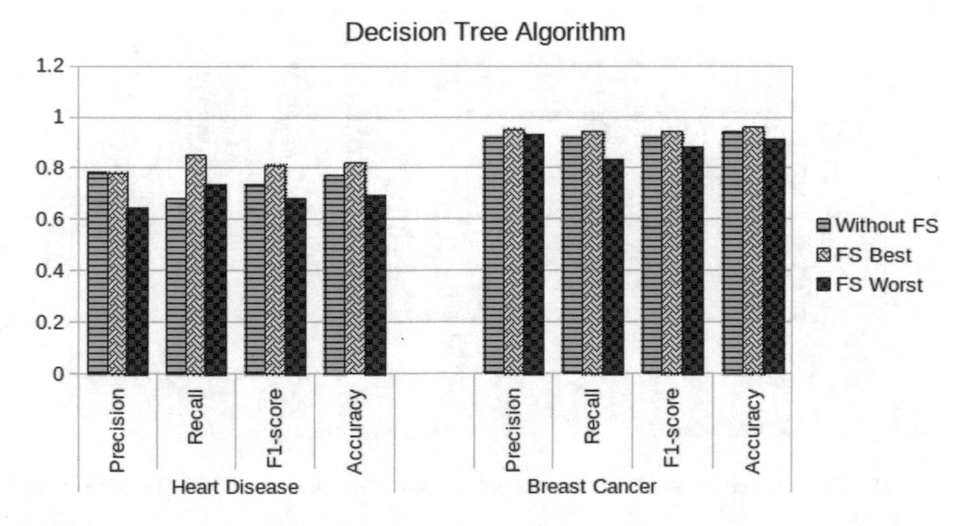

Fig. 10. Comparison of the Performance of Feature Selection Versus Without Feature Selection

disease dataset. The recall score of the worst performance is higher than the one without feature selection on the heart disease dataset. While the precision score of the worst performance is slightly higher than the one without feature selection on the breast cancer dataset.

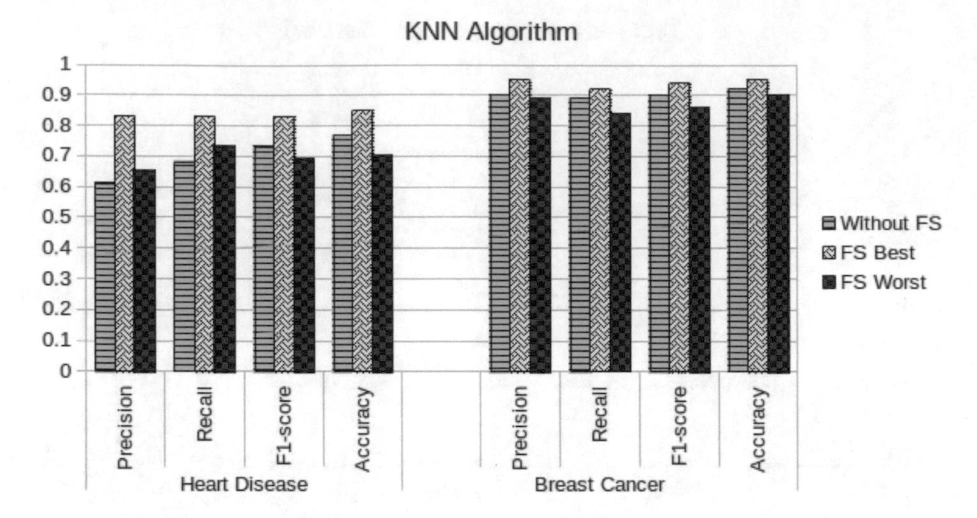

Fig. 11. Comparison of the Performance of Feature Selection Versus Without Feature Selection

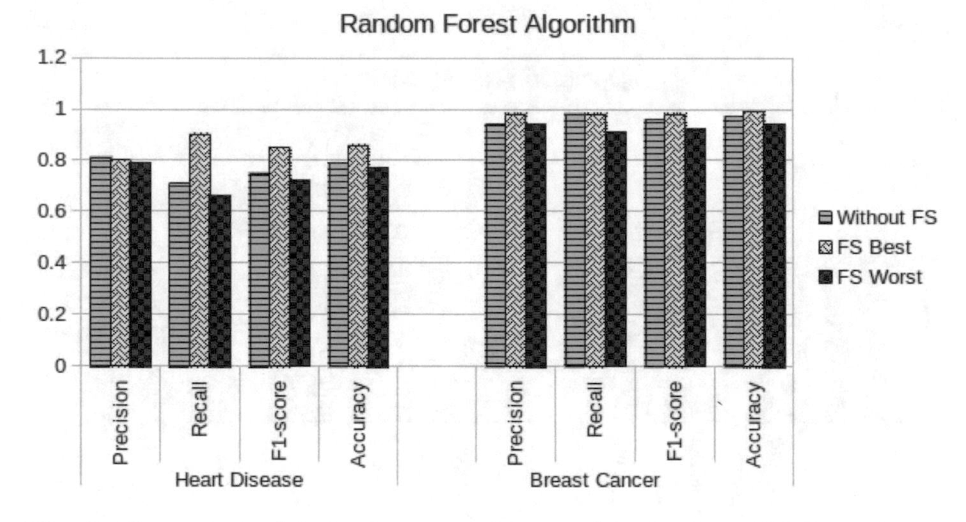

Fig. 12. Comparison of the Performance of Feature Selection Versus Without Feature Selection

In the result for the KNN algorithm (Fig. 11), these exceptions can be seen in the precision and recall scores of the worst performance being higher than those without feature selection on the heart disease dataset. In the result for RF algorithm (Fig. 12), the precision score without feature selection is slightly higher than the best performance on the heart disease dataset. And the recall score of the best performance and the one without feature selection are the same on the breast cancer dataset.

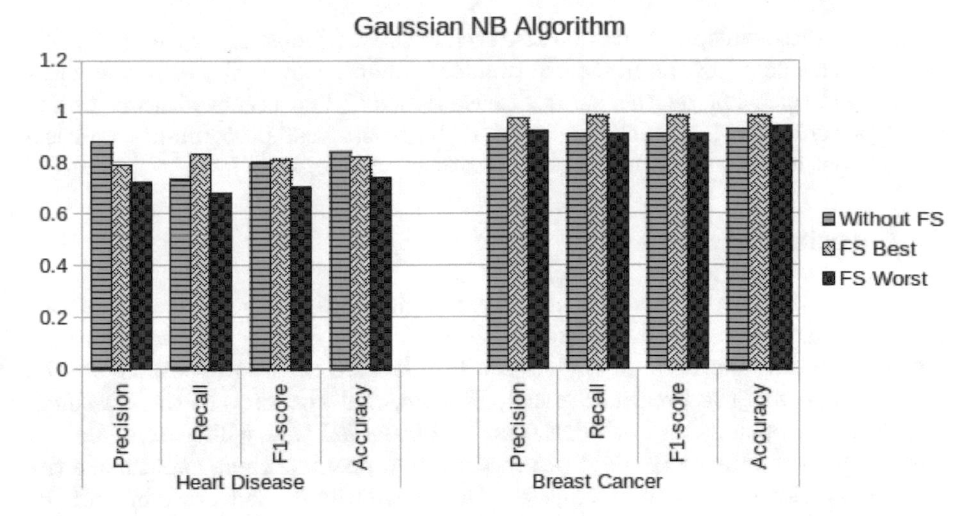

Fig. 13. Comparison of the Performance of Feature Selection Versus Without Feature Selection

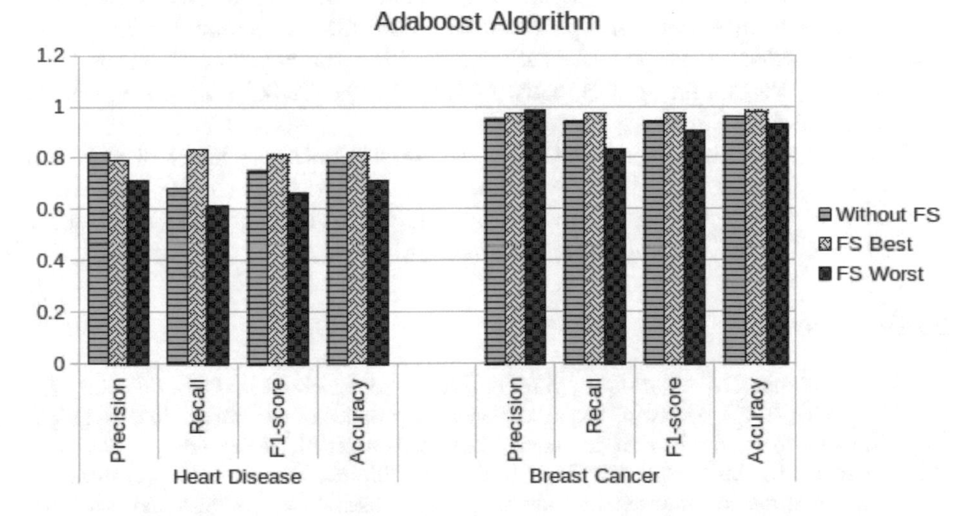

Fig. 14. Comparison of the Performance of Feature Selection Versus Without Feature Selection

In the result for NB algorithm (Fig. 13), the precision score without feature selection is apparently higher than that of the best performance on the heart disease dataset. While the accuracy score without feature selection is slightly higher than the accuracy of the best performance still on the heart disease dataset. The precision and accuracy scores of the worst performance are slightly higher than those without feature selection in the breast cancer dataset. The recall and f1-score scores of the worst performance are the same with the ones without feature selection on the same dataset.

Last of these exceptions can be also seen in the Adaboost algorithm (Fig. 14). The precision score of the best performance is slightly lower than the one without feature selection on the heart disease dataset. The precision score of the worst performance is slightly higher than both the best performance and the one without feature selection on the breast cancer dataset.

7 Conclusion

Summarily, this paper tried to improve the classification performance of two healthcare datasets - heart disease and breast cancer - using genetic algorithm feature selection method. Applying the GA feature selection method on the dataset improved the precision, recall, f1-score, and accuracy of the classifiers with a few exceptions on both datasets. It is observed that while some selected features from the datasets could produce optimal results, others could make the classifiers perform lower than expected. Hence, we would like to recommend the use of suitable algorithm for the selection of relevant features to the dataset at hand. It is important to use several feature selection methods to see which one works best for your problem. Specifically, since GA uses a natural heuristic search for its feature selection approach, we would like to recommend the use of it for feature selection steps. GA is efficient and has the capacity to escape local minima while searching an expansive feature spaces [22,30] unlike some other search optimization methods.

The best algorithm, from this study, to classify both the heart disease and breast cancer datasets, with and without feature selection, is the RF algorithm. In future work, we will try to do more parameter tuning for the GA used for feature selection and for the classification algorithms.

References

1. Abu Khurma, R., Aljarah, I., Sharieh, A., Abd Elaziz, M., Damaševičius, R., Krilavičius, T.: A review of the modification strategies of the nature inspired algorithms for feature selection problem. Mathematics **10**(3), 464 (2022)
2. Alhenawi, E., Al-Sayyed, R., Hudaib, A., Mirjalili, S.: Feature selection methods on gene expression microarray data for cancer classification: a systematic review. Comput. Biol. Med. **140**, 105,051 (2022)
3. Ali, N., Neagu, D., Trundle, P.: Evaluation of k-nearest neighbour classifier performance for heterogeneous data sets. SN Appl. Sci. **1**(12), 1–15 (2019). https://doi.org/10.1007/s42452-019-1356-9
4. Batra, M., Agrawal, R.: Comparative analysis of decision tree algorithms. In: Panigrahi, B.K., Hoda, M.N., Sharma, V., Goel, S. (eds.) Nature Inspired Comput. AISC, vol. 652, pp. 31–36. Springer, Singapore (2018). https://doi.org/10.1007/978-981-10-6747-1_4
5. Bolón-Canedo, V., Sánchez-Maroño, N., Alonso-Betanzos, A.: A review of feature selection methods on synthetic data. Knowl. Inf. Syst. **34**(3), 483–519 (2013)
6. Bommert, A., Welchowski, T., Schmid, M., Rahnenführer, J.: Benchmark of filter methods for feature selection in high-dimensional gene expression survival data. Briefings Bioinform. **23**(1), bbab354 (2022)

7. Calzolari, M.: manuel-calzolari/sklearn-genetic:sklearn-genetic 0.3.0 (2020). https://doi.org/10.5281/zenodo.4081754. https://doi.org/10.5281/zenodo.4081754
8. Chandrashekar, G., Sahin, F.: A survey on feature selection methods. Comput. Electr. Eng. **40**(1), 16–28 (2014)
9. Charbuty, B., Abdulazeez, A.: Classification based on decision tree algorithm for machine learning. J. Appl. Sci. Technol. Trends **2**(01), 20–28 (2021)
10. Chen, S., Webb, G.I., Liu, L., Ma, X.: A novel selective naïve bayes algorithm. Knowl.-Based Syst. **192**, 105,361 (2020)
11. Cutler, A., Cutler, D.R., Stevens, J.R.: Random forests. In: Ensemble machine learning, pp. 157–175. Springer (2012). https://doi.org/10.1007/978-1-4419-9326-7_5
12. Fund, W.C.R.: Breast cancer statistics. https://www.wcrf.org/dietandcancer/breast-cancer-statistics/ (2018)
13. Ganesan, K., Acharya, U.R., Chua, C.K., Min, L.C., Abraham, K.T., Ng, K.H.: Computer-aided breast cancer detection using mammograms: a review. IEEE Rev. Biomed. Eng. **6**, 77–98 (2012)
14. García-Martínez, C., Rodriguez, F.J., Lozano, M.: Genetic algorithms. In: Handbook of heuristics, pp. 431–464. Springer (2018)
15. Haldurai, L., Madhubala, T., Rajalakshmi, R.: A study on genetic algorithm and its applications. Int. J. Comput. Sci. Eng. **4**(10), 139 (2016)
16. Halim, Z., et al.: An effective genetic algorithm-based feature selection method for intrusion detection systems. Comput. Security **110**, 102,448 (2021)
17. Joel, L.O., Doorsamy, W., Paul, B.S.: Artificial intelligence and machine learning for health risks prediction. In: Marques, G., Kumar Bhoi, A., de la Torre Díez, I., Garcia-Zapirain, B. (eds.) Enhanced Telemedicine and e-Health. SFSC, vol. 410, pp. 243–265. Springer, Cham (2021). https://doi.org/10.1007/978-3-030-70111-6_12
18. Joel, L.O., Doorsamy, W., Paul, B.S.: A review of missing data handling techniques for machine learning. Int. J. Innov. Technol. Interdisc. Sci. **5**(3), 971–1005 (2022)
19. Kaggle: Breast cancer wisconsin (diagnostic) data set. https://www.kaggle.com/uciml/breast-cancer-wisconsin-data (2016)
20. Kaggle: Heart disease uci. https://www.kaggle.com/ronitf/heart-disease-uci (2018)
21. Katoch, S., Chauhan, S.S., Kumar, V.: A review on genetic algorithm: past, present, and future. Multimedia Tools and Applications, pp. 1–36 (2020)
22. Krömer, P., Platoš, J., Nowaková, J., Snášel, V.: Optimal column subset selection for image classification by genetic algorithms. Ann. Oper. Res. **265**(2), 205–222 (2016). https://doi.org/10.1007/s10479-016-2331-0
23. Kumar, S., Jain, S., Sharma, H.: Genetic algorithms. In: Advances in Swarm Intelligence for Optimizing Problems in Computer Science, pp. 27–52. Chapman and Hall/CRC (2018)
24. Louppe, G.: Understanding random forests: From theory to practice. arXiv preprint arXiv:1407.7502 (2014)
25. Murphy, S.L., Xu, J., Kochanek, K.D., Arias, E.: Mortality in the united states, 2017 (2018)
26. Nadimi-Shahraki, M.H., Zamani, H., Mirjalili, S.: Enhanced whale optimization algorithm for medical feature selection: a covid-19 case study. Comput. Biol. Med. **148**, 105,858 (2022)
27. Onesmus, M.: Introduction to random forest in machine learning. (2020) https://www.section.io/engineering-education/introduction-to-random-forest-in-machine-learning/

28. Organization, W.H.: Breast cancer (2021). https://www.who.int/news-room/fact-sheets/detail/breast-cancer
29. Sarker, I.H.: Machine learning: Algorithms, real-world applications and research directions. SN Comput. Sci. **2**(3), 1–21 (2021)
30. Schulte, R., Prinsen, E., Hermens, H., Buurke, J.: Genetic algorithm for feature selection in lower limb pattern recognition. Front. Robot. AI **8**, 710,806 (2021). https://doi.org/10.3389/frobt.2021.710806
31. Scikit-learn: An introduction to machine learning with scikit-learn (2017-2021). https://scikit-learn.org/stable/tutorial/basic/tutorial.html
32. Sharma, H., Kumar, S.: A survey on decision tree algorithms of classification in data mining. Int. J. Sci. Res. (IJSR) **5**(4), 2094–2097 (2016)
33. Song, J., Lu, X., Wu, X.: An improved adaboost algorithm for unbalanced classification data. In: 2009 Sixth International Conference on Fuzzy Systems and Knowledge Discovery, vol. 1, pp. 109–113. IEEE (2009)
34. Subbiah, S.S., Chinnappan, J.: Opportunities and challenges of feature selection methods for high dimensional data: A review. Ingénierie des Systèmes d'Information **26**(1) (2021)
35. Sun, B., Chen, H.: A survey of nearest neighbor algorithms for solving the class imbalanced problem. Wireless Commun. Mobile Comput. **5520990**, 12 (2021)
36. Sun, B., Chen, S., Wang, J., Chen, H.: A robust multi-class adaboost algorithm for mislabeled noisy data. Knowl.-Based Syst. **102**, 87–102 (2016)
37. Tang, J., Alelyani, S., Liu, H.: Feature selection for classification: A review. Data classification: Algorithms and applications p. 37 (2014)
38. Vembandasamy, K., Sasipriya, R., Deepa, E.: Heart diseases detection using naive bayes algorithm. Int. J. Innov. Sci., Eng. Technol. **2**(9), 441–444 (2015)
39. Wang, R.: Adaboost for feature selection, classification and its relation with svm, a review. Phys. Procedia **25**, 800–807 (2012)
40. Wikipedia contributors: Naive bayes classifier (2021). https://en.wikipedia.org/w/index.php?title=Naive_Bayes_classifier&oldid=1053686606
41. Yildirim, P.: Filter based feature selection methods for prediction of risks in hepatitis disease. Int. J. Mach. Learn. Comput. **5**(4), 258 (2015)
42. Ying, C., Qi-Guang, M., Jia-Chen, L., Lin, G.: Advance and prospects of adaboost algorithm. Acta Automatica Sinica **39**(6), 745–758 (2013)
43. Zhang, Y., et al.: Research and application of adaboost algorithm based on svm. In: 2019 IEEE 8th Joint International Information Technology and Artificial Intelligence Conference (ITAIC), pp. 662–666. IEEE (2019)
44. Zhang, Z.: Introduction to machine learning: k-nearest neighbors. Ann. Trans. Med. **4**(11) (2016)

BSS in Underdetermined Applications Using Modified Sparse Component Analysis

Anil Kumar Vaghmare(✉)

Department of ECE, Chandigarh College of Engineering and Technology, Chandigarh 160019, India
anil2siddu@gmail.com
https://ccet.ac.in/ECE-faculty.php

Abstract. The number of sources in an underdetermined blind source separation (UBSS) is more than the observed mixed signals. The UBSS involves two stages. In the first stage, the mixing matrix is estimated, and in the second, the source separation is performd. Researchers have proposed a number of methods based on clustering, time-frequency, and sparse component analysis to address UBSS. The performance in source separation as well as the estimated mixing matrix coefficients utilising older methods differed from the actual mixing matrix. In this approach, source signals are recovered by using the series of least square problem which is known as modified sparse component analysis. The suggested method estimates the mixing matrix by considering the one dimensional subspace, associated with time-frequency points of mixtures combined with hierarchical clustering. The one dimensional subspace is considered only for active source in which more energy exists compared to other sources in the mixing matrix estimation process. The proposed method's performance is contrasted with that of approaches created using single source points. According on experimental findings, the suggested methodology performs better than other common approaches.

Keywords: Mixing matrix estimation · hierarchical clustering · sparse component analysis

1 Introduction

The practise of separating source signals from their mixtures without any prior knowledge of the sources or the mixing procedure is known as blind source separation (BSS). The BSS concept is first given by J. Herault and C. Jutten in the year 1986. Blind source separation (BSS) is essential in signal and image processing for reconstructing source signals from mixes without knowing the input source signals [1–3]. Numerous uses [4,5] have been found in a variety of fields, including passive sonar and antenna arrays for wireless communication [6,7], as well as speech processing, seismic signal processing, data communication, and biological signal processing. Apart from this BSS is also used in speech enhancement in which

H. Sharma et al. (Eds.): ICIVC 2022, PALO 17, pp. 375–387, 2023.
https://doi.org/10.1007/978-3-031-31164-2_31

noise is removed [8]. There are three different types of mixing processes for blind source separation; if there are less source signals than sensors, the mixing process is called overdetermined blind source separation (OBSS) [9–13]. In the determined conditions [14,15], there are an equal number of sources and sensors. It is an underdetermined blind source separation (UBSS) mixing process if there are more source signals than sensors. Many methods have been given [16–24] to solve the UBSS. In [24], UBSS is solved based on single source points which are identified with improving the sparsity of the signal based on dynamic data field clustering. Recently, when only one source is active, the time-frequency (TF) points are found by taking into account the sparse constraint in the source signals, and these identified points are referred to as a single source point (SSP) [22]. The identified SSPs are clustered and mixing matrix is estimated. The performance of the methods given in [16,20,25–27] depends upon the directions of SSPs in clustering algorithm. In these methods, mixing matrix is estimated depending upon the ratio between different signals of mixtures at TF point. Due to the sensitivity of these ratios to noise, performance suffers. To reconstruct the original source signals based on the ratios of the time-frequency distributions of the data, a degenerate unmixing estimation technique (DUET) was suggested in [16] using time frequency distributions (TFD). In [16], a solution for n disjoint orthogonal signals is put forth. This approach can only be used when the sources are orthogonally disjoint and performance is deteriorating from source overlapping. Even though the methods given in [16,25] can work for small overlapping among sources in time-frequency domain, but performance is degraded due to increase of TF points [27]. [28] Identifies the time-frequency (TF) points that each source's single source occupies. The identification procedure compares the normalised real and imaginary sections of the mixed signals' TF coefficient vectors to find the single source points. Use of the [29] method, which sequentially estimates the mixing matrix, is advised. By using the rank-1 detecting device, the difficulty of blind identification is reformulated as a restricted optimization issue. As a result, the task of identifying one mixing matrix column is reduced to optimization, for which a successful iterative approach is suggested. Then, using a generalised eigenvalue decomposition-based deflation technique, the other columns of the mixing matrix are identified.

According to the suggested strategy, the issue is resolved by employing an approach based on a sparse linear combination of each TF vector with other time-frequency mixing vector by applying ℓ_1-regularization. Here, sparse coding is used to identify few TF mixture vector from the subspace. The obtained sparse coding values identify the SSPs at each time-frequency points of mixture, which are located in different one dimensional subspace (1-D). To estimate the mixing matrix, these SSPs are grouped, and several least squares formulations are made. The sources are reconstructed by solving the least square problem. The time-frequency distributions [16,25] are taken into account in this work as the preliminary work for identifying an active source at any time from the time-frequency (TF) domain.

The proposed method is used for UBSS in this study and contrasted with the current approaches. This paper is organised as follows: Sect. 2 discusses current methods, Sect. 3 presents the modified sparse component analysis methodology, Sect. 4 examines the UBSS results, and Sect. 5 discusses the conclusion and future directions.

2 Existing Methods

This section deals with the existing methods which are used to separate the sources in UBSS condition. The linear statistical model for instantaneous mixing in a noisy environment can be represented as

$$\mathbf{x}(t) = \mathbf{A}\mathbf{s}(t) + \mathbf{N} \tag{1}$$

where $\mathbf{x}$, $\mathbf{A}$, $\mathbf{s}$ and $\mathbf{N}$ represents the p dimension vector of observations, the $p \times n$ mixing matrix, the n dimensional source vector and an additive noise respectively. The instantaneous mixing model is defined specifically for the underdetermined BSS problem in this study. When the propagation time between sources and sensors is ignored, the instantaneous mixing model is particularly useful. The term "UBSS" is used when the number of source signals exceeds the number of sensors. The block diagram for two sensors and five sources is shown in Fig. 1. The matrix form of representation for the 5 sources and 2 sensors is given in Eq. (2)

$$\begin{bmatrix} \mathbf{x}_1(t) \\ \mathbf{x}_2(t) \end{bmatrix} = \begin{bmatrix} a_{11} & a_{12} & a_{13} & a_{14} & a_{15} \\ a_{21} & a_{22} & a_{23} & a_{24} & a_{25} \end{bmatrix} \times \begin{bmatrix} \mathbf{s}_1(t) \\ \mathbf{s}_2(t) \\ \mathbf{s}_3(t) \\ \mathbf{s}_4(t) \\ \mathbf{s}_5(t) \end{bmatrix} \tag{2}$$

where,

$$\mathbf{A} = \begin{bmatrix} a_{11} & a_{12} & a_{13} & a_{14} & a_{15} \\ a_{21} & a_{22} & a_{23} & a_{24} & a_{25} \end{bmatrix};$$

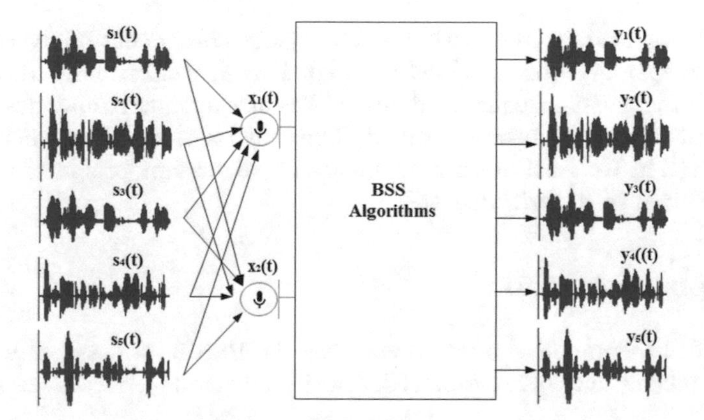

Fig. 1. Underdetermined BSS with 5 sources and 2 sensors

$$\mathbf{x}(t) = \begin{bmatrix} \mathbf{x}_1(t) \\ \mathbf{x}_2(t) \end{bmatrix}; \quad \mathbf{s}(t) = \begin{bmatrix} \mathbf{s}_1(t) \\ \mathbf{s}_2(t) \\ \mathbf{s}_3(t) \\ \mathbf{s}_4(t) \\ \mathbf{s}_5(t) \end{bmatrix}$$

2.1 Hierarchal Clustering

[22] provides the hierarchical clustering approach for estimating mixing matrices. By contrasting the actual and fictitious components of the visible mixtures, single source locations are initially located. The specific TF point is used as SSPs in this method. Technique 1 outlines a hierarchical clustering algorithm and its specific steps.

Algorithm 1. Hierarchical clustering

1: The given observation matrix $\mathbf{x}(t)$ is converted to domain of frequency by applying DFT, which is represented by $\mathbf{X}(t,\ k)$.

2: Check the following conditions
$\left| \frac{R\{\mathbf{X}(t,\ k)\}^T I\{\mathbf{X}(t,\ k)\}^T}{\|R\{\mathbf{X}(t,\ k)\}\| \|I\{\mathbf{X}(t,\ k)\}\|} \right| > cos(\Delta\theta)$
where R is a real part and I is the imaginary part and k is the kth frequency bin.

3: If the condition in step 2 is satisfied, then the single source point is taken as $\mathbf{X}(t,k)$ and mixing matrix is estimated. If the condition is not satisfied, discard the time frequency point.

4: rRepeat steps 2 and 3 for each point in the time-frequency (TF) plane until there are enough single source points for calculating the mixing matrix to be achieved.

2.2 Sparse Component Analysis

Blind source separation frequently employs sparse component analysis. In order to obtain a sparse representation, the signals that sensors have detected are converted into the time-frequency domain. The mixing matrix matches the single source points that have been identified. The signal $\mathbf{s}(t)$ is sparse when the each column of $\mathbf{s}(t)$ has at least one zero element. The algorithm for sparse component analysis is given in Algorithm 2 [18].

3 Proposed Method

The provided technique is used to conduct the BSS in two steps. In the first stage, the mixing matrix is estimated, and in the second, sources are separated

Algorithm 2. Mixing matrix estimation using sparse component analysis

1: The given observation $\mathbf{x}(t)$ is converted into the domain of frequency by applying DFT, which is represented by $\mathbf{X}(t,\ k)$.
2: All zero columns of $\mathbf{X}(t,\ k)$. are removed to obtain a sparse matrix of $\mathbf{X}(t,\ k)$. $\in R^{m\times n_1}$
3: The columns of $\mathbf{X}_i(t,k),\ i=1,2,3.....n_1$ are normalized and denote it as $\mathbf{z}_i = \dfrac{\mathbf{X}_i(t,\ k)}{\|\mathbf{X}_i(t,\ k)\|}$, and multiply each column by -1 if the 1st element of $\mathbf{z}_i$ is negative.
4: Do clustering for $\mathbf{z}_i,\ i=1,2,3.....n_1$ in $N-1$ groups $G_1, ...G_{N+1}$ such that for any $i=1,2,3.....N,\ \|\mathbf{X}(t,\ k)-z\|<\varepsilon,\ \forall\ \mathbf{X}(t,\ k),\ \mathbf{z}\in G_i$ and $\|\mathbf{X}(t,\ k)-z\|\geq \varepsilon,$ for any $\mathbf{X}(t,\ k),\ \mathbf{z}\in$ different groups.
5: Chose any $\mathbf{z}_i\in G_i$ and put $\mathbf{a}_i=\mathbf{z}_i$. The matrix $\mathbf{A}$ with columns $\{\mathbf{a}_i\}_{i-1}^{N}$ is an estimate of the mixing matrix, till the permutation and scaling.

3.1 Procedure for Mixing Matrix Estimation

The observed signals at sensors, mathematically represented as $\mathbf{x}(t) = \mathbf{A}\mathbf{s}(t)$. The measured signals at the sensors are converted into frequency domain using the short time Fourier transform (STFT), where they are mathematically expressed as

$$\mathbf{X}(t,\ k) = \mathbf{A}\mathbf{S}(t,\ k) \tag{3}$$

where, t is a time instant and k is a frequency bin. The following assumptions are taken into consideration for mixing matrix identification [30].

Assumption 1: Each source $\mathbf{S}_i$ has some time-frequency (TF) points (t,k) if $\mathbf{S}_i$ is active or dominant, that is, $|\hat{S}_i(t,\ k)| >> |\hat{S}_j(t,\ k)|,\ \forall\ j\neq i$.

Assumption 2: The columns of mixing matrix $\mathbf{A}$ should be linearly independent. Compared to the constraints given in [16,19], assumption 1 is more relaxed in this approach. In [16], the degenerate unmixing estimation technique (DUET) is put forward under the presumption that only one source is active at a given time-frequency point. However, the major source occurs in nearby windows according to the technique based on the time-frequency ratios of mixtures (TIFROM) [19]. In the proposed method as per assumption 1 the dominant source can occur at any instant of time-frequency point. Clearly, it indicates that the single source points (SSPs) are distributed arbitrarily in the time-frequency plane, it is called as single source point condition [31]. To recover the sources, assumption 1 is considered with probability one for any random mixing matrix $\mathbf{A}$.

Based on the assumption 1 and 2, we represent a time-frequency point $(\mu,\ \nu)$ where one source $\mathbf{S}_i$ is dominant and it can be written as

$$\mathbf{X}(\mu,\ \nu) = \hat{S}(\mu,\ \nu)\mathbf{a}_i \tag{4}$$

Equation (4) represents the product of a single source point $\hat{S}(\mu,\ \nu)$ and ith column vector $\mathbf{a}_i$ is equal to time-frequency point of observed vector $\mathbf{X}(\mu,\ \nu)$. Hence column vectors of mixing matrix is estimated at these single source points.

We can observe that same single dominant source lie in the one dimensional subspace (1-D) with respect to time-frequency mixture vector. It can be represented linearly by another time-frequency vector $\mathbf{X}(\psi, \omega)$ in the subspace as given in Eq. (5).

$$\mathbf{X}(\mu, \nu) = \alpha\mathbf{X}(\psi, \omega) \tag{5}$$

where α is a real number. The Eq. (5) is useful to transform the single source points to 1-D subspace in the time frequency mixture vectors. The modified sparse component analysis can identify these low dimensional subspace. In this each time-frequency mixture vector is coded such that it should be a linear combination of another few TF mixture vectors. From Eq. (5) it is clear that, the sparse coding of time-frequency mixture vector has only one nonzero element which lies in the one dimensional subspaces with probability equal to one (as per assumption 2). The sparse coding solution for the observed $\mathbf{X}(\mu, \nu)$ is explained as follows:

Let $\mathbf{X}_1(t, k)$, $\mathbf{X}_2(t, k)$, $\mathbf{X}_3(t, k)....\mathbf{X}_P(t, k)$ be the TF mixture vectors. Here, each time-frequency mixture vector can be coded as a linear combination of other time-frequency mixture vector given in (6).

$$\mathbf{X}_i(t, k) = \hat{\mathbf{X}}\mathbf{C}_i \;\; \text{s.t.} \;\; C_{i,i} = 0 \tag{6}$$

where, $\hat{\mathbf{X}}$ is a TF vector and $\mathbf{C}_i$ is the coding coefficient vector, which are defined in Eq. (7) and Eq. (8) respectively.

$$\hat{\mathbf{X}} \triangleq [\mathbf{X}_1(t, k), \; \mathbf{X}_2(t, k), \; \mathbf{X}_3(t, k)....\mathbf{X}_P(t, k)] \tag{7}$$

$$\mathbf{C}_i \triangleq [C_{i1}, \; C_{i2}, \; C_{i3}....C_{iP}] \tag{8}$$

The constraint $C_{i,i} = 0$ nullify the trivial solution of time-frequency mixture vector by itself. The solution of $\mathbf{C}_i$ can be obtained by sparse coding and nonzero entries of $\mathbf{C}_i$ represents the time-frequency mixture vectors from the same subspace as $\mathbf{X}_i(t, k)$. Actual sparse coding problem can be solved using the ℓ_0-norm minimization given in Eq. (9).

$$\min\|\mathbf{C}_i\|_0 \;\; \text{s.t.} \;\; \mathbf{X}_i(t, k) = \hat{\mathbf{X}}\mathbf{C}_i \; , \quad C_{i,i} = 0 \tag{9}$$

where $\|\mathbf{C}_i\|_0$ is the ℓ_0-norm which is equal to nonzero component in $\mathbf{C}_i$. It is a nondeterministic polynomial hard (NP-hard) problem [32] because generally it is a combinational optimization problem. Using compressed sensing theory, if the answer to $\mathbf{C}_i$ is sparse then Eq. (9) is equal to ℓ_1-norm of [32] and it can be represented as

$$\min\|\mathbf{C}_i\|_1 \;\; \text{s.t.} \;\; \mathbf{X}_i(t, k) = \hat{\mathbf{X}}\mathbf{C}_i \; , \quad C_{i,i} = 0 \tag{10}$$

where $\|\mathbf{C}_i\|_1$ represents ℓ_1-norm. The solution of Eq. (10) it is solvable using [32] in polynomial period. In real time applications the mixed signal observations are corrupted by noise. Hence, to improve the robustness the error term is introduced in the sparse coding as given in Eq. (11).

$$\mathbf{J}(\mathbf{C}_i \; ; \; \lambda) = \lambda\|\mathbf{C}_i\|_1 + \frac{1}{2}\|\mathbf{X}_i(t, k) - \hat{\mathbf{X}}\mathbf{C}_i\|_2^2 \;\; \text{s.t.,} \quad C_{i,i} = 0 \tag{11}$$

where, $\lambda > 0$ which balances the compromise between the sparsity and reconstruction error. It is a convex optimization problem and it is solved by ℓ_1-Homotopy [33]. This method recovers the solution with Q non zeros in $O(Q^3 + P)$ here P represents number of columns in $\hat{\mathbf{X}}$ [33].

Each time-frequency point's solution to sparse coding has one nonzero component, just as the single source points (SSPs). The obtained SSPs are clustered using the methods that are available [21,22,34,35]. Algorithm 3 provides the suggested approach for estimating the mixing matrix.

Algorithm 3. Mixing matrix estimation using proposed method

1: The given observation $\mathbf{x}(t)$ is converted into domain of frequency by applying STFT, which is represented by $\mathbf{X}(t,\ k)$. with n sources.
2: All the mixture vectors are normalized with unit norm.
3: Compute the sparse coding for each time-frequency mixture vector by minimizing Eq.(16) with ℓ_1-Homotopy [33]
4: Add all the nonzero components of the sparse coding coefficient of the clustering group of time-frequency mixture vector.
5: Use method of clustering on SSPs to group its components into n clusters.
6: Centre of these n clusters is calculated for the columns of the mixing matrix.
7: The mixing matrix is estimated.

3.2 Procedure for Source Separation

The calculated mixing matrix is employed for reconstruction of original sources as it is an irreversible problem [23]. The following definition 1 and assumption 3 are considered for recovering the sources.

Definition 1: Set $\mathbb{A}$ contains all the submatrices $m \times (m-1)$ of the matrix $\mathbf{A}$ i.e.,

$$\mathbb{A} = \{\mathbb{A}_i | \mathbb{A}_i = [\mathbf{a}_{\theta 1}, \mathbf{a}_{\theta 2}, ..., \mathbf{a}_{\theta m-1}]\} \tag{12}$$

$\mathbb{A}$ contains C_n^{m-1} components.

Assumption 3: Mostly $m-1$ sources are active among n sources at each time-frequency point.
The condition given in assumption 2 and 3 is easily satisfied in BSS problems when the sources are speech signals [27] and also supported by the following theorems.

Theorem 1: For any given time-frequency mixture vector $\mathbf{X}(t,\ k)$ there exist a submatrix $\mathbb{A}_* = [\mathbf{a}_{\phi 1}, \mathbf{a}_{\phi 2}, ..., \mathbf{a}_{\phi m-1}]$ in the set $\mathbb{A}$ with condition given in Eq. (13),

$$\mathbf{X}(t,\ k) = \mathbb{A}_* \mathbb{A}_*^{\dagger} \mathbf{X}(t,\ k) \tag{13}$$

here, $\mathbb{A}_*^\dagger$ is the pseudo inverse of $\mathbb{A}_*$. Using Theorem 1, the sources can be reconstructed with n-dimensional vector $\hat{\mathbf{S}}_j(t,\ k)$ for j^{th} component as

$$\hat{\mathbf{S}}_j(t,\ k) = \begin{cases} \mathbf{e}_i & \text{if } j = \phi_i \\ 0 & \text{otherwise} \end{cases} \tag{14}$$

where, $\mathbf{e} = [e_1, e_2, ..., e_{m-1}]^T = \mathbb{A}_*^\dagger \mathbf{X}(t,\ k)$. Based on Theorem 1 we can obtain the Theorem 2.

Theorem 2: The reconstructed vector $\hat{\mathbf{S}}_i(t,\ k)$ is equal to the original source vector $\mathbf{S}_i(t,\ k)$ with probability one.

Theorem 2 gives theoretical basics to estimate the time-frequency representation $\hat{\mathbf{S}}_i(t,\ k)$ of the original sources. In case if noise is present, it is not possible to find submatrix $\mathbb{A}_*$ satisfying Eq. (13). Hence $\mathbb{A}_*$ is obtained from following criterion:

$$\mathbb{A}_* = \underset{\mathbb{A}_i \subset \mathbb{A}}{\arg\min} \|\mathbf{X}(t,\ k) - \mathbb{A}_i \mathbb{A}_i^\dagger \mathbf{X}(t,\ k)\|_2 \tag{15}$$

The complete procedure for reconstruction of sources is given in Algorithm 4.

Algorithm 4. Source separation by proposed method

1: The given observation $\mathbf{x}(t)$ is converted into the frequency domain by applying STFT, which is represented by $\mathbf{X}(t,\ k)$. with n sources.
2: Find the corresponding sub-matrix $\mathbb{A}_*$ for each time-frequency vector using Eq.(15).
3: Find the estimation of source with time-frequency representation using Eq.(14).
4: Add all the nonzero components of the sparse coding coefficient of the clustering group of time-frequency mixture vector.
5: Use clustering method on SSPs to group its elements into n clusters.
6: Convert the estimated source of frequency domain to time domain by applying inverse STFT.
7: Finally, represent the recovered sources.

4 Results and Discussion

In this section the results are carried out for underdetermined BSS (5 sources and 2 mixtures) for speech samples taken from NOIZEUS [36] database and real environmental data which is recorded in the laboratory environment. The block diagram of considered environment is given in Fig. 1. The normalized mean square error (NMSE) [22] and performance index [30] are used to measure the effectiveness of the suggested technique for estimating the mixing matrix. The NMSE values are obtained using Eq. (16).

$$\text{NMSE} = 10\ \log_{10}\left(\frac{\sum_{i,j}(\hat{\mathbf{a}}_{ij} - \mathbf{a}_{ij})^2}{\sum_{i,j}\mathbf{a}_{ij}^2}\right)(dB) \tag{16}$$

where $\hat{\mathbf{a}}_{ij}$ is the i, j^{th} component of calculated mixing matrix $\hat{\mathbf{A}}$ and a_{ij} is the i, j^{th} component of mixing matrix $\mathbf{A}$. The performance index value is calculated by using Eq. (17).

$$\text{Perfomance index} = \frac{1}{n}\sum_{j=1}^{n}\left(1 - \frac{\mathbf{a}_j^T\hat{\mathbf{a}}_j}{\|\mathbf{a}_j\|\|\hat{\mathbf{a}}_j\|}\right) \tag{17}$$

where $\hat{\mathbf{a}}_j$ is the estimated value of the steering vector $\mathbf{a}_j$ of jth element of mixing matrix and n represents the number of sources. The performance of the method for mixing matrix estimation is better while it has lower performance index value. Source to interference ratio (SIR), source to distortion ratio (SDR), and source to artifact ratio are used to test the source separation performance [14,15].

(i) *Source to interference ratio* (SIR): It measures the quality of separation, the higher the value of SIR better will be the separation. It is defined as follows.

$$\text{SIR} = 10\ \log_{10}\frac{\|\mathbf{s}_{\text{target}}\|^2}{\|\mathbf{e}_{\text{interference}}\|^2}(dB) \tag{18}$$

(ii) *Source to distortion ratio* (SDR) : It is a valid global performance measure given in Eq. (19) and its unit is the decibel (dB). The higher value of the SDR is needed for better performance in BSS.

$$\text{SDR} = 10\ \log_{10}\frac{\|\mathbf{s}_{\text{target}}\|^2}{\|\mathbf{e}_{\text{interference}} + \mathbf{e}_{\text{noise}} + \mathbf{e}_{\text{artifact}}\|^2} \tag{19}$$

The performance of the proposed method is compared with hierarchical clustering [22] and sparse component analysis [18] which are the state of the art methods for underdetermined BSS.

4.1 Performance Analysis for Underdetermined BSS

The five signals are mixed with two sensors by instantaneous mixing with the following mixing matrix $\mathbf{A}$.

$$\mathbf{A} = \begin{bmatrix} 0.258 & 0.707 & 0.965 & 0.965 & 0.707 \\ -0.965 & -0.707 & -0.258 & 0.258 & 0.707 \end{bmatrix} \tag{20}$$

The estimated mixing matrices using various algorithms as follows:
1. Using hierarchical clustering

$$\hat{\mathbf{A}}_{1s} = \begin{bmatrix} 1.123 & 3.668 & 4.576 & 5.152 & 4.745 \\ -4.010 & -3.710 & 1.304 & -1.527 & 4.833 \end{bmatrix}$$

2. Using sparse component analysis

$$\hat{\mathbf{A}}_{2s} = \begin{bmatrix} -0.6211 & 0.659 & -0.317 & -0.691 & 0.599 \\ 0.226 & -0.111 & -0.613 & -0.202 & -0.260 \end{bmatrix}$$

3. Using proposed method

$$\hat{\mathbf{A}}_{3s} = \begin{bmatrix} 0.308 & 0.926 & 0.967 & 0.741 & 0.203 \\ -0.951 & -0.376 & 0.253 & 0.671 & 0.979 \end{bmatrix}$$

The performance index values obtained for various methods are tabulated in Table 1. From the obtained performance index values it is observed that the proposed method giving lower index value. The obtained NMSE values for various methods are illustrated in Fig. 2. Estimated mixing matrix element values are approximately near to considered mixing matrix element values with proposed method and it is reflected in the Fig. 2. The obtained SIR and SDR values are tabulated in Table 2.

Table 1. The performance index values for various methods

Method	Error
Hierarchical clustering	0.5
Sparse component analysis	0.40
Proposed method	**0.09**

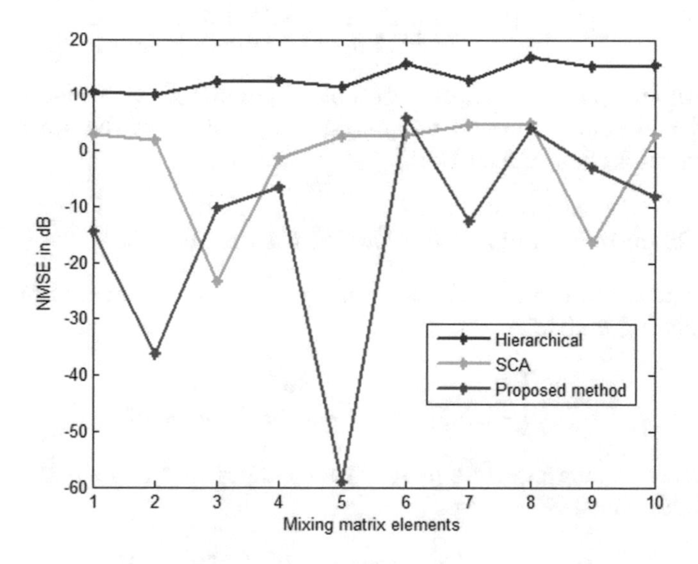

Fig. 2. Normalized mean square error values for various methods

Table 2. Obtained SIR values using various algorithms

Separated Source	Hierarchical	SCA	Proposed method
Source signal 1	–1.08	**10.65**	8.98
Source signal 2	0.57	–7.22	**5.46**
Source signal 3	–0.51	–5.02	**7.69**
Source signal 4	–6.78	4.48	**20.10**
Source signal 5	**–1.95**	–2.37	–6.52

Table 3. Obtained SDR values using various algorithms

Separated Source	Hierarchical	SCA	Proposed method
Source signal 1	–1.08	3.71	**4.45**
Source signal 2	0.57	–8.31	**2.66**
Source signal 3	–0.51	–5.08	**2.09**
Source signal 4	–6.78	3.34	**10.95**
Source signal 5	**–1.95**	-2.97	–12.13

From the Table 2 and Table 3, it is clear that the proposed method is giving higher values of SIR and SDR for more number of sources in the separation process. From the obtained SDR values, the proposed method gives less distortion in separation of sources in UBSS. The higher SIR values show less interference from other sources while reconstruction of source signals. From the Table 2 and Table 3, the proposed method effectively eliminates the interference from other sources.

5 Conclusion

This work is focused on UBSS in audio processing applications. The underdetermined BSS problem is solved using modified sparse component analysis. The degraded performance is found with existing methods which are developed based on SSPs in UBSS. In the proposed method, by exploiting the one dimensional subspaces (1-D) which are associated with time-frequency points in which only one source is active or dominant source. The identified dominant sources are grouped in the subspaces using clustering algorithm and mixing matrix is estimated. The sources are reconstructed using an estimated mixing matrix by solving the least square problems. The performance of the mixing matrix estimation is analyzed in terms of performance index and NMSE values. The source separation performance is analyzed by SIR and SDR values. From the simulated results It is clear that the proposed technique performs better when the BSS is underdetermined. The performance of future studies can be enhanced by taking into account the source signal's direction of arrival.

References

1. Comon, P., Jutten, C.: Handbook of Blind Source Separation: Independent component analysis and applications. Academic press (2010)
2. Yu, X., Hu, D., Xu, J.: Blind source separation: theory and applications. John Wiley & Sons (2013)
3. Belouchrani, A., Abed-Meraim, K., Cardoso, J.-F., Moulines, E.: A blind source separation technique using second-order statistics. IEEE Trans. Signal Process. **45**(2), 434–444 (1997)
4. Malik, H., Iqbal, A., Joshi, P., Agrawal, S., Bakhsh, F.I. (eds.): SCI, vol. 916. Springer, Singapore (2021). https://doi.org/10.1007/978-981-15-7571-6
5. Naik, G.R., Wang, W. (eds.): SCT, Springer, Heidelberg (2014). https://doi.org/10.1007/978-3-642-55016-4
6. Mansour, A., AI-Falou,A.: Performance indices for real-world applications. In: European Signal Processing Conference, pp. 1-5 (2006)
7. Comon, P., Jutten, C., Herault, J.: Blind Separation of Sources, Part II: Problem Statement. Signal Process. **24**(1), 11–20 (1991)
8. Anil Kumar, V., Rama Rao, C.h.V.: Unsupervised noise removal technique based on constrained NMF. IET Signal Process. **11**(7), 788–795 (2017)
9. Wang, L., Reiss, J.D., Cavallaro, A.: Over-determined source separation and localization using distributed microphones. IEEE/ACM Trans. Audio Speech Lang. Process. **24**(9), 1569–1584 (2016)
10. Joho, M., Mathis, H., Lambert, R.H.: Overdetermined blind source separation: Using more sensors than source signals in a noisy mixture. In: Proceedings of ICA, pp. 81-86 (2000)
11. Souden, M., Affes, S., Benesty, J.: A new approach to blind separation of two sources with three sensors. In: Proceedings of IEEE Vehicular Technology Conference, pp. 1-5 (2006)
12. Duarte, L., Ando, R.A., Attux, R. Deville, Y., Jutten, C.: Separation of sparse signals in overdetermined linear-quadratic mixtures. In: Latent Variable Analysis and Signal Separation, pp. 239-246 (2012)
13. Osterwise, C., Grant, S.L.: On over-determined frequency domain BSS. IEEE/ACM Trans. Audio Speech. Lang. Process. **22**(5), 956–966 (2014)
14. Anil Kumar, V., Rama Rao, C.h.V., Anirban, D.: Performance analysis of blind source separation using canonical correlation. Circ. Syst. Signal Process. **37**(2), 658–673 (2018)
15. Anil Kumar, V., Rama Rao, C.h.V., Anirban, D.: Blind speech separation using canonical correlation and performance analysis. In: 2nd International Conference on Communication Systems, Computing and IT Applications (CSCITA), pp. 37-41 (2017)
16. Jourjine, A., Rickard, S., Yilmaz, O.: Blind separation of disjoint orthogonal signals: Demixing N sources from 2 mixtures. In: Proceedings of IEEE International Conference on Acoustics, Speech, and Signal Processing, vol. 5, pp. 2985-2988 (2000)
17. Bofill, P., Zibulevsky, M.: Underdetermined blind source separation using sparse representations'. Signal Process. **81**(11), 2353–2362 (2001)
18. Georgiev, P., Theis, F., Cichocki, A.: Sparse component analysis and blind source separation of underdetermined mixtures. IEEE Trans. Neural Netw. **16**(4), 992–996 (2005)

19. Abrard, F., Deville, Y.: A time-frequency blind signal separation method applicable to underdetermined mixtures of dependent sources. Signal Process. **85**(7), 1389–1403 (2005)
20. Li, Y., Amari, S.-I., Cichocki, A., Ho, D., W.C., Xie, S.: Underdetermined blind source separation based on sparse representation. IEEE Trans. Signal Process. **54**(2), 423-437 (2006)
21. Luo, Y., Wang, W., Chambers, J.A., Lambotharan, S., Proudler, I.: Exploitation of source nonstationarity in underdetermined blind source separation with advanced clustering techniques. IEEE Trans. Signal Process. **54**(6), 2198–2212 (2006)
22. Reju, V.G., Koh, S.N., Soon, I.Y.: An algorithm for mixing matrix estimation in instantaneous blind source separation. Signal Process. **89**(9), 1762–1773 (2009)
23. Peng, D., Xiang, Y.: Underdetermined blind separation of nonsparse sources using spatial time-frequency distributions. Digit. Signal Process. **20**(2), 581–596 (2010)
24. Guo, Q., Ruan, G., Nan, P.: Underdetermined mixing matrix estimation algorithm based on single source points. Circ. Syst. Signal Process. 1–15 (2017)
25. Linh Trung, N., Belouchrani, A., Abed-Meraim, K., Boashash, B.: Separating more sources than sensors using time-frequency distributions. EURASIP J. Appl. Signal Process. 2828–2847 (2005)
26. Belouchrani, A., Amin, M.G.: Blind source separation based on time-frequency signal representations. IEEE Trans. Signal Process. **46**(11), 2888–2897 (1998)
27. Aissa-El-Bey, A., Linh-Trung, N., Abed-Meraim, K., Belouchrani, A., Grenier, Y.: Underdetermined blind separation of nondisjoint sources in the time-frequency domain. IEEE Trans. Signal Process. **55**(3), 897–907 (2007)
28. Dong, T., Lei, Y., Yang, J.: An algorithm for underdetermined mixing matrix estimation. Neurocomputing **104**, 26–34 (2013)
29. Zhang, M., Yu, S., Wei, G.: Sequential blind identification of underdetermined mixtures using a novel deflation scheme. IEEE Trans. Neural Netw. Learn. Syst. **24**(9), 1503–1509 (2013)
30. Zhen, L., Peng, D., Yi, Z., Xiang, Y., Chen, P.: Underdetermined blind source separation using sparse coding. IEEE Transac. Neural Netw. Learn. Syst. **28**(12), 3102–3108 (2017)
31. Fadaili, E.M., Moreau, N.T., Moreau, E.: Nonorthogonal joint diagonalization/zero diagonalization for source separation based on time-frequency distributions. IEEE Trans. Signal Process. **55**(5), 1673–1687 (2007)
32. Donoho, D.L.: For most large underdetermined systems of linear equations the minimal ℓ_1-norm solution is also the sparsest solution. Commun. Pure Appl. Math. **59**(6), 797–829 (2006)
33. Asif, M.S., Romberg, J.: Sparse recovery of streaming signals using ℓ_1-homotopy. IEEE Trans. Signal Process. **62**(6), 4209–4223 (2014)
34. Peng, X., Yi, Z., Tang, H.: Robust subspace clustering via thresholding ridge regression. In: AAAI Conference on Artificial Intelligence (AAAI), pp. 3827–3833 (2015)
35. Peng, X., Tang, H., Zhang, L., Yi, Z., Xiao, S.: A unified framework for representation-based subspace clustering of out-of-sample and large-scale data. IEEE Trans. Neural Netw. Learn. Syst. **27**(12), 2499–2512 (2016)
36. https://ecs.utdallas.edu/loizou/speech/noizeus/
37. Vincent, E., Gribonval, R., Févotte, C.: Performance measurement in blind audio source separation. IEEE Trans. Audio Speech Lang. Process. **14**(4), 1462–1469 (2006)

Optimal Reservoir Operation Policy for the Multiple Reservoir System Under Irrigation Planning Using TLBO Algorithm

Bhavana Karashan Ajudiya[1(✉)] and Sanjay Madhusudan Yadav[2]

[1] Civil Engineering Department, Marwadi University, Rajkot, Gujarat, India
bhavana.ajudiya@gmail.com

[2] Civil Engineering Department, SVNIT, Surat, Gujarat, India
smy@ced.svnit.ac.in

Abstract. This paper presents a novel Teaching Learning Base Optimization (TLBO) model to derive optimal reservoir policies under irrigation planning for a multiple reservoir system in a semi-arid river basin. In the present study, the objective function of the model is maximizing net benefits from the command area subjected to constraints of land allocation, storage continuity, water allocation, evaporation, and overflow. The TLBO model is developed and applied to the multiple reservoir system namely, the Aji-2, Nyari-2, and Aji-3 reservoir systems in Gujarat, India. A 75% dependability level of inflow into the reservoir is considered in model development. The optimal multiple reservoir operation policies are derived based on results obtained from the proposed TLBO model. The optimal policy is validated with the results of the LP model. The result reveals that the TLBO model performed better. Optimal reservoir operation policy provides a guideline to multiple reservoir managers to decide on the amount of water released throughout the operation.

Keywords: TLBO · Irrigation Planning · LP · Multiple Reservoir System · Optimal Reservoir Operation Policy

1 Introduction

The reservoirs play a significant role in the advancement and growth of the nation. Major reservoirs are constructed to provide multiple services such as water supply, irrigation, flood control, hydropower, navigation, recreation, etc. In the arid and semi-arid regions, excess water in monsoon and its deficiency in the summer season pose imitative challenges to efficient reservoir operation. The aim of attention must focus on improving the operational effectiveness and efficiency of the existing reservoir system for maximizing the beneficial uses of large-scale water storage projects in a developing country stated by Labadie [17], Andrieu et al. [1], Ajudiya et al. [3]. The assessment of developed policies was carried out using irrigation indices proposed by Jothiprakash et al. [9]. In the context of reservoir management for irrigation planning, Haouari and Azaiez [8] state that a large number of models have been developed to identify optimal operating

H. Sharma et al. (Eds.): ICIVC 2022, PALO 17, pp. 388–400, 2023.
https://doi.org/10.1007/978-3-031-31164-2_32

policies for a given horizon, such as conventional optimization models proposed by Lee et al. [18], Jothiprakash et al. [10], and Ajudiya et al. [4] as well as metaheuristic algorithm models such as Genetic Algorithm developed by Nageshkumar et al. [21], Jothiprakash et al. [9], Parmar et al. [23]); and Fuzzy Algorithm models proposed by Kamodkar and Regular [11]. In addition to it, the optimal multi-purpose reservoir operation planning is derived using particle swarm optimization by Nageshkumar and Reddy [22]. The GA-NLP hybrid approach was adopted to find a global solution for multipurpose reservoir operation by K. Leela et al. [12, 13]. K. Leela et al. [14] state that the optimal reservoir operation policy helps a manager to decide on the operation of the multipurpose reservoir system.

Numerous literature is available on conventional optimization techniques of those Linear Programming proposed by Wurbs [29], Non-Linear Programming proposed by Carvallo et al. [6]; Stochastic Linear Programming proposed by Lee et al. [18]; Dynamic Programming proposed by Nagesh Kumar and Baliarsigh [20], etc. Linear programming techniques were widely employed in the optimization of reservoir operation problems in the past time. According to Kumar and Yadav [15], conventional techniques have their own drawback despite the several advantages in solving the optimization problem.

Several Metaheuristic algorithms were used for reservoir operation optimization. Bozorg-Haddad et al. [5] state that metaheuristic algorithms have their superior capacity over conventional optimization techniques in providing globally optimal solutions. Based on the consequences, other metaheuristic algorithms are developed from the individuality of biological and physical systems in nature. In addition, they have gained a reputation among methods that are used for solving numerous complex optimization problems occurring in the field of sciences and engineering.

However, the tuning of the parameters of these algorithms is a limitation that influences the performance of the optimization problem, according to Rao [27]. Therefore, it is essential to approach a new algorithm that does not require any tuning of algorithm-specific parameters. The TLBO algorithm requires only the common control parameters like population size and the number of iterations for its working by Rao et al. [24]. A novel TLBO model was developed to derive optimal reservoir operation by Kumar and Yadav [15]; and optimal cropping patterns by Varade and Patel [28] as well as Kumar and Yadav [16] for a single reservoir system. Moreover, the application of TLBO algorithms to derive optimal cropping patterns was observed for multiple reservoir systems by Ajudiya et al. [2]. However, the application of the TLBO algorithms to derive optimal reservoir operation policy under irrigation has the potential to become a novel contribution. The main objective of the present study is to derive optimal multiple reservoir operation policies under irrigation planning.

2 Study Area

The Aji-2 and Nyari-2 irrigation schemes are located in Rajkot and Paddhari taluka of Rajkot district respectively while the Aji-3 irrigation scheme is located in the Dhrol taluka of Jamnagar district. The Nyari-2 and Aji-3 irrigation schemes envisage irrigation facilities in the command area, situated in Rajkot and Jamnagar districts respectively. These schemes provide services to a drought-prone area of the state that experiences scarcity and recurring drought conditions. This study evaluates the possible benefits of irrigation in drought-prone areas by integrated operation of multiple reservoirs i.e. Aji-2, Nyari-2, and Aji-3. The Aji-2 reservoir has a gross storage capacity is 22.09 MCM, with 1.33 MCM as a dead storage capacity. It has a left bank main canal (LBMC) system providing irrigation facilities to a culturable command area of 2384 ha and its discharge capacity is 2.92 m^3/s. The Aji-3 reservoir has a gross storage capacity is 61.95 MCM, with 4.75 MCM as a dead storage capacity. It has a LBMC and a right bank main canal (RBMC) system, providing composite irrigation facilities to a cultivable command area of 6635 ha. Consequently, it has been named to be a composite bank main canal (CBMC) system. Its discharge capacity is 4.5 m^3/s of LBMC and 2.27 m^3/s of RBMC. The Nyari-2 reservoir has a gross storage capacity is 12.25 MCM, with 0.75 MCM as a dead storage capacity. It has a LBMC system providing irrigation facilities to a cultivable command area of 1695 ha. The LBMC discharge capacity is 3.06 m^3/s.

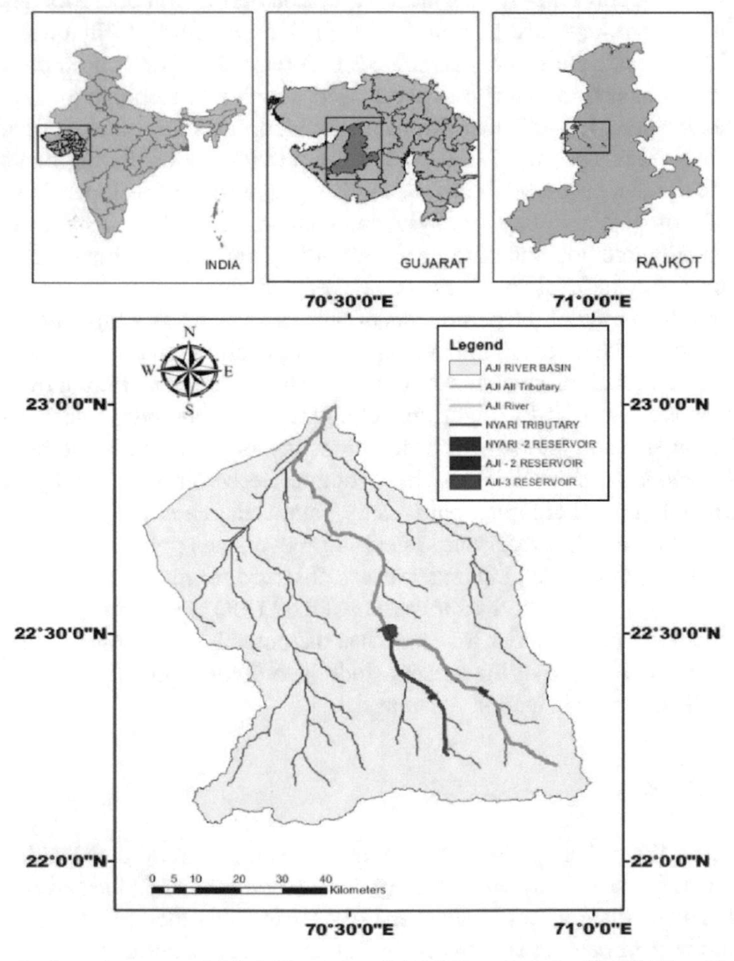

Fig. 1. Location map of Aji-2, Nyari-2 and Aji-3 Reservoir in the Aji River Basin.

There are eight, seven, and eighteen villages that fall under the command area of Aji-2, Nyari-2, and Aji-3 reservoirs respectively, which have a total population of around 89221 souls as per the census 2011 and are expected to increase the population to 105280 souls as of the year 2020. Additionally, data on storage and overflow of the all three reservoirs shows that the two reservoirs namely Aji-2 and Nyari-2 reservoir have overflowed every year. While, the Aji-3 reservoir has hardly overflowed. Suppose the Aji-2 and Nyari-2 reservoirs are operated in an integrated way with the Aji-3 reservoir. In that case, the surplus water of the Aji-2 reservoir and Nyari-2 reservoir can be utilized for irrigation purposes in the command area of the Aji-3 reservoir, and the agricultural yield can be increased. With this consideration and due to the non-availability of data on other reservoirs of the Aji River basin, the Aji-2 reservoir, Nyari-2 reservoir, and Aji-3 reservoir are only included in this integrated operation of the reservoir system, which is portrayed in Fig. 1. The necessary data of the all three reservoirs were collected from the Rajkot irrigation division for the years 2005–06 to 2016–17.

3 Methodology of TLBO Algorithm

The TLBO algorithm is a mimic T-L phenomenon seen in the class which was proposed by Rao et al. [25, 26]. The algorithm has distinct two phases of learning: one is the teacher phase and the other is the learner phase. The objective function f(y) is to be maximized or minimized. The population size, iteration Number, and elite size (zero) are decided. The initial solution is randomly generated as per Eq. (1). From the list of a solution, the best solution is identified as the maximum and minimum for the maximization and minimization functions, respectively.

$$\text{Randomly Generated Populations} = [\text{L} + \text{r}(\text{U} - \text{L})] \tag{1}$$

where; U = variable's Upper limit; L = Variable's lower limit; r = random number between [0, 1].

In the teaching phase, the teacher takes an effort to their knowledge to improve the results of students by the difference mean method and mathematically express as per below.

$$\text{Difference mean} = \text{r}(\text{Ybest} - \text{Ymean}) \tag{2}$$

where Y_{best} = the teacher's solution; Y_{mean} = the mean of all the student's solution; r = random number between [0, 1]. Now, a modified solution is found based on old solutions using Eq. (3)

$$\text{Ynew} = (\text{Yold} + \text{Difference mean}) \tag{3}$$

where Y_{old} = old solution; Y_{new} = modified solutions; the better solution of Y_{new} is accepted and replaced with the previous (Y_{old}).

In the learning phase, learners use knowledge taught by a teacher and interact each other's randomly to improve overall knowledge. Suppose, $Y_{old,\,i}$ and $Y_{old,j}$ as two random learners, where $Y_{old,\, i}$ interacts with $Y_{old,j}$. Then two possibilities will occur, which can be mathematically expressed as per Eq. (4) and Eq. (5).

$$\text{If } f(Yold, i) < f(Yold, j) : \text{Then } Ynew.i = Yold, i + r1(Yold, j - Yold, i) \quad (4)$$

$$\text{If } f(Yold, i) > f(Yold, j) : \text{Then } Ynew, i = Yold, i + r1(Yold, i - Yold, j) \quad (5)$$

According Eq. (4) & (5), find the $Y_{new.i.}$ The better functional value of $Y_{new.i,}$ was obtained then accepted and replaced with the previous Y_{new}. The first iteration was completed up to the previous steps. If the maximum number of iterations is accomplished than the algorithm stopped and displays the optimum solution. Else, it repeats itself from the teaching phase.

4 Methodology for Model Formulation

In the presented study, the objective function is to obtain the maximum net benefits (NBs) from the command area of a multiple reservoir system over the year, which is mathematically expressed as per Eq. (6).

$$f(Y) : \text{Maximize NBs} = \sum_j^1 (\sum_{i=1}^3 C_i ak_i + \sum_{i=1}^6 C_i ar_i + \sum_{i=1}^2 C_i at_i) + \sum_j^2 (\sum_{i=1}^2 C_i ak_i + \sum_{i=1}^6 C_i ar_i + \sum_{i=1}^2 C_i at_i) + \sum_j^3 (\sum_{i=1}^2 C_i ak_i + \sum_{i=1}^7 C_i ar_i + \sum_{i=1}^2 C_i at_i) \quad (6)$$

Table 1. The Coefficient of NBs for the Crops Sown in the Command Area of Aji-2, Aji-3, and Nyari-2 Reservoir

Sr. No	Crop	Seed cost (Rs/ha)	Fertilizer (Rs/ha)	Khed (Rs/ha)	Pesticides (Rs/ha)	Irrigation water charges (Rs/ha)	Total Expenditure (Rs/ha)	Yield (kg/ha)	Selling price (Rs/Kg)	Gross benefit (Rs/ha)	Net benefit coefficient (Rs/ha)
1	Castor	1000	3000	1500	4000	2560	12060	3750	44.8	168000	155940
2	Juvar/Bajri/ Makai	750	3000	1200	2000	640	7590	3750	14.93	56000	48410
3	Cotton	2300	6000	1800	7000	2560	19660	3750	58.3	218625	198965
4	Ground nut	14000	4000	1200	6000	800	26000	2500	50	125000	99000
5	Wheat	5000	12000	1200	4000	800	23000	6250	15	93750	70750
6	Cumin	2500	4000	1200	17000	480	25180	1250	149.6	187000	161820
7	Gram	6000	4000	800	7000	480	18280	2500	44.8	112000	93720
8	Vegetables	9500	8000	800	1500	480	20280	1250	40	50000	29720
9	Others Crop(Til)	1000	4000	800	8000	480	14280	1250	50	62500	48220
10	Onion/ Garlic	13125	9375	6250	9375	800	38925	10000	15	150000	111075

where NBs are the net benefits from the command area of the multiple reservoir system in million rupees; j is the index of reservoir i.e. 1, 2, 3; i is the index of Crop area i.e.1, 2,3,.....n; ak_i is the Kharif crop area allocation under i^{th} crop (ha); ari is the Rabi crop area allocation under i^{th} crop (ha); at_i is the biannual crop area allocation under i^{th} crop (ha); C_i is the coefficient of net benefit for the ith crop (Rs/ha). The coefficient of net benefit of each crop sown in the study area is computed which is presented in Table 1. The necessary information for finding out the NBs coefficient is collected from the office of agronomist Rajkot. As per year 2015–16, the selling rate of crop yield was collected from the marketing yard, Rajkot. The procedure followed in finding NB co-efficient is as follows: (a) The rate of application of seeds per hectare is calculated and after which, the multiplication with the price of seeds gives the cost of the seeds (b) The total fertilizer requirement of each crop per hectare is calculated and after which, the multiplication with a price of fertilizer gives a cost of fertilizers (c) The cost of cultivation for each crop per hectare is calculated (d) The total cost of pesticides for each crop per hectare is calculated (e) The total cost of irrigation water is calculated for each crop (f) The sum of seed, fertilizer, cultivation, pesticides, and irrigation water charges give total expenditure for each crop per hectare (g) The yield of each crop per hectares and the selling price for each crop is multiplied for computing gross benefits (h) After deducting total expenditure from gross benefits, the co-efficient of NBs is computed for each crop per hectare.

Subjected to a several constraints functions which are discuss as per below section.

4.1 The Land Allocation

The total crop area allocated during the Kharif or Rabi season must be less than or equal to the CCA_j at site j^{th}. The constraints' function is mathematically expressed as per Eq. (7) and Eq. (8) for the Kharif and the Rabi season, respectively.

$$\text{For Kharif Season, } g_{1j}(t) = \sum_{i=1}^{n} ak_{ji} + at_{ji} - CCA_j \leq 0, \text{ and} \tag{7}$$

$$\text{For Rabi Season } g_{2j}(t) = \sum_{i=1}^{n} ar_{ji} + at_{ji} - CCA_j \leq 0 \tag{8}$$

where ak_{ji} is the Kharif crop area allocation under i^{th} crop at site j^{th} (ha); ar_{ji} is the Rabi crop area allocation under i^{th} crop at site j^{th} (ha); CCA_j is the culturable command area at site j^{th} (ha).

4.2 The Storage Continuity

This constraint function expresses the link of the month to next month storage of multiple reservoir systems during an operating period 't' for the jth reservoir. The constraint functions are mathematically expressed as per Eq. (9), Eq. (10), and Eq. (11). According to Ajudiya et al. [2], the Aji-2 and Nyari-2 reservoirs were operated in parallel while at the same time, the Aji-2 and the Nyari-2 reservoirs together were operated in series with the Aji-3 reservoir.

$$g_{3j}(t) = S_t^1 + I_t^1 - RI_t^1 - E_t^1 - Ovf_t^1 - S_{t+1}^{j=1} = 0, (j = 1 \text{ for Aji} - 2) \tag{9}$$

$$g_{3j}(t) = S_t^2 + I_t^2 - RI_t^2 - E_t^2 - Ovf_t^2 - RW_t^2 - S_{t+1}^{j=2} = 0, \ (j = 2 \text{ for Nyari} - 2) \tag{10}$$

$$g_{3j}(t) = S_t^3 + I_t^3 - RI_t^3 + Ovf_t^1 + Ovf_t^2 - E_t^3 - Ovf_t^3 - RW_t^3 - Sj_{t+1}^{=3} = 0, \ (j = 3 \text{ for Aji} - 3) \tag{11}$$

where t is the operating period i.e. 1,2,….12; S_t is the initial storage at the period 't' at site j^{th} (MCM); S_{jt+1} is the final storage at period't' at the jth site (MCM); RI_t is the irrigation release at period 't' at site j^{th}, I_{jt} is the dependable Inflow for the jth reservoir at period 't' (MCM); Ovf^j_t is the overflow from the jth reservoir at period 't' (MCM); RW^j_t is the release for water supply from the jth reservoir at period 't' (MCM); E_{jt} is the evaporation at the jth reservoir at period 't'(MCM).

4.2.1 Dependable Inflow

The uncertainty in inflows arising due to variations in rainfall and continuous change in characteristics of the catchment area is tackled through dependable level flow. Nine years of historical monthly inflow data is used to compute the probability of exceedance of inflow. The monthly inflow is arranged in descending order of magnitude. The Weibull method is used, proposed by Chow et al. [7]; and mathematically expressed as Eq. (12).

$$p = \left[\frac{m}{n+1}\right] * 100 \tag{12}$$

where p is the probability of occurrence of month (%); n is the numbers of the year of available data of monthly inflow; m is the rank of monthly inflow arranged in descending order of magnitude.

Table 2. 75% Dependability level of inflow for multiple reservoir system

Month	Aji-2 Reservoir (MCM)	Nyari-2 Reservoir (MCM)	Aji-3 Reservoir (MCM)
Jan	0.39	0.00	0.00
Feb	0.25	0.00	0.00
March	0.55	0.00	0.00
Apr	0.43	0.00	0.00
May	0.05	0.00	0.00
Jun	0.55	0.00	0.00
July	3.96	2.74	0.28
Aug	10.77	4.49	8.21
Sep	10.74	5.27	21.95
Oct	1.30	0.83	0.68
Nov	0.88	0.03	0.00
Dec	0.49	0.00	0.00

Compute the probability exceedance of inflow at a different level. Then interpolation method is used to project the magnitude of inflow at the various probability of exceedance of inflow such as 90%, 85%, 80%, 75%, 70%, 65%, 60%, 55%, 50%; and named as dependability level of inflow. In the presented study, a 75% dependability inflow level was used which is presented in Table 2.

4.3 Water Allocation

Total water allocation for growing crops at period 't' should be less than or equal to irrigation release by the canal during the period 't'. The constraint function is mathematically expressed as per Eq. (13).

$$g4j(t) = \left(\sum_{i=1}^{n} \mathrm{NIR}_{jit} a_{ji}/\eta_s\right) - \mathrm{RI}_{jt} \leq 0 \tag{13}$$

where NIR_{jit} is the net irrigation requirement of the ith crop at the jth reservoir during period 't'; ηs is the efficiency of the surface water system at the jth reservoir. Moreover, the data on NIR for each crop and ηs were collected from the report of the office of Agronomist, Valsad.

4.4 Evaporation

The monthly evaporation loss from the reservoir is computed using the equation reported by Jothiprakash et al. [9]. The constraint function is mathematically expressed as per Eq. (14).

$$g_{5j}(t) = C_{jt} + m_{jt}\left[\left(S_{jt} + S_{j(t+1)}\right)/2\right] - E_{jt} = 0 \tag{14}$$

where E_{jt} is the evaporation loss at the jth reservoir at period't' (MCM); C_{jt} and m_{jt} is the regression coefficient at the jth reservoir at period 't'. Actual monthly evaporation volume and average storage were used to develop a linear regression model. The required parameters exported from the regression models which were used in Eq. (14) to compute monthly evaporation loss from the reservoir.

4.5 Overflow

Surplus water can spill over at period 't' from the reservoir through the given overflow constraint function in the model, which is mathematically expressed as per Eq. (15).

$$g_{6j}(t) = S_{jt} + I_{jt} - R_{jt} - E_{jt} - Kj - Ovf_{jt} >= 0 \tag{15}$$

where K_j is the storage capacity at rule level for the jth reservoir (MCM).

If the objective function does not meet the constraints presented through Eqs. (7) to Eq. (11) and Eqs. (13) to Eqs. (15), then the solution is infeasible and a penalty function is applied. The penalty function, $h_{1j}(t)$, $h_{2j}(t)$, $h_{3j}(t)$, $h_{4j}(t)$, $h_{5j}(t)$, and $h_{6j}(t)$ applied to the corresponding constraints function of Eqs. (7) to Eqs. (11) and Eqs. (13) to Eqs. (15) is expressed as follows:

$$h_{1j}(t) = \mathrm{sum}\left(10 * \mathrm{abs}\left(g_{1j}(t)\right)^2\right) \tag{16}$$

$$h_{2j}(t) = \text{sum}\left(10 * \text{abs}\left(g_{2j}(t)\right)^2\right) \tag{17}$$

$$h_{3j}(t) = \text{sum}\left(10 * \text{abs}\left(g_{3j}(t)\right)^2\right) \tag{18}$$

$$h_{4j}(t) = \text{sum}\left(10 * \text{abs}\left(g_{4j}(t)\right)^2\right) \tag{19}$$

$$h_{5j}(t) = \text{sum}\left(10 * \text{abs}\left(g_{5j}(t)\right)^2\right) \tag{20}$$

$$h_{6j}(t) = \text{sum}\left(10 * \text{abs}\left(g_{6j}(t)\right)^2\right) \tag{21}$$

where $g_{1j}(t)$ to $g_{6j}(t)$ are the respective constraint function at the jth reservoir at period 't' expressed through Eqs. (7) to Eq. (11) and Eqs. (13) to Eqs. (15). However, the penalized objective function is rewritten, and mathematically expressed as per the following:

$$\begin{aligned} f'(Y)\text{MaximizeNB} = f(Y) - \textstyle\sum_{j=1}^{3}[(\sum_t^{12} h_{1(t)}) + (\sum_t^{12} h_{2(t)}) + \\ (\textstyle\sum_t^{12} h_{3(t)}) + (\sum_t^{12} h_{4(t)}) + \left(\sum_t^{12} h_{5(t)}\right) + \left(\sum_t^{12} h_{6(t)}\right)] \end{aligned} \tag{22}$$

The TLBO model contains several function files which included initialization, implementation, objective, TLBO run, etc. The upper and lower limit of the variable given to the initialization of TLBO algorithms. The population size, iteration numbers, and numbers of variables are given to implement the algorithms. The optimization model was developed in an objective function file. There were 32 decision variables out of 212 as total variables involves in the TLBO model. The TLBO algorithm code algorithm and LP model were run in MATLAB R2015b and LINGO 18 unlimited constraint version [19].

5 Results and Discussion

In the present study, the optimal reservoir operating policies and optimal rule curves of Aji-2, Nyari-2, and Aji-3 reservoirs of multiple reservoir systems at 75% dependability levels were derived by LP and TLBO algorithm approach using LINGO 18 unlimited constraints version and MATLAB software respectively. The TLBO and LPM75 models were developed as an objective function to derive maximum NBs in the command area of multiple reservoir systems subjected to a constraint on land allocation, storage continuity, water allocation, evaporation, and overflow. The TLBO algorithm was developed using different combinations of population size such as 25, 50, 75, 100, 150, 160, and 200 and numbers of iterations such as 100, 150, 160, and 200. The accuracy of the TLBO algorithm was tested based on the LP result. The NBs resulting from the TLBO model over 24 independent runs are analyzed. The maximum NBs of Rs. 1283.67×10^6 were derived using a combination of 160 population size and iteration numbers. The optimal storage and release resulted from the TLBO model with a combination of 160 population sizes and iteration numbers used to derive optimal operation policy.

5.1 Optimal Rule Curve

The optimal storages resulted from the TLBO and LP model of multiple reservoir systems are portrayed in Fig. 2(a), 2(b), and 2(c). As shown in Fig. 2(a), the optimal storage in the reservoir is more with the TLBO model compared to the LP. Optimum storage resulting from the TLBO model is observed as the highest and equal to rule-level storage in September. Optimum storage in July and August is observed as less than the rule curve storage. It means that starting at the operation period surplus water is released to the downstream reservoir.

As shown in Fig. 2(b), the optimum storage is better with the TLBO model compared to the LP model. The Optimal storage resulted from the TLBO model is observed as the highest and equal to rule-level storage in September. Also, it is equal to the maximum observed storage in November, December, February, and March. Optimal storage in July and August is observed as less than the rule curve storage. It means that starting at the operation period surplus water is released to the downstream reservoir. As shown in Fig. 2(c), the optimal storage is better with the TLBO model compared to the LP model. Optimal storage resulted from the TLBO model is observed as less than rule-level storage as well as the maximum observed storage in August, September, and October. The maximum observed storage was higher than the rule level storage in August.

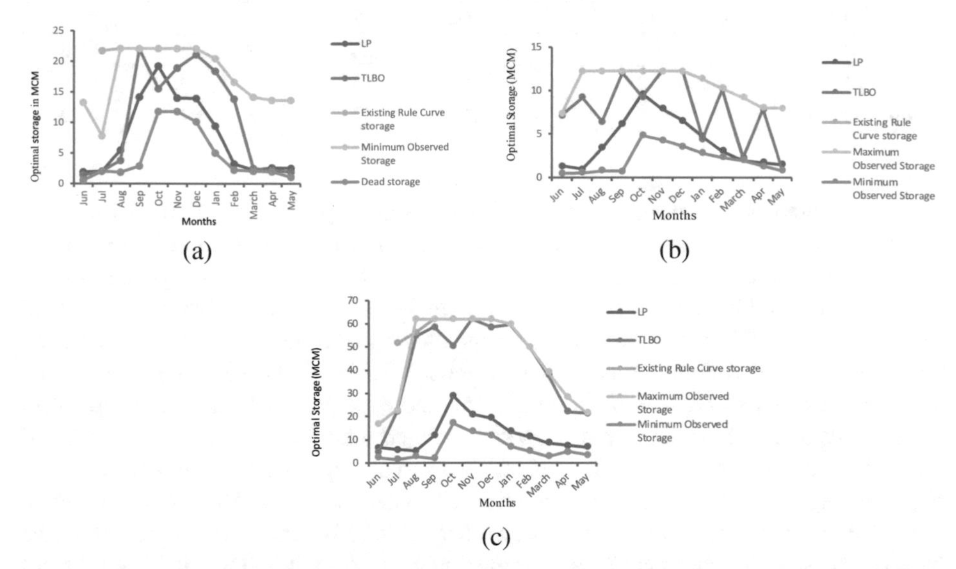

Fig. 2. **(a)** Optimal rule curves for the Aji-2 reservoir of multiple reservoir system **(b)** Optimal rule curves of the Nyari-2 reservoir of multiple reservoir system **(c)** Optimal rule curves of the Aji-3 reservoir of multiple reservoir system

So, it is advisable to maintain the rule-level storage for August. Moreover, it is advisable to maintain optimal storage during the operation period in all three reservoirs to derive maximum NBs in the command area of multiple reservoir systems.

5.2 Optimal Irrigation Release

Table 3. Optimal releases resulted from LP and TLBO models for multiple reservoir system

Month	Aji-2 Reservoir Operation			Nyari-2 Reservoir Operation			Aji-3 Reservoir Operation		
	Optimal release (LP)	Optimal release (TLBO)	Actual release as of the year 2009–10	Optimal release (LP)	Optimal release (TLBO)	Actual release as of the year 2009–10	Optimal release (LP)	Optimal release (TLBO)	Actual release as of the year 2009–10
Jun	0.23	0.18	1.19	0.08	2.05	0.55	0.23	6.43	0.00
Jul	0.18	1.1	0.00	0.09	1.99	0.00	0.22	6.75	0.00
Aug	1.49	1.35	2.24	0.52	0.18	0.18	0.97	4.51	0.00
Sep	4.75	0.18	0.00	1.26	0.37	0.37	3.9	0.01	1.90
Oct	5.71	0.01	0.00	2	0.18	0.18	7.33	0.23	4.27
Nov	0.34	2.71	2.76	0.09	0.23	0.55	0.54	5.43	7.77
Dec	3.78	4.76	5.23	1.57	1.22	1.10	6.02	2.18	10.10
Jan	2.98	3.41	5.82	1.42	1.27	1.07	5.27	6.71	8.83
Feb	1.12	0.17	3.76	0.9	1.45	0.79	2.44	2.54	4.93
Mar	0.36	0.24	0.51	0.01	0.92	0.12	0.76	2.14	0.92
Apr	0	0.02	0.00	0	0.36	0.00	0	1.01	0.00
May	0	0.19	0.10	0	0.23	1.10	0	3.11	0.00
Total	**20.94**	**14.32**	**21.61**	**7.94**	**10.45**	**6.01**	**27.68**	**41.04**	**38.72**

To find a bitterness of model for the optimal operating policy, the optimal irrigation release resulted from the TLBO model as well as the LP model of multiple reservoir systems is compared with the actual release made from the reservoir for the year 2009–10, which is presented in Table 3. As shown in Table 3 Aji-2 reservoir, it has been observed that optimum irrigation releases are high from November to January in the Rabi season with the TLBO model compared to the LP model. The optimal releases resulted from the TLBO model from November to January are more relevant to the actual release, also. This shows that if the reservoir is properly operated then the irrigation demand could be alleviated. As shown in Table 3 Nyari-2 reservoir, optimal irrigation releases are observed high during the Jun-July of the Kharif season with the TLBO model compared to the LP model. This shows that irrigation releases should be increased at the start of the Kharif season to increase net benefits in the command area. As shown in Table 3 Aji-3 reservoir, optimal irrigation releases are observed high in the month of Kharif as well as Rabi season with the TLBO model compared to the LP model. It means the surplus water was released from Nyari-2 and Aji-2 reservoirs during the operation period that is utilized for irrigation purposes in the command area of the Aji-3 reservoir. Therefore, the NBs were better with the TLBO model compared to the LP model.

6 Conclusion

In the presented study, the novel TLBO algorithm is applied to determine the optimal rule curve and optimal releases for Aji-2, Nyari-2, and Aji-3 reservoirs when it is operated in an integrated manner in multiple reservoir systems. The results of the TLBO model were compared with those of the LP model, which was purposefully generated in the study. The optimal policy is validated with the results of the LP model. The result reveals that the TLBO model performed better. In the Aji-3 reservoir, the maximum storage was observed higher than the existing rule curve storage in August for the year 2006–07. Therefore, it is advisable to maintain the existing rule storage value in August to avoid the emergency operation of gates. The optimal reservoir operation policy provides a guideline to multiple reservoir managers to decide on the amount of water released for irrigation over operation periods. As a scope of work, the integrated reservoir operation policy should be determined, including a greater number of reservoirs in the Aji river basin such as the Aji-4, Demi-2, and Demi-3 reservoirs. Also, the flexibility in the formulation allows the model to be modified to include the conjunctive use of groundwater.

Disclosure Statement. The authors declare that they have no conflict of interest

References

1. Andrieu, L., Henrion, R., Römisch, W.: A model for dynamic chance constraints in hydropower reservoir management. Eur. J. Oper. Res. **207**(2), 579–589 (2010)
2. Ajudiya, B.K., Yadav, S.M., Majumdar, P.K.: Optimization of cropping pattern in the command area of multiple reservoir system using TLBO algorithm. ISH J. Hydraulic. Eng. **28**(3), 271–280 (2021)
3. Ajudiya, B.K., Yadav, S.M., Majumdar, P.K.: Optimization of water supply using linear programming monthly model. In: Presented in 22nd International Conference on Hydro-2017, LDCE, Ahmadabad, 21–23 December 2017 (2017)
4. Ajudiya, B.K., Majumdar, P.K., Yadav, S.M.: Deriving operation policy for multiple reservoir system under irrigation using LPM model. Int. J. Recent. Technol. Eng. **8**(4), 4487–4496 (2019)
5. Bozorg-Haddad, O., Karimirad, I., Sifollahi-Aghamiuni, S., Loaiciga, H.A.: Development and application of the bat algorithm for optimizing the operation of reservoir systems. J. Water Res. Plng. Mgmt. **141**(8), 040140971–0401409710 (2015)
6. Carvallo, H.O., Holzapfel, E.A., Lopez, M.A., Marino, M.A.: Irrigated cropping optimization. J. Irrig. Drain. Eng. **124**(20), 67–72 (1998)
7. Chow, V.T., Maidment, D.R., Mays, L.W.: Applied Hydrology. McGraw-Hill, New York (1988)
8. Haouari, M., Azaiez, M.N.: Optimal cropping patterns under water deficit. Eur. J. Oper. Res. **130**, 133–146 (2001)
9. Jothiprakash, V., Ganeshanshanthi, J., Arunkumar, R.: Development of operational policy for multiple reservoir systems in india using genetic algorithm. Water Res. Manag. **25**(10), 2405–2423 (2011)
10. Jothiprakash, V., Arunkumar, R., Ashok Rajan, A.: Optimal crop planning using a chance-constrained linear programming model. IWA Water Policy **13**, 734–749 (2011)

11. Kamodkar, R.U., Regular, D.G.: Multiple reservoir operating policies: a fully fuzzy linear programming approach. J. Agr. sci. Tech **15**(2013), 1261–1274 (2013)
12. Krishna, K.L., Mahesh, N.V.U., Srinivasa, P.A.: Optimal multipurpose reservoir operation planning using Genetic Algorithm and Non-Linear Programming (GA-NLP) hybrid approach. ISH J. Hydraulic Eng. **24**(2), 258–265 (2018)
13. Krishna, K.L., Mahesh, N.V.U., Srinivasa, P.A.: Optimal crop water allocation coupled with reservoir operation by Genetic Algorithm and Non-Linear Programming (GA-NLP) hybrid approach. J. Phys.: Conf. Ser. **1344**(2019), art no. 012006, 1–12 (2019)
14. Karnatapu, L.K., Annavarapu, S.P., Nanduri, U.V.: Multi-objective reservoir operating strategies by genetic algorithm and nonlinear programming (ga–nlp) hybrid approach. J. Inst. Engineers (India): Series A **101**(1), 105–115 (2019). https://doi.org/10.1007/s40030-019-00419-2
15. Kumar, V., Yadav, S.M.: Optimization of reservoir operation with a new approach in evolutionary computation using TLBO algorithm and jaya algorithm. Water Res. Manag. **32**, 4375–4391 (2018)
16. Kumar, V., Yadav, S.M.: Optimization of cropping patterns using elitist-jaya and elitist-TLBO algorithms. Water Resour. Manag. **33**, 1817–1833 (2019)
17. Labadie, J.W.: Optimal operation of multiple reservoir systems: state-of-the-art Review. J. Water Res. Plng. Mgmt. ASCE, **130**(2), 93–111 (2004)
18. Lee, Y., Kim, S.-K., Ko, I.-H.: Multistage stochastic linear programming model for daily coordinated multi-reservoir operation. J. Hydro Inf. **10**(1), 23–41 (2008)
19. LINGO version 18.0, Lindo System, Chicago
20. Nageshkumar, D., Baliarsingh, F.: Folded dynamic programming for optimal operation of multiple reservoir system. Water Res. Manag. **17**, 337–353 (2003)
21. Nageshkumar, D., Raju, K.S., Ashok, B.: Optimal reservoir operation for irrigation of multiple crops using genetic algorithms. J. Irrig. Drain. Eng. ASCE **132**(2), 123–129 (2006)
22. Nageshkumar, D., Reddy, M.J.: Multipurpose reservoir operation using particle swarm optimization. J. Water Res. Plng. Mgmt. ASCE **133**, 192–201 ((2007))
23. Parmar, N., Parmar, A., Raval, K.: Optimal reservoir operation for irrigation of crops using genetic algorithm: a case study of sukhi reservoir project. Int. J. Civ. Eng. Technol. **6**(4), 23–27 (2015)
24. Rao, R.V., Savsani, V.J., Vakharia, D.P.: Teaching–learning-based optimization: a novel method for constrained mechanical design optimization problems. Comput. Aided Des. **43**, 303–315 (2011)
25. Rao, R.V., Savsani, V.J., Vakharia, D.P.: Teaching learning based optimization: an optimization method for continuous nonlinear large-scale problems. Inf. Sci. **183**, 1–15 (2012)
26. Rao, R.V., Savsani, V.J., Vakharia, D.P.: Teaching learning based optimization algorithm for unconstrained and constrained real parameter optimization problems. Eng. Optim. **44**(12), 1447–1462 (2012)
27. Rao, R.V.: Teaching Learning Based Optimization Algorithm and its Engineering Application. Springer, Switzerland (2016). https://doi.org/10.1007/978-3-319-22732-0_2
28. Varade, S., Patel, J.N.: Determination of Optimum Cropping Pattern using Advanced Optimization Algorithms. J. Hydrol. Eng. **23**(6), 05018010 (2018)
29. Wurbs, R.: Reservoir system simulation and optimization models: a state-of-the-art review. J. Water Resour. Plng. Mgmt. ASCE **116**(1), 52–70 (1993)

Feature Reduction Based Technique for Network Attack Detection

Somya Agrawal[1](✉), Astha Joshi[1], Rajan Kumar Jha[1], and Neha Agarwal[2]

[1] Department of CSE, Jaipur Engineering College and Research Centre, Jaipur, India
{somyaagrawal.cse,asthajoshi.cse,rajanjha.cse}@jecrc.ac.in
[2] Department of CSE, Vivekanand Global University, Jaipur, India
neha.agarwal@vgu.ac.in

Abstract. Day by day naive powerful attack methods have come up to break the security of a single host or a network system. Typically, attacks are a type of action aimed to compromise the confidentiality, integrity, and accessibility of a single host network. Hence, in the real time state of affairs the network security issues have attracted the researchers and become a hot research field of the present time. On the other hand, since the age of the network system a variety of exclusive attack detection algorithm have been proposed by number of researchers and a lot of research work is going in direction to design an efficient algorithm for providing additional safety to the single host to the network security system with high detection rate of attacks in real time. However, these approaches improve the level of security but the majority of accessible approaches are designed for detecting only identified network attacks, trained by takeout exact patterns from before known attacks. Proposed approach achieves improved results against the accessible security approaches in conditions of high exposure rate, precision and reduction of false alarms concurrently, required parameters of a successful intrusion detection system.

Keywords: Security Attack · NSL KDD · DoS · R2L · U2R · Probe Attack

1 Introduction

Many attackers have routinely introduced naive and powerful techniques to compromise the security of the network system. Yet, from the time of the age of the network, many researchers and industry communities have developed various proprietary attack detection algorithms to protect in sequence from such threats. Each of these algorithms, however, has its own character. Performance problem. Most available techniques use a basic signature algorithm to identify only the beforehand identified types of attacks, detect new attacks, and generate a large number of attacks.

False alarms are therefore not suitable for broadband networks. These problems severely limit the usefulness of the deterrent system and are addressed in this work by the introduction of a new technique that generates low false positives with an increased detection rate of known and abnormal attacks on the network.

H. Sharma et al. (Eds.): ICIVC 2022, PALO 17, pp. 401–408, 2023.
https://doi.org/10.1007/978-3-031-31164-2_33

2 Types of Network Attacks

In general, an attack is an event designed to circumvent security settings such as the secrecy, integrity, ease of use of a stand-alone computer system or network. Fig. 1 is showing the types of attacks. An attack may arrive in a variety of shapes as well as hacks, virus, worms, Steganography dictionary and denial of service attacks etc. to harm the system or a network. The whole attacks can be categorized in one of the following categories.

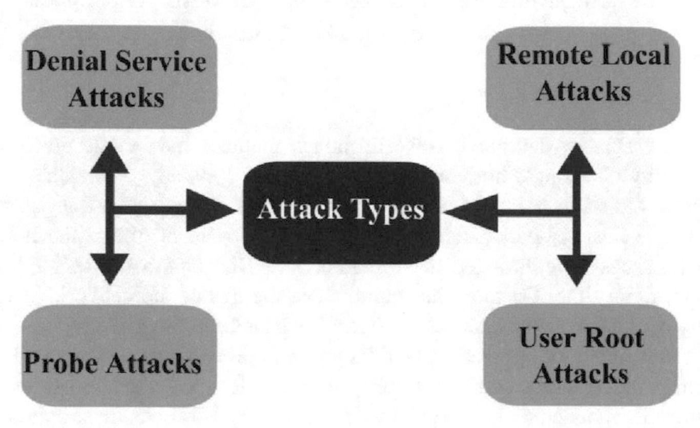

Fig. 1. Types of Attacks over a host or Network System

2.1 Denial of Service Attacks (Dos)

This type of attack, bandwidth or computer network connectivity has been deliberately attacked by flooding a large volume of traffic or link request to avoid legitimate user of the right from using the requested service. Types of DoS Attacks [1, 2].

2.2 Remote to Local Attacks (R2L)

This kind of attack allows the assailant to change the in order or access the host's resources not including having an account on a victim's computer. Ftp write, and Phf are common The latest R21L attacks contain an NT PowerPoint Macro Attack a Frame Spoof Attack, a remote administration tool installed by a Trojan (Net bus), and a Trojan Trojan S S H Server Trojan) and a version of a Linux FTP file way in utility with an mistake that permit distant commands to run on a nc ftp.

2.3 Probe Attacks

This kind of attack automatically searches for open ports on a computer network to avoid legitimate users from accessing information or services. The attacker uses an extended procedure on another network or is surrounded by a single host for services to examine the target network or a single computer for open ports. This process is called probe attack [3]. Ipsweep, Portsweep, Nmap, Satan is an attack type attack.

2.4 User to Remote Attacks (U2R)

This type of attack serves to unauthorized gain root privileges on a single host or network system. Attackers are granted permission to access special ex files with a common application such as Mail or FTP [4]. Load Module, Perl, Buffer Overflow are types of U2R attacks. Table 1 is showing the attacks associated with the NSL KDD dataset.

Table 1. Attacks Associated with NSL KDD Dataset

Attack Type	Attack Name
DOS	spy, warezclient, warezmaster Smurf, Neptune, Back, Teardrop, Pod, land, syslogd, ftp write, imap, phf, guess passwd
PROBE	Portsweep, Nmap Satan, Ip sweep, Mscan
R2. L	Guess passwd, Warezclient, Imap, ftp write, Guest, Netbus, Multihop, Phf, Spy Warezmaster
U2. R	Perl, Multihop Overflow, Buffer, Rootkit, Load Module

3 Issues with AS Attack Detection

In the area of higher bandwidths and ease of connectivity of wireless and mobile devices the conventional detection systems take more time for analysis and are often computationally exhaustive.

a. Produce low accuracy.
b. Majority of conventional systems are tuned to detect only known major service level attacks.
c. Major attack recognition systems demonstrate superior accuracy in detecting convinced classes of attacks while performing inadequately for the additional classes.
d. Unable to detect incongruity attacks over the system or network.
e. Always required updating processes by humans.
f. Generate huge false alarms.

4 Proposed Approach

Proposed approach selects an only small set of features capable of detecting anomaly and hybrid type of attacks with high accuracy and generation of low false alarms. However, human input is essential to keep the accuracy of the system. In general, the framework of projected move toward consists of the following parts.

(i) Gathering. Data
(ii) Preprocessing and
(iii) Classification

The DoS and Probe attacks are in the right place to the bulk class, while U2R and R2L belong to the minority class, also called the rare attack class [10]. The selected dataset contains 45927 DoS, 11656 probes, 995 R2L, 52 instances of U2Rattack. Below is Fig. 2 which is showing the percentages of attack instances associated with selected dataset.

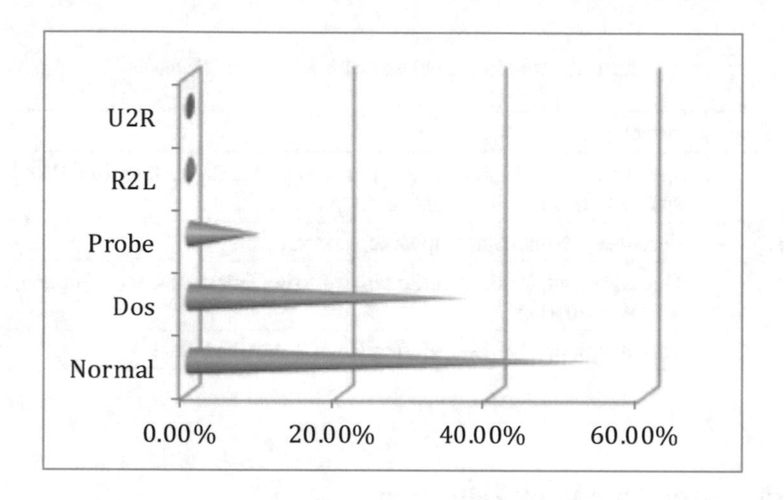

Fig. 2. Percentage of attack Instances Associates with Selected Dataset

Proposed approaches get better the power of existing attack discovery systems but suffer from their own limitations such as in case when an attack acts as a normal data a number of detection approaches fail to detect such events which harms the system and sometimes may play a more dangerous role. To improve the detection rate of an attack the proposed approach is implemented in a layered format in which each input data packet examines for an attack activity in a sequential manner.

5 Algorithm Used for Proposed Approach

Figure 3 is showing the alogirthm for feature reduction used in propsed approach. Here detection rate will be calculated.

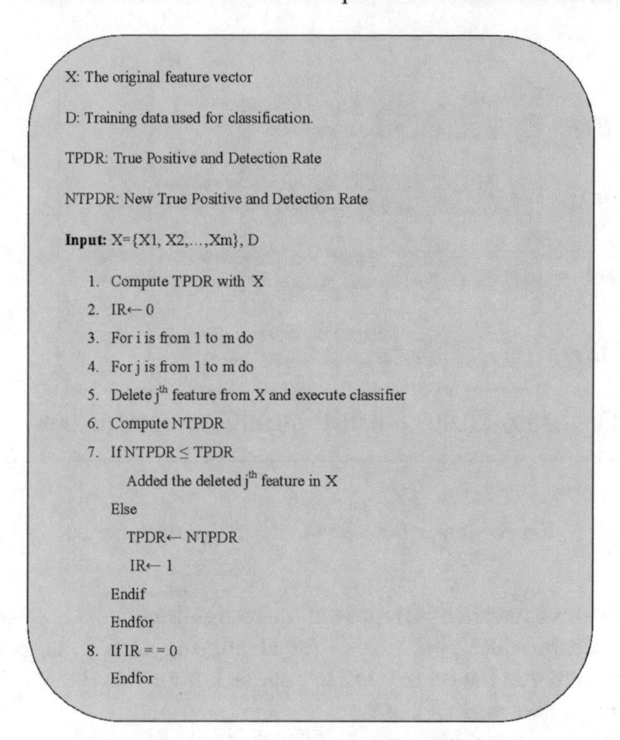

Fig. 3. Feature Reduction Method in Proposed Approach

6 Result Analysis

After select the categorization technique, the experiment was execute with the reduced set of features selected. Table 2 is showing the result analysis.

Table 2. Comparison between Nb and Proposed Approach Classification Results

S. No	Attack Type	categorization Technique	Attack Detection Rate (%)
1.	DoS	NB attribute Set	93.90
		Proposed Model	99.10
2.	Probe	NB attribute Set	98.15
		Proposed Model	97.70
3.	R21L	NB attribute Set	96.39
		Proposed Model	97.10
4.	U2R	NB attribute Set	64.34
		Proposed Model	67.30

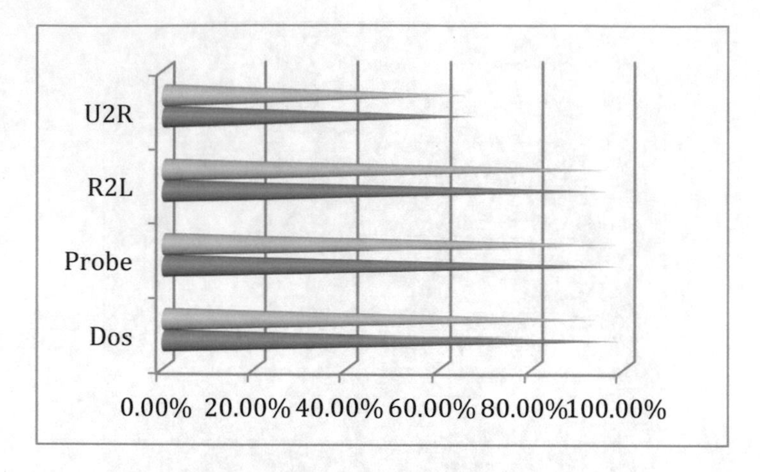

Fig. 4. Attack Detection Performance of Proposed

The planned move toward has improved the recognition rate. On the other hand, the proposed approach provides poor results for identifying an R2L class attack, but the dissimilarity is small and the on the whole approach improves the recognition rate for the Major and negligible attack classes.

As Shown in Fig. 5 the effectiveness of the planned approach in shortening the time required to build the model. This demonstrates that the proposed real-time approach is effective and efficient in detecting a range of attacks.

Table 3. Total Times Taken by Proposed Approach to Build Model.

S.No	Categorization Technique	Time Taken
1	N B with whole Feature Set	6.61
2	D T with whole Feature Set	43.2
3	Proposed Model	2.71

The result also compares the generation of false alarms. Figure 4 shows the apparent effectiveness of the proposed approach measure up to the existing classification technique of Naive Bayes. Table 3 is showing the total times taken by proposed approach to build the model.

The experimental result of the proposed approach presented in the previous section demonstrates the relevance of the proposed approach to existing attack detection approaches. In this division, the results of the proposed were compared with those of the different approaches offered by the number of researchers.

This result of the proposed approach is compared in two ways depending on the precision of the attack recognition and the false alarm generation rate. In both cases, the proposed approach resulted in significantly improved comparative results compared

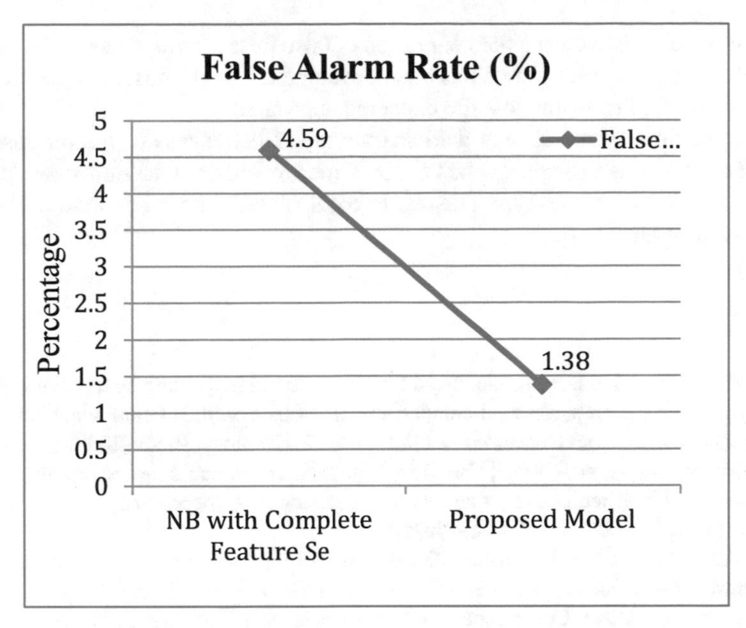

Fig. 5. N B and Proposed by False Alarms Generate

Table 4. Detection Accuracy rate of proposed Approach over the other offered Algorithms

Approaches	Correctly Classified Instances (%)	False Alarms Rate
Proposed Approach	**98.62**	**1.38**
Hybrid layered intrusion detection system [5], Feb. 2015	92.13	7.87
improved PSO-BP neural network algorithm [6], 2015	96.73	3.27
DTNB Based Approach [7], June 2014	97.14	2.86
Dimension reduction Based Approach [8], July 2013	75.79	24.21

to other available approaches. Table 4 is showing the detection accuracy raote of the proposed approach over the other offered algorithms.

7 Conclusion

There are modern approaches to securing information on the unsecured channel, but with the regular introduction of naïve attack techniques, it is not possible to create a fully secure system in reality. In addition, many accesses to the recognition method generate huge rates of false alarms [9]. Therefore, to improve the precision of the attack recognition mechanism during the generation of false false alarms, a new approach based on the selection of characteristics has been developed. The proposed approach uses the NSLKDD record to train and test the expected approach.

The evaluation results clearly demonstrate the effectiveness of the proposed algorithm over other existing approaches to get superior attack detection rates, primarily root-level attacks & remote-type attacks. to local places where rates are higher. to be lowered by false alarms.

References

1. Bahrololum,M., Salahi, E., Khaleghi, M.: Machine learning techniques for feature reduction in intrusion detection systems: a comparison. In: 2009 Fourth International Conference on Computer Sciences and Convergence Information Technology. IEEE (2009)
2. Khraisat, Ansam, Gondal, Iqbal, Vamplew, Peter, Kamruzzaman, Joarder: Survey of intrusion detection systems: techniques, datasets and challenges. Cybersecurity **2**(1), 1–22 (2019). https://doi.org/10.1186/s42400-019-0038-7
3. Chou ,T.S., Chou, T.N.: Hybrid classifier systems for intrusion detection. In: Seventh Annual Communications Networks and Services Research Conference. IEEE (2009)
4. Jin, S., Yeung, D.A.: Covariance analysis model for DDoS attack detection. In: IEEE International Communication Conference (ICC 2004), vol. 4, pp. 20–24 (2004)
5. Yao, X.H.: A network intrusion detection approach combined with genetic algorithm and back propagation neural network. In: International Conference on E-Health Networking Digital Ecosystems and Technologies (EDT). IEEE (2010)
6. Fu, Y., Zhu, Y., Yu, H.: Study of neural network technologies in intrusion detection systems. In: 5th International Conference on Wireless Communications, Networking and Mobile Computing. IEEE (2009)
7. Dartigue ,C., Jang, H.I., Zeng, W.: a new data-mining based approach for network intrusion detection. In: Seventh Annual Communication Networks and Services Research Conference. IEEE (2010)
8. Kargl, F., Maier, J., Schlott S., Weber, M.: Protecting Web Servers from Distributed Denial of Service Attacks, 1-58113-348-0/01/0005. ACM (2001)
9. Sharma , N., Mukherjeeb, S.: A novel multi-classifier layered approach to improve minority attack detection in ID. In: 2nd International Conference on Communication, Computing & Security [ICCCS-2012]. Elsevier Ltd. (2010)
10. Jones, A.K., Sielken, R.S.: Computer system intrusion detection: a survey. Int. J. Comput. Theory Eng. **2**(6) (2010)

Machine Learning Based Solution for Asymmetric Information in Prediction of Used Car Prices

Amit Kumar(✉)

Government of Andhra Pradesh, Visakhapatnam, India
amitkumar@andhrauniversity.edu.in

Abstract. It is common for people in all countries to buy and sell used cars. The car sellers and buyers usually determine the price of a car by comparing the price of the same type of car in some advertisements. In reality, demand and supply of the cars determine the real market prices for cars. Also used car markets are characterized by information asymmetries and thus a buyer lack trust while buying a car. The purpose of this paper is to examine markets characterized by asymmetric information about the price of used cars. Thus, by using machine learning techniques, price of used cars is predicted. The dataset which was used in this study was initially not cleaned and hence it was pre-processed by handling the outliers and missing data, and thus it was finally prepared for machine learning algorithms. Exploratory data analysis was performed on the dataset and the most important features were identified which had a maximum impact on the price of a car. To estimate the car price, four machine learning algorithms were used which included Linear Regression, Random Forest, Extra Tree Regressor and Extreme Gradient Boosting Regression. Extra Tree Regressor outperformed all the other algorithms in this study with highest testing R^2 of about 95%. Finally, a cloud based application was developed and deployed that provides an estimation of the price of a particular car, giving users helpful guidance in identifying their car's value.

Keywords: Car Price Prediction · Machine Learning · Linear Regression · Random Forest · Extra Tree Regressor · Extreme Gradient Boosting Regression

1 Introduction

The world's automobile markets are ruled by mistrust and uncertainty. Particularly, the used car market is heavily affected by these factors. G.A. Akerlof, Nobel Laureate and author of "The Market for 'Lemons': Quality Uncertainty and the Market Mechanism" offered insight into why the used car market is plagued by distrust among the buyers and the sellers of cars [1]. The primary reason for uncertainty is that the buyers and the sellers have different knowledge about the quality of the car. Having a better understanding of the quality of the

H. Sharma et al. (Eds.): ICIVC 2022, PALO 17, pp. 409–420, 2023.
https://doi.org/10.1007/978-3-031-31164-2_34

car, the seller will try to sell it at a higher price, creating a mistrust between the buyers and sellers that leads to the failure of the market.

In many markets, buyers use market statistics to judge the quality of prospective purchases. There is an incentive for sellers to market poor quality goods in this case, since better returns are gained by the entire group in this case rather than the individual seller. As a result, the quality of goods and the size of the market tend to decline. New vehicle prices are at all-time highs, therefore many buyers will turn to buy used cars. Nevertheless, it can be difficult for buyers to determine whether they are getting a good deal or not. Despite careful examination, the buyer still may not know everything the seller knows. There is asymmetric information when one party knows more about the product than the other.

Objectives of the study

The main objectives of this study are as follows.

- To develop a prediction model i.e., a fair price mechanism system for estimating the sale price of used cars based on details such as the car's model, the age of the vehicle, the type of fuel it consumes, the distance the vehicle has already travelled etc.
- Based on the car's features, this paper will help to determine the selling price of a used car and thus it will reduce consumer and seller risks.
- Develop and launch a cloud based computer application (app) which can be used by anyone to predict the price of used cars.

In order to accomplish this task, this study uses four machine learning algorithms i.e., Linear Regression, Extra Tree Regressor, Random Forest and Extreme Gradient Boosting Regression. The rest of the paper is organized as follows. Section 2 discusses the literature review. Section 3 and Sect. 4 describes the dataset, exploratory data analysis of the dataset and methodologies used in this study respectively. Section 5 contains the results and comparison of models. Finally, Sect. 6 summarizes and concludes the findings.

2 Literature Review

The work on estimating the price of used cars is very recent and is also very sparse, with results varying from accuracy of 50% to 95%. [2] examined how reverse selection operates in the second hand car market. It has been observed that new car dealers (new and second hand cars) differ greatly from those who mostly deal in second hand cars in the wholesale market. Based on reverse selection models, it is likely that the vendor type that sells a higher percentage of trade in the wholesale market will sell higher quality cars and receive a higher price in return. To test this estimate, a survey form was used to analyze the wholesale behaviors of the dealers and the prices were collected at the wholesale auction. However, inverse selection seemed to be weakly supported.

[3–7] used Hedonic price model to estimate the price of cars using price differences between the various car models in terms of the differences in the car characteristics. [4] compared Fuzzy Hedonic model predictions and normal model estimates to find the best way to inform customers at a high level. In light of expectations theory, [7] developed a Hedonic price model to address the price structure of the used automobile market. According to the results, consumers avoided the risk when the second hand car's reliability was below the expected reference value and above the expected reference value. Furthermore, the model showed how automobile quality affects residual values and how buyers evaluate used cars.

[8,9] used K-nearest neighbours algorithm to predict prices of used cars. [8] used different machine learning algorithms like K-nearest neighbours, multiple linear regression, naive Bayes and decision trees and built a model which would predict the price of used cars in Mauritius. Results were 60%–70% accurate, with naive Bayes and decision trees having the greatest weakness as they require discretizing and classifying prices, which leaves inaccuracies in their results. [9] achieved an accuracy of about 85% by using K-Nearest Neighbour on the data which contained 14 different attributes. [10] used multiple linear regression, extreme gradient boosting regression and random forest on data from a German e-commerce website which contained 11 different attributes. Extreme gradient boosting regression outperformed all the other algorithms used in their study with least mean absolute error (MAE) of about 0.28.

[11,12] used Artificial Neural Network (ANN) methods for used car price prediction. [11] evaluated the performance of Neural Networks in predicting used car prices. However, the predicted values were not very close to the actual prices, particularly for expensive cars. They concluded that support vector machine regression outperformed neural networks and linear regression in predicting used car prices. The BP neural network and nonlinear curve fit were combined for optimizing the model [12]. The neural network was trained on schemed data and it reached the training goal. A nonlinear curve fit was performed using the schemed data and NN outputs, which had the highest accuracy. [13] proposed the use of the optimized BP neural network algorithm for online used car price evaluation. In order to optimize hidden neurons, they developed a new optimization method called Like Block-Monte Carlo Method (LB-MCM). When compared to the non-optimized model, the optimized model yielded higher accuracy.

A total of three machine learning techniques were used to predict the price of used cars by [14]. After pre-processing the data scraped from a local Bosnian website for used cars, the total number of samples remained 797 observations and the machine learning models which they used in their study included support vector machine, random forest and artificial neural network. When each algorithm was used individually, the accuracy of the results were less than 50%, but when all the models were combined with random forest algorithm, the results increased to 87.38%. Based on the results, SVM and multiple linear regression were more accurate when evaluating leased car prices in large dataset with high levels of dimensionality [15,16] used Random Forest algorithm to predict used car prices using Kaggle's dataset. Approximately 500 decision trees were trained using random forest. The accuracy of the training dataset and testing dataset was 95.82% and 83.63% respectively. [17] conducted a study and developed a system that consists primarily of three different parts i.e., acquisition of a data system, an algorithm which would forecast the price and finally a model which would analyze the performance of the model. A conventional Artificial Neural Network (ANN) with a back-propagation network was compared to the proposed ANFIS due to its adaptive learning capability. The ANFIS incorporates qualitative fuzzy logic approximation as well as adaptive neural network capabilities. In the experiment, ANFIS performed better in predicting used car prices than other expert systems. Multiple linear regression model was used by [18,19] to predict the price of used cars. [18] used a dataset from a Pakistan website which contained about 1699 observations. After data pre-processing and reducing the dimensionality of the data, they achieved an accuracy of about 98% on the testing dataset.

3 Dataset and Its Exploratory Data Analysis

The dataset used in this study was downloaded from an online commercial website. As shown in the Table 1, there were six attributes in the dataset namely name of the car company, model of the car, year of manufacturing, total kilometres driven, type of fuel that car utilizes and price of the used car. Initially there were 892 observations and 6 columns. But after pre-processing the data, there were about 816 observations as shown in Fig. 1. The data was unclean with many missing values and strings in between the numbers and hence it needed cleaning while pre-processing it. Conversion of few attributes from objects to integer value was also done.

Table 1. Dataset description

Sl.No	Variable name	Description	Value
1	company	the name of the company	Categorical value
2	name	model of the company	Categorical value
3	year	year of manufacturing	Integer value
4	kms_driven	car's overall distance travelled	Continuous value
5	fuel_type	car's fuel type based on Petrol, Diesel or LPG	Categorical value
6	Price (dependent variable)	the selling or the buying price of the car	Continuous value

3.1 Company vs Price

Figure 2 shows the relationship between the car company and its prices. Prices are shown in y-axis while the car company is shown in x-axis. The median price of few car companies like Audi, Mercedes and Jaguar is very high when compared to other companies. Figure 2 also shows that there are few outliers in the dataset.

Out[44]:

	name	company	year	Price	kms_driven	fuel_type
0	Hyundai Santro Xing	Hyundai	2007	80000	45000	Petrol
1	Mahindra Jeep CL550	Mahindra	2006	425000	40	Diesel
2	Hyundai Grand i10	Hyundai	2014	325000	28000	Petrol
3	Ford EcoSport Titanium	Ford	2014	575000	36000	Diesel
4	Ford Figo	Ford	2012	175000	41000	Diesel
...	...	...	...	...	...	...
811	Maruti Suzuki Ritz	Maruti	2011	270000	50000	Petrol
812	Tata Indica V2	Tata	2009	110000	30000	Diesel
813	Toyota Corolla Altis	Toyota	2009	300000	132000	Petrol
814	Tata Zest XM	Tata	2018	260000	27000	Diesel
815	Mahindra Quanto C8	Mahindra	2013	390000	40000	Diesel

816 rows × 6 columns

Fig. 1. Total number of observations after cleaning the data

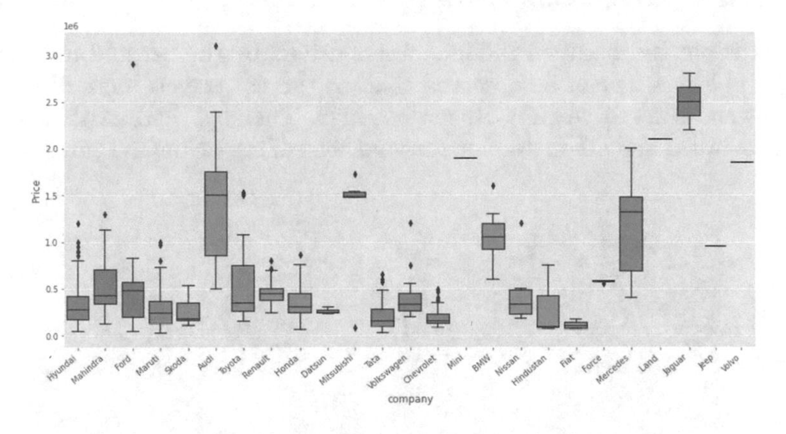

Fig. 2. Boxplot of Car company's price

3.2 Relationship of Price with Year

As shown in Fig. 3, a swarmplot is plotted between the year and the price of the cars. It shows that, as the year gets increased, the price of the cars also gets increased which can be also interpreted as the car gets old, the price of the car falls and if the car is new, its price gets increased.

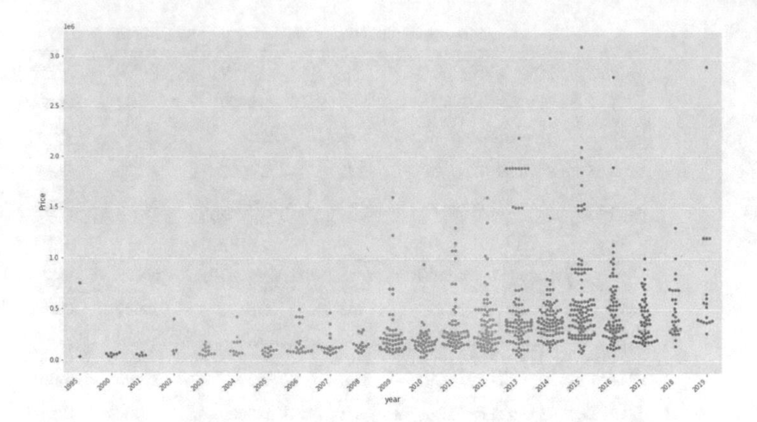

Fig. 3. Swarmplot of price with year

3.3 Price vs Kilometers Driven

A relplot is plotted between the kilometers driven by the car and its price and is shown in Fig. 4. It can be observed that, as the car travels more distance, its price falls and if its driven less, its price is high. Thereby, there exists a negative relation between the kilometers driven and the selling or buying price of a car.

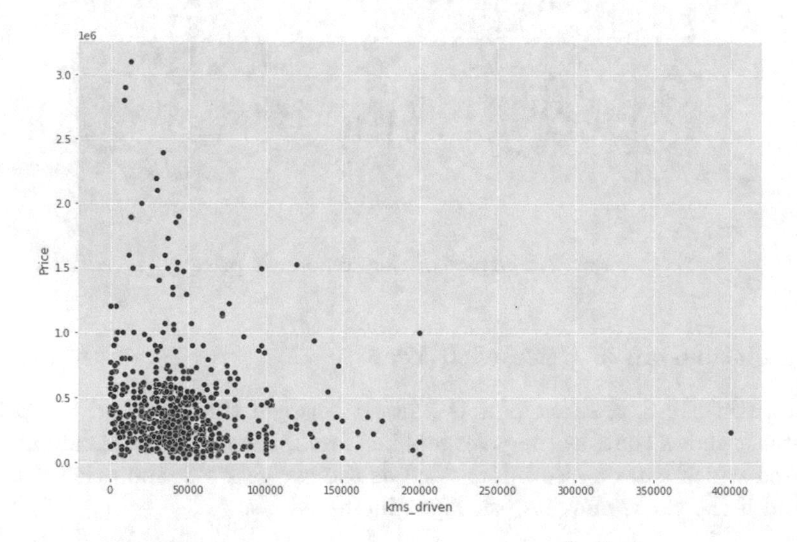

Fig. 4. Relationship between the distance travelled (kms) and Price of a car

3.4 Fuel Type vs Price

Figure 5 shows the relation between the Fuel type used by the car and its price. The median price of diesel type cars is highest and the median price of Petrol and LPG cars is almost equal.

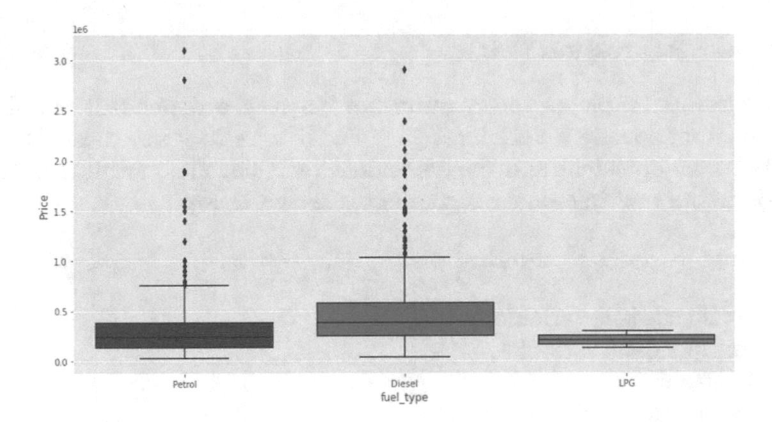

Fig. 5. Boxplot of fuel type car's price

3.5 Relationship of Price with Fuel Type and Year

Figure 6 shows the relationship of price with fuel type and the manufacturing year of a car. For companies like Hyundai, Audi, BMW and Jaguar, the price of petrol type cars is high when compared to diesel and LPG type of cars. While in other car company, the price of diesel type of cars is high. It can also be seen that the larger bubble has high price which means the recent year car's manufacturing has more price when compared to the older one.

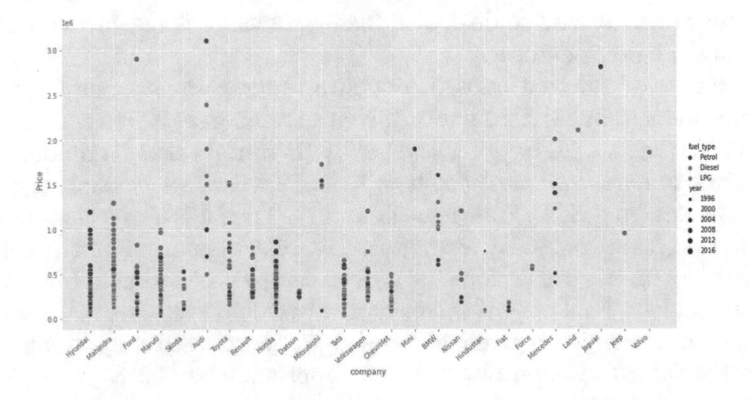

Fig. 6. Relationship of Price with Car's year and fuel type

4 Methodologies

Before applying the machine learning algorithms on the clean dataset, the dataset was divided into training and testing dataset. 80% of the data was used for training dataset and 20% of the data was used for testing dataset.

4.1 Linear Regression (LR)

Linear regression explains the relationship between a dependent variable and one or more independent variables. Thereby, it tries to find a best fit straight line between independent and the dependent variable. The mathematical representation of the multiple linear regression is shown in Eq. 1.

$$y = \alpha + \beta_1 x_1 + \beta_2 x_2 + ... + \beta_n x_n + \varepsilon \tag{1}$$

where, y = dependent variable,
x_i = independent variables,
α = intercept,
β_i = slope,
ε = residual (error).

The R^2 obtained after running linear regression on this dataset was about 92%.

4.2 Random Forest (RF)

Random Forest (RF) are ensembles of decision trees where it has a set of decision trees called Random Forest. Firstly, RF tries to make the subsets of the original dataset. Then the row sampling and feature sampling with replacement is done and later it creates subsets of the training dataset. In the next step, it creates an individual decision tree for each subset and then it will give an output. Decision tree algorithm faces an issue of low bias in training dataset and high variance in testing dataset. But, Random forest eliminates this drawback by using bagging technique which tries to deal with bias-variance problem by decreasing the variance in the testing dataset.

In order to train the tree learners, bootstrap aggregation technique is used for the training dataset by random forest. Given a training set X = x_1, ..., x_n and its outputs as Y = y_1, ..., y_n, bagging repeatedly (B times) selects a random sample and fits trees to these samples. For b = 1, ..., B, with replacement, it creates n training samples from X, Y, known as X_b and Y_b. Then it trains a classification or regression tree f_b on X_b,Y_b. By averaging the predictions from all the individual regression trees on x', predictions for unseen samples x' is made. Since this is a regression problem, final output is considered based on average of all the outputs of the decision tree which is given as $\widehat{f} = \frac{1}{B}\sum_{b=1}^{B} f_b(\text{x}')$. After applying Random forest on the dataset, the obtained R^2 was approximately 93%.

4.3 Extra Tree Regressor (ETR)

As an extension of the Random Forest (RF) model, [20] suggested the Extra Tree Regression (ETR) approach. A decision tree or regression tree constructed by ETR is similar to the normal top-down approach [20]. In the random forest model, bootstrapping and bagging are used for regression. From a set of randomly selected training data, a set of trees are bootstrapped to produce a set of decision trees. An ensemble of decision tree nodes is constructed by bagging, where subsets of training data are previously picked at random. Thus by selecting the best subset and its value, the decision-making process is achieved [21].

RF models were viewed as a series of decision trees, where G(x,θ_r) provided the Gth predicting tree, whilst 'θ' was interpreted as a uniform independent distribution vector before the tree shape was determined [22]. An ensemble of trees (forming a forest) of G(x) is constructed using Breiman's equation as shown in Eq. 2.

$$G(x, \theta_1, ., \theta_r) = \frac{1}{R}\sum_{r=1}^{R} G(x, \theta_r) \tag{2}$$

RF and ETR models differ in two important ways. Firstly, the nodes are randomly divided by the ETR and secondly it tries to achieve minimal bias by utilizing all the learning samples of the data [20]. The splitting process is controlled by two parameters, namely, 'k' and 'n_{min}', where 'k' is the number of features that are randomly chosen for the node, and 'n_{min}' is the minimum sample size required to separate the nodes. Moreover, 'k' and 'n_{min}' estimate the selected attributes strength and the average output noise strength. They both help to reduce overfitting and improve the precision of the ETR model [23,24]. The R^2 obtained after applying ETR on this dataset was about 95%.

4.4 Extreme Gradient Boosting Regression (XGBoost)

The Extreme Gradient Boosting (XGBoost) approach implements gradient boosting in an efficient manner which can be applied to regression predictive modeling. This algorithm is used when large dataset is available to make a prediction with high accuracy. Boosting is actually an ensemble of learning algorithms that combine the prediction of several estimators in order to enhance robustness over a single estimator [25,26]. To build a strong predictor, it combines several weak or average predictors. The main difference between this technique and the other gradient boosting methods is the objective function as shown in Eq. 3, which basically consists of two components. The first one is the training loss and the second one is the regularization term.

$$\mathcal{L}(\varnothing) = \sum_{i=1}^{n} \ell(\widehat{y_i}, y_i) + \sum_{k} \Omega(f_k) \tag{3}$$

$$where, \Omega(f) = \gamma T + \frac{\|w\|^2}{2}\lambda \tag{4}$$

The training loss is a measure of how predictive the model is based on the training data. Regularization improves generalization by controlling the complexity of the model. In most cases, the training loss is calculated based on the Mean Square Error. XGBoost expands the polynomial loss function using a Taylor expansion up to the second order. The R^2 obtained after applying the XGBoost on this dataset was about 88%.

5 Results and Accuracy Comparison of Models

Table 2 shows the accuracy comparison of all the models based on their obtained R^2. It can be seen that the R^2 of XGBoost is least i.e., 88% while the R^2 of Extra Tree Regressor is highest with 95% accuracy. The R^2 of Linear Regression and Random Forest were 92% and 93% respectively. Since the accuracy of Extra Tree Regressor was highest, it was used to build the cloud based app which can be used by anyone to predict the price of used cars [27].

Table 2. Accuracy of all the models

Sl.No	Model	R^2
1	Linear Regression (LR)	92%
2	Random Forest (RF)	93%
3	Extra Tree Regressor (ETR)	95%
4	Extreme Gradient Boosting Regression(XGBoost)	88%

6 Conclusion

By using four different machine learning algorithms, this paper has made an attempt to predict the price of used cars. An exploratory data analysis of the dataset was done and relationship with the dependent variable and the independent variables were plotted. The data wasn't a clean data, like Price had commas in its prices, kms_driven had object values with kms written after the numbers, year had many non-year values, fuel_type had nan values etc. So, the data was cleaned and pre-processed. Finally, the cleaned data was used for testing and training purpose. Based on the results, the Extra Tree Regressor outperformed all the models with highest R^2 value of about 95%. Finally, a web based application was made and deployed on cloud which predicts the price of used car. Also, the performance of the models can also be increased if we use large dataset or use sophisticated models or tune the hyper parameters and create some new attributes in the models.

References

1. Akerlof, G.A.: The market for "lemons": Quality uncertainty and the market mechanism. In: InUncertainty in Economics, pp. 235–251. Academic Press (Jan 1 1978)
2. Genesove, D.: Adverse selection in the wholesale used car market. J. Polit. Econ. **101**(4), 644–645 (1993)
3. Murray, J., Sarantis, N.: Price-quality relations and hedonic price indexes for cars in the United Kingdom. Int. J. Econ. Bus. **6**(1), 5–27 (1999)
4. Pazarlioglu, M.V., Gunes, M.: The hedonic price model for fusion on car market. In: Proceedings of the Third International Conference on Information Fusion 2000, Jul 10, vol. 1, pp. TUD4-13. IEEE (2000)
5. Erdem, C., Şentürk, İ.: A hedonic analysis of used car prices in Turkey. Int. J. Econ. Perspect. **3**(2), 141–149 (2009)
6. Matas, A., Raymond, J.L.: Hedonic prices for cars: an application to the Spanish car market, 1981–2005. Appl. Econ. **41**(22), 2887–2904 (2009)
7. Prieto, M., Caemmerer, B., Baltas, G.: Using a hedonic price model to test prospect theory assertions: The asymmetrical and nonlinear effect of reliability on used car prices. J. Retail. Consum. Serv. **1**(22), 206–212 (2015)
8. Pudaruth, S.: Predicting the price of used cars using machine learning techniques. Int. J. Inf. Comput. Technol. **4**(7), 753–764 (2014)
9. Kumar, K.S.: Used Car Price Prediction using K-Nearest Neighbor Based Model
10. Monburinon, N., Chertchom, P., Kaewkiriya, T., Rungpheung, S., Buya, S., Boonpou, P.: Prediction of prices for used car by using regression models. In: 2018 5th International Conference on Business and Industrial Research (ICBIR), pp. 115–119. IEEE (17 May 2018)
11. Peerun, S., Chummun, N.H., Pudaruth, S.: Predicting the price of second-hand cars using artificial neural networks. In: The Second International Conference on Data Mining, Internet Computing, and Big Data (BigData 2015), p. 17 (29 Jun 2015)
12. Gongqi, S., Yansong, W., Qiang, Z.: New model for residual value prediction of the used car based on BP neural network and nonlinear curve fit. In: 2011 Third International Conference on Measuring Technology and Mechatronics Automation, vol. 2, pp. 682–685. IEEE (6 Jan 2011)
13. Sun, N., Bai, H., Geng, Y., Shi, H.: Price evaluation model in second-hand car system based on BP neural network theory. In: 2017 18th IEEE/ACIS International Conference on Software Engineering, Artificial Intelligence, Networking and Parallel/Distributed Computing (SNPD), pp. 431–436. IEEE (26 Jun 2017)
14. Gegic, E., Isakovic, B., Keco, D., Masetic, Z., Kevric, J.: Car price prediction using machine learning techniques. TEM J. **8**(1), 113 (2019)
15. Listiani, M.: Support vector regression analysis for price prediction in a car leasing application. Unpublished (Mar 2009). https://www.ifis.uni-luebeck.de/moeller/publist-sts-pw-andm/source/papers/2009/list09.pdf
16. Pal, N., Arora, P., Kohli, P., Sundararaman, D., Palakurthy, S.S.: How much is my car worth? a methodology for predicting used cars' prices using random forest. In: Arai, K., Kapoor, S., Bhatia, R. (eds.) FICC 2018. AISC, vol. 886, pp. 413–422. Springer, Cham (2019). https://doi.org/10.1007/978-3-030-03402-3_28
17. Wu, J.D., Hsu, C.C., Chen, H.C.: An expert system of price forecasting for used cars using adaptive neuro-fuzzy inference. Expert Syst. Appl. **36**(4), 7809–7817 (2009)

18. Noor, K., Jan, S.: Vehicle price prediction system using machine learning techniques. Int. J. Comput. Appli. **167**(9), 27–31 (2017)
19. Kuiper, S., College, G.: Introduction to multiple regression: how much is your car worth. JournalStat. Educ. **16**(3) (2008)
20. Geurts, P., Ernst, D., Wehenkel, L.: Extremely randomized trees. Mach. Learn. **63**(1), 3–42 (2006)
21. Sharafati, A., Asadollah, S.B., Hosseinzadeh, M.: The potential of new ensemble machine learning models for effluent quality parameters prediction and related uncertainty. Process Saf. Environ. Prot. **1**(140), 68–78 (2020)
22. Breiman, L.: Random forests. Mach. Learn. **45**(1), 5–32 (2001)
23. John, V., Liu, Z., Guo, C., Mita, S., Kidono, K.: Real-time lane estimation using deep features and extra trees regression. In: Bräunl, T., McCane, B., Rivera, M., Yu, X. (eds.) PSIVT 2015. LNCS, vol. 9431, pp. 721–733. Springer, Cham (2016). https://doi.org/10.1007/978-3-319-29451-3_57
24. Mishra, G., Sehgal, D., Valadi, J.K.: Quantitative structure activity relationship study of the anti-hepatitis peptides employing random forests and extra-trees regressors. Bioinformation **13**(3), 60 (2017)
25. Chen, T., Guestrin, C.: Xgboost: A scalable tree boosting system. In: Proceedings of the 22nd acm Sigkdd International Conference on Knowledge Discovery and Data Mining, pp. 785–794 (13 Aug 2016)
26. Friedman, J.H.: Greedy function approximation: a gradient boosting machine. Ann. Stat. **1**, 1189–232 (2001)
27. https://second-hand-car-price-app.herokuapp.com/

Semantic Aware Video Clipper Using Speech Recognition Toolkit

Adishwar Sharma, Karanjot Singh(✉), Keshav Dubey, Amit Kumar, and Prajakta Ugale

School of Computer Engineering and Technology, MIT Academy of Engineering, Pune, India
{adishwarsharma,kjsingh,kkdubey,agkumar,pvugale}@mitaoe.ac.in

Abstract. The Internet learning framework was presented as not a single one of us could meet up like previously, during the pandemic. Most understudies go to their web-based classes on portals like Google Meet, Microsoft Groups, Zoom. Students having some doubt regarding a topic, or they somehow missed the lesson, to cover the classes they can go through the recordings, but in these recordings, students are only interested in watching the content where the teacher was teaching a particular concept/topic, but the video also has a lot of the content which is not so useful, i.e., unrelated to the content being taught and also want to skip the part where nothing is happening. So, to eliminate these problems, we propose Semantic Aware Video Clipper (SAVC), a software to clip the irrelevant stuff and the part where no activity is there (silence) and make these videos/recordings relevant to the topic and hence save the priceless time of users, with respect to learning. SAVC will automatically cut some video fragments and then join these fragments together and this entire process will be completely automatic, without human intervention.

Keywords: Video Clipper · Semantic · Automatic · Silence

1 Introduction

This pandemic has been a curse and blessing to us all. Let's keep the negativity aside and let's discuss the blessing. As a blessing, it introduced an online teaching system in our Indian education system, which was not at all entertained beforehand, a new change for the education System. Nowadays, all institutions are using platforms like teams, zoom for online teaching, but after missing a lecture, students need to go through a recording to get along with the class. But the recording contains various things such as the part where no activity is there (silence) and the starting time of the class where the teacher is waiting for the students to join, attendance, which is not relevant to the topic and watching all this will be time consuming.

Many advanced video editing techniques are present to eliminate this problem. But video editing itself is always time-consuming. Removing unnecessary video fragments is not a difficult, but time-consuming task. One must thoroughly review the video (possibly multiple times!), select all necessary fragments, join them, and then render the video for

H. Sharma et al. (Eds.): ICIVC 2022, PALO 17, pp. 421–430, 2023.
https://doi.org/10.1007/978-3-031-31164-2_35

an extended period. An hour-long video usually takes more than three hours to edit. So, SAVC (semantic aware video clipper) proposes to trim the video automatically where no activity is taking place (silence), hence will help the students in saving their precious time while watching the recordings. In our future endeavors, we will be working on trimming stuff irrelevant to the topic being taught and make the video to the point.

2 Literature Survey

Sergey Podlsenyy et al. [1] have proposed new features based on the state vector to dramatically improve the whole quality of the videos. It also provides an automated way of editing video footage from multiple cameras to create a meaningful story of events present in any video. The steps involved in selecting the most valuable parts from the video in terms of visual as well as audio quality. The importance of the action planned was about cutting the footage into a meaningful story that would be interesting to watch. The extracts being used here were from ImageNet-trained convolutional neural networks.

Tushar Sahoo and Sabyasachi Patra [2] have suggested a technique for composite silence removal under short time energy and statistical methods. The performance of the proposed algorithms was good as compared to STE and the statistical method. The proposed algorithm was applied in the first stage of the speaker identification system and a 20% silence removal is identified after the final comparison.

Takuya Furukawa and Hironobu Fujiyo [3] have discussed a technique for editing personal videos and sensor information automatically (i.e., without human convention). The proposed technique uses CRIM (continuous rank increase measure and MC (motion correlation) values which generally capture any motion-change, acceleration, etc. from sensors to cut the scenes including object movement. Higher quality results were obtained using the suggested method.

Goutam Saha, S.S. Sandipan Chakroborty and Suman Senapati [4] have proposed a new silence removal and endpoint detection algorithm for speech-recognition applications. Their proposed algorithm used Probability Density Function (PDF) of the background noise and a linear pattern classifier for identifying the silence part of the video. The proposed method also performs better end point detection and the silence removal than ZCR and STE methods.

A. Adjila, M. Ahfir and D. Ziadi [5] have suggested a method for detection and removal of silence from the audio. This method working is based on the continuous average of the speech signals. The technique used was beneficial to highlight the overall performance and its accuracy. But the audio from only three languages including English, Arabic and French, performed better in noisy areas.

Dharmik Timbadia1 and Hardik Shah [6] have developed a straightforward computation for making the tests of the first audio records. Their suggested calculation generalizes the clear edges and clamour by changing over sign. They have utilized MATLAB to generate substantial results.

Frederico Pereira et al. [7] have presented a web-based speed-to-term recognition approach for providing an easier interaction with baseline functionalities by using a blend of techniques like Voice Activity Detection (VAD), Automatic Speech Recognition (ASR) and NLP (Natural Language Processing).

Prerana Das, Kakali Acharjee, Pranab Das and Vijay Prasad [8] have designed and implemented a voice recognition system for speech to text conversion. Their proposed system has two main parts i.e., first one for processing acoustic signal and the second one for interpreting. They have used the Hidden Markov Model (HMM) for building letters.

Babu Pandipati and Dr. R.Praveen Sam [9] have researched and evaluated the methods used in STT conversions. The method built on the interactive voice response was proposed. They explored various methods of speech-to-text conversion for utilizing it in an e-mail system entirely based on voice. They have also suggested a model which uses both ANN and HMM techniques for STT conversion.

Brezeale et al. [10] proposed various text, audio, and visual video classification techniques. They have suggested that only audio and visual feature extraction are used in various applications while being equally important. As a result, if they employ their combination, the results will be extremely accurate and precise. They have worked upon various automatic video classification techniques to assist viewers for finding best interest in viewing. They have also surveyed and discovered some features that are drawn from text, audio and visual modalities and furthermore investigated a wide range of feature and classification combinations.

Kranthi Kumar Rachavarapu et al. [11] have addressed the major problems in data driven cinematography and video retargeting using gaze. An algorithm for content adaptation and automating the process of video creation was also proposed which edits with cut, pan and zoom operations by optimizing the path of a cropping window with the original video. They have also tried to preserve the crucial information. And they have focused on two algorithms i.e., one for efficient video content adaptation and the other for automating the whole process of video content creation. A novel approach was also presented by them in order to tackle the problems of efficient video content adaptation to optimally retarget the videos for varied displays with totally different aspect ratios and also by preserving important scene content.

Abdelkader Outtagarts et al. [12] have presented a cloud-based collaboration and automatic video editing technique with an approach to automatic video editing from audio transcription to text using transcripts that are selected and automatically concatenated video sequences to create a new video. Their approach was based on extraction of keywords from audio-embedded in videos. They have proposed a keyword-based mashup modeling approach in which a video-editing testbed was designed and implemented. They have used keywords to automatically edit videos to create different mashups. It's been also stated that video editing is the process of selecting segments from a set of raw videos and then concatenating all of them by adding audio, effects and different transitions to create a new video or a type of mashup. Their proposed software wants to enable the automatic keyword-based video editing, deployed in the cloud with a preview web-client. Different videos which came from different videosharing platforms or from webcams were stacked and also viewed on the fly. They have also suggested that the video editor is also a research testbed for the study of automatic video editing and summarization based on the statistical and textual data and its metadata.

Edirlel Soares de Lima et al. [13] have presented a real-time editing method for interactive storytelling. Their proposed method generates the most appropriate shot transitions, swaps video fragments to avoid jump cuts, and also creates ample looping scenes. Their research focuses on the concept of video-based storytelling to use pre-recorded videos with real actors. In their approach, automatic video editing was generally required for developing interactive narratives with the quality of feature films. But that was also their critical issue which was not fully addressed.

Favlo Vazquez et al. [14] have suggested an idea to fix annoying silence in videos and all related errors. So, Python Jump cutter is used here to edit videos. Jump cutter is a program written in Python that automatically cuts silent parts of videos. The purpose method facilitates any post-recording work. It has one limitation in the form of very low video quality.

Joseph C. Tsai et al. [15] have proposed a new motion-in-painting technique which allows the users to change the dynamic texture used in a video background for production of special effects. Motions estimations of global and local textures are used by them. Furthermore, they have used video blending techniques in conjunction.

In video editing, segments are selected from a set of raw videos and concatenated by adding audio, effects, and transitions to create a new video or mashup. Most of the video editing tools available are time-consuming and require specific knowledge. We discovered how automatic editing of multi-camera video footage works to create a coherent narrative of an event, where the components and training used are semantic data extraction. And model training for automatic editing. The steps to select the most valuable footage in terms of visual quality and action shot importance. Regarding video classification, different methods such as text, audio and video are used, but in different applications only visual and acoustic feature extraction are used, although they are equally important. We found that features are extracted from three modalities (text, audio, and image) and that a variety of feature and classification combinations were examined, while cropping windows within the original video while attempting to retain important information and follow cinematic principles.

3 System Design and Preliminaries

3.1 System Architecture

The Fig. 1 explains the overall architecture, Services and requests over the system. It acts as a blueprint that shows how to manage all the crucial service and requests over the system.

The below Fig. 2 explains the overall flow of our project and shows the overall design of new systems and also provides a major overview of all the system components, key participants and the rest relationships of the proposed system. In this diagram, firstly, the user has to import the video in the given format i.e., avi, yuv, mkv, flv, wmv then the user will have gone to processing stage where the synchronization of video takes place and the video is automatically edited and once processing is completed, the file will get be ready to export.

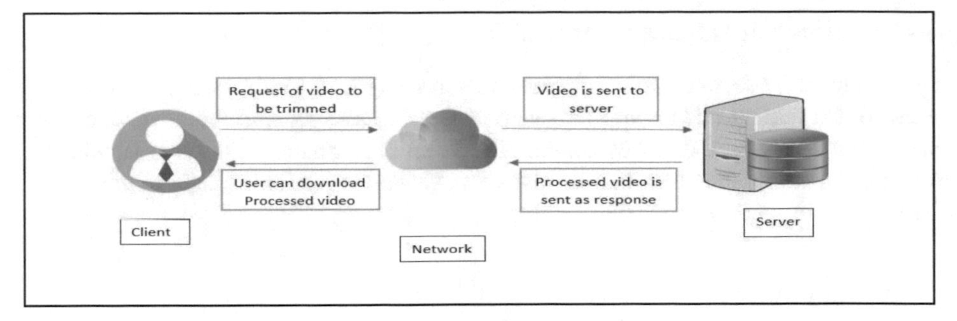

Fig. 1. Proposed System

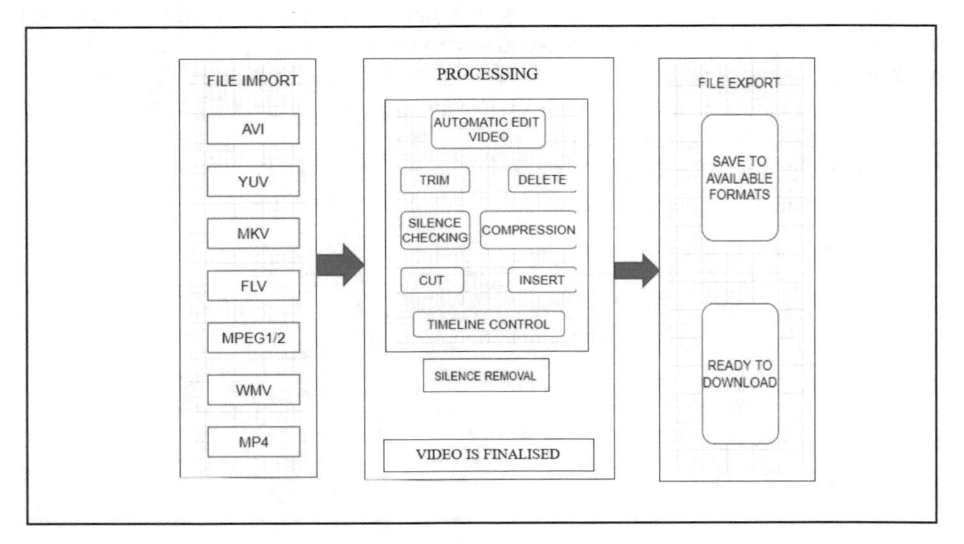

Fig. 2. Block Diagram

3.2 Preliminaries Vosk API

Vosk is a speech recognition toolkit that supports more than 20 languages (e.g., English, German, Hindi, etc.) and dialects. It works offline and even on lightweight devices like Raspberry Pi. Its portable models are only 50 Mb each. However, there are also much larger models.

3.3 Silence Removal

This method cuts moments when silence (no activity happens) lasts longer than a certain threshold (for example, 2 seconds). This approach is fully automated and requires no human intervention during or after video recording. Just enter the video path and you will get the video without any silent moments.

3.4 Control Words Removal Algorithm

Control word is a set of commonly used words in a language. Control words are primarily used in text mining and natural language processing and generally remove words that are so frequently used that they contain little useful information. This approach is used to remove parts of the video that are not very relevant to the topic taught in the lecture. Especially for recorded university lectures.

4 Methodology

By using the software, user will drop/upload the video recording of the lecture afterwards it will process to backend in which it will iterate the length of the whole video and will store the intervals with no conversations/no audio (longer than 3 s duration) will be detected using Machine Learning/Artificial Intelligence algorithm's and once the iteration is complete, it will clip out /trim the noted interval's using MoviePy library in python. And after the processing is completed, it will compress the video using High Efficiency Video Coding (HEVC) algorithm so that the user can download the video while saving the data (Fig. 3).

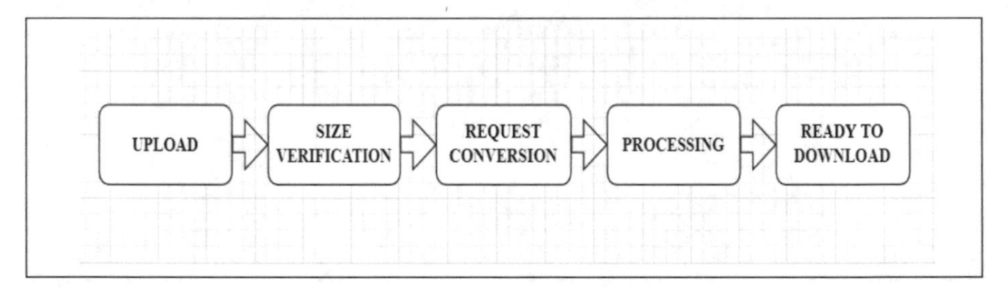

Fig. 3. General processing of the software

4.1 Techniques Used

The proposed system will automatically cut video into fragments in two ways:

A) Identifying long moments of silence, where no activity happens.

According to the proposed technique as described in Fig. 4. The video will be imported first on the server, following it, video-to-audio conversion will be done with the help of VOSK API. Now, it will iterate the audio file converted from the video, following by the conditions that if the time between the current word of a sentence and the first word of the adjacent sentence(next word) is greater than the threshold time set, then storing and clipping of the video segments, according to timestamps will be done and using MoviePy, trimming will be done according to the timestamp stored in the segments. Otherwise, it will keep iterating the audio file, until it finishes/ends.

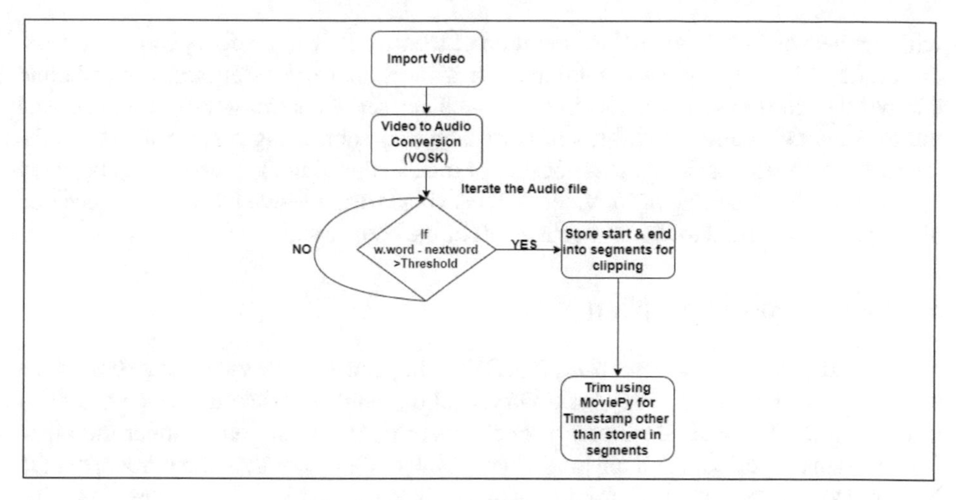

Fig. 4. Automatic Silence Removal

B) Recognizing control word

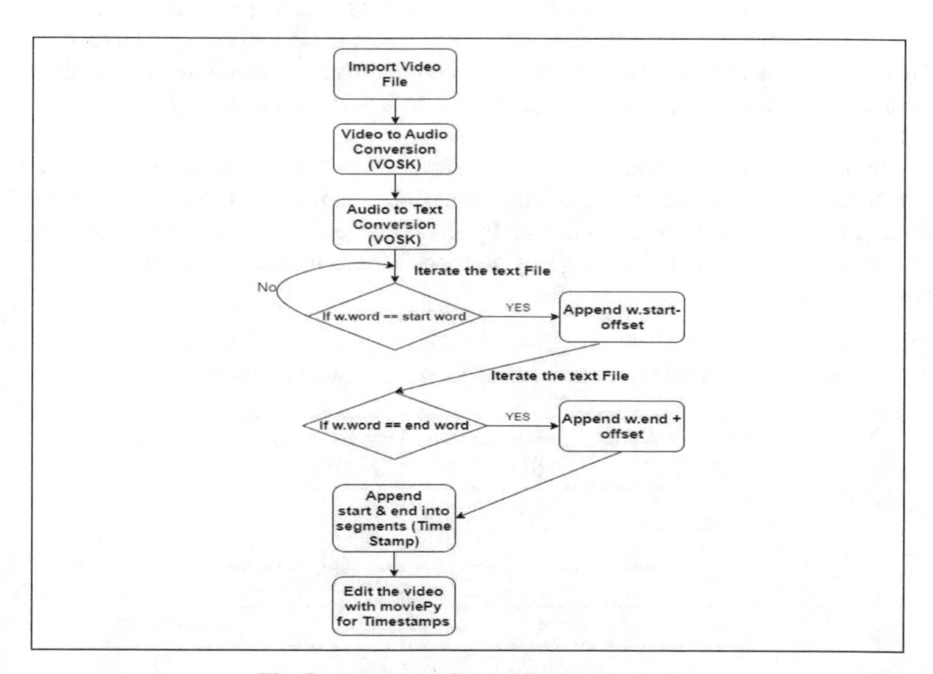

Fig. 5. Automated Control Words Removal

This method will be the future endeavor of the proposed system. With the proposed technique as described in Fig. 5, the video will be imported first in the server, following to it, video-to-audio conversion followed by the audio-to-text conversion will be done with the help of VOSK API. Now, iteration of the text file will be done, converted from the audio, following by the conditions that if the start word of current sentence and end word of current sentence lies in the trained dataset of control words, it will append the updated timestamps in the list of timestamps by subtracting offset value from the respective timestamps, hence the timestamps to be included in the video would be ready by this time. Now, editing of I/P video will take place using Moviepy, and the processed video will be returned to the user as a result of the request sent.

5 Results and Discussion

Video editing is always time-consuming. Removing unnecessary video fragments is not a difficult, but time-consuming task. One must thoroughly review the video (possibly multiple times!), select all necessary fragments, join them, and then render the video for an extended period. An hour-long video usually takes more than three hours to edit. The major targeted audience for the software is students, as to help them by saving their precious time while watching lecture recordings. As stated in the problem statement, the proposed software will automatically cut all video fragments and then join these fragments together and this entire process will be completely automatic, without human intervention. The suggested algorithm is very simple, as the various tasks involved are divided into 3 subtasks: · Edit videos using moviepy library. · Recognize control words/silence and their timestamps · Connect these two components together This approach is fully automated and does not require any human intervention during or after video recording. One must specify the path to the video and get the processed video.

Table 1 shows the data collected during the testing phase of our project which was conducted throughout the month of May 2022 (on more than hundreds of videos), and these are some of the results, followed by the visual plotting of the values obtained. The testing was conducted by our group, to check various functionalities of the software being developed.

Table 1. Results among various processed videos

Original Video (IN SEC)	Silence (IN SEC)	Processed Video (IN SEC)
26	15	11
6	3	3
77	58	19
350	60	290
600	2	600
660	50	610

Comparison and testing of more than 100 videos have been done and, in the following table and the chart as in Fig. 6, a total of seven videos are considered. If the amount of silence part is more than the threshold value then only the silence part gets removed and if it is less than the threshold value then the silence part doesn't get removed from the original video. And by testing, we have concluded on a note that the more the length of the video, the more time it will take to completely process the video.

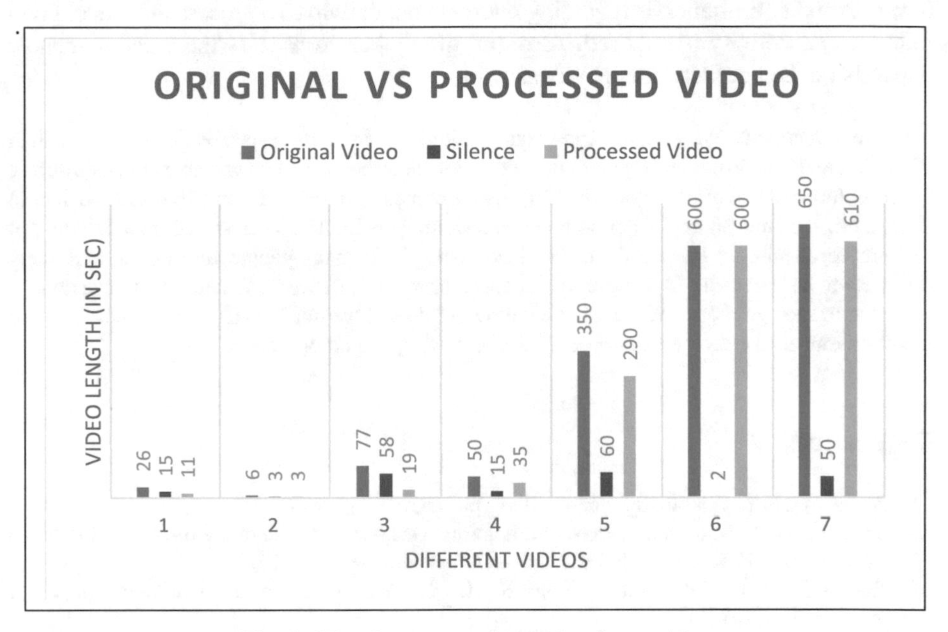

Fig. 6. Visual representation of the above table

6 Conclusion

Online learning has been a trend in the last decade as anyone can learn at their own place and whenever they like. Since the pandemic, online learning has become one of the key factors in the field of education. Nowadays, the majority of users are using platforms like teams, zoom for online teaching, but if users miss a lecture, they need to go through a recording to get along with the class. So, the recording material should be short and up to the mark. The software focuses on using the available techniques to achieve many of these problems. The proposed software not only trims the silence part (no audio) of the recording but is also capable of reducing the memory size of the recording while retaining the original quality of the recording. Therefore, it is very feasible for users as they can directly access the relevant content and is considerable as a time-saving component to some extent. By examining all the above-mentioned features and benefits of the software, it has a lot of room for improvement, especially as India moves closer to becoming an educationally advanced country and more ideas and implementation in the field of online teaching and learning are needed. In our future endeavors, our goal will be to trim the stuff irrelevant to the topic being taught and make the video to the point.

The proposed software has some limitations as well, first being the time to process video. It totally depends on the length of video, the more the length of video, the more the time it takes to process the video. For processing a 10-min video, it takes around 7 to 8 min. Another limitation is regarding internet connectivity. To process the video, good internet connectivity is required. Also, the time taken to process the video somehow depends on the internet connectivity as well.

Acknowledgement. We want to express our gratitude towards our respected faculty Dr. Vaishali Wangikar and Mrs. Vinodini Gupta for their constant encouragement and valuable guidance during the completion of this project work. We also want to express our gratitude towards respected School Dean Mrs. Ranjana Badre for her continuous encouragement. The success and final outcome of this project required a lot of guidance and assistance from many people, and we are extremely fortunate to have got this all along the completion of my project work. Whatever we have done is only due to such guidance. We would be failing our duty if we don't thank all the other staff and faculty members for their experienced advice and evergreen co-operation.

References

1. Podlesnyy, S.Y.: Automatic video editing, pp. 155–191 (2021)
2. Brezeale, D.C.: Automatic video classification: a survey of the literature,pp. 416–430 (2008)
3. Rachavarapu, K.K.: Towards Data-Driven Cinematography (2019)
4. Tsai, J.C., Shih, T.K., Wattanachote, K., Li, K.: Video editing using motion inpainting, pp. 649–654 (2012)
5. Outtagarts, A., Mbodj, A.: A cloud-based collaborative and automatic video editor. In: 2012 IEEE International Symposium on Multimedia, pp. 380–381. IEEE (2012)
6. Lima, E.S.D., Feijó, B., Furtado, A.L., Ciarlini, A., Pozzer, C.: Automatic video editing for video-based interactive storytelling. In: 2012 IEEE International Conference on Multimedia and Expo, pp. 806–811. IEEE (2012)
7. Sahoo, T., Patra, S.: Silence removal and endpoint detection of speech signal for text independent speaker identification. In: 2014 International Journal of Image, Graphics and Signal Processing, pp. 27–35 (2014)
8. Saha, G., Chakroborty, S.S., Senapati, S.: A new silence removal and end- point detection algorithm for speech and speaker recognition applications (2005)
9. Furukawa, T., Fujiyoshi, H.: A cut method for cutting and editing personal videos using st-patches and sensor information. J. Inst. Image Inf. Television Eng., 93–100 (2012)
10. Adjila, A., Ahfir, M., Ziadi, D.: silence detection and removal method based on the continuous average energy of speech signal. In: 2021 International Conference on Information Systems and Advanced Technologies (ICISAT), pp. 1–5 (2021)
11. Pereira, T., et al.: A web-based voice interaction framework proposal for enhancing information systems user experience. Procedia Comput. Sci. **196**, 235–244 (2021)
12. Das, P., Acharjee, K., Das, P., Prasad, V.: Voice recognition system: speech-to-text. J. Appl. Fund. Sci. **1**, 2395–5562 (2015)
13. Merabti, B., Christie, M., Bouatouch, K.: A virtual director using hidden markov models. In: Computer Graphics Forum. Wiley (2015). https://doi.org/10.1111/cgf.12775.Hal-01244643
14. Timbadia, D., Shah, H.: Removing silence and noise using audio framing, pp. 118–120 (2021)
15. Vosk. https://alphacephei.com/vosk/adaptation. Accessed 12 Nov 2022

Feature-Rich Long-Term Bitcoin Trading Assistant

Jatin Nainani(✉), Nirman Taterh, Md Ausaf Rashid, and Ankit Khivasara

K. J. Somaiya College of Engineering, Vidyavihar 400077, India
{jatin.nainani,nirman.t,mdausaf.r,ankit.khivasara}@somaiya.edu

Abstract. For a long time predicting, studying and analyzing financial indices has been of major interest for the financial community. Recently, there has been a growing interest in the Deep-Learning community to make use of reinforcement learning which has surpassed many of the previous benchmarks in a lot of fields. Our method provides a feature rich environment for the reinforcement learning agent to work on. The aim is to provide long term profits to the user so, we took into consideration the most reliable technical indicators. We have also developed a custom indicator which would provide better insights of the Bitcoin market to the user. The Bitcoin market follows the emotions and sentiments of the traders, so another element of our trading environment is the overall daily Sentiment Score of the market on Twitter. The agent is tested for a period of 685 d which also included the volatile period of Covid-19. It has been capable of providing reliable recommendations which give an average profit of about 69%. Finally, the agent is also capable of suggesting the optimal actions to the user through a website. Users on the website can also access the visualizations of the indicators to help fortify their decisions.

Keywords: Bitcoin · Cryptocurrency · Reinforcement Learning · Sentiment Analysis · Technical Analysis · Trend trading

1 Introduction

Cryptocurrency is a virtual currency which is secured by cryptography which uses cryptographic functions to facilitate financial transactions and form a system to store & transfer value. It leverages blockchain technology in order to be decentralized. Its most important feature is that it's not controlled by a central authority like the government or bank to uphold or maintain it. Bitcoin, which was released in 2009 is the first decentralized cryptocurrency.

While there are numerous solutions and trading bots that try to generate profits by trading and through short term patterns, there is a need for an investing assistant which tries to maximise profits by considering long term trends and real-time user sentiments.

It helps give the investor a perspective of the overall market situation and accordingly enables them to plan their investment and allocate an appropriate percentage of their financial portfolio to cryptocurrencies and Bitcoin.

H. Sharma et al. (Eds.): ICIVC 2022, PALO 17, pp. 431–442, 2023.
https://doi.org/10.1007/978-3-031-31164-2_36

It is important to note that more than 98% percentage of short-term traders do not make a profit net of fees. This is because it is complicated to time the market and make the right trading decision every time. This is because there are too many variables involved in predicting the short-term price of any asset, more so in the case of Bitcoin since it is one of the most volatile financial assets in the market.

The critical difference between a trading bot and an investing assistant is: In the case of the trading bot, the final trade decision is made by the bot itself, while in the investing assistant, the final investment decision is on the investor.

There are numerous challenges involved in building a robust and accurate reinforcement learning model that's a good enough representation of a cryptocurrency market's environment. Cryptocurrency price actions are extremely volatile and depend on a large amount of real-world and statistical factors. We will start by outlining the overall flow and architecture of the model. We then move forward to discuss the implementation of various features like Twitter sentiments, technical indicators and the custom index. Finally, all these features will be utilized in the reinforcement learning model. The model will be evaluated using robust evaluation metrics like entropy loss, policy loss and value loss. The final product is hosted on a website for the user to get the recommendations and gain insights from the visualizations.

2 Literature Survey

Behavioural economists like Daniel Kahneman and Amos Tversky have established that decisions, including financial consequences, are impacted by the emotions and not just value alone [1]. Insights from these researchers open up prospects to find advantages through various tools like sentiment analysis as it shows that demand for goods, and hence their price, may be impacted by more than its economic fundamentals. Panger et al. found that the Twitter sentiment also correlated with people's overall emotional state [2].

Similar work has been done on cryptocurrencies after their introduction to see if such methods effectively predict cryptocurrency price changes. In the paper by Stenqvist et al. [3], the authors describe how they collected tweets related to Bitcoin. First, tweets were cleaned of non-alphanumeric symbols (using "#" and "@" as examples of symbols removed). Then the tweets which were not relevant were removed from the analysis. The authors then used VADER (the Valence Aware Dictionary and Sentiment Reasoner) to finally analyze the sentiment of each tweet and hence classify it as negative, neutral, or positive. Only tweets that could be considered positive or negative were kept in the final analysis . The sentiment analysis done in this project builds off everything above but is unique, and we solve the problem of giving the tweets a weight.

Sattarov et al. [4] created a recommendation system for cryptocurrency trading with Deep Reinforcement Learning. Data was taken from cryptodatadownload and focused on three currencies Bitcoin (BTC), Litecoin (LTC), and Ethereum (ETH). The environment was responsible for accounting for stock assets, money management, model monitoring, buying and holding or selling stocks and finally calculating the reward for actions taken. The experiment on Bitcoin via DRL application shows that the investor got 14.4% net profits within one month. Similarly, tests when carried out on Ethereum and Litecoin also finished with 74% and 41% profit, respectively.

Liu et al. [5] regarded the transaction process as actions. The returns were regarded as awards and prices are considered to be states to align with the idea of reinforcement learning. A Deep Reinforcement Learning Algorithm - Proximal Policy Optimization (PPO) was used for high-frequency bitcoin trading. They first compared the results between advanced machine learning algorithms like support vector machine (SVM), multi-layer perceptron (MLP), long short-term memory (LSTM), Transformers and temporal convolutional network (TCN), by applying them to real-time bitcoin price and the experimental results demonstrated that LSTM outperforms. It was then decided to use LSTM as the policy for the PPO algorithm. The approach was able to trade the bitcoins in simulated environment paired with synchronous data and obtained a 31.67% more return than that of the best benchmark, improving the benchmark by 12.75%.

Borrageiro et al. [6] created an agent that learns to trade the XBTUSD (Bitcoin versus US Dollars) perpetual swap derivatives contract on BitMEX on an intraday basis. The cryptocurrency agent realized a total return of 350%, net transaction costs over five years, 71% of which is down to funding profit. The annualized information ratio that it achieves is 1.46. The echo state network provides a scalable and robust feature space representation.

We can conclude a few points from the above literature survey. There is a distinct lack of a consistent environment, leading to some really restrictive while others are too free and ill-defined. Most of the agents are restricted to a single type of market. The variety of preprocessing techniques used led to the question of whether the improvement in the metric was the result of the model or the data fed. Most models are treated as a complete black box with a lot of hyperparameter tuning. Perhaps some form of explainable AI might find some use here to convince the investors and help them understand on what basis our model recommends actions.

3 Approach

3.1 Block Diagram

The block diagram is shown in Fig. 1. It describes the flow of the project.

3.2 Data Description

Bitcoin Historical Data: We obtained the data on past prices from Cryptocompare as it regularly reviews exchanges, carefully monitoring the market to deliver the most accurate pricing indices. The main difference between BTC prices and usual stock prices is that the BTC prices change on a much larger

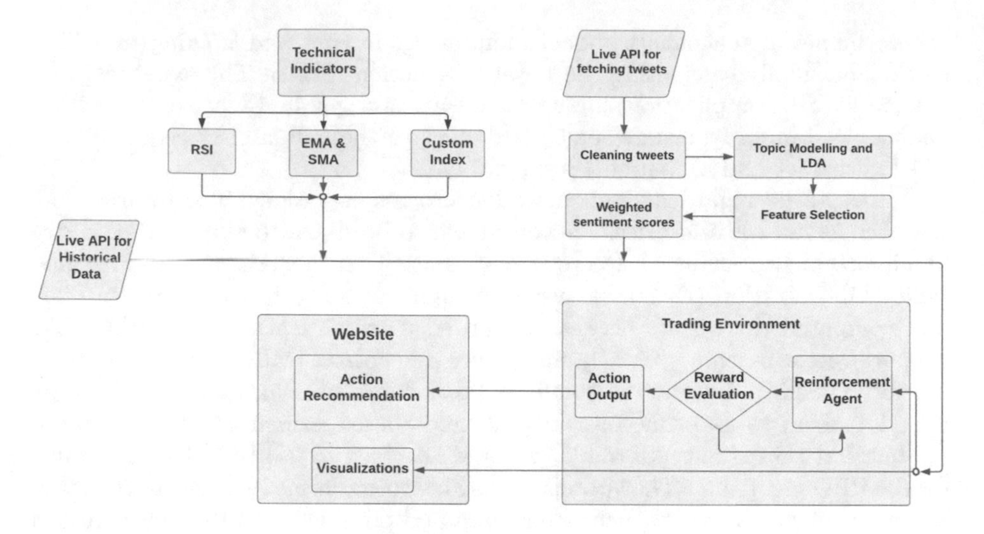

Fig. 1. Block Diagram

scale than the local currencies. The overall data collection period is from January 2016 to April 2021 on a daily fidelity which gives us a total of 1918 values. This dataset consists of six different attributes, namely, Date, High, Low, Open, Close, and Volume.

Bitcoin Related Tweets: Twitter is a minefield of information due to the volume and variety of users. This has resulted in Crypto Twitter being significantly influential on cryptocurrency trading decisions. Thus, we collected tweets to perform sentiment analysis on. We used the Twitter API to query tweets by applying to the developer role. bitcoin, BTC, #Bitcoin, and #BTC were used as the keyword filters for fetching tweets for a period of 2016/1 to 2021/04. Also, all tweets were fetched in the English language. This gave us a total of 4,265,266 tweets to work with.

3.3 Technical Analysis

Trend Indicators

Simple Moving Average: SMA is just the mean price over the specified period. The mean average is called "moving" because it is plotted on the chart bar-wise, which forms a line that moves along the chart with respect to the average value. When it comes to our project, we aim at maximizing profits in the long run. So, we took a longer time span of 21 weeks to calculate our SMA to make sure that it is not sensitive to all the little changes in price but also does not overlook the major price changes.

$$SMA = \frac{(A_1 + A_2 + ... + A_n)}{n} \tag{1}$$

A_n is the price of Bitcoin at period n
n is the number of total periods

Exponential Moving Average: EMA is used to determine trend direction and to trade in that direction. When the EMA rises buying when prices dip near the EMA is a good move. When the EMA falls, selling when prices rally towards EMA is a good choice. Testing for different values, we found the value of 2 to give just enough credibility to recent data. The EMA has a shorter delay than the SMA within the same period. Hence we decided to go for a period of 20 weeks for EMA calculations.
The formula for EMA at period n is:

$$EMA_n = \left(A_n * \left(\frac{S}{1+n}\right)\right) + EMA_{n-1} * \left(1 - \left(\frac{S}{1+n}\right)\right) \tag{2}$$

where, A_n is the price of Bitcoin at period n
S is the smoothing factor
n is the number of total periods

Relative Strength Index: The RSI is a momentum oscillator that measures the change of price movements. Traditionally the RSI is considered to be overbought when its value is above 70 and oversold when the value is below 30. RSI is also helpful to identify the general trend. Usually 14 periods are used to calculate the initial RSI value. Once 14 periods of data is available, the second part of the RSI formula is to be done which smooths the results.
Value of RSI is calculated with a two steps that starts with the below formula:

$$RS = \frac{Avg.Gain}{Avg.Loss} \tag{3}$$

$$RSI = 100 - \frac{100}{1 + RS} \tag{4}$$

Custom Indicator-BMSB: BMSB index is a Bollinger band that helps us understand the current valuation of bitcoin against USD, vis., undervalued, overvalued, or fair valued. It ranges from -100 to 100, with the negative side corresponding to extreme undervaluation and the positive side indicating extreme overvaluation with respect to recent price movements.

Usually, when the price of Bitcoin falls below BMSB, there is a bearish momentum in the market, and when the price of Bitcoin holds support or stays above the BMSB, it indicates the onset of a bull market. That being said, being over-extension (+70 to +100 range) does not necessarily imply a bull run, and under-valuation (–70 to -=100 range) does not necessarily imply a bearish movement. In fact, it can indicate the end of a bull run and bear market, respectively. Therefore, in our project, when the BMSB index tends to -100, we show it as a buying opportunity, and when it tends to +100, we show it as a time to take profits.

Algorithm: Let us introduce a few of the crucial variables that will enable us to formulate the algorithm for the BMSB Index: We define Combined Average (μ) as the average of the current 21 weeks EMA and 20 weeks SMA. We define K_p as the Price Coefficient. The price coefficient decides the extent of the price range from which the Bitcoin price is considered to be normal. The normal price range varies from $\mu(1 - K_p)$ to $\mu(1 + K_p)$.

Algorithm 1: BMSB Indicator

Inputs: Price of Bitcoin (p), Exponential Moving Average (EMA), Simple Moving Average (SMA), Price Coefficient (K_P) and Scaling Coefficient (K_S)
Outputs: BMSB Index (I)

START

Combined Average (μ) = Average(SMA, EMA)
Upper Bound (B_U) = $\mu(1 - K_P)$
Lower Bound (B_L) = $\mu(1 + K_P)$

IF $p > B_L$ && $p < B_U$ **then**

$$\mathrm{I} = \left(\frac{p - \mu}{\mu}\right) \frac{K_S}{K_P} \times 100$$

ELSE IF $p < B_L$ **then**

$$\mathrm{I} = \left[\frac{p(1 - K_S)}{\mu(1 - K_P)} - 1\right] \times 100$$

ELSE $p > B_U$

$$\mathrm{I} = \left[1 - \frac{\mu\,(1 + K_P)\ (1 - K_S)}{p}\right] \times 100$$

end IF

STOP

K_s is defined as the scaling coefficient. The scaling coefficient decides the extent on the BMSB Index scale for the normal price range. It varies from 100(0-K_s) to 100(0+K_s) and corresponds to the $\mu(1 - K_p)$ to $\mu(1 + K_p)$ price range. We calculate the BMSB Index as shown in Algorithm 1.

3.4 Sentiment Analysis

Preprocessing

Cleaning: After scraping the tweets, we had to drop some columns that we deemed irrelevant for the sentiment analysis. We kept the columns that we deemed useful, which include the date, username, number of replies, retweets, favorites, and the text of the tweet. Before we are able to start doing any form of sentiment analysis, the tweets collected have to be cleaned.
Sample tweet before cleaning:

"Bitcoin Climbs Above 11,316.7 Level, Up 6% https://yhoo.it/2YV6wKZ #bitcoin #crypto #blockchain #btc #news #cryptocurrency pic.twitter.com/mPv3rd4kLn"

These tweets contain a large amount of noise. Using regex expressions, all these noises were removed. Preprocessing is a very important aspect of sentiment analysis; if we were to pass these raw tweets to our analyzer, chances are it will perform very poorly and take up much more time
Sample Tweet after cleaning:

Bitcoin Climbs Above Level Up

Next, we set all the characters to lower cases and also removed stopwords in our tweets. Stop words are commonly used words such as a, the, and as, which provide no valuable information in Sentiment analysis. We made use of the NLTK to get rid of them. This was enough for VADER as it was specially tuned for social media content.

Sentiment Analysis: We then compare between VADER and TextBlob, which are lexicon-based approaches for sentiment analysis. We selected a few tweets to compare if VADER or TextBlob performs better. We noticed that the sentiment score for advertisements and spam tweets was mostly quite positive. Hence, we filtered out such tweets, which has been described in detail in the Topic Handling section. From the tests we conducted on the selected few tweets, VADER works better with things like slang, emojis, etc., which is quite prevalent on Twitter, and it is also more sensitive than TextBlob. So, we choose to go with VADER.

After the sentiment of each Tweet is computed, we take the daily average sentiment, but taking the average sentiment results in the loss of information. For example, a prominent user's positive tweet (score of +1.0) may have greater influence over public opinion as compared to a common users' negative tweet (score of 0). To address this issue, we decided to use weighted Sentiment analysis scores.

Table 1. Established Weight Rules

Rule	Weight Scores
Prominent Users: List of prominent Crypto Twitter users is generated.	If the Tweet is made by user on the list, +1 to the weight score
Bitcoin Keyword List: List of keywords used in Bitcoin are generated.	Add weight +1 if at least one or more keywords are used.
Number of Retweets, Replies and Favourites: The larger the number, the greater the impact and reach of the tweet.	Added the round(log(no of retweets + replies + favourites) to the weight score. We took log to normalise the data.
Time: Tweets made nearer to closing price are more reflective of the current public opinion and carry more weight.	For daily, if the tweet is made from 13:00 to 23:59 and for hourly, if the tweet is made in the last 30 mins, +1 is added to the weights.
Advertisement Score: We tag each tweet with an ad score based on topic clustering.	The Advertisement Score weight is the ad score obtained during data pre-processing

Feature Selection and Weighted Sentiment Score: Our collection of features consists of Tweets text, Time of Tweet, Number of Retweets, Number of Replies, Number of Favorites, Advertisement score, Tweet Volume, and Sentiment Analysis Score (VADER). To prevent the loss of information by taking the average sentiment, we have generated weights to create weighted sentiment scores. Each tweet will have a weighted score. Table 1 shows the Weight Rules we have established.
The final weights equation is as follows:

Weight = [weight(prominent user) + weight(keyword) + weight(# of retweets) + weight(# of replies) + weight(# of favorites) + weight(time)] * ad_score

Weighted Sentiment per tweet = sentiment score of tweet * Weight

Weighted Sentiment per day = sum(Weighted Sentiment per tweet)/ # of tweets in one day

To address the issue with VADER assigning ads high polarity scores, we multiply the ad_score to the other weights. As ad_score for ads are assigned as 0 and so Weighted Sentiment will be 0. As the weighted sentiment of ads are now neutral, we have effectively filtered out the ads as it will not affect our model prediction. Hence these newly generated Weighted Sentiment Scores are then passed onto the RL model as a feature.

3.5 Reinforcement Learning

Environment

Positions: Positions of the market describe the amount of Bitcoin a particular user holds at that moment. In our environment, we have considered long and short positions. These suggest the two potential directions of the price required to maximize profit. The Long position wants to buy shares when prices are low and profit by sticking with them while their value is going up, and the Short position wants to sell shares with high value and use this value to buy shares at a lower value, keeping the difference as profit. The environment assigns a value of 0 to the agent if the position is discovered to be Short and 1 if the position is discovered to be Long. When starting a new environment, the position of the user is considered short as default.

Actions: The environment assigns a value of 0 to the agent if the recommended action is Sell and 1 if recommended action is Buy. However, performing a trade on every interval does not produce reliable results for a real-life situation. To bridge the gap between simulation and reality and maximize long-term profits, the algorithm accepts the recommendation from the model and the user's current position in the market to provide the final verdict. To do that, it refers to the position of the environment. When you make a short trade, you are selling a borrowed asset hoping that its price will go down and you can buy it back later for a profit. So our algorithm will recommend the user to Buy after the short trade is satisfied and Sell after the long trade is satisfied. If the above conditions are not satisfied, then no trade occurs (Holding) because our conditional logic suggests that profit can be maximized just by holding in such a case.

Algorithm 2: Final Recommendation

Inputs: Recommended Action, Current Position
Outputs: Final Recommendation

IF Recommended Action = Buy && Current Position = Short **then**
Final Recommendation = BUY

ELSE IF Recommended Action = Sell && Current Position = Short **then**
Final Recommendation = SELL

ELSE,
Final Recommendation = HOLD

end IF

STOP

In the end, our algorithm can now recommend three independent actions depending on the situation of the market, namely, Buy, Hold or Sell.

Learning Algorithm

Model Specifications: The Asynchronous Advantage Actor-Critic method (A3C) has been very influential since the paper [7] was published. The algorithm combines allows shareing of layers between the policy and value function and asynchronous updates.

However, Synchronous A2C performs better than asynchronous implementations, and we have not seen any evidence that the noise introduced by asynchrony provides any performance benefit. The host of this algorithm was the 'stable baselines 3' library on Python. The library offers multiple learning policies depending on the type of input data.

We have used the MLP (Multi-Layer Perceptron) policy which acts as a base policy class for actor-critic networks allowing both policy and value prediction. It provides the learning algorithm with all the base parameters like the learning rate scheduler, activation functions, normalization, feature extraction class, and much more. This model is learned with a sliding window approach.

3.6 Training of Proposed Model

The training of the RL agent was done on the Google Colab GPU, with *NVIDIA-SMI 460.67 Driver Version: 460.32.03 CUDA Version: 11.2.* The model was tested for multiple scenarios, and the below gave the most consistent results. The time steps were set to be 50000, which is equivalent to the number of iterations of training. The 1918 data instances were present, and 1233 were used for training the model with a window size of 30. The remaining 685 instances were used for testing.

3.7 Results and Discussion

After training for 50000 time steps, the entropy loss of 0.0209 was obtained. The learning rate was optimized at 0.007. Finally, the testing was performed on the latest 685 d in the data. Specifically, the testing period started from 17th May 2019 and ended on 1st April 2021. This period is highly significant as it captures the peak points of Covid-19 and its effects on the market. Given these extreme scenarios, the model was able to generate a total reward of 39,412 and a profit value of 1.69486, which is equivalent to a 69.486% increase over 685 d.

The green dots in the Fig. 2 indicate a long position, whereas the red dots indicate a short position. As discussed before, the agent itself recommends only Buy or Sell with the aim to capture every penny. Our algorithm then reviews the current situation/position of the user in the market and either recommend the agent action or asks them to hold their assets.

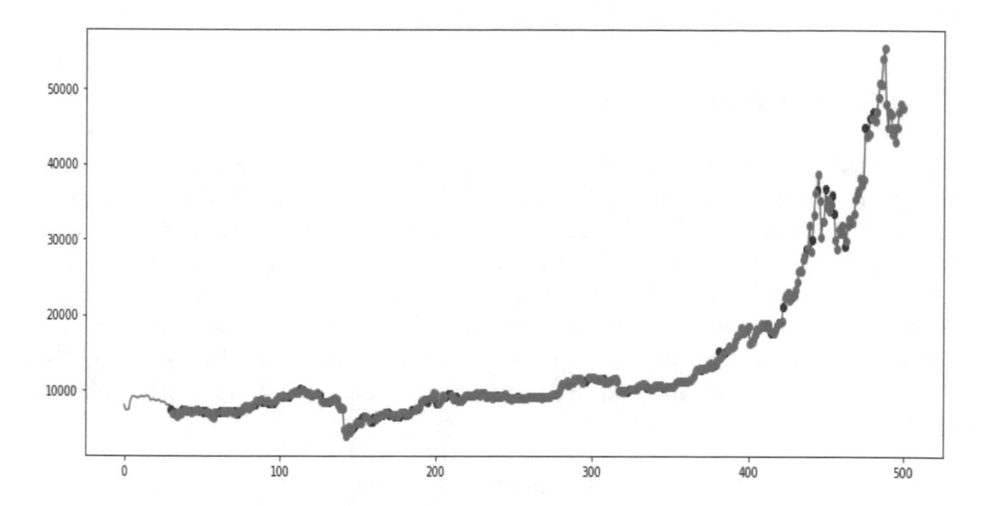

Fig. 2. Graph of recommendations

4 Conclusion

Investing in the crypto market is a tedious task and requires a lot of effort to make good trading strategies. Moreover, analyzing real-time data is a more difficult task. Traders make several trading decisions and keep on learning from them. During this process, they improve their decision-making skills. This is the major idea behind implementing a Reinforcement Learning agent, which trains against the environment to adapt and make better decisions. Thus, we can say that trading can be approached as a Reinforcement Learning Problem.

Our plan was to allow the learning agent to learn in such a feature-rich learning environment. Specifically, we have used some of the most popular and tested long-term trend indicators like SMA, RSI, and EMA. We have also crafted our technical indicator, which is used by the agent to find opportunities for trade action. Another huge factor that impacts the flow of prices is the overall sentiment of the market. By including this factor in the environment, the agent was better able to understand the market situation.

The value of the added features was demonstrated as without these, the average profit of the model was not always positive. In addition to the features, we were able to get consistently positive profits over a long period of testing. With these handcrafted features, our model was able to analyze the situation of the user and market and recommend smart decisions to the user.

We were able to provide an interface to the user for their better understanding of the current market situation through visualizations of the important indicators and sentiment scores. The interface was able to host the agent to provide its final recommendation to the user. The results show great potential for the approach, but the bitcoin markets are quite large, complex and volatile, so the modelling of this environment still presents a lot of challenges.

5 Further Work

Many experts in the field of cryptocurrency and stock trading utilize trend analysis by identifying the popular patterns in the price action. Each of these patterns helps in the analysis of the price changes to occur in the future. The ascending and descending triangle pattern leads to a continuity in the trend of prices. Because of their impact, the recognition of these patterns becomes of most significance. However, current pattern matching algorithms fail to work for different pattern spans. This problem is highly significant as even though we have an idea of what pattern we are looking for, most patterns occur at significantly different intervals.

Acknowledgments. We would like to express my very great appreciation to Ankit Khivasara Sir for their valuable and constructive suggestions during the planning and development of this project. Their willingness to give their time so generously has been very much appreciated. We would also like to extend our gratitude to the entire team of OpenAI as well as that of Stable Baselines3 for implementing and maintaining their framework, without which this project would not be possible.

References

1. Kahneman, D., Tversky, A.: Prospect theory: An analysis of decision under risk. Handbook of the fundamentals of financial decision making: Part **I**, 99–127 (2013)
2. Panger, G.T.: Emotion in social media. University of California, Berkeley (2017)
3. Stenqvist, E., Jacob, L.: Predicting Bitcoin price fluctuation with Twitter sentiment analysis (2017)
4. Sattarov, O., et al.: Recommending cryptocurrency trading points with deep reinforcement learning approach. Appli Sci. **10**(4), 1506 (2020)
5. Liu, F., et al.: Bitcoin transaction strategy construction based on deep reinforcement learning. Appli. Soft Comput. **113**, 107952 (2021)
6. Borrageiro, G., Firoozye, N., Barucca, P.: The recurrent reinforcement learning crypto agent. IEEE Access **10**, 38590–38599 (2022)
7. Mnih, V., et al.: Asynchronous methods for deep reinforcement learning. In: International Conference on Machine Learning. PMLR (2016)

PHR and Personalized Health Recommendations Using Rule-Based Approach

Tatukuri Charishma(✉), Shaik Fathimabi, Yarlagadda Mohana Bhargavi, and Sindhu priya Gude

Velagapudi Ramakrishna Siddhartha Engineering College, Vijayawada, Andhra Pradesh, India
charishmatatukuri@gmail.com, fathimabi@vrsiddhartha.ac.in

Abstract. Personal health record (PHR) systems have been increasingly popular among patients and consumers in recent years. Nowadays, the interest in following a healthy life style has increased. The needs for health information and prevention of disease have also been gaining people's interest. A personal health record (PHR) is now a patient-centric method for exchanging medical data. In this study, we provide a user-centered approach to rule-based reasoning for designing a tailored healthcare assistance system. The suggested system performs five main tasks, including modifying PHR log data and analyzing user profiles, visualization of user profile, self-tracking and recommendation of the proper diet suggestions for each user. The proposed system was implemented as a web application. Ultimately, Various alternative menus are used to give nutritional diet recommendations, ensuring that each menu is a part of a healthy, balanced diet. This web application is useful in real time to keep track of one's health related information and feedback the results to enable individual self-health care. A rule-based technique is used to construct a personalized recommendation system that aids the end user in structuring diet recommendations in accordance with numerous personal criteria, such as food ideas, nutritious diet, and deficits.

Keywords: PHR · Personalized Recommendation · Rule Based Reasoning · Web Application · Self-tracking

1 Introduction

As the world becomes technologically advanced people are interested to store their sensitive information in a centralized manner through cloud, emails, personal health records (PHR) etc. Physicians must adjust to this new data source as PHRs have become widely adopted in recent years. When properly interpreted, a PHR can offer important data points to evaluate the clinical course prior to the presentation and offer a level of fidelity that cannot be obtained by using only conventional patient interviews [15]. PHR services have increased the effectiveness of medical data storage, retrieval, and sharing by enabling patients to create, manage, and control their personal health data in a single location via web-enabled devices [16]. Systems for keeping personal health records combine information, expertise, and software tools to help individuals take an active

H. Sharma et al. (Eds.): ICIVC 2022, PALO 17, pp. 443–455, 2023.
https://doi.org/10.1007/978-3-031-31164-2_37

role in their own care. They go beyond being static stores for patient data [9]. Having a PHR can truly be a lifesaver such as in an emergency the vital health information can readily made available to the first responders such as diseases you are being treated for, medications you take and any allergies related to drug. Through a defined and proper PHR one can prevent health hazards and increase the quality of healthy life [1]. In today's integrated world people are susceptible to many health issues due to the sessile and inactive lifestyle. Majority of the diseases are caused due to the unhealthy eating habits. Hence it is crucial to maintain a healthier lifestyle.

We have developed a web application, we intend to provide the management of user's health related information as a personal health record (PHR) stored in MySQL database [2]. There exists a self-tracking system that enables users to monitor, analyze and interpret personal performance on health data. Data visualization is carried out using graphs and charts [3]. A personalized recommendation system is developed to generate nutritional diet recommendations based on the user health profile and PHR log data through the implementation of rule-based reasoning.

Our research is unique in that it combines all three of the respective modules into a single web application. The patient can upload, update, and track their own health using this application. The patient should not always seek a doctor's advice if any type of nutrition diet is necessary. He can easily track their health data using this website application and receive the necessary nutritional diet advice based on their own health.

The remainder of the essay is structured as follows. The basic vocabulary used in the article is explained in Sect. 2. The existing relevant works are included in Sect. 3. The details of the algorithm used is described in Sect. 4. The system's architecture is described in Sect. 5, and the implementation and assessment of the suggested approach are covered in Sect. 6. The conclusion and upcoming work are discussed in Sect. 7.

2 Preliminaries

The basic terms and concepts used in this paper are explained in this section.

2.1 PHR

The Public Health Operational Group defines PHR as an electronic function as: a private, secure, and confidential environment through which individuals can access, handle, and communicate both their own health information and that of others for whom they are certified. A set of internet technologies known as the Personal Health Record (PHR) [12] enables people to link, synchronize, and create pertinent elements of their lifetime health information that are made accessible to those who require it. PHRs give an extensive and integrated picture of health data, including information from patients (such as warning signs and prescription use), clinicians (such as analysis and test results), pharmacies, insurance companies, and other sources.

2.2 Personalized Recommendation System

A personalized diet recommendation system takes the user's specific health profile and preferences into account while making meal recommendations [13]. The suggested dietary items must not increase the user's disease if they are a patient of any ailment [3]. The recommendations are meant to be useful, adaptive, and of interest to the user.

2.3 Self-tracking

A self-tracking system enables people to keep track of the information about themselves collected knowingly and purposively, which they review and consider the application of changes in their lives. This is done on daily basis. The personal information is usually maintained and is tracked using digital tools.

2.4 Rule Based Reasoning

Rule-based reasoning is one way to develop knowledge-based intelligent decision support systems. Using plausible rules, most frequently if-then rules, rule-based reasoning represents both common sense knowledge and domain-specific domain expertise.

2.5 Natural Language Processing

Natural Language Processing is referred to as NLP [14]. It is a subfield of artificial intelligence that aids in the comprehension, interpretation, and manipulation of human language by computers. The OCR algorithm is implemented in this module to extract text from the image that will be used as input. The three steps in a standard OCR algorithm are to eliminate background complexity [8], manage various lighting situations, and remove noise from the image. To identify the text in a picture and draw a bounding box around the text-containing area, text detection algorithms are needed. It will also be possible to detect this objection using standard methods.

3 Literature Study

Many researchers have presented a variety of systems that provide the users with wellness recommendations which might improve their health condition. These recommendations were generated based on the users' considering their health status and the information learned from users who are similar to them. According to earlier studies, collaborative filtering, machine learning, and rule-based algorithms were used to construct the majority of recommendation systems. This section primarily focuses on the references that improved our knowledge of the algorithms behind recommendation systems.

Shadi Alian et al. [1] for American Indian diabetes patients, it was suggested to use an ontology-enhanced recommendation system to deliver real-time, individualized recommendations. The system was created using a knowledge base built on an ontology and a collection of semantic rule sets. The ontology knowledgebase comprises patient profiles that include details about their health, preferences, cultures, socioeconomic statuses, and the environment in which they reside. Based on the information from the ontology knowledge base and the semantic rules from the rule set, a reasoning engine can come to conclusions and offer specialized recommendations to a particular user.

Sarah et al. [2] developed a web-based application that enables users to periodically submit their personal health records, such as blood glucose readings or notes about their physical activity, diet, and other factors. Patients are given the chance to systematically set goals and monitor their advancements during the intervention. It incorporates into the system the recommendations and treatment objectives of doctors. Methodically connects digitized and personal health records with the web-based application.

Sharvil et al. [3] presented a tool that organizes the user's smartphone's calendar of health-related activities and generates personalized health advice from various providers. SVM and fast Text are used to train the suggested model on a server, and the learned model is then used to make predictions on a client device. Using calendar-classified occurrences, the method is used to rerank health-related recommendations gathered from outside health service providers, assessing the usefulness of each advice for the user. As a result, it can produce recommendations for your own health. The model achieved decent accuracy for the recommendation system and classifier module.

Gouthami et al. [4] proposed a model that provides nutrition recommendations based on BMI calculations in an informative and user-friendly environment. In this food recommendation application, they focused on daily diet plan and nutrition needed. According to user food preferences and consumption they generated suggestions, food nutrition's, deficiencies and kept track of user's food habits history. In this application memory-based Content-Based Filtering and Collaborative Filtering methods are used to get users choice of his food recommendation for the daily nutrition with the help of USDA dataset and grocery data.

Archenaa et al. [5] explained how big data analytics are used in healthcare. Additionally, the diabetes prediction and recommendation system was researched. Machine learning algorithms are used to build predictions and recommendations. The features are divided into three categories using Bayesian classifiers: diabetes, pre-diabetic, and not diabetic. Using a confusion matrix, the prediction's accuracy is evaluated. Based on how closely their health profiles resemble those of other people, recommender systems can help users by making recommendations for alternative treatments, health insurance, clinical pathway-based treatment options, and diagnoses. Content filtering, Rule-based, and Case based filtering algorithms were employed in the hybrid recommender system filtering method. To increase the recommendations' accuracy, a hybrid filtering strategy was applied.

Roehrs et al. [6] attempted to define the taxonomy and open questions in order to investigate the most recent research on PHRs. Furthermore, this study tried to clearly identify data kinds, standards, profiles, goals, methodologies, functionalities, and architecture with reference to PHRs. Regarding coverage and localization, they gave a comparison of various architecture types and PHR implementation styles. Last but not least, the difficulties and problems relating to security, privacy, and trust are described and are expected to be the subject of future studies.

Sumithra et al. [7] developed an innovative method for restricting access to PHRs, and the RSA algorithm is utilized to encrypt the PHR files for each patient. For the safe exchange of individual health records in the cloud, a novel structure was put forth. To fully actualize the patient-centric idea, partially trustworthy cloud servers are taken into consideration and discussed. Encrypting their PHR data will provide patients total control over their own privacy and allow for fine-grained access. In comparison to other research, the framework greatly decreases the complexity of key management while improving privacy guarantees. It accomplishes this by solving the specific problems raised by numerous PHR users and owners. In order to provide access from both personal users and a variety of users from the public domains with a range of professional positions, qualifications, and affiliations, the PHR data is encrypted using RSA.

4 Description of Algorithm

This NLP technique is used to extract text from images that users have contributed during the record upload process. Based on the pictures that the user uploads, this back-end process tries to understand the individual and his experiences.

Algorithm for NLP Technique: TextExtraction_from_images (filename, path)
Require:

- Filename: Name of the file uploaded
- Path: The path of the file it is uploaded to the storage
- Import cv2,pytesseract

Step 1: Initially the file has to be imported through the specified path.
Step 2: The image has to be read from the specified path and should be stored in the local variable.
Step 3: Initialize a new variable and apply pytesseract.image_to_string() function to the image and store the output in the created variable.
Step 5: Normalize the output in order to fit into the recommendation algorithm.
Step 6: Store the data in the database.

5 Architecture Diagram

Figure 1 System Architecture describes the following steps.

1. First the user will register into the personal health record management.
2. After registration he/she will login into the personal health record management.
3. The patient will update his health details like BP, Sugar, Height, Weight, and lab reports.
4. All the data will be preprocessed & stored in SQL Database.
5. Based on the reports uploaded and vitals updated, metrics are to be generated.
6. These metrics visualizes the total health analysis of the user.
7. Based on the lab reports of the user, nutritional diet recommendations are generated.

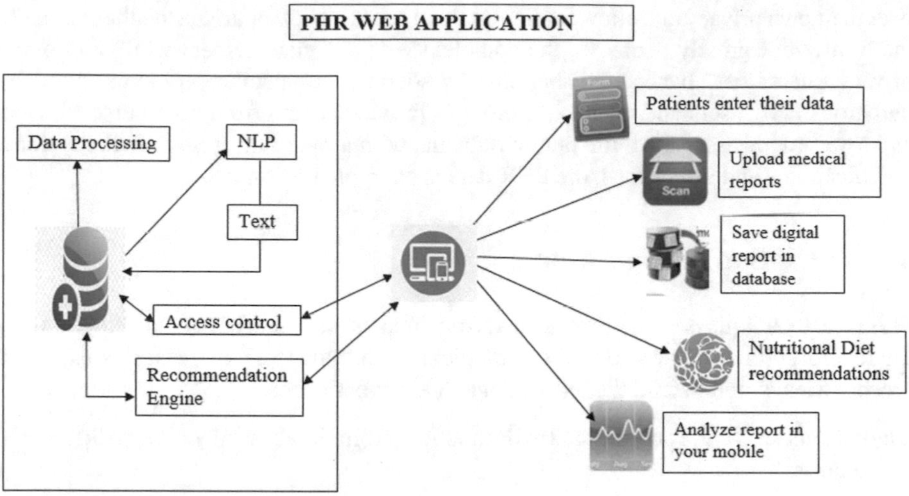

Fig. 1. System Architecture

6 Design Methodology

In this paper, we have developed a web application to store the user's health information data according to the survey items prepared in advance, and generate real time personalized diet recommendations for the user. To suggest our methodology, we cited the publication [1]. Rule-based reasoning and domain knowledge are both used in our suggested method for producing suggestions. The rule-based reasoning approach has the advantage that it overcomes the data shortage that collaborative filtering- and machine learning-based approaches have.

The system framework typically consists of a database which stores the PHR/log life data provided by the user. When the user is prompted to upload a file, the data can be extracted using a certain algorithm, and health analysis can be performed using the data gathered [11]. The recommendation system consists of a knowledge base and a set of semantic rules. The knowledge base contains users' profiles including the health, preferences which the user provides or obtained from log data in the database [10]. Based on medical literature, the expert knowledge of doctors, and a variety of additional sources, a semantic rule set is created.

The acquired knowledge is converted into logic-based rules. A reasoning engine is used to draw conclusions and offer the user tailored recommendations based on the data from the knowledge base and the semantic rules from the rule set.

To represent the rules in the form that computers can understand, we have adopted the basic if-then clauses. The if-then rules can be understood as if (condition) then (conclusion). A group of logical phrases linked together by logical operators that relate the user profiles make up the condition (such as BMI, health status). The "if" clause evaluates each condition to determine if it is true or false. A conclusion will follow if all of the preconditions are determined to be true. The user's recommendations are included in the conclusion.

7 Implementation of Results

Our proposed system consists of three components a PHR module, a recommendation module and an administrator module. The PHR module stores the health-related data and documents from the user in a knowledge base database. The recommendation module receives the data from the knowledge base and performs rule-based approach over the data acquired and generates recommendations. The administrator module is responsible for managing the entire system.

The software used to implement the proposed system were shown in Table 1. PHP, MySQL, and the Apache web server have all been used. JavaScript, PHP, HTML, and CSS were used to create the web application.

Table 1. Development Environment

OS	Windows 10
Programs	PHP7, Apache2
Database	MySQL

The steps that follow demonstrate how the suggested system is implemented as a web application.

7.1 Registration and Authentication

The user registers onto the website for the first visit providing the email and password. By registering an account is created for the user. If the user already has an account, he/she can directly log on to the website by providing the login credentials. After the verification of the entered details with the existing data in the database is done, he/she is able to log on to their profile.

7.2 PHR Module

After the successful login, the user is able to access the data provided by him/her. In PHR module, the user data such as health related information, personal details, preferences, electronic documents are stored in a knowledge base database.

7.3 Personalized Recommendations

The recommendation module fits the user's profile and applies rule-based reasoning to generate rules and inferences that suit the user requirements. The user profile consists of the health condition of the user such as height, weight, BP, Diabetes, Pulse rate and any other disease they are suffering from. The BMI, Diabetes and BP are taken as the primary elements in the generation of the recommendations. These rules thus generated are asserted as personalized recommendations unique to each user profile.

7.4 Administrator Module

The administrator module performs user management and content management. Predefined rule sets and generated rules can be viewed from this module. Any new rules inferred or generated can be adjoined to the knowledge base through the administrator module.

7.5 Knowledge Base

The knowledge base is developed by the acquisition and access of the knowledge related to food and nutrition as well as user health status. The construction of the knowledge base depended on knowledge from domain experts, nutrition guidelines and clinical practices. The knowledge base typically consisted of rule-based knowledge pertaining to the decision model used in the generation of recommendations.

7.6 Results

The signup page is used for signing the user for the first time. The details required for the signup are as displayed in Fig. 2. After the successful signup the user is able to login to the platform using his/her login credentials.

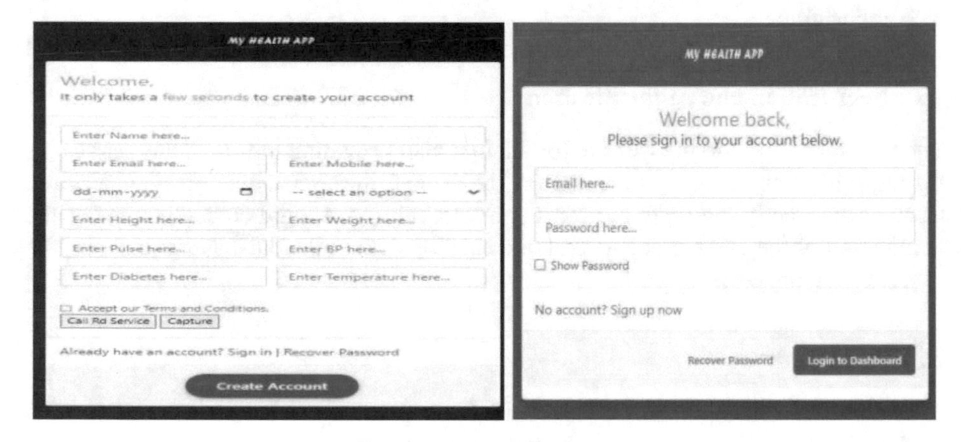

Fig. 2. Register and Login

Figure 3 describes about that user is able to upload the health-related documents on this page. The uploaded documents are stored in the database and can be viewed timely.

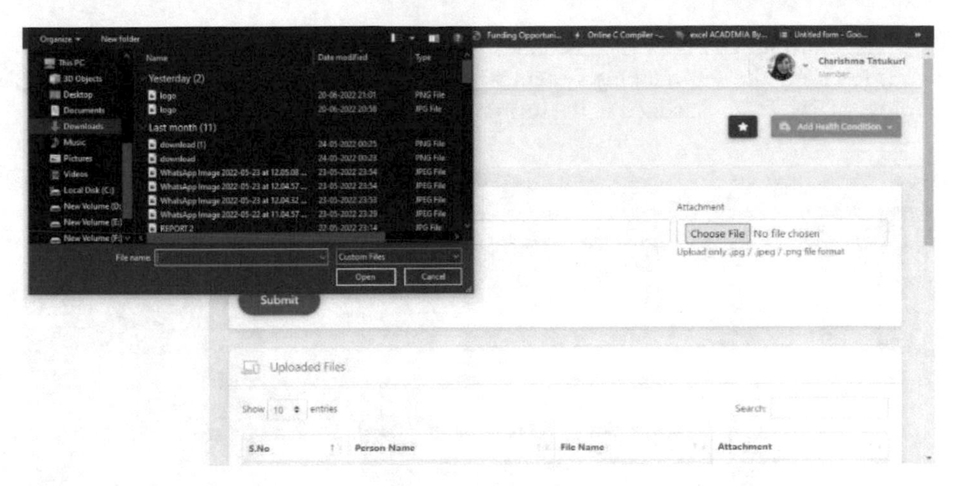

Fig. 3. Upload file

Figure 4 reports about the user profile is created by providing the medical details such as Height, Weight, Pulse Rate, BP and Diabetes. This information is stored in the database can be viewed in a timely fashion.

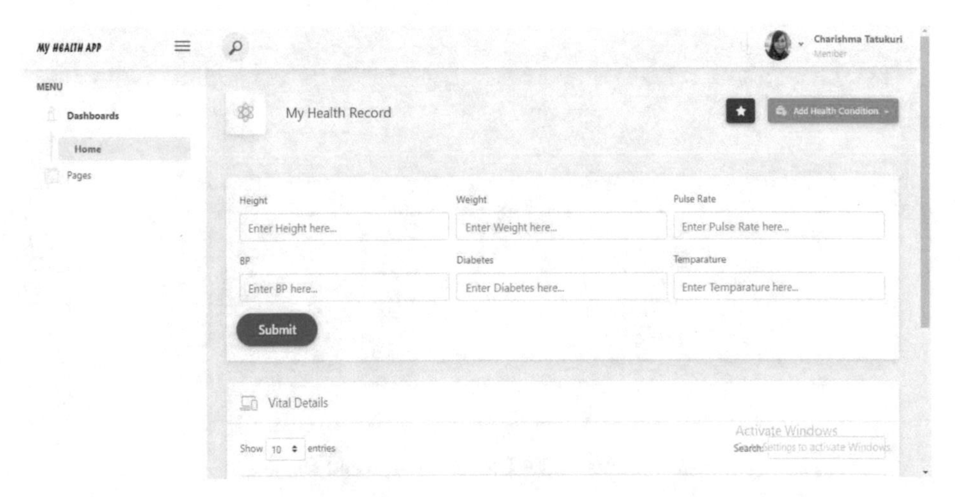

Fig. 4. Update Profile

Figure 5, Fig. 6, Fig. 7 details about that the user is able to analyze their health condition using the self-tracking system where they can find insights about their health demographics. It is visualized in the form of graphs and charts.

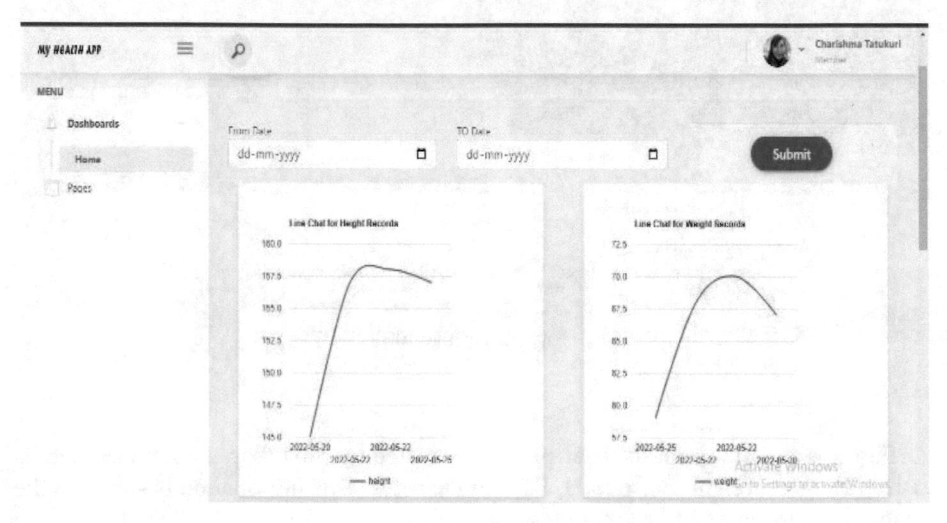

Fig. 5. Statistics for height and weight

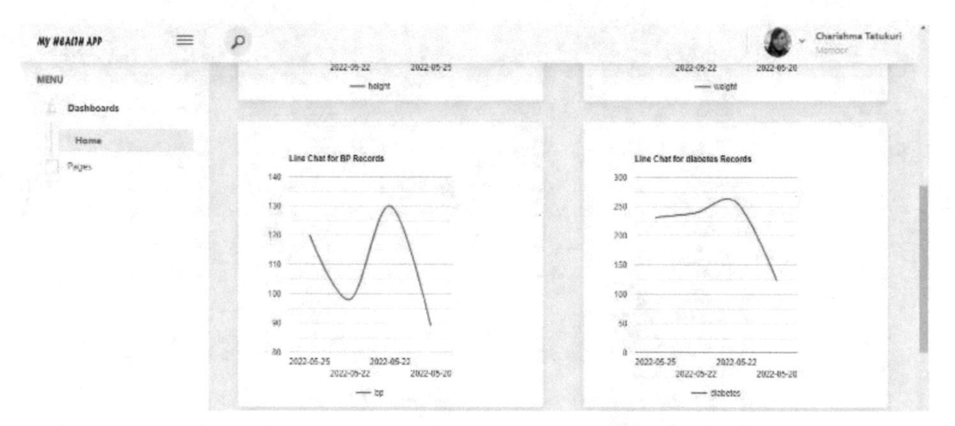

Fig. 6. Statistics for Blood Pressure and Diabetes

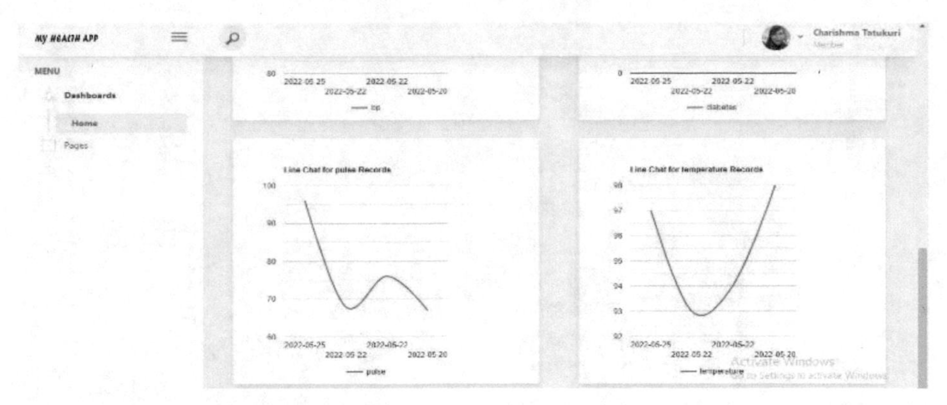

Fig. 7. Statistics for Pulse and Temperature

Figure 8 outlines about the recommendations are generated by fitting the user profile with the existing rules. The user is provided with the status of their health condition and are given suggestions for the food intake.

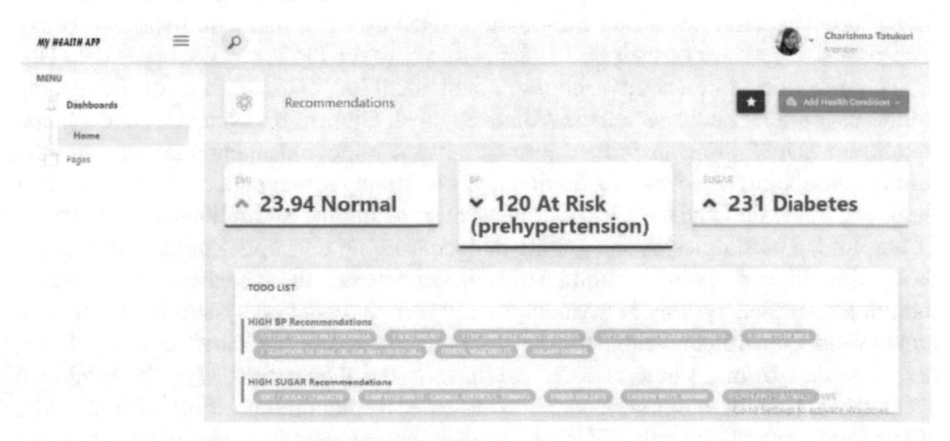

Fig. 8. Suggested food intake

Figure 9. Trace about the Administrator module that concerns with the management of the users registered on to the platform and the management of the content of recommendations.

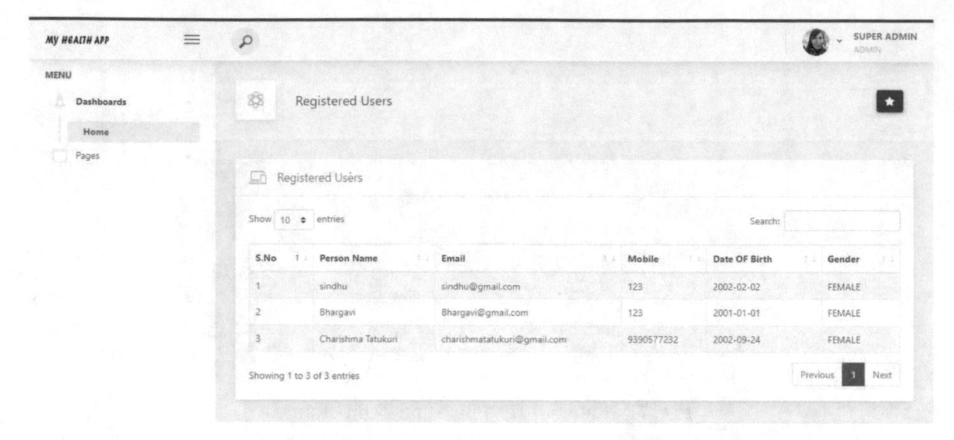

Fig. 9. Registered Users in admin module

8 Conclusion

We described a user-friendly interactive platform in this research for storing and managing individualized healthcare information. Each user can maintain their own health records and keep track of their health status thanks to the PHR module it provides. This assists in learning more about each user's individual health circumstances. Depending on the user's selections, personalization is enabled. Clinical best practises and general recommendations are transformed into rule-based logic and integrated with the recommender's knowledge base. Additionally, our strategy integrates advice from other fields regarding diet and exercise. Our integration technique exemplifies the real benefit of employing ontologies by integrating and exchanging concepts from the knowledge bases of multiple domains, creating new linkages between these concepts, and merging them into a unified system. A reasoning-based recommendation system is used to generate tailored recommendations for the users with the aid of the knowledgebase. Using real-time data from a range of users, we have tested the usefulness of the suggested system. The breadth of this system can be increased by examining various recommender algorithms, strengthening the PHR's security resolution, and investigating the data that is kept in the system and providing pertinent knowledge through big data analysis.

9 Future Work

In the future, large amounts of data can be stored in the future by integrating Big Data. By developing new recommender algorithms, we may also enhance a variety of other aspects, which can then be ingrained within a specific organization to produce superior outcomes.

References

1. Alian, S., Li, J., Pandey, V.: A personalized recommendation system to support diabetes self-management for American Indians. IEEE Access **6**, 73041–73051 (2018)

2. Mantwill, S., Fiordelli, M., Ludolph, R., Schulz, P.J.: EMPOWER-support of patient empowerment by an intelligent self-management pathway for patients: study protocol. BMC Med. Inform. Decis. Mak. **15**(1), 1–7 (2015)
3. Katariya, S., Bose, J., Reddy, M.V., Sharma, A., Tappashetty, S.: A personalized health recommendation system based on smartphone calendar events. In: Mokhtari, M., Abdulrazak, B., Aloulou, H. (eds.) ICOST 2018. LNCS, vol. 10898, pp. 110–120. Springer, Cham (2018). https://doi.org/10.1007/978-3-319-94523-1_10
4. Gangappa, M., Gouthami, B.: Nutrition diet recommendation system using user's Interest. Int. J. Adv. Res. Eng. Technol. **11**(12), 2910–2919 (2020). http://iaeme.com/Home/issue/IJARET?Volume=11&Issue=12
5. Iyamu, T.: A framework for selecting analytics tools to improve healthcare big data usefulness in developing countries. South African J. Inform. Manage. **22**(1), 1–9 (2020)
6. Alex, R., Da Costa, C.A., da Rosa Righi, R., De Oliveira, K.S.F.: Personal health records: a systematic literature review. J. Med. Internet Res. **19**(1), e5876 (2017)
7. Sumithra, R., Jayalakshmi, V., Priya, R.: Personal health record using cloudcomputing technology. Indian J. Public Health Res. Develop. **11**(3) (2020)
8. Ray, S.: An overview of the Tesseract OCR engine. In: Ninth International Conference on Document Analysis and Recognition (ICDAR 2007), vol. 2, pp. 629–633. IEEE (2007)
9. Xiong, C., Yuan, S., Mu, B.: App engine-based personal health records system. In: 2012 International Conference on Information Engineering and Applications (IEA) Proceedings, pp. 653–659. London-based Springer (2013)
10. Latha, K., Ambika, M.: A patient-centered framework for an intelligence-based recommendation system for healthcare. In: International Conference on Cutting-Edge Engineering and Technologies, CAASR (2015)
11. Mallik, S., Pradhan, C., Mishra, B.S.P., Barik, R.K., Das, H.: are among the authors. Mechanism for making health recommendations based on intelligence and big data analytics for efficient management of healthcare utilising big data analytics, pp. 227–246. Academic Press (2019)
12. Ash, J.S., Bates, D.W., Overhage, J.M., Sands, D.Z. Tang, P.C.: Definitions, advantages, and methods for removing adoption barriers related to personal health records. J. American Med. Inform. Assoc. **13**(2), 121–126 (2006)
13. Alison, G., Threlkeld, R.J.: Nutritional recommendations for individuals with diabetes. (2015)
14. The use of natural language processing to change the information in health records. A. Roberts. Psychiatry in Europe 30.S1, 1–1 (2015)
15. David, K., Pan, E.C.: The value of personal health record (PHR) systems. In: AMIA Annual Symposium Proceedings, vol. 2008, p. 343. American Medical Informatics Association (2008)
16. Norman, A., Fevrier-Thomas, U., Lokker, C., Ann McKibbon, K., Straus, S.E.: Personal health records: a scoping review. J. American Med. Inform. Assoc. **18**(4), 515–522 (2011)

A Novel Remote Sensing Image Captioning Architecture for Resource Constrained Systems

Nipun Jain(✉), Medha Wyawahare, Vivek Mankar, and Tanmay Paratkar

Vishwakarma Institute of Technology, Bibwewadi, Pune 411037, India
{nipun.jain18,medha.wyawahare,vivek.mankar18,
tanmay.paratkar18}@vit.edu

Abstract. Machine learning, specifically deep learning, has proven to be successful in many aspects along with groundbreaking results in various fields. One such area where the applications of Deep Learning are yet to show differences is in Resource Constrained Environment. In such conditions, not only the accuracy but also the feasibility with which one can process computations on the machines play an important role. The biggest challenge demanded by such systems is the simple yet accurate application that can operate under such challenging conditions. In this paper, we present a novel architecture for Image Captioning on remote sensing images, which delivers information based on satellite images with decent accuracy and can work efficiently in a resource-constrained atmosphere. The model is trained and tested on standard Remote Sensing Datasets such as UCM-Captions and Sydney Captions. Our proposed model has achieved the BLUE-1 score of 71.2 with only 8.28M parameters which is significantly less when compared to other benchmark models in Remote Sensing Image Captioning.

Keywords: Deep Learning · Image Captioning · Resource-Constrained Systems

1 Introduction

Image captioning gives machines the ability to generate texts which can describe the contents of the image. Image Captioning [1] models generally use a framework that consists of an encoder and a decoder network. The encoder is a model network that reads the input image and encodes it using internal representation. The decoder is another network model which decodes the encoded image and generates captions that describe the image. Various techniques have been proposed for deep-learning base image description generation systems. Image captioning methods have two aspects, image representation, and sentence generation. For image representation, a Deep Convolutional Network [2] is used, whereas, for text generation, Deep Recurrent Networks [3] are used since convolutional networks can handle special data and recurrent networks can handle sequential data. Remote sensing is a method of observing and processing image data of an area from a distance usually captured with the help of a satellite. Remote Sensing [4] is used in various areas like weather forecasting, forestry, agriculture, maps, and disaster management. Deep neural networks have become popular in remote sensing image captioning.

H. Sharma et al. (Eds.): ICIVC 2022, PALO 17, pp. 456–465, 2023.
https://doi.org/10.1007/978-3-031-31164-2_38

Influenced by the current progress of artificial satellites, remote-sensing images have received massive attention. Remote sensing images are used for scene classification, target detection, and segmentation. Resource Constrained Environment poses a challenge to the system as it is not helpful when it comes to the execution and implementation of the models. The paper talks about the development of an architecture that produces top results with minimal possible parameters.

2 Literature Survey

There are several approaches that have been made to generate natural sentences in the domain of remote sensing image captioning. Some of the relevant approaches that focus on image caption generation for remote sensing data in resource-constrained systems are discussed in this section.

Chowdhery et al. [5] demonstrated the development of a model for microcontroller systems by presenting a unique dataset named "Visual Wake Words". This dataset focuses on a general vision use-case of recognizing whether the point of interest is available in the image or not, and acts as a benchmark for remote sensing vision models. The model has low memory footprints which is very helpful when it comes to deploying a neural network on microcontrollers. Along the same lines, a study by Liberis et al. [6] showed concerns about neural network deployment in memory-limited systems and proposed a solution by modifying the default implementation strategy of a neural network. They showed that with a little overhead cost, a defragmentation strategy is an achievable step and helps improve the performance of lightweight neural network models.

Fedorov et al. [7] presented a way that can automatically design CNNs which are tiny enough to fit with the memory limitations of microcontrollers (MCUs). This method combines the techniques of neural architecture development and network pruning that is able to learn from popular datasets of IoT. Another method by Sumbul et al. [8] demonstrates a three-step approach for image captioning. It generates the descriptions of an image by using standard CNN-LSTM-based architecture. Then, it summarizes the ground truth captions of each image used in training with the help of an ANN and removes the redundancies witnessed in the training dataset. In the penultimate stage, the adaptive weight of the remote sensing image is defined automatically, which uses the semantic content of the image to combine summarized and standard captions.

There is also some work that focuses on model development using smaller datasets for remote sensing image captioning. Hoxha and Melgan [9] have provided an alternative RSIC system on the basis of SVM. Extraction of features and descriptions are created by pre-trained CNN and Support Vector Machines respectively. The proposed IC model when ran on small RSICD gives promising results and can be taken as an alternative to other deep learning frameworks.

Huang et al. [10], proposed a mechanism that aggregates multiscale features along with the denoising operations during image feature extraction This method results in the denoising multiscale feature representation. They have applied their system to two well-known datasets in remote sensing i.e. UCM - captions and Sydney captions and have set out a comparative analysis of the results of each dataset. Zheng et al. [11] explained how accurate and flexible captions can be generated for the RSICD. They have provided instructions that describe RSIC based on general characteristics. They have also developed a wide-ranging dataset of aerial images for generating textual descriptions of these images. This helps in the development of models that can fully leverage the contents of RSIC images.

Han et al. [12] proposed a unique attention framework called attentive linear transformation (ALT). It learns different feature abstractions like visual dependence and spatial and channel-wise attention. They have also proposed a regression-based threshold that is able to produce the attention probabilities values for image regions. While Shen et al. [13] proposed a Variational Autoencoder and Reinforcement Learning based Two-stage Multi-task Learning Model (VRTMM) for RSIC-based applications. Initially, they set the CNN along with a Variational Autoencoder. Furthermore, with the help of spatial as well as semantic features the corresponding captions are generated by using a transformer.

Shen et al. [14] proposed a model which involves a transformer to predict accurate captions. A few factors such as residual connections, drop-out layers etc., makes the transformer much more adaptive to RSIC tasks. Afterward, to improve the quality of textual descriptions generated, Reinforcement Learning is applied.

A caption generation approach by Zhao et al. [15] generates corresponding words in sequence to sequence manner and marks the attention areas in the image according to the weights assigned. The textual descriptions are generated by common attention-based methods on the basis of unstructured attention units which fail to identify the semantic relationships between the objects present in the RSICD. Yang [16] presented a unique cross-modal feature fusion retrieval model. This system gives a modified cross-modal common feature space than previous models and thus improves performance and accuracy. The model adds sentence information to the audio feature for image retrieval in remote sensing. Hence the image captioning model generates accurate textual descriptions with the help of intra-modality semantic discrimination to solve voice description tasks.

The study proposed by Ramos and Martins [17] shows how the nature of the output can affect the generated image descriptions. The authors proposed a generation function that leads to output sequences of word embeddings rather than word probability. They are exempt from the idea that continuous output models have better chances of mapping the relationship between the generated description and input images based on semantic similarity. Their theory includes a comparison of representations for a single token and for complete captions against image representations. Bejiga et al. [18] focus on extracting ancient textual scripts on the basis of geography, civilizations, and equivalent remote sensing images by using generative adversarial networks (GANs).

A study by Hoxha and Melgani [19] have given a unique network architecture of SVMs to convert image details to caption generation. The proposed Image Captioning method works especially in environments where a limited number of training resources are available. They performed on various datasets which proves that the quality of captions generated is maintained throughout and highly correlates with image information. Another network architecture by Assef Jafar et al. [1] combines the techniques of typical encoder-decoder model architecture with object detection model architecture to develop a new architecture with improved performance. On the other hand, P. R. Devi et al. [20] proposed a unique metric containing a reinforcement learning-based reward system that combines BLEU and CIDE metrics to generate the score. This approach has shown superior performance when compared on Flickr8k and Flickr30 datasets.

3 Methodology

The proposed encoder-decoder architecture achieves superior results over traditional encoder-decoder architects with less size and computation requirements. Generally, the encoder network draws out relevant details from the image and the decoder network uses that image information to generate the output sentences [21]. The quality of output sentences relies on the information extracted from the image by the encoder. If we increase the size of the encoder output, the decoder will have more information to process. This improves the chances of generating better results. However, This technique increases the number of floating point operations in the model and makes it difficult to deploy in resource-constrained applications. We present a novel compression network that focuses on reducing the size of the encoder output while maintaining the quality of extracted information.

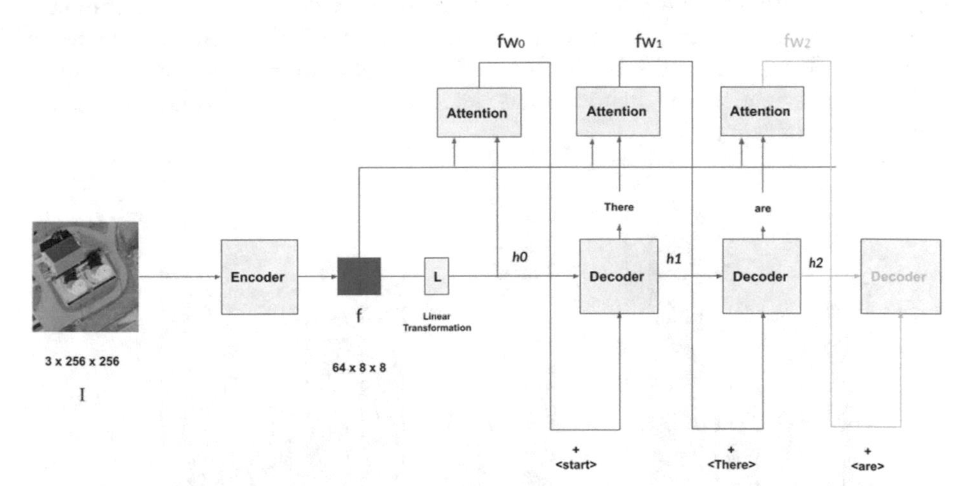

Fig. 1. Enhanced lightweight encoder-decoder architecture with attention mechanism

Figure 1 shows the proposed architecture. A three-channeled RGB image is first passed to the encoder block with the compressor network. The output of this block is then fed to the decoder network. The decoder network uses an attention mechanism [22] to improve information extraction.

3.1 Encoder Block

As shown in Fig. 2, the encoder block mainly contains two networks. The first network is a feature extractor network. A pre-trained mobile net v2 [23] model is chosen as a feature extractor because of its small size and better performance. In this network, the residual connections are placed between bottleneck layers to avoid the loss of information. The extracted features are then passed to the CNN compressor network. The job of this network is to learn to process the extracted features in such a way that the feature scale is reduced by a factor of 20 at the same time the information quality is preserved.

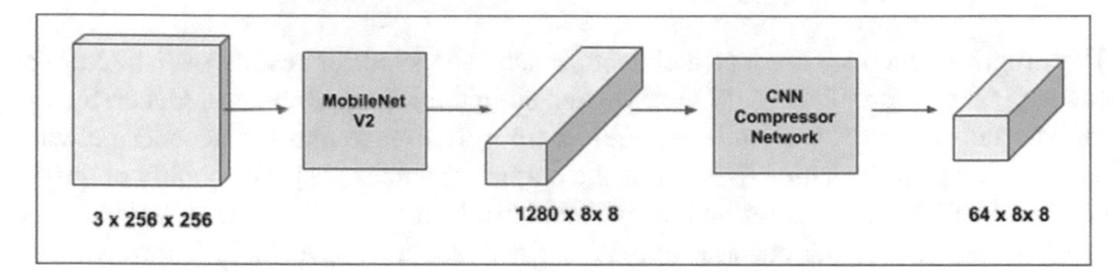

Fig. 2. Encoder Block with feature extractor and compressor networks

The CNN compressor network uses a combination of convolution and pooling layers. The architecture is described in the following Fig. 3. The output of the feature extractor model is passed through a global average pooling layer, then batch normalization is applied on top of it. After that, a special dropout layer is used. Furthermore, a couple of convolution layers are applied on top of it with ReLU activation. Next, the initial set of pooling, normalization, and dropout layers is applied again on the CNN layers output to enhance the efficiency.

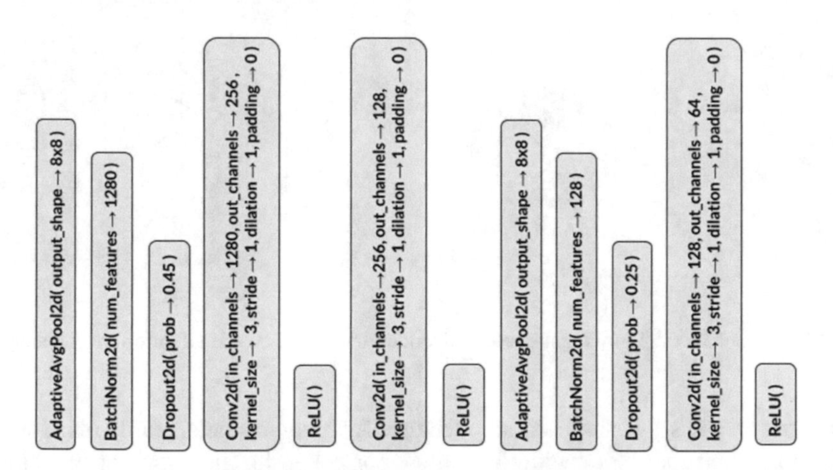

Fig. 3. CNN Compressor Network Layers

In the last, a final convolutional layer is applied with ReLU activation to produce the updated feature map. Since the mobile net network is pre-trained, all of the encoder block learnable parameters are present in the compressor network. During training, the compressor network updates its weights to efficiently reduce the size of the mobile net output while preserving the quality of information.

3.2 Attention Block

The attention mechanism uses image information and the current state decoder output to generate the weighted attention encoding. This attention-weighted encoding helps the decoder to focus on the necessary image information while producing the next words. The following Fig. 4 shows the structure of the attention network. The weighted encodings are generated by performing element-wise multiplication of the attention weights and the image encoding. The attention weights are calculated by using the information from the encoder as well as the decoder output. First, a linear transformation is applied to the encoder and decoder output, then a summation operation is performed on the transformed vectors.

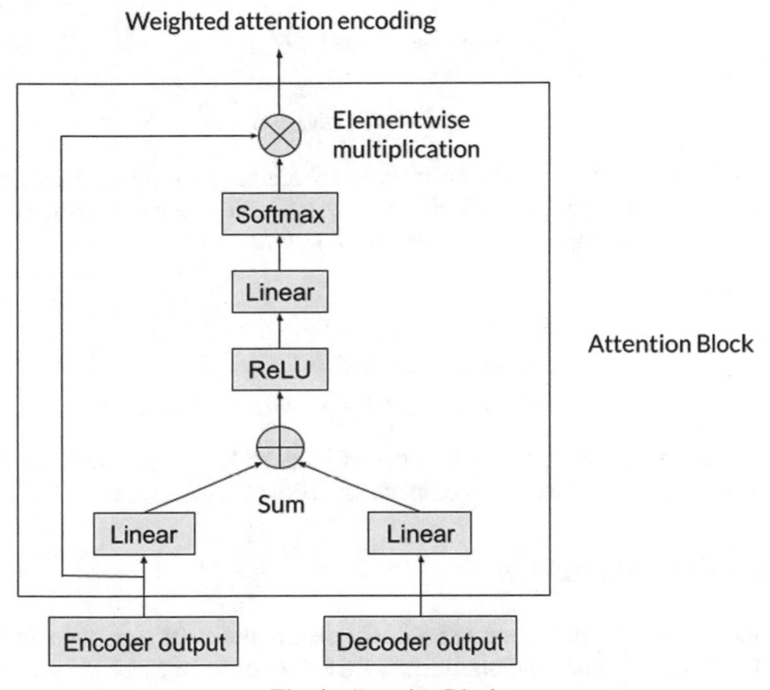

Fig. 4. Attention Block

The resultant vector is passed through the ReLU activation layer and a linear layer. Finally, the output vector is passed through the softmax activation to get the attention weights.

3.3 Decoder Block

The decoder block uses a Long and Short Term Memory(LSTM) [24] Network to generate the output words. As shown in Fig. 1, the compressed image encoding (f) are used as an initial hidden state(h_0) for the LSTM decoder after applying a linear transformation to it. The decoder block uses an attention mechanism to improve its performance. First, the image encoding (f) and initial hidden state(h_0) are passed to the attention block to get the weighted attention encoding (fw). This (fw) and the $< start >$ token are fed to the decoder network to generate the probability vector of the next word. Afterward, the same process is repeated till either the sentence limit is reached or the $< end >$ token is generated by the decoder.

The complete flow of the data through the architecture is described in the following Eqs. 1–4.

$$e = MobileNetV2(I) \tag{1}$$

$$f = CNNCompressorBlock(e) \tag{2}$$

$$fw_t = AttentationBlock(f, h_t) \tag{3}$$

$$h_{t+1}, w_{t+1} = Decoder(fw_t + we_t, h_t) \tag{4}$$

Where I is the input image, e is the encoded image features, f is the compressed image features, h_t is the decoder hidden state at instance t, fw_t is the weighted attention encoding at instance t, we_t is the input word embedding at instance t.

At $i = 0$,

$$\begin{aligned} h_0 &= Linear(f) \\ fw_0 &= AttentationBlock(f, h_0) \\ we_0 &= Embedding\ of\ the < start >\ token \end{aligned}$$

This architecture leverages the benefits of transfer learning as well as attention mechanism to improve the performance at the same time the size of the model is reduced.

4 Results and Discussion

The model is trained on an Ubuntu 18.04.6 LTS distribution, with a single-core Intel(R) Xeon(R) CPU @ 2.30 GHz and one Tesla T4 GPU. In order to check the performance of the developed model two most common RSICDs were used: the Sydney Captions dataset [25] and the UCM Captions dataset [25]. Sidney Captions dataset is a special image captioning dataset which contains satellite images taken from Sydney city in Australia. It is derived from the Sydney Data Set [26]. In this dataset, each image is provided with 5 captions. These images are of different regions in Sydney city including residential areas, industrial areas, rivers, etc. UCM Captions dataset is a dataset similar to the Sydney captions dataset. It is derived from the Merced Land Use Dataset [27]. This dataset is bigger with respect to the Sydney captions dataset and contains more variety of regions in the images. The performance of the model is compared with some of the previous implementations for remote sensing image captioning. The model is evaluated using standard image captioning metrics such as BLUE [28], CIDEr [29], and ROUGE [30]. The model performance analysis on UCM captions and Sydney captions dataset is given below in Table 1 and Table 2 respectively.

Table 1. Comparative performance analysis of our model on UCM dataset

Model			Results	
	BLUE-1	BLUE-4	CIDEr	ROUGE
CSMLF [31]	43.6	12.1	22.2	39.2
Multimodal [25]	70.8	45.9	292.5	66.1
Baseline_ref [8]	76.1	51.7	169.0	65.7
Attention_with_CNN_Compressor	71.2	51.1	195.4	**66.3**

Table 2. Comparative performance analysis of our model on Sydney dataset

Model			Results	
	BLUE−1	BLUE−4	CIDEr	ROUGE
CSMLF [31]	59.9	34.3	75.5	50.1
Multimodal [25]	69.6	54.3	220.2	63.4
Baseline_ref [8]	74.8	53.8	312.2	69.5
Attention_with_CNN_Compressor	71.2	**54.3**	251.4	67.3

From the above table, we can say that the proposed model provides reasonable accuracy despite having smaller architecture and much lesser number of parameters. The proposed architecture only contains 8.28 million parameters which is significantly less even compared to the first encoder-decoder approach proposal [21] in image captioning.

5 Conclusion

In this paper we present a Remote Sensing Image Captioning model architecture that can be used to develop intelligent software for resource-constrained systems. The goal of this study was to develop an architecture with a reduced number of parameters and a good score. We have modified the typical encoder-decoder architecture by using a specially designed compressor network along with attention techniques.

The study was mainly conducted using simple neural network architecture so that the model size can be limited. However, further enhancement in sophisticated deep learning architectures like transformers may lead to the development of small-sized specialized models that can provide state-of-the-art accuracy with fewer system requirements. Such development will not only revolutionize the deep learning applications in the constrained system but also become a go-to resource for most deep learning deployments.

References

1. Al-Malla, M.A., Jafar, A., Ghneim, N.: Image captioning model using attention and object features to mimic human image understanding. J. Big Data **9**, 20 (2022)

2. Arora, D., Garg, M., Gupta, M.: Diving deep in deep convolutional neural network. In: 2020 2nd International Conference on Advances in Computing, Communication Control and Networking (ICACCCN), pp. 749–751 (2020). https://doi.org/10.1109/ICACCCN51052.2020.9362907
3. Mun, J., Ha, S., Lee, J.: Automotive radar signal interference mitigation using RNN with self attention. In: ICASSP 2020 - 2020 IEEE International Conference on Acoustics, Speech and Signal Processing (ICASSP), pp. 3802–3806 (2020). https://doi.org/10.1109/ICASSP40776.2020.9053013
4. Xu, K., Huang, H., Deng, P.: Remote sensing image scene classification based on global–local dual-branch structure model. IEEE Geosci. Remote Sens. Let. **19**, 1–5 (2022). https://doi.org/10.1109/LGRS.2021.3075712
5. Chowdhery, A., Warden, P., Shlens, J., Howard, A., Rhodes, R.: Visual wake words dataset. arXiv preprint arXiv:1906.05721 (2019)
6. Liberis, E., Lane, N.D.: Neural networks on microcontrollers: saving memory at inference via operator reordering. ArXiv, abs/1910.05110 (2019)
7. Fedorov, I., Adams, R.P., Mattina, M., Whatmough, P.: Sparse: Sparse architecture search for CNNs on resource-constrained microcontrollers. Adv. Neural Inform. Process. Syst. **32** (2019)
8. Sumbul, G., Nayak, S., Demir, B.: SD-RSIC: summarization-driven deep remote sensing image captioning. IEEE Trans. Geosci. Remote Sens. **59**(8), 6922–6934 (2021). https://doi.org/10.1109/TGRS.2020.3031111
9. Hoxha, G., Melgani, F.: Remote sensing image captioning with SVM-based decoding. In: IGARSS 2020 - 2020 IEEE International Geoscience and Remote Sensing Symposium, pp. 6734–6737 (2020). https://doi.org/10.1109/IGARSS39084.2020.9323651
10. Huang, W., Wang, Q., Li, X.: Denoising-based multiscale feature fusion for remote sensing image captioning. IEEE Geosci. Remote Sens. Lett. **18**(3), 436–440 (2021). https://doi.org/10.1109/LGRS.2020.2980933
11. Lu, X., Wang, B., Zheng, X., Li, X.: Exploring models and data for remote sensing image caption generation. IEEE Trans. Geosci. Remote Sens. **56**(4), 2183–2195 (2018). https://doi.org/10.1109/TGRS.2017.2776321
12. Ye, S., Han, J., Liu, N.: Attentive linear transformation for image captioning. IEEE Trans. Image Process. **27**(11), 5514–5524 (2018). https://doi.org/10.1109/TIP.2018.2855406
13. Shen, X., Liu, B., Zhou, Y., Zhao, J., Liu, M.: Remote sensing image captioning via Variational Autoencoder and Reinforcement Learning
14. Shen, X., Liu, B., Zhou, Y., Zhao, J.: Remote sensing image caption generation via transformer and reinforcement learning. Multimedia Tools Appl. **79**(35–36), 26661–26682 (2020). https://doi.org/10.1007/s11042-020-09294-7
15. Zhao, R., Shi, Z., Zou, Z.: High-resolution remote sensing image captioning based on structured attention. IEEE Trans. Geosci. Remote Sens. **60**, 1–14 (2022). https://doi.org/10.1109/TGRS.2021.3070383
16. Yang, R.: Cross-modal feature fusion retrieval for remote sensing image-voice retrieval. In: 2021 IEEE International Geoscience and Remote Sensing Symposium IGARSS, pp. 2855–2858 (2021). https://doi.org/10.1109/IGARSS47720.2021.9554533
17. Ramos, R., Martins, B.: Using neural encoder-decoder models with continuous outputs for remote sensing image captioning. IEEE Access **10**, 24852–24863 (2022). https://doi.org/10.1109/ACCESS.2022.3151874
18. Bejiga, M.B., Melgani, F., Vascotto, A.: Retro-remote sensing: generating images from ancient texts. IEEE J. Select. Top. Appl. Earth Observ. Remote Sens. **12**(3), 950–960 (2019). https://doi.org/10.1109/JSTARS.2019.2895693

19. Hoxha, G., Melgani, F.: A novel SVM-based decoder for remote sensing image captioning. IEEE Trans. Geosci. and Remote Sens. **60**, 1–14 (2022). https://doi.org/10.1109/TGRS.2021.3105004
20. Devi, P.R., Thrivikraman, V., Kashyap, D., Shylaja, S.S.: Image captioning using reinforcement learning with BLUDEr optimization. Pattern Recognit. Image Anal. **30**(4), 607–613 (2020). https://doi.org/10.1134/S1054661820040094
21. Vinyals, O., Toshev, A., Bengio, S., Erhan, D.: Show and tell: Lessons learned from the 2015 mscoco image captioning challenge. IEEE Trans. Pattern Anal. Mach. Intell. **39**(4), 652–663 (2016)
22. Anderson, P., et al.: Bottom-up and top-down attention for image captioning and visual question answering. In: IEEE Conference on Computer Vision and Pattern Recognition (CVPR) (2018)
23. Sandler, M., Howard, A., Zhu, M., Zhmoginov, A., Chen, L.C.: Mobilenetv2: inverted residuals and linear bottlenecks. In Proceedings of the IEEE Conference on Computer Vision and Pattern Recognition, pp. 4510–4520 (2018)
24. Hochreiter, S., Schmidhuber, J.: Long short-term memory. Neural Comput. **9**, 1735–1780 (1997). https://doi.org/10.1162/neco.1997.9.8.1735
25. Qu, B., Li, X., Tao, D., Lu, X.: Deep semantic understanding of high resolution remote sensing image. In: International Conference on Computer, Information and Telecommunication Systems, pp. 124–128 (2016)
26. Zhang, F., Du, B., Zhang, L.: Saliency-guided unsupervised feature learning for scene classification. IEEE Trans. Geosci. Remote Sens. **53**(4), 2175–2184 (2015)
27. Yang, Y., Newsam, S.: Bag-of-visual-words and spatial extensions for land-use classification. In: ACM SIGSPATIAL International Conference on Advances in Geographic Information Systems, pp. 270–279 (2010)
28. Papineni, K., Roukos, S., Ward, T., Zhu, W.J.: Bleu: a method for automatic evaluation of machine translation. In: Proceedings of the 40th Annual Meeting of the Association for Computational Linguistics, pp. 311–318 (2022)
29. Vedantam, R., Lawrence Zitnick, C., Parikh, D.: CIDEr: Consensus-based Image Description Evaluation. In: CVPR (2015)
30. Lin, C.Y.: Rouge: a package for automatic evaluation of summaries. In: Text summarization branches out, pp. 74–81 (2004)
31. Wang, B., Xiaoqiang, L., Zheng, X., Li, X.: Semantic descriptions of high-resolution remote sensing images. IEEE Geosci. Remote Sens. Let. **16**(8), 1274–1278 (2019). https://doi.org/10.1109/LGRS.2019.2893772

Determination of Burnout Velocity and Altitude of N-Stage Rocket Varying with Thrust Attitude Angle

Paras Bhatnagar(✉) and Rupa Rani Sharma

Department of Mathematics, G. L. Bajaj Institute of Technology and Management, Greater Noida, UP, India
paras.bhatnagar@glbitm.ac.in

Abstract. In the performance of step rockets, optimum staging plays a dominant role from two considerations, either to derive maximum performance with pre-assigned resources or to achieve the required performance with minimum resources. Weisbord, Subotowicz and Hall and Zambelli have investigated the general solution for minimum gross weight for non-homogeneous stages holding good for the arbitrary number of stages. In all the above investigations, the variation of thrust attitude angles with stages is neglected and a field-free space is assumed. In this paper, an optimum staging program to find burnout velocities and the burnout altitudes for step rockets of an arbitrary number of stages having different specific impulses and mass fractions with stages is derived. Variation of thrust attitude angle with stages and effect of gravity is taken into consideration. The results are in general so far as specific impulses and mass fractions may have different values with stages. The altitude constraint is relaxed and the problem reduces to the optimization problem where variations of thrust attitude angle with stages and gravity effect are included and the velocity equation is the sole restrictive condition.

Keywords: Physics of the rocket · Rocket Equation · Lagrangian undetermined multipliers · optimization analysis

1 Introduction

Many authors have discussed the multiple staging rocket optimization with a minimum gross weight of the rocket. Malina and Summerfield's [14] discussed the optimization for homogeneous stages with minimum gross weight to achieve a given burnout velocity. Relaxing the restrictions of constant specific impulse and structural factors in Malina and Summerfield problem, Goldsmith [8] derived a relation for two-stage rockets when the structural weights were proportional to the fuel weight and the powerplant weights were proportional to the stage gross weight. The results in Weisbord [24], Subotowicz [21], Hall and Zambelli [11] investigations are derived as special cases constraint optimization problems of this velocity. The analytical approach to determine the optimal condition

H. Sharma et al. (Eds.): ICIVC 2022, PALO 17, pp. 466–475, 2023.
https://doi.org/10.1007/978-3-031-31164-2_39

for burnout velocity and optimum altitude with the optimum payload ratio of n stage rocket is discussed.

The lifting force of gravity produced and operated on a flying model of a rocket in the air is aerodynamic force [12]. Lifting actions are in line with the direction of air movement [4]. As shown in Fig. 1, drag D works in the same way as the associated airflow [9] and weight W is the energy produced by the gravitational force of a rocket, oriented in the earth's center. The resultant of the forces depends on the fuel amount, loading charge and the size of rocket parts [10]. Some of the scientific studies with a rocket are simple, inexpensive and faster as compared with a satellite [18, 22]. Many countries are running rocket programs [6, 7] and developing technologies to exploit the advantages [3, 15].

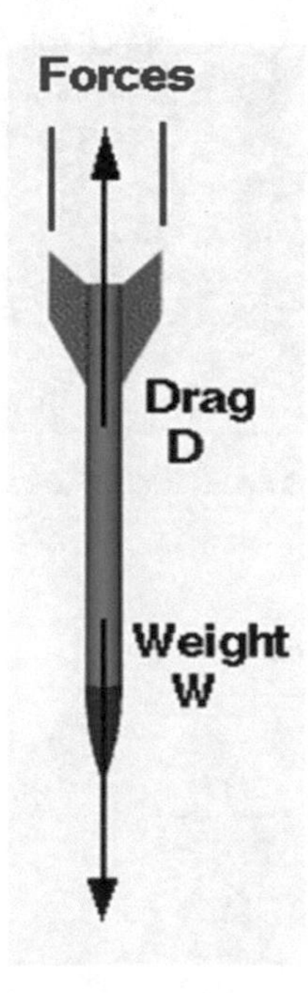

Fig. 1. Forces on a Rocket [16]

Mass Ratios of Rocket [23]

Empty Mass $= me = md + ms$

Full Mass $= mf = md + ms + mp$

$$mf = me + mp$$

Propellant Mass Ratio $= MR = \dfrac{mf}{me}$

$$MR = 1 + \frac{mp}{me}$$

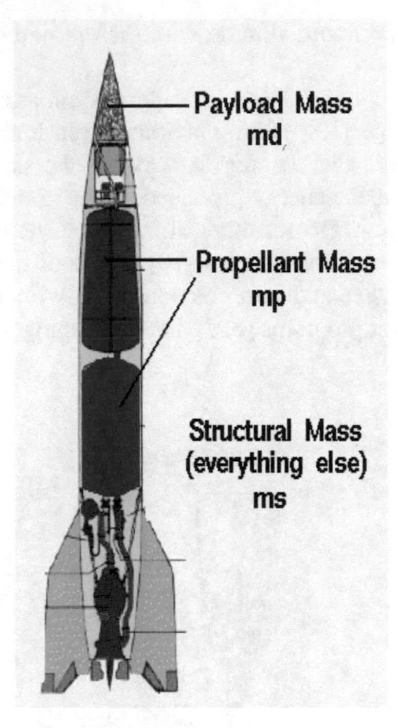

Fig. 2. Mass Ratios of Rocket [17]

$$\text{Payload Ratio} = \lambda = \frac{md}{mf - md}$$

$$\lambda = \frac{md}{mp + ms}$$

$$\text{Structural Coefficient} = \varepsilon = \frac{ms}{mp + ms}$$

$$MR = \frac{1 + \lambda}{\varepsilon + \lambda}$$

2 Mathematical Formulation

In this paper, we discuss optimum staging [20] for the minimum gross weight of a rocket of an arbitrary number of stages to achieve a given burnout velocity at an assigned altitude measured vertically from the point of location of the rocket firing. The results are in general so far as specific impulse and mass fraction may have different values with stages [5]. Then altitude constant is relaxed and the problem reduces to the optimization problem where variations of thrust attitude angle [13] with stages and gravity effect are included and the velocity equation is the sole restrictive condition. The structural mass, payload mass and propellant mass are distributed as shown in Fig. 2. The propellant mass ratio, payload ratio and structural coefficient given in [23] are used as standard results.

The thrust attitude angle ψ_1, the burnout velocity V_{1b} and the burnout altitude h_{1b} measured vertically from the launch point for a single stage, given by Seifert [19, 25] is

$$V_{1b} = V_0 + gI_1 + log\left[\frac{M_f}{M_e}\right] - gt_1 sin\psi_1 \tag{1}$$

$$h_{1b} = V_0 t_1 sin\psi_1 + gI_1 t_1 sin\psi_1 \left[1 - \frac{log\left(\frac{M_f}{M_e}\right)}{\frac{M_f}{M_e} - 1}\right] - \frac{1}{2} g t_1^2 sin\psi_1 \tag{2}$$

Where $t_1 = \frac{gI_1(M_f - M_e)}{F}$,

Here I_1 is the specific impulse in the first stage, M_f is the initial mass of the rocket and M_e is the final mass, V_0 is the initial velocity of the rocket and F is the rocket thrust [1].

Let r_{01} be the initial thrust-to-weight ratio, then

$$r_{01} = \frac{F}{M_f g}.$$

Therefore $F = r_{01} M_f g$, so that

$$t_1 = \frac{I_1}{r_{01}}\left(1 - \frac{M_e}{M_f}\right) \tag{3}$$

$$\frac{M_f}{M_e} = \frac{\lambda_1}{\lambda_1(1 - \mu_1) + \mu_1},$$

where λ_1 is the gross weight payload ratio and μ_1 is the mass fraction of the first stage.

$$\frac{M_f}{M_e} - 1 = \frac{\mu_1(\lambda_1 - 1)}{\lambda_1(1 - \mu_1) + \mu_1} \text{ and}$$

$$1 - \frac{M_f}{M_e} = \frac{\mu_1(\lambda_1 - 1)}{\lambda_1},$$

Using these values Eqs. (1), (2) and (3) can be written as

$$V_{1b} = V_0 + gI_1 + log\left[\frac{\lambda_1}{\lambda_1(1 - \mu_1) + \mu_1}\right] - gt_1 sin\psi_1 \tag{4}$$

$$h_{1b} = V_0 t_1 sin\psi_1 + gI_1 t_1 sin\psi_1 \left[1 - \frac{log\left(\frac{\lambda_1}{\lambda_1(1-\mu_1)+\mu_1}\right)}{\frac{\mu_1(\lambda_1-1)}{\lambda_1(1-\mu_1)+\mu_1}}\right] - \frac{1}{2} g t_1^2 sin\psi_1 \tag{5}$$

Where

$$t_1 = \frac{I_1}{r_{01}} \frac{\mu_1(\lambda_1 - 1)}{\lambda_1} \tag{6}$$

For a two-stage rocket whose thrust attitude angle of the first stage is ψ_1 and that of the second stage is ψ_2, the burnout velocity V_{2b} of the rocket will be given by the sum of the component of the burnout velocity after the first stage in the direction of ψ_2, that is $V_{1b}cos(\psi_2 - \psi_1)$ and the velocity contribution due to second stage, that is V_2. The burnout altitude h_{2b} of the two-stage rocket will be the sum of the altitude attained due to the first stage, that is h_{1b} and due to the second stage h_2. Hence assuming the rocket starts from the rest that is $V_0 = 0$,

Therefore, we have

$$V_{2b} = V_{1b}cos(\psi_2 - \psi_1) + V_2 = V_1 cos\varepsilon_1 + V_2 \tag{7}$$

Where

$$V_2 = gI_2 + log\left[\frac{\lambda_2}{\lambda_2(1-\mu_2)+\mu_2}\right] - gt_2 sin\psi_2,$$

$$\varepsilon_1 = \psi_2 - \psi_1, t_2 = \frac{I_2}{r_{02}}\frac{\mu_2(\lambda_2-1)}{\lambda_2}$$

And

$$h_{2b} = h_{1b} + h_2,$$

Where

$$h_2 = V_{1b}cos\varepsilon_1 sin\psi_2 t_2 + gI_2 t_2 sin\psi_2\left[1 - \frac{log\left(\frac{\lambda_2}{\lambda_2(1-\mu_2)+\mu_2}\right)}{\frac{\mu_2(\lambda_2-1)}{\lambda_2(1-\mu_2)+\mu_2}}\right] - \frac{1}{2}gt_2^2 sin^2\psi_2 \tag{8}$$

Therefore using (5) and (8)

$$h_{2b} = gI_1 t_1 sin\psi_1\left[1 - \frac{log\left(\frac{\lambda_1}{\lambda_1(1-\mu_1)+\mu_1}\right)}{\frac{\mu_1(\lambda_1-1)}{\lambda_1(1-\mu_1)+\mu_1}}\right] - \frac{1}{2}gt_1^2 sin^2\psi_1 + V_{1b}cos\varepsilon_1 sin\psi_2 t_2 + gI_2 t_2 sin\psi_2\left[1 - \frac{log\left(\frac{\lambda_2}{\lambda_2(1-\mu_2)+\mu_2}\right)}{\frac{\mu_2(\lambda_2-1)}{\lambda_2(1-\mu_2)+\mu_2}}\right] - \frac{1}{2}gt_2^2 sin^2\psi_2 \tag{9}$$

Or

$$h_{2b} = V_1 cos\varepsilon_1 sin\psi_2 t_2 + \sum_{n=1}^{2} gI_n t_n sin\psi_n\left[1 - \frac{log\left(\frac{\lambda_n}{\lambda_n(1-\mu_n)+\mu_n}\right)}{\frac{\mu_n(\lambda_n-1)}{\lambda_n(1-\mu_n)+\mu_n}}\right] - \frac{1}{2}\sum_{n=1}^{2} gt_n^2 sin^2\psi_n$$

Similarly, for a three-stage rocket, we have

$$\begin{aligned} V_{3b} &= V_{2b}cos\varepsilon_2 + V_3 \\ &= V_1cos\varepsilon_2cos\varepsilon_1 + V_2cos\varepsilon_2 + V_3 \\ &= V_1\prod_{k=1}^{2}cos\varepsilon_k + V_2\prod_{k=1}^{2}cos\varepsilon_k + V_3 \\ &= \sum_{n=1}^{2}V_n\prod_{k=n}^{2}cos\varepsilon_k + V_3 \end{aligned} \tag{10}$$

Where

$$V_3 = gI_3log\left[\frac{\lambda_3}{\lambda_3(1-\mu_3)+\mu_3}\right] - gt_3sin\psi_3,$$

$$t_3 = \frac{I_3}{r_{03}}\frac{\mu_3(\lambda_3-1)}{\lambda_3}$$

And $h_{3b} = h_{2b} + h_3$.
Where

$$h_3 = V_{2b}cos\varepsilon_2sin\psi_3t_3 + gI_3t_3sin\psi_3\left[1 - \frac{log\left(\frac{\lambda_3}{\lambda_3(1-\mu_3)+\mu_3}\right)}{\frac{\mu_3(\lambda_3-1)}{\lambda_3(1-\mu_3)+\mu_3}}\right] - \frac{1}{2}gt_3^2sin^2\psi_3 \tag{11}$$

Using Eq. (7)

$$\begin{aligned} V_{2b}cos\varepsilon_2sin\psi_3t_3 &= (V_1cos\varepsilon_1 + V_2)cos\varepsilon_2sin\psi_3\ \ t_3 \\ &= (V_1cos\varepsilon_2cos\varepsilon_1 + V_2cos\varepsilon_2)sin\psi_3\ \ t_3 \\ &= (V_1\prod_{(k=1)}^{2}cos\ \varepsilon_k + V_2\prod_{(k=1)}^{2}cos\ \varepsilon_k)sin\psi_3\ t_3 \end{aligned} \tag{12}$$

Using (9), (11) and (12) we have

$$\begin{aligned} h_{3b} = {} & V_1cos\varepsilon_1sin\psi_2\ t_2 + (V_1\prod_{k=1}^{2}cos\varepsilon_k + V_2\prod_{k=1}^{2}cos\varepsilon_k)cos\psi_3\ t_3 \\ & + \sum_{n=1}^{3}gI_nt_nsin\psi_n\left[1 - \frac{log\left(\frac{\lambda_n}{\lambda_n(1-\mu_n)+\mu_n}\right)}{\frac{\mu_n(\lambda_n-1)}{\lambda_n(1-\mu_n)+\mu_n}}\right] - \frac{1}{2}\sum_{n=1}^{3}gt_n^2sin^2\psi_n \end{aligned}$$

Or

$$\begin{aligned} h_{3b} = {} & \sum_{n=1}^{3}(V_1\prod_{k=1}^{n-1}cos\varepsilon_k + V_2\prod_{k=2}^{n-1}cos\varepsilon_k)sin\psi_n\ t_n \\ & + \sum_{n=1}^{3}gI_nt_nsin\psi_n\left[1 - \frac{log\left(\frac{\lambda_n}{\lambda_n(1-\mu_n)+\mu_n}\right)}{\frac{\mu_n(\lambda_n-1)}{\lambda_n(1-\mu_n)+\mu_n}}\right] - \frac{1}{2}\sum_{n=1}^{3}gt_n^2sin^2\psi_n. \end{aligned}$$

Thus, proceeding further, the burnout velocity V and burnout altitude H of N stage rocket with varying thrust attitude angle from stage to stage is given by

$$V = \sum_{n=1}^{N-1} (V_n \prod_{k=1}^{N-1} cos\varepsilon_k) + V_n \tag{13}$$

Where $\varepsilon_k = \psi_{k+1} - \psi_k, k = 1, 2, 3, \ldots\ldots, (N-1)$

$$V_n = gI_n log\left[\frac{\lambda_n}{\lambda_n(1-\mu_n)+\mu_n}\right] - gt_n sin\psi_n$$

And

$$H = \sum_{n=1}^{N-1} (V_1 \prod_{k=1}^{n-1} cos\varepsilon_k + V_2 \prod_{k=2}^{n-1} cos\varepsilon_k + \ldots + V_{n-1} cos\varepsilon_{n-1}) \times sin\psi_n\, t_n$$
$$+ \sum_{n=1}^{N} gI_n t_n sin\psi_n \left[1 - \frac{log\left(\frac{\lambda_n}{\lambda_n(1-\mu_n)+\mu_n}\right)}{\frac{\mu_n(\lambda_n-1)}{\lambda_n(1-\mu_n)+\mu_n}}\right] - \frac{1}{2} \sum_{n=1}^{N} gt_n^2 sin^2\psi_n \tag{14}$$

Where $t_n = \frac{I_n}{r_{0n}} \frac{\mu_n(\lambda_n-1)}{\lambda_n}$.

3 Results and Discussion

With the help of burnout velocity V_{1b} and burnout altitude h_{1b} for a single-stage given by Seifert in Eqs. (1) and (2) and proceeding step by step, burnout velocity V and burnout altitude H for N-stage rocket have been derived in Eqs. (13) and (14). These are nonlinear algebraic equations that can be solved by iterative methods, giving preassigned values of specific impulse, mass fractions, initial thrust, weight ratios and average thrust attitude angle.

Now we discuss the analytical approach in optimal analysis to determine the optimal condition for burnout velocity and optimum altitude with the optimum payload ratio of n stage rocket.

4 Optimization Analysis

The relationship between initial gross weight W_T, payload W_{PL} and the gross weight payload ratio λ, of the rocket, can be given by

$$\frac{W_T}{W_{PL}} = \lambda_1\lambda_2 \ldots\ldots\ldots\lambda_n \tag{15}$$

Let the restrictive conditions for the payload ratios λ be given by

$$V = V_c \tag{16}$$

$$H = H_c \tag{17}$$

where V and H are given by the Eqs. (13) and (14) respectively. If α and β be the Lagrangian undetermined multipliers, the optimization equation can be written as

$$\frac{\partial}{\partial\lambda_n}\left(\frac{W_T}{W_{PL}}\right)+\alpha\frac{\partial V}{\partial\lambda_n}+\beta\frac{\partial H}{\partial\lambda_n}=0 \tag{18}$$

Reduction of optimization equations:
From Eq. (15)

$$log\left(\frac{W_T}{W_{PL}}\right)=log\lambda_1+log\lambda_1+\ldots\ldots+log\lambda_n$$

Differentiating partially with respect to λ_n

$$\frac{1}{\left(\frac{W_T}{W_{PL}}\right)}\frac{\partial}{\partial\lambda_n}\left(\frac{W_T}{W_{PL}}\right)=\frac{1}{\lambda_n}$$

Or

$$\frac{\partial}{\partial\lambda_n}\left(\frac{W_T}{W_{PL}}\right)=\frac{1}{\lambda_n}\left(\frac{W_T}{W_{PL}}\right) \tag{19}$$

The number of non-linear algebraic equations giving optimum payload ratios can be reduced to N only by the elimination of Lagrangian multipliers [2]. From relation (19) Eq. (18) can be written as.

$$\alpha\lambda_n\frac{\partial V}{\partial\lambda_n}+\beta\lambda_n\frac{\partial H}{\partial\lambda_n}=-\left(\frac{W_T}{W_{PL}}\right), n=1,2,3,\ldots\ldots\ldots,N \tag{20}$$

From Eq. (20) we have the relations

$$\alpha\lambda_{n+1}\frac{\partial V}{\partial\lambda_{n+1}}+\beta\lambda_{n+1}\frac{\partial H}{\partial\lambda_{n+1}}=-\left(\frac{W_T}{W_{PL}}\right) \tag{20a}$$

$$\alpha\lambda_{n+2}\frac{\partial V}{\partial\lambda_{n+2}}+\beta\lambda_{n+2}\frac{\partial H}{\partial\lambda_{n+2}}=-\left(\frac{W_T}{W_{PL}}\right) \tag{20b}$$

Subtracting (20a) from (20) and (20b) from (20a) we get

$$\alpha\left(\lambda_n\frac{\partial V}{\partial\lambda_n}-\lambda_{n+1}\frac{\partial V}{\partial\lambda_{n+1}}\right)=-\beta\left(\lambda_n\frac{\partial H}{\partial\lambda_n}-\lambda_{n+1}\frac{\partial H}{\partial\lambda_{n+1}}\right) \tag{20c}$$

$$\alpha\left(\lambda_{n+1}\frac{\partial V}{\partial\lambda_{n+1}}-\lambda_{n+2}\frac{\partial V}{\partial\lambda_{n+2}}\right)=-\beta\left(\lambda_{n+1}\frac{\partial H}{\partial\lambda_{n+1}}-\lambda_{n+2}\frac{\partial H}{\partial\lambda_{n+2}}\right) \tag{20d}$$

Dividing (20c) by (20d) we get

$$\left(\lambda_n\frac{\partial V}{\partial\lambda_n}-\lambda_{n+1}\frac{\partial V}{\partial\lambda_{n+1}}\right)\left(\lambda_{n+1}\frac{\partial H}{\partial\lambda_{n+1}}-\lambda_{n+2}\frac{\partial H}{\partial\lambda_{n+2}}\right)=$$

$$\left(\lambda_n \frac{\partial H}{\partial \lambda_n} - \lambda_{n+1} \frac{\partial H}{\partial \lambda_{n+1}}\right)\left(\lambda_{n+1} \frac{\partial V}{\partial \lambda_{n+1}} - \lambda_{n+2} \frac{\partial V}{\partial \lambda_{n+2}}\right) \quad (21)$$

Where $n = 1, 2, 3, \ldots\ldots\ldots, (N-2)$. Equation 21 together with Eqs. (16) and (17) will give optimum values of λ. From Eq. (13) we have

$$V = \sum\nolimits_{n=1}^{N-1} (V_n \prod\nolimits_{k=1}^{N-1} cos\varepsilon_k) + gI_n log\left[\frac{\lambda_n}{\lambda_n(1-\mu_n)+\mu_n}\right] - g\frac{I_n}{r_{0n}} \frac{\mu_n(\lambda_n - 1)}{\lambda_n} sin\psi_n$$

Differentiating with respect to λ_n, we find

$$\frac{\partial V}{\partial \lambda_n} = g\frac{I_n}{r_{0n}} \frac{\mu_n}{(\lambda_n)^2}\left[\frac{r_{0n}\lambda_n}{\{\lambda_n(1-\mu_n)+\mu_n\} sin\psi_n} - 1\right] sin\psi_n \quad (22)$$

And

$$\frac{\partial H}{\partial \lambda_n} = \frac{\mu_n I_n sin\psi_n}{r_{0n}\lambda_n^2}[g(sin\psi_{n+1} t_{n+1} cos\varepsilon_n + sin\psi_n$$

$$+ \prod_{n+1}^{k=n} cos\varepsilon_k + \ldots\ldots + sin\psi_N t_N \prod_{N-1}^{k=n} cos\varepsilon_k)$$

$$+ \left(V_1 \prod_{n-1}^{k=1} cos\varepsilon_k + V_2 \prod_{n-1}^{k=2} cos\varepsilon_k + \ldots\ldots + V_{n-1} cos\varepsilon_{n-1} + V_n\right)]$$

5 Conclusion

Proceeding step by step the burnout velocity V and burnout altitude H of N stage rocket with varying thrust attitude angle from stage to stage is calculated in Eq. (13) and the optimum value of payload can be calculated from Eq. (22).

Future scope

With the help of these results, the shape and design of a rocket can be improved such that the given burnout velocity at an assigned altitude can be achieved and it will validate and enhance the aerodynamic performance.

References

1. Peressini, A.L.: Lagrange multipliers and the design of multi-stage rockets. The UMAP J. **7.3** (1986)
2. Bhatnagar, P., Rajan, S., Saxena, D.: Study on optimization problem of propellant mass distribution under restrictive conditions in multistage rocket. In: IJCA Proceedings on International Conference on Advances in Computer Applications 2012, by IJCA Journal ICACA - Number 1 (2012)
3. Chern, J.S., Wu, B., Chen, Y.S., Wu, A.M.: Suborbital and low thermos-spheric experiments using sounding rockets in Taiwan. Acta Astronaut. **70**, 159–164 (2012)

4. Coleman, J.J.: Optimum stage weight distribution of multistage rockets. ARS J. **31** (1961)
5. Dief, T.N., Yoshida, S.: System identification and adaptive control of mass-varying quad-rotor. Evergreen Joint J. Novel Carbon Resource Sci. Green Asia Strat. **4**(1), 58–66 (2017)
6. Dougherty, K.: Upper atmospheric research at Woomera: the Australian built sounding rockets. Acta Astronaut. **59**, 54–67 (2006)
7. Eberspeaker, P.J., Gregory, D.D.: An overview of the NASA sounding rocket and balloon programs. In: 19th ESA Symposium on European Rocket and Balloon Programs and Related Research, Bad Reichenhall, Germany, 7–11 June 2009
8. Goldsmith, M.: On the optimization of two-stage rockets. Jet Propulsion **27**(April), 415–416 (1957)
9. Muhammad, H.I., Rahman, A., Prihantini, N.B., Deendarlianto, N.: The application of poly-dispersed flow on rectangular airlift photobioreactor mixing performance. Evergreen **7**(4), 571–579 (2020). https://doi.org/10.5109/4150508
10. Halawa, A.M., Elhadidi, B., Yoshida, S.: Aerodynamic performance enhancement using active flow control on du96-w-180 wind turbine airfoil. Evergreen Joint J. Novel Carbon Resource Sci. Green Asia Strat. **5**(1), 16–24 (2018)
11. Hall, H.H., Zambelli, E.D.: On the optimization of multi-stage rockets. Jet Propulsion, p. 463 (1958)
12. Ismail, N.I. Mahadzir, M.M, Zurriati Ali M.: Computational aerodynamics study on neo-ptero micro unmanned aerial vehicle. Evergreen Joint J. Novel Carbon Resource Sci. Green Asia Strat. **8**(2), 438–444 (2021–06)
13. Maitra, S.N.: Analytical solutions to rocket performance with an arbitrary thrust program having constant thrust inclination with respect to the horizon in a vacuum. Sadhana **12**, Part 4, 339–352 (1988)
14. Malina, F.G., Summerfield, M.: The problem of escape from the earth by rocket. J. Aeron. Sci. **14**(471), 480 (1947)
15. NASA sounding rocket program overview. https://rscience.gsfc.nasa.gov
16. National Aeronautics and space administration. https://www.grc.nasa.gov/www/k12/rocket/termvr.html
17. National Aeronautics and space administration.https://www.grc.nasa.gov/www/k12/rocket/rktwtp.html
18. Sanz-Aranguez, P., Calero, J.S.: Sounding rocket developments in Spain. Acta Astronaut. **64**, 850–863 (2009)
19. Seifert, H.S., Mills, M.M.: Summerfield: "The Physics of Rocket". Amer. J. Phys. (1947)
20. Sohurmann, E.H.: Ernest: "Optimum staging technique for multi- staged rocket vehicles." Jet Propul. **27**(August), 863–865 (1957)
21. Subotiwicz, M.: The optimization of the N-step rocket with different construction parameters and propellant specific impulsion in each stage. Jet Propul. 460 (1958)
22. Thomson, W.T.: Introduction to Space Dynamics. Willey, New York, London (1961)
23. Vertregt, M.: A method of calculating the mass ratio of step rockets. J. British Inter Planet. Soc. **15**, 95 (1956)
24. Weisbord, L.: A generalized optimization procedure for N-staged missiles. Jet Propul. 164 (1958)
25. Khin, Z.H.K., Ohnmar, M.: Mathematical Techniques in rocket motion. World Acad. Sci. Technol. **46** (2008)

Emotion Detection Using Deep Fusion Model

Ashwini Raddekar, Akash Athani, Akshata Bhosle, Vaishnavi Divnale, and Diptee Chikmurge(✉)

MIT Academy of Engineering, Alandi, Maharashtra 412105, India
diptee.chikmurge@mitaoe.ac.in

Abstract. Extracting personal emotions from a given input format such as an image has been one of the most powerful and challenging research activities in social networking. Deep learning is the trending technology in the market which provides a good working model for emotion recognition. So Deep Learning algorithms will definitely perform better than traditional methods of image processing. Automatic emotion detection can help robots communicate intelligently with humans to provide better services. To increase the accuracy of the model proposed here, we defined the emotion recognition process using the CNN fusion method, which links two deep convolutional neural networks. The overall result of this experiment is based on the dataset FER 2013 We used a fusion method consisting of two image classification models VGG 16 and ResNet 50. The feature extraction using deep residual network ResNet-50 and VGG 16 combining convolutional neural networks for facial emotion recognition. It is based on developing a detailed emotion recognition model by reading the input images to provide the recognised output according to the emotions. The developed emotion recognition model is mainly used in the healthcare industry to recognise patients' emotions. Our model would be able to detect all the emotions of the patient, which is helpful for the doctors to treat the patient accordingly.

Keywords: Emotion detection · Deep learning · CNN fusion · VGG 16 · ResNet 50 · Feature Extraction

1 Introduction

Emotions are feelings expressed by the environment, mood or connectedness with others and are also a part of human communication [1]. Nowadays, human emotion recognition is an important part of human-to-human interaction to develop intelligent models for appropriate recognition. Emotions have a fundamental impact not only on human interactions, but also on human-computer interactions [1]. Since a person's emotional state can affect his or her concentration, ability to solve tasks, and make decisions, the vision of affective computing is to enable systems to recognize human emotions and influence them to increase productivity and efficiency when working with computers. [1, 17]. Emotion recognition using deep fusion models was introduced specifically for the medical domain. Our deep fusion model will be able to diagnose the emotional state of a patient so that medical professionals can monitor the patient's health condition and

H. Sharma et al. (Eds.): ICIVC 2022, PALO 17, pp. 476–487, 2023.
https://doi.org/10.1007/978-3-031-31164-2_40

accordingly provide appropriate treatment to the patient. Facial expressions indicate the movement of facial muscles, which represent the emotions shown by the patient. It gives information about the patient's attitude. Emotions are a state of mind that a patient is currently having [16]. It is what the person feels internally in response to things happening around him [16]. In any case, details about a patient's mental health state depend on his facial expression. Moreover, robots can intelligently communicate with humans to provide better services, and much more [7]. Emotion recognition in conversations can also be used to extract a large number of participants' opinions from very large conversation data in social networks. Face recognition is used to emotionally evaluate and process a given system based on a person's face [4].

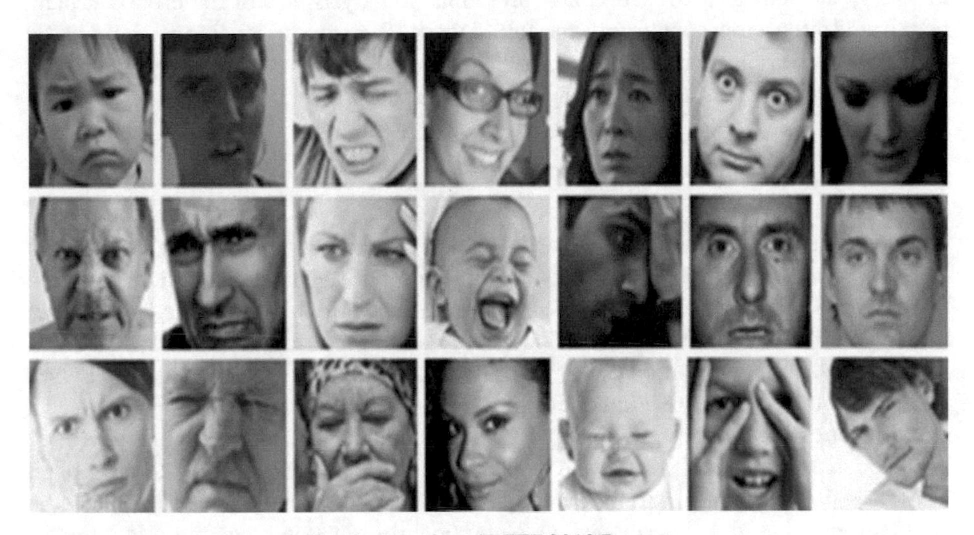

Fig. 1. Emotions in FER2013 Dataset

The goal of this study is to create an emotion recognition system that can divide an image into six different categories of emotions. These emotions are combined into a single dataset, the FER2013 dataset, which we see in Fig. 1 [6]. Emotional awareness provides benefits for many organizations or sectors and for different aspects of one's life. It is important and very useful for health and safety purposes. Moreover, it is important to easily know a person's feelings about the detection of their emotions at a given time without having to ask or look at them all the time. In healthcare today, facial recognition is certainly being used. They use it to find out if a patient needs medication or doctors (specialists) so they know who to treat first. In this project, we developed an emotion-based face recognition system in real-time image frames with deep learning algorithms [9]. Unlike similar applications, the system achieves high accuracy under different conditions or at any location with different models.

We developed a CNN view of an integrated database from different databases, which led to this research on higher verification accuracy and recognition level compared to existing models. Our goal is to improve the accuracy of emotion recognition in still images [3]. We proposed a convolutional fusion architecture (CFN) that could provide favorable insights into the judicious use and combination of multi-level features in deep neural networks. We decided to use CFN models to solve the image-level classification task [1]. The deep learning algorithms of fusion convolutional neural networks are very powerful as they are used to better understand other emotion-related behaviors, such as mental disorders. In this work, the combined fusion of two deep CNNs will be evaluated to achieve higher accuracy in emotion recognition than single models [3]. This study will explain how to build, train, develop and evaluate a fusion CNN. It will enable us to understand and know how to improve the ability to extract from a given dataset or database and how to explain the results by analyzing the training performance model. The remaining sections of this paper include related emotional activities, overview of the study method, data collection and preliminary analysis, use of the proposed plan, results and discussion, conclusion, and future work. The proposed fusion model is based on the analysis of facial information to detect six emotion categories, namely anger, fear, happiness, sadness, surprise, disgust, and neutral state.

2 Related Work

In recent years, many researchers have developed facial emotion models using the deep learning approach using the open library "Keras" for human facial emotion recognition. Here, a robust CNN network was used for image recognition. The images were taken from a provided dataset that contains facial expressions for different emotions like angry, sad, happy and many more and they also identified the expressions using an emotion model built by a CNN network model using Deep Learning. Gupta et al. [6] proposed a method consisting of convolutional layers with residual connections. An attention block is placed in the middle of the first and dense layers to give more sensitivity to the model. This model showed good results for the dataset FER and could be used for real-time applications. Li et al. [7] developed a CNN model for emotion recognition in humans, which consists of CNN and LSTM and transfers the parameters of CNN to present an improved model. The emotion recognition system based on this neural model detects six different facial emotions.

Abdullah et al. [7] proposed a model for recognizing facial expressions in videos whose methodology consists of CNN training, spatial feature observation, and long and short term memory. The developed method shows that using a better trained CNN followed by RNN training is almost as effective for identifying facial expressions in videos as for another task of the same type. Mutegeki et al. [8] used CNN and LSTM based techniques for human activity identification, training datasets, creating CNN and LSTM layers and obtaining the required output. Their advantage is to improve the accuracy of activity detection by exploiting the robustness in obtaining features of this network, which takes advantage of the process work and LSTM model for time series prediction, classification, and ranking. Shi et al. [9] have developed a method in which preprocessing of input images and feature extraction of expression images are first performed. Each feature is obtained by different network parts and then combined together (fusion), and this step improves the feature extraction capacity of the model MBCC-CNN. Finally, global or universal mean pooling is applied to the feature maps to average them in the last layer, and then the output is sent to the next SoftMax layer for identification. Ullah et al. [10] recommended a system that is a systematic literature review for human activity recognition (HAR), whose primary goal is to provide a foundation for novices in human activity recognition research. In human activity identification, the different activities are detected using different sensors/identifiers, but here in this paper, we restrict ourselves to the video-based ones because they are more applicable in the real world.

SI MIAO et al. [11] constructed the SHCNN model without deep temporal architectures such as 3D convolution and LSTM. There are two categories of facial expressions, namely static and micro expressions, and the proposed shallow model SHCNN is capable of understanding both. The testing of this algorithm is done on some public datasets. The algorithm gives the number of categories of facial expressions. It converts the available video images into grayscale images during preprocessing. If the expression category is micro expressions, the voting method is used to classify the video. Davide Anguita et al. [12] has performed some experiments to obtain the HAR dataset. Li et al. [13] developed a human expression recognition model consisting of CNN through ResNet50 which is an improved model. The emotion recognition system based on this neural model captures six different facial emotions and is a better multiple classification model. Enrique Correa et al. [14] provided a framework for their preferred model that includes convolutional frames or layers with residual connections. They also announced an attention block in the middle of the convolutional layer and the thick layer, through which the visual clarity of the model is increased. It gives finer results on the FER dataset and we can use it in real applications.

Liliana et al. [15] have developed a system using CK and a database trained with different data sizes for facial expression recognition and they have used CNN architecture. By using all these things, they have tried to classify/identify facial emotions based on 8 classes. In all 8 classes, there are some misclassifications which indicate that the system needs further improvement. Diah Anggraeni Pitalokaa et al. [16] have worked on a basic emotion classification system for humans. They used some preprocessing steps such as face detection, cropping the image, resizing the image, adding noise and normalizing the data. This model works efficiently at 32×32 or 64×64 resolution and requires maximum capacity at large data resolutions.

Wu et al. [17] developed a facial expression identification method based on an improved pre-trained VGG16 network model, which targets problems with poor representability and very short feature data when traditional expression identification methods are considered with these crisp applications. The VGG16 model network is modified by using a large convolutional filter instead of a small convolutional kernel and minimizing some fully connected layers to minimize the model complexity, values and parameters. The high-dimensional abstract phenomena output using the modified VGG16 are then fed into a convolutional neural network to recognize emotion expression types with greater accuracy. The expression identification or recognition method combined with the greatly improved modified VGG16 model and the CNN will be used for the human-computer interaction of the NAO robot. The robot performs different interactive actions depending on the emotion or expression. Mou et al. [18] developed an attention-based convolutional neural network (CNN) and a long-short-term memory (LSTM) model to combine non-invasive data such as eye data, vehicle data, and environmental data. The proposed model can then automatically extract features from each modality separately and pay different attention to features from different modalities through self-observation mechanisms.

3 Methodology

The selection of an appropriate algorithm that will help our system (CNN algorithm). Classification of images using CNN algorithm plays a very important role in emotion recognition. Here, we use a fusion CNN to combine multiple layers of features and provide a suitable output. Our method uses a deep CNN as the basis for our fusion model. The deep CNN models we combine are VGG-16 and ResNet-50, these models for image classification. The deep VGG-16 network with 16 layers uses small (3×3) convolutional layers throughout the network [3, 17], and the ResNet-50 (residual network) is used to build a deeper neural network. [3]. Using these models for image classification. We performed feature fusion and found the model with the best accuracy for our system. We compared these two models and found out which model has better accuracy. In doing so, we compared some terms of the two models such as accuracy, f1 score, precision, and recall.

The following steps were performed in this study.

1. A set of six basic sensory data was created from images found in online social searches, e.g., Kaggle.
2. This face detection process is applied to photos using the Viola Jones algorithm to detect faces in photos and crop them to create new image files.
3. The database is divided into two parts: the training database and the testing database. The Fusion-CNN method is trained with a training database and the performance of the network is tested with a test database.
4. Finally, the resulting network is used to separate facial expressions into image files.

4 Proposed Work

The internal process of the CNN algorithm is that our input in the form of an image goes through all these layers, the features are extracted, and a fully connected layer finally classifies the image and provides the output. In our system, we used an image as input, then it was passed to data preprocessing, after which we added two models for image classification, namely VGG 16 and ResNet 50, and we used these models to extract the features of the CNN algorithm. Then, we concatenate the CNN of these two models and compare them to achieve better accuracy; this complete process is called fusion. The flowchart below (see Fig. 2) gives an overview of the fusion classification model.

Data set: data consists of 48x48 pixel grayscale face images. The faces were automatically registered so that the face is more or less centered and occupies approximately the same space in each image. The task is to classify each face into one of seven categories based on the emotion in the facial expression. The training set consists of 28,709 examples and the public test set consists of 3,589 examples.

CNN algorithm: the Convolutional Neural Network (CNN) model is one of the components of Deep Learning Emotional Networks used to analyze images. The CNN architecture consists of three layers:

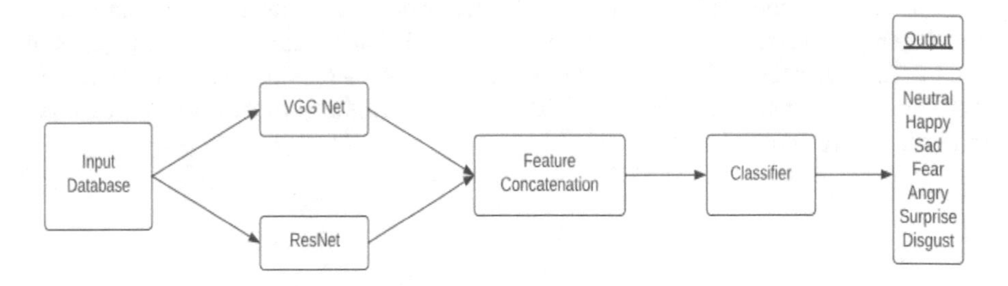

Fig. 2. Fusion based classification model

1) Convolutional layers: It helps to remove features from the input
2) MAX Pooling
3) The fully connected (dense) layers: Uses the data from the convolutional layers to obtain the output [15]

The first layer, i.e., the convolutional layer, essentially removes some of the features such as edges. This output is passed to the next layer up, which detects complex and difficult features such as corners. The filter/kernel size of (3 × 3 matrices) was used for the input image to improve the integrated feature [16]. This integrated feature is passed to the pooling layer. The pooling layer has the task of reducing the base size of the modified feature. Pooling can be done by average pooling and max pooling [16], but we choose max pooling here because it consists of the maximum number of pixels from a portion of an image that is overlapped by a kernel. It is also a noise suppressor. It also eliminates noise activation and provides noise removal and size reduction. On the other hand, central blending provides definition of all values from the part of the image covered by the kernel. Medium blending simply reduces the size as a way to compress the audio. Therefore, we can say that the performance of Max Pooling is much better than Average Pooling. That is why we used Max Pooling. After that, the output of the last and each of the pooling and convolution layers will be a 2D matrix from which all values must be subtracted into a vector; this method is called flattening. Then we get our integrated feature and it is transferred to the next layer, the fully connected layer. A fully connected layer (FC) has the weights, biases, and neurons used to connect neurons between two different FC layers. These layers are usually present before the last layer and form some previous layers of the CNN architecture. In this case, the input image from the previous layers is flattened and transferred to the layer FC. Then, the apartment vector gets a few additional FC layers, where mathematical operations are often performed, and when the process is complete, we get our final CNN output [16]. In this complete process, we used some basic terminologies, i.e. We used padding and strides in the convolutional layer, activation functions after the convolutional layer, and flattening before the fully connected layer. This mapping is repeated throughout the layer. When the fully linked layers are passed, the output layer uses SoftMax as the activation function instead of ReLU and is therefore used to invalidate the input possibilities into the appropriate categories.While implementing the model we imported necessary libraries and gave directories of train and test dataset as parameter when using ImageDataGenerator. We have 6 classes in the FER 2013 dataset. So we had chosen categorical as class_mode. Our batch_size is 32 and the number of epochs is 30. Input shape for all images is (48, 48, 1). This overall CNN implementation process is shown below (see Fig. 3). It shows the whole basic architecture process of the CNN algorithm from the convolutional layer to the last fully connected layer.

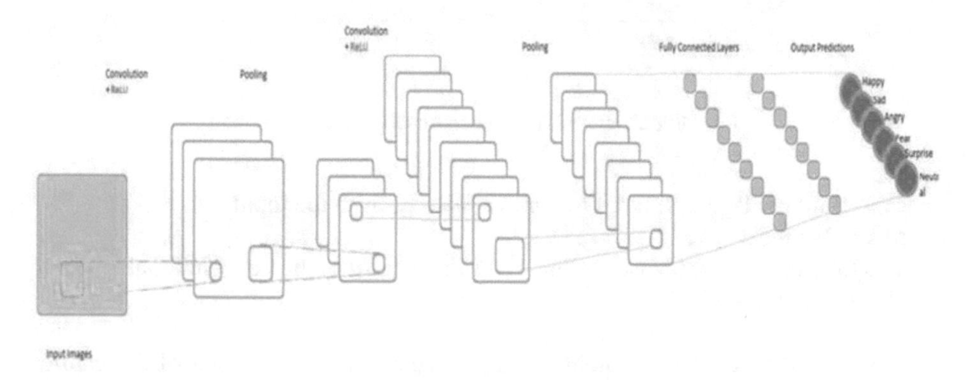

Fig. 3. Basic architecture of CNN

The last flat vector connects to smaller or more fully connected frames or layers and performs the similar mathematical operations! As given in below Eq. (1). In each layer of the Artificial Neural Network, the following operations are performed.

$$g(wx + b) \tag{1}$$

where,

x—the input vector of **[p_l, 1]**
W—the weight matrix of **[p_l, n_l]**
b—the bias vector of **[p_l, 1]**
g—ReLU activation function

This architectural flow is repeated throughout feature extraction. As the fully linked layers are traversed, the output layer uses SoftMax as the activation function instead of ReLU and is therefore used to invalidate the input features into the appropriate categories.

VGG-16 model: we used the pre-trained VGG-16 model, which has multiple convolutional layers and was developed by deep learning researchers who mainly focused on increasing the filters of the network [17, 20]. Second, VGG-16 uses only 3 × 3 kernels in each of the convolutional layers to perform the mentioned convolutional operation. The architecture of VGG-16 consists of 5 blocks. Each of the first two blocks contains two convolutional layers and a max-pooling layer. The remaining three blocks of the network contain three convolutional layers and one max-pooling layer. Third, after the last block of the mesh, three fully connected layers are also included: Layer one and layer two have 4096 neurons and layer three has 1000 neurons for the classification task in ImageNet [17]. Therefore, the VGG-16 is a very comprehensive network designed for the deep learning community. The final layer of the network is the soft-max layer [5].

The purpose of the VGG-16 is: 1) to increase the resolution of the relu activation function by including more layers (i.e. two or three layers as opposed to one for a receptive field of 7 × 7) [3], 2) a giant convolution is used to reduce the dimensionality directly on a higher order feature map that does not generate many computations, and 3) by extension, the continuous large size of the convolution kernel replaces the small size of the convolution, to subtract the complications of the model, further compress the number of parameters and reduce the proportion of the fully connected layer, it will not disturb the expression of the feature layer, but reduce the number of important parameters [17]. The 16 in VGG16 itself expresses that it has 16 layers that contain weights [20].

ResNet 50 model: this is a kind of convolutional neural network that is 50 layers deep and is one of the deepest neural networks [19]. The model, an abbreviation for ResNet, can be a classical neural network used for many tasks in computer vision. The main advantage of ResNet was that it allows us to train very deep neural networks [2]. The execution of the residual network uses stack normalization (BN) to reduce the internal drift of the covariates [2, 3]. Since a deep neural network uses stochastic gradient descent, BN solves the problem by standardizing each mini-batch [3]. This is supported by two intuitions:-

1) When we go deeper into implementing an oversized number of layers, we should always ensure that we do not reduce the accuracy as well as the error rate. This is solved by identity mapping.
2) Continue to examine the residuals to map the predicted with the original. These are the residual network functions. Below (see Fig. 4) is the detailed working principle of the ResNet 50 model with the mathematical understanding.

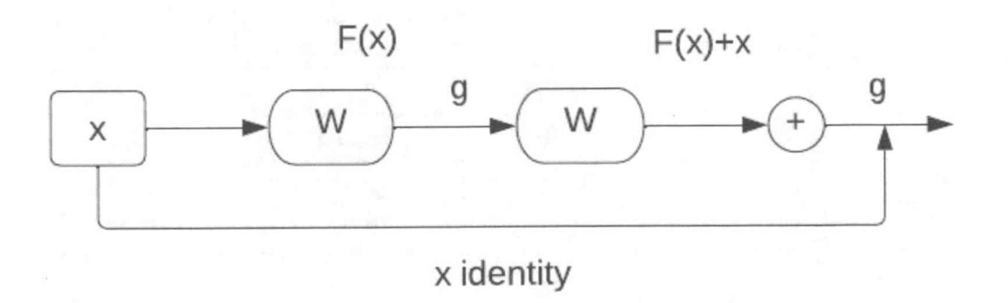

Fig. 4. Mathematical modeling of ResNet 50

Here [15] is what the paper says about Identity Function and it's implemented.

$$y = F(x, \{Wi\}) + x \quad (2)$$

x and y are the input and output vectors values of the layers taken. The function F(x, {Wi}) represents the residual mapping. For the instance in Fig. 4 that has two layers,

$F = W2\sigma(W1x)$, in which σ denotes ReLU

$$y = F(x, \{Wi\}) + Wsx \quad (3)$$

We can also use the matrix Ws(2). However, it can be shown by experiments that the identity mapping is more satisfactory to resolve the degradation issue and thus Ws is simply used when comparing dimensions [18].

5 Results and Discussion

We proposed only onc CNN model consisting of 5 convolutional layers and obtained an output that was not satisfactory for test accuracy. To avoid this error, we moved to the VGGNET 16 model and trained our dataset. We obtained a good accuracy, but when testing the model there were some unsatisfactory parameters. So we decided to work with the ResNet 50 model and trained the dataset. Again, the accuracy was good enough, but the model was not able to give the correct results. Therefore, we had the idea to link the two models VGGNET 16 and ResNet 50 in the form of a fusion CNN. We trained our dataset using the fusion CNN model. We proposed the CNN model for emotion recognition, facing problems related to accuracy. The training accuracy was good, but the testing accuracy was not satisfactory, so the model experienced overfitting. In addition, the emotions were not detected correctly. Therefore, we need to choose the Fusion-CNN because it contains multiple CNN algorithms. This could help us to overcome the overfitting and improve the accuracy. As a result, we obtained better accuracy and also the expected output of classified emotions.

The following Table 1 shows that comparing the validation results of the different models, we compared the test accuracy, F1 score, precision and recall to get a better accuracy of our model.

Table 1. Comparison of validation result

Model	Accuracy	F1 Score	Precision	Recall
CNN Model	64%	0.6661	0.6812	0.6222
VGG 16	85%	0.7011	0.7021	0.7121
ResNet 50	85.71%	0.7111	0.7122	0.7132
Fusion CNN	88%	0.7521	0.7535	0.7522

To test our model we have given an input image as shown in Fig. 5. Which is a happy emotion picture. As a result the following graph in Fig. 6 is showing the proper output, that is the input image belongs to the class happy.

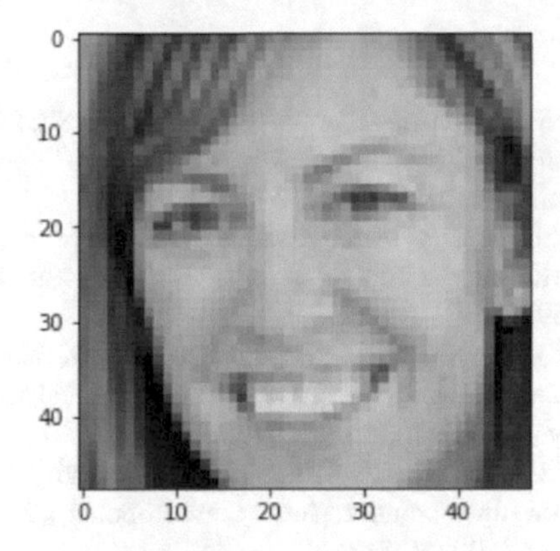

Fig. 5. Input Image

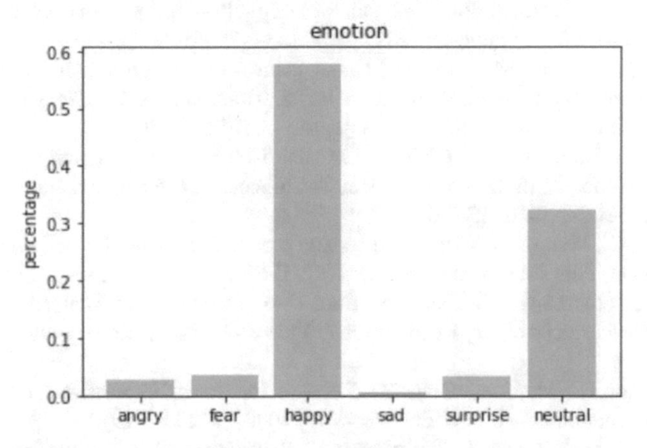

Fig. 6. Output Graph

6 Conclusion

We know that this program automatically identifies emotions. Therefore, with this automated program, we can create intelligent healthcare facilities that help diagnose patients' stress and depression, and we can start treatment early. With this system, we are able to make the most effective processing result using a real-time face-to-face separation of learning faculty according to CNN the experimental work using experimental data and according to CNN algorithm we use Fusion CNN obtained. In the Fusion CNN, we used VGG 16 and ResNet 50 models. This is a method of combining source images obtained from the same scene and concatenating the final output of these two fusion models. In this overall system, it was recognized that a pre-trained network with an extended training database provides effective and consistent results under changing light and environmental conditions. In the results, we can see examples of separating the facial emotions of anger, fear, joy, neutrality, sadness, disgust, and surprise in real-time in image frames.

References

1. Do, L.-N., Yang, H.-J., Nguyen, H.-D., Kim, S.-H., Lee, G.-S., Na, I.-S.: Deep neural network-based fusion model for emotion recognition using visual data. J. Supercomput. **77**(10), 10773–10790 (2021). https://doi.org/10.1007/s11227-021-03690-y
2. Liu, Y., Guo, Y., Georgiou, T., Lew, M.S.: Fusion that matters: convolutional fusion networks for visual recognition. Multimedia Tools Appl. **77**(22), 29407–29434 (2018). https://doi.org/10.1007/s11042-018-5691-4
3. Aza, M.F.U., Suciati, N., Hidayati, S.C.: Performance study of facial expression recognition using convolutional neural network. In: 2020 6th International Conference on Science in Information Technology (ICSITech), pp. 121–126. IEEE (2020)
4. Ercolano, G., Rossi, S.: Combining CNN and LSTM for activity of daily living recognition with a 3D matrix skeleton representation. Intel. Serv. Robot. **14**(2), 175–185 (2021). https://doi.org/10.1007/s11370-021-00358-7
5. Jaiswal, A., Raju, A. K., Deb, S.: Facial emotion detection using deep learning. In: 2020 International Conference for Emerging Technology (INCET), pp. 1–5. IEEE (2020)
6. Gupta, A., Arunachalam, S., Balakrishnan, R.: Deep self-attention network for facial emotion recognition. Procedia Comput. Sci. **171**, 1527–1534 (2020)
7. Abdullah, M., Ahmad, M., Han, D.: Facial expression recognition in videos: an CNN-LSTM based model for video classification. In: 2020 International Conference on Electronics, Information, and Communication (ICEIC), pp. 1–3. IEEE (2020)
8. Mutegeki, R., Han, D.S.: A CNN-LSTM approach to human activity recognition. In: 2020 International Conference on Artificial Intelligence in Information and Communication (ICAIIC), pp. 362–366. IEEE (2020)
9. Shi, C., Tan, C., Wang, L.: A facial expression recognition method based on a multibranch cross-connection convolutional neural network. IEEE Access **9**, 39255–39274 (2021)
10. Ullah, H.A., Letchmunan, S., Zia, M.S., Butt, U.M., Hassan, F.H.: Analysis of Deep Neural Networks For Human Activity Recognition in Videos–A Systematic Literature Review. IEEE Access (2021)
11. Miao, S., Xu, H., Han, Z., Zhu, Y.: Recognizing facial expressions using a shallow convolutional neural network. IEEE Access **7**, 78000–78011 (2019)
12. Anguita, D., Ghio, A., Oneto, L., Parra Perez, X., Reyes Ortiz, J.L.: A public domain dataset for human activity recognition using smartphones. In: Proceedings of the 21th International European Symposium on Artificial Neural Networks, Computational Intelligence and Machine Learning, pp. 437–442 (2013)
13. Li, B., Lima, D.: Facial expression recognition via ResNet-50. Int. J. Cogn. Comput. Eng. **2**, 57–64 (2021)
14. Correa, E., Jonker, A., Ozo, M., Stolk, R.: Emotion recognition using deep convolutional neural networks. Technical Report IN4015 (2016)
15. Liliana, D.Y.: Emotion recognition from facial expression using deep convolutional neural networks. J. Phys. Conf. Ser. **1193** (1), 012004 (2019). (IOP Publishing)
16. Pitaloka, D.A., Wulandari, A., Basaruddin, T., Liliana, D.Y.: Enhancing CNN with preprocessing stages in automatic emotion recognition. Procedia Comput. Sci. **116**, 523–529 (2017)
17. Wu, S.: Expression recognition method using improved VGG16 network model in robot interaction. J. Robot. (2021)
18. Mou, L., et al.: Driver stress detection via multimodal fusion using attention-based CNN-LSTM. Expert Syst. Appl. **173**, 114693 (2021)

19. Kale, S., Shriram, R.: Suspicious activity detection using transfer learning based ResNet tracking from surveillance videos. In: Abraham, A., et al. (eds.) SoCPaR 2020. AISC, vol. 1383, pp. 208–220. Springer, Cham (2021). https://doi.org/10.1007/978-3-030-73689-7_21
20. Mathavaraj, S., Padhi, R.: Performance comparison. In: Satellite Formation Flying, pp. 139–146. Springer, Singapore (2021). https://doi.org/10.1007/978-981-15-9631-5_8

GTMAST: Graph Theory and Matrix Algorithm for Scheduling Tasks in Cloud Environment

Iqbal Gani Dar(✉) and Vivek Shrivastava

Department of Computer Science, International Institute of Professional Studies, Devi Ahilya Vishwavidyalaya, Indore 452001, India
{iqbal.gani.dar-rs,vivek.shrivastava}@iips.edu.in

Abstract. Mapping users' tasks to virtual machines (VMs) is an NP-Complete problem. Researchers have tried to optimize various performance indicators of the tasks while taking into consideration certain attributes of the tasks. The number of tasks completed successfully by the algorithm, turnaround time, throughput, response time, execution time, and other characteristics are examples of single or combinations of parameters that can be used in optimization. In this work, we present an efficient algorithm GTMAST for scheduling heterogeneous tasks in the cloud environment. The proposed algorithm aims at maximizing the number of tasks executed and the throughput while minimizing the turnaround time. The algorithm has been tested in an open-source simulation tool, CloudSim 3.0.1. The performance of this algorithm outperforms the well-known algorithms such as Elimination Et Choice Translating REeality (ELECTRE) and Technique for Order Preference by Similarity to Ideal Solution (TOPSIS) in terms of the number of tasks executed, the average turnaround time and the average throughput.

Keywords: Cloud Computing · Task Scheduling · GTMAST · ELECTRE · TOPSIS

1 Introduction

Over the last fifty years, numerous computing models have been proposed. One such model is cloud computing (CC). CC is a distributed computing paradigm [1] that offers a framework for hosting and providing on-demand services to users over the Internet. CC is becoming popular due to its many features such as virtualization, pay-per-use pricing policy, scalability and elasticity. However, the dynamic nature of the cloud gives rise to many challenges also. Scheduling tasks submitted by multiple users at once to a cloud service provider (CSP) is one of the major challenges. Task scheduling problem (TSP) as shown in Fig. 1 involves the efficient placement of users' tasks on VMs for execution, satisfying certain constraints imposed by the user as well as the CSP. These constraints can be Quality of Service (QoS) parameters specified by the users in the Service Level Agreement (SLA) or to maximize revenue from the CSPs' point of view.

H. Sharma et al. (Eds.): ICIVC 2022, PALO 17, pp. 488–499, 2023.
https://doi.org/10.1007/978-3-031-31164-2_41

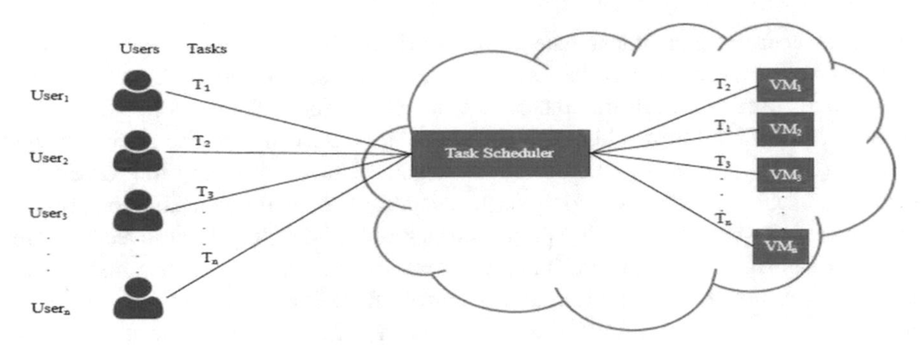

Fig. 1. Task Scheduling Problem.

The aims of this paper are as follows:

- Design an efficient task scheduling algorithm for scheduling users' tasks onto VMs.
- Evaluate the performance of this algorithm in terms of performance metrics such as the number of tasks executed, the average turnaround time and the average throughput.
- Compare its performance with well-known algorithms, ELECTRE and TOPSIS.

The rest of this paper is organized as follows. The related work is discussed in Sect. 2. Section 3 presents the problem definition and describes the proposed algorithm. Section 4 shows the results of experiments done to check the performance of the proposed algorithm. Finally, the conclusion and future work are provided in Sect. 5.

2 Related Work

Task scheduling entails the efficient distribution of resources among distinct tasks while taking into account several performance measures. To improve these performance indicators, several single-criteria and multi-criteria decision analysis (MCDA) methods have been suggested. Considering one scheduling criterion will result in better outcomes for that criterion alone, but maintaining balance across all parameters is crucial. Decision-making problems with numerous criteria are addressed using a number of Multi-Criteria Decision Analysis (MCDA) or Multi-Criteria Decision Making (MCDM) techniques [2]. The MCDM procedure chooses the best alternatives from a set of available alternatives. Multiple criteria are used to do the computation primarily. MCDM evaluates many alternatives under a finite number of conflicting criteria [3]. The use of mathematical approaches for handling TSPs, such as MCDM methods, is becoming more popular. Alla et al. [4] proposed the MCPTS algorithm to address the priority issue in both users' tasks and CSPs' resources. For evaluation and task prioritization, differential evolution and ELECTRE III are proposed. To prevent task scheduling conflicts in the backfilling algorithm for carrying out deadline-based tasks in CC, Nayak et al. [5] incorporated the MCDM techniques: AHP (Analytical Hierarchy Process), VIKOR (VIseKriterijumska Optimizacija I Kompromisno Resenje) and TOPSIS (Technique for Order Preference by Similarity to Ideal Solution). Using the BWM and the TOPSIS, Khorsand et al. [6]

devised an energy-efficient task scheduling algorithm. The makespan, energy consumption, and resource usage are the performance indicators used in their work, and the BWM procedure is used to assign the importance of weights for each criterion. Kumar et al. [7] introduced a workflow scheduling algorithm based on TOPSIS. The entropy weight method (EWM) is used to determine the weights of the criteria and four scheduling attributes including schedule length, cost, reliability and energy consumption are considered in their work. Mahmoud et al. [8] introduced the Task Scheduling-Decision Tree (TS-DT) algorithm for task allocation and execution. Load balance parameters such as makespan, power consumption and resource utilization are used for performance evaluation of the proposed task scheduling strategy. The proposed algorithm reduces makespan, cost, and energy consumption while increasing reliability. EWM is applied to determine the input weight for the attributes of schedule length, cost, reliability, and energy consumption. Singh et al. [9] proposed the Cuckoo Search optimization (CSO) algorithm. Modified Best Fit Decreasing (MBFD) algorithm is used for VM placement. With this information, the security level is obtained using a neural network (NN) for final task scheduling. The evaluation metrics used are the number of SLA violations and energy consumption. A multi-objective optimization model that reduces energy usage and increases span was introduced by Shukla et al. [10] for the effective scheduling of tasks. Additionally, a method based on the Non-dominated Sorting Genetic Algorithm (NSGA) was proposed to address the optimization of the multi-objective task allocation problem. To enhance work scheduling in the cloud, Shariat et al. [11] suggested the HATMOG technique, which was based on the intelligent hybrid multiple criterion decision-making algorithms of AHP-TOPSIS and Non-Dominated Sorting Genetic Algorithm (NSGAII). Chraibi et al. [12] proposed three modifications to the bin-packing algorithm. The modified bin packing task scheduling (MBPTS) reduced waiting time and makespan, and improved resource utilization in comparison to the particle swarm optimization (PSO) and first come first serve (FCFS). Swain et al. [13] suggested Interference Aware Workload Scheduling (IAWS). The two prediction models in IAWS are used to execute time-sensitive tasks. IAWS minimizes interference for a better quality of service and improves task and priority guarantee ratios, while improving resource utilization as compared to other well-known methods. Shrivastava et al. [14] proposed EBTASIC to balance resource utilization and revenue generation. Various performance metrics such as actual response time, execution time, turnaround time, throughput, average response time, resource utilization, revenue generation and deadline are calculated using EBTASIC and compared with baseline algorithms FCFS and Earliest Deadline First (EDF). Table 1 provides the related work in a summarized manner. However, none of the surveyed literature used Graph Theory and Matrix approach to schedule the tasks in the CC environment. The novelty in this work is the application of graph theory and matrix approach in CC for TSP. In this work, we have proposed the novel GTMAST. This work aims to extract the priority from the attributes of the tasks based on their relative importance.

Table 1. Related work in a summarized manner.

Author (Year)	Techniques used	Performance Metrics used
Kaur et al. [1] (2021)	FCFS, Round Robin (RR), PSO	makespan
Alla et al. [4] (2021)	MCPTS, Random, Earliest Deadline First, Shortest Job First	makespan, resource utilization, response time, Degree of Imbalance (DI)
Nayak et al. [5] (2020)	Backfill, MCDM techniques AHP, VIKOR, TOPSIS	VM slot utilization, VM slot wastage and average task acceptance ratio
Riehaneh Khorsand and Mohammadreza Ramezanpour [6] (2020)	MCDM techniques BWM, TOPSIS	makespan, energy consumption, and resource utilization
Kumar et al. [7] (2021)	MCDM techniques EWM, TOPSIS	makespan, cost, energy consumption and reliability
Mahmoud et al. [8] (2022)	TS-DT, Heterogeneous Earliest Finish Time (HEFT), TOPSIS-EWM, and combining Q-Learning with HEFT (QL-HEFT)	makespan, resource utilization, load balancing
Mandeep Singh and Shashi Bhushan [9] (2022)	CSO, MBFD, NN	SLA violation, energy consumption
Shukla et al. [10] (2021)	NSGA	makespan, energy consumption
Sahar Samsam Shariat Behrang Barekatain [11] (2022)	HATMOG based on MCDM techniques of AHP-TOPSIS and NSGAII	makespan, resource utilization
Chraibi et al. [12] (2022)	MBPTS, FIFO, PSO	makespan, waiting time, resource utilization,
Chinmaya Kumar Swain and Aryabartta Sahu [13] (2021)	IAWS	task guarantee ratio, priority guarantee ratio, resource utilization
Vivek Shrivastava and Yasmin Shaikh [14] (2022)	EBTASIC (An Entropy-Based TOPSIS Algorithm for Task Scheduling in IaaS Clouds)	actual response time, execution time, turnaround time, throughput, average response time, resource utilization, revenue generation, deadline

3 Proposed Work

3.1 Problem Definition

In this work, the proposed methodology consists of a finite set of N, N = $\{T_1, T_2, \ldots T_n\}$ heterogeneous users' tasks. These N tasks from the global queue are mapped onto a finite set of K heterogeneous VMs, R = $\{VM_1, VM_2, \ldots, VM_k\}$ for execution. The problem is to find a schedule S that maximizes the number of tasks executed by the scheduler, minimizes the turnaround time and maximizes the throughput, subject to certain constraints imposed by both users and CSPs. We have used graph theory and matrix approach to find an efficient schedule. Figure 2 shows the proposed methodology.

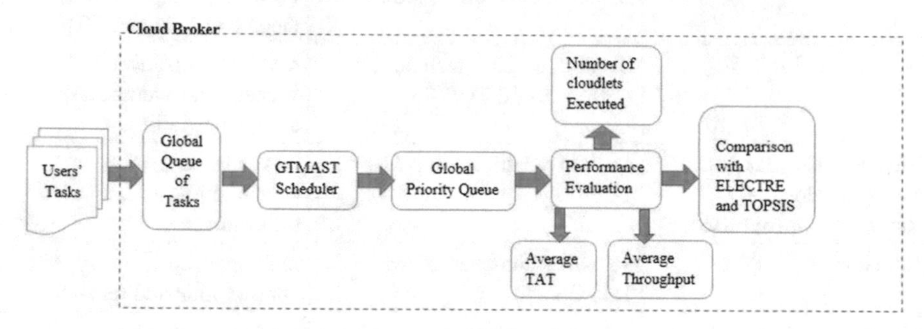

Fig. 2. Proposed Methodology.

3.2 Proposed Algorithm

Graph theory is used to study graphs in mathematics. Graphs are mathematical structures used to model the pairwise relationship between objects. In the proposed algorithm GTMAST, the vertices represent the attributes such as length, deadline and number of resources. The edges represent the pairwise relationship among the attributes. The performance of GTMAST is evaluated using different performance metrics such as the number of cloudlets executed, the average turnaround time (TAT) and the throughput. The various steps in GTMAST are as follows:

Step 1: Identification and selection of attributes and alternatives involved in the problem.
Step 2: Formation of a directed graph (or digraph) of the attributes. Each vertex in the digraph represents an attribute. In our problem, the digraph has three vertices, namely, length, deadline and number of resources. These three vertices are connected via six edges representing the relationship between vertices in the digraph as shown in Fig. 3.

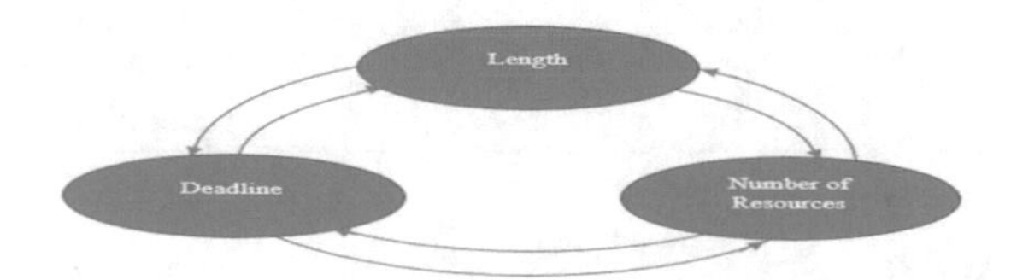

Fig. 3. Attribute digraph.

Step 3: Formation of the permanent matrix (PM) for the attributes of the digraph. This matrix will be an nxn matrix where n is the number of attributes. In our problem, the 3x3 matrix is as follows:

$$\begin{array}{c} \\ \text{Length} \\ \text{Deadline} \\ \text{PesNumber} \end{array} \begin{array}{c} \begin{array}{ccc} \text{Length} & \text{Deadline} & \text{PesNumber} \end{array} \\ \begin{pmatrix} \boldsymbol{A1} & a_{12} & a_{13} \\ a_{21} & \boldsymbol{A2} & a_{23} \\ a_{31} & a_{32} & \boldsymbol{A3} \end{pmatrix} \end{array}$$

Where A_i is the measure of the particular alternative for the ith attribute. In the PM, A_i, i.e. A_1, A_2, A_3 takes the place of diagonal elements. A_{ij} is the relative importance (RI) value of one attribute over another. For example, a_{21} represents the RI of deadline over length i.e. how important is deadline over length. The values of RI are taken from Table 2 in which equal importance is assumed a value of 0.5 and exceptionally more important is assumed a value of 1.0. The decision maker needs to give the value of a_{ij}. The value of a_{ji} is calculated as $a_{ji} = 1 - a_{ij}$.

Table 2. RI values of one attribute over another.

Description	Relative Importance	
	a_{ij}	$a_{ji} = 1 - a_{ij}$
Equally important	0.5	0.5
Slightly more important	0.6	0.4
Strongly more important	0.7	0.3
Very strongly important	0.8	0.2
Extremely important	0.9	0.1
Exceptionally more important	1.0	0.0

Assumptions: In this work, we have taken the following assumptions.

Deadline is extremely important over pesNumber and strongly important than length. Length is strongly more important than PesNumber.

Using the above assumptions, the matrix of step 3 takes the form:

$$
\begin{array}{r} \text{Length} \\ \\ \text{Deadline} \\ \\ \text{PesNumber} \end{array}
\begin{array}{c} \begin{array}{ccc} \text{Length} & \text{Deadline} & \text{PesNumber} \end{array} \\ \begin{pmatrix} \boldsymbol{A1} & 0.3 & 0.7 \\ & & \\ 0.7 & \boldsymbol{A2} & 0.9 \\ & & \\ 0.3 & 0.1 & \boldsymbol{A3} \end{pmatrix} \end{array}
$$

Step 4: Substitute the values of A_i and find the value of the PM. To find the value of the PM, we first normalize the decision matrix (DM). We have performed linear normalization. The formula for linear normalization is given below:

For beneficial attributes:

$$\underline{X}ij = \frac{Xij}{Xj^{Max}} \tag{1}$$

For non-beneficial attributes:

$$\underline{X}ij = \frac{Xj^{Min}}{Xij} \tag{2}$$

To compute the value of the PM we substitute the value of A_i from the DM.

For Alternative$_1$, PM_1, $Per(A_1)$=

0.917926	0.3	0.7
0.7	0.520833	0.9
0.3	0.1	1

Next, we calculate the permanent of A_1. The value of $Per(A_1)$ equals 1.010075. Similarly, we find the PM and $Per(A_n)$ value for A_n.

Step 5: Prioritize the alternatives. Finally, we prioritize the alternatives. The alternative with the highest permanent value gets priority. Figure 4 shows the workflow of GTMAST. Table 3 shows the pseudocode of GTMAST.

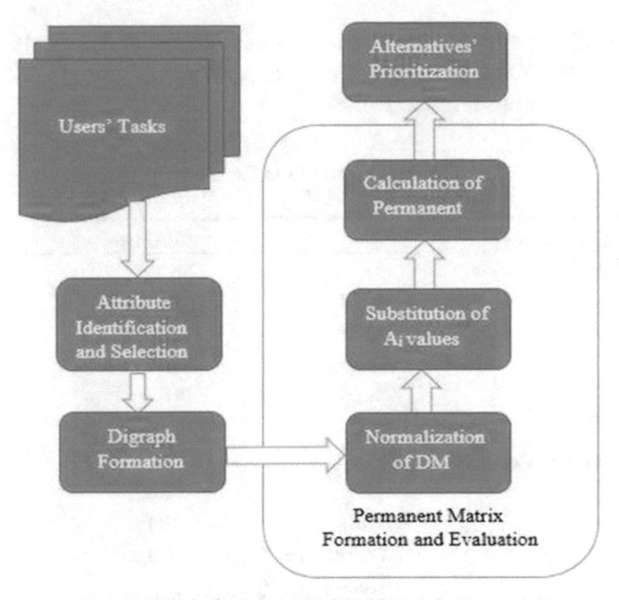

Fig. 4. Workflow of GTMAST.

Table 3. Pseudo code of GTMAST.

```
Algorithm: GTMAST
Input:  A global queue of N tasks, T1, T2,…., Tn.
Output: Efficient Schedule of N tasks, T1, T2,…., Tn.
while workload in cloud broker is running do
        Obtain attribute values from the tasks
end
Normalize Decision Matrix.
Substitute Ai values from Decision Matrix.
Calculate the Permanent of T1, T2,…, Tn.
Prioritize the tasks T1, T2,…Tn.
Calculate the performance metric values.
Compare with ELECTRE and TOPSIS.
```

4 Results and Discussion

In this section, the simulation tool used for experimentation as well as the results of the experiments has been provided.

4.1 Simulation Setup

The performance of the proposed algorithm has been evaluated in CloudSim 3.0.1 toolkit. Various parameters that include data centers, hosts, cloudlets and VMs are used in the simulation. The simulation is done using 2 data centers and 2 hosts, with different parameter values for VM and cloudlet used as shown in Table 4. A dataset of 2400 cleaned tasks has been used in this work.

Table 4. Simulation Parameters

Parameters		Values
VM Parameters		
	RAM	512 MB
	MIPS	1000
	Bandwidth	1000
	Number of CPUs	1
	VMM	Xen
Cloudlet Parameters		
	Length	1000
	File size	300
	Output size	300
	Number of CPUs	1

4.2 Performance Evaluation and Comparison

Three performance metrics have been used to evaluate the performance of GTMAST. We also compute the performance improvement rate (PIR). PIR indicates the performance improvement in percentage of various performance metrics. It is calculated using the following formula.

$$PIR(\%) = \left(\frac{Metric(Proposed\ Algorithm) - Metric(Old\ Algorithm)}{Metric(Old\ Algorithm)}\right) * 100 \quad (3)$$

where Metric (Proposed Algorithm) is the value of the performance metrics of GTMAST and Metric (Old Algorithm) is the value of the performance metrics of TOPSIS and ELECTRE.

4.2.1 Number of Cloudlets Executed

The more the number of cloudlets executed by the algorithm the better it is. The proposed algorithm GTMAST executes 1801 cloudlets in comparison to 1656 by ELECTRE and 1691 by TOPSIS. The performance improvement rate (PIR) of GTMAST is 8.75% more than TOPSIS and 6.56% more than ELECTRE. Figure 5 shows the comparison.

4.2.2 Average Turnaround Time

We define turnaround time as the total time between the submission of a cloudlet and its completion. The proposed algorithm GTMAST has a less average turnaround time of 6.30 as compared to 6.53 in TOPSIS and 6.45 in ELECTRE. The PIR of GTMAST is 3.58% more than TOPSIS and 3.87% more than ELECTRE. Figure 6 shows the average turnaround time comparison.

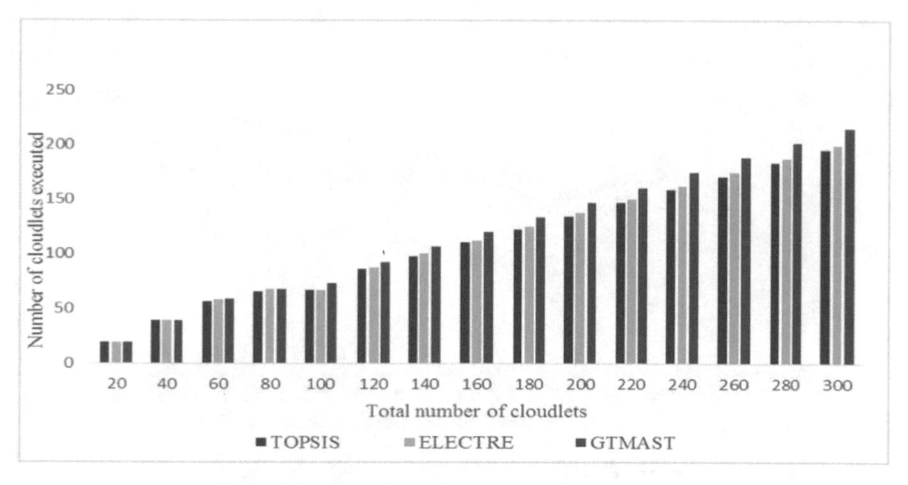

Fig. 5. Number of cloudlets executed comparison.

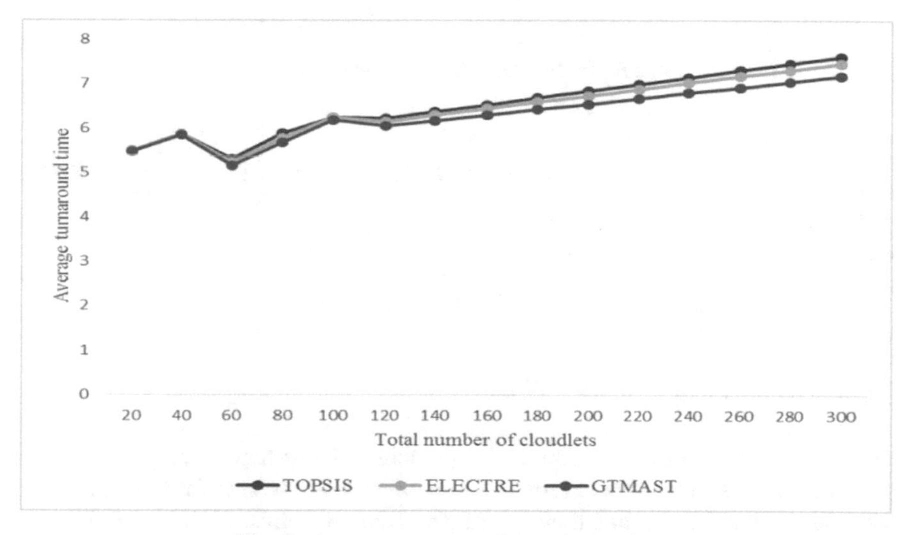

Fig. 6. Average turnaround time comparison.

4.2.3 Average Throughput

The throughput is the number of cloudlets executed per unit of time. The average throughput of GTMAST is 2.62 as compared to 1.92 in TOPSIS and 2.13 in ELECTRE. The PIR of GTMAST is 14.90% more compared to TOPSIS and 8.90% more compared to ELECTRE. Figure 7 shows the average throughput comparison. Table 5 shows the performance metrics and comparison of GTMAST with TOPSIS and ELECTRE. The table also shows the PIR of GTMAST over TOPSIS and ELECTRE.

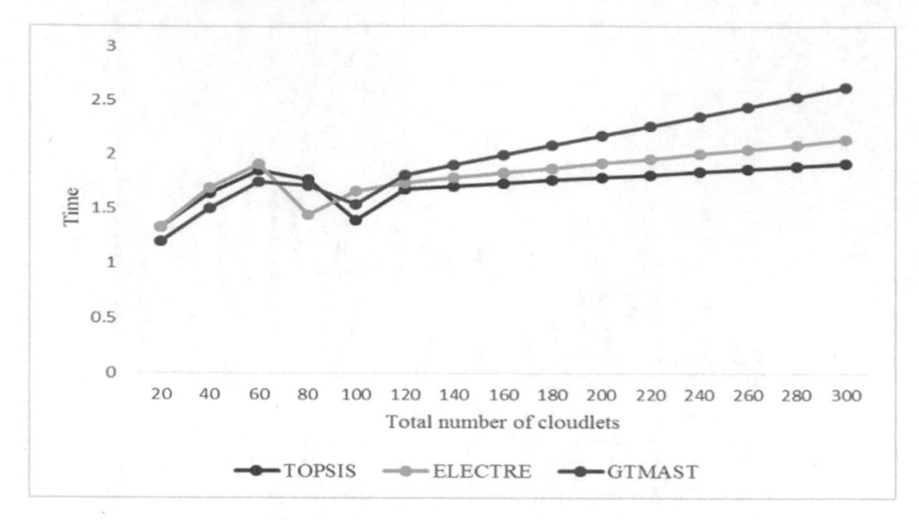

Fig. 7. Average throughput comparison.

Table 5. Performance Metrics and Comparison.

Performance Metrics	Comparison			PIR (%)
	TOPSIS	ELECTRE	GTMAST	TOPSIS ELECTRE
Number of Cloudlets Executed	1656	1711	**1801**	8.75 6.56
Average Turnaround Time	6.53	6.51	**6.30**	3.58 3.87
Average Throughput	1.92	1.81	**2.62**	14.90 8.90

5 Conclusion and Future Work

Cloud computing is a popular computing paradigm that offers users computing as a service and allows them to use resources in a pay-per-use policy. Currently, more and more organizations are shifting their businesses from on premise to the cloud. However, this expansion has also led to a more difficult task scheduling and resource provisioning process. Numerous research in this area ignores multiple criteria that are present in task scheduling in the real world and instead concentrate only on one. This does not meet the multi-criteria QoS requirements outlined in the SLA from the perspectives of cloud users and providers. To address this issue, we have proposed GTMAST based on MADM. Another direction of research can involve expanding the attributes that are used in the scheduling decision. Moreover, in future, a comparative study of more MADM scheduling methods can be done.

References

1. Kaur, R., Laxmi, V., Balkrishan: Performance evaluation of task scheduling algorithms in virtual cloud environment to minimize makespan. Int. J. Inf. Technol. (Singapore), **14**(1), 79–93 (2022). https://doi.org/10.1007/s41870-021-00753-4
2. Cinelli, M., Kadziński, M., Miebs, G., Gonzalez, M., Słowiński, R.: Recommending multiple criteria decision analysis methods with a new taxonomy-based decision support system. Eur. J. Oper. Res. **302**(2), 633–651 (2022)
3. Gyani, J., Ahmed, A., Haq, M.A.: MCDM and various prioritization methods in AHP for CSS: a comprehensive review. IEEE Access **10**, 33492–33511 (2022). https://doi.org/10.1109/ACCESS.2022.3161742. https://doi.org/10.1016/j.ejor.2022.01.011
4. Ben Alla, H., Ben Alla, S., Ezzati, A., Touhafi, A.: A novel multiclass priority algorithm for task scheduling in cloud computing. J. Supercomput. **77**(10), 11514–11555 (2021). https://doi.org/10.1007/s11227-021-03741-4
5. Nayak, S.C., Parida, S., Tripathy, C., Pati, B., Panigrahi, C.R.: Multicriteria decision-making techniques for avoiding similar task scheduling conflict in cloud computing. Int. J. Commun. Syst. **33**(13), 1–31 (2020). https://doi.org/10.1002/dac.4126
6. Khorsand, R., Ramezanpour, M.: An energy-efficient task-scheduling algorithm based on a multi-criteria decision-making method in cloud computing. Int. J. Commun. Syst. **33**(9), 1–17 (2020). https://doi.org/10.1002/dac.4379
7. Kumar, M.S., Tomar, A., Jana, P.K.: Multi-objective workflow scheduling scheme: a multi-criteria decision making approach. J. Ambient. Intell. Humaniz. Comput. **12**(12), 10789–10808 (2021). https://doi.org/10.1007/s12652-020-02833-y
8. Mahmoud, H., Thabet, M., Khafagy, M.H., Omara, F.A.: Multiobjective task scheduling in cloud environment using decision tree algorithm. IEEE Access **10**, 36140–36151 (2022). https://doi.org/10.1109/ACCESS.2022.3163273
9. Singh, M., Bhushan, S.: CS optimized task scheduling for cloud data management. Int. J. Eng. Trends Technol. **70**(6), 114–121 (2022). https://doi.org/10.14445/22315381/IJETT-V70I6P214
10. Shukla, D.K., Kumar, D. Kushwaha, D.S.: Task scheduling to reduce energy consumption and makespan of cloud computing using NSGA-II. Mater. Today Proc. (2021). https://doi.org/10.1016/j.matpr.2020.11.556
11. Samsam Shariat, S., Barekatain, B.: HATMOG: an enhanced hybrid task assignment algorithm based on AHP-TOPSIS and multi-objective genetic in cloud computing. Computing **104**(5), 1123–1154 (2022). https://doi.org/10.1007/s00607-021-01049-y
12. Chraibi, A., Alla, S. B., Ezzati, A.: An efficient cloudlet scheduling via bin packing in cloud computing. Int. J. Electr. Comput. Eng. **12**(3), 3226–3237(2022). https://doi.org/10.11591/ijece.v12i3.pp3226-3237
13. Swain, C.K., Sahu, A.: Interference aware workload scheduling for latency sensitive tasks in cloud environment. Computing **104**(4), 925–950 (2021). https://doi.org/10.1007/s00607-021-01014-9
14. Shrivastava, V., Shaikh, Y.: EBTASIC : an entropy-based TOPSIS algorithm for task scheduling in IaaS. Indian J. Sci. Technol. **15**(37), 1850–1858 (2022)

Dynamic Priority Based Resource Scheduling in Cloud Infrastructure Using Fuzzy Logic

Kapil Tarey(✉) and Vivek Shrivastava

International Institute of Professional Studies, DAVV, Indore, India
kapiltareycs@gmail.com

Abstract. Cloud computing is a viable solution for infrastructure services in distributed applications with the primary goal of optimizing resource utilization. Cloud infrastructure services may offer thousands of resources to complete a single operation. Appropriate resource scheduling results in optimal usage and an improvement in the service provider's QoS. In this work, a fuzzy logic-based technique is used, and an efficient algorithm for scheduling resources is introduced. Implementation of the proposed approach and its efficacy is tested through the Cloudsim and MATLAB tools. The proposed technique's waiting and turnaround times were compared with prominent algorithms SJF and FCFS. The simulation findings are remarkable when compared with the turnaround time and waiting time of existing algorithms. The proposed approach has an improvement of 80% in waiting time and 17% in turnaround time.

Keywords: Cloud computing · Fuzzy Logic · Scheduling · Waiting time · Turnaround time

1 Introduction

Cloud computing enables a global, accessible, and need-based group of configurable resources such as networking, workstations, memory, services, and applications. Cloud facilities can be accomplished quickly and with little effort. The most typical cloud computing strategy is allocating a resource for a specified time in response to a specific demand. Several popular cloud services use this strategy in exchange for hourly resource usage. The cloud environment consists of many resources that are allotted based on their tasks and provide services to various customers [1]. Scheduling resources in a cloud-based environment is a challenging task. Scheduling is the process of allocating resources appropriately to tasks [2]. Every consumer may choose from hundreds of resources to complete a task in a cloud environment. However, resources cannot be manually distributed to jobs. Cloud computing solutions are designed to reduce resource costs and maximize service delivery to applications [3, 4]. In resource scheduling, the additional complexity introduced by users' imprecise requirements can be resolved using fuzzy logic theory. With precise, logical reasoning, fuzzy logic represents the human mind and creates a "human-friendly system" [5]. This fuzzy logic setup can be dynamically modified based on user demands and current resource performance. This research proposes

H. Sharma et al. (Eds.): ICIVC 2022, PALO 17, pp. 500–508, 2023.
https://doi.org/10.1007/978-3-031-31164-2_42

an algorithm based on fuzzy logic for resource scheduling. For designing the resource scheduling algorithm, fuzzy logic is applied. The system dynamically prioritizes tasks depending on three inputs: cost, length, and power consumption, and this priority-based allocation of resources improve QoS.

2 Literature Review

This section focuses on several resource scheduling techniques and discusses their benefits and drawbacks. FIFO is an experimental scheduling method utilized as the first optimization technique [6]. A new job is added to the queue considering the arrival time. This approach is simple to design and implement, but it may take longer to run on machines with lesser throughput [7]. In Min-Min scheduling, the technique assigns a task to a processor with the least predicted time to completion [8]. Min-Min reviews all unassigned jobs at every turn. In this paper, Max-Min is also considered that works similarly to the Min-Min method. The Max-Min process takes a number of all unassigned jobs and estimates the shortest remaining time for each activity in the set. The distinction between Max-Min and Min-Min is that the scheduling activities in Max-Min are accomplished and mapped to the machine in the least amount of time. The scheduled task is then deleted from the collection. The method is repeated until all tasks have been scheduled [9, 10]. The SHEFT algorithm was introduced as an alternative scheduling technique in [11]. Experimental results of SHEFT reveal that the algorithm achieves better regarding the number of scheduled jobs, workflow runtime efficiency, flexibility, and sustainability. An algorithm for allocating resources based on the method described in [12] takes two parameters for prioritizing: resource prioritization and task prioritization. The cost of the resources determines resource prioritization. For priorities of the user's requests, only those requests that require the resource for a smaller amount of time are selected. In [13], Kulkarni et al. presented a method that hierarchically handles the requests and allocates available resources. A method for allocating requests to resources to minimize overall run time is described in [14], A cost matrix is built based on credits consumed for each task assigned to a resource. Each assignment is more likely to be more reliable regarding resource allocation. In [15], M. Hosseini et al. developed a scheduling method for the cloud environment that examines jobs in groups and assigns distinct costs and mathematical operations to resources. The classification operation directs computational resources and improves communication for larger workloads. An effective load-balancing technique utilizing fuzzy logic has been presented by Li et al. [16]. This scheme proposed a new load-balancing algorithm with two input variables for fuzzification: allocated load and processor speed. After fuzzification, 12 rules are provided in the fuzzy inference system to obtain the defuzzified result. In [17], Siddiqi et al. described a job scheduling algorithm based on fuzzy neural networks in a cloud environment. Fuzzy logic converts input factors such as bandwidth, memory, and time to linguistic variables. It maps system resources to jobs, and then fuzzy logic is used to translate linguistic variables to crisp values. A hybrid task scheduling system has been developed based on two artificial intelligence technologies: genetic algorithms and fuzzy logic, by Logeswaran et al. [18]. The researchers employed fuzzy logic in the genetic algorithm. The resources were allocated according to the duration of the job to avoid resource waste. In [19], Tajvidi et al. presented a fuzzy-based improved multi-queue work scheduling algorithm that uses the load balancing approach by assigning jobs in ascending order based on burst duration. The load is distributed among nodes, and load movement is performed via fuzzy logic. The use of fuzzy logic lowers the waiting and response time. A fuzzy logic-oriented resource scheduling technique was applied by Akbar et al. in [20] to improve cloud computing's reliability. Bhardwaj et al. [21], the authors optimized the operational costs while ensuring efficiency and meeting deadline constraints by considering parameter uncertainty and heterogeneity and assessing all three available pricing plans. One of the critical characteristics of cloud computing that distinguishes it from traditional IT approaches is resource pooling, as described by Agarkhed et al. [22]. The combination of resource pooling, virtualization, and resource sharing results in dynamic cloud behavior. Chen et al. [23] proposed an efficient approach to increasing resource utilization. The suggested fuzzy logic approach improved the performance of the planning phase, allowing for more efficient decision-making. Siddiqi et al. [17], the authors presented a novel elasticity controller for the autonomic resource provisioning controller model. This controller computes the required number of processor core(s) based on information about the application's internal workload and virtual machine resource utilization. The servers store many shared data, which authorized users can access. The probabilistic C means module (PCM) is broadly used for image processing and analysis. As presented by Zhu et al. [23], PCM cannot extract knowledge from large amounts of heterogeneous data. Therefore, this paper proposes fuzzy logic-based data provisioning via cloud computing to solve the problem. This study aims to assess numerous implementations of fuzzy logic to address the above issues and improve the waiting and turnaround time for prioritizing cloud computing-based resource demands.

3 Fuzzy Logic

The classical logical systems do not represent the vague meaning in natural language, but fuzzy logic helps make decisions with uncertainty and vagueness [16]. When inputs which are given to any system are not clear, uncertain, and not so accurate, then fuzzy logic suggests a way of decision-making.

Fuzzy logic works in the following five steps-

- Fuzzification: For finding the related membership degrees of Cost, Power Consumption, and Length.
- Relate Fuzzy Operations: Using the "AND" and the "OR" operator degree of membership, 27 rules are designed for mapping.
- Implication: Fuzzy membership for all the above three inputs is measured (Table 1).
- Aggregation: The aggregate output for all fuzzy sets with every rule in terms of "Priority" is calculated.
- Defuzzification: Any defuzzification formula can transform all the accumulated fuzzy sets into a crisp value.

Figure 1 depicts the structural design of fuzzy logic. A mathematical function determines membership having values in [0, 1] interval. The primary goal of fuzzy logic is to

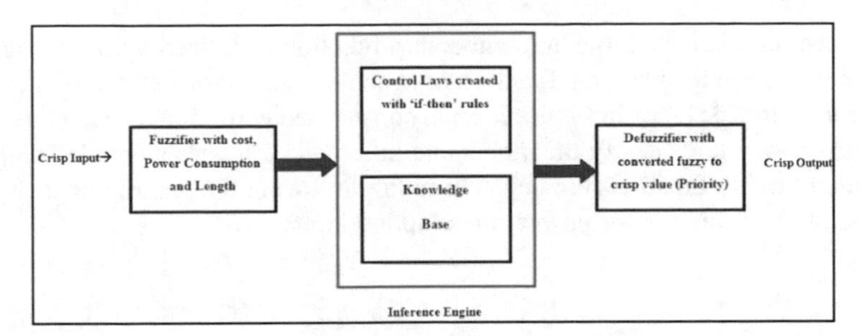

Fig. 1. Representative structural design of the fuzzy logic system

describe algorithms for control laws based on 'if-then' rules [17]. Generally, these rules are created by experts' knowledge. It may have various rules created with AND/OR logic. The strategy modeled in this paper shows how the distinct values of the given parameters can affect cloud performance.

4 The Proposed Model

This model uses three fuzzy parameters: cost, length, and power consumption. Before applying the suggested fuzzy system, parameters must be converted to phase space. The fuzzy measure is shown in Table 1.

Table 1. Phase space of fuzzy system parameters

Verbal Parameters	A	β	γ
Low	[0 1 50]	[0 1 50]	[0 1 50]
Medium	[1 50 100]	[1 50 100]	[1 50 100]
High	[50 100 125]	[50 100 125]	[50 100 125]

The range of each variable's verbal output is presented in Table 2.

Table 2. Fuzzy system output parameters

Parameters	Output
Very Low	[−25 0 15]
Low	[15 25 35]
Medium	[30 50 60]
High	[60 75 85]
Very High	[85 100 125]

For each linguistic variable, a membership function is defined by fuzzy logic. The adjustment of these membership functions can achieve an optimal effect. In this experiment, Mamdani-type fuzzy inference logic is constructed with twenty-seven rules. Trimf is used in this work to fuzzify the inputs and output. The suggested model's input and output membership functions are depicted in the following figures. Figure 2 shows the membership function plot for power consumption input.

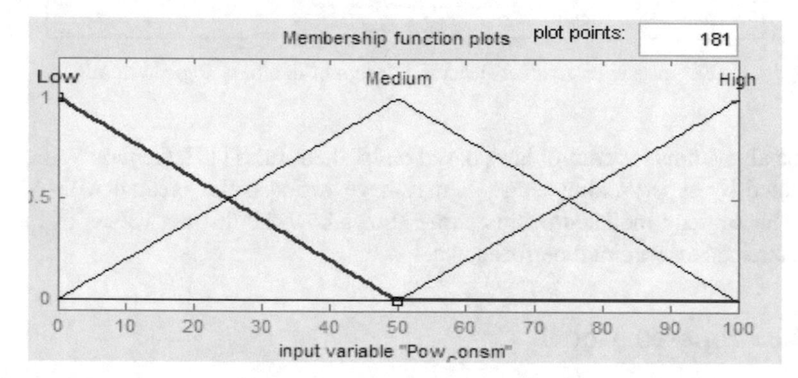

Fig. 2. Power Consumption (Input)

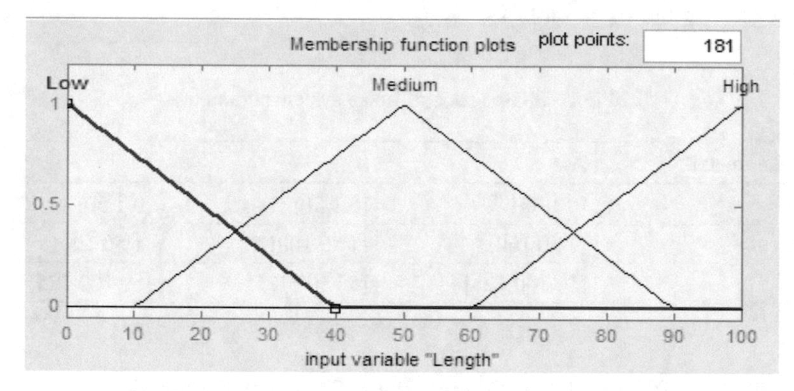

Fig. 3. Length (Input)

Membership functions plots for the other two inputs, Length, and Cost, are depicted in Fig. 2, Fig. 3, and Fig. 4, respectively. Figure 5 illustrates the output membership function plot as a dynamic priority. The system's output is to select priority based on cost, length, and power consumption. As illustrated in Fig. 6, the suggested model requires three inputs ((cost, power consumption, and length) and produces a single output (i.e., dynamic priority). Since each of the three parameters contains three membership functions, this model employs 27 rules. The proposed scheduling method allocates resources to tasks under fuzzy dynamic priority.

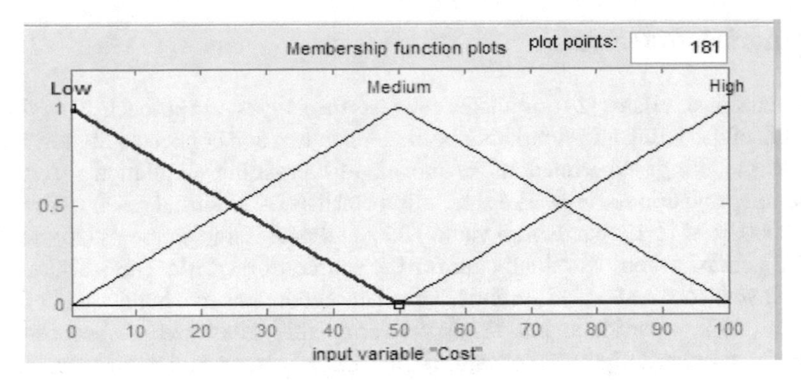

Fig. 4. Cost (Input)

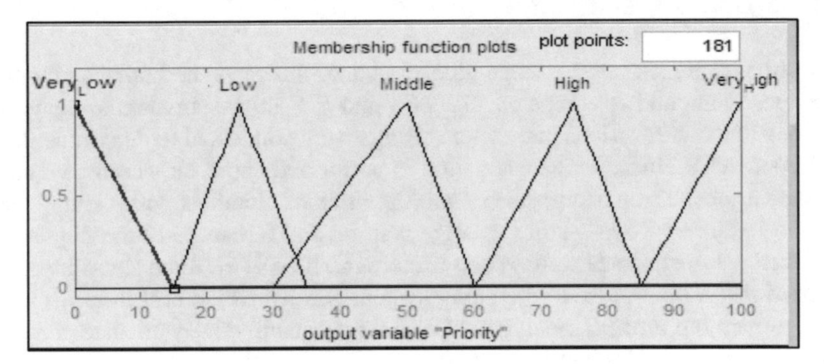

Fig. 5. Dynamic Priority (Output)

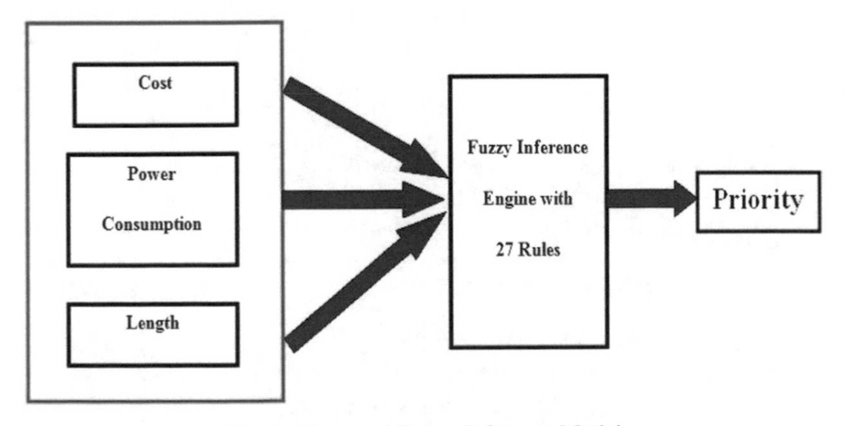

Fig. 6. Proposed Fuzzy Inference Model

5 Experimental Setup

The CloudSim simulator [24] simulates and tests the suggested approach. This simulator is capable of modeling and simulating both independent and connected clouds. It allows seamless modeling, experimentation, and cloud computing simulation services with application provision settings. The simulation consists of 20 virtual machines with CPU MIPS 1000, RAM 512, bandwidth value 1000, and processing elements (No. of CPUs) is 1. We get tasks from workload data for the source of the task. The workload traces are employed from real-world systems. This dataset contains five hundred records, each associated with a particular job. After delivering this dataset to the proposed fuzzy system, the proposed algorithm's waiting time and turnaround time are calculated.

6 Results and Discussion

This section examines the suggested algorithm and illustrates the findings graphically. The proposed method is compared to FCFS and SJF in the simulation experiments. Numerous trials with various parameter settings were conducted to determine the suggested approach's efficacy. The evaluation is performed based on workload data from independent jobs. The performance of multiple sets of cloudlets with varying sizes is examined in terms of two metrics average waiting and turnaround time. Figure 7 and Fig. 8 illustrate the average waiting and turnaround time of all three algorithms. It was concluded that a fuzzy-based scheduling system performed the best in terms of waiting and turnaround times. The suggested technique can effectively allocate cloudlets to resources in the most efficient manner.

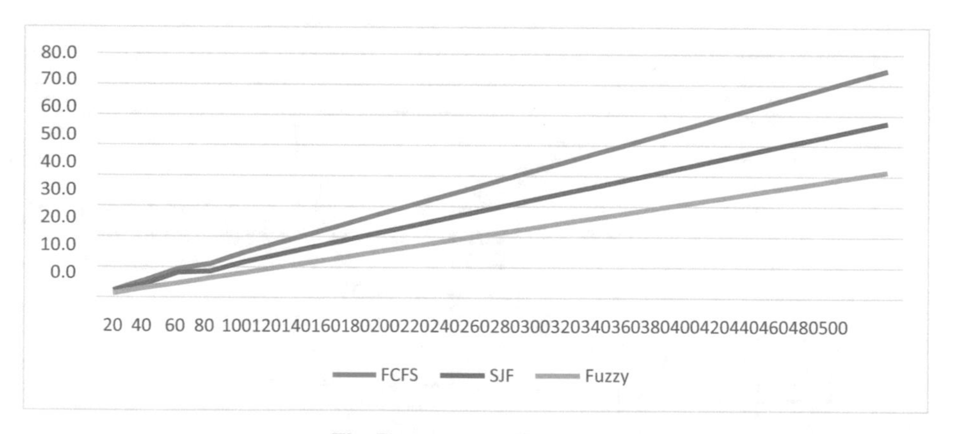

Fig. 7. Average Waiting Time

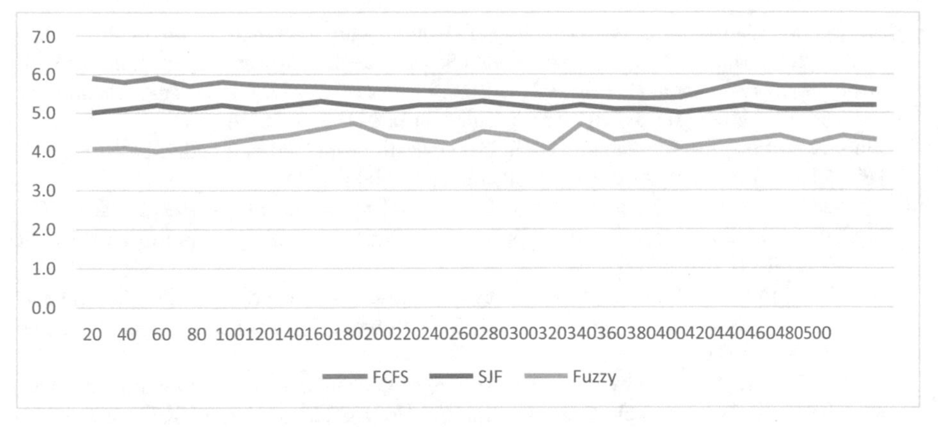

Fig. 8. Average Turnaround Time

7 Conclusion

Scheduling is a significant challenge in cloud computing. The influence of cloudlets number and task precedence on job/task completion time was studied. The proposed scheduling algorithm outperformed FCFS and SJF. The experiment results reveal that the fuzzy-based technique minimizes waiting and turnaround times. This strategy can be improved in the future by including more parameters to help with scheduling decisions.

References

1. Vamshikrishna, S.: Efficient resource allocation and job scheduling for improving the throughput of the SDN-A review (preprint) (2019)
2. Bhagwan, J., Kumar, S.: Independent task scheduling in cloud computing using meta-heuristic HC-CSO algorithm. Int. J. Adv. Comput. Sci. Appl. **12**(7) (2021)
3. Priya, S.S., Mehata, K.M., Banu, W.A.: Ganging of resources via fuzzy Manhattan distance similarity with priority tasks scheduling in cloud computing. J. Telecommun. Inf. Technol. **1**(2018), 32–41 (2018)
4. Shrimali, B., Bhadka, H., Patel, H.: A fuzzy-based approach to evaluate multi-objective optimization for resource allocation in cloud. Int. J. Adv. Technol. Eng. Explor. **5**(43), 140–150 (2018)
5. Liu, Y.: Adaptive job-scheduling algorithm based on queuing theory in a hybrid cloud environment. Int. J. Perform. Eng. (2019)
6. Alenizi, F., Rana, O.: Minimising delay and energy in online dynamic fog systems. Comput. Sci. Inf. Technol. (CS & IT) (2020)
7. Prakash, V., Bawa, S., Garg, L.: Multi-dependency and time based resource scheduling algorithm for scientific applications in cloud computing. Electronics **10**(11), 1320 (2021)
8. Ben Alla, H., Ben Alla, S., Ezzati, A.: A priority based task scheduling in cloud computing using a hybrid MCDM model. Ubiquitous Netw., 235–246 (2017)
9. Jin, M., et al.: Research on fuzzy scheduling of cloud computing tasks based on hybrid search algorithms and differential evolution. Fractals **30**(02) (2022)
10. Tang, X.: Reliability-aware cost-efficient scientific workflows scheduling strategy on multi-cloud systems. IEEE Trans. Cloud Comput., 1 (2021)

11. Krämer, M.: Capability-based scheduling of scientific workflows in the cloud. In: Proceedings of the 9th International Conference on Data Science, Technology and Applications (2020)
12. Kulkarni, M.R.A., Patil, D.S.H., Balaji, D.N.: Efficient fuzzy based real time scheduling of workflows on cloud. Int. J. Eng. Adv. Technol. **8**(6), 221–225 (2019)
13. Guo, X.: Multi-objective task scheduling optimization in cloud computing based on fuzzy self-defense algorithm. Alex. Eng. J. **60**(6), 5603–5609 (2021)
14. HosseiniShirvani, M., NoorianTalouki, R.: A novel hybrid heuristic-based list scheduling algorithm in heterogeneous cloud computing environment for makespan optimization. Parallel Comput. **108**, 102828 (2021)
15. Rai, K., Vemireddy, S., Rout, R.R.: Fuzzy logic based task scheduling algorithm in vehicular fog computing framework. In: 2021 IEEE 18th India Council International Conference (INDICON) (2021)
16. Li, W., Tang, Z., Qi, F.: A hybrid task scheduling algorithm combining symbiotic organisms search with fuzzy logic in cloud computing. In: 2020 IEEE 23rd International Conference on Computational Science and Engineering (CSE) (2020)
17. Siddiqi, M.H., Alruwaili, M., Ali, A., Haider, S.F., Ali, F., Iqbal, M.: Dynamic priority-based efficient resource allocation and computing framework for vehicular multimedia cloud computing. IEEE Access **8**, 81080–81089 (2020)
18. Logeswaran, L., Bandara, H.M.N.D., Bhathiya, H.S.: Performance, resource, and cost aware resource provisioning in the cloud. In: 2016 IEEE 9th International Conference on Cloud Computing (CLOUD) (2016)
19. Tajvidi, M., Essam, D., Maher, M.J.: Deadline-constrained stochastic optimization of resource provisioning, for cloud users. In: Proceedings of the 8th International Conference on Cloud Computing and Services Science (2018)
20. Akbar, M.A., Tehreem, T., Hayat, S., Mateen, M.: A self-adaptive resource provisioning approach using fuzzy logic for cloud-based applications. Int. J. Comput. Digit. Syst. **9**(3), 341–348 (2020)
21. Bhardwaj, T., Sharma, S.C.: Fuzzy logic-based elasticity controller for autonomic resource provisioning in parallel scientific applications: a cloud computing perspective. Comput. Electr. Eng. **70**, 1049–1073 (2018)
22. Agarkhed, J., Kodli, S.: Fuzzy logic based data provisioning using cloud computing. In: 2018 International Conference on Information, Communication, Engineering and Technology (ICICET) (2018)
23. Chen, Z., Zhu, Y., Di, Y., Feng, S.: A dynamic resource scheduling method based on fuzzy control theory in cloud environment. J. Control Sci. Eng. **2015**, 1–10 (2015)
24. Calheiros, R.N., Ranjan, R., Beloglazov, A., De Rose, C.A., Buyya, R.: CloudSim: a toolkit for modeling and simulation of cloud computing environments and evaluation of resource provisioning algorithms. Softw. Pract. Exp. **41**(1), 23–50 (2010)

Identification of Social Accounts' Responses Using Machine Learning Techniques

Medha Wyawahare, Rahul Diwate, Agnibha Sarkar(✉), Chirag Agrawal, Ankita Kumari, and Archis Khuspe

Department of Electronics and Telecommunication, Vishwakarma Institute of Technology, Pune, India
{medha.wyawahare,chirag.agrawal20,ankita.kumari20, archis.khuspe20}@vit.edu, diwate.rahul@gmail.com, agnibha10@gmail.com

Abstract. Twitter bots, or "zombies," are automated Twitter accounts run by bot software. Bots carry out tasks like sending tweets, following other users in bulk, and tweeting and retweeting material on a vast scale to achieve predetermined objectives. The proliferation of social bots—automated agents typically exploited for nefarious purposes—on social media sites like Twitter is a significant issue. These include using the impact of these accounts to influence a community on a particular issue, disseminate false information, enlist individuals in criminal organisations, control people's behaviour in the stock market, and extort people into disclosing their data. This work has investigated how bots react to unexpected political events compared to human accounts, explained the general prevalence of political bots on Twitter, and created and implemented a model to recognise them only based on user profiles.

Keywords: automation · accounts · bots · classifiers · followers · spam · twitter

1 Introduction

Bots are computer programs that use the Internet to carry out automated tasks. On social media platforms like Twitter, bots have taken over and automated exchanges with other users. Twitter bots that are focused on politics tend to focus on elections, crises, and policy issues. Research has focused on detecting a bot's presence on social media platforms. Not much research is available on the bot's dynamic response to political events and the overall presence of political bots. Political bots have mostly been investigated in relation to their involvement in organised political events, such as elections. The differential sentiment analysis by Shukla et al. focuses on how bots try to sway people's attitudes inside their network [1]. The size of the tree and each decision node are visualised in the decision trees. When the tree is larger, the model fits the learning algorithm better, as researched on by Galindo et al. [2]. Additionally, it was noted by Fazzolari et al. [3] that, to the exclusion of the 2 datasets, bot accounts typically do not link to well-known online news websites but verified accounts do so more frequently. Dukić

H. Sharma et al. (Eds.): ICIVC 2022, PALO 17, pp. 509–521, 2023.
https://doi.org/10.1007/978-3-031-31164-2_43

et al. [4] stated that using deep learning models is redundant compared to shallow models. Mutlu et al. [5] conclude that bots have a high friend-to-follower ratio, little follower growth, and most screen names with eight digits. When dealing with noisy data, supervised techniques were not very reliable at identifying temporal data patterns. Instead of examining users' interactions, supervised algorithms depend on statistical features according to Barhate et al. [6]. To find the overlap between the two spam and real fuzzy groups, a fuzzy C-Means technique is applied. Web crawlers and chatbots are considered good bots, while malevolent bots that imitate human behaviour are considered bad bots, according to Schnebly and Sengupta [11]. Social bot accounts have gotten increasingly good at imitating human behaviour. The research community must urgently create tools to detect social bots to prevent malpractice, rightfully mentioned by González et al. [12]. Benigni et al. [14] have also stated that human users need to understand whom they are engaging with to be able to become genuine users of social media; hence, Twitter needs to classify human and spambot accounts. Shi et al. [13] focused on building precise machine learning classifiers to find and identify bots in social networks and conducted extensive research. Their recent research on social bots focuses on the roles, locations, and collective presence of the bots.

2 Literature Survey

In paper [1], Shukla et al. have made a substructure to identify Twitter bots using the repository of accounts on Twitter. They have initiated an extensive, qualified analysis to choose ideal feature encoding, feature choice, and grouping methods. Galindo et al. [2] studied the detection of bots and analysed their actions in the elections held in Spain in 2019. Fazzolari et al. [3] examined tweets made by various Twitter profile categories, comparing the emotion relayed by automated profiles with the sentiments conveyed by verified accounts. Dukić et al. [4] presented a bot detection model. Deep Learning models were concluded to be redundant compared to shallow models. As conclusive evidence suggests according to Mutlu et al. [5], i) when compared to humans, bot-managed accounts use a powerful vocabulary. ii) human-generated data exhibits more consistent behaviour, as seen in the moral outcomes of human-generated content, as opposed to the haywire patterns displayed by bots; iii) human-generated content exhibits greater predictability when compared to bots. The bot probability of the user is determined using a feature named bot_score. A supervised approach to selected features obtained a precision of 0.9743. The hashtag #StudentsLivesMatter had a high percentage of bots. The bot's weekly and hourly tweet counts were analysed and compared with verified users. Barhate et al. [6] proposed the LA-MSBD algorithm by recognising trustworthy (verified) users. It recognises the end users as genuine or hostile bots based on URL features. The combined algorithm is 2%–3% more precise than the LA-MSBD, which has been put forward with only direct trust. Rout et al. [7] conclude that bots have low follower growth, a high followers-to-friends ratio, and an 8-digit screen name. Bot systems may improve over time, so classifier training on old, existing data may not be fruitful. Continuous monitoring of changes in tweeting characteristics is important. According to Chu et al. [8], the actions of humans and bots on Twitter have been measured and distinguished and the research planned a computerised classification system based on four crucial sectors: machine learning, profile properties, entropy, and decision-making. Based on the findings of Kouvela et al. [9], a modern AI technique was developed to tackle specific issues related to Twitter bots. A new dataset was used on which the machine-learning model was trained. The method used by Bello and Heckel [10] is geolocation analysis to find the location, and similar hashtags are used for data collection. In this paper, new studies have been discovered regarding the bots' behaviour and role in the Brexit debate. They showed how to discover the tactics executed by the bots by reverse engineering those bots using an unconventional method. Schnebly and Sengupta [11] introduced a classifier to detect Twitter bots. The dataset used in this work was taken from IIT in Italy. The accuracy of classification results in this study, increases by combining variational pattern-based categorisation and Random Forest. Better results are obtained using combined models. It has been discussed that differentiating Twitter users as bots and humans is necessary, which will help take legal action against the responsible account holders, according to González et al. [12]. The proposed method includes semi-supervised clustering and clickstream sequences that transition to probability-based feature extraction. The behavioural clickstream's probability-based detection is 12.8% more accurate than quantitative analysis methods. Shi et al. [13] used only a tiny count of labelled users and promptly detected malicious social bots. Benigni et al. [14] analysed the issue of automated engineering on Twitter, and mentioned a particular class of SIBNs, or social influence bot networks. SIBNs that alter topics have been introduced by creating an artificial core group in the mentions and pages of Twitter co-mentions. An RNN BiLSTM model has been presented by Wei and Nguyen [15], with word implants to differentiate genuine accounts from bots that rely only on tweets and do not need advanced feature engineering. Battur and Yaligar [16] proposed a method for detecting bots on Twitter. A technique called 'Bag of Words' was discovered as the upfront grasping framework with a precision of 96.70% for trained information and 96.65% for tested information. Apart from these algorithms, the random forest, multinomial naive Bayes, and decision tree algorithms were also implemented. Similarly, Alothali et al. [17] reviewed bot detection methods in recent Twitter social networks. Bello et al. [18] summed up the major observations in the literature review in four main subsections: classifier, dataset, performance considerations, and analysed features. In this paper, the work proposed is the reverse engineering of Twitter bots. They suggested ways to acknowledge the actions of the bots with greater detail. Their study makes the activity of bots easier to understand and broader, which assists in reaching a possible cause. Wald et al. [19] employed six classifiers to construct models for forecasting the relationship between a bot and a genuine user. The number of friends, Klout score, status, and several profile followers indicate the chances of that user interacting with a bot. In this case, the foremost classifier was that of Random Forest. These outcomes help determine which users are most susceptible to interacting with bots. According to Badawy et al. [20], the consequences of a manipulation drive were studied by analysing the users who shared posts on Twitter with bot accounts from Russia. On Twitter, false information was mostly shared by conservatives as compared to liberals. Aswani et al. [21] used numerous testing methods by merging the data from social media to model various Twitter spammers using bio-inspired computing. The laid-out stance gives 97.98% precision by considering 13 important components after an analytical test. Kantepe and Ganiz [22] collected data of 1800 specific Twitter users, extracted 62 features from each of them, and considered and labelled the suspended Twitter accounts as 'bots' and others as 'normal users'. They obtained an 86% accuracy and 83% F1 score with gradient-boosted trees (GBT), which is the best result among all the other used algorithms. According to Paudel et al., [23] social media bots are becoming more sophisticated. Also, they pose great challenges for their detection. Studying these Twitter bots increases the data on the research that has been performed on them, including natural interactivity and developmental features compared to the bots that have already been studied. Meanwhile, as researched by Dickerson et al., [24] SentiBot depends on the tweet's semantics, syntax, user actions, and network-centric user data. Automatic buzzer detection was performed on a particular Twitter dataset by Ibrahim et al. to eliminate bots, financed profiles, and extremist users that generally produced clamour in their data allotments.

This paper focuses on the comprehensive analysis of all the documents mentioned in the survey and the algorithms implemented. The algorithms are chosen based on their use cases, and their accuracies have been noted for comparisons. The algorithm with the best score has been further implemented in a Flask-based web application to make the process of obtaining results interactive for the user.

3 Methodology

The dataset was initially cleaned, and missing data were handled, following which data analysis was done. Feature extraction and engineering techniques have been performed using a collection of specific words (bag of words). Multiple algorithms have been applied to the datasets to get maximum accuracy. The results and predictions are displayed on the GUI. Figure 1 depicts the entire flow of this study, right from collecting data, finding discrepancies in the data, removing unnecessary information, dropping missing values, standardising/normalising the data, splitting data into train/test data, running the selected machine learning algorithms for finding the one which fits best over the given data, and making necessary optimizations for the specified machine learning algorithms by tweaking their parameters to avoid bias/underfitting or variance/overfitting.

3.1 Dataset

Kaggle was used to obtain the Twitter-Bot detection dataset. The dataset contains a total of 3461 observations and 20 features. The features include the account ID, followers count, friends count, listed count, favourites count, screen name, location, and account status (verified or not). The target variable is a binary value, with 1 being assigned to a bot user and 0 for a human user. The dataset consists of 1321 observations, classified as bot accounts, and 1476 observations, classified as non-bot accounts, making it a balanced dataset. 70% of the training data is used to build the model. The model is tested using 30% of the data.

3.2 Data Pre-processing

There may be instances where data is missing owing to various factors, such as corrupted data, information that cannot be imported, or even improper feature extraction. One of

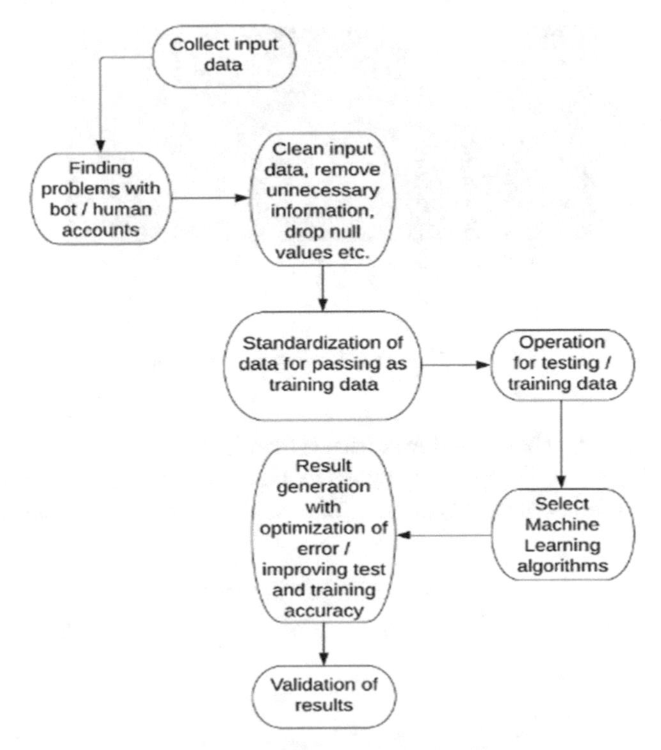

Fig. 1. Methodology flowchart.

the most challenging tasks for data professionals is managing this missing information. Therefore, making the appropriate choice to either drop or fill those missing values with that row or column's mean enables the analyst to generate robust data models. Data pre-processing is, therefore, one of the most important steps through which information and data can be extracted to the fullest.

3.3 Identifying the Missing Values in the Dataset

Figure 2 represents a heatmap, a data visualisation approach that uses colour to depict the number of missing values in the feature columns. The more missing values there are, the more yellow cells there are. All the missing values from the dataset and the imbalance in records have been identified in the following heatmap in Fig. 2. Here, the features like location, description, URL, and status have the most significant number of yellow cells, which indicates that these features have a lot of missing values and therefore were dropped as they do not make an impact on the classification process.

3.4 Data Analysis

The dataset, as mentioned before, contains the follower count and friend count of the Twitter accounts. When comparing bots' friend-to-follower ratio to that of non-bots,

Fig. 2. Heatmap showing missing data.

Fig. 3 demonstrates that bots have fewer friends than followers in general. This is because bots follow other accounts in large numbers to gain traction. On the other hand, human users naturally have more friends in their followers' lists than bot accounts.

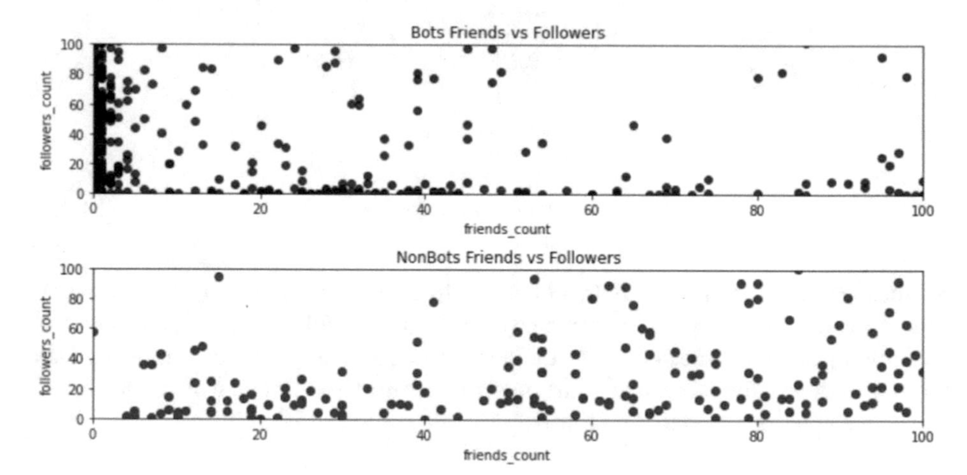

Fig. 3. Friends vs. Followers Count.

3.5 Feature Independence

Spearman Correlation was used to determine how the input (independent variables) and the target (dependent variables) related to one another. From the heatmap (as shown in Fig. 4), no correlation was found between the 'id,' 'statuses_count,' 'default_profile,' 'default_profile_image' variables and the target variable ('bot'), whereas the 'Verified,' 'listed_count,' 'friends_count,' 'followers_count' variables were found to be strongly correlated to the target variable ('bots'). Those features where the correlation was not found have been eliminated from the dataset.

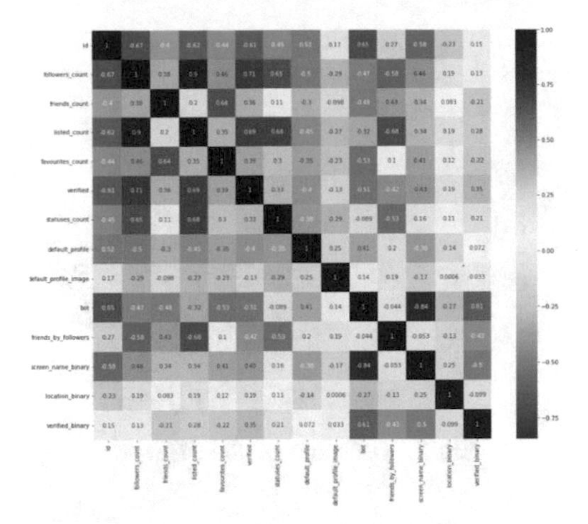

Fig. 4. Feature Correlation Matrix.

3.6 Feature Extraction

Observed datasets in this technique are reduced to smaller, more manageable groups for further processing. If the datasets are too large and contain many variables, they would have very high computational costs. Based on the analysis performed above in Sects. 3.3, i.e., missing values, and Sect. 3.5, i.e., the independence of features from each other, the following features have been extracted: 'screen_name_binary, 'name_binary,' 'description_binary,' 'status_binary,' 'verified,' 'followers_count,' 'friends_count,' 'statuses_count,' 'listed_count_binary.'

3.7 Feature Engineering

The raw data is transformed into features that can better represent the problem to the predictive model, further improving the accuracy of the given data. A list of frequently used words by bots are selected for feature engineering like 'tweet me,' 'follow me', 'update,' 'every,' 'prison,' 'paper,' 'fake,' 'droop,' and 'free' (Bag of Words).

3.8 Machine Learning Classifiers

Decision Tree: The nodes of the decision tree represent features in the dataset. The decision node makes any decision, and the leaf node gives an output of that decision. The tree starts with the root node and asks the question whether the answer/result is 'yes' or 'no' for all the selected features; it further splits into subtrees, branches, and sub-branches and reaches the leaf node for the final output result. In Fig. 5, the test AUC probability obtained is 0.94, whereas the train AUC is 0.96, which is a very good score, and thereby fitting very well over the given dataset.

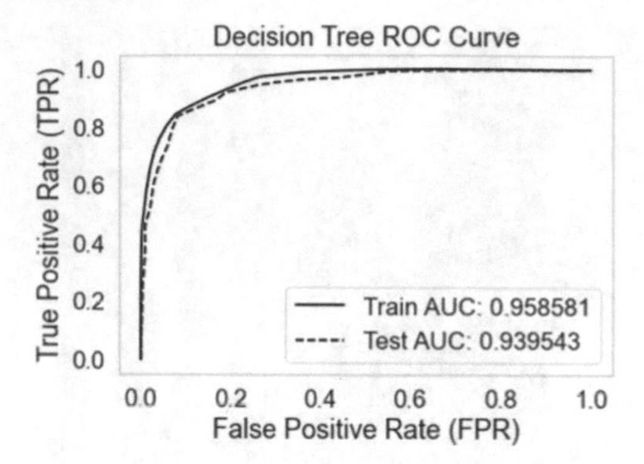

Fig. 5. Decision tree ROC Curve.

Random Forest Classifier: The Random Forest classifier is capable of handling high-dimensional and large datasets. Random Forest uses many decision trees that run on different data subsets and independently predict the output. Then the prediction with the largest number is selected, which becomes the model's final prediction. In Fig. 6, the test AUC obtained is 0.93, whereas the train AUC is 0.94, which is almost at par (slightly lower) with the results obtained by using a single decision tree. This classifier also fit perfectly on the given data, as there is no major difference between the train and test AUC.

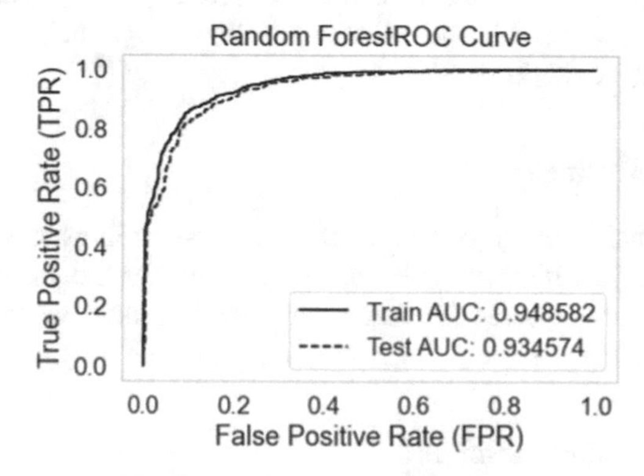

Fig. 6. Random Forest ROC Curve.

Multinomial Naive Bayes: This algorithm deals with probabilities and the Bayes theorem. It strongly assumes that each feature contributes independently to the likelihood of classification. As depicted in Fig. 7, for Multinomial Naïve Bayes, the AUC of training data is 0.70, and that of testing data is 0.69.

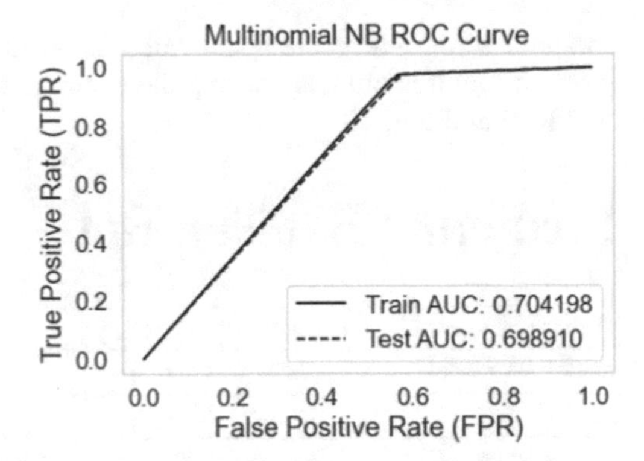

Fig. 7. Multinomial NB ROC Curve.

Support Vector Machine Classifier: SVM differentiates the dimensional space into classes by drawing boundary lines. In the ROC curve graph, as shown in Fig. 8, 0.61 test AUC is obtained, whereas the train AUC is 0.99, which confirms that overfitting takes place when the SVM classifier is used for training using the given dataset.

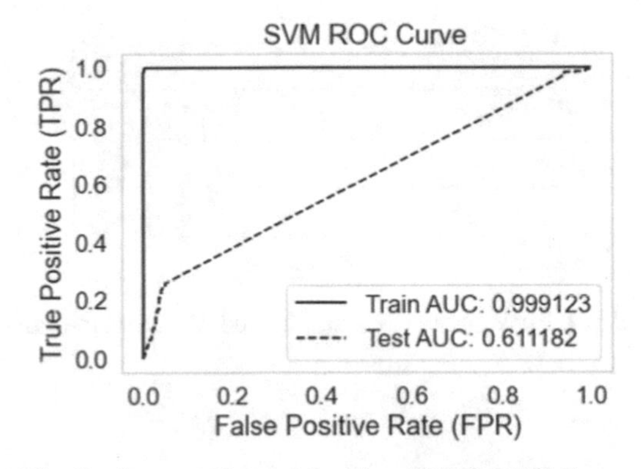

Fig. 8. Support Vector Machine (SVM) ROC Curve.

3.9 Building and Integrating Model Using a Flask Application

When the accuracy scores of all four algorithms were compared, the decision tree classifier emerged as the algorithm with the highest accuracy score. The model was therefore built using this algorithm, with nine input features: 'screen_name_binary,' 'name_binary,' 'description_binary_n,' 'status_binary_n,' 'verified_n,' 'followers_count_n,' 'friends_count_n,' 'statuses_count_n' and 'listed_count_binary_n.' The target variable or output is a binary value, with 1 indicating a bot account and 0 indicating a human user. Backend Python commands and libraries were used to generate the results. On hitting the 'Predict' button, the output is generated, and the verdict on the account is provided. Flask API has been used for integration with a web application, where the results are displayed in an HTML file along with the other plots and figures obtained during the exploratory data analysis of the dataset, as shown in Fig. 9 and Fig. 10.

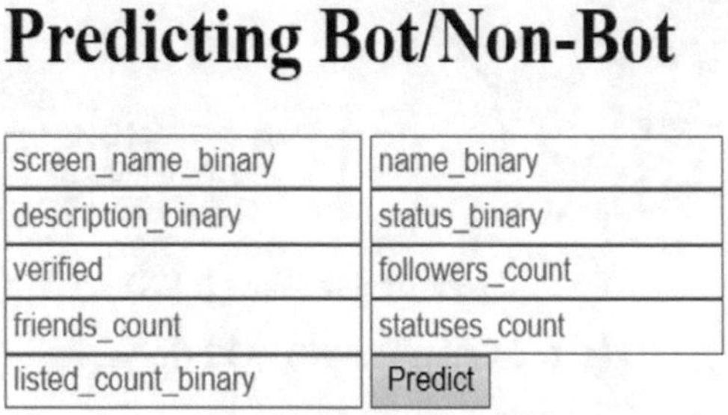

Predicting Bot/Non-Bot

screen_name_binary	name_binary
description_binary	status_binary
verified	followers_count
friends_count	statuses_count
listed_count_binary	Predict

Fig. 9. Snapshot of the Flask application where user inputs values for the specific Twitter account for testing.

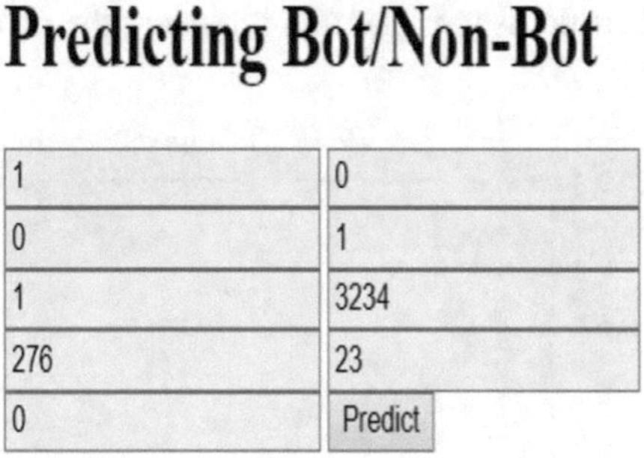

Predicting Bot/Non-Bot

1	0
0	1
1	3234
276	23
0	Predict

Fig. 10. Values entered by the user in the Flask application. Based on the values, the model outputs human.

4 Results

Four machine-learning classifiers have been used to classify Twitter users as bots or non-bots. The Decision Tree classifier has an accuracy score of 0.8785, which is the highest. The multinomial Naive Bayes classifier gives an accuracy score of 0.6976, clearly performing poorly and not being a good choice as the train's AUC is just 0.556 and the test is 0.555. The Random Forest classifier, whose accuracy score is 0.8608, and the support vector machine give an accuracy score of 0.5369, which is very low. The decision tree, therefore, has the highest accuracy among all. A comparison is made for all the algorithms and their accuracy, precision, recall, and F1 scores, which are mentioned in Table 1. It was found that the decision tree classifier is the best classifier for Twitter bot detection, with an accuracy of 87.85%, precision of 91.12%, recall of 87.23%, and an F1 score of 89.14%. The prediction model was then built with the best and most accurate algorithm, i.e., the Decision Tree classifier, which, on receiving the input features, outputs the correct predicted binary output value in the form of 'Bot' or

Table 1. Comparison of classifier outputs.

	Decision Tree	Random Forest	Multinomial Naïve Bayes	Support Vector Machine	Best Score
Accuracy	**0.88**	0.86	0.69	0.53	**Decision Tree**
Precision	**0.91**	0.89	0.62	0.80	**Decision Tree**
Recall	**0.87**	0.85	0.76	0.16	**Decision Tree**
F1 Score	**0.89**	0.87	0.68	0.26	**Decision Tree**

'Human', with the value of 1 indicating the presence of a bot account and 0 for a human user.

The performance of various algorithms used was analysed in this work. The naive Bayes algorithm depends on the premise that all features are independent or unrelated, which is not the case in real-time case studies. So, it cannot model the relationship accurately and has proven to be less accurate in this work. When the data set contains target classes that overlap, SVM does not perform very well. Also, for this, choosing the proper kernel function is vital. A linear kernel was used in this classifier, resulting in lesser accuracy. Varying the kernel function and exploring the effect on accuracy can be more helpful and a part of our future work. Random Forest consists of a cluster of decision trees, which work similarly to decision trees but work better in the case of overly complex data. In the case of this dataset, though, as the dataset is not that complicated, the results are very close for both the classifiers, with decision trees obtaining a slightly higher score as compared to random forests, and they were chosen based on the average obtained after multiple runs.

5 Conclusion

Twitter has become one of the biggest platforms on social media today, where people can present their opinions, thoughts, and ideas in front of the world. However, many social bot accounts (created by humans, nonetheless) negatively influence people by spreading fake news and creating hatred. Therefore, identifying those accounts of people impersonating humans is necessary in today's world. Bot activities have been studied, and multiple algorithms have been tried and tested to find the most accurate one, which was concluded to be the decision tree classifier, which was then finally used to build a model that could detect and differentiate between bots and humans on social media (Twitter) with good accuracy.

6 Future Work

Apart from the four machine learning algorithms mentioned in this study, deep learning-based artificial neural networks can be used, which can vastly improve the model's accuracy and ability to detect and differentiate bots and human users on any social media platform.

References

1. Shukla, H., Jagtap, N., Patil, B.: Enhanced Twitter bot detection using ensemble machine learning. In: 2021 6th International Conference on Inventive Computation Technologies (ICICT), pp. 930–936. IEEE, January 2021
2. Pastor-Galindo, J., et al.: Spotting political, social bots in Twitter: a use case of the 2019 Spanish general election. IEEE Trans. Netw. Serv. Manag. **17**(4), 2156–2170 (2020)
3. Fazzolari, M., Pratelli, M., Martinelli, F., Petrocchi, M.: Emotions and interests of evolving Twitter bots. In: 2020 IEEE Conference on Evolving and Adaptive Intelligent Systems (EAIS), pp. 1–8. IEEE, May 2020
4. Dukić, D., Keča, D., Stipić, D.: Are you human? Detecting bots on Twitter using BERT. In: 2020 IEEE 7th International Conference on Data Science and Advanced Analytics (DSAA), pp. 631–636. IEEE, October 2020
5. Mutlu, E.Ç., Oghaz, T., Tütüncüler, E., Garibay, I.: Do bots have a moral judgment? The difference between bots and humans in moral rhetoric. In: 2020 IEEE/ACM International Conference on Advances in Social Networks Analysis and Mining (ASONAM), pp. 222–226. IEEE, December 2020
6. Barhate, S., Mangla, R., Panjwani, D., Gatkal, S., Kazi, F.: Twitter bot detection and their influence in hashtag manipulation. In: 2020 IEEE 17th India Council International Conference (INDICON), pp. 1–7. IEEE, December 2020
7. Rout, R.R., Lingam, G., Somayajulu, D.V.: Detection of malicious social bots using learning automata with URL features in the Twitter network. IEEE Trans. Comput. Soc. Syst. **7**(4), 1004–1018 (2020)
8. Chu, Z., Gianvecchio, S., Wang, H., Jajodia, S.: Who is tweeting on Twitter: human, bot, or cyborg? In: Proceedings of the 26th Annual Computer Security Applications Conference, pp. 21–30, December 2010
9. Kouvela, M., Dimitriadis, I., Vakali, A.: Bot-detective: An explainable Twitter bot detection service with crowdsourcing functionalities. In: Proceedings of the 12th International Conference on Management of Digital Ecosystems, pp. 55–63, November 2020
10. Bello, B.S., Heckel, R.: Analysing the behaviour of Twitter bots in post Brexit politics. In: 2019 Sixth International Conference on Social Networks Analysis, Management and Security (SNAMS), pp. 61–66. IEEE, October 2019
11. Schnebly, J., Sengupta, S.: Random forest Twitter bot classifier. In: 2019 IEEE 9th Annual Computing and Communication Workshop and Conference (CCWC), pp. 0506–0512. IEEE, January, 2019
12. Loyola-González, O., Monroy, R., Rodríguez, J., López-Cuevas, A., Mata-Sánchez, J.I.: Contrast pattern-based classification for bot detection on Twitter. IEEE Access **7**, 45800–45817 (2019)
13. Shi, P., Zhang, Z., Choo, K.K.R.: Detecting malicious social bots based on clickstream sequences. IEEE Access **7**, 28855–28862 (2019)
14. Benigni, M.C., Joseph, K., Carley, K.M.: Bot-ivistm: assessing information manipulation in social media using network analytics. In: Agarwal, N., Dokoohaki, N., Tokdemir, S. (eds.) Emerging Research Challenges and Opportunities in Computational Social Network Analysis and Mining. LNSN, pp. 19–42. Springer, Cham (2019). https://doi.org/10.1007/978-3-319-94105-9_2
15. Wei, F., Nguyen, U.T.: Twitter bot detection using bidirectional long short-term memory neural networks and word embeddings. In: 2019 First IEEE International Conference on Trust, Privacy, and Security in Intelligent Systems and Applications (TPS-ISA), pp. 101–109. IEEE, December 2019

16. Battur, R., Yaligar, N.: Twitter bot detection using machine learning algorithms. Int. J. Sci. Res. (IJSR) (2018)
17. Alothali, E., Zaki, N., Mohamed, E.A., Alashwal, H.: Detecting social bots on Twitter: a literature review. In: 2018 International Conference on Innovations in Information Technology (IIT), pp. 175–180. IEEE, November 2018
18. Bello, B.S., Heckel, R., Minku, L.: Reverse engineering the behaviour of Twitter bots. In: 2018 Fifth International Conference on Social Networks Analysis, Management and Security (SNAMS), pp. 27–34. IEEE, October 2018
19. Wald, R., Khoshgoftaar, T.M., Napolitano, A., Sumner, C.: Predicting susceptibility to social bots on Twitter. In: 2013 IEEE 14th International Conference on Information Reuse & Integration (IRI), pp. 6–13. IEEE, August 2013
20. Badawy, A., Ferrara, E., Lerman, K.: Analysing the digital traces of political manipulation: the 2016 Russian interference Twitter campaign. In: 2018 IEEE/ACM International Conference on Advances in Social Networks Analysis and Mining (ASONAM), pp. 258–265. IEEE, August, 2018
21. Aswani, R., Kar, A.K., Ilavarasan, P.V.: Detection of spammers in Twitter marketing: a hybrid approach using social media analytics and bio-inspired computing. Inf. Syst. Front. **20**(3), 515–530 (2018)
22. Kantepe, M., Ganiz, M.C.: Pre-processing framework for Twitter bot detection. In: 2017 International Conference on Computer Science and Engineering (UBMK), pp. 630–634. IEEE, October 2017
23. Paudel, P., Nguyen, T.T., Hatua, A., Sung, A.H.: How the tables have turned: studying the new wave of social bots on Twitter using complex network analysis techniques. In: Proceedings of the 2019 IEEE/ACM International Conference on Advances in Social Networks Analysis and Mining, pp. 501–508, August 2019
24. Dickerson, J.P., Kagan, V., Subrahmanian, V.S.: Using sentiment to detect bots on Twitter: Are humans more opinionated than bots? In: 2014 IEEE/ACM International Conference on Advances in Social Networks Analysis and Mining (ASONAM 2014), pp. 620–627. IEEE, August 2014
25. Ibrahim, M., Abdillah, O., Wicaksono, A.F., Adriani, M.: Buzzer detection and sentiment analysis for predicting presidential election results in a Twitter nation. In: 2015 IEEE International Conference on Data Mining Workshop (ICDMW), pp. 1348–1353. IEEE, November 2015

Pesticide and Quality Monitoring System for Fruits and Vegetables Using IOT

R. S. Keote(✉), Pranjal Rewatkar, Prachi Durve, Vaishnavi Domale, and Sharvari Buradkar

Yeshwantrao Chavan College of Engineering, Nagpur, India
rashmikeote@gmail.com

Abstract. The risk of adverse health impacts in people should increase dramatically with the use of pesticides, steroids, and fertilizers. Dangerous pesticides enter the human body through fruits and vegetables so it is necessary to monitor the pesticide present and also check the quality. The outcomes of the hardware and software design are precise and timely. The prototype for the system to gather data regarding the presence of pesticides is created using the MQ2 sensor, Arduino, and Wi-Fi ESP-01 module. The market's fruits and vegetables are inspected for pest presence using an Arduino and a pest detection sensor. Additionally, the LCD is utilized to display the presence of pests in fruits and vegetables. The issue is displayed on an LCD to highlight the percentage of fruits and vegetables that are present.

Keywords: MQ2 sensor · Arduino · Wi-Fi ESP-01 Module

1 Introduction

At various stages of cultivation and during the post-harvest storage of crops chemical substances like pesticides are applied to crops. The use of pesticides is intended to prevent the destruction of food crops by controlling agricultural pests or unwanted plants and to improve plant quality. Fruit diseases are the cause of crop degradation and economic losses in agriculture Through the use of fertilizers, pesticides, and high quality fruit pesticides is the reason, an increase in adverse effects on humans due to the uncontrolled level of pesticides in those fruits or vegetables, so we should develop an appropriate solution to diagnose the disease and pesticides. Hardware and software design are designed to get the right result. A system prototype is created which combines MQ2 sensor, Arduino, Wi-Fi module ESP-01 to obtain information on the presence of pesticides in fruits and vegetables. And have we seen in the software program Tested by diagnosing the disease the discovery by dividing it into three categories using the end-to-end processing method required with image separation, the first stage is RGB to gray conversion, followed by media filtering, edge detection, and morphological functions. In the case of a second-degree exit feature both domains are compared to extract the feature and a separate third-step image using a different kernel on a vector support machine.

H. Sharma et al. (Eds.): ICIVC 2022, PALO 17, pp. 522–531, 2023.
https://doi.org/10.1007/978-3-031-31164-2_44

The maximum amount of pesticide that both humans and animals may consume legally is set down in the Embedded Program. If pesticides are found in a fruit in a range above or below the threshold level, the fruit is said to have pesticides. The market's fruits and vegetables are inspected for pest presence using an Arduino and a pest detection sensor. Additionally, the LCD is utilized to display the presence of pests in fruits and vegetables. The issue is displayed on an LCD to highlight the percentage of fruits and vegetables that are present. The pesticide content and the readings from each sensor are shown via IOT on Things of Speak, an IOT-based platform. Since the majority of nations have defined maximum residue levels (MRL) for pesticides in food items. Monitoring of pesticides in fruit and vegetable samples has grown in recent years. The procurements of vegetables and fruits are most in markets and supermarkets due to advancement of urbanization construction. In most of the markets and supermarkets there is no provision of pesticide residues detection devices. Traditional analytical techniques like Gas chromatography (GC), liquid chromatography (LC) or combinations (GC-MS or LC-MS/MS) are used for identification and quantity determination of pesticides residues. The primary goal is to control the amount of pesticide used in fruits and vegetables so that they reduce diseases that are caused in our human system. There is a way to control & reduce the pesticide by using vinegar, baking soda, and powdered turmeric in fruits and vegetables. IOT technology can make monitoring agricultural food products easier, automatic, and effective. The establishment and application of agricultural products quality and safety system is based on IOT technology, will provide the whole process of tracking and detecting the food products and meet the public needs of high-quality and safe agricultural products. The application of this pesticides detection device has been performed on real samples.

2 Related Work

Crop quality is impacted by the overuse of herbicides and fertilizer. Additionally, it poses a risk to human health. A variety of methods are suggested as a solution to this issue, including a sensor network-based advisory system that recommends how much fertilizer and insecticide should be used on a given crop under specific weather and soil conditions. This paper [1] develops a system to monitor crop fields using sensors like soil moisture, temperature, humidity, light and automate the irrigation system. Sensor data is sent to the web server using wireless transmission to encode the server data in JSON format. The developed system is more efficient than the conventional approach. Usage of green energy and smart technology provide better productivity in the agriculture sector. Also the Internet of Things (IOT) plays an important role in smart farming.

A remote-controlled vehicle can be used for a variety of agricultural tasks like spraying, cutting, and weeding in both automatic and manual modes. The controller continuously checks the temperature, humidity, and soil quality, and provides the field with water as needed [2]. Global Positioning System (GPS), Wireless Sensor Networks (WSN), and Radio Frequency Identification (RFID) are employed in this study for sensing, tracking, etc. in environmental management. The market's fruits and vegetables are inspected for pest presence using an Arduino and a pest detection sensor. Additionally, the LCD is utilized to display the presence of pests in fruits and vegetables. The issue is displayed on an LCD to highlight the percentage of fruits and vegetables that are present [3]. The most ideal solution, presented in this study, eliminates the drawbacks of the previous ones, such as cost, accuracy, and complexity. Both the pesticide content and the disease identification are carried out in this work whereas only one of the two tasks is carried out by the present approaches in terms of efficiency. The suggested system is the greatest option available since it is comprehensive, precise, and real-time. After some testing, it was discovered that CNN performed significantly better than SVM. Although CNN produced very accurate results, its training and testing procedures took 6.5 min longer than those of SVM [4, 7]. In Paper [5], a novel method for identifying pesticide damage to fruits and vegetables is developed utilizing the photosensitive effect and the Internet of Things. As a result, the consumer will be confident in the food's safety. A smart phone application and portable, accurate, and user-friendly gadget called Fresh-O-Sniff were created to assist consumers choose premium fruits and vegetables as a solution to this problem. This innovative method predicts the presence of pesticides in fruits and vegetables. The general population will be the main gainer from the solution. This will contribute to the general public's health being improved, which would in turn speed up India's digitization process. IoT, mobile applications, and food technology are all combined in this technology. This paper [6] presents a prototype of the system with the use of four sensors, a Node MCU microcontroller and to get the information about the presence of pesticides. The high level of pesticide that can be accepted legally to be consumed by animals and humans is given by the Embedded C Program. If a fruit is detected to belong in a range above or below the threshold level then it is said to contain pesticides. Through IoT, the pesticide content and the values obtained from each sensor are shown in Blynk APP.

Paper [7, 10] proposed work is divided primarily into two modules, the first of which uses CNN's image processing network to identify the fruit. 360 photos of fruits are used in total to train CNN using these images. Cameras are used to record and classify input for CNN. In module 2, pesticides in fruits and vegetables are detected by calculating their NDVI in three different methods, utilizing an IR sensor, a gas sensor, and comparing the results. The screen will show the detection information. And a plot of the result graph is made. Out of these techniques, gas sensors provide more accuracy for spotting pesticide residue on the outside of fruits and vegetables, as demonstrated above.

The portable nature of the device will increase the effectiveness of the food safety authorities and guarantee quality [8, 9, 11].

3 Methodology

The system's IoT components, which include gas sensors and other devices for fruit quality and pesticide detection, make up the system architecture. And finally, a computer-connected Arduino microcontroller. The user's device displays the output that the Arduino provides. The output data from the Arduino is transferred to the MATLAB ThingSpeak cloud server. This data can be sent to the server using a Wi-Fi module linked to the internet, and it subsequently provides information about the presence of pesticides that is presented using an application displayed on the user's device. A ThingSpeak channel dedicated to pesticide detection has been launched.

3.1 Hardware and Software System Description

Hardware Components

1. *MQ3(Gas Sensor)-*
   The MQ sensor series frequently uses the Metal Oxide Semiconductor (MOS) type sensor, designated MQ3. This kind of sensor is called a chemiresistor, and it detects alcohol by changing the resistance of the detecting material. Alcohol concentrations can therefore be determined by integrating it into a straightforward voltage divider network.
2. *Arduino-*
   An open source programmed circuit board called Arduino can be used in a wide range of makerspace projects, both straightforward and intricate. This board has a microprocessor that may be designed to recognize and manage physical things. The Arduino is able to communicate with a wide range of outputs, including LEDs, motors, and displays, by responding to sensors and inputs. Arduino has grown to be a very popular option for makers and makerspaces looking to construct interactive hardware projects because of its adaptability and inexpensive price. The Arduino Uno is one of the most widely used Arduino boards. Although it wasn't the first board to be produced, it is still the one that is used the most frequently and has the most documentation.
3. *ESP8266X*
   The ESP8266EX provides a full and independent option for offloading Wi-Fi networking functions from one application processor to another. The ESP8266EX launches the application straight from an external flash when it serves as the host is shown in Fig. 1. The system's performance in these applications has improved. Alternatively, any micro controller-based design with straightforward connectivity (SPI/SDIO or I2C/UART interface) can be enhanced with wireless internet access by acting as a Wi-Fi adaptor.

Fig. 1. ESP-01 ESP8266 Serial WIFI Transceiver Module

4. *ESP-01 WIFI MODULE*
   Features of the Fig. 2. ESP8266-01 includes its affordability, portability, and power; its current consumption is 100 mA; its maximum I/O voltage is 3.6 V; and its maximum I/O source current is 12 mA. (Max)

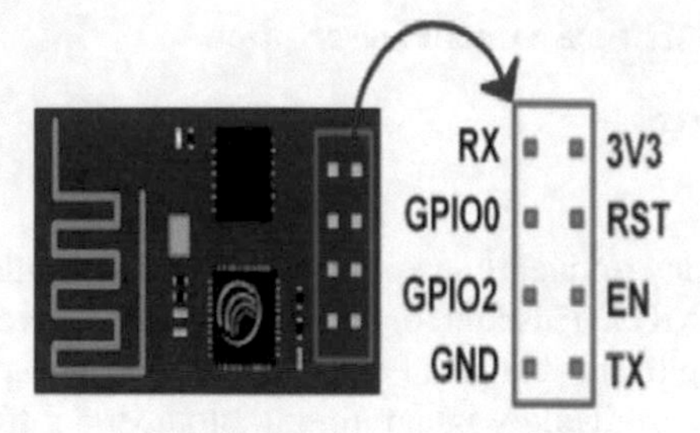

Fig. 2. ESP8266X PINOUT

5. *WIFI INTERFACING*

- Join the VCC/3.3V/Power Pin and Enable Pin (both red wires) of the ESP to the +3.3 V power pin of the Uno.
- Join the Ground/GND Pin (Black Wire) on the ESP to the Ground/GND Pin on the Uno.
- Attach the green wire from the ESP's TX to Uno's Pin 6. Connect the ESP's RX (blue wire) to the Uno's Pin 7 and then to a 1K resistor.
- Join the 1K resistor and the Uno's GND pin with the ESP's RX (blue wire) (Fig. 3).

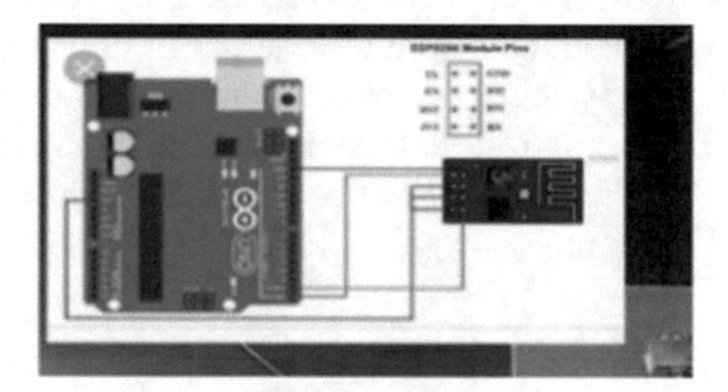

Fig. 3. WIFI Interfacing

Software Implementation

The TCP Client sample makes use of the ThingSpeak server. Everyone may monitor and analyze real-time data from their sensor devices on the open IOT platform ThingSpeak. Additionally, using MATLAB code embedded in ThingSpeak, data analysis may be done on data submitted by remote devices. ESP8266 Response. Checking ESP8266 answers at the client end is necessary. On a PC or laptop's serial interface, we can verify it. As illustrated in Fig. 4 below, attach the ESP8266 module's transmit pin (TX) to the Arduino UNO's receive pin (RX) and the USB to serial converter's receive pin (RX). Connect the PC or laptop to the USB to serial converter. To see the ESP8266's response to the AT instruction delivered from the Arduino UNO, open the serial terminal on your computer or laptop.

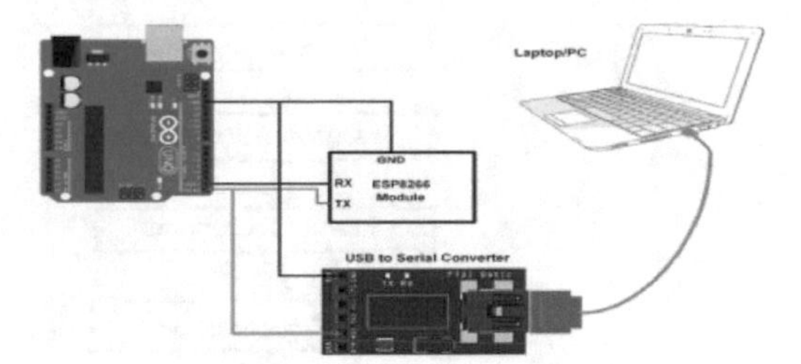

Fig. 4. ESP8266 Response

3.2 Working

The proposed Circuit diagram for Pesticides detection monitoring system in fruits and vegetables is shown in Fig. 5 below. An Arduino microcontroller connected to the computer and shown on the user's device in this case produces the output. The output data from the Arduino is transferred to the cloud server.

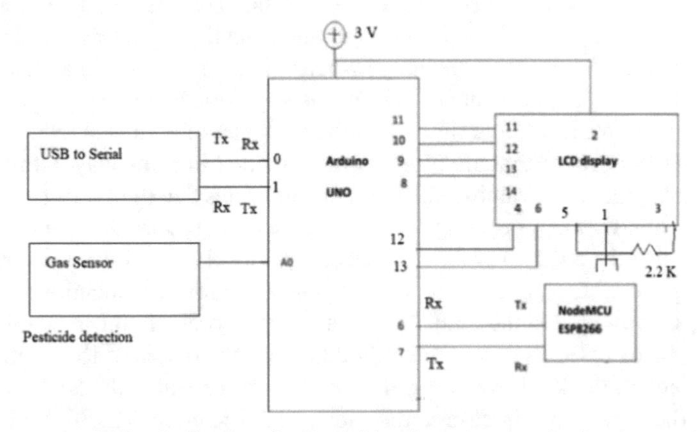

Fig. 5. The proposed Circuit diagram for Pesticides detection monitoring system in fruits and vegetables

This data is sent to the server by using the Wi-Fi module which gives information about pesticides presence. A channel created in ThingSpeak for the detection and monitoring of pesticides. Flowchart for identification and classification of diseases monitoring is shown in Fig. 6 below

Three levels—high level, mid-level, and low level—are present in the recorded input image that was used to identify this situation. Here, high level is nothing, but this high level image restoration allows for the recovery of any missing features, while midlevel images are used for picture segmentation and low level images are used for image enhancement in accordance with the workings of the image. And moving on to the feature extraction, the image's foreground and background are included in this feature extraction. The image's foreground is made up of the image's data values, which are represented by pixels, while the image's background is made up of the image's colour

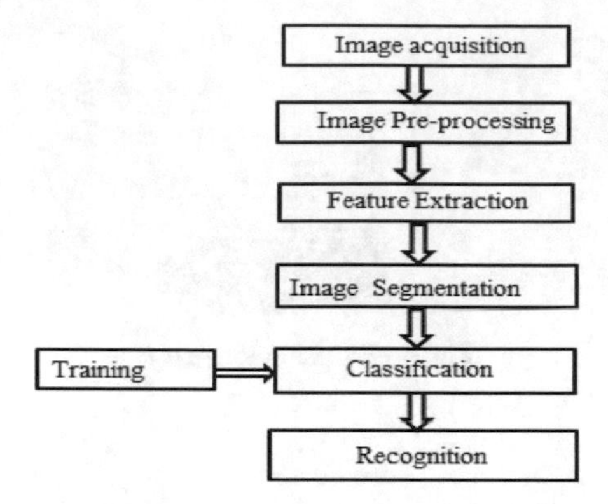

Fig. 6. Flowchart for identification and classification of diseases monitoring.

values, or RGB (red, green, and blue) values. How much data needs to be calculated in the image's foreground is done here. The image segmentation process is then processed. A picture is simply divided or subdivided into sections during segmentation. Because we do not need to process the complete image when identifying diseases, we just need to consider the required portion of the image when detecting diseases in a fruit image. We employ the clustering method known as k-means clustering in the segmentation process. Simply said, a cluster is nothing more than the grouping of elements, which divides an image into smaller groups that are related to one another in some way. Similar refers to the same colour, pixel, etc. values. The number k indicates that the dataset would be split into k unique, non-overlapping subgroups by the clustering method. After clustering, we require these data in order to compare them to previously saved datasets. We actually have training and testing; during the testing procedure for classification, the obtained input image is compared to the image that is already recorded in the available database. The many sorts of diseases can be identified here when compared to a stored diseases dataset. To foretell the kind of disease that may harm the fruits, the SVM algorithm is employed. The kernel in this case is the radial basis function (RBF). A piece of data could have one dimension, two dimensions, three dimensions, or infinite dimensions. RBF is used to select the location of the threshold that facilitates data splitting in infinite dimensions.

4 Result

The hardware for the pesticide detection Monitoring system is shown below in Fig. 7 which shows result on LCD display depends upon the values detected. After the execution of the program data is sent to the LCD Display. Messages of pesticide detection and quality of fruit is displayed on the display (Fig. 8).

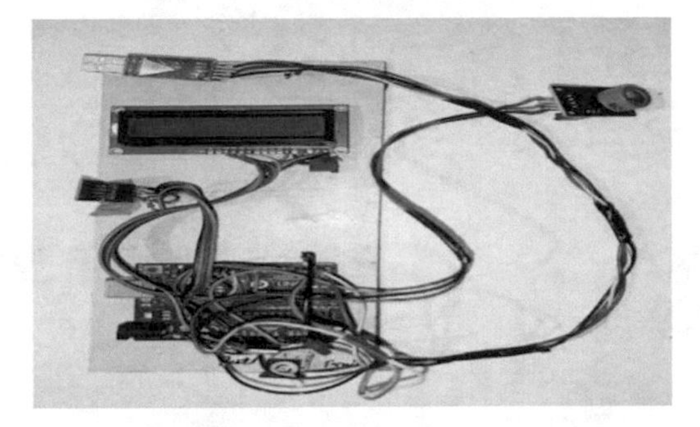

Fig. 7. Hardware setup of Pesticide detection monitoring system

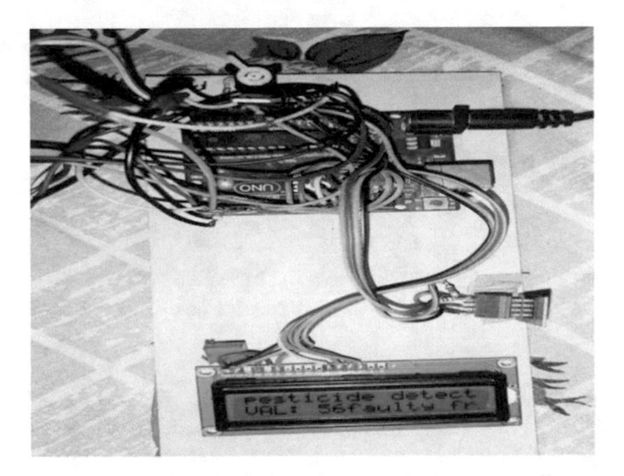

Fig. 8. Pesticide detection

The pesticide detection graphs are shown in Fig. 9 and 10 on the ThingSpeak website.

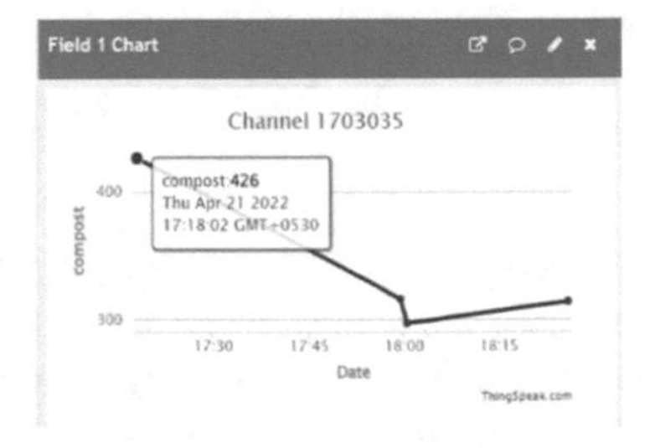

Fig. 9. Pesticide Detection Graph1

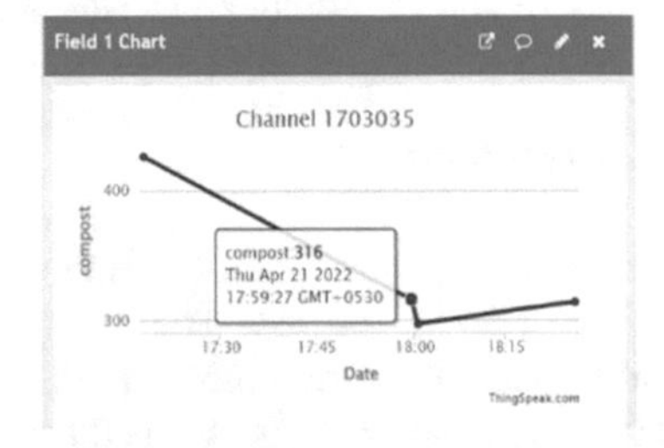

Fig. 10. Pesticide Detection Graph2

When the value is 100, the fruit is said to be faulty and the result of quality of fruit is shown on the LCD. Faulty Fruit Detection with graph on ThingSpeak for Apple & Banana is shown in Fig. 11 and 12 below.

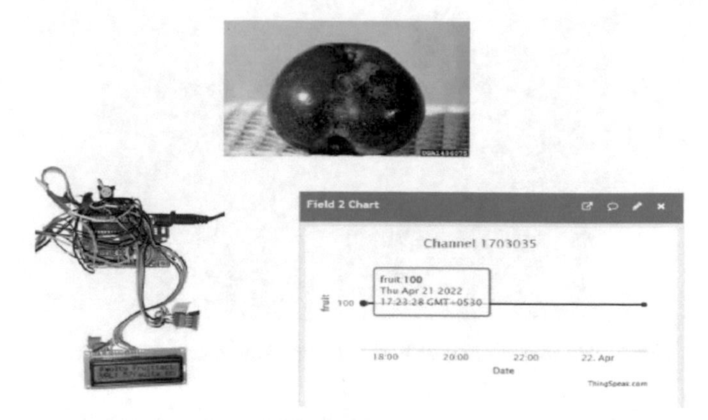

Fig. 11. Faulty Fruit Detection with graph on ThingSpeak for Apple

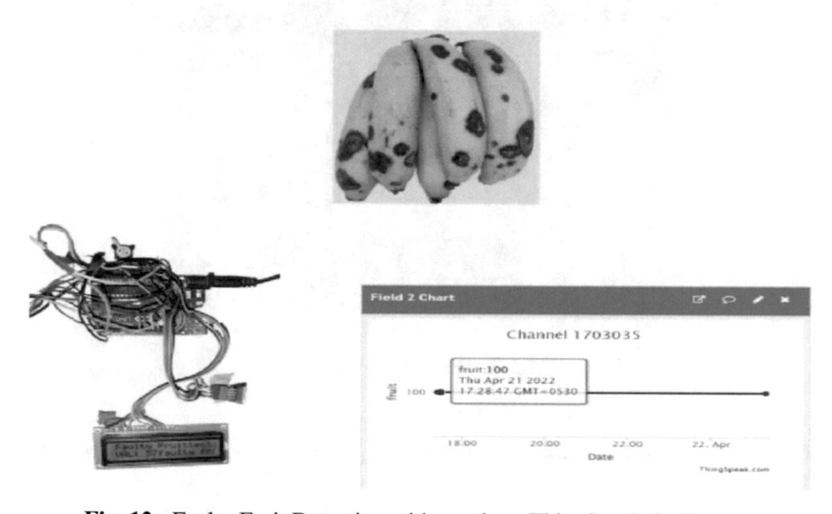

Fig. 12. Faulty Fruit Detection with graph on ThingSpeak for Banana

5 Conclusion

The Quality & pesticide monitoring of Fruits & vegetables are done by using machine learning of SVM and IoT. The most efficient, precise, and sophisticated solutions are possible with the current methodology. To detect the fruit's quality by using the sensor and the Internet of Things, this is the best method for detecting pesticides. According to the literature, the color, texture, and morphological traits are all suited for the condition. SVM algorithms and neural networks are utilized for classification and comparison; the KNN method is used to identify fruit illnesses. While neural networks are used for comparison with training data pixel by pixel to identify the type of fruit disease, SVM and NN are both utilized for classification. The suggested system outperforms other current systems in terms of reliability, real-time functionality, and output quality. The performance is quite precise.

References

1. Rajalakshmi, P., Devi Mahalakshmi, S.: IOT based crop-field monitoring and irrigation automation. In: 10th International Conference on Intelligent Systems and Control (ISCO), 7–8 January 2016. IEEE Xplore, November 2016
2. Bhattacharjee, A., Das, P.: Smart farming using IoT. In: International Conference on Intelligent Systems and Control (ISCO), 2017. IEEE Xplore (2017)
3. Nikitha, R., Pavithra, T.K., Pavithra, V., Kavyashree, M., Vijayalakshm: IoT based solution for monitoring of pollution through pesticide in fresh fruits and vegetable available in market. Int. J. Sci. Eng. Res. **10**(3) (2019). ISSN 2229-5518
4. Devi, D., Anand, A., Sophia, S.S., Karpagam, M., Maheswari, S.: IoT- deep learning based prediction of amount of pesticides and diseases in fruits. In: 2020 International Conference on Smart Electronics and Communication (ICOSEC) (2020)
5. Kanmani, R., Maheswar, R., Sureshkumar, A., Praveena, V., Mugilan, A.: Non-destructive approach to detect pesticides in fruits and vegetables using IoT technology. In: 2020 International Conference on Computer Communication and Informatics (ICCCI - 2020), Coimbatore, India, 22–24 January 2020 (2020)
6. Kanmani, R., Yuvashree, M., Sneha, S., Sahana Reshmi, V., Pavithra, R.: IoT based pesticides monitoring system in food products. Int. Res. J. Eng. Technol. (IRJET) **08**(03) (2021). e-ISSN: 2395-0056, p-ISSN: 2395-0072
7. Aradhana, B.S., Raj, A., Praveena, K.G., Joshy, M., Reshmi, B.S.: Quality and pesticides detection in fruits and vegetables. J. Emerg. Technol. Innov. Res. (JETIR) **8**(5) (2021)
8. Nagajyothi, D., Hema Sree, P., Rajani Devi, M., Madhavi, G., Hasane Ahammad, S.k.: IOT-based prediction of pesticides and fruits diseases detection using SVM and neural networks. Int. J. Mech. Eng. **7**(1) (2022). ISSN: 0974-5823
9. Sangewar, S., Peshattiwar, A.A., Alagdeve, V., Balpande, R.: Liver segmentation of CT scan images using K mean algorithm. In: Proceedings of the 2013 International Conference (2013)
10. Devi, D., Lashmi Sruthy, D., Madhuvarsni, T., Mohana Ranjini, V: IOT based root rot detection system. In: 2022 8th International Conference on Advanced Computing and Communication System (ICACCS) (2022)
11. Madankar, A., Patil, M., Khandait, P.D.: Automated waste segregation system and its approach towards generation of ethanol. In: 2019 5th International Conference on Advanced Computing & Communication System (ICACCS) (2019). https://doi.org/10.1109/ICACCS.2019.8728358Corpus. ID: 174817427

Forecasting Crop Yield with Machine Learning Techniques and Deep Neural Network

B. G. Chaitra, B. M. Sagar, N. K. Cauvery, and T. Padmashree(✉)

Department of ISE, RVCE, Bangalore, India
{sagarbm,cauverynk,padmashreet}@rvce.edu.in

Abstract. The Crop yield forecast is an essential agricultural problem in the modern world. According to statistics 60% of the Indian economy depends on agriculture whereas 10% farmers are educated, only 2% use modern agricultural techniques. The paper discusses the significance of deep learning in predicting the yield from the parameters relevant to the Indian scenario. Crop yield prediction using deep learning deals with developing a system that leverages deep learning technology to predict the yield. It also compares the performance of the deep learning model with other machine learning techniques. The system has a user interface through which a user can do the real time prediction by passing the required parameters. The work involves using three algorithms which are used to build an ensemble approach for better accurate prediction of the yield. The selected algorithms are Random Forest (RF), Extreme Gradient Boosting (XGBoost) and Deep Neural Network (DNN). The results for accuracy obtained are as follows respectively, 92.9%, 93.3% and 96%. The prediction will enable farmers to have the best knowledge to know the reason behind the poor yield as well as determine which method can help grow the crops in the best way and gain high profits in return based on particular conditions.

Keywords: Deep Neural Network · Machine Learning · Crop yield · Ensemble

1 Introduction

Farming plays a critical function in the economic situation across the globe. With the development of the human population around the world with varying cultures, crop yield is central to dealing with food security difficulties and lowering the influences of climate changes. Some of the major parameters that create impact include area, crop, year, temperature, precipitation and wind speed.

India's economy is driven by agriculture as a major factor. Agricultural activities in the nation has 54.6% of the total employment according to a survey conducted by Census 2011 and constitutes 17.8% of the country's Gross Value Added (GVA) for the year 2019–20. The Land Use Statistics states in 2016–17 that out of the total geographical area of the country is 328.7 million hectares, 139.4 million hectares is the reported net sown area and 200.2 million hectares is the total area which is cropped.

The cropping intensity is seen to be 143.6%. The total area where sowing is accomplished is said to be around 42.4% of the total area. The Annual National Income as

H. Sharma et al. (Eds.): ICIVC 2022, PALO 17, pp. 532–542, 2023.
https://doi.org/10.1007/978-3-031-31164-2_45

released by Central Statistics Office (CSO), Ministry of Statistics & Programme Implementation, have depicted that the agricultural domain has majorly been 17.8% of India's GVA during 2019–20, and it is comparatively higher than in 2015–16 which was 17.7% [13].

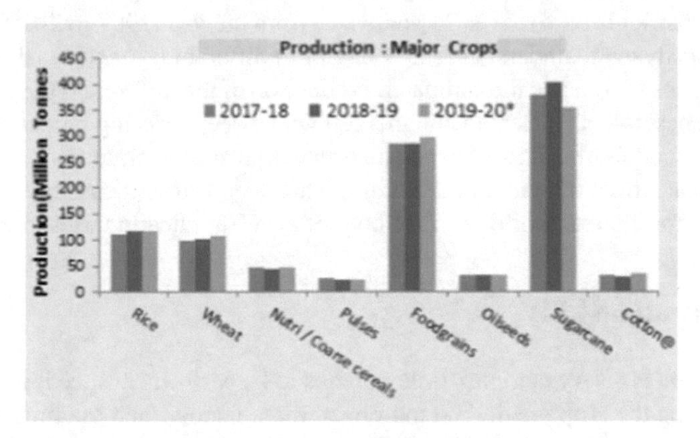

Fig. 1. Production: Major Crops grown

Figure 1 shows the Production of major crops grown. India GDP (Gross Domestic Product) from agriculture: It is estimated that India's agriculture sector accounts only for around 14% of the country's economy but for 42% of total employment. As around 55% of India's arable land depends on precipitation, the amount of rainfall during the monsoon season is very important for economic activity.

Figure 2 shows the GDP from Agriculture in India and it decreased to 5688.80 INR Billion in the first quarter of 2022 from 6626.88 INR Billion in the fourth quarter of 2021 [14].

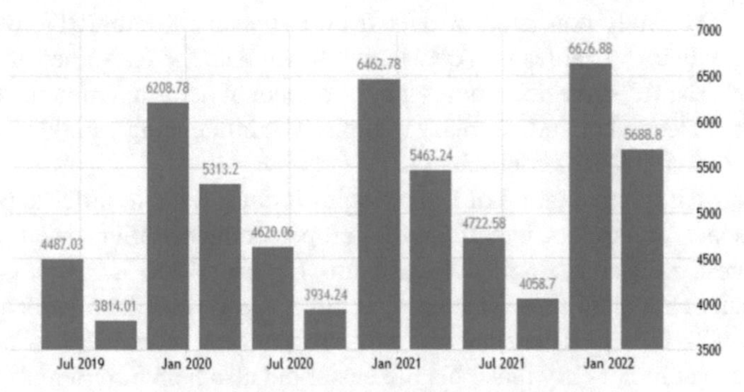

Fig. 2. GDP from Agriculture in India

Food consumption varies significantly across the globe, but the standard facilities for human sustenance remain the same. Usually all of us consume a great variety of vegetables, fruits and various grains. In this work, deep learning is employed for predicting the most familiar Indian crops, making use of existing available data from the Ministry of Agriculture and Farmers Welfare of India. The data is surveyed across the Indian subcontinent at district level for over 23 years i.e., from the year 1997 to 2019. The data is studied and analyzed to understand the behavior of different types of Machine Learning (ML) algorithms to forecast the estimated production of the different types of crops. Out of which the best two are selected and merged with a deep learning model for obtaining the ensemble model. The algorithm can also state how the supernatural factors like rain and climate can affect the yield of the crops. This deep learning ensemble algorithm is implemented for the real world with the help of a web application user interface.

2 Background Study

Yield prediction is a very complex task which is to be performed since it involves many factors affecting the crop yield. And moreover, these factors such as climate conditions cannot be controlled by human limits. These factors have a huge effect on the yield production. To have an efficient algorithm to perform the task will require a simulated environment where factors that cannot be controlled can be applied to test and determine the effect on crop yield. These predictions can be helpful to understand the adverse effect of any factor being excess.

Many ways to predict the crop yield have already been proposed but not yet efficiently implemented. There is a huge gap that has to be filled in research and development. One of such studies involves the use of a random forest algorithm to forecast the yield production of cotton in the region of Maharashtra. The study involved the research at three different time periods before the actual harvest season [2]. This helped to determine a few aspects such as integration and processing of some large number of inputs that were collected from different satellite modalities, un-scaled and non-uniform ground-based information. The study consisted of data from the year 2001 to 2017, where the co-linearity of predictor variables was considered to validate the R2 value for the Random Forest model. The R2 value determines the coefficient of determination in the final yield for September, December and February months. The influencing variable threshold was kept near to 69% to 39% of coefficient [2].

An effective implementation of technologies is considered in order to provide maximum assistance to farmers in the area of crop recommendation. Methods used are Random Forest, XGBoost, Logistic Regression, Decision Tree are given to data of soil nutrients and weather by these machine learning algorithms. The work predicted an accuracy of 98% to help determine the right crop to grow. If the farmer adapts to this technology it would not only make his life easier but also help him in making decisions in accordance with the environment [1].

Other than the fundamental Random Forest algorithm, few of the deep learning algorithms are also under research as mentioned by Chaoya Dang [3]. Here the redundancy analysis technique was employed to carry out feature selection and explanatory factors. This study gives a clear outcome on how deep neural networks tend to perform better as compared to Support Vector Machine (SVM) and Random Forest and Deep Neural Network models. For real time prediction it can be said that the ensemble approach on the crop yield prediction scenario can give much better results than having a single algorithm running in the background [4]. Also a web framework built with flask is deployed and hosted live in the cloud.

A new concept of digital farming is arising from the traditional way of farming. This involves using an information mining approach; where the calculation employed is to improve the harvest suggestion by making use of Support Vector Machine [5].

Earlier the production for crops was determined by the experience that a farmer has. Today, the same experience is available in terms of data gathered from surveys and other methods. Hence implying the existing experience to the model will tend to give a similar outcome as predictions done by an experienced farmer. The models trained are capable of recognizing the pattern in the data efficiently and hence grouping such similarities on the current real-life scenario will give a favourable outcome [6].

Implementation of Internet of Things (IoT) and other such latest technologies have already started to make its way in the agriculture domain. Not just individual farmers but also the Government organizations are also pushing forward to ensure that the areas such as crop, ground, water, yield and weed management can be brought into picture [7]. New techniques include usage of satellite images of different agricultural zones. These zones differing in multiple aspects were having a different set of problems that impact the crop yield production. It is found that the solution to a specific problem cannot be the same in all the zones [8].

To address all such scenarios Long Short-Term Memory (LSTM) and Recurrent Neural Network (RNN) models have already been implemented but the results were very specific with a given set of data. The data that was used had very few factors that have been put under consideration where a huge number of parameters were found to be left out [9].

Hence to develop an ideal model the understanding of the real time environment is very much required to be understood. As mentioned above, the solutions need not be the same for specific kinds of problems where the zones are different with respect to the super natural factors. The model trained will be accurate only if the data collected will be localized. This is done by simplifying the approaches in two modules, one that consists of the data with factors carrying supernatural factors like weather, temperature and humidity. The other module considers data related to the crop production methods and techniques [10].

Deep Learning has the following advantages over Machine learning models.

1. Feature Generation Automation: Deep learning can perform complex tasks that often require extensive feature engineering.
2. Works Well With Unstructured Data: Train deep learning networks with unstructured data and with appropriate labelling.
3. Better Self-Learning Capabilities: Deep learning's performance is directly proportional to the volume of training datasets. The larger the datasets, the more accuracy.
4. Cost Effectiveness: Deep learning models can be cost-intensive, once trained [11].

5. Improve performance with Algorithm tuning: Tuning neural network algorithms includes Weight Initialization, Activation Functions, Batches and epochs, Early Stopping [12].

Based on the extensive literature survey, it can be concluded that when the model is trained using ML techniques, adding more features will give more accurate results and is reliable. Random Forest and Deep learning techniques give better results compared to other models and most of the studies are focusing on particular regions and crops.

3 Objectives

A significant and challenging issue in agriculture is the forecast of crop yield. Agricultural yield primarily depends on basic factors like weather conditions such as temperature, rainfall and wind speed etc. Accurate data about previous yield is essential for making decisions by using deep learning to make agricultural risk management and future predictions simple. Hence, the objectives of the crop yield prediction are as follows:

- To gather the data, that identifies the features influencing the crop yield.
- To Pre-process the gathered data by changing the raw data into a clean data set.
- To train a deep learning model for top crops grown in different parts of the Indian sub-continent to predict their yield.
- To develop a web application where the inputs of the factors affecting the crop yield can be provided to get appropriate predictions.
- To estimate the crop yield accurately for the given features in less time that act as a recommendation system for farmers.

4 Methodology

The following steps were implemented in order to achieve the results:

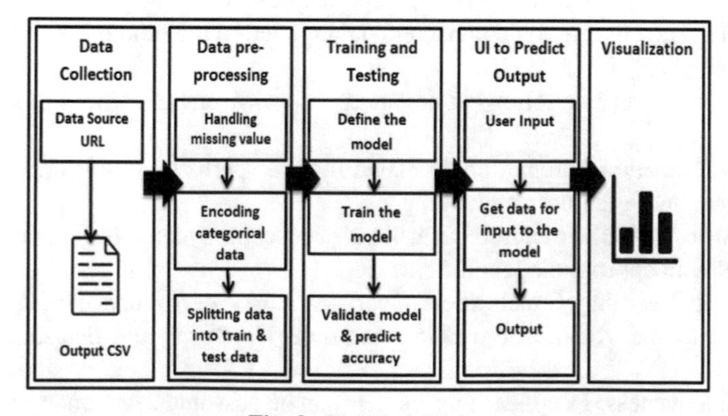

Fig. 3. Detailed Design

Figure 3 shows the Detailed Design of the implementation of the proposed work.

Data Collection

The data gathering process incorporates determining different features that influence the harvest of a particular crop. The previous yield data which is openly available is gathered from the Ministry of agriculture and farmer's welfare development of India.

The features include area, crop, year and weather information like temperature, wind speed and rainfall is gathered from weather tower data from weather.com.

Data Pre-processing

This is a crucial phase which involves bifurcating the data into training and testing, managing missing data, putting the categorical data in the format required etc.

Management of values which are not available is accomplished by eliminating the row, or deciding on a particular constant value to fill in the row, or by adding a known mean value which is decided based on the feature selected.

The categorical values which are present as string constants are converted into integer values to enable the model to interpret the values seamlessly.

Bifurcation of data into training and testing is the standard 80:20 ratio where 80% of data is considered for training the model and 20% for testing.

Training and Testing

The phase consists of using the training data for executing the model using the test data for verification of the model for its prediction. 80% of data is considered for training the model and 20% for testing. The work carried out considers the following algorithms: Random Forest, XG Boost and Deep Neural Network.

Random Forest: The working of Random forest algorithm is based on creating decision trees and the given data. Every tree disseminates predictions and the algorithm works to decide on the optimal solution by the process of voting. This algorithm follows supervised learning. Since the results here are averaged out before deciding on the solution, it is always more efficient than a single decision tree output as it reduces the over-fitting.

XG Boost: Generally, for data in tabulated form, XGBoost algorithm is used. This algorithm is mostly used where speed and performance matters the most. Decision trees which are gradient boosted are used in the design of this algorithm.

Deep Neural Network: The basic component of a Deep Neural Network (DNN) is a model that associates weights for inputs (neuron) and the output is generally an activation function. Activation functions used are Relu, softmax. At each layer the input is transformed and no direct association can be found between layers which are not consecutive. Back propagation and other GD methods are used to optimize the network parameters. This includes Adam. Each update of the network is done for the complete training dataset called an epoch. The epoch count used is 10 and is trained until the validation accuracy saturates or starts dropping in order to avoid overfitting. Results are discussed in the later sections.

Web Based Interface to Predict the Output for Real Time Attributes
Once the model is set up, it should be deployed to be used in real-time. Hence a simple web page is developed.

The user (essentially the farmer/authority interested to know the Crop Yield) begins with the registration process on the portal followed by login as depicted in Fig. 4.

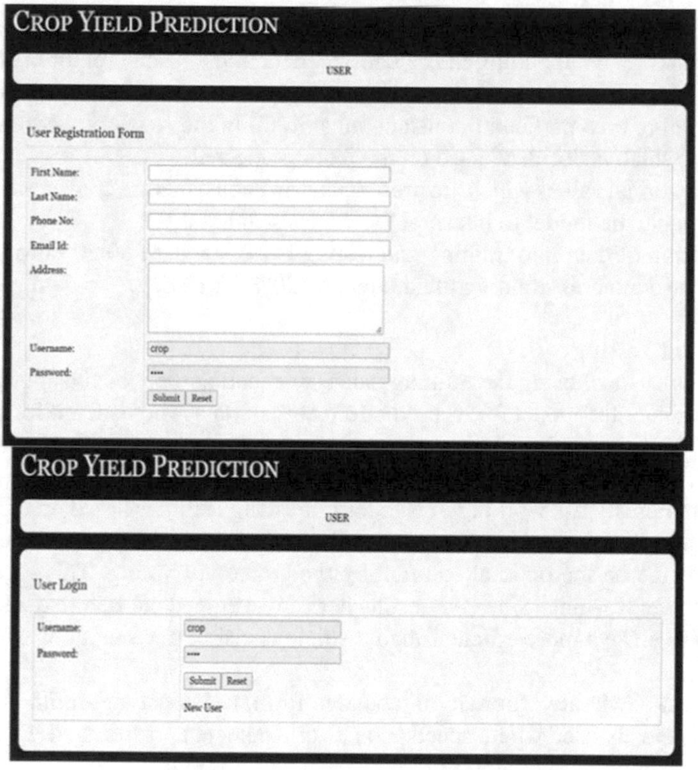

Fig. 4. UI for Registration and Login

Once the login process is successful user can enter inputs like area (Hectare) and crop. Depending on the input, weather information for the particular attributes is retrieved and the model depicts the prediction results on the screen. Result is the yield prediction in Tonnes. Figure 5a and 5b are screenshots of the User Interface (UI) which shows the yield prediction.

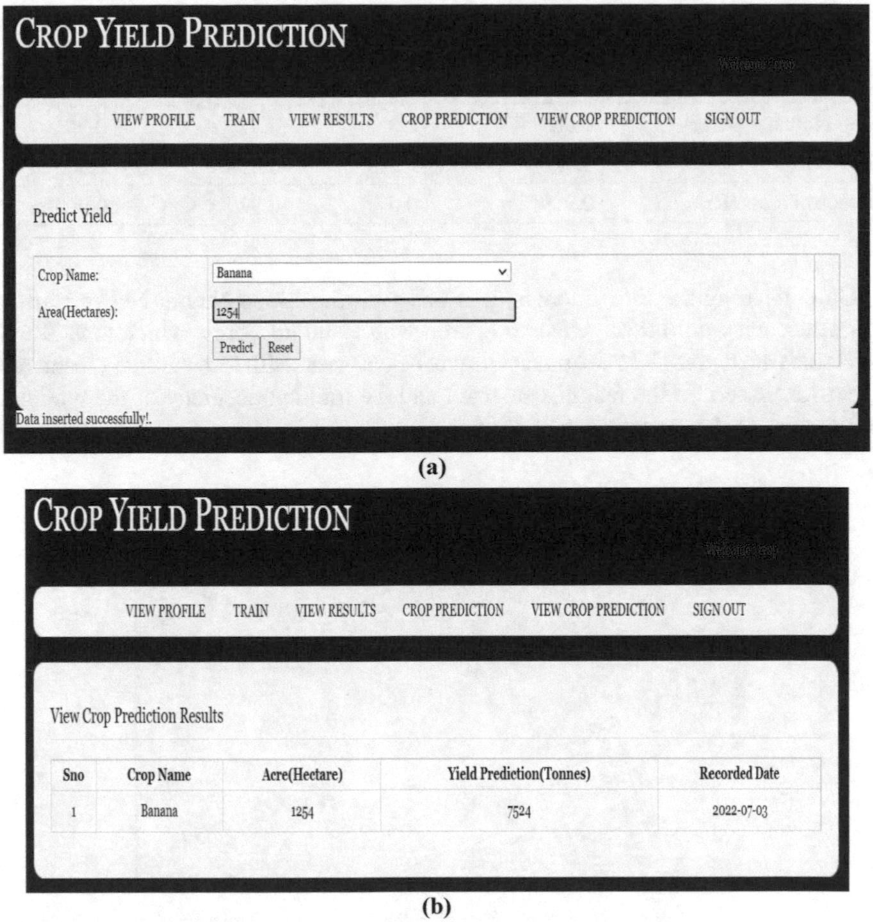

(a)

(b)

Fig. 5. (a) UI for Data selection (b) UI for Yield Prediction

Visualization

Data visualization is the graphical representation of the information. In this work, visualization involves the graphical representation of the Deep Neural Network model for accuracy and loss.

5 Results

The models considered for training using the dataset are Random Forest, Extreme Gradient Boosting and Deep learning model. Results obtained from training are depicted here.

The metric values for each of these models are as shown in below Table 1.

Table 1. Metric Values

Model/Metrics	Accuracy	Loss	Precision	Recall
Deep Neural Network	0.96	0.3	0.97	0.95
XG Boost	0.933	0.511	0.958	0.965
Random Forest (RF)	0.929	0.5	0.929	0.963

On comparing the three models, it is observed that Deep Neural Network has the highest accuracy with 96% when compared with Random Forest which is 92.9% and XG Boost which gives 93.3% and results can be observed with loss, prediction and recall as metrics. Based on the features selected and the training accuracy of the model, the performance of the model can be determined.

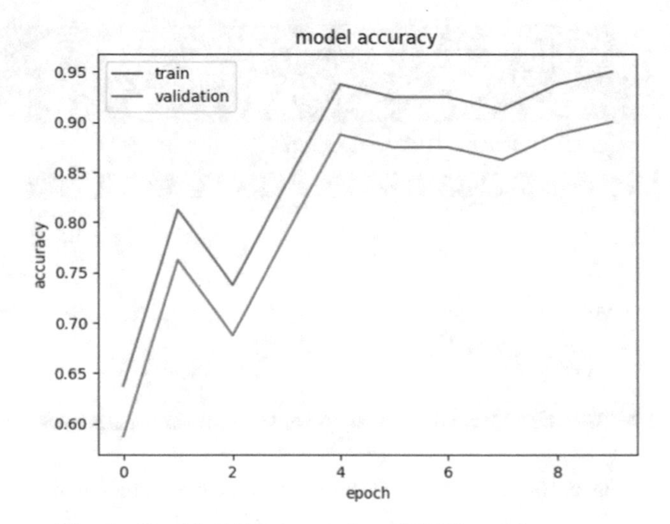

Fig. 6. Graphical Representation of DNN model accuracy

The performance of the model can be visualized in the graphical representation of model accuracy as shown in Fig. 6. The blue line indicates the training accuracy and the orange line represents the validation accuracy. Number of epochs is plotted on the X-axis and on the Y-axis the accuracy values are plotted. As visualized, the accuracy increases rapidly up to 10 epochs, indicating that the network is learning fast and flattens in the end indicating that not too many epochs are required to train the model further. The training data accuracy is increasing while the validation data accuracy is below which encounters over-fitting and indicates that the model is started to memorize the data.

Figure 7 shows the graphical representation of the DNN model loss. The blue line is the training loss and the orange line represents the validation loss. On the X-axis, the number of epochs is plotted and on the Y-axis the loss values are plotted. While making the predictions, the weights and biases stored at every last epoch is used. So the model is trained completely till the specified number of epochs, which is 10 epochs. And the

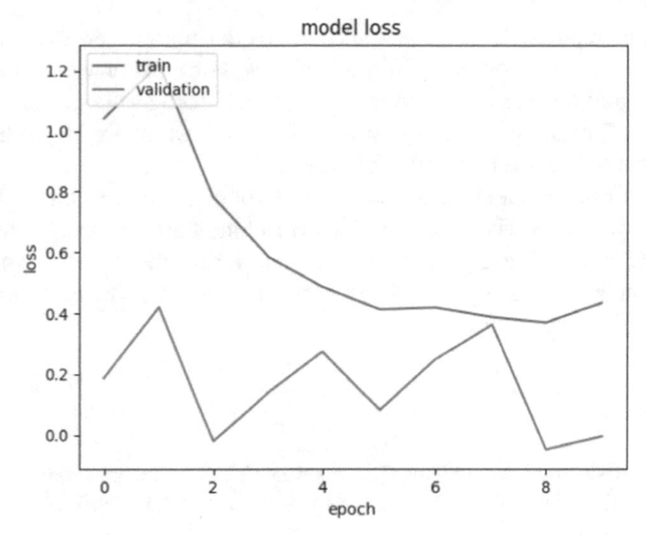

Fig. 7. Graphical Representation of DNN model loss

parameters learned during the last epoch are used in order to make the predictions. For 10 epochs, the training loss, training accuracy, validation loss, and validation accuracy is used for predictions.

With respect to Loss as visualized, the loss on the training set decreases rapidly up to 8 epochs. For the validation data, the loss does not decrease at the same rate as the training set, but remains almost flat for 10 epochs. Thus the model is generalizing well for the unseen data.

6 Future Scope

The current work considers a subset of features to build a prediction model. In future, additional features of soil attributes, the looseness variant that determines the pressure in the soil, images which are retrieved from satellite communication can also be considered to build a much more efficient model.

Limited crops are being considered and therefore in addition to this the model can be trained to include more variety of crops. The web application concentrates on the English language but in reality the work carried out is more appropriate to stakeholders who prefer to use regional languages. Such an application would be deployed on a larger scale to reach out to additional sections of people. Another extended application of this could be plant disease classification using CNN. The model can also be trained to give possible remedies to the identified disease.

7 Conclusion

The problem for prediction is difficult even today as there is no particular solution completely reliable. Farmers as of now are not equipped to that involve novel techniques in their cultivating practices. Hence, the work carried out helps farmers to use technological solutions rather than depending on old techniques. For a better yield, it requires a lot of information and knowledge into numerous aspects such as the temperature, rainfall, wind speed etc. Educating farmers on usage of technology involving prediction models can lead to even better results in real time scenarios.

The deep learning model works better in terms of performance and prediction results with the accuracy of 96% when compared with machine learning models such as Random Forest and XG Boost. These predictions can help farmers to have the knowledge that can multiply their earnings by increasing production yield of the crops and also know the best suitable crop to be grown.

References

1. Jadhav, A., Riswadkar, N., Jadhav, P., Gogawale, Y.: Crop recommendation system using machine learning algorithms. Int. Res. J. Eng. Technol. (IRJET), **09**(04) (2022). ISSN 2395-0056
2. Prasad, N.R., Patel, N.R., Danodia, A.: Crop yield prediction in cotton for regional level using random forest approach. Spat. Inf. Res. **29**(2), 195–206 (2020). https://doi.org/10.1007/s41324-020-00346-6
3. Dang, C., Liu, Y., Yue, H., Qian, J., Zhu, R.: Autumn crop yield prediction using data-driven approaches:-support vector machines, Random Forest, and deep neural network methods. Can. J. Remote. Sens. **47**(2), 162–181 (2021)
4. Shetty, S.A., Padmashree, T., Sagar, B.M., Cauvery, N.K.: Performance analysis on machine learning algorithms with deep learning model for crop yield prediction. In: Jeena Jacob, I., Shanmugam, S.K., Piramuthu, S., Falkowski-Gilski, P. (eds.) Data Intelligence and Cognitive Informatics. AIS, pp. 739–750. Springer, Singapore (2021). https://doi.org/10.1007/978-981-15-8530-2_58
5. Suresh, G., Kumar, A.S., Lekashri, S., Manikandan, R.: Efficient crop yield recommendation system using machine learning for digital farming. Int. J. Mod. Agric. **10**(1), 906–914 (2021)
6. Janmejay Pant, R.P., Pant, M.K., Singh, D.P., Singh, H.P.: Analysis of agricultural crop yield prediction using statistical techniques of machine learning. Mater. Today: Proc. **46**, 10922–10926 (2021)
7. Nathgosavi, V.: A survey on crop yield prediction using machine learning. Turkish J. Comput. Math. Educ. (TURCOMAT) **12**(13), 2343–2347 (2021)
8. Sharifi, A.: Yield prediction with machine learning algorithms and satellite images. J. Sci. Food Agric. **101**(3), 891–896 (2021)
9. Agarwal, S., Tarar, S.: A hybrid approach for crop yield prediction using machine learning and deep learning algorithms. In: Journal of Physics: Conference Series, vol. 1714, no. 1, pp. 012012. IOP Publishing (2021)
10. Kutsenogiy, P.K., Kalichkin, V.K., Pakul, A.L., Kutsenogiy, S.P.: Machine learning as a tool for crop yield prediction. Russ. Agric. Sci. **47**(2), 188–192 (2021). https://doi.org/10.3103/S1068367421020117
11. www.width.ai/post/advantages-of-deep-learning. Accessed 10 Nov 2021
12. www.machinelearningmastery.com/improve-deep-learning-performance/
13. https://agricoop.nic.in/sites/default/files/Web%20copy%20of%20AR%20%28Eng%29_7.pdf
14. http://tradingeconomics.com/india/gdp-from-agriculture. Accessed 06 Aug 2019
15. LNCS Homepage. http://www.springer.com/lncs. Accessed 21 Nov 2016

Semantic Segmentation on Land Cover Spatial Data Using Various Deep Learning Approaches

Rashmi Bhattad[1](✉), Vibha Patel[2], and Samir Patel[3]

[1] Gujarat Technological University, Ahmedabad, India
rashmibhattad26@gmail.com
[2] Vishwakarma Government Engineering College, Ahmedabad, India
[3] Department of Computer Science and Engineering, Pandit Deendayal Energy University, Gandhinagar, India

Abstract. After AlexNet's huge success at the ImageNet Large Scale Visual Recognition Competition, 2012, deep learning techniques are arguably the most ground-breaking work in the computer vision/deep learning community in the last few years. Later UNet took the attention starting from biomedical images. Which has predicted pixel-wise data by contracting and expanding the image back to the original. Deep neural networks are popular due to their automatic feature extraction capability while deep semantic segmentation networks are popular due to their encoder and decoder architecture. In our experiment, we are using an ultra-high-resolution DeepGlobe dataset to predict granularity in the images. For low-level pixels, an information prediction encoder is used while for high-level semantic information identification decoder architecture is used. This decoder architecture also enhances encoded input feature maps to original input size feature maps by using skip connections. These skip connections add encoded feature maps with targets to make sure that there is no loss of information. This way we can use spatial information effectively. Our experimentation lies around various models such as attention networks, residual networks, and SegNet keeping on UNet as its base structure. Also, skip connections are generated in a novel way. Additionally, a comparison of each of these models is provided. Our aim is not only to give accurate results but also to compare the behaviour of various methodologies and parameters affecting the segmentation prediction.

Keywords: Semantic Segmentation · Land Cover Data · Spatial Data · UNet · ResNet · Attention module · SegNet

1 Introduction

Deep learning techniques are providing excellent results not only in "computer vision" but also in the domain of remote sensing image processing [26–28]. Semantic segmentation is a step in image processing where each pixel value

H. Sharma et al. (Eds.): ICIVC 2022, PALO 17, pp. 543–555, 2023.
https://doi.org/10.1007/978-3-031-31164-2_46

is compared with available pixel classes to determine which one best matches it. Pixel wise segmentation is required as land cover maps do not have particular fixed size shape like in object detection and hence for accurate prediction and anomaly detection it is mandatory to go with semantic segmentation. While doing this, it's also crucial to analyse the data contained in the image; to do this, you may either analyse the data immediately around the image, which is class classification, or you can look at the pixel alignment, which may also include an object's forms or shapes. This is referred at localization of image data. This data mapping is done with the help of respective image masks. Traditional models were focusing mainly on object localization where even if data is dislocated it should identify shape appropriately. While many recent segmentation algorithms solves these problems hand in hand. But this might decrease the classification performance.

The deep learning techniques basically gives solutions to these problems via deep encoder-decoder networks [33]. And we have experimented with many variations of this encoder-decoder design. Additionally, it has been tested with a variety of settings including batch normalisation, dropout, max pooling, activation functions, picture cropping, and rescaling. As in land cover dataset there are no fixed shape boundaries to improve localization. Hence, we have used the transfer learning approach introduced by [22,23] to understand and categories the boundaries appropriately in our residual structure. Residual structure is used in ResNet module. These are like pre-trained weights which are used to solve boundary delineation problems. This enhancement is done in network itself and is trained end to end. Quite different approach to this is explained in CRF post processing [41].

In these cases, if a network is not learning any weight or feature then it will act as a shallow layer which could then be compared with skip connections. Hence in our work as well, primary focus is on comparing ResNet with UNet. As one of them uses identity layers while the other uses skip connections. As ResNet is based on the addition phenomenon, skip connections are based on the concatenation. We are using ImageNet pre-trained weights for this purpose in ResNet, which is also known as transfer learning. Using transfer learning into a model has proven very effective in various algorithms such [22,23]. And proposed some enhancement in skip connections.

In this research work performance of various deep learning methodologies from scratch is represented. The effectiveness of different deep learning approaches, including UNet, Attention Net, ResNet, and SegNet, is proved in this study paper. We have created scratch models of each of these strategies to help us achieve this. Section 2 describes work done by various authors for various techniques related to the field. While Sect. 3 describes various methodologies with proposed method. Section 4 describes the main experiment and results that we have derived, using the UNet, Attention Net, SegNet, and ResNet architecture. Section 5 gives Discussion and conclusion and finally, Sect. 6 concludes with a future scope.

2 Existing Work

From feature extraction to learning weight and again from back propagation to relearning deep learning techniques has given significantly good results compared to traditional machine learning methods [3]. To analyze satellite-based remote sensing images semantic segmentation plays a very important role, that may be to extract road data or change in land usage or any geographical mapping. Though there are numerous solutions given by lot many researchers in the domain still there are places to achieve accurate predictions [26–29] some of them are due to their class imbalance problem and a few because of the type of data structures [30–32] here we are dealing with the spatial type of data. For classification, traditional techniques such as support vector machines, random forests, and decision trees have proven good results but for semantic segmentation deep learning has given better results in comparative less computational efficiency [4,6,34]. For this purpose, lots of experiments are done from a fully convolutional network (FCN) to the convolutional neural network (CNN) [35].

Then UNet [4] came into the picture and has given promising results by using multi scale information. Then SegNet took that popularity by introducing innovation in max pooling indices with its various variants [7,19,20]. Along with SegNet, DSSN [6] proposed graph-based networks which were compared with deep Lab V3+ [6] and have done extensive work to improve the results for semantic segmentation. An expansion of this attention mechanism came into the picture [11,36] to highlight useful features and avoid not-so-important ones. To reduce the complexity of the attention network and to improve accuracy new concept called squeeze-and-excitation block was introduced [37] which again enhanced further for spatial type of data [36,38] which has also focused on the channel information. As UNet was compared with every other model to check its performance more work got done on UNet [39] considering its post-processing parameters. However, given the difficulties associated with dealing with spatial data when it comes to land cover data, one additional biological vision perspective model [40] was created. This model produced noteworthy outcomes in the field of medical pictures. Numerous parametric tests have been conducted on this [41]. Along with the dice coefficient, mean IoU plays a very important role in semantic segmentation, and to improve it graph based connection networks [6,10,21] have also been experimented with extensively.

3 Deep Learning Approaches and Proposed Methodology

3.1 UNet

To deal with semantic segmentation in remote sensing images we need to extract pixel-level information and need to consider the overall patch of that region. To get better results considering both the above conditions we need to extract low-level pixel information and high-level semantic information. To achieve this, we have U-shaped encoder-decoder architecture with skip connections named as

UNet. This UNet architecture fine-tunes feature maps by extracting low-level pixel information from the encoder while high-level semantic information from the decoder [1]. Skip connections fine tunes the feature maps learned by encoder and these learned weights are given as input to decoder, to effectively understand weights and to get a dense classification [4] this also helps to retrieve original image back at the end.

The Skip connection in UNet gives an advantage of processing multi-scale information in images. As the name says it skips a few connections and passes the output to the next ones. This can also be seen in Fig. 2. Where output of encoder batch normalization is added to cropping of decoder. Sometimes shallow layered model performs well over deep layers. This might be due to any reason such as random weight initialization but over fitting might not be the issue here. However, the appropriate use of batch normalization helps to achieve better results with few layers as well [9]. In the case of deeper networks, a few additional layers which do not contribute to learning can be replaced by skip connections.

3.2 ResNet-UNet

The deeper the network goes gives better results but at the same time it may also create a vanishing gradient problem and to solve this we have residual blocks to be added. Based on how deep a residual model is the number of times it uses the stacked layers in each encoder. This lets high level gradients to be back propagated through short networks to learn better features without or with fewer gradient problems [5]. Where residual function calculates residual and added with an identity function of input parameter, based on the concept of matrix addition. This residual is then passed through rectified linear unit (ReLU) activation function. To make sure that the model is considering useful parameters. Like, as we implement 50-layered Resnet the stacked layers in first to fourth encoders are [3, 4, 6, 14].

While decoding the output of the last layer of an encoder in a block is fed as input to the respective decoder block, to get concatenated with the output of the previous transposed upsampling output. So that, we can retrieve the original input image without loss of information. After doing lots of experiments on various layers of convolution layers with and without residual function, this has been proven that using residual functions reduces validation error [14]. This allows upper gradients to be directly back-propagated via shorter connections to learn better features. Unlike traditional convolutional networks in which the network goes in a forward direction one after the another, this network adds the output of a one base model to the next model by maintaining the dimensions as well. The base model is a two-layer convolution module. When the module goes to the next encoding layer, a dimension mismatch occurs.

Solving these residual functions could be added in two ways one is by padding zeros to match the higher dimensional data. Other is to project data with some parameter to map the change when mapping from one dimension to another.

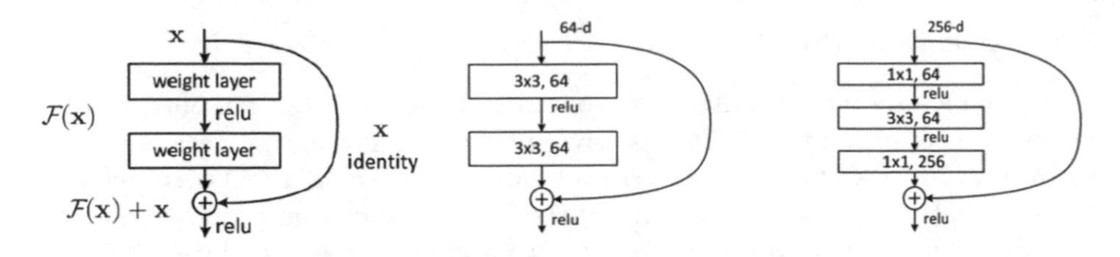

Fig. 1. Left: Residual function in general, Middle: Weight wise residual function, Right: Residual function for deeper layers

This second approach can also be extended by projecting data to all the layers wherever we are using residual mapping. This strategy, however, adds to the time and computational complexity without making a meaningful impact. Identity mappings are chosen above other mapping approaches for the reasons listed above.

$$\begin{aligned} (y_l) &= h(x_l) + F(x_l, W_l) \\ x_{l+1} &= f(y_l) \end{aligned} \tag{1}$$

working of residual function is explained in Eq. (1) where x_l is the input weight of first layer which is added with weight of respective layer via $\&F(.)$, which is nothing but the residual function. Output of this multiplication is then added with identity mapping function $h(.)$ which plays very important role in residual function. Output generated via this concatenation is then passed to activation function $f(.)$. Finally current output is then added with previous layer identity function. When this transmission occurs between different dimensional data then to handle such situations we have different ways, they are mentioned below. Where x_{l+1} is the final output of the layer.

As our residual network is 50 layers deep here instead of using two-layer base convolution as shown in Fig. 1 left, three-layer convolution is used. Where 1×1 convolution is used to reduce time complexity as shown in Fig. 1 right, as the first 1×1 convolution reduces feature map size, the second layer performs the 3×3 convolution as the last layer increases the dimension to the original as the first one. In this way even if the network goes deeper its time and computational complexity will not make the model complex and will also solve the deprecation issue. This is also referred to as feature re-usability or preserving learned features. Due to this replacement, our ResNet can generate results in the same time as that of UNet. Though it is 50 layers deep. Rectified linear activation unit is used as an activation function that activates only useful neurons. This also helps to reduce gradient descent problems. We have seen above as the residual network goes deeper; we also preserve learned features which help to solve the degradation problem up to a certain extent. This is also one of the reasons why residual networks give better results without making the module complex and are popularly used as well.

3.3 Attention Model

The attention module can deal with spatial and channel attention both simultaneously. As remote sensing images have highly complex spatial data compared to natural images our first focus is to study spatial images and abstract data from them. By adjusting the weights of the features, it can suppress and enhance useless and useful features in it. Features generated via the convolution layer are fed as input to the attention layer which then performs both spatial attention and channel attention in parallel. Spatial attention performs $1 \times 1 \times 1$ convolution and sigmoid activation is applied to the input feature to learn those spatial features. This shows the usefulness of different spatial positions and then, multiplies this spatial attention map with input to strengthen the expression of spatial semantic information.

Channel attention first performs global average pooling on input and then, two convolutions along with the sigmoid activation function calculate the channel attention mask. To calculate the usefulness of each channel its channel-wise multiplication is done with its mask and their weights will give its usefulness. Finally, the addition of both spatial and channel attention modules will be the output of the attention module for semantic segmentation. Encoding is to extract useful features for prediction, but it also reduces the input image size by applying various functions to it. Later while decoding the way, one extracts the original image back by appropriate decoding technique making it different than others. The Attention module, it is based on various gating mechanisms like the input gate, output gate, and forget gate these gates help it predict both spatial and channel data simultaneously. It also helps to deal with spatiotemporal data, but here we are dealing with spatial data only hence, it may not give that accurate results compared to other models which are designed for spatial data only.

In this model convolution output generated in the encoder is added with an output of the UNet gating module. After then, the attention gating module is used on the data. This was subsequently reversed and the initial data added back. From this data, the weights which are below the threshold are omitted to avoid overfitting and then again convoluted to generate a feature map. More caution must be used while working with the attention module during the bridge and initial layer of decoding.

3.4 SegNet

SegNet architecture is based on VGG16 network architecture, where the first 13 convolution layers are the same as VGG16. After studying various approaches for pixel-based semantic segmentation [17–20] few researchers [21] have concluded that max-pooling operation done during encoding part of an image reduces feature map resolution. They have thus developed a unique architecture for un-pooling while decoding the visual data in order to address this issue. While this experiment was performed on road scene classification data, and we have implemented it for land cover classification on the DeepGlobe dataset. As max pooling is applied at every encoder various layer of it produces more sparse data. More translational invariance is produced by this strong categorization approach and this can result in the delineation of boundaries. In solution to this [21] storing values before and after sub-sampling is required.

Instead of fetching feature map from max pooling the whole data is stored in max pooling indices format and retrieved during decoding. In some ways, this can also be called as an alternative to skip connection. Because feature maps are generated during encoding and indices are retrieved during decoding. While dealing with spatial data, to segment data based on pixel color may be challenging in some cases like the barren land and forest land. In these situations, the SegNet technique handles border delineation and provides simple end-to-end training [7].

Novelty is in the way it upsamples its low-resolution input feature maps. While decoding data instead of concatenating feature maps SegNet uses max pooling indices to retrieve the encoded feature map. These indices should match with corresponding layers in the network. Like last of the encoder should be mapped with the first layer of the decoder and henceforth. This way it also reduces the task of using additional parameters and layers to be learned for upsampling of encoded data [15]. However, SegNet model saves the indices of max pooling data to make it more effective and less complicated. Which retrieves the data accurately during the decoding or up sampling process. This approach of data retrieval makes SegNet different than others. This way it solves boundary a delineation issue as well. According to [19,21] there are two more variants of SegNet but SegNet basic which we have implemented also gives almost similar results.

3.5 Proposed Methodology

In addition to the primary UNet architecture described earlier here at last of every convolution layer additional batch normalization layer is added, as shown in Fig. 2. Dropout ignores parameters which are contributing below threshold instead of this batch normalization is used which focus more on parameters that are learning more. Though both things sound similar, we have tried both approaches and we have got a significant difference in the performance of both models. Also, this data is added with up sampled batch normalized data. This is like taking a feature map from an encoder to get the original image size back with properly learned weights. In our experiment, we handle the same by doing pixel-wise rescaling and special care is taken while resizing the data. The detailed flow of our work is shown in the Fig. 4.

Usage of batch normalization does not only avoid overfitting but also with proper weight normalization [8] avoids gradient descent as well. Concatenated data after cropping is then given for up sampling and the decoding process starts. We have added batch normalization layer because using this adds a regularization effect which sometimes reduces usage of dropout function [12]. And reduces the random weight initialization problem and helps in faster training. This ultimately

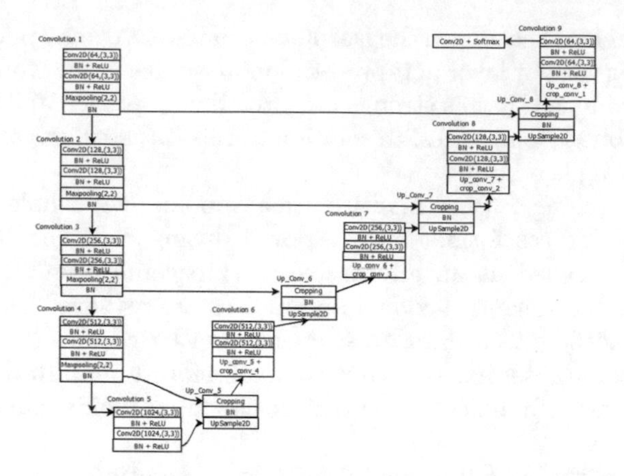

Fig. 2. Architecture of Proposed UNet

improves the accuracy and training time required for data processing. While it also helps to use higher learning rates.

As with an increasing number of layers in down sampling, channel features get doubled and input image size parameters get halved, and vice versa for up sampling. The same concept is applicable in each UNet-based structure. In our experiment during up sampling, instead of giving transposed data directly for convolution, we are applying batch normalization and then cropping, following that the cropped data is again concatenated with the output of the respective encoded batch normalized data. The reason for doing so is to make sure that data is not getting lost during image resizing and will generate a more accurate pixel prediction.

4 Experiments and Results

For land cover classification, a rich and large dataset of satellite images is given by DeepGlobe 2018 challenge at Kaggle [2]. This ultra high-resolution image data is very challenging and covers area of 1,716.9 km^2. For better visualization and a clear understanding of the deep learning technique's performance on a spatial dataset, we have done all experimentation on scratch models. The findings are presented in Table 1 after extensive experimentation on various parameters. Results are also compared with the work of certain other researchers.

From Table 1, we can clearly see that our proposed UNet has improved dice-coefficient significantly over the base UNet model. Also, it is outperforming the SegNet and Attention-based UNet models. Though here this one hot mean IoU has not shown any outperforming results when the same has been checked over the class-wise performance of a selected image they have given good results, this can be seen in Fig. 3. As shown in Table 2 our modified UNet is giving more

Table 1. Result comparison of our work with various other methodologies

Model Name	Dice co-eff	mIoU
Attention-UNet	75.66	20.56
Attention-UNet [6]	85.70	FWIoU
Attention-UNet [25]	82.60	56.91
SegNet	86.50	60.85
SegNet [6]	81.97	FWIoU
SegNet [25]	80.22	52.94
UNet-Base	78.45	42.85
UNet-proposed	**90.66**	**43.07**
UNet [6]	79.77	FWIoU
UNet [24]	88.79	Hausdorff
UNet [25]	81.83	55.70
ResNet50	92.27	57.93
Att-ResUNet	85.70	FWIoU

accurate prediction over the urban and agriculture classes. As this dataset is having lots of imbalanced data, Where agricultural data is 57.88% and urban data is 10.88% only. And hence, our results are close to those class values. Though here in land cover spatial data ResNet is outperforming where there are various medical image-related problems [4,8,24] where this UNet outperforms.

Table 2. Class wise mean IoU

Model	Urban	Agricultural	Rangeland	Forest	Water	Barren	Unknown
Attention UNet	23.41	77.12	0	0	0	0	0
ResNet	11.08	78.38	0	0	0	0	10.53
SegNet	3.37	76.11	0	0	0	0	0
UNet Modified	13.7	75.97	0.0026	0.016	0.0068	0.0038	10.28
UNet Base	0	90.51	0	0	0	0	9.48

To consider appropriate weight initialization done by batch normalization we have used ReLU [14] activation function. We have used Adam as an optimizer as it is stable when it comes to reaching local minima [11]. The actual segmentation of these high-resolution data is also shown in Fig. 5. To check performance in semantic segmentation two popularly used metrics are mean Intersection over Union (IoU) [9] and dice coefficients [10]. Mean IoU is calculated as given in an Eq. (2), which needs truly predicted positives, false positives, and false negative predicted pixels. Considering pixels of unknown class gives variation in segmentation.

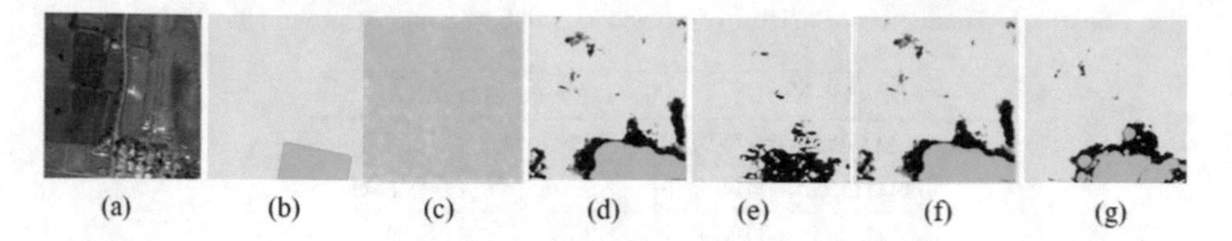

Fig. 3. Segmented images of all experimented models are given in following sequence (a) Original Image, (b) Ground truth, (c) attention, (d) ResNet, (e) UNet base, (f) SegNet, (g) UNet modified

The last segmented map of Fig. 3 clearly says that even though mean IoU is near to half it is predicting urban and agriculture data well. The pixel values with black color are unknown class pixels, these pixels are predicted as unknown due to change in class where spatio-temporal data is required. Instead of predicting pixel into wrong category our model has kept them unknown. The values which we have seen in Table 2 are very justifiable now. And also, the dice coefficient from Table 1. This does not mean that our classifier is performing to the mark, it is just that one of the classes is with the maximum occupied area, and hence others are getting divided into it.

$$mIoU = \sum_{c=1}^{6} IoU_c, IoU_c = TP_c/(TP_c + FP_c + FN_c) \tag{2}$$

The value of mean IoU closer to 1 means prediction says multi class prediction is more accurate. While on the other hand, dice coefficient is calculated as given in Eq. (3) gives the ratio of total intersected data as per predicted and mask data to whole data in predicted and actual mask.

$$DiceCoefficient = 2 * TP/(2 * TP + FP + FN) \tag{3}$$

In dice coefficient, truly positive predicted data are multiplied twice, many researchers claim to give similar results like the mean IoU itself [10]. Classifiers may get confused with small green grass as algae in images. To solve this type of problem we need 3D data or spatiotemporal data to get more textual information. Though high pixel quality gives good results, data should also be balanced.

5 Discussion and Conclusion

By looking at the images shown in Fig. 3 one might think that these predictions are not good enough as it is not exactly matched to the ground truth. However, if you look closely at the data, the black pixels are wherever there is a change in the patch or area around it. This shows spatial data can not precisely identify change and there is need for spectral data. Class imbalanced data could also one of the reasons behind these varied predictions.For class imbalance issue may be resolved by concentrating on one or two categories at once, based on the idea of the weighted average or by leveraging the previously described spatiotemporal data.

In case of Resnet, convolutional neural networks perform well but adding residual networks makes ease convergence at early stages and better results with comparatively fewer layers [14]. Like 34 layers plain network will perform somewhat like 18 layered residual networks. Hence, adding residual function in layers and that too in deep layers always performs better and gives more accurate results. This reduces validation error and greatly reduces training error too.

6 Future Scope

Though there are enormous enhancements getting done in the domain of remote sensing with respect to deep learning as well there are still some challenges that need to be resolved. Analysing an exact pixel colour is also one of the challenges when it comes to deep learning with remote sensing images. As accurate pixel identification becomes difficult when we deal with spatial data only. When batch normalization is used for both down sampling and up sampling data still there is convergence in results which leads to training error, this must be studied more [8].

Predicting classes from imbalanced data is one of the challenges in spatial type of data. A solution could be provided for this if we limit number of categories to two or three. Also, some extended usage of batch normalization in ResNet might give even more enhanced prediction.

References

1. Sahayam, S., Nenavath, R., Jayaraman, U., Prakash, S.: Brain tumor segmentation using a hybrid multi resolution U-Net with residual dual attention and deep supervision on MR images. Biomed. Signal Process. Control **78**, 103939 (2022)
2. Demir, I., et al.: DeepGlobe: a challenge to parse the earth through satellite images. In: CVPR Workshop Open Access. IEEE Xplore (2018)
3. Camps-Valls, G., Tuia, D., Bruzzone, L., Benediktsson, J.A.: Advances in hyperspectral image classification. IEEE Signal Process. Mag. **31**, 45–54 (2014)
4. Ronneberger, O., Fischer, P., Brox, T.: U-Net: convolutional networks for biomedical image segmentation. In: Proceedings of the International Conference on Medical Image Computing and Computer Assisted Intervention, Munich, Germany, 5–9 October, pp. 234–241 (2015)
5. He, K., Zhang, X., Ren, S., Sun, J.: Deep residual learning for image recognition. In: Proceedings of the IEEE Conference on Computer Vision and Pattern Recognition, Las Vegas, NV, USA, 27–30, June 2016
6. Ouyang, S., Li, Y.: Combining deep semantic segmentation network and graph convolutional neural network for semantic segmentation of remote sensing imagery
7. Badrinarayanan, V., Kendall, A., Cipolla, R.: SegNet: a deep convolutional encoder-decoder architecture for image segmentation. arXiv:1511.00561, October 2016. https://doi.org/10.48550/ARXIV.1511.00561
8. He, K., Zhang, X., Ren, S., Sun, J.: Deep residual learning for image recognition. Microsoft Res. (2015). arXiv:1512.03385v1

9. Rakhlin, A., Davydow, A., Nikolenko, S.: Land cover classification from satellite imagery with U-Net and Lovasz-Softmax loss. In: IEEE Conference on Computer Vision and Pattern Recognition (2018)
10. Daudta, R.C., Le Sauxa, B., Boulcha, A., Gousseaub, Y.: Multitask learning for large-scale semantic change detection, August 2019. arXiv:1810.08452v2
11. Zhang, J., et al.: Why adam beats SGD for attention models. ICLR 23, October 2020. arXiv:1912.03194v2
12. Ioffe, S., Szegedy, C., Google Inc.: Batch normalization: accelerating deep network training by reducing internal covariate shift, March 2015. arXiv:1502.03167
13. Santurkar, S., Tsipras, D., Ilyas, A., Madry, A.: How does batch normalization help optimization? arXiv:1805.11604
14. Hinton, G.E., Srivastava, N., Krizhevsky, A., Sutskever, I., Salakhutdinov, R.R.: Department of Computer Science, University of Toronto: improving neural networks by preventing co-adaptation of feature detectors, July 2012. arXiv:1207.0580v1
15. Noh, H., Hong, S., Han, B.: Learning deconvolution network for semantic segmentation. In: Proceedings of the IEEE International Conference on Computer Vision, pp. 1520–1528 (2015)
16. Szegedy, C., et al.: Going deeper with convolutions. In: Proceedings of the IEEE Conference on Computer Vision and Pattern Recognition, pp. 1–9 (2015)
17. Badrinarayanan, V., Handa, A., Cipolla, R.: SegNet: a deep convolutional encoder-decoder architecture for robust semantic pixel-wise labelling (2015). arXiv:1505.07293
18. Höft, N., Schulz, H., Behnke, S.: Fast semantic segmentation of RGB-D scenes with GPU-accelerated deep neural networks. In: Lutz, C., Thielscher, M. (eds.) KI 2014. LNCS, vol. 8736, pp. 80–85. Springer, Cham (2014). https://doi.org/10.1007/978-3-319-11206-0_9
19. Badrinarayanan, V., Kendall, A., Cipolla, R.: SegNet: a deep convolutional encoder-decoder architecture for image segmentation. IEEE Trans. Pattern Anal. Mach. Intell. **39**(12) (2017)
20. West, J., Ventura, D., Warnick, S.: Spring research presentation: a theoretical foundation for inductive transfer. Brigham Young University, College of Physical and Mathematical Sciences (2007)
21. George Karimpanal, T., Bouffanais, R.: Self-organizing maps for storage and transfer of knowledge in reinforcement learning. Adapt. Behav. 111–126, S2CID 53774629 (2019). https://doi.org/10.1177/1059712318818568
22. Jiang, Z., Ding, C., Liu, M., Tao, D.: Two-stage cascaded U-Net: 1st place solution to BraTS challenge 2019 segmentation task. In: Crimi, A., Bakas, S. (eds.) BrainLes 2019. LNCS, vol. 11992, pp. 231–241. Springer, Cham (2020). https://doi.org/10.1007/978-3-030-46640-4_22
23. Li, R., Zheng, S., Duan, C., Wang, L., Zhang, C.: Land cover classification from remote sensing images based on multi-scale fully convolutional network. Geo-Spat. Inf. Sci. (2021). https://doi.org/10.1080/10095020.2021.2017237
24. Mountrakis, G., Li, J., Lu, X., Hellwich, O.: Deep learning for remotely sensed data. J. Photogramm. Remote Sens. **145**, 1–2 (2018)
25. Ma, L., Liu, Y., Zhang, X., Ye, Y.: Deep learning in remote sensing applications: a meta-analysis and review. J. Photogramm. Remote Sens. **152**, 166–177 (2019)
26. Li, Y., et al.: Accurate cloud detection in high-resolution remote sensing imagery by weakly supervised deep learning. J. Photogramm. Remote Sens. Environ. **250**, 112045 (2020)

27. Li, Y., Zhang, Y., Huang, X., Yuille, A.L.: Deep networks under scene-level supervision for multi-class geospatial object detection from remote sensing images. J. Photogramm. Remote Sens. **146**, 182–196 (2018)
28. Zhu, X.X., et al.: Deep learning in remote sensing. IEEE Geosci. Remote Sens. Lett. **5**, 8–36 (2017)
29. Li, Y., Chao, T., Yihua, T., Ke, S., Jinwen, T.: Unsupervised multilayer feature learning for satellite image scene classification. IEEE Geosci. Remote Sens. Lett. **13**, 157–161 (2016)
30. Li, Y., Ma, J., Zhang, Y.: Image retrieval from remote sensing big data: a survey. Inf. Fusion **67**, 94–115 (2021)
31. Basaeed, E., Bhaskar, H., Al-Mualla, M.: Supervised remote sensing image segmentation using boosted convolutional neural networks. Knowl. Based Syst. **99**, 19–27 (2016)
32. Long, J., Shelhamer, E., Darrell, T.: Fully convolutional networks for semantic segmentation. In: Proceedings of the IEEE Conference on Computer Vision and Pattern Recognition, Boston, MA, USA, 7–12, pp. 3431–3440 (2015)
33. Noh, H., Hong, S., Han, B.: Learning deconvolutional network for semantic segmentation. In: Proceedings of the IEEE International Conference on Computer Vision, Santiago, Chile, 11–18 December, pp. 1520–1528 (2015)
34. Roy, A.G., Navab, N., Wachinger, C.: Concurrent spatial and channel 'squeeze & excitation' in fully convolutional networks. In: Proceedings of the International Conference on Medical Image Computing and Computer-Assisted Intervention, Granada, Spain, 16–20, September 2018
35. Hu, J., Shen, L., Albanie, S., Sun, G., Wu, E.: Squeeze-and-excitation networks. In: Proceedings of the IEEE Conference on Computer Vision and Pattern Recognition, USA, pp. 18–22 (2018)
36. Li, H., et al.: SCAttNet: semantic segmentation network with spatial and channel attention mechanism for high-resolution remote sensing images. IEEE Geosci. Remote Sens. Lett. (2020)
37. Alirezaie, M., Längkvist, M., Sioutis, M.: Semantic referee: a neural-symbolic framework for enhancing geospatial semantic segmentation. Semant. Web. **10**, 863–880 (2019)
38. Yong, L., Wang, R., Shan, S., Chen, X.: Structure inference net: object detection using scene-level context and instance-level relationships. In: Proceedings of the IEEE Conference on Computer Vision and Pattern Recognition, USA, 18–22, pp. 6985–6994 (2018)
39. Scarselli, F., Gori, M., Tsoi, A.C., Hagenbuchner, M., Monfardini, G.: The graph neural network model. IEEE Trans. Neural Netw. **20**, 61–80 (2009)
40. Veličković, P., Cucurull, G., Casanova, A.: Graph attention networks. In: Proceedings of the International Conference on Learning Representations, BC, Canada, May 2018
41. Maggiolo, L., Marcos, D., Moser, G., Tuia, D.: Improving maps from CNNs trained with sparse, scribbled ground truths using fully connected CRFs. In: International Geoscience and Remote Sensing Symposium, pp. 2103–2103. IEEE (2018)

A Framework for Identifying Image Dissimilarity

Rahul Jamsandekar(✉), Milankumar Jasanil, and Alwin Joseph

CHRIST (Deemed to Be University), Pune Lavasa Campus, Pune, India
rahul.jamsandekar@msds.christuniversity.in

Abstract. Images are captured to keep memories of events in and around a person's life, documents and many such use cases. After capturing images, comparing the images captured over time and finding the differences between similar captured images requires significant effort. The proposed framework aims to distinguish between two comparable photographs and computes the dissimilarity features. The model will also assist the user in recognizing the local terrain from an image and help the user take a second image identical to the first reference image. The similarity between the two images is based on various factors, including the facial structure, camera orientation, terrane, focus object on the image, background, face points, and boundaries. The proposed framework is designed to identify the similarity between two images from possible known categories like groups, portrait, and landscape images. The framework can identify the differences between the first shot and the second reference shot of the image using the extracted features from both images and will be able to display the affecting factors for dissimilarity.

Keywords: Image Similarity · Dissimilarity · Edge Detection · Face Coordinates · Detection · Recognition · Image Comparison

1 Introduction

People capture images to keep memories and gather data for various use cases. Many image processing methods and models are designed to process the images. These models are widely used by commercials, developers, multimedia applications, and other domains to process and analyze images. Some applications require an image similarity algorithm to complete their functionalities. Comparing images is one of the complex problems in image processing. Identifying the similarity between images is always a challenging task. The similarity of images is calculated using different extraction methods. These methods include Facial image extraction, background removal, and deep learning techniques. Various similarity methods are combined to obtain the overall similarity score between two images [1], and it is a standard process to combine multiple techniques to produce an efficient result.

This model can create a similarity-based framework that focuses on the dissimilarity of the images and gives the output as the affecting factors responsible for the dissimilarity present in the images. The pre-existing models on image similarity discuss the

H. Sharma et al. (Eds.): ICIVC 2022, PALO 17, pp. 556–567, 2023.
https://doi.org/10.1007/978-3-031-31164-2_47

similarity between the images. Our proposed model will be able to compute the dissimilarity between the images by performing different similarity models and give the factors affecting the similarity. The proposed model uses a category-based approach that is not commonly used in pre-existing models. Also, we are using different models used on different categories of images.

The similarity/dissimilarity identification helps to categorize the images into different groups and will be able to show the relationship between those images. Various methods work for a specific type of image, but a multimodal system where the framework that can process different types of images is limited. This model aims to help the users take a similar image to the previous one and is highly challenging. With the help of the proposed model, we could generate a framework that will bridge some issues for identifying dissimilarities from the previous images.

2 Related Work

For developing similarity algorithms, we lack the labelled data for training purposes and require highly configured systems for processing the images. In this situation, implementing a deep learning approach is always challenging. However, researchers are trying to propose and develop various frameworks for image similarity analysis. Block Similarity Evaluation is a framework designed on deep learning approaches for image Forensics [2]. It has three main parts: feature extraction, similarity, and refined network. It takes the colour image as input; the first layer crushes the image's content. Attention mechanisms are placed between convolutional layers, significantly increasing the computing process. The Siamese network, a widely used network for parallel processing, is used to extract two features that determine the similarity between two images.

Many face recognition techniques and approaches to human face simulation and recognition have been proposed to achieve high success rates for accuracy, identification confidence, and authentication in security systems. One of the most significant steps was presented [1], where the images are compared, and the image similarity value is returned, which tells how much the images are similar; if it is 0, it is identical, and if it is 1, it is similar. The Oriented Fast Rotated Brief ORB algorithm is used, which measures the similarity, and other similarity methods are used like Structural Similarity Index Measure SSIM, pixel-based similarity, Brute Force Matcher (BF-Matcher), and earth mover's distance used to obtain the similarity score.

The process is initialized by taking two resized and normalized input images fixed in size defined by the height and width parameter for smooth processing tasks of the images. Then, histograms of a grayscale 8-bit image of a vector of 256 units are obtained, which are used for comparing the images. Here, the histogram is obtained from the CDF (Cumulative Distribution Function) called a cumulative histogram. The total values for pixel intensity are up to the new bin container. Now, the similarity measure is obtained by giving the path of images as arguments that return the float value output. Then by providing the file path of the images as arguments, the similarity score between two images is computed, either 1 or 0. 1 signifies that the two images are more likely to be similar, and 0 indicates that the two images are not much alike. Using ORB, features are detected, and then the similarity is measured by finding the image key points and descriptors. BF Matcher is used to compute the match similarity between two images.

The method deals with a single person's face from a group image using a skin detection algorithm to recognize faces and extract the image's features. Binary distance is used to guide bounding boxes which signifies the limit. The input image can be created using various colour information, which plays a crucial role in extracting images; this critical role is used for segmenting regions. It is a process that divides the image into a set of areas. Colour spaces like RGB (Red, Green, Blue), HSV (Hue, Saturation, Value), and YCrCb (Luminance, red, Chromaticity, and Blue Chromaticity) play an essential role in giving better position in the skin detection process [3]. Morphological Operations like Dilation and Erosion fill the blank regions using pixel processing in the input image. Using this operation, a bounding box is acquired, which detects the boundary of the image. The face is tracked and projected by a square frame by specifying the limits of bounding boxes. After extracting the face, the GLCM approach is used for texture feature extraction, which can extract face properties projected as a square frame in the group image. These collected properties are compared with the properties from the database to identify the person from the group image. This method's advantage applies to manufacturing industries, surveillance systems, criminal identifications, and intelligence.

The model used for the real-time face detection system through the videos with the addition of detecting eyes and relation threads between two successive images [4]. Among the other different face detection approaches, a problem still exists in determining contextual information and cue combinations. In the case of two working nodes in the face detection system, the first one is After no detection, which starts after no individual has been detected in the viewing field. In this first case, two window shift detectors based on general object detection are used, which achieves better recognition rates for the lower-resolution images. For any face detected by the system, facial features that are frontal faces will be assumed, and it verifies some geometric restrictions. For recent detection(s) for every detected face, the system will store position and size, average normalized colour space, patterns of the eyes, and the whole face. Eye tracking is also applied to the area surrounded by previously detected eyes which is only possible if available. Faces recorded recently are also searched in the area covered by the recent detections, which helps reduce tracking shift problems. The main advantage of this approach is that it considers multiple face detections since no restrictions are placed on it. But this approach is still not able to detect robust facial features and non-frontal poses.

A method is developed to search for images with the target pose described by the user's graphical input [5]. When we compare the two images and find the difference, this paper mainly describes the numerous movements and activities in the photographs. A computer vision algorithm is used to examine their diverse positions and viewpoints. The SVM, Random Forest, and GaussianNB models compare the matching poses. By creating a user interface and dragging the pose the user wants to be found in the database, that drawable figure is taken as an image input. A web SRII function launches an HTTP request to the web server hosting the pre-trained model. The web server then calls the model, which was trained using the dataset of all preprocessed photos. The model returns the image that is the closest and then goes back to the web server. Following the request's interpretation by the web server, a request is made to query the database on our database server, where we store and make accessible permanent data (our preprocessed photos).

After receiving the request, the database server sends the requested image back to the web server. The client browser can then render the page to display the associated image when the web server has constructed the answer and sent it back through an HTTP response. So here is the result: the SVM algorithm pre-built model gives a more accurate image than the Random Forest and GaussianNB algorithms.

Measuring the similarity between visual entities (images or videos) based on matching self-similarities is one of the most critical processes [6], using internal self-similarities to identify similar images and determine the similarity between images. A compact local "self-similarity descriptor" measured densely across the image or video at various scales efficiently captures these internal self-similarities when regional and global geometric distortions are considered. This leads to the ability to match complex visual input, including handling textured objects without clear borders, recognizing complex actions in crowded video footage, and detecting objects in real-world chaotic photographs using simple hand sketches.

In our approach, both photos are taken as input and, using geometric terminology, identified the coordinates before determining the difference between the two coordinates to find similarities between one image and the second image. The similarities inside each image can be rapidly discovered using a straightforward SSD (Sum of Square Difference), which creates local self-similarity descriptors that can now be compared between images. However, it is not required to do so to identify the similarities existing inside each image. Measuring similarities between photos can be overly complicated. Local terminology links the surrounding image region to the image patch centre. The relative geometric positions of each local descriptor are preserved in a single global "ensemble of descriptors" made up of all the local descriptors in the tiny template. A successful match is indicated by discovering an ensemble of descriptors in a great template, like the small templates, in terms of their descriptor values and relative geometric positions.

Face Detection and Recognition Using OpenCV [8]. This paper presents the main OpenCV modules and features based on Python. The paper also presents common OpenCV applications and classifiers used in these applications, like image processing, face detection, face recognition, and object detection. Finally, we discuss some literary reviews of OpenCV applications in the fields of computer vision, such as face detection and recognition, or recognition of facial expressions such as sadness, anger, happiness, or recognition of the gender and age of a person. This study discusses one-millisecond face alignment with an ensemble of regression trees [9]. This paper The issue of face alignment for a single picture. We demonstrate how to estimate the landmark positions of the face directly from a sparse subset of pixel intensities, resulting in super-realtime performance and high-quality forecasts. We provide a broad framework based on gradient boosting for learning an ensemble of regression trees that naturally accommodates missing or partially labelled data and minimises the sum of square error loss. We demonstrate how effective feature selection is made possible by applying priors that take advantage of the structure of picture data.

Recognizing the indoor scene based on the object detection dataset to train a scene dataset [7]. The proposed scene recognition technique is evaluated with various models trained to assess which performs better for a new dataset specialized in indoor space. High accuracy is required to apply scene recognition in indoor situations. This approach uses convolutional neural networks (CNN) that are 50 layers deep, transferring the learning with the new dataset's training. The new dataset consists of 15 files containing 5,000 training photos and 100 validation images. These 15 categories include the following: elevator, corridor, chemical lab, computer lab, classroom, conference room, lobby, gym, library, locker room, office, and cafeteria. The "ImageNet" dataset [10] is an input, followed by some transformation. Then, split into the training and testing datasets and apply the CNN Model. In this model, use the CNN algorithm and continue working at a learning rate with each model until the minimum loss is found. When the model finds the minimum loss, it gives the label classifier in the image file. After that, the following process is sending a fingerprint map for indoor localization. The architecture comprises the cluster map selection, name matching, and query name to produce the fingerprint map. Getting the scene name and comparing it to the setup list is known as a "Query name", and later initialize the kNN, Euclidean distance estimation is used to estimate the error distance between the user RSSI and the database RSSI.

3 Methodology

The proposed methodology for the identification of similarities and showing the dissimilarity can be categorized into three steps:

a) To apply image classification on both the images that are "Base Image" (BI) and the live video feed which is used while capturing the "Second Reference Image" (SRI). Categories a variety of images can be divided into three categories which are – 1) Portrait images, 2) Group images 3) Scene-based images.
b) To apply computer vision technologies and statistical methods to classify the images according to their categories.
c) To display the factors which determine the similarity between two images. The affecting factors can be as follows:

   a. Image Alignment
   b. The object is missing
   c. Persons are mission
   d. Full body poses mismatch
   e. Face alignment mismatch
   f. New objects are added
   g. Objects misplaced
   h. Mismatch of patterns in people.
   i. A new face is added
   j. Exposure mismatch.

We can initiate the model by providing an image for which we have to detect the image similarity. Then the image will be classified into one of the three categories, as shown in Fig. 1. And the live feed will be provided to the model to detect the factors affecting the similarity. The initial step for the model is to apply an image classification algorithm. When the model receives the first image, then that image will be classified as the predicted category. When we click for the SRI, the model will recognize the live feed from the camera and based on that, will display the classification categories in real time.

For BI classification, the model will take the image as an input and use computer vision-based libraries to detect objects and faces in the photo. Initially, the model will detect if there are any faces inside the photo. Suppose there is only one face inside the photo. The model will again classify the image as a portrait with the pose estimation coordinates, which will be extracted using machine learning algorithms. And suppose the single face is not detected. In that case, the model will detect that all possible objects in the image the detected objects are indoor scene objects like cupboards, chairs, dining tables, or other pre-defined indoor scenery objects. In that case, it will be classified as an indoor scenery image with its respective object position coordinates, which will be highlighted using a bounding box. And the pose estimation coordinates of the human and facial coordinates in the portrait image using machine learning methods. And if detected objects in the image are like trees, plants, and natural pre-defined outdoor objects which the help of machine learning libraries, then the photo will be classified as an outdoor scenery image with the object position coordinates using computer vision methods and the pose estimation coordinates of the human, and face will also be stored for affecting measures estimation.

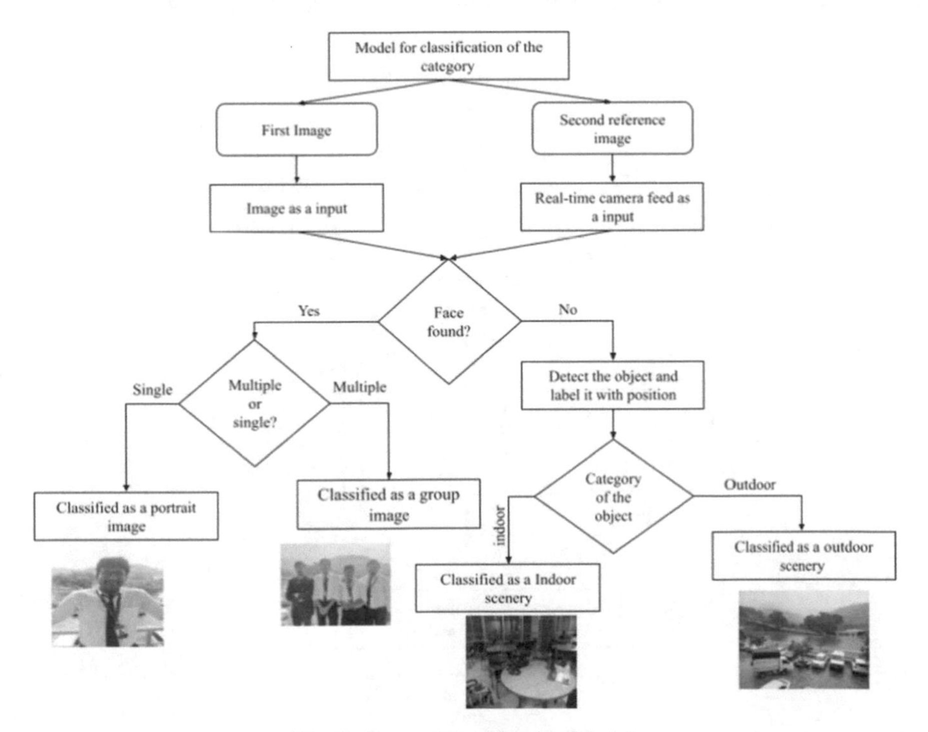

Fig. 1. Image Classification Model.

For SRI classification, our model proposes to give the classification in the live feed whenever the camera's feed changes, and then it will display the classification category. If the model finds one face, it will guide with the classifier name as a portrait image. If suddenly two or more faces come into the camera feed, the model will change the classification to a group photo, like the real-time BI classification method. Suppose there is no face in the camera feed. In that case, the model detects the object in real-time, and if detected objects are like plants, trees, roads, or any other outdoor scene objects, the classification name in the feed is changed to an outdoor scene, and if detected objects are like dining tables, bedrooms, tables, chairs, or any other indoor scene object, the classification name in the feed is changed to an indoor scene. Object classification can be obtained using predefined machine learning libraries.

If the model finds two or more faces in the photo, then the model will enable an eye-tracking algorithm to detect eyes from FI. For this eye tracking purpose, the model will first take the face image and extract the eye image, then invert it, which will convert it to grayscale and apply the Erosion Transform. Then the model will use a binary filter taking the threshold value of 220, and then find the most significant object with its center point coordinates and height value, and in the end, it will highlight that circle which will detect the pupil from the eye. From the pupil's position, the model can detect whether the person is watching towards or away from the camera. If the multiple faces are looking toward the camera, the model will predict that the image is a group image.

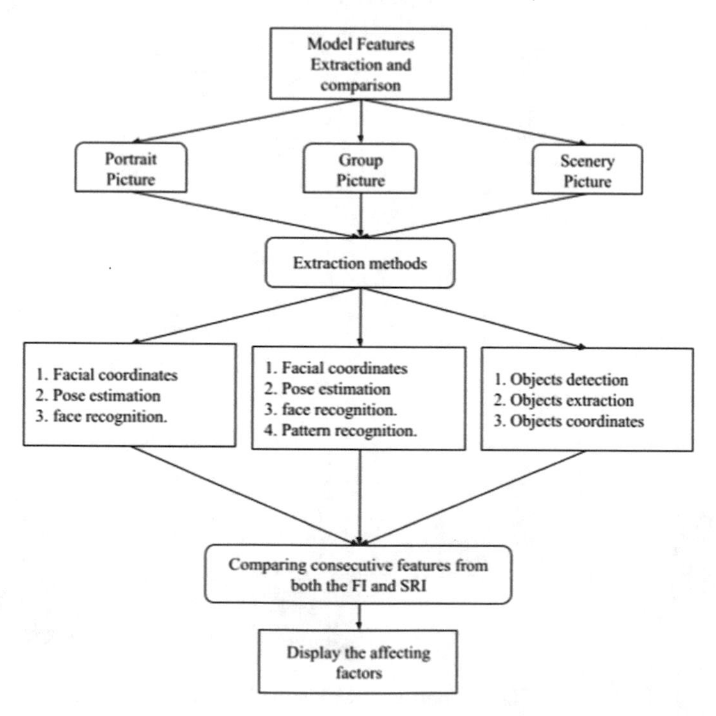

Fig. 2. Feature Extraction Model.

If the model cannot find the face inside the photo or if the face is detected. Still, if that person's pupil is not facing the camera, the model will detect all the objects in the scene with their position coordinates. First, the model will detect every object inside the captured photo. The model will use the computer vision library to recognize every object. After detecting every object, the model will label every object inside that image. Suppose detected objects are indoor scene objects like cupboards, chairs, dining tables, or other objects. In that case, it will be classified as an indoor scenery image category. If detected objects in the image are like trees, plants, and natural outdoor objects, which the machine learning libraries predefine, then the photo will be classified as an outdoor scenery image; after classifying the image, the model will detect the position or location of every object. The model will use the geometric phrases (Faces, Edges, and Vertices) concept to find the position or location of every object in the photo. This is how the model will work for all three categories and extract the features from the images with the help of the features extraction method described in Fig. 2 (Table 1).

Table 1. Models applied in the framework for image extraction and processing.

Image Category	Models
Portrait image	Mediapipe: Google created Mediapipe. This cross-platform framework offers incredible pre-built ML solutions for computer vision problems [11]
	OpenCV: A collection of Python bindings called OpenCV-Python was created to address issues with computer vision [12]
	OpenPose: The first real-time multi-person system, OpenPose, can jointly identify 135 critical spots on a single picture, including crucial points on the human body, hands, facial features, and feet. Researchers from Carnegie Mellon University suggested [13]
Group image	Face recognition: A facial recognition system is a technology that can compare a human face in a digital photo or video frame to a database of faces [14]
	Euclidean distance: A weighted Euclidean distance measure between row (or column) profiles in a table, where each squared difference between profile elements is divided by the corresponding element of the mean profile [15]
Scenery image	Yolo v7: The YOLO v7 algorithm outperforms all existing real-time object identification methods in terms of accuracy [16]
	TensorFlow: You can implement best practices for data automation, model tracking, performance monitoring, and model retraining with the help of the TensorFlow platform. Success depends on using production-level tools to automate and monitor model training over a good, service, or business process [17]
	Pillow: You can extract some statistics from an image using the Pillow module's histogram method and then use those statistics for automatic contrast amplification and statistical analysis [18]

In Portrait Image, our main task is extracting the facial and body pose coordinates, so using the mediapipe library, which was built by google, we can extract all the coordinates from the image to recognise the person. We use the OpenCV, and OpenPose libraries for the body pose to identify whether a person is the same and extract all the information for the portrait image. In group images, our main aim is to identify all the faces and pattern extraction. We use the face recognition algorithm to recognise all the faces and find the pattern created in the photo, so we extract it through the Euclidean distance. For the scenery image, our main aim is first to detect all objects inside the image and get their position coordinates. For detection of the object, we use the pre-trained model yolov7 and get the list of object labels in terms of tensor. We manipulate that tensor using the TensorFlow framework and compare it with the second image object tensor with the help of the pillow library to find the difference in the position of the objects. After detecting the live feed category as a Portrait image, the model can extract three features from the coming live feed. The first feature is to detect the human face and get the facial coordinates, the second is to get the human pose estimation, and the third feature is the face extraction from the live feed. The same three features can be extracted using this same model and stored independently. All these tasks can be done with the help of computer vision technologies and algorithms. The model will get the facial coordinates from both images and calculate the Euclidean distances between two consecutive points. The average of such Euclidean distances of points will be obtained and scaled into 0 to 100 values. Then the similarity scale will be defined. For example, if the similarity scaled value between BI and SRI coming from the live feed is above 90, then the model will declare that the SRI is almost similar. If the similarity scaled value is below 90, it will display the factors affecting the similarity. If the model detects human pose estimation from the BI, then the model will try to find the pose estimation from SRI from a live feed. Then the model will store the geometric pattern from the BI by storing the points, their distance between each other, and the angles between every pair of consecutive edges. After storing these parameters, the model will extract the patterns from every SRI coming from the live feed and compare the distances between the points from both BI and SRI using the Euclidean distance method, which will be scaled from 0 to 100. The lower the Euclidean scaled distance value more remarkable the similarity. The angles between the edges will be compared, and the matching percentage can be calculated for every consecutive angle of BI and SRI, which can contribute to calculating the average percentage, which will be used to calculate the overall scaled similarity between human pose patterns in range values 0 to 100. The third extracted feature is to store faces from both SRI and BI, and the model will use the machine algorithm to find the image similarity and give values that can be 0 or 1. This method is like the facial recognition algorithm [7].

If the model detects live feed SRI as a group image category, then the model will extract three features from the BI and SRI. The first feature is detecting and extracting all the images' faces. The second feature is to extract the pattern between the faces in an image, and the third is to extract the pose estimation of every person in the group. We can extract the faces and use face recognition for every person in the group to recognize whether every person exists in SRI. The human position coordinates can be stored with the essential reference to the person's face, making it easy to compare with the SRI.

Then we can find the pattern between the faces extracted using the second feature. The pattern can be drawn by creating the bounding box for every face and then connecting the centroid of every such bounding box using an edge. This pattern can be extracted by storing the coordinate values.

In the same way, we can extract the coordinates from every SRI coming from the live feed by comparing consecutive coordinates from BI as well as SRI by calculating the Euclidean distance between them and by taking an average of all such Euclidean distances and scaling that value in between 0 to 100 will give a scaled similarity estimated value. After extracting the third feature, which will provide all the coordinates from the human pose estimation, we can compare the human pose of every person from BI and SRI using the SPH method. This will help to detect whether any person present in the SRI does not match the pose from BI.

For the third category, image similarity, when the scenery image is detected, three features will be extracted using machine learning methods: object detection, object extraction, and object coordinate extraction. If the extracted object image from BI is not present in SRI or if the object is misplaced, it can be detected using object coordinates that are extracted using computer vision algorithms. A computer vision library can see extra edges by subtracting SRI from BI. All detected object's faces and coordinates can be extracted using a computer vision model and python geometric libraries. Then we can display the affecting factors due to which similarity has been decreased.

4 Results and Discussion

Fig. 3. Sample of results from the prototype of the proposed model.

The proposed model will produce output according to the models running on different image categories. The output for a portrait image will give three results which can be – a) There is a face mismatch or not, b) Face alignment is correct or not. c) Position of the person (If any) is correct. According to these results, the model will display the factors affecting the portrait image similarity. Influencing factors can be like "Face alignment mismatch", "Full body pose mismatch", and "Face mismatch". The output for the group image will give three results which are – a) There is a mismatch of pattern in people or not, b) a new face is added or not c) Faces are missing or not. Affecting factors for these results can be "Persons are missing", "New person is in the frame", and "Mismatch of pattern in people". For the scenery category, the model can display the following categories – a) the Object's coordinates are not matching, b) New image pixels are added, and c) Object geometric faces and coordinates are not matched between the BI and SRI. Using this results model can display the affecting factors like "New object is added", "Object is missing", and "Object is misplaced". The sample output for group image dissimilarity can be displayed as described in Fig. 3. The main advantage of using this model is to find the reason behind image dissimilarity and give assistance to increasing the image similarity value using machine learning models. Unlike pre-existing models, the proposed model uses a selective model for the image categories rather than the same similarity model for all images.

5 Conclusion

Based on the image classification with the finding of dissimilarity between two images in our model process discussed above, we conclude that our proposed algorithm may outperform the existing models for finding the dissimilarity between two images. Our work discusses the reasons for the distinction between the two images. Our model can run different algorithms based on the category of images obtained by the image classification model and will find the affecting factors responsible for the dissimilarity between BI and SRI. Multiple models running at one time may take more computational processes, so it can take more time to process. The proposed model lacks cloth colour detection, human age gap difference, exposure detection, person identification failure while capturing images in crowded places, and object tracking and identification. There is a possibility that more affecting factors are present, but these are considered the possible elements in this model. In the future, improvements can be made to the model to identify more factors, decrease the computational process, and give more accurate results with the application-based model.

References

1. Appana, A.V., Guttikonda, T.M., Shree, D., Bano, S., Kurra, H.: Similarity score of two images using different measures. In: 2021 6th International Conference on Inventive Computation Technologies (ICICT), pp. 741–746 (2021). https://doi.org/10.1109/ICICT50816.2021.9358789
2. Wang, H.-T., Su, P.-C.: Deep-learning-based block similarity evaluation for image forensics. In: 2020 IEEE International Conference on Consumer Electronics - Taiwan (ICCE-Taiwan), pp. 1–2 (2020). https://doi.org/10.1109/ICCE-Taiwan49838.2020.9258247
3. Kumar, K.R., Praneeth, K.R., Rumani, S., Lavanyamani, A.N.S.: Department of Electronics and Computer Engineering, K L University, Guntur - 522502, Andhra Pradesh, India
4. Castrillón-Santana, M., Lorenzo-Navarro, J., Déniz-Suárez, O., Isern-González, J., Falcón-Martel, A.: Multiple face detection at different resolutions for perceptual user interfaces. In: Marques, J.S., Blanca, N., Pina, P. (eds.) Pattern Recognition and Image Analysis, pp. 445–452. Springer, Heidelberg (2005). https://doi.org/10.1007/11492429_54

5. Wyld, D.C., et al. (eds.): DMML, SEAS, ADCO, NLPI, SP, BDBS, CMCA, CSITEC – 2022, pp. 187–194 2022. CS & IT - CSCP (2022). https://doi.org/10.5121/csit.2022.120716
6. Shechtman, E., Irani, M.: The Weizmann Institute of Science. Matching Local Self-Similarities across Images and Videos. 76100 Rehovot, Israel (2008)
7. Labinghisa, B., Lee,D.M.: A deep learning based scene recognition algorithm for indoor localization. In: 2021 International Conference on Artificial Intelligence in Information and Communication (ICAIIC), pp. 167–170 (2021). https://doi.org/10.1109/ICAIIC51459.2021.9415278
8. Khan, M., Chakraborty, S., Astya, R., Khepra, S.: Face detection and recognition using OpenCV. In: 2019 International Conference on Computing, Communication, and Intelligent Systems (ICCCIS), pp. 116–119 (2019). https://doi.org/10.1109/ICCCIS48478.2019.8974493
9. Kazemi, V., Sullivan,J.: One millisecond face alignment with an ensemble of regression trees. In: 2014 IEEE Conference on Computer Vision and Pattern Recognition, pp. 1867–1874 (2014). https://doi.org/10.1109/CVPR.2014.241
10. Deng, J., Dong, W., Socher, R., Li, L.-J., Li, K., Fei-Fei,L.: ImageNet: a large-scale hierarchical image database. In: 2009 IEEE Conference on Computer Vision and Pattern Recognition, pp. 248–255 (2009). https://doi.org/10.1109/CVPR.2009.5206848
11. mediapipe: PyPI. https://pypi.org/project/mediapipe/. Accessed 10 Nov 2022
12. OpenCV-python: PyPI. https://pypi.org/project/opencv-python/. Accessed 10 Nov 2022
13. Gandhi, N.: Real-time pose estimation in webcam using OpenPose: Python 2/3 & OpenCV, Pixel-wise (2018). https://medium.com/pixel-wise/real-time-pose-estimation-in-webcam-using-openpose-python-2-3-opencv-91af0372c31c. Accessed 10 Nov 2022
14. Face-recognition: PyPI. https://pypi.org/project/face-recognition/. Accessed 10 Nov 2022
15. Euclidean Distance. https://doi.org/10.1016/B978-0-08-044894-7.01317-8. Accessed 10 Nov 2022
16. Bandyopadhyay, H.: YOLO: real-time object detection explained (2022). https://www.v7labs.com/blog/yolo-object-detection. Accessed 11 Nov 2022
17. TensorFlow: TensorFlow. https://www.tensorflow.org/. Accessed 11 Nov 2022
18. Pillow. https://pillow.readthedocs.io/en/stable/index.html. Accessed 11 Nov 2022

Work from Home in Smart Home Technology During and After Covid-19 and Role of IOT

Abdalhafeez Alhussein(✉), Baki Kocaballi, and Mukesh Prasad

University Technology of Sydney, 15 Broadway, Ultimo, NSW 2007, Australia
abdalhafeezalhussein@gmail.com

Abstract. In the modern world that has become highly digitalized, more individuals prefer to work remotely rather than perform tasks at the office. As a result, the Working from Home (WFH) concept has experienced shifts, facilitating the process of incorporating more smart home innovations. This study aims to use the systematic review approach to study the application of the smart home technologies, specifically Internet of Things (IoT), in the context of working from home during the COVID- 19 pandemic. It was found that device manufacturers, utility companies, and house planners try to market their products and services as techniques to simplify life and save time. As a result, employees can access all corporate features, incorporate sustainable approaches, and digitalize the household. Though specific challenges make smart home systems vulnerable, the benefits of such a concept are found to optimize work and operations for both employees and businesses.

Keyword: COVID19 · remote work · teleworking · home based telework · smart home

1 Introduction

As remedies to COVID-19 have been developing, technology and data collide in unprecedented. It is becoming evident that the outbreak of coronavirus has not only transformed the way many companies work but has hastened the demise of countless firms worldwide [3]. This paper explores the literature on the evolution of digitalization in light of the COVID-19 pandemic, which has resulted in the widespread adoption of smart homes [28] and is expected to continue. Furthermore, several technologies that have become especially important under these circumstances will be emphasized.

From a critical perspective, the post-COVID-19 era has hastened the trans-formation from traditional work modalities to remote working. The paradigm shift has been occasioned by the technological advancements witnessed over the years. More specifically, working from home [5] has shifted from the traditional perception of domestic workers and caregivers to more advanced professions such as banking, teaching, and engineering, among others. Though subtly, several research studies have revealed the importance of working from home [24]. In their explanations, different studies have tried to explain the benefits of employees' rising need to work from home.

H. Sharma et al. (Eds.): ICIVC 2022, PALO 17, pp. 568–579, 2023.
https://doi.org/10.1007/978-3-031-31164-2_48

By evaluating different articles about working from home in the current era, it can be deduced that home working has its advantages as well as disadvantages. Working from home in a fully smart home has introduced new possibilities for how organizations operate in addition to structuring themselves [17]. With the breakout of COVID-19, working from home has offered some employers the flexibility they require to continue their business operations while prioritizing the health and safety of their workers [4]. Indeed, before the coronavirus pandemic, there was a rising trend of working from home because various organizations had established the advantages it could bring to their businesses.

Within unforeseeable conditions, organizations are forced to digitize their working practices regardless of their resources and skills. The pandemic has provided an opportunity to investigate the sustainability of remote work [19] in a new setting where employees are confined to their homes [25]. Such circumstances, on the one hand, put pressure on both firms and employees. On the other hand, the case has blurred the lines between home and work activities and locations. The usage of smart homes has expanded tremendously as work and home areas have merged into a hybrid environment.

Thus, this study aims to explore the details of applying smart home [1] technologies within the work from home setting in the context of coronavirus-associated restrictions. The research has great potential because the technologies have gained wider popularity during the pandemic and will continue within the upward trend due to the changes in overall work-life balance. Exploring both advantages and disadvantages of the technologies will allow for thc more effective implementation of them in practice during the post-COVID-19 era.

2 Theoretic Framework

Remote working or teleworking are terms that are interchangeably used but are related. Pioneer research was fuelled at a time when working from home meant working away from the workplace, basically utilizing technological communication substitute for commuting [8]. Currently, teleworking is defined as a form of flexible work that involves remote work, which is dependent on the use of information and communication technologies. However, modern teleworking processes entail three major forms of working.

Important forms of teleworking are home-based telework, teleworking from remote offices, and mobile telework. Home-based telework is a category of remote working that involves low or high-skilled duties that are done at home. On the other hand, teleworking from remote offices refers to working in which jobs are done far away from the main office or headquarters using communication technologies. Mobile telework incorporates the completion of assignments that are ordinarily done by those who routinely work away from their working base. Interestingly, the COVID-19 pandemic changed the notion of teleworking to mean working from home, or as it is referred to as remote working.

Today, smart homes make people's lives easier and more convenient day-to-day. Many manual tasks have become automated or considerably improved with the help of smart devices. The smart home concept entails the use of Internet-connected solutions that can be operated remotely from the smartphone or tablet [15]. The COVID-19 pandemic had a drastic influence on the future of smart home gadgets, resulting in their rapid adoption because people have begun transforming their living environments into places where they can work and study effectively [20]. The move Many organizations have been a critical factor in the development of smart houses. This has led to a surge in the creation and spread of smart home technologies, which are strongly linked to the new working circumstances.

3 Methodology

Published papers from various peer-reviewed journals linked to the development of smart homes and digitalization in the age of COVID-19 were summarized as part of the search for relevant material. Various internet databases and Google Scholar were used to conduct the literature search. Jstor (April 2022)-Two result, Science direct (April 2022) – Four result, ACM digital library (April 2022) – three results, Scopus (May 2022) – Three result, Elsevier (May 2022)- Two result, Sci-Hub (May 2022) - three results, UTS Library (March 2022) - three results, and the Free Library were among the databases used. The search was made using a specific combination of phrases and keywords (refer to the abstract). To evaluate eligibility and gather information about the study, each of the publications was examined. Since the research issues discussed are extensive, the majority are narrative studies.

3.1 Research Questions

A total five research questions (RQ) were formulated for this study. They are the following: RQ1: How the smart house technologies been adopted to serve their purpose in users' homes? RQ2: What are the characteristics of the smart home environment? RQ3: What is the role of cyber-security technologies in smart houses? RQ4: What is the relevance of Internet of Things (IoT) in the context of smart home technologies? RQ5: How the smart home context changed due to the social shifts brought by the COVID-19 pandemic?

3.2 Research Approach

A systematic review evaluation approach was used for this literature review. Its goal was to find papers that described the issue in question. The narrative review is a conventional method of examining the literature that focuses on a qualitative interpretation of past knowledge [27]. This review style may be beneficial for gathering and summarizing the body of literature on a particular topic area. Its primary goal is to present thorough information to the reader to grasp existing knowledge and underline the necessity of new research [27]. Descriptive reviews can assist researchers in generating new ideas by finding gaps or inconsistencies in the corpus of knowledge, thus allowing to construct research questions or hypotheses.

There is a small set of literature on the usage of smart home technology. Most prior studies have focused on smart house technologies from a technological standpoint. In contrast, another area of studies has explored the services that smart home products can deliver. While the research on smart homes suggests that technology can enhance living circumstances and housekeeping duties, there is little evidence that smart home shave an in fluence on work-related results. The study used a systematic approach to analyse and synthesize materials relevant to the smart home phenomenon. Furthermore, the given research discusses the concept of the smart home, its benefits to remote work, and possible threats. The study's objective is to identify future trends of smart homes and their influence on remote work.

4 Results

The literature exclusively dealing with the issue of smart home development and its relation to the future of work-from-home settings is relatively scarce. However, several articles discuss this issue within the broader topic of digitalization and the Internet of Things [13] (IoT). Several studies touch upon the issue of possible dangers of further digitalization of working conditions and its direct link to the development of smart homes, cities, and innovative building concepts. Overall, the way in which smart home technologies will help people work from home [23] in the future relies on several sets of factors, which are directly or indirectly discussed in the allocated research literature. The working environment is considered crucial in this regard.

4.1 Adoption of Smart House Technologies

There are still two basic ways to categorize and understand smart houses, based on whether they rely on living spaces or power systems [21]. Establish the first category by identifying them as homes with Information and Communication Technology (ICT). Waleed et al. (2018) pointed to the increasing cost-effectiveness of a smart home; hence, a wider range of people can enjoy the benefits of smart home and pursue remote working. Currently, numerous technologies containing sensors can be embedded in smart homes. [26] emphasize the importance of the constant development of the smart home to be up to date with the technological growth taking place over the years. This paper thus analyses the future trends of smart homes, such as their simplicity, cost-effectiveness, multi-purposefulness, and flexibility, especially since they entail the use of a smartphone. The system design is user-friendly and successfully tested and assessed by various users. However, there is a need for the integration of high-resolution cameras and more security features to enable the ubiquitous adoption of this cost-effective system for those working remotely [31]. The system should be expanded in order to be used by several people; for example, everyone in the house who owns an iPhones.

The justification for adopting smart homes is matching the supply and demand in real-time and reducing peak time, as well as assisting in integrating more distributed renewable generation into electricity systems. Scholars specializing in energy industry issues assume that energy efficiency is directly related to a home's integration of smart solution. However, this assumption is limited by the idea that a smart home is not occupied by people who can significantly affect energy outcomes [12].

Researchers have emphasized the importance of proper level automation in increasing the adoption of smart homes and using these spaces for remote work. Despite the smart homes' growing popularity and advantages, people have not vastly responded to the trend due to limited consumer demand, high device prices, and long device replacement cycles that inhibit smart home diffusion [32]. Thus, a significant challenge is linked to the disregard of social aspects and issues integrating and distributing smart home services. Researchers have previously analysed smart homes but ignored user attributes and their environments. Most approaches focus on technology entirely and experiments; hence they do not consider what an individual would need from as mart home to enjoy working and living there. According to Yang controllability, reliability, and interconnectedness significantly affect users' satisfaction. Interestingly, automation is not a defining factor because people generally seek more effective and safer options of remote feature management instead of highly automated services. Thus, users feel much safer when they can control the devices hence the need for a limited form of automation. Apartment residents prioritize inter connected ness and controllability, while house owners prefer reliability and automation. Men are more likely to be drawn to interconnectivity of devices, while women choose reliability over other traits because they are more risk-averse [32]. However, the study is limited because its main findings are based on South Korean data. Thus, future studies need to be geographically and ethnically diverse to generalize the results [32]. The study sheds light on the specific factors that will ensure the adoption of smart home services and their customization based on such characteristics as gender, age, and living conditions. The study is a foundation for customizing smart home services for remote work according to the different user preferences.

The commercial smart home platforms often limit their multi-user support, giving house residents different levels of information, as evidenced by Smart-Things [10]. Consequently, tensions and challenges arise between remote workers sharing a similar smart home. Researchers have con-ducted a mixed-methods study that included 18 subjects, primarily those pushing to adopt smart devices in their homes [10]. The re-searchers observed an outsized role of the installer of the smart home devices in terms of selection, control, and fixing of these devices. There were tensions in device selection and installation, users not knowing what to do when things go wrong, regular device usage, and the long-term changes in homes [10]. The central themes included power, agency, technical interest, and skill difference between active and passive users working in a smart home. The drivers get more functionality from working remotely, while the in- active users depend on them for their comfort [16] According to the study, the co-occupants of the smart home minimally voiced their privacy concerns. However, this forms part of the study's limitations as they had a bias in their selection of subjects, with the subjects mainly being smart home drivers [10]. They may be unaware of the severe concerns that passive users may have. From the preceding, smart home designers should include multi-user functionalities during the account creation process to streamline the process for all the residents, especially those working from home.

4.2 Relevance of Internet of Things (IoT)

IoT technologies are expected to play a significant role in the post-pandemic environment. IoT has swiftly become one of the most well-known commercial and technical terms in the light of the COVID-19 pandemic [33]. The Internet of Things, when combined with other technologies such as cloud computing and the embedding of actuators and smart sensors, makes it easier to interact with smart devices, allowing access from multiple locations, increasing data exchange efficiency, and improving storage and computing power [29]. The research by [29]. Showed that these improvements are likely to influence remote work habits and are ideally suited to make it simpler to maintain the work environment.

However, while IoT-based technology has the potential to revolutionize the way we live beyond COVID-19, further research and validation is required before widespread use (see Fig. 1). Security and privacy problems are critical obstacles to installing smart infrastructures and have been the focus of much research. Home automation attacks, for example, have grown widespread, with cyber criminals using Internet of Things (IoT) devices to gain access to a more extensive network [15]. The move to a digital office employing smart houses is being hampered by factors such as organizational inflexibility, cash-based company practices, and the digital divide in the case of small businesses.

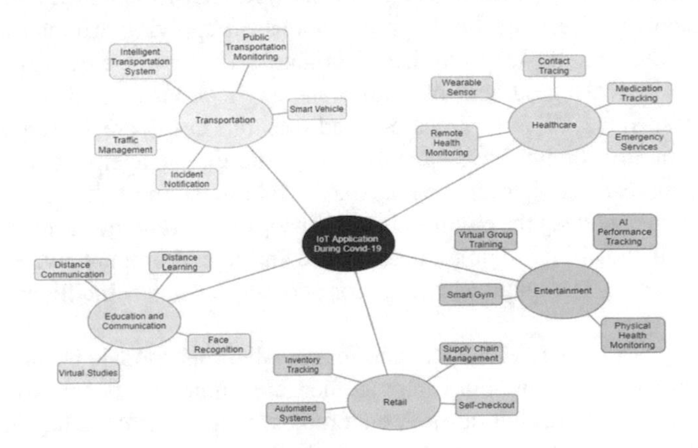

Fig. 1. IoT application: recent trends [33].

There are differences between research and the practical deployment of artificial intelligence in smart homes. Although research has been conducted to investigate artificial intelligence in smart homes, there is little information and material on how to integrate the products [14]. Evaluated smart home goods, technological trends, and the link between smart home items and literature. They perform product and literature reviews to identify the responsibilities and tasks of artificial intelligence in smart homes and examine their relevance (see Fig. 2). According to their research, the six critical tasks of artificial intelligence in smart homes are data proc processing, activity identification, picture recognition, speech recognition, projection, and decision making [14]. Activity recognition involves recognizing human activity and analysing sensor data to detect actions and alert residents of abnormal activity. Data processing entails analysing data to extract information from various sources and recognize intrinsic relationships [18]. Voice recognition is used in the famous Alexa and employs voice- driven technologies that allow the interaction between people and artificial intelligence through conversation. Image recognition can analyse the physical attributes of humans, while decision-making involves deciding which action is the most appropriate in consideration of the input data [18]. In terms of forecasting, sensors create data gathered by the computer network and archived for processing and producing trends, patterns, and fore- casts [18]. These functions are responsible for the trends in adopting smart home technology, as shown in the diagram below.

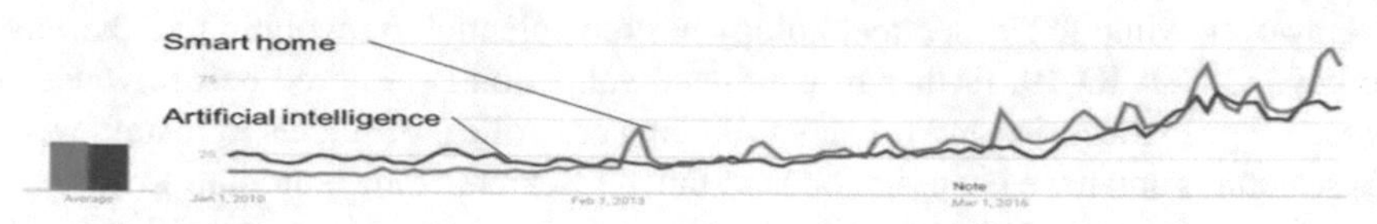

Fig. 2. Interest over time in smart home and artificial intelligence [14]

Researchers found that artificial intelligence products have six functions in smart homes: entertainment, energy management, personal robots, healthcare, intelligent interaction, and security. There are limited publications on applying artificial intelligence products, with the existing literature based primarily on new research. However, one can observe an increasing number of studies and applications of artificial intelligence in smart homes. Since 2015, healthcare-based research has decreased steadily, while intelligent interaction studies have increased [14]. Energy management is another field with intensifying research; hence, one may deduce that in the future, the trend will focus on the interactions between the environment and people, smart home customization, and increasing the sustainability of smart homes. The findings of product reviews indicated that decision-making is a popular reason for adopting artificial intelligence in smart homes.

Interestingly, artificial intelligence functions are disproportionate between the products and literature distribution. For example, there are limited studies on voice recognition and image recognition, although they encompass most artificial intelligence products [14]. Notably, when the technology matures in the coming decades, more products will utilize the various ranges of technology that artificial intelligence offers [14]. The multiple functions of artificial intelligence are expected to meet more consumer needs and enable them to interact with the technology in their everyday routines seamlessly; hence, it will positively influence remote work.

[9] address the security issues at smart homes by proposing an efficient and anonymous authentication scheme for secure communications in smart homes. They emphasize the resource constraints of IoT devices and their deployment in unmonitored and insecure environments [9]. Device identity and mechanisms for authentication help in securing IoT infrastructure. However, conventional authentication schemes require large computations that IoT devices cannot undertake due to limited resources. In addition, many IoT devices lack the processing capability and storage capacity needed to implement current authentication methods that rely on costly cryptographic techniques [2]. These mechanisms need user intervention regarding provisioning and authentication. Using password-based methods is no longer an option, especially after the Mirai IoT botnet de- strayed a considerable portion of the network in a distributed denial-of-service assault. Passwords are easily cracked, and many IoT systems lack a password authentication mechanism. As a result, there is a need for a new approach and protocol to robustly authenticate IoT agents while catering to the IoT ecosystem's environmental and architectural constraints.

Although authorization techniques for remote smart home access have been proposed, they are neither safe nor easy enough to accommodate a smart house's IoT resource-constrained components. They ignore physical context knowledge, such as geolocation and transaction history, which are critical in preventing known authentication assaults such as the Mirai attack [9] The authors of the study used anonymous authentication and integrated situational awareness with transaction records in their concept for remote access to smart homes. The proposed system's advantages are that it avoids clocking synchronization problems and does not store verification data [9]. The study revealed that the suggested strategy was more cost-effective than the others regarding bits and messages sent. However, it was less cost-effective than one of the previously proposed ones because they had more safety features and functionality verification between user and gateway and stolen smartphone threat, password reset attack, mutual authentication between customers. And gateway, and physical situational awareness [9]. Furthermore, employers can be involved with their workers' smart homes. According to [30]), the Internet of Things (IoT) technology is a dependable and organized technique for handling healthcare services delivery features such as telehealth and health care Fog-computing programs run on the edge of network devices rather than in cloud computing data centres [30]. It generates enormous amounts of data that cloud computing can process. It collects health data from various medical and IoT sensors and notifies one of any unfolding adverse events. At the network's edge, this architecture employs sophisticated technologies and services such as distributed memory, integrated data mining, and notification systems [7]. The research results on this issue show that the model is highly accurate and has a high response time another classification algorithms. The data helps companies make decisions based on the employee's real-time health care data, enhancing the proposed system's value.

5 Discussion

The literature review results indicate that the smart homes involve using various digital equipment and promoting sustainability and security. Most scholars agree that a smart home is beneficial for remote employees because it enhances efficiency and productivity [22]. However, researchers claim that smart homes involve technological vulnerability that can be abused by cyber attackers who can obtain private data for personal gain.

Nowadays, a series of events led industries and numerous businesses to adapt to the new reality, exploring innovative ways of employment. Mean-while, employees explore the concept of a smart home that utilizes in-built technologies that can be controlled remotely as some digital equipment does not need human monitoring. The benefit of such an approach is a significant increase in productivity. Thus, remote work and the smart home concept might be considered a promising combination that can gain more popularity in the future.

The current discussion about working from home is grounded on the technological advancements in the communication, science, engineering, and mathematics sector. Smart working is currently embraced by the advanced professions that were traditionally office-dependent because of the COVID-19 and guidelines that are presented against the disease. Compared to any other phenomenon, working from home has advantages and disadvantages as shown in Table 1.

Table 1. Advantages and disadvantages of working remotely from home.

Advantages of Remote Working Flexibility and activeness	Disadvantages of Remote Working Does not suit everyone
Employee retention Competitive advantage Increased productivity Increased staff motivation	Social alienation
Improved health and social welfare Reduced costs	Difficulty monitoring performance
Convenience for clients Work-life balance Increased efficiency	Possible burnouts (Lacey and Spector 2021) Initial high costs
	No staff development Information security Low staff morale
	Inconvenient to hire new workers Concerns over cybersecurity

6 Limitations

During the pandemic, various institutions advised their employees to work from home [8]. As a challenge, this paper explored information security as a challenge that bedevils smart workers. The increase in the problem of cyber-attacks is attributed to the rise in the number of those who are working from their homes [6]. Clearly, the COVID-19 pandemic period acted as the turning point for a paradigm shift in work modalities.

Since the review depended heavily on research conducted amid the COVID-19 pandemic, there are limitations regarding the reliability of the information. Ideally, people tend to be attracted to surveys if there is a monetary gain attached to them. Therefore, considering the special conditions of isolation and a challenging economic situation due to COVID-19 lockdowns, the reliability of the studies might be compromised [11]. Few samples were collected for some surveys as few individuals came forward for the survey. As such, the error is cascaded further to this review.

7 Conclusion

Incorporating smart home concepts is predicted to increase due to helpful and accessible features. Employees who work from home can obtain such features as remote network access, network security, web filtering, and a corporate management dashboard. The smart home concept provides time- efficient features that facilitate the successful implementation of work tasks. Moreover, it is beneficial to companies since smart homes for employees and remote work decrease business' costs. Smart working eliminates the need for one to travel over some distances to go to a workplace. The pros are derived from the organizations which then dependable, effective, and responsive. The benefits that both the employer and employee receive are correlational because they depend on others, and a benefit enjoyed by one party ignites the rise of another advantage.

However, the model of smart working has its short falls. A notable weakness of working from home is that the model does not match every one because of the variation in people's needs, abilities, and preferences. With disregard for the uniqueness of each individual, it is most likely that home working creates social detachment in an employee. An additional challenge to the smart working paradigm is the difficulty that arises when more than one worker is to be monitored from a control centre by the employer. Other disadvantages of working from smart working entail disruption from the home environment, the likelihood of burnout, high costs for establishing home working stations, and the inability to maintain career development.

References

1. Abdi, N., Zhan, X., Ramokapane, K.M., Such, J.: Privacy norms for smart home personal assistants. In: Proceedings of the 2021 CHI conference on human factors in computing systems, pp. 1–14 (2021)
2. Ali, O., Ishak, M.K., Bhatti, M.K., et al.: A comprehensive review of internet of things: technology stack, middlewares, and fog/edge computing interface. Sensors **22**, 995 (2022). https://doi.org/10.3390/s22030995
3. Amankwah-Amoah, J., Khan, Z., Wood, G., Knight, G.: Covid-19 and digitalization: the great acceleration. J. Bus. Res. **136**, 602–611 (2021). https://doi.org/10.1016/j.jbusres.2021.08.011
4. Birimoglu Okuyan, C., Begen, M.A.: Working from home during the COVID-19 pandemic, its effects on health, and recommendations: the pandemic and beyond. Perspect. Psychiatr. Care **58**, 173–179 (2021). https://doi.org/10.1111/ppc.12847
5. Bolisani, E., Scarso, E., Ipsen, C., Kirchner, K., Hansen, J.P.: Working from home during COVID-19 pandemic: lessons learned and issues. Manag. Mark. Challenges Knowl. Soc. **15**(s1), 458–476 (2020)
6. Butt, U.J., Richardson, W., Nouman, A., Agbo, H.-M., Eghan, C., Hashmi, F.: Cloud and its security impacts on managing a workforce remotely: a reflection to cover remote working challenges. In: Jahankhani, H., Jamal, A., Lawson, S. (eds.) Cybersecurity, Privacy and Freedom Protection in the Connected World. ASTSA, pp. 285–311. Springer, Cham (2021). https://doi.org/10.1007/978-3-030-68534-8_18
7. Debauche, O., Mahmoudi, S., Manneback, P., Assila, A.: Fog IOT for health: a new architecture for patients and elderly monitoring. Procedia Comput. Sci. **160**, 289–297 (2019). https://doi.org/10.1016/j.procs.2019.11.087
8. Diab-Bahman, R., Al-Enzi, A.: The impact of covid-19 pandemic on conventional work settings. Int. J. Sociol. Soc. Policy **40**, 909–927 (2020). https://doi.org/10.1108/ijssp-07-2020-0262
9. Fakroon, M., Alshahrani, M., Gebali, F., Traore, I.: Secure remote anonymous user authentication scheme for smart home environment. Internet of Things **9**, 100158 (2020). https://doi.org/10.1016/j.iot.2020.100158
10. Geeng, C., Roesner, F.: Who's in control? In: Proceedings of the 2019 CHI Conference on Human Factors in Computing Systems (2019). https://doi.org/10.1145/3290605.3300498
11. Georgiadou, A., Mouzakitis, S., Askounis, D.: Working from home during COVID-19 crisis: a cyber security culture assessment survey. Secur. J. **35**, 486–505 (2021). https://doi.org/10.1057/s41284-021-00286-2
12. Gram-Hanssen, K., Darby, S.J.: "Home is where the smart is"? evaluating smart home research and approaches against the concept of home. Energy Res. Soc. Sci. **37**, 94–101 (2018). https://doi.org/10.1016/j.erss.2017.09.037

13. Guan, A.L.C., Manavalan, M., Ahmed, A.A.A., Azad, M.M., Miah, S.: Role of internet of things (IoT) in enabling productive work from home (wfh) for environmental volatiles. Acad. Mark. Stud. J. **26**(1), 1–11 (2022)
14. Guo, X., Shen, Z., Zhang, Y., Wu, T.: Review on the application of artificial intelligence in smart homes. Smart Cities **2**, 402–420 (2019). https://doi.org/10.3390/smartcities2030025
15. Gupta, D., Bhatt, S., Gupta, M., Tosun, A.S.: Future smart connected communities to fight covid-19 outbreak. Internet of Things **13**, 100342 (2021). https://doi.org/10.1016/j.iot.2020.100342
16. Hargreaves, T., Wilson, C.: Introduction: smart homes and their users. In: Hargreaves, T., Wilson, C. (eds.) Smart Homes and Their Users. Human-Computer Interaction Series, pp. 1–14. Springer, Cham (2017). https://doi.org/10.1007/978-3-319-68018-7_1
17. Ipsen, C., van Veldhoven, M., Kirchner, K., Hansen, J.P.: Six key advantages and disadvantages of working from home in Europe during COVID-19. Int. J. Environ. Res. Public Health **18**(4), 1826 (2021). https://doi.org/10.3390/ijerph18041826
18. Kopytko, V., Shevchuk, L., Yankovska, L., et al.: Smart home and artificial intelligence as environment for the implementation of new technologies. Path Sci. **4**, 2007–2012 (2018). https://doi.org/10.22178/pos.38-2
19. Lacey, K., Gray, C., Spector, P.E.: Remotely stressed: investigating remote work stressors, employee burnout, and supervisor support. In: Academy of Management Proceedings, vol. 2021, no. 1, p. 15896. Academy of Management, Briarcliff Manor (2021)
20. Maalsen, S., Dowling, R.: Covid-19 and the accelerating smart home. Big Data Soc. **7**, 205395172093807 (2020). https://doi.org/10.1177/2053951720938073
21. Marikyan, D., Papagiannidis, S., Alamanos, E.: A systematic review of the smart home literature: a user perspective. Technol. Forecast. Soc. Chang. **138**, 139–154 (2019). https://doi.org/10.1016/j.techfore.2018.08.015
22. Mustajab, D., Bauw, A., Rasyid, A., et al.: Working from home phenomenon as an effort to prevent COVID-19 attacks and its impacts on work productivity. TIJAB (Int. J. Appl. Bus.) **4**, 13 (2020). https://doi.org/10.20473/tijab.v4.i1.2020.13-21
23. Purwanto, A., et al.: Impact of work from home (WFH) on Indonesian teachers' performance during the Covid-19 pandemic: an exploratory study. Int. J. Adv. Sci. Technol. **29**(5), 6235–6244 (2020)
24. Giedre Raišiene, A., Rapuano, V., Varkuleviciute, K., Stachová, K.: Working from home—who is happy. A survey of Lithuania's employees during the COVID-19 quarantine period. Sustainability **12**, 5332 (2020)
25. Rana, O., Papagiannidis, S., Ranjan, R., Marikyan, D.: Working in a smart home-office: exploring the impacts on productivity and wellbeing. In: Proceedings of the 17th International Conference on Web Information Systems and Technologies (2021). https://doi.org/10.5220/0010652200003058
26. Schieweck, A., Uhde, E., Salthammer, T., et al.: Smart homes and the control of indoor air quality. Renew. Sustain. Energy Rev. **94**, 705–718 (2018). https://doi.org/10.1016/j.rser.2018.05.057
27. Snyder, H.: Literature review as a research methodology: an overview and guidelines. J. Bus. Res. **104**, 333–339 (2019). https://doi.org/10.1016/j.jbusres.2019.07.039
28. Strengers, Y., Nicholls, L.: Convenience and energy consumption in the smart home of the future: Industry visions from Australia and beyond. Energy Res. Soc. Sci. **32**, 86–93 (2017)
29. Umair, M., Cheema, M.A., Cheema, O., et al.: Impact of covid-19 on IOT adoption in healthcare, Smart Homes, smart buildings, smart cities, transportation and industrial IOT. Sensors **21**, 3838 (2021). https://doi.org/10.3390/s21113838
30. Verma, P., Sood, S.K.: Fog assisted-IOT enabled patient health monitoring in smart homes. IEEE Internet Things J. **5**, 1789–1796 (2018). https://doi.org/10.1109/jiot.2018.2803201

31. Waleed, J., Abduldaim, A.M., Hasan, T.M., Mohaisin, Q.S.: Smart home as a new trend, a simplicity led to revolution. In: 2018 1st International Scientific Conference of Engineering Sciences - 3rd Scientific Conference of Engineering Science (ISCES) (2018). https://doi.org/10.1109/isces.2018.8340523
32. Yang, H., Lee, W., Lee, H.: IOT smart home adoption: the importance of proper level automation. J. Sens. **2018**, 1–11 (2018). https://doi.org/10.1155/2018/6464036
33. Yousif, M., Hewage, C., Nawaf, L.: IOT technologies during and beyond COVID-19: a comprehensive review. Future Internet **13**(5), 105 (2021)

Using Genetic Algorithm for the Optimization of RadViz Dimension Arrangement Problem

Samah Badawi[1(✉)], Hashem Tamimi[2], and Yaqoub Ashhab[3]

[1] College of IT, Hebron University, Hebron, Palestine
samahb@hebron.edu
[2] Department of IT, PPU, Hebron, Palestine
[3] Department of Biotechnology, PPU, Hebron, Palestine

Abstract. Visualization of high dimensional data aims to eliminate the difficulties and efforts of working with tabular and abstract data forms. One of the critical challenges for visualization methods is the dimension arrangement problem; where the final result of visualization is completely affected by the set and the order of the dimensions along the visualization anchors. According to the nature of this problem it is treated as an NP-complete problem; where optimization tools are required for solving such a problem. In this study, Researchers implemented the genetic algorithm (GA) to be used for the dimension arrangement optimization of radial coordinate visualization tools. During the testing of GA we work with a dataset of proteomic data to preserve the pairwise structural relations of the dataset instances as much as possible. We compared the result obtained using our GA optimization with some solutions obtained without optimization, and we found that our result was close to the optimal solution 4 times more than non-optimized solution.

Keywords: High-Dimensional Data · NP-Complete Problem · Genetic Algorithm (GA) · Radial Coordinate Visualization (RADVIZ)

1 Introduction

Working with large and high dimensional data sets becomes one of the critical research issues. This issue is related to the fact of the complexity and difficulty of understanding the overall behavior and content of large data sets, also because if we rely on abstract tabular and raw data form, this will be very exhausting. Therefore, we are looking for a method that represents these data sets in a more meaningful, representative, and useful way than its raw format, In order to help researchers easily gain insight into the data, analyze, and understand their content. This leads to a large number of visualization methods that are developed during last years, some of these methods are; Scatter Plot Matrix [1], Parallel Coordinates [2], Andrews Curves [3], Space Filling Curve [4], Chernoff Faces [5], Permutation Matrix [6], Self Organizing Map [7], RADVIZ [8]. The clear difference between these different visualization methods, related to the way each method represents discrete samples, the way they project high dimensional space into lower space, and the way they distribute its dimension anchors within visualization space.

H. Sharma et al. (Eds.): ICIVC 2022, PALO 17, pp. 580–588, 2023.
https://doi.org/10.1007/978-3-031-31164-2_49

Most of the visualization methods, suffer from the problem of instability of the final visualization results of a dataset; where the subset and the order of dimensions within one subset can completely affect the final result of visualization; according to their distribution over the visualization anchors, which leads to an important research question, which is how to choose the subset and its dimensions order along the visualization anchors, so to enhance the result of the visualization as much as possible, which is referred as dimension arrangement problem DA [9].

In this study, we introduce new implementation of GA to be used during the optimization of DA problem within the context of RadViz visualization tool, where we propose the complete custom definition for the GA functions, with the goal of finding a good solution based on the subset selection and the order of its dimension to minimize the required fitness function.

2 RADVIZ Visualization Tool

RadViz mainly relies on the concept of Hooke's law for mechanical spring forces; it works by treating each attribute as spring, with one end attached to a perimeter point on a circle, and the other end attached to one data sample. Each sample is set at an exact location so that the summation of spring forces on that point is equal to zero. More formally, Hooks law says that the restoring force f of a given stretched spring, is proportional to its elongation e, and this force acts in the opposite direction to its elongation; this is stated by Eq. (1).

$$f = e \times k \tag{1}$$

where k is the spring constant, and e is the elongation of that spring. The way that dimensions around RadViz visualization anchors interact with each sample is shown in Fig. 1, which represents data set of 6 dimensions with one sample, where (x1,x2, ..., x6) values represent the elongation of corresponding springs over sample p, and the final result of sample p projection into two dimension, is defined as x and y values, that computed as set by Eq. (2).

$$x = \frac{\sum_{i=1}^{d} a_i cos\theta_i}{\sum_{j=1}^{d} a_j} y = \frac{\sum_{i=1}^{d} a_i sin\theta_i}{\sum_{j=1}^{d} a_j} \tag{2}$$

where d is the number of dimensions, a_i is the value of attribute i for the current sample, θ_i is the angle of attribute j on the circle.

2.1 Dimension Arrangement Problem

One of the important challenges for RadViz, as well as all other similar visualization methods, is the problem of instability of the final visualization results for same data set. The subset and the order of the dimensions that are distributed over visualization space anchors can completely affect the final result of the visualization. This is called the Dimension Arrangement problem (DA). To clarify the problem in RadViz, we use a simple example for one sample with four features and to show the variations of the point location with respect to the dimension orders around the visualization anchors, we give three of the variations of dimension order and the change of the sample location as shown in Fig. 2.

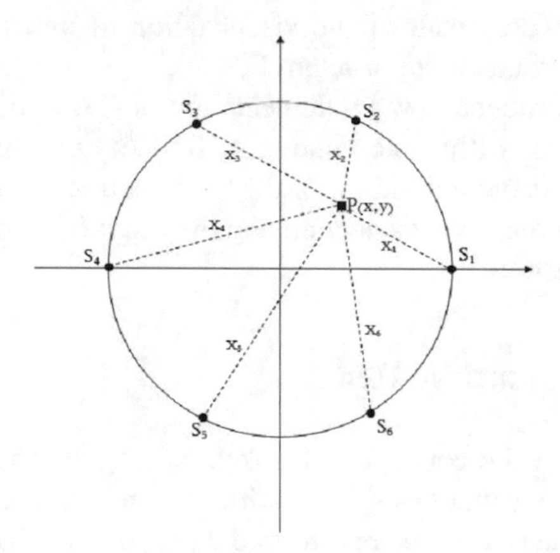

Fig. 1. RadViz with 6 dimensions over one sample.

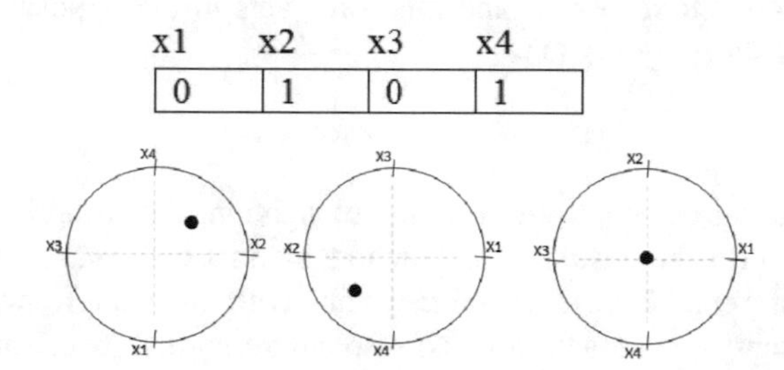

Fig. 2. RadViz visualization for different dimension arrangement of proposed sample example.

Due to the nature of the dimensional arrangement problem and to the fact that DA problem is NP-complete problem as approved in 1998 by Ankert, et al. [9]. Then to check all possible solutions according to Eq. (3) and with a large number of dimensions we have to work with a huge number of possibilities.

$$Number\ of\ solutions = \sum\nolimits_{r=3}^{p} \frac{p!}{(p-r)!} \tag{3}$$

To solve such a type of problem there are different attempts, some are exhaustive searches, when the number of dimensions that we work with is low. Other methods work with greedy and local search: In 2006 by Olivette et al. proposed a greedy method called Similarity based attribute arrangement (SBAA) [10]. This method is based on computing similarity between all possible pair of attributes. Then, starting with the highest similar pair as starting axes and continue until all dimensions are ordered according to their similarity. Another method was proposed in 2009 by Albuquerque et al. [11]. They are based on a weighted sort, where all dimensions are weighted and then sorted according to their weights..

3 Proposed Genetic Algorithm

3.1 Individual Representation

To solve the DA problem of RADVIZ in this paper we proposed a custom definition of the GA to be used to solve this problem. The individual representation is presented as a vector of length equal two times the number of dimensions (l = 2d), that is mainly divided into two equal parts. The first part is used to represent the order of the d dimensions over visualization anchors, the second part is used to represent the selected and not selected features. The representation of individuals is as shown in Fig. 3. In the initial population, the individuals are filled with random permutation of d numbers, also the second part is filled with random binary values.

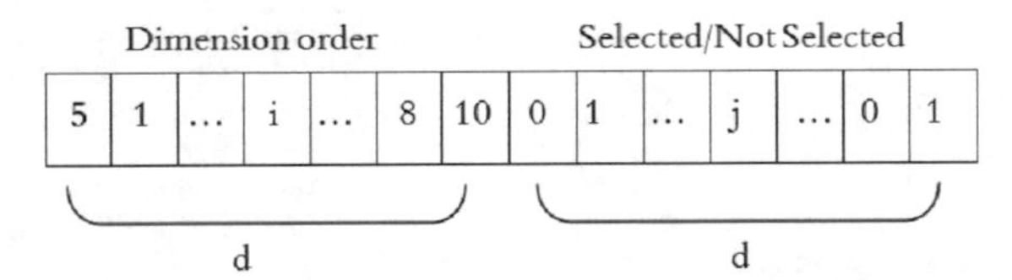

Fig. 3. Our system GA individual representation.

In Fig. 3, i is an integer number that must be greater than or equal 1 and less than or equal to d, and its value must be unique within the individual, j is a binary value that can be 0 or 1, when 1 means that the corresponding dimension in the dimension order half is applied within the current set of dimensions, and when 0 do not. To clarify the interpretation of one individual we introduce an example for data set of 5 dimensions (x1, x2, x3, x4, x5), and one example of possible individuals is as shown in Fig. 4, where it means that the current dimension arrangement is to take feature {5, 3, 4, and 1} according to their corresponding values that equal 1 in the second half of the individual that related to the selected or not selected part, and leave dimension 2 because it is not selected according to its corresponding 0 value in the second half. The order of these dimensions will be in the same order as they appear in the first half of the individual.

3.2 Cross Over Operation

In our system we need a careful definition of the crossover operation. We must force the generated children to have no duplication of the selected features. In our system the general idea for crossover is to take two random values that represent the start and the end of a block from one parent, which passes to its children, with the rest of the children cells filled from the other parent, in a way to ensure that we have no feature duplicated twice times. The process of one child generation from its parents is shown in

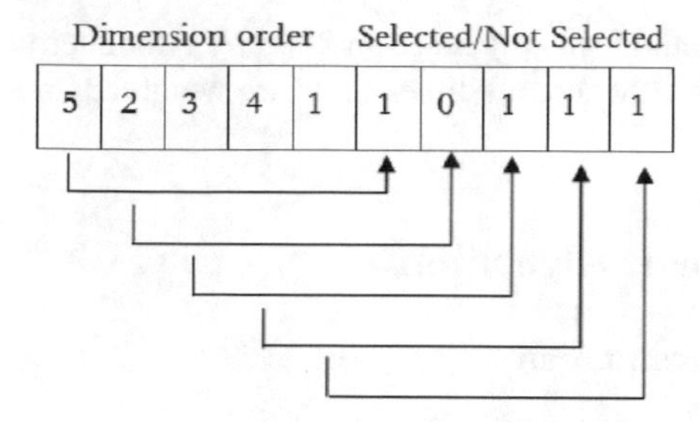

Fig. 4. One Individual representation for 5 dimensions dataset.

Algorithm 1. We take the first block from one parent (P1) to be the same in the children, then fill the rest of the children from the other parent (P2) in the way that we ensure all dimensions to be included only once. Figure 5 represents an example of crossover operation, with two parents and their two generated children, with their interpretation as RadViz DA solutions.

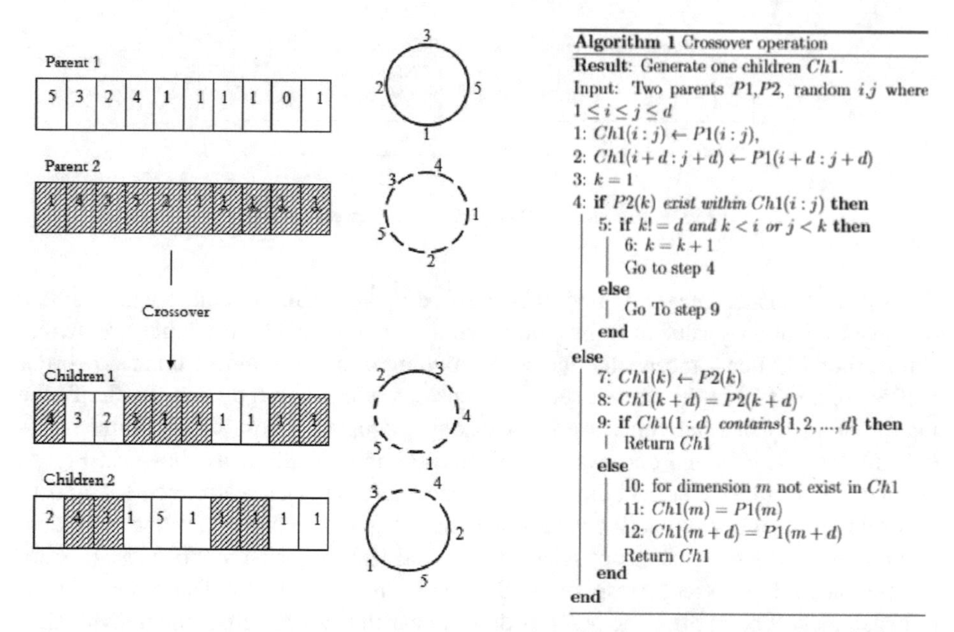

Fig. 5. Crossover operation over two parents as used in our system, with their interpretation as RadViz DA solutions.

3.3 Mutation Operation

In our system, we make a custom mutation operation that is a combination between exchange and swap operation, where the swap operation is applied on the first part of the individual that related to the dimension order objective with the corresponding selected or not selected pair of values, and the exchange operation is applied on the second half of the individual that related to the selected and not selected objective, that switches from 0 to 1 or from 1 to 0. Figure 6, shows two examples of the possible mutation over two individuals.

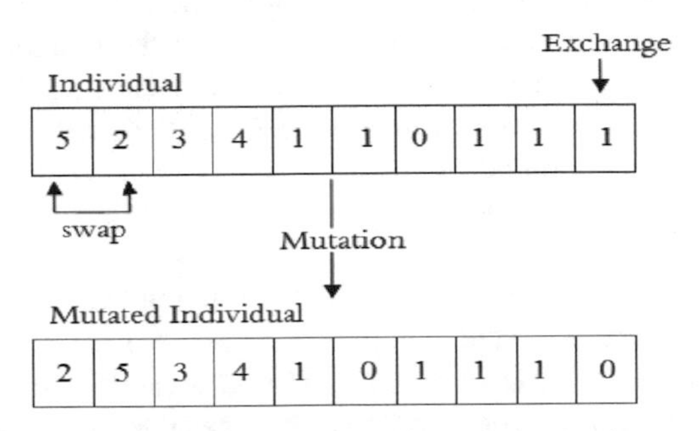

Fig. 6. An example related to the mutation operation in our system.

3.4 GA Fitness Function Definition

This is one of critical parts that directly related to our problem definition, and in GA it represents the goal that we want to minimize, so to have a good solution, In our study and for testing we work with proteomic data set and our goal is to visualizes the data instances in RadViz 2D space, where the pairwise distances is close as much as possible to their 3D Structural relations. And the following function represents the cost function as shown by Eq. (4).

$$C = \sum\nolimits_{I=1}^{n} \sum\nolimits_{j=1}^{n} \| RadViz(i,j) - 3DStructur(i,j) \| \tag{4}$$

where n is the number of data set samples, RadViz2D(i,j) is the 2D Euclidean distance between sample i, and sample j in RadViz space that is represented by (x,y) pair obtained with a specific DA solution. 3DStructure(i,j) is the structural similarity between sample i, and sample j as measured from STRUCTAL tool.

4 Experiments and Results

This section covers the experiments and the results that are related to our work. The experiments are divided mainly into three phases. The first phase is related to the GA parameters tuning. The second phase is related to the system set up phase. And the third phase is related to the testing of our system general applicability.

4.1 Experimental Settings

During the set up phase, we used the CATH500 database with 5-fold cross validation during training and testing. We used Matlab implementation of the GA with experimental tuning of GA parameters and custom implementation of the initial population, crossover, mutation, and fitness functions. We used local sequence alignment score as an indicator to measure the performance of our developed system versus direct sequence relations.

4.2 GA Parameters Tuning

During optimization we based on Matlab GA optimization tool, and tuning its related parameters experimentally that chosen as follow; first we fix all parameters except elite number, and study the effect of elite number on the percentage of early convergence before reaching the optimal solution, also we measure the relation between the elite number and the fitness fall interval, where the result was as shown in Fig. 7, and according to these results we choose the elite number to be 2% of the population.

For crossover and mutation percentage we fix all parameters except cross over and mutation, then allow the GA algorithm to operate on the selected parameters and with cross over ratio vary from 0 to 1 of steps 0.2, then we show the result for the effect of different cross over ratios over the average fitness function which is as shown in Fig. 8, also we show the result for the relation between the best fitness value, iteration number, and the cross over ratio which is as shown in Fig. 9. So according to these results, we found that the crossover ration is preferred to be between 0.8 and 0.7, so we use 0.75:0.25 crossover to mutation ratio.

For the population size parameter we fix all parameters except population size, that are tested to be 10,100,1000,10000, the result for the relation between the population size and the best fitness value, at what generation the best fitness is reached, and the time it takes to finish, which is as shown in Fig. 10. And according to the results, we found that the population size is preferred to be around 100.

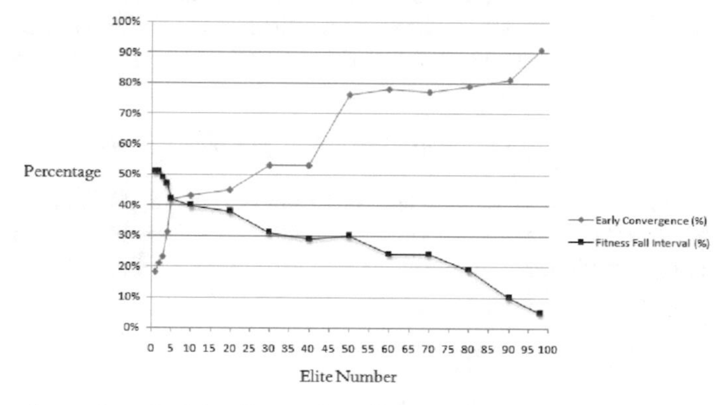

Fig. 7. Experimental result of the elite number effects over the early convergence percentage, and the fitness fall interval.

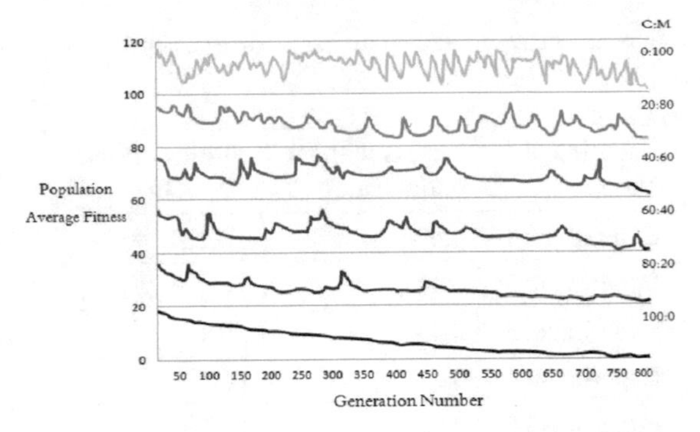

Fig. 8. Experimental result of the relation between $C:M$ ratio and the average fitness value.

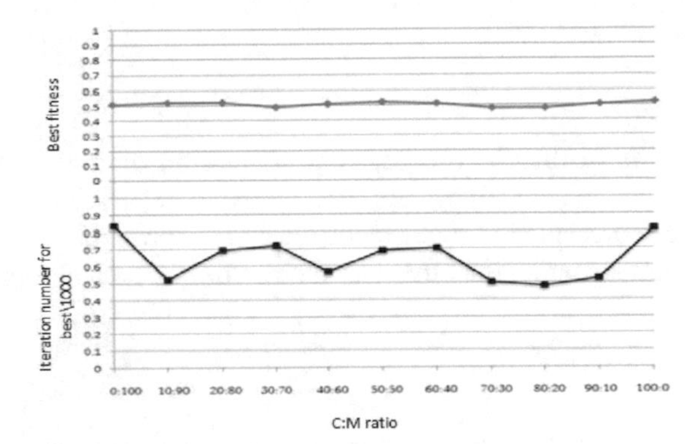

Fig. 9. Experimental result of the relation between $C:M$ ratio and the best fitness value and iteration number it reached.

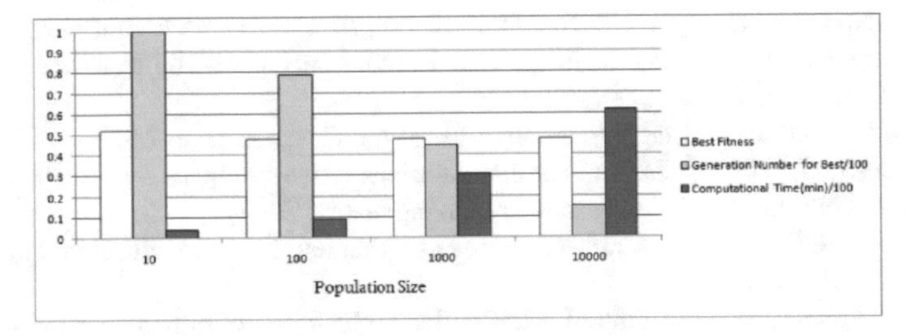

Fig. 10. Experimental result of the relation between population size and best fitness, generation, computational time.

4.3 Performance Evaluation

For testing our system performance, we perform 5 experiments, each time we compare the distance of the generated solution based on our final GA optimization process, with the result obtained from non-optimized solution, where the result is as shown in Table 1.

Table 1. Preservation of distance to the goal based on our optimized solution versus non-optimized solution.

Experiment Number	Fitness without optimization	Fitness using our GA
Experiment 1	0.47	0.13
Experiment 2	0.34	0.11
Experiment 3	0.49	0.12
Experiment 4	0.39	0.1
Experiment 5	0.46	0.09
Average	0.43	0.11

5 Conclusion

Our custom definition of the GA provides a good optimization algorithm for the DA problem to be searched, where our system visualizes data relations that in average close to the 3D structure representation by distance which is approximately 4 times closer to the 3D structure representation than that based on solutions without optimization.

References

1. Friendly, M., Denis, D.: The early origins and development of the scatterplot. J. Hist. Behav. Sci. **41**, 103130 (2005)
2. Rundensteiner, E.A., Fua, Y.-H., Ward, M.O.: Hierarchical Parallel Coordinates for Exploration of Large Datasets. pp. 43–50 (2000)
3. Spencer, N.H.: University of Hertfordshire. Investigating data with hernof plots (2003)
4. Wattenberg, M.: A note on space-filling visualizations and space-filling curves, pp. 181–186. IEEE (2005)
5. Spinelli, J.G., Zhou, Y.: Mapping quality of life with Chernoff faces (2009)
6. Hinterberger, H.: The visulab: an instrument for interactive, comparative visualization (2010)
7. Shyam M. Guthikonda. Kohonen self-organizing maps. (2005)
8. Hoffman, P., Grinstein, G., Marx, K., Grosse, I., Stanley, E.: DNA visual and analytic data mining. IEEE (1997)
9. Keim, S., Ankerst, D.A., Berchtold, M.: Similarity clustering of dimensions for an enhanced visualization of multidimensional data. In: INFOVIS, vol. 39, pp. 52–59 (1998)
10. Levkowitz, C., Olivette, H., Cristina, A.: Enhanced high dimensional data visualization through dimension reduction and attribute arrangement. In: 10th International Conference on Information Visualization (2006)
11. Lehmann, M., Theisel, D., Mangor, H., Albuquerque, M., Eisemann, G.: Quality-based visualization matrices. In: VMV, pp. 341–349 (2009)

COLEN-An Improvised Deep Learning Model for Plant Disease Detection Using Varying Color Space

Subham Divakar[1], Rojalina Priyadarshini[2](✉), Surendra Kumar Nanda[2], Debendra Muduli[2], and Ramchandra Barik[2]

[1] Persistent Technology, Pune, India
[2] C. V. Raman Global University, Bhubaneswar, Odisha, India
priyadarshini.rojalina@gmail.com

Abstract. Plant disease detection is a rapidly evolving area of research with new techniques being discovered and proposed almost every year. Most work done in this field focuses upon increasing the accuracy of the classifiers, however when it comes to real images taken from fields, the accuracy drops drastically due to varying light intensity and background of the images. In this paper a novice approach has been used to improvise the performance of the plant disease detection model. It has been seen from the tremendous experimentation that, mostly in Red Green Blue (RGB) input images, while detecting the diseased leaf images, the green channel in healthy part dominates the blue channel in the diseased part of the same leaf which is found to be one of the other factors which is reducing the accuracy. Hence, here eight widely used color spaces are considered to measure the impact of the performance of the model by changing the color space. For conducting the experimentation on the proposed idea, a public dataset is being used in varying color spaces against RGB space. From the experimentation results it has been observed that for the majority of the classifiers, for the Hue Saturation and Lightness (HSL) color space the accuracy was more than the accuracy for RGB or any other color space. Furthermore, to improvise the performance, an ensemble algorithm named as COLEN has been proposed to choose the color space with the maximum accuracy and F1 score.

Keywords: Plant Disease Detection · Deep Learning · Color Space · Ensemble Learning

1 Introduction

Farming is a process that is carried from generations in many families. In the current time period farming has seen tremendous growth in its methodology. With the latest machinery and advanced techniques, farmers are now equipped with the latest technologies and methods to boost the growth of soil. Mankind today has produced various species of plants and crops through genetic engineering that produce twice the old species. Soil monitoring and quick on the go analysis also helps farmers take quick decisions regarding

H. Sharma et al. (Eds.): ICIVC 2022, PALO 17, pp. 589–599, 2023.
https://doi.org/10.1007/978-3-031-31164-2_50

the spread of diseases in plants. The use of machine learning and artificial intelligence in analyzing the data is now adopted as a general procedure. Computer Vision or the use of images to identify specific patterns in the image is also becoming popular when it comes to applying for farming.

There have been lots of recent developments done in the area of plant disease detection. Many people have proposed deep learning classifiers and architectures that increase the accuracy of the classifiers. However, in most of the work authors have talked about the need of reducing false predictions and increasing the accuracy of the classifiers as there is shortage of diseased leaf data for any crop with lots of variations. Also, when the proposed works are tested against real-field images, the accuracy drops drastically. In our previous work [1], we proposed a novel ensemble algorithm which aims at increasing the accuracy and reducing the false predictions in case of real-field images. We also did a vast analysis of the RGB channel distribution of the dataset and observed the uneven channel distribution in case of diseased leaf images. Carry forwarding the approach and observations, the main contribution of the papers are:

1. An in-depth analysis has been carried out with respect to the effect of RGB channel distribution over the input images towards plant disease detection.
2. The impact of the use of different color space as inputs are experimented and their effect against the performance of the disease detection model is being analyzed. The performance matrix which is considered are F1 score and accuracy.
3. A novel ensemble algorithm named as COLEN has been proposed which automatically selects the best color space suited for the dataset and assigns it to the classifier.
4. A comparison has been done among the proposed models' performance with some other existing state of art methods and approaches.

2 Related Works

In this section a literature survey has been done to understand the contribution done recently in the field of plant disease detection and understand the techniques used widely. In one of the author's contributions in previous work [1], they have used Synthetic Minority Oversampling Technique (SMOTE) to balance the imbalanced dataset and proposed a novel ensemble algorithm which takes as input list of classifiers and outputs the best classifier or the classifier with the best accuracy and f1 score together. In this work, the author's main objective is to reduce the false predictions and their algorithm showed EfficientNetB7 as the best classifier. Authors have also done vast visualization of the dataset which forms the backbone of our paper as the visualization in [1] revealed the uneven spread of the RGB channel values for the first time and also it was observed that healthy leaf images have more green channel values and diseased leaf images have more blue channel values. In [2] authors have used GAN to solve the problem of shortage of dataset and proposed PDNet, a deep learning classifier whose accuracy is 93.67%. In their work authors have also outlined the drop in classification accuracy when the models trained using transfer learning approaches are tested on real field images.

Authors in [3] used a transfer learning approach to train a few deep learning models on their vast PlantVillage Dataset and achieved an accuracy of 99.7%. However, their work does not talk about testing the classifiers on real field images which is also a major drawback. In [4], authors have used transfer learning on a huge dataset of images to detect diseases in leaf images. They compared 10 pre-trained deep learning models like GoogleNet, AlexNet, VGG16 to train the bottleneck layers and compare the results. This work showed the drop in the training time of models and also highlights the importance of transfer learning approach when we have shortage of dataset or with imbalanced classes. However, they did not analyze the field images or real images with lots of variations in light intensity and background. One main concern raised by the authors was the lack of suitable dataset of the field images. In [7] authors have proposed a novel deep learning architecture ReTs to detect disease in the leaf images. ResTS is a tertiary adaptation of formerly suggested Teacher/Student architecture and authors achieved a good accuracy and f1 score close to 99%. Similarly in [8, 9] and [10] authors have proposed various machine learning and deep learning architectures and also achieved high accuracy but when it comes to real field images, the accuracy drops [11, 12].

After going through this literature, it has been observed that almost all have used only the RGB images to detect the diseases in the plants. However, in [1] authors did say about the further investigation needed to explore the RGB channel content variation which would allow them to explore other color spaces which could increase the accuracy. Though the use of transfer learning approach yields classifiers with good accuracy, when it comes to real field images, the accuracy drops drastically, but when it comes to tackling the real images most are found to be noisy due to varying background and light intensity. Further one major factor addressed in almost all the papers, the issue of false predictions. Thus, in this work the main objective is to take a different approach by experimenting and analyzing the RGB channel distribution of input images, which can enhance the accuracy and reduce the false predictions using the effect of various color spaces.

3 Proposed Work

The proposed work has divided into the following sections discussed below.

3.1 Data Collection

Two datasets were collected for our work. First is the widely available public dataset, PlantVillage which contains 38 categories of diseased and non-diseased plant leaf images. For our paper we have taken only the Apple Leaf Images belonging to 4 categories. Second dataset is from the Kaggle Competition "Plant Pathology 2020-FGVC7" [8], which contains images belonging to apple leaf diseases. It contains images belonging to 4 category of images "healthy", "multiple_diseases", "scab" and "rust". "scab" and "rust" refers to the apple scab and apple rust disease and "multiple_diseases" refer to the images having multiple diseases. In our earlier work [1] we have used dataset [6] to train a deep learning model using Transfer Learning approach.

3.2 Data Pre-processing and Visualization

The dataset [5] contains Apple Leaf Images belonging to 4 categories of "Healthy", "Apple Scab", "Apple Rust", "Apple Black Rot". An analysis has been done on the RGB channel values to see the distribution and understand the effect of the RGB channels.

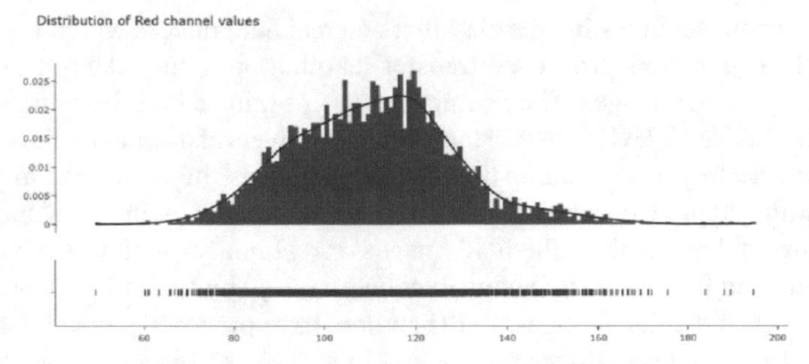

Fig. 1. Distribution of red channel in the dataset

Figure 1 shows the distribution of Red channel values across the dataset. From this it can be observed that the Red channel values are concentrated at 120 and that the graph is rightly skewed. This indicates that most of the images in the dataset are having Red channel content below 120.

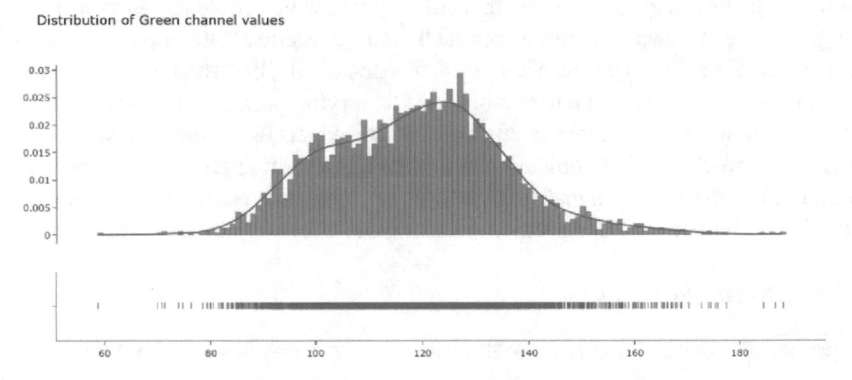

Fig. 2. Distribution of green channel in the dataset

Figure 2 shows the distribution of Green channel values from which it can be observed that the graph is slightly right skewed and concentrated at 130. This indicates that most of the images have less than 130 Green channel content.

Then an analysis has been done on the Blue channel content in Fig. 3 and found that the graph is almost normally distributed as compared to the Red and Green channel values. Also this graph is concentrated at 110.

From above observations it can be claimed that Red and Green channels are not normally distributed as compared to the Blue channel, thus while performing image classification images with more Red and Green channel values will be dominating each other. For this let's see the combined channel graph Fig. 4 from which we can observe that each channel has different concentration points and below 120, most of the images have blue as dominating channel and above 120 we have green as the dominating channel.

It is evident from the visualization that mean channel value spread of the dataset in Fig. 5, in which we can see the mean channel to be concentrated at 120. After observing

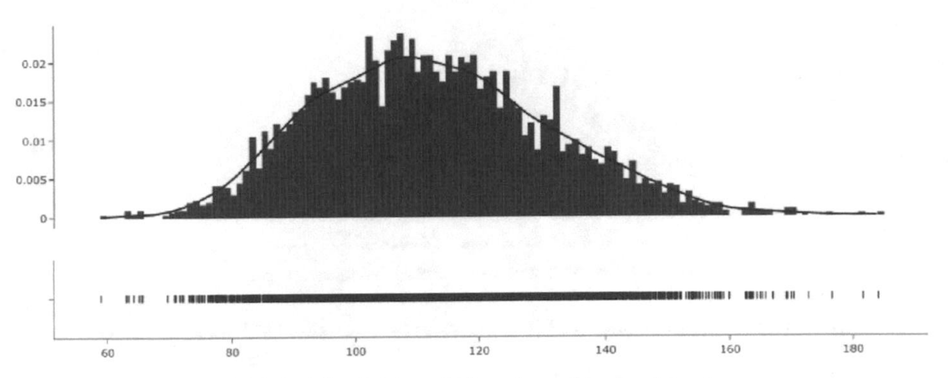

Fig. 3. Distribution of blue channel in the dataset

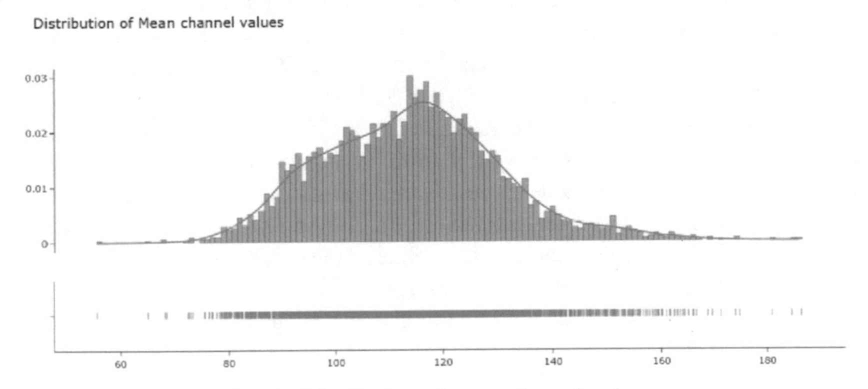

Fig. 4. Distribution of mean channel values

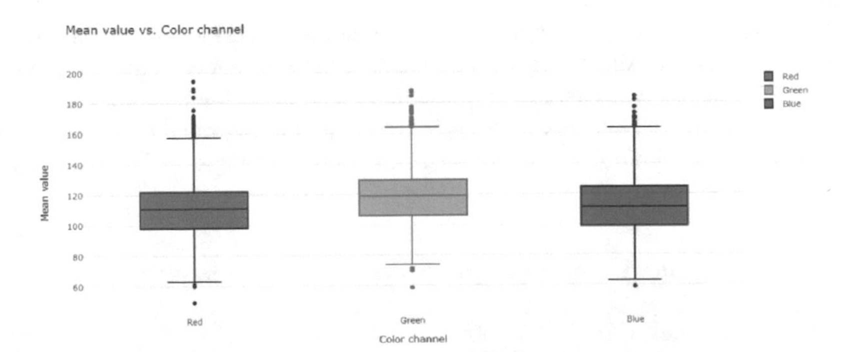

Fig. 5. Showing mean channel value versus color channel plots

the channel values from the above graphs it could be noticed that due to so much uneven spread of Blue and Green channel values, performing image classification in just RGB color spaces would not be sufficient as in some images Blue channel is dominating and in

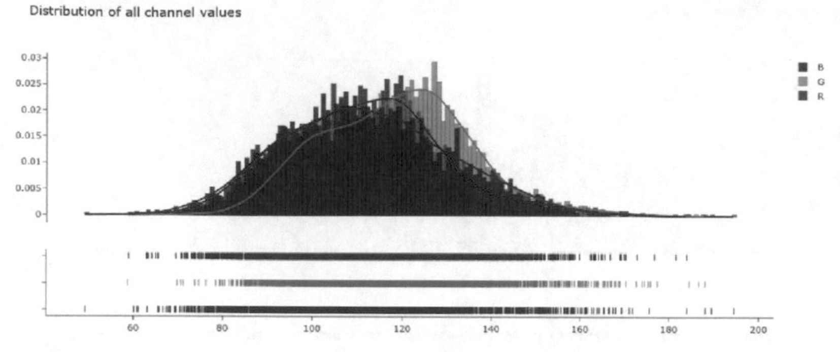

Fig. 6. Combined distribution of red, green and blue channel values in the dataset

some Green channel is dominating. The experiment has also been done to take the RGB channel value of the healthy and diseased part of a diseased leaf which are presented in Table 1 and found that the healthy leaf part of images dominates over the diseased parts which is also evident from the mean channel vs the RGB graph given in Fig. 6. The Green channel value 137 dominated over Red and Blue in the case of the healthy part and the Blue channel dominated over the Red and Green channels in the diseased part. This is in line with the observation that we did from the graphs and also if we compare the Dominating channels in both the leaf parts, we found that the Green channel dominated over the Blue channel of the diseased part. Hence, it is a fact that changing the default RGB images to different color spaces where either Blue or Green dominates would help in more precise detection of diseased leaf images.

Two findings which are drawn from the above visualization and experimentation that,

1) Green channel values dominate as seen from the combined RGB channel graph and Table 1.
2) Changing the color spaces with more Blue and Red channels could improve the accuracy of the classifier as this would increase the dominance of the diseased part. This becomes the vital aspect of our proposed work as the leaf images obtained from the fields have lots of variations in the form of backgrounds and varying light intensity. Thus with different color spaces, the effect of these variations could also be reduced.

Table 1. Comparison of RGB values of diseases leaves

Leaf Part	Red	Green	Blue
Healthy Part	48	137	31
Diseased Part	45	99	120
Healthy Part	48	137	31

3.3 Novel Ensemble Algorithm

For measuring the effect of the changed color space the ensemble learning classifier is being chosen. Ensemble techniques use multiple models and combine them to predict the output. Methods like Bagging and Boosting have been widely used in classification tasks to improve the classifier's performance. Max Voting, Averaging, Bagging and Boosting are some widely used ensemble approaches. So a novel ensemble algorithm-(COLEN) has been proposed which selects the color space with the highest score. Its main objective is to consider all color spaces and then predict results using classifiers trained on each. This is done to reduce the effect of background and light intensity in images. This algorithm takes as input the classifier to be trained, dataset and color spaces on which the classifier will be trained. Then the classifier is trained on all the color spaces and then predictions are done for each. For our classifier we will have 'n' predictions where n is the number of color spaces given to the algo. Out of these n predictions, the one with the maximum score will be considered and the color spaces corresponding to it will be considered as the best color space suited for the dataset. The detailed steps of the algorithm are given below.

Proposed Algorithm *COLEN*
Input – M: Best model/classifier
I-$\{I_1, \ldots\ldots\ldots, I_m\}$ – Set of image formats, m: no of image formats taken
D – set of training images in RGB format
Output – Set P1-$\{(C_i, A_i, F1_i), \ldots\ldots\}$ – Sorted in decreasing order of A_i where Ci is the set of classifiers, Ai and F1i are the accuracy and F1 score of ith classifiers.

1. Set C_i – where Ci is a set of Classifiers and '*i*' ranges from 1 to n, n= number of classifiers
2. For i=1 to m, do
3. Convert D into D*i* where D_i is the transformed images into "ith" format and i ={1 to 8}
4. Call (Max(C1,C2, ….Mn))
5. Return *Mi where Mi is the model which gives highest accuracy and F1 score*
6. Train M_i on D_i.
7. Evaluate the model.
8. Store $(M_i, A_i, F1_i)$ in P1.
9. Evaluate the test data, calculate the max(Mi) where max(Mi) returned the predictions of the model M trained on all image formats.
10. X=Input for Testing
11. For i = 1 to m:
12. Predict X. and store the greater of the Result and prediction.
Result = ($M_i.predict(X)>=$ Result) ?$M_i.predict(X)$: Result
13. Finally Result contains the highest-scoring color space and its score.

3.4 Color Spaces

Color spaces are mathematical models that show how a particular color can be represented. When we talk about different color spaces like RGB, HSV or any other we mean to specify the amount of Red, Green and Blue content in them. There are a variety of color spaces available, RGB or BGR being the most common one. In this paper we choose eight different color spaces for our experiment.

1. RGB OR BGR – RGB or BGR is the default color space used most widely for image classification tasks. It defines color space using the values of Red, Green and Blue, the three primary colors ranging from 0 to 255. It is an additive model as the primary colors are combined with one another to create non-primary colors.
2. HSV – HSV also known as HSB, aims at defining colors through hue, saturation and brightness. It is often used as an alternative to the RGB color model. In our work, we converted the RGB images to HSV. The RGB to HSV conversion was done using the rgb_to_hsb() which takes 'red', 'blue' and 'green' as values.
3. HSL – It aims at defining the color space through Hue, Saturation and Lightness. Hue is the color type ranging from 0 to 360 degree. Unlike RGB, in HSL, both the saturation and lightness of a color can change.
4. LAB – This color space gives us a way to locate and communicate colors. L*a*b stands for Lightness, Red/Green value, Blue//Yellow Value.
5. YCRCB -YCRCB represents colors as two color difference signals and it represents color as brightness whereas RGB represents color as three basic colors. It is better than RGB as it can separate luminance from chrominance more effectively than RGB color space.
6. LUV – L*u*v color space is widely used in computer graphics dealing with colored lights.
7. XYZ – This color space encompasses color sensations that are visible to a person with average eyesight.
8. YUV – YUV color spaces are divided into packed formats and planar formats in a single array. It is typically used as a part of the color image pipeline. YUV color spaces are more efficient coding and reduce the bandwidth more than RGB capture can.

4 Results and Discussion

After performing the data analyzation in Section 4B, we concluded that we needed more color spaces in which the diseased parts could be more precisely detected by the classifiers as the green channel values in the healthy part of the diseased leaf dominates over the blue channel value of the diseased part of the leaf. To further validate this point we decided to train four image classifiers using the Transfer Learning approach on the eight color spaces which are RGB, HSV, HSL, LAB, YCRCB, LUV, XYZ, YUV as mentioned in Section 4E. Initially dataset [5] already has images in RGB or BGR format, hence we converted the images from RGB or BGR to the remaining seven types and each classifier was trained on all the eight color spaces and their accuracy and F1 score is presented in Tables 2 and 3.

Table 2. Comparison of the Accuracy of various deep learning models trained on 8 different color spaces.

Model Name	BGR	HSV	HSL	LAB	YCRCB	LUV	XYZ	YUV
Inception	0.8646	0.8898	0.9024	0.8787	0.8394	0.9276	0.8283	0.8898
Renset50	0.8247	0.7699	0.7753	0.7342	0.7342	0.7479	0.7425	0.6932
EfficientNet	0.9221	0.9102	0.9381	0.8841	0.8921	0.8763	0.9105	0.9074
VGG16	0.669	0.681	0.726	0.655	0.645	0.678	0.65	0.691

Table 3. Comparison of the F1 Score of various deep learning models trained on 8 different color spaces.

Model Name	BGR	HSV	HSL	LAB	YCRCB	LUV	XYZ	YUV
Inception	0.8625	0.8913	0.9016	0.8772	0.8391	0.9281	0.8281	0.8904
Renset50	0.8251	0.7544	0.7608	0.7173	0.7304	0.7274	0.7384	0.6638
EfficientNet	0.9101	0.9011	0.931	0.8911	0.9012	0.8712	0.921	0.912
VGG16	0.6335	0.68	0.719	0.641	0.642	0.671	0.641	0.691

From Table 2 we can observe that the accuracy of the Inception model is highest for the LUV color spaces which is 92%. For Resent50 the accuracy is highest for the RGB or BGR color spaces. For Efficient the highest accuracy is for the HSL color spaces and the same is observed for the VGG 16 and VGG19. Similarly the F1 score also follows the same pattern as shown in Table 3. For the Inception model, the highest F1 score is achieved for the LUV color spaces. For resnet50 highest F1 score is achieved for the RGB or BGR color spaces and for the EfficentNet, VGG16 and VGG19, highest F1 score is achieved for the HSL color spaces. These observations point to the fact that there are color spaces other than RGB which could help in increasing the accuracy and F1 Score of the classifiers. Hence we proposed a novel ensemble algorithm: COLEN mentioned in Section 4D which aims at finding the best color spaces in which a classifier performs to its best and produces low false predictions.

So, we used our novel ensemble algorithm(COLEN), which takes as input the dataset [5] having images in RGB or BGR format, list of color spaces and classifier which is to be trained. Its main objective is to train the classifier on all the color spaces and then choose the one with maximum accuracy and f1 score to predict the color space with the highest score. It does predictions on all color spaces and the one with the maximum value is chosen as the final output. Applying COLEN to the data input, results obtained in Tables 2 and 3, we achieved better test accuracy and f1 score as compared to the one received in RGB color spaces. Since we got the highest accuracy for RGB images using the EfficientNet classifier in our previous work [1] and also it is in line with the results from Tables 2 and 3, we took the results and compared them in Table 4 along with some other results from various works. Results obtained in Table 4 are on the test images taken from dataset [6] which was not used while training.

Table 4. Comparison of the performance of various works.

Serial Number	Model	Accuracy	F1 Score	Color space
1	EfficientNet [1]	92.21%	91.10%	RGB
2	PDnet [2]	93.67%	-	RGB
3	SPMohanty [3]	84%	-	RGB
4	ReTs [7]	91%	92.6%	RGB
5	Adarshet al. [4]	91.49%	88.8%	RGB
6	COLEN	93.81%	93.1%	HSL

We also compared the results obtained using the proposed algorithm COLEN with the existing state of the art methods and architectures. The results obtained by our approach (accuracy-93.81% and F1 score-93.1%)was better than our previous work [1] (accuracy-92.21% and F1 score-91.10%) in which EfficientNet was trained on RGB images only. Results obtained by us were also better than the approach mentioned by the authors in [2] where they achieved an accuracy of 93.67%. Similarly authors in [3] quoted an accuracy of 99.6% on training images but when we tested the same on test images from dataset [6], it showed only 84% accuracy. Similarly authors in [4] used modified transfer learning approach and achieved 99% accuracy but when tested on the test images, the accuracy dropped to 91.49%, which is lesser than ours. Thus our proposed algorithm COLEN outperforms many existing works and increases the accuracy and f1 score of the classifiers and also proves that color spaces like HSL, are also important for the classifiers as often highest accuracy is not always attained by training on RGB images.

5 Conclusion and Future Scope

In this paper we have presented a novel ensemble algorithm which trains the classifiers on various color spaces. The accuracy and f1 score yielded in this manner proved to be better than the existing state of the art methodologies and deep learning architectures as mentioned in Table 4. We performed vast data visualization on the dataset which we used for training purposes and found that the green channel values were dominant with respect to the blue channel values in the diseased leaf image. Hence choosing only RGB color spaces to train the classifiers will definitely reduce the accuracy and F1 score. So, we decided to experiment with the 8 color spaces as mentioned in Section 4E and trained various deep learning models on them and presented the results in Table 2 and 3. From Table 2 and 3 we concluded that for the majority of models HSL was the model with best accuracy and f1 score and for the resnet50 and inception model it was different. Thus we concluded that instead of choosing one color space, we should choose an ensemble approach to consider the color spaces which perform best. Thus, using COLEN, we achieved better results than our previous work and it also outperformed deep learning architectures proposed in existing literature. Also, it outperformed the results obtained by authors using Transfer Learning techniques on RGB color spaces. Our results are in line with the main objective of increasing the accuracy as well as decreasing the false predictions as with different color spaces, the background, light intensity and other factors affecting images are reduced. The Future scope of this paper lies in making the classifier more robust to deal with the variations in the images.

References

1. Divakar, S., Bhattacharjee, A., Priyadarshini, R.: Smote-DL: a deep learning based plant disease detection method. In: 2021 6th International Conference for Convergence in Technology (I2CT), pp. 1–6. IEEE, April 2021
2. Arsenovic, M., Karanovic, M., Sladojevic, S., Anderla, A., Stefanovic, D.: Solving current limitations of deep learning based approaches for plant disease detection. Symmetry **11**(7), 939 (2019)
3. Prasanna Mohanty, S., Hughes, D., Salathe, M.: Using Deep Learning for Image-Based Plant Disease Detection. arXiv e-prints, pp. arXiv-1604 (2016)
4. Pratik, A., Divakar, S., Priyadarshini, R.: A multi facet deep neural network model for various plant disease detection. In: 2019 International Conference on Intelligent Computing and Remote Sensing (ICICRS), pp. 1–6. IEEE, July 2019
5. Mohanty, S.: spMohanty/PlantVillage-Dataset. GitHub (2016). Accessed 10 June 2022
6. Thapa, R., Zhang, K., Snavely, N., Belongie, S., Khan, A.: The Plant Pathology Challenge 2020 data set to classify foliar disease of apples. Appl. Plant Sci. **8**(9), e11390 (2020)
7. Shah, D., Trivedi, V., Sheth, V., Shah, A., Chauhan, U.: ResTS: residual deep interpretable architecture for plant disease detection. Inf. Process. Agric. **9**(2), 212–223 (2022)
8. Khirade, S.D., Patil, A.B.: Plant disease detection using image processing. In: 2015 International Conference on Computing Communication Control and Automation, pp. 768–771. IEEE, February 2015
9. Singh, V., Sharma, N., Singh, S.: A review of imaging techniques for plant disease detection. Artif. Intell. Agric. **4**, 229–242 (2020)
10. Shah, J.P., Prajapati, H.B., Dabhi, V.K.: A survey on detection and classification of rice plant diseases. In: 2016 IEEE International Conference on Current Trends in Advanced Computing (ICCTAC), pp. 1–8. IEEE, March 2016
11. Shruthi, U., Nagaveni, V., Raghavendra, B.K.: A review on machine learning classification techniques for plant disease detection. In: 2019 5th International Conference on Advanced Computing & Communication Systems (ICACCS), pp. 281–284. IEEE, March 2019
12. Shrivastava, V.K., Pradhan, M.K., Minz, S., Thakur, M.P.: Rice plant disease classification using transfer learning of deep convolution neural network. Int. Arch. Photogramm. Remote Sens. Spat. Inf. Sci. **3**(6), 631–635 (2019)

Tuberculosis Detection Using a Deep Neural Network

Dipali Himmatrao Patil(✉) and Amit Gadekar

Sandip University, Nashik, India
patil.dipali07@gmail.com, amit.gadekar@sandipuniversity.edu

Abstract. The infectious lung illness tuberculosis (TB) has been a major global killer for decades. Chest X-rays (CXR) are now the gold standard for tuberculosis (TB) screening because of their high sensitivity, low cost, and convenience of use. By using automated CXRs for TB detection and pinpointing locations that may display TB, the quality of TB diagnosis has the potential to dramatically improve. Researchers are working on a computer-aided detection method to help clinicians (and radiologists) make correct TB diagnoses using patient CXRs. Therefore, our study suggests a self-sufficient technique for TB diagnosis using deep neural networks (DNNs). A CXR image with large patches of black can be misleading to machine learning algorithms and is therefore of little diagnostic use. The proposed method utilizes segmentation networks to isolate the region of interest in multimodal CXRs. These segmented images are subsequently fed into deep neural network (DNN) models. For the sake of the subjective evaluation, we use explainable NN to represent the TB-infected regions of the lungs. Using public-domain CXR datasets, we compare the accuracy of different Deep neural network (DNN) models for categorization. Compared to raw CXR pictures, the proposed method may be more effective at segmenting the lungs.

Keywords: Tuberculosis (TB) · X-ray · Deep neural network (DNN) · CXR · Accuracy

1 Introduction

Tuberculosis (TB) is a very contagious lung disease that is the leading cause of death globally. Over ninety-five percent of TB patients are located in low-income nations with inadequate healthcare systems and resources.

Computed tomography (CT) is currently the most widely used method for TB detection. Because of its low radiation exposure, low cost, ease of availability, and ability to identify unexpected pathologic alterations among TB detection modalities, chest X-rays (CXRs) are the preferred method of confirmation for the vast majority of early cases. For many years, scientists have tried to figure out how to utilize computer-aided detection (CAD) software and medical imaging to make a preliminary diagnosis of tuberculosis-related disorders. Selecting and extracting helpful disease traits from photos can provide significant quantitative insights in the early stages of computer-aided design (CAD) [1, 12, 13].

H. Sharma et al. (Eds.): ICIVC 2022, PALO 17, pp. 600–608, 2023.
https://doi.org/10.1007/978-3-031-31164-2_51

As the field of deep learning has advanced, convolutional neural networks (CNNs) have regularly surpassed more conventional methods of recognition in the categorization and recognition of images. Because of its remarkable ability to automatically extract substantial information from the underlying features of data, CNN is the best option for addressing challenging medical situations. Past implementations of deep learning algorithms in CAD systems have yielded a wide variety of high-quality diagnostic solutions for the diagnosis of medical illnesses, but have also tended to draw attention to worrisome facets of the process [15–19].

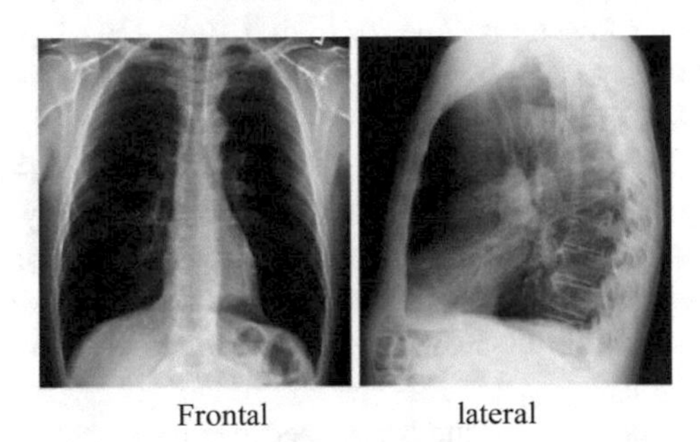

Fig. 1. Chest x-ray

As a result, in a favorable environment, the TB bacteria might grow and result in symptoms. As Fig. 1 shows chest CXRs captured in two ways frontal and lateral. If the TB bacteria proliferate, patients could develop the disease. Doctors are compelled to treat individuals with active TB as a result. Early diagnosis of tuberculosis allows for early treatment commencement, shortens the incubation period, and enhances patient outcomes [2]. Getting the proper diagnosis is crucial because lively occurrences might be fatal. The medical history of a patient will be reviewed to identify TB susceptibility. Usually, doctors are aware of TB symptoms [3].

2 Background

In this publication, a number of noteworthy contributions are mentioned [4]. The construction of a unique automated TB discovery system using CXR pictures is the aim of the proposed effort [5]. To enhance presentation across a number of objective and arbitrary TB diagnostic criteria, separation and DNN representations will be used. The suggested method would use a patient's CXR to determine whether or not they had tuberculosis (TB) [5, 6, 14].

Software with a range of learning techniques and algorithms must be used if computers are to learn. Data is fed into NNs. Thanks to the usage of AI and CAD structures, academics now have more possibilities for producing representations, and NNs may now choose the goal of learning machinery [3, 5, 6]. The medical industry will leverage big data and multimedia information analysis to provide outcomes in a number of ways, including by giving clinicians another choice for decision-making and enhancing their ability to diagnose patients. Additionally, these outcomes might be reached faster and cheaper [9].

2.1 CNN-Based Transfer Learning

In order to deal with this growing issue, the authors of [10] use a deep residual learning strategy. They need criticism in order to assess the merits of their proposed design and demonstrate the growing flaw in the concept. Also, ResNets with 18 and 34 layers were compared in [11]. As soon as a picture is received, it goes through a series of transformations, including scaling, grouping, enhancement, and resampling. For instance, whereas an 18-layer plain network displays a greater working error during training, their respective explanation spaces are contiguous. For residual learning, however, the tables are flipped, and the 34-layer ResNet is superior to its 18-layer predecessor (2.8 percentage points). Because of its superior performance and its special relevance to authentication data, the 34-layer [14–19].

ResNet [13] stands out. Using a system of counting "Top-1-err" and "Top-5-err" phrases, the authors ranked the proposed structure. The frequency with which the classifier (ResNet) committed an error when analyzing the data is represented by the top-1 error. Top-5 error is a metric for evaluating how often a classifier fails to select the proper class given the top 5 probabilities. ResNet has a Top-1 error rate of 21.43% and a Top-5 error rate of 5.71% [4].

In light of the theoretical background presented in [5], an assessment of its correctness and generalizability was called for. Disturbances in a network's architecture can be fixed to make it better. In terms of the representational size of the input data and the representational size of the output data, there are four primary schools of thought [16]. Higher-dimensional representations increase local processing in photo categorization while a network is being trained. CNN's activation rate must be increased in order to distinguish more features.

In [7], the authors demonstrated a novel construct for a convolutional neural network (CNN) that allowed for the depth-based partitioning of convolution layers during training. Authors claim that by using mapping techniques, the primary goal can be more easily attained. Theoretically, the correlation mappings between channels and between channels and space in CNN feature maps can be rather different. They offered a structural [8] that they dubbed "Extreme Inception," or "Xception" for short. For feature extraction, the Xception architecture relies on a whopping 36 convolutional layers. Image categorization was an important part of this effort [9]. A linear tangle of depth-divided complexity layers; remaining Xception linkages simplified. The Xception architecture transfers information via an incoming stream (299299 bytes), a circulating stream (8 iterations), and an outgoing stream. To keep tabs on any and all complications, we use batch standardization to keep track of data, and we divide up the convolution layers as necessary. Total difficulty is calculated by multiplying each level of difficulty by 1. (No depth development). Results from the Xception framework were superior, with a Top-1 accuracy of 79% and a Top-5 accuracy of 94.5%. The study of compact and powerful neural networks has been increasingly popular in recent years. Academic research has recently placed a greater emphasis on the creation of efficient neural networks with a minimal resource footprint.

The authors of [2, 10] indicate that a paradigm with dormancy and size constraints may be used if networks were created with variable capacity. Mobile Net places an emphasis on dormancy to achieve size balance across the system. The authors also presented several strategies for creating small systems that incorporate the functionality of bigger ones, such as Xception and Squeeze net. All point-wise layers, the authors claim, are monitored by batch normalization using a nonlinear activation function during training, with the exception of the final entirely linked layer, which is linearly stimulated and feeds a SoftMax layer for classification (a rectified linear unit, ReLU).

By analyzing the relationship between the width and depth of CNN representations, Tan et al. [11] outline an approach for developing extremely efficient CNN models. The authors aimed to reduce the number of parameters used by CNN models while increasing their classification accuracy. The term "Efficient Net CNN models" describes the process through which these models were developed.

Contrasting the results of this study with those of others that have examined the utility of CXR pictures in the diagnosis of tuberculosis is presented in Table 1. It's worth noting that while some of these projects relied on openly available data, others relied on confidential information [1–19].

Table 1. Comparison with other recent similar works

Author	Year	Images	DataSet Type	Method	Results
Lakhani et al.	2017	1007	1. Montgomery 2. Shenzhen 3. Belarus	Two CNN models using normal CXRs Used augmentation compression	Accuracy is 98%
Becker et al.	2017	–	Mulago National Hospital	Detecting patterns in photographs	Accuracy is 98%
Liu et al.	2017	4248	Partners in Health Peru	CNN transfer learning	Accuracy is 85.68%
Stirenko et al.	2018	800	1. Montgomery 2. Shenzhen	CNN segmented lung CXRs	Accuracy lower than 85%
Hwang et al.	2019	4559	Seoul National Hospital	Develop deep learning	Accuracy is 97.1%
Nguyen et al.	2019	1032	Seoul National Hospital	Five CNN models using transfer learning	Accuracy is 99%
Heo et al.	2019	800	1. Montgomery 2. Shenzhen	Five CNN pre-trained models	Accuracy is 92.13%

3 Problem Definition

Computed tomography (CT) is currently the most widely used method for TB diagnosis. Among TB detection methods, chest X-rays (CXRs) are most commonly used for confirmation because to their low radiation dose, low cost, easy availability, and ability to identify unexpected clinical alterations. For decades, scientists have sought to perfect a computer-aided detection (CAD) system that could use medical imaging to make a preliminary diagnosis of TB-related disorders. For early quantitative insights, CAD chooses and extracts important pathologic features from photos. These methods, while effective, are time-consuming and reliant on fabricated data pattern extraction. The manifestation of many diseases typically takes up only a tiny fraction of the image, which quickly complicates the feature detection process. In addition, the CAD system's inability to make an informed, high-accuracy judgment in light of amassed medical image data and evolving disease mutations have been hindered by issues with uneven performance in regard to freshly created data and poor data transferability between datasets.

4 Datasets and Preprocessing

We analyzed data from 662 frontal posteroanterior CXR pictures of varying sizes, of which 326 were deemed to be from healthy individuals and 336 from those with TB symptoms. The NIH data set is among the largest available CXR datasets in the public domain. The collection includes 112,120 CXR images with 14 thoracic diseases in the posteroanterior and anteroposterior planes and was taken from the clinical PACS database at the NIH Clinical Center. Sickness information and labels for CXRs must be text-mined with an accuracy of >90% using natural language processing techniques, as the original radiology report is unlikely to be made public. As a result of its large size, rich annotations, and breadth of thoracic disorders covered, this dataset has been extensively used by deep-learning researchers exploring the diagnosis of thoracic diseases.

5 Methodology

In this study, we provide a method for rapidly identifying TB in CXR pictures. In this method, segmented CXR images of the lungs are used in conjunction with pre-trained CNN models. Multiple datasets and cross-dataset scenarios are investigated. There are several building blocks that make up the system. Given that each block in the proposed model relies on the one before it, the system as a whole is quite sturdy. This graphic depicts the system block diagram for the future tuberculosis discovery system.

Figure 2 shows a visual broken down into phases of the system, with each block representing one of these phases. Automatic tuberculosis detection in CXR pictures is a goal of this project, and DNN models will be employed to get there. We take a closer look at a variety of cutting-edge CNN detection techniques. Our research technique includes the following procedures to help us accomplish our aims:

Building systems, analyzing data, and utilizing deep neural networks Comparisons made utilizing a variety of presentational measures and formats.

In this research, we introduce a TB detection system that uses DL and DNN techniques for automated analysis. The first of the three fundamental elements of the proposed system is the radiographer-taken chest radiographs of the patient. Thanks to their availability in public databases, the input CXR images can be used in CAD system research and development, allowing universities to carry out their experiments and reach their objectives. TB diagnosis frequently makes use of photographic archives in the medical

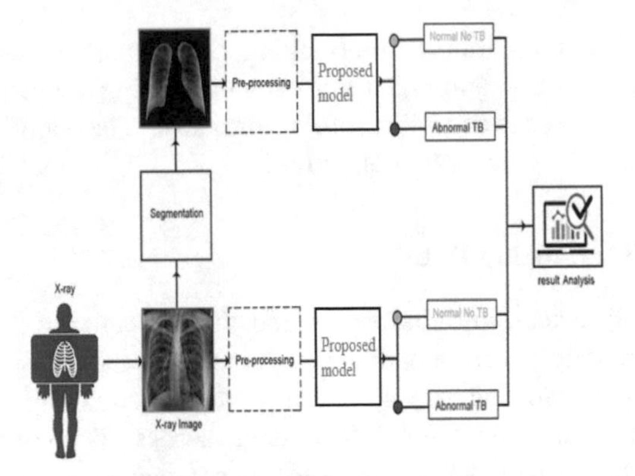

Fig. 2. TB diagnosis pipeline.

field. The primary goal of our system is to enhance detection performance and lower mistake rates through the use of enhanced input photographs and subsequent use of image processing methods.

Image borders can be filtered to make them look better. It's possible that filters could be used to successfully extract spatial and temporal visual properties. Therefore, it would be useful to conduct a more thorough exploratory information analysis and preprocessing with lung group filtering for people of different ages, sexes, and geographical areas. When applied to an image, preprocessing can either improve it or reorganize its structure. These pictures can be utilized once the text and the blocks have been taken out. Lung masks can be seen in the CXRs of the Shenzhen and Montgomery datasets that are available to the public. The mask's minimalist black-and-white style is immediately recognizable. As a result of applying the appropriate mask to each image, we are left with segmented images that only contain the lungs [4, 5]. To extract the ROI from CXR images, a preprocessing technique called lung field subdivision must be applied [5]. Several new segmentation methods have been developed specifically for creating segmented photographs. Subdivision methods vary per system, but some examples include multi-resolution pixel categorization and dynamic appearance representations [9]. In the past, segmentation has been used on numerous types of medical images. We can take the following measures to protect ourselves from tuberculosis: (TB).

This research proposes a number of different tuberculosis diagnostic modules based on the use of machine learning techniques (TB).

5.1 Collection of Datasets

The suggested method begins with the collection of datasets, which is also its most crucial step. Since a flawed dataset would result in misleading conclusions, this is the most crucial stage. Before beginning data collection, we must determine the type of datasets to use.

The next step in the data collection process is deciding on a classification scheme from which to pull primary data, and then settling on a specific technique to use in gathering that data.

Surveys, online tracking, transaction tracking, online market analytics, social media monitoring, and subscription and registration data gathering are just a few of the many approaches that can be used to compile useful information. The majority of these methods aren't appropriate for the data we're collecting.

5.2 Sorting and Organizing Data

Gathering raw data on learning challenges and then organizing it into a meaningful order is an example of data sorting, which is done to make the data easier to understand, analyze, and visualize. Sorting is a common method for making research data more comprehensible through visualization. Raw data (across all records) and aggregated data can be sorted in the same way (in a table, chart, images, or some other aggregated or summarized output).

A machine learning algorithm is used to sort the data, split it into a train and test set, and then learn from it. After all, the pre-processing stage of prediction involves selecting an appropriate training and testing method before applying it to the collected datasets.

Both the weights and the bias are used to learn (determine) positive values from a subset of samples during model training. For the purpose of developing a supervised learning model, a machine learning algorithm looks at many examples and attempts to select one that minimizes loss.

5.3 Feature Extraction from the Dataset

After all, the pre-processing phase of the prediction process involves picking the right training and testing techniques before applying them to the collected datasets. Since utilizing several prediction algorithms on inconsistent data sets may lead to decreased accuracy and performance, selecting the appropriate datasets is vital.

A feature is a salient piece of information extracted from an image or signal that allows for in-depth familiarity with the dataset. To be disabled is to have a distinguishing property of the signals in a certain amount of data. Therefore, classification machine learning models can be put through their paces using accuracy prediction methods. The Holdout process is one such method, in which the data is split into a training set and a test set (usually with a 2/3 training set and a 1/3 test set designation) and the trained model's results are evaluated on the test set.

5.4 Testing of Data Using the Proposed Method

To evaluate a model's efficacy with a certain performance indicator, analysts consult its efficacy on a test set of data. The assessment set must not contain any findings that were previously used in the training set. If the test set comprises examples from the training set, it will be difficult to evaluate if the algorithm has learnt to generalize from the training set while we are using other datasets, each with their own numerous values.

After adequate adaptation, the software should be able to carry out a process with the updated data. Therefore, training becomes the most important part of our TB prediction system. Software that learns from its training data by memorizing an overly complex model may be able to accurately predict the values of the response variable within the training set, but it would be unable to do so for new instances. As a means of making reliable predictions, we would put the TB dataset through its paces with the help of the best possible proposed DL model.

5.5 Showing the Prediction and Storing the Result

We have arrived at metrics like forecasts on the number of persons who would contract tuberculosis once the forecasts have been created, losses and errors have been minimized, the necessary protections have been implemented, and the predictions' accuracy has been raised (TB). Using this prognosis, we can better understand how to mitigate these threats to people's health. Once the forecasts have been shown, the data should be carefully stored for use in the future.

6 Conclusion

In summary, it has been proposed to make standard changes to the structure of the deep learning model and then to tweak it to perfection. Our proposed methodology had been tested using a number of deep learning models, each with its own unique module topology and set of layers.

To automatically detect TB in CXRs, this research proposed a strategy based on deep neural networks and transfer learning (NNs). We picked original segmented input photographs that had the essential part of the creative image. Afterward, trained DNN models were used to extract the images' respective attributes. Finally, the models' common ability to identify TB in CXRs allows for a wide range of performance evaluations.

References

1. Chandra, T.B., et al.: Automatic detection of Tuberculosis related abnormalities in chest X-ray images using hierarchical feature extraction scheme. Expert Syst. Appl. **158**(15), 113514 (2020)
2. Rajpurkar, P., et al.: CheXpedition: investigating generalization challenges for translation of chest X-Ray algorithms to the clinical setting. In: ACM Conference on Health, Inference, and Learning, Ontario, Canada (2020)
3. Rahman, T., et al.: Reliable Tuberculosis detection using chest X-Ray with deep learning, segmentation and visualization. IEEE Access **8**, 191586–191601 (2020)
4. Muhammad, G., Hossain, M.S.: COVID-19 and non-COVID19 classification using multi-layers fusion from lung ultrasound images. Inf. Fusion **72**, 80–88 (2021)
5. Muhammad, G., Alqahtani, S., Alelaiwi, A.: Pandemic management for diseases similar to COVID-19 using deep learning and 5G communications. IEEE Netw. **35**(3), 21–26 (2021)
6. Altaheri, H., et al.: Deep learning techniques for classification of electroencephalogram (EEG) Motor Imagery (MI) signals: a review. Neural Comput. Appl. (2021). https://doi.org/10.1007/s00521-021-06352-5
7. Alshehri, F., Muhammad, G.: A comprehensive survey of the Internet of Things (IoT) and AI-based smart healthcare. IEEE Access **9**, 3660–3678 (2021)

8. Razzak, M.I., Imran, M., Xu, G.: Efficient Brain Tumor segmentation with multiscale two-pathway-group conventional neural networks. IEEE J Biomed. Health Inform. **23**(5), 1911–1919 (2019)
9. Khan, T.M., et al.: Width-wise vessel bifurcation for improved retinal vessel segmentation. Biomed. Sig. Process Control **71**, 103169 (2022)
10. Muhammad, G., Alshehri, F., Karray, F., et al.: A comprehensive survey on multimodal medical signals fusion for smart healthcare systems. Inf. Fusion **76**, 355–375 (2021)
11. Tan, M., Le, Q.: Efficient Net: rethinking model scaling for convolutional neural networks. In International Conference on Machine Learning, Long Beach, CA, USA, pp. 6105–6114 (2019)
12. Lakhani, P., Sundaram, B.: Deep learning at chest radiography: automated classification of Pulmonary Tuberculosis by using convolutional neural networks. Radiology **284**(2), 574–582 (2017)
13. Liu et al.: TX-CNN: detecting tuberculosis in chest X-ray images using convolutional neural network. 2017 IEEE International Conference on Image Processing (ICIP), Beijing, pp. 2314–2318 (2017)
14. Nguyen et al.: Deep learning models for Tuberculosis detection from chest X-ray images. In: 26th International Conference on Telecommunications (ICT), Hanoi, Vietnam, pp. 381–385 (2019)
15. Norval, M., Wang, Z., Sun, Y.: Pulmonary Tuberculosis detection using deep learning convolutional neural networks. In: ICVIP 2019: Proceedings of the 3rd International Conference on Video and Image Processing, Shanghai, China, pp. 47–51 (2019)
16. Hwang, E.J., et al.: Development and validation of a deep learning-based automatic detection algorithm for active pulmonary tuberculosis on chest radiographs. Clin. Infect. Dis. **69**(5), 739–747 (2019)
17. Heo, S.J., et al.: Deep learning algorithms with demographic information help to detect Tuberculosis in chest radiographs in annual workers' health examination data. Int. J. Environ. Res. Public Health **16**, 250 (2019)
18. Rajaraman, S., Antani, S.K.: Modality-specific deep learning model ensembles toward improving TB detection in chest radiographs. IEEE Access **8**, 27318–27326 (2020)
19. Chang, R.-I., Chiu, Y.-H., Lin, J.-W.: Two-stage classification of tuberculosis culture diagnosis using convolutional neural network with transfer learning. J. Supercomput. **76**(11), 8641–8656 (2020). https://doi.org/10.1007/s11227-020-03152-x

Capacity Enhancement Using Resource Allocation Schemes

A. Geetha Devi, M. Roja(✉), and J. Ravindra Babu

P.V.P Siddhartha Institute of Technology, Kanuru, Vijayawada, Andhra Pradesh, India
muvvalaroja@gmail.com

Abstract. Multi-user OFDM (MU-OFDM), also known as Orthogonal Frequency Division with Multiple Access (OFDMA), is an orthogonal multiple access technology that is based on OFDM. Numerous users with sets of sub-carriers that are mutually exclusive can share an OFDM symbol because of OFDMA. Two names are used to refer to two different kinds of multi-user resource allocation problems: rate adaptive (RA) and margin adaptive (MA). While considering the data rates of customers and bit error rates, margin adaptive seeks to lower the overall transmit power i.e., BER. Rate adaptive systems are designed to increase overall system capacity under a total power restriction. At this point, the resource allocation for each base station is dispersed. Different techniques have been used during the past ten years to eliminate interference as noise, only in Single-Input Single-Output (SISO) condition. The major drawback in all of these strategies is that the bandwidth required must rise proportionately to the user base in order to grant each user a minimal generalized degree of freedom. Additionally, interference channel separate coding is virtually always ineffective. Another drawback of those approaches is the adaptive margin. The plan has been presented to stop those problems. The suggested approach makes use of complete transmitter channel status data. This method employs iterative power allocation and a power control of distributed convergence. The margin adaptive (MA) problem is modified into a convex optimization problem, which solves a variety of antenna configurations at real-world SNR and outage capacity levels. The suggested approach cancels power divergence issues at any load, making it more effective over iterative water-filling.

Keywords: MU-OFDM · SISO · SNR · Multiuser channel · CDMA

1 Introduction

Inter-symbol interference is one of the main obstacles to sending broadband data over wireless channels (ISI). OFDM systems are adaptable and bandwidth-effective modulation techniques that may minimise ISI because they break a wide-band channel profile into a number of narrow-band orthogonal sub channels, each bearing a Quadrature Amplitude Modulation (QAM) symbol. Since channel fading of each sub-carrier differs, it is possible to manage each sub-carrier separately using an adaptive power allocation and modulation approach. Given the increasing interest for wireless data traffic, it has

H. Sharma et al. (Eds.): ICIVC 2022, PALO 17, pp. 609–618, 2023.
https://doi.org/10.1007/978-3-031-31164-2_52

been recognized that future wireless networks must use adaptive resource allocation to fully utilize their limited power and bandwidth resources.

"Orthogonal Frequency Division with Multiple Access," or multiuser OFDM (MU-OFDM), is a type of orthogonal multiple access based on OFDM (OFDMA). Multiple users with sub-carrier sets that are mutually exclusive can share an OFDM symbol thanks to OFDMA. Since the fading that each user experiences is unrelated to the fading that another user experiences, multi-user diversity benefits can be realised by applying a channel-sensitive resource allocation method that increases the total amount of capacity available.

Rate adaptive (RA) and margin adaptive (MA) multi-user resource allocation issues fall into these two groups [3, 4, 5, 6, and 7]. The goal of margin adaptive is to reduce the overall transmit power while taking into account the user's constraints on data speed and bit error rate (BER). When there is a total power restriction, the rate adaptive seeks to increase system capacity. In order to address the proportionality restriction, Wong et al. [1, 2, 7 and 15] expanded the rate maximization issue by incorporating a preset set of priority criteria.

For flexible charging and the difference of quality of service (QOS) levels, the proportionality restriction is essential. The method described in [7] is nearly perfect, but it is challenging to utilize since numerous nonlinear power distribution equations must be solved. Considering that each user ought to receive an amount of sub-carriers proportionate to its predetermined priority index, [1] simplifies the preceding procedure. For the distribution of power among the customers, this supposition generates linear equations, greatly simplifying the problem. However, the most current technique does not ensure that proportionality will be precisely met.

2 Multiple Access Techniques

The most effective method for sharing the allocated spectrum among multiple mobile users is the implementation of various access strategies. Because there is a finite amount of the spectrum, sharing is necessary to boost cell or area capacity by allowing numerous users to simultaneously utilize the available bandwidth, and it needs to be done in a way that maintains the high caliber of the current user's experience with the service. In networks of wireless communication, it is typically beneficial to permit subscribers to transmit and receive data from the base station simultaneously. A cellular network is separated into cells, each of which contains a mobile phone with base station connectivity. Increasing the channel's capacity is the major goal while developing a cellular network in order to support as many calls as possible within a given bandwidth while maintaining an acceptable level of service quality. There are several possibilities to grant access to the channel. The following are the main ones:

1) Frequency Division Multiple Access (FDMA)
2) Time Division Multiple Access (TDMA)
3) Code Division Multiple Access (CDMA)

2.1 Frequency Division Multiple Access

This was the first method of multiple access that gave each user a pair of frequencies to use for making or receiving calls. The same frequency pair is used for both uplink and downlink. The allotted frequency pair is blocked throughout the call in the same cell or in cells nearby thanks to frequency division duplexing (FDD), which lowers co channel interference. To increase share capacity when they are idle, inactive FDMA channels cannot be accessed by other users. After the voice channel selection, BS and MS both start transmitting constantly and simultaneously.

FDMA systems employs many strategies to address the near-far issue. In this instance, the worst-case situation would be a weak signal recovering at a frequency range close to a strong signal. The stronger signal is used to determine the gain at the input of a gain stage when as a composite both signals are present at once; the weaker signal may be lost in the noise floor.

2.2 Time Division Multiple Access

Digital systems don't need constant transmission because users don't constantly consume the available bandwidth. In some situations, TDMA serves as an additional access mechanism to FDMA. Global Systems for Mobile Communications (GSM) employs the technique of TDMA. The complete bandwidth is temporarily available to TDMA users, but it is restricted. For the reason that users share the frequency domain bandwidth and are given time slots to use the entire channel capacity, the number of channels that can use the available bandwidth is frequently subdivided than with FDMA. There are fewer channels and nearly no inter channel interference. In TDMA, particular time periods are employed for transmission and reception. Time division duplexing (TDD) is the name for this particular sort of duplexing. In TDMA, several users share a single carrier frequency while maintaining non-overlapping time periods. A number of factors, such as the modulation technique and the available bandwidth, among others, affect how many time slots are allocated to each frame. Instead of continuously transmitting data, TDMA does so in bursts. Battery usage is decreased as a result of the variable OFF mode of subscriber side transmitter. The subscriber unit can listen to other base stations while it is waiting for a handoff because TDMA employs discontinuous transmission, which makes the handoff process simpler. Due to the necessity for separate time slots for transmission and reception, it does not require duplex transmission.

2.3 Code Division Multiple Access

Numerous radio communication techniques use the code division multiple access channel access technology (CDMA). Numerous transmitters can simultaneously transfer data across a single communication channel in multiple access systems like CDMA. This enables multiple users to share a frequency band. Spread-spectrum technology and a special coding method used by CDMA to reduce user interference allow for this (where each transmitter is assigned a code). Numerous mobile phone standards, such as CDMA One and CDMA2000, rely on the CDMA access technology. Spread-spectrum multiple access is one method (CDMA). For the same quantity of transmitted power, spread spectrum technology divides the data's bandwidth evenly. Unlike regular narrow pulse codes, a spreading code has a limited ambiguity function. In CDMA, locally generated codes run at a pace that is significantly faster than the data being sent. Before transmission, data is bit-wise XOR (exclusive OR) concatenated using the quicker coding. Spread spectrum signal production is shown in the figure. The code signal and the data signal, both with pulse durations of Tb and Tc, are XORed (symbol period and chip period). (Remember that bit timc, or 1/T, and bandwidth have an inverse relationship.) As a result, the bandwidth of a spread spectrum signal is reduced to 1/Tc, while the bandwidth of data transmission remains at 1/T band. The spread spectrum signal has a significantly wider bandwidth since Tc is much smaller over Tb. The processing gain, sometimes referred to as the spreading factor or ratio Tb/Tc, determines the most concurrent users that a base station can support.

2.4 Orthogonal Frequency Division Multiple Access (OFDMA)

Users are provided with sub-carriers dynamically in various time slots in OFDMA, which combines FDMA and TDMA. In terms of effective multi path suppression and frequency diversity, OFDMA has an advantage over single user OFDM. Additionally, OFDM is a versatile multiple access method, that supports a substantial user base with a variety of applications, data speeds, and QoS needs. The use of multiple access, which is flexible and efficient, allows for the allocation of digital bandwidth (before the IFFT process). To best serve the user population, this permits the development of complex temporal and frequency domain scheduling algorithms. In the section that follows, we'll discuss a number of these algorithms. When compared to OFDM, OFDMA offers the opportunity to decrease transmit power while also reducing the ratio of peak-to-average power issue (PAPR) [11, 12]. The PAPR problem is specifically significant in the uplink since power efficiency and power amp cost are such delicate variables. Each MS only uses a small

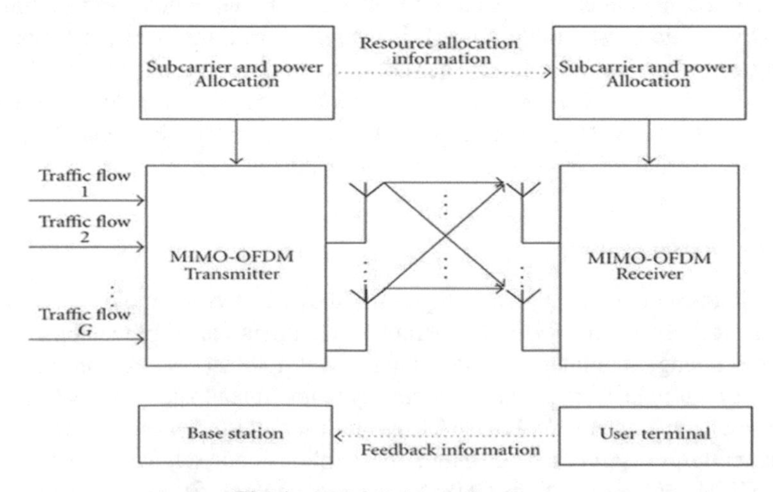

Fig. 1. OFDMA block diagram.

portion of the sub-carriers because the cell's multiple MSs Share the available bandwidth among themselves. Each MS transmits with significantly lower overall power and PAPR because it doesn't have to use the entire bandwidth (recall that PAPR rises with the number of sub-carriers). Compared to single user OFDM, TDMA, or CSMA, OFDMA handles bursty traffic and slower data rates far better. Below Fig. 1 depicts the process of OFDMA.

3 Channel Side Information at the Transmitter (CSIT)

Finding a best input signal is necessary to determine the capacity using dynamic CSIT. Optimal input for memory less channels when the receiver receives perfect channel information, is referred to as a Gaussian distribution with a mean of zero. Therefore, obtaining its optimal covariance still stands as the objective [8]. The covariance Eigen vectors indicate the transmit beam directions, while the Eigen values describe the beam power. The capacity optimization problem for dynamic CSIT with a non-zero effective channel mean and a non-trivial effective channel covariance requires the application of a non-central Wish art distribution expectation analysis. Only a small number of scenarios have been thoroughly addressed, including mean CSIT and covariance CSIT, where the covariance is the identity matrix but the channel mean is non-zero and the channel covariance is non-trivial respectively. The ideal beam directions that can be computed analytically using the Eigen vectors of the mean or covariance matrix occasionally do not coincide with the power distribution. Iterative covariance CSIT [10] and mean CSIT algorithms are effective tools for numerical optimization, which is typically required by the latter.

The capacity is first defined using dynamic CSIT, following which the asymptotic examination of the ideal capacity gain and input covariance from the CSIT for both low and high SNRs. In-depth research has been done on asymptotic MIMO capacity [9]. We provide new research on the optimal mode-dropping at high SNRs in addition to simplifying some previous findings in the context of capacity increase. When the SNR is low, the ideal input typically transitions to simple single-mode beam formation, where the capacity gain is very multiplicative. The ratio of receiving and broadcasting antennas with high SNRs determines the optimum strategy. Systems having an equal number of broadcast and receive antennas, or less, the capacity gain is close to zero, although the ideal input is close to equi-power. For others, however, the CSIT is crucial in identifying the appropriate input as well as the capacity expansion. Unlike equi-power, perfect inputs in systems with more send antennas than receive antennas may lose modes at high SNRs. We describe the prerequisites for mode dropping on typical channels with high K factors or substantial transmit antenna correlation. These circumstances help us comprehend intuitively what happens when, given the CSIT, just a subset of the available Eigen-modes are optimally engaged at all SNRs. In such situations, CSIT offers a high SNR additive capacity gain. Because it is a convex problem, capacity optimization using dynamic CSIT may fortunately be well handled numerically [13]. Using the programme in [14, 15] we investigate the effects of channel mean, transmit covariance, CSIT quality, and K factor on non-asymptotic capacity. The programme also aids in evaluating a suboptimal input covariance that tightens Jensen's capacity constraint and finds the circumstances under which adopting this covariance enables a simple, analytical capacity approximation and offers a tight lower-bound on the capacity.

4 Channel Side Information at the Transmitter at any Load

Sub-carrier Allocation attempts to tighten the standards (this here is comparable to raising the direct channel's maximum singular value). Here we consider two scenarios in which each user's desired data rate is $R_{target =}$ 256 or 384 kbps. The statistics acquired for performance evaluation includes a proportion of clients rejected (relative to all clients connected to the central cell) and a proportion of active sub-carriers (with respect to SC). If the user intended data rate cannot be reached during the time interval, which consists of 1, 2, or 3 user-orthogonalized time slots and 1 time slot using the suggested approach and iterative water-filling. An allocated sub-carrier is active if it has a transmit power greater than zero and is assigned to a user. In contrast, repeatedly water-filling both scenarios results in a sharp increase in the proportion of clients who are denied, while our suggested technique avoids this tendency. When power divergence occurs, the number of active sub-carriers also rapidly decreases with repeated water filling. Some sub-carriers are dormant due to low load (less than 64 users per cell), as not all of the authorized sub-carriers are used to give appropriate data rate due to power control. Due to admission control turning away many users whose requested power is too high, the proportion of active sub-carriers drops at high loads using our suggested method and at medium loads using iterative water-filling. The entire CSIT MIMO convergence criterion is approximate. Therefore, with this criterion, we cannot essentially guarantee that no power divergence scenarios will occur. The numerical outcomes, however, demonstrate that, both at the low and high load, our recommended technique is substantially more effective than repeat water-filling. The convergence criterion is not too demanding when applied to light loads.

5 Simulation Results

The research network is composed of two rings of omnidirectional cells with equal inter-site spacing of 1.212 km. The shadowing has a log-normal distribution with a standard deviation of 7 dB and follows the Okumura-Hata path-loss model with (d) = 137.74 + 35.22 log (d). 174 dBm/Hz is the spectral density of thermal noise. The maximum permitted transmit power per BS is 43 dBm. Each cell transmits using OFDMA with an available LSC = 256 sub-carriers and a 10 MHz network bandwidth. The total number of antennas is given by $n_t = n_r = 2$. We compare our suggested approach to iterative water-filling and a conventional orthognalisation procedure using Monte-Carlo simulations. To achieve orthognalisation, neighbouring cells with a Frequency Reuse Factor (FRF) of 3 transmit on varied groups of LSC 3 sub-carriers. When there are between LSC 3 and 2LSC 3 users per cell, a TDMA technique with two time slots is employed. At a targeted data rate of 2Rtarget, 50% of the clients are served at each time slot. A three-time slot TDMA strategy is employed when there are between 2LSC 3 and LSC users. Before all power control systems, the same sub-carrier allocation method is used. Each of the 32,

64, 224, etc., sites have an equal number of users. To avoid unintended repercussions, statistics are weighted according to the total number of central BS users.

To use criteria E, the requirement = 103 must be met. Criteria E will improve by using sub-carrier allocation (which is here comparable by increasing the direct channel's highest singular value). We examine two cases where the intended data rate for each participant is R_{target} R_{target} = 256 or 384 kbps. The percentage of consumers turned away (regarding all customers treated by the central cell) and the percentage of active sub-carriers are used to assess performance (compared to LSC). If a user is unsuccessful in achieving the necessary data rate R_{target} during the time interval, which consists of 1, 2, or 3 time slots with user orthognalisation and 1 time slot utilising the suggested approach and iterative water-filling. If a sub-carrier is assigned to a user and has the transmit power greater than zero, it is said to be active.

While the approach we recommend avoids this propensity, repeated water-filling abruptly increases the proportion of clients who are denied in both scenarios. When power divergence occurs, the quantity of active sub-carriers likewise rapidly decreases with repeated water filling. Some of the allocated sub-carriers are idle at low load because they are not all required to produce the appropriate data rate during power control (less than 64 users per cell). Because admittance control forbids human users whose necessary power is too high, the proportion of active sub-carriers drops for medium loads with iterative water-filling and big loads with our suggested technique. This suggested method also has less interference from inside cells. Regardless of the outage likelihood, the findings demonstrate that the suggested method has a lower rejection rate than iterative water-filling. In general, between 33 and 35% fewer rejections were made when comparing the three specified cases to iterative water-filling. The suggested method avoids situations where there is a power imbalance and admission control rejects a large number of applicants. As depicted in Fig. 2, the rate of rejection and the proportion of active sub-carriers are statistical measures rather than step functions as in the entire CSIT scenario, we are dealing with outage data rates.

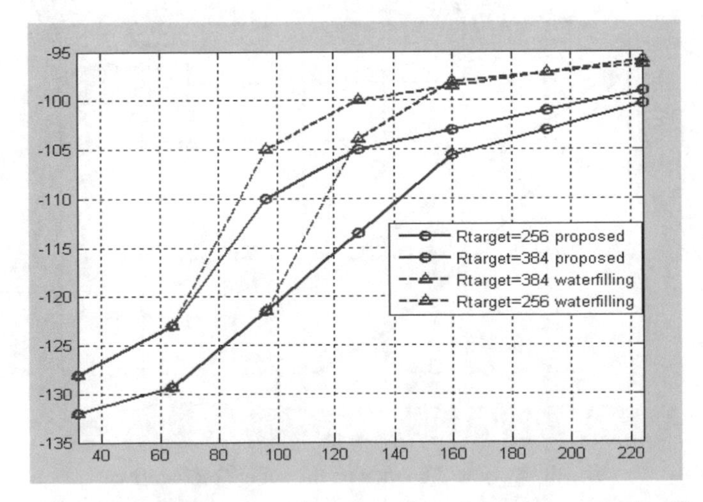

Fig. 2. Inter-cell interference per active sub-carriers, depending on the load for CSIT.

Above Fig. 2 illustrates the Inter-cell interference in CSIT depending on the load for different number of active sub-carriers in the system.

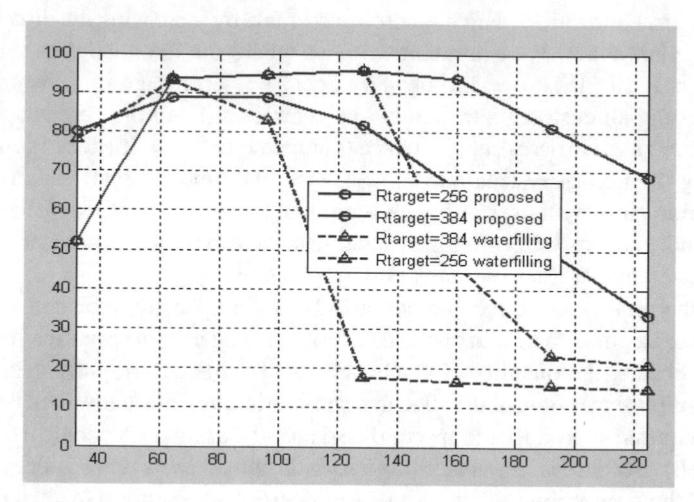

Fig. 3. Percentage of active sub-carriers depending on the load for CSIT.

The recommended number of active sub-carriers to be imposed on the system depending on the CSIT load is shown in the pictorial representation of Fig. 3 above.

The outage probability of $n_{min} = 2$ is illustrated in below Fig. 4 where P_{out} denoting the actual outage probability. P_{min} and P_{max} represents the possible minimum and maximum probability bounds respectively. In Fig. 4, blue colour curve indicates the P_{out}, green colour line represents the P_{min}, while P_{max} is represented with red line.

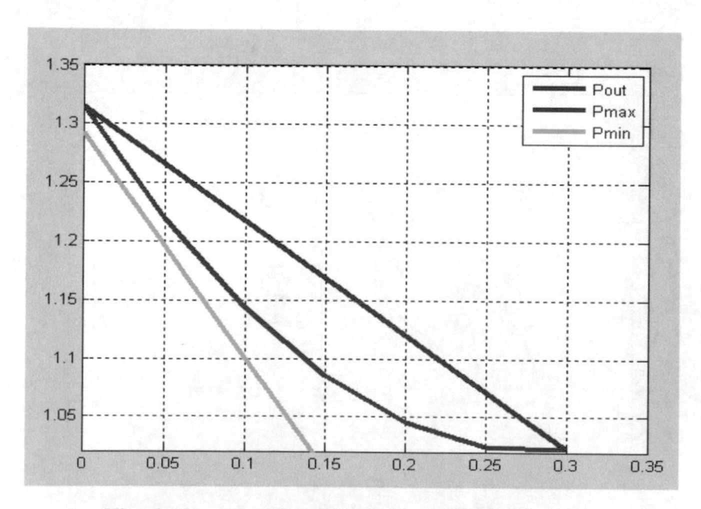

Fig. 4. $N_{min} = 2$ bounds on the outage probability.

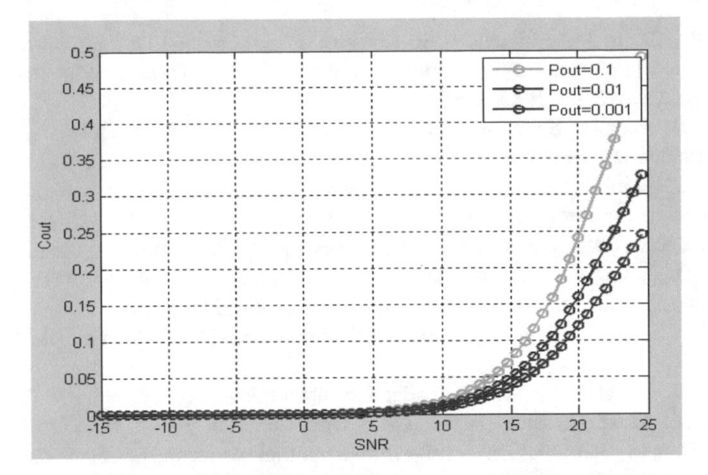

Fig. 5. For $n_t = n_r = 2$, varying P_{out}, accuracy of the approximation.

Possible approximation values for $n_t = n_r = 2$, the variation in P_{out} is clearly explained with the help of above Fig. 5.

6 Conclusions

In this study, the margin adaptation problem in MIMO study can only be solved during transmission if the channel's statistical characteristics are known and resource allocation algorithms and power control in OFDMA are computed. A convergence criterion has been devised for both iterative power control and sub-carrier allocation. The outage capacity has been roughly analytically expressed using the outage probability and the SNR. Based on the Signal-to-Noise Ratio and the outage probability, a preliminary analytical expression for the outage capacity in the CSIT condition is obtained. For various antenna designs, it converts the MA problem into a convex optimization problem that is suitable for the current Signal-to-Noise Ratio and outage capacity values. Because at any load, they restrict power divergence situations, and the suggested solutions handle the margin adaptive (MA) issue problem more effectively than iterative water filling.

References

1. Wong, I.C., Shen, Z., Evans, B., Andrews, J.: A low complexity algorithm for proportional resource allocation in OFDMA systems. In: Proceedings of the IEEE Workshop on Signal Processing Systems, pp. 1–6, October 2004
2. Shen, Z., Andrews, J., Evans, B.: Optimal power allocation in multiuser OFDM systems. In: Proceedings of the IEEE Global Communications Conference, San Francisco, CA, pp. 337–341, December 2003
3. Wong, C.Y., Cheng, R.S., Latief, K.B., Murch, R.D.: Multiuser OFDM with adaptive subcarrier, bit, and power allocation. IEEE J. Sel. Areas Commun. **17**(10), 1747–1758 (1999)
4. Jang, J., Lee, K.B.: Transmit power adaption for multiuser OFDM systems. IEEE J. Sel. Areas Commun. **21**(2), 171–178 (2003)

5. Rhee, W., Cioffi, J.M.: Increasing in capacity of multiuser OFDM system using dynamic subchannel allocation. In: Proceedings of the IEEE International Vehicular Technology Conference, vol. 2, pp. 1085–1089 (2000)
6. Kivanc, D., Liu, H.: Subcarrier allocation and power control for OFDMA. In: The 34th Asilomar Conference on Signals, Systems, and Computers, vol. 1, pp. 147–151 (2000)
7. Wong, I.C., Shen, Z., Evans, B., Andrew, J.: Adaptive resource allocation in multiuser OFDM systems with proportional fairness. IEEE Trans. Wirel. Commun. http://www.ece.utexas.edu/bevans/papers/2005/multiuserOFDM/AdaptResAllocMUOFDM.pdf
8. Kivanc, D., Li, G., Liu, H.: Computationally efficient bandwidth allocation and power control for OFDMA. IEEE Trans. Wirel. Commun. **2**(6), 1150–1158 (2003)
9. Li, G., Liu, H.: On the optimality of the OFDMA network. IEEE Commun. Lett. **9**(5), 438–440 (2005)
10. Zhang, Y.J., Latief, K.B.: Multiuser adaptive subcarrier-and-bit allocation with adaptive cell selection for OFDM systems. IEEE Trans. Wirel. Commun. **3**(4), 1566–1575 (2004)
11. Mohan Ram, C., Bhashyam, S.: A sub-optimal joint subcarrier and power allocation algorithm for multiuser OFDM. IEEE Commun. Lett. **9**(8), 685–687 (2005)
12. Chow, J., Tu, J., Cioffi, J.: A discrete multitone transceiver system for HDSL applications. IEEE J. Sel. Areas Commun. **9**(6), 895–908 (1991)
13. Chow, P., Bingham, J.: A practical discrete multitone transceiver loading algorithm for data transmission over spectrally shaped channels. IEEE Trans. Commun. **43**(2/3/4), 773–775 (1995)
14. Cendrillon, R., Moonen, M., Verliden, J., Bosteon, T., Yu, W.: Optimal multiuser spectrum management for digital subscriber lines. In: Proceedings of the IEEE International Conference on Communications, vol. 1, pp. 1–5, June 2004
15. Yu, W., Ginis, G., Cioffi, J.: Distributed multiuser power control for digital subscriber lines. IEEE J. Sel. Areas Commun. **20**(5), 1105–1115 (2002)

A Method for Price Prediction of Potato Using Deep Learning Techniques

Savya Sree Adudotla[1,2](✉), Prathyusha Bobba[1,2], Zakiya Pathan[1,2], Tejaswi Kata[1,2], C. C. Sobin[1,2], and Jahfar[1,2]

[1] SRM University, Andhra Pradesh, Mangalagiri, India
{adudotla_savya,bobba_prathyusha,kata_tejaswi, sobin.c}@srmap.edu.in, pathan_zakiya@smap.edu.in

[2] IIIT Hyderabad, Hyderabad, Telangana, India

Abstract. The Indian agricultural sector contributes to 19.9% of the total economy. Agricultural growth can end extreme poverty in a country like India where 70% of the rural population depends on agriculture for their livelihood. It is important to make timely decisions to grow and sustain profitably. Among agricultural items, vegetables have the most supply and price variations. Vegetable prices play a major role in the national economy. It's tough to keep the supply and prices of vegetables stable because they're cultivated outside and their yields fluctuate a lot depending on the weather. Despite the government's efforts to stabilize vegetable supply and pricing, frequent meteorological shifts in recent years have resulted in unpredictable vegetable supply and price swings. Potatoes are one of India's most popular vegetables. Variation in potato output over time leads to wide price fluctuations, putting growers in a high-risk situation. Therefore, accurate forecasting of prices is critical. The current study tried to investigate variations in potato arrivals and prices in Uttar Pradesh's biggest potato-producing district, Agra. The data set consists of potato prices from the Achenra market of Agra for the years 2018–2022. The prediction was made and the RMSE value is compared between three different methods namely: Auto-Regressive Integrated Moving Average Model (ARIMA), Artificial Neural Network (ANN) and Long-Short Term Memory (LSTM). When these Rmse values were compared ANN had a lower Rmse value depicting that ANN was best suited for this dataset.

Keywords: Price prediction · Deep Learning

1 Introduction

Agriculture is an important part of the Indian economy. It has been found throughout the country for thousands of years. Agriculture directly or indirectly employs two-thirds of the population. It is a way of life as well as a source of money. It has changed with time, and modern technology and equipment have largely replaced all traditional farming methods. Furthermore, because of their shortage of resources to employ modern technologies, some small farmers in India continue to use outdated conventional agricultural

H. Sharma et al. (Eds.): ICIVC 2022, PALO 17, pp. 619–629, 2023.
https://doi.org/10.1007/978-3-031-31164-2_53

methods. Agricultural products are necessary for people's daily existence. They are crucial for economic development, and the price prediction of agricultural commodities has a significant impact on market economy stability. Price swings in agricultural goods have been more severe in recent years, with detrimental consequences for society. Large price variations will raise the uncertainty of output for farmers, increasing the number of risks that must be managed.

As a result, precise price forecasting of agricultural commodities is crucial for agricultural authorities to implement scientific judgments and ensure that the social economy flows efficiently. Vegetables are essential for human nourishment. They are an important source of vitamins and minerals for maintaining a healthy lifestyle. India is the world's second-largest vegetable producer, trailing only China [1]. Most vegetables have an erratic supply, owing to their short growth time and high perishability. The price of vegetables is dynamic and changes rapidly, having a significant impact on residents' daily lives and serving as the root cause of rapid changes in market prices of commodities as well as changes in people's living and consumption indexes. As a result, governments at all levels should pay close attention to changes in vegetable prices. Because it is difficult to foresee market information for vegetables, scientifically recording the changing trend of vegetable prices and precisely projecting market price trends are critical.

Potatoes are one of the country's most important commercial crops. It is one of the most popular goods in both the rich and poor Indian markets. India is the world's second-largest producer of potatoes. Potato marketing is a big concern for farmers due to the price volatility of the crop. Inadequate marketing infrastructure and an increase in the number of middlemen between the producer and the customer result in high marketing costs, which reduce farmers' profit. These farmers confront the issue of choosing which agricultural commodities to cultivate since they lack the tools required to determine which agricultural products will have the best prices. Uttar Pradesh cultivates the largest quantity of potatoes in India, followed by the states West Bengal and Bihar. Prices will remain stable as long as there is sufficient supply. This also ensures that the cost of living and incomes remain stable.

Agricultural production has an impact on the price level. For example, due to crop damage caused by unexpected rainfall in the important agricultural regions of Uttar Pradesh and West Bengal, potato prices soared to Rs 14–15/kg for a week after falling to Rs 9–10 at the beginning of March [4]. The average retail price of all important food products has risen in 2019, with potatoes seeing the largest increase of 92% [5]. Between 2019 and 2020, wholesale potato prices increased by 108 percent, from one thousand seven hundred rupees per quintal to three thousand six hundred per quintal in domestic markets. Potatoes are a common household vegetable, their price rise has a substantial impact on trade, exports, and household budget allocation [2]. Due to specific challenges, potato production is not always profitable for farmers. Farmers are gradually reducing potato cultivation or changing to other 'less-risky' crops due to the hazards inherent in potato farming and the risk of large losses. The area under potato production in Uttar Pradesh has come down by 8–10% for the new crop [3]. Farmers in certain parts of the state have shifted to crops like garlic, mustard, and others [3].

2 Related Work

STL-Attention-based LSTM is a method proposed by Yin et al. [5] and is a cutting-edge approach for forecasting agricultural product prices. STL-ATTLSTM uses various sources of data to anticipate monthly vegetable prices in their original article, including vegetable prices, and meteorological data. Removal of the trend and seasonality components and send the remaining elements into LSTM network components. Yin et al. [5] incorporate the STL method containing an attention layered attention mechanism into LSTM, with promising performance.

BV et al. [6] attempted to forecast the prices and demand for tomatoes. The MapReduce framework is used to create a similar version of Holt Winter's and regression models in this paper. Holt Winter's model, simple linear regression, and multiple linear regression models were all compared in this study. Multiple Linear Regression is effective in forecasting commodity demand and supply. In comparison to the other two models, Holt Winter's price predicting model produces better outcomes. They surmised that seasonality was a deciding variable because Holt Winter's model, that takes seasonality into account, had the greatest results.

To anticipate cabbage, hot pepper, cucumber, kidney bean, and tomato prices, Xiong, Li, and Bao [7] first apply the STL-based technique to analyze the price series. They take into account the seasonal properties of vegetables and use these attributes to preprocess time-series price data. In this attempt, a hybrid approach which combined EML and STL is proposed for seasonal forecasts of vegetable price series, coined STL-ELM. Their experiments show that their method is effective.

The Auto-Regressive Integrated Moving Average (ARIMA) forecasting model was employed by Ashwini et al. [8] to anticipate future cotton prices in India's key producing states. Monthly cotton price time-series data was acquired for the study to anticipate prices for the Kharif 2017–18 harvest months. This prediction is based on past data and models, and the actual market price may differ from what is predicted. Cotton prices are expected to rise in the next few years, as there is demand for the crop, according to the ARIMA model's predictions.

With the weekly retail prices of eggs from January 2008 to December 2012 in China, this paper [9] develops a short-term prediction model of weekly retail prices for eggs based on chaotic neural networks. Finally, this model is used to forecast retail egg prices and compared to ARIMA. According to the results, the chaotic NN has a stronger accuracy and nonlinear fitting ability in predicting the weekly retail price of eggs. Because of its greater prediction accuracy and ideal fitting effect, Weng et al. [9] concluded that the chaotic neural network model developed in this study is a good tool for short-term forecasting of nonlinear time series data.

The BP neural network prediction model of the vegetable market price is established in this paper [10]. This paper talks about the weekly and monthly prediction discoveries. The outcomes suggest that implementing a neural network to predict the market price of vegetables using non-linear time series is a viable option. The final output of the Backpropagation NN indicates the absolute error percentage of weekly and monthly vegetable price prediction and analyzes the price prediction accuracy percentage.

An ANN-based method was utilized in this work [11] to assess the relative importance of predictor factors for corn and soybean yield prediction, as well as the potential of ANNs for corn and soybean yield prediction. As crop production predictor variables, many satellite-derived vegetation indices, and slope data were employed, with the hypothesis that different vegetation indices reflect distinct crop and site characteristics. The corn-specific model outperformed the soybean-specific model. The findings suggest that satellite photos can be utilized to estimate yield as long as models for specific crop kinds are developed.

A deep learning system is presented in this study [12] for accurately predicting future product prices based on past pricing and volume patterns. PECAD solves shortcomings seen in previous machine learning techniques. PECAD presents a new deep and wide NN algorithm made up of two CNN models. PECAD beats existing state-of-the-art baseline approaches by obtaining a 25% lower coefficient of variation, according to simulation data.

Yuan et al. [13] developed software with price predicting characteristics to help farmers maximize their profits from farming. This study explores advanced ML algorithms, such as LSTM, ARIMA, Support Vector Regression, Prophet, and Extreme Gradient Boosting, and then conducts an experiment to determine which technique gives the maximum accuracy. The accuracy of each model is compared based on their mean-square error. LSTM had the lowest MSE among the models in the second experiment, followed by XGBoost, Prophet, ARIMA, and finally SVR. LSTM is a preferred pick than ARIMA, since it produces great results in the initial experiment attempt, because of its superior ability to handle larger datasets, with respect to software maintainability and scalability.

Gupta, et al. [14] used machine learning and numerical methods to analyze Minimum Support Price (MSP) prediction for crops. Different ML algorithms, as well as numerical techniques, were used in this article to anticipate the exact value of MSP. The data set for the experiment was obtained from the Indian government's official website. The performance of the various approaches was evaluated for predicting Paddy crop minimum support prices. The experimental results show that Newton's interpolations and Lagrange's interpolation methods are superior. Furthermore, for the prediction of minimal support price data, Lagrange's interpolation produces the best results.

Zhang et al. [15] proposed a multi-layer artificial neural network architecture for agricultural planting prediction. They put the framework to the test in Lancaster, Nebraska, in the United States of America. They created a ML system to forecast the spatial distribution of yearly crop planting maps. The multi-layer ANN architecture is used in the prediction model. They implemented the forecasting model at a county level and analyzed the forecast result's uncertainty. Then, for the Corn Belt in the United States, they forecasted and confirmed annual crop planting maps. They concluded that the ASD-level machine-learned crop planting map will achieve 88 percent relevance with future CDL by scaling up the suggested technique to the US Corn Belt.

Below Fig. 1 shows a brief timeline view of the above mentioned research work by various authors for various commodities in price prediction. As explained in detail below Fig. 1 is a brief timeline view of some of the research works done in the price prediction field in the years 2013–2021.

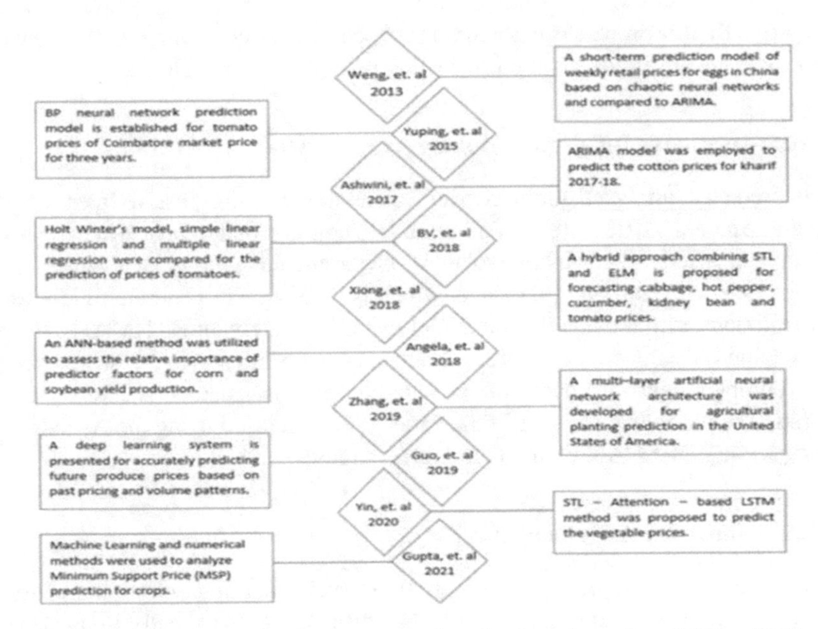

Fig. 1. Timeline of all the research works done by the authors from 2013 to 2021

3 Proposed System

This section encompasses the techniques used in this study for forecasting the temperature. The reason for using these models is to compare the traditional (ARIMA, LSTM) methods with ANN. These models are extensively used in time series forecasts and can estimate complex non-linear relations between the data points unlike the traditional methods.

3.1 Artificial Neural Network (ANN)

A neural network is made up of numerous layers, where each serves a specific purpose. As the number of layers increases, the complexity of the algorithm also rises. For this reason, they are called multi-layer perceptrons. A neural network has three layers in its most basic form: an input, a hidden, and an output layer. Signals are taken as input in the input layer and forwards them to the hidden layer, which then sends the final prediction to the output layer. The NN's, for instance, ML methodologies, have to be trained with data before being used to solve a specific problem. The input layer receives data initially, then passes it on to the hidden layers. Weights are assigned to each random input and the interconnection between these two layers assigns weights to each input at random at first. After bias is added to every neuron input, the total weight, that is a mixture of bias and weights, is sent through the activation function. The Activation Function determines which node to fire for feature extraction after that it calculates the result. Forward propagation is this whole process termed. The weights in the nodes are corrected using BP to reduce the inaccuracy when actual value and expected value are compared in BP this process is iterated till the specified epoch count. Finally, the weights in the model are adjusted, and the forecasting procedure is completed.

3.2 Auto-regressive Integrated Moving Average Model (ARIMA)

Price is predicted, analyzed, and forecasted using the Auto-Regressive Integrated Moving Average model. ARIMA (p, d, q) is a standard notation that indicates that the ARIMA model is being used right away by replacing the parameters with integer values. A series of replies is identified in the first stage, which is then used to generate time series and autocorrelations with the identity statement using the statement ESTIMATE, the previously specified variables, as well as the parameters, are estimated in this stage. This stage entails running diagnostic checks on the variables and parameters that were previously collected. The predictive values of time series are Forecasted using the ARIMA model and the phrase FORECAST, which are future values.

3.3 Long Short Term Memory (LSTM)

LSTM is a development of recurrent neural networks that adds more interactions per module to overcome the aforementioned RNN disadvantages (or cell). Long short-term memory is a type of recurrent neural network that can learn the long-term dependencies and remember information for lengthy periods. Long short-term memory model is structured in the form of a chain. The repeating module, on the other hand, has a different structure. It features four interacting layers with a unique form of communication, rather than a single neural network like a normal RNN. A typical long short term memory network is made up of cells which acts as a memory block for that network. The following cell receives both the cell state and the concealed state. The cell state is the most important link in the data flow chain because it allows data to pass forward almost unchanged. However, linear transformations are possible. Data can be added or removed from the cell state using sigmoid gates.

4 Results

Data necessary for this study was acquired from Data.gov.in. The potato data set of the Agra district is used for the study. Agra district of Uttar Pradesh produces nearly 14 lakh tons of potatoes (as per the 2021 report). This district also owns a large cold storing capacity with almost 2.2 million tons which is 7% of the total cold storage capacity of our country. So, we have purposefully selected this district because of its large production capability. We have considered the data set For the five consecutive years (2018–2022). This data consists of prices from seven different markets (Achenra, Agra, Fatehabad, Fatehpur Sikri, Jagnair, Jarar) in Agra from which the Achenra market was chosen.

The data set consists of different properties of potato-like the type, max price, min price, and modal price for 1 quintal of potato on their respective dates. The missing values in the data set were pre-processed by considering the average of previous and next values. The models were trained on 2018–2021 values and the prediction was done on 2022 data. The above (Table 1) shows the head of the data set that was used in this

Table 1. Head of the database

District	Market	Commodity	Variety	Min Price	Max Price	Modal Price	Date
Agra	Archenra	Potato	Desi	430	470	450	01-jan-18
Agra	Archenra	Potato	Desi	440	480	460	02-jan-18
Agra	Archenra	Potato	Desi	400	420	410	03-jan-18
Agra	Archenra	Potato	Desi	400	420	410	04-jan-18
Agra	Archenra	Potato	Desi	400	420	410	05-jan-18

study. Below Fig. 2 is a graphical representation of the data set. The x-axis in this graph shows the dates from 2018-to 2022 (17th April) and the y-value depicts the modal price which is the average of minimum and maximum prices of a particular day.

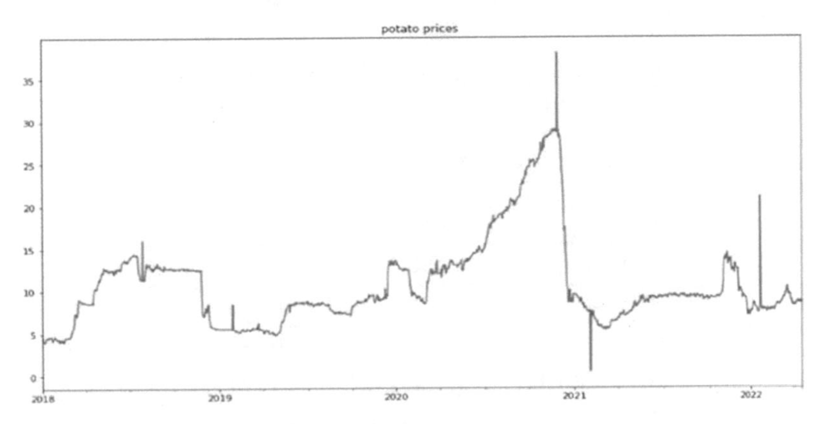

Fig. 2. Graph showing the modal prices of respective data

ARIMA

In this study, we trained the model with 2018–2021 data and later the forecasting was done on 2022 data. When the dickey fuller test was performed on the data it showed that the p-value is 0.15(which is more than 0.05), which indicated that the data is non-stationary. The seasonality in data was removed by performing first-order differentiation. And p-value after the first differentiation gave the value 2.159e−11 (<0.05) indicating that the data got stationary. And the auto-correlation lag value obtained for this data was0.98.

The above Fig. 3 shows the trend of the data over the period and through the seasonality graph, we can see that this data is not stationary and the residual analysis which shows the difference between actual and predicted output data tells us that ARIMAs predictions are not precise for this particular data. The ARIMA which gives a minimum AIC value is chosen. In general $AIC = -2Log(L)+2(p+q+k+1)$ where L is the likelihood of data and p, q, k are values of the ARIMA.

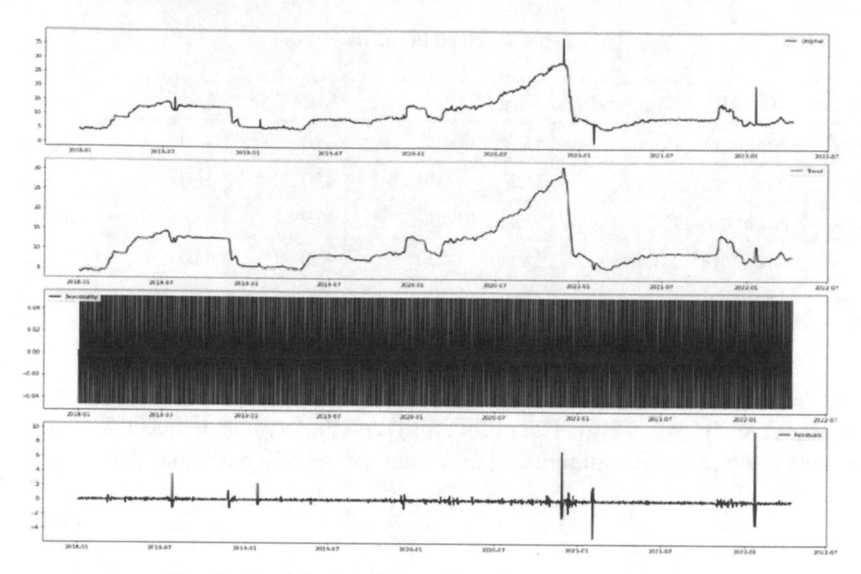

Fig. 3. Tend Seasonality and Residual analysis of data

Table 2. ARIMA Models with AIC value

Model	AIC	Time
ARIMA(0,1,1)	3142.298	0.14 s
ARIMA(0,1,0)	3377.501	0.37 s
ARIMA(1,1,0)	3150.607	0.39 s
ARIMA(2,1,1)	3137.493	1.10 s
ARIMA(2,1,3)	3105.44	6.09 s
ARIMA(0,1,2)	3132.853	0.23 s
ARIMA(1,1,2)	3101.254	1..80 s

As depicted by the values in table (Table 2), the ARIMA (1,1,2) was chosen as it has the minimum AIC value. This shows that the lag value of the auto regression function is set to 1 and this one tells that the data is not stationary and can become stationary when first-order differentiation is performed on the data. Finally, best fit parameters are used to fit the model and sort out the coefficients of regression. The RMSE value obtained for this data is 1.44. Figure 4 shows the graphical representation of predicted values and actual values of the test data set.

LSTM

The Lstm methodology was implemented with the same data set that was used on ARIMA. The training data set contained modal prices of potatoes on their respective arrival dates from 2018–2021 and then the model was tested with 2022 data. When

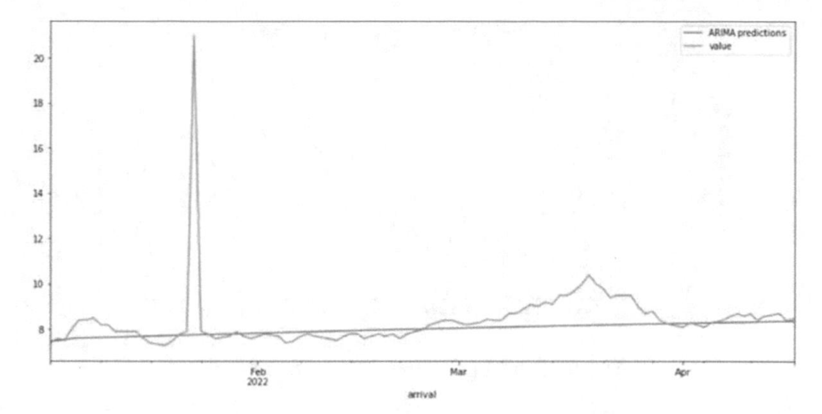

Fig. 4. Modal price prediction using ARIMA

compared to ARIMA, LSTM showed a better result. It had an RMSE value of 0.78 for training data and 0.31 for the test data set. The below Fig. 5 shows the graphical representation of forecasted values on seen and unseen data and the original values vs the arrival time.

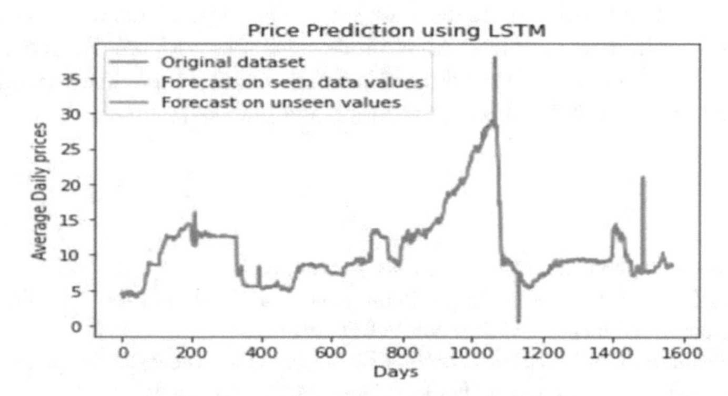

Fig. 5. Modal Price Predictions using LSTM

ANN

The same dataset is utilized to train and test the ANN model. For this study, we trained the model for 25 epochs because after nearly 10 epochs the loss value became stable. An epoch that has the lowest mean squared error was chosen and an adam optimizer was used. ANN showed a better result than ARIMA and LSTM with its RMSE value of 0.73 for training data and 0.19 for test data. Below Fig. 6 is the graphical representation of both train and test data.

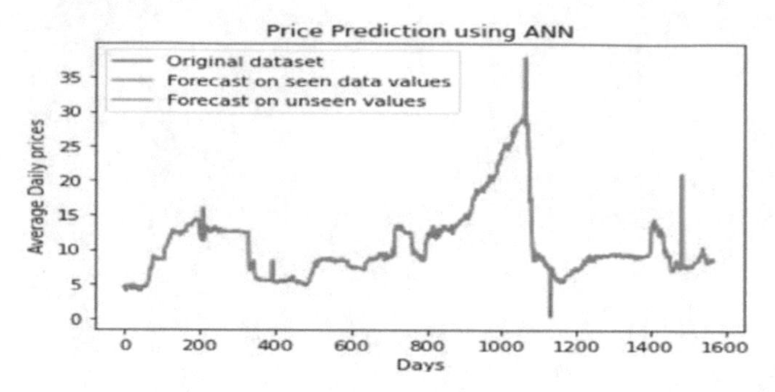

Fig. 6. Modal price Prediction using ANN

5 Conclusion

Price forecasting has always been a difficult area to study because many factors influence the price of any product. In this study, we have tried to predict the price of potatoes. To forecast potato prices in the Achenra market of Agra district, this study uses ANN, LSTM, and ARIMA models. These models were chosen because of their ability to do precise prediction with both stationary and nonstationary data. With an RMSE of 0.74 in the train data set and 0.19 in the test data set, ANN outperforms these models. When compared to the other two models, ANN predictions are more precise according to the Rmse value. Hence the use of hybrid models which can find complex dependencies and patterns would be more helpful for an accurate and precise result.

References

1. https://www.atlasbig.com/en-in/countries-by-potato-production
2. Premi, S., Premi, B.R.: Onion Supply Chain Analysis: Constraints and Way Forward. Rural Pulse, Issue: XXI May-June, 2017. NABARD (2017)
3. https://economictimes.indiatimes.com/markets/commodities/news/potato-prices-unlikely-to-drop-to-8-10-a-kg-this-year-experts/articleshow/73149385.cms
4. https://economictimes.indiatimes.com/news/economy/agriculture/potato-price-gais-in-india-due-to-rains/articleshow/74704243.cms?from=mdr
5. Yin, H., Jin, D., Gu, Y.H., Park, C.J., Han, S.K., Yoo, S.J.: STL-ATTLSTM: vegetable price forecasting using STL and attention mechanism-based LSTM. Agriculture **10**, 612 (2020)
6. Balaji Prabhu, B.V., Dakshayini, M.: Performance analysis of the regression and time series predictive models using parallel implementation for agricultural data. Proc. Comput. Sci. **132**, 198–207 (2018)
7. Xiong, T., Li, C., Bao, Y.: Seasonal forecasting of agricultural commodity price using a hybrid STL and ELM method: evidence from the vegetable market in China. Neurocomputing **275**, 2831–2844 (2018)
8. Darekar, A., Reddy, A.A.: Cotton price forecasting in major producing states. Econ. Aff. **62**, 373–378 (2017)
9. Li, Z.M., et al.: Prediction model of weekly retail price for eggs based on chaotic neural network. J. Integr. Agric. **12**, 2292–2299 (2013)

10. Lu, Y., Yuping, L., Weihong, L., Qidao, S., Yanqun, L., Xiaoli, Q.: Vegetable price prediction based on PSO-BP neural network. In: 2015 8th International Conference on Intelligent Computation Technology and Automation (ICICTA), pp. 1093–1096 (2015). https://doi.org/10.1109/ICICTA.2015.274
11. Kross, A., et al.: Evaluation of an artificial neural network approach for prediction of corn and soybean yield. In: Proceedings of the 14th International Conference on Precision Agriculture, Montreal, QC, Canada (2018)
12. Guo, H., Woodruff, A., Yadav, A.: Improving lives of indebted farmers using deep learning: predicting agricultural produce prices using convolutional neural networks. In: AAAI (2020)
13. Yuan, C.Z., Ling, S.K.: Long short-term memory model based agriculture commodity price prediction application. In: Proceedings of the 2020 2nd International Conference on Information Technology and Computer Communications (2020)
14. Gupta, S., Agarwal, A., Deep, P., Vaish, S., Purwar, A.: Analysis of minimum support price prediction for Indian crops using machine learning and numerical methods. In: Gupta, D., Khanna, A., Bhattacharyya, S., Hassanien, A.E., Anand, S., Jaiswal, A. (eds.) International Conference on Innovative Computing and Communications. AISC, vol. 1166, pp. 267–277. Springer, Singapore (2021). https://doi.org/10.1007/978-981-15-5148-2_24
15. Zhang, C., et al.: Machine-learned prediction of annual crop planting in the US Corn Belt based on historical crop planting maps. Comput. Electron. Agric. **166**, 104989 (2019)

Author Index

H. Sharma et al. (Eds.): ICIVC 2022, PALO 17, pp. 631–633, 2023.
https://doi.org/10.1007/978-3-031-31164-2

n the United States
: Taylor Publisher Services